AF776228

Crossed Products of C^*-Algebras

Mathematical Surveys and Monographs

Volume 134

Crossed Products of C^*-Algebras

Dana P. Williams

American Mathematical Society

EDITORIAL COMMITTEE

Jerry L. Bona　　Ralph L. Cohen
Michael G. Eastwood　　Michael P. Loss
J. T. Stafford, Chair

2000 *Mathematics Subject Classification.* Primary 46L55, 46L05, 22D25, 22D30, 46L45, 54H15.

For additional information and updates on this book, visit
www.ams.org/bookpages/surv-134

Library of Congress Cataloging-in-Publication Data

Williams, Dana P., 1952–
Crossed products of C^*-algebras / Dana P. Williams.
p. cm. — (Mathematical surveys and monographs, ISSN 0076-5376 ; v. 134)
Includes bibliographical references and index.
ISBN-13: 978-0-8218-4242-3 (alk. paper)
ISBN-10: 0-8218-4242-0 (alk. paper)
1. C^*-algebras. 2. Operator algebras. I. Title. II. Mathematical surveys and monographs : no. 134.

QA326.W55 2007

2006047931

Copying and reprinting. Individual readers of this publication, and nonprofit libraries acting for them, are permitted to make fair use of the material, such as to copy a chapter for use in teaching or research. Permission is granted to quote brief passages from this publication in reviews, provided the customary acknowledgment of the source is given.

Republication, systematic copying, or multiple reproduction of any material in this publication is permitted only under license from the American Mathematical Society. Requests for such permission should be addressed to the Acquisitions Department, American Mathematical Society, 201 Charles Street, Providence, Rhode Island 02904-2294, USA. Requests can also be made by e-mail to reprint-permission@ams.org.

© 2007 by the American Mathematical Society. All rights reserved.
The American Mathematical Society retains all rights
except those granted to the United States Government.
Printed in the United States of America.

∞ The paper used in this book is acid-free and falls within the guidelines
established to ensure permanence and durability.
Visit the AMS home page at http://www.ams.org/

10 9 8 7 6 5 4 3 2 1　　12 11 10 09 08 07

For V, O and H.

Contents

Introduction

This book is meant to provide the tools necessary to begin doing research involving crossed product C^*-algebras. Crossed products of operator algebras can trace their origins back to statistical mechanics, where crossed products were called covariance algebras, and to the group measure space constructions of Murray and von Neumann. Now the subject is fully developed with a vast literature. Crossed products provide both interesting examples of C^*-algebras and are fascinating in their own right. Simply put, a crossed product C^*-algebra is a C^*-algebra $A \rtimes_\alpha G$ built out of a C^*-algebra A and a locally compact group G of automorphisms of A. When the C^*-algebra A is trivial, that is, when A is the complex numbers, then the crossed product construction reduces to the group C^*-algebra of harmonic analysis. In fact, the subject of crossed product C^*-algebras was introduced to me as the harmonic analysis of functions on a locally compact group taking values in a C^*-algebra. This is a valuable analogy and still serves as motivation for me and for the approach in this book.

The subject of crossed products is now too massive to be covered in a single volume. This is especially true for this treatment as I have tried to write at a level suitable for graduate students who are just beginning to search out research areas, and as I also want to make the treatment reasonably self-contained modulo a modest set of prerequisites (to be described below). As a result, it has been necessary to leave out many important topics. A list of what is covered is given below in the "Reader's Guide". (A brief discussion of what is not covered is given under "Further Reading" below.) In choosing what to include I have been guided by my own interests and bounded by my ignorance. Thus the central theme of this book is to uncover the ideal structure of crossed products via the Mackey machine as extended to C^*-algebras by Rieffel and to crossed products by Green. Crossed products, just as group C^*-algebras, arise via a subtle completion process, and a detailed structure theorem determining the isomorphism classes of such algebras is well out of reach. Instead, we try to understand these algebras via their representations. Thus a key objective is to determine the primitive ideals of a given crossed product and the topology on its primitive ideal space. (We settle for primitive ideals in the general case, since it is not practicable to try to collect the same sort of information about irreducible representations of non-type I algebras.) If we have an abelian C^*-algebra $B = C_0(X)$, then the spectrum and primitive ideal space coincide with X as *topological* spaces. Thus, a description of X and its topology contains everything there is to know about $C_0(X)$. If B is arbitrary, then the primitive ideal space

$\operatorname{Prim} B$ is a modest invariant, but it still contains a great deal of information about B. For example, we can recover the lattice of ideals of B. (Hence the term "ideal structure".)

The basic paradigm for studying $\operatorname{Prim}(A \rtimes_\alpha G)$ is to recover the primitive ideals via induction from the algebras $A \rtimes_{\alpha|_H} H$ where H is a closed subgroup of G. Loosely speaking, this process is the "Mackey machine" for crossed products. The centerpiece for this is the GRS-Theorem ("GRS" stands for Gootman, Rosenberg and Sauvageot), and we devote a fair chunk of this book to its proof (Chapter 9 and several appendices).

Errata

There are mistakes and typographical errors in this book. I only wish that the previous sentence were one of them. As I become aware of typographical errors and other mistakes, I will post them at `http://math.dartmouth.edu/cpcsa`. If you find a mistake that is not listed there, I would be grateful if you would use the e-mail link provided there to send me a report so that I can add your contribution to the list.

Prerequisites and Assumptions

I have tried to keep the required background to a minimum in order to meet the goal of providing a text with which a graduate student in operator algebras can initiate an investigation of crossed products without having to consult outside sources. I do assume that such a student has had the equivalent of a basic course in C^*-algebras including some discussion of the spectrum and primitive ideal space. For example, the first few chapters of Murphy's book [110] together with Appendix A of [139] should be enough. Since it is a bit harder to pick up the necessary background on locally compact groups, I have included a very brief introduction in Chapter 1. Of course, this material can also be found in many places, and Folland's book [56] is a good source. Rieffel's theory of Morita equivalence is assumed, and I will immodestly suggest [139] as a reference. Anyone interested in the details of the proof of the GRS-Theorem will have to sort out Borel structures on analytic Borel spaces. While some of that material can be found in the appendices, there is no better resource than Chapter 3 of Arveson's beautiful little book [2]. Of course, I also assume a good deal of basic topology and functional analysis, but I have tried to give references when possible.

I have adopted the usual conventions when working in the subject. In particular, all homomorphisms between C^*-algebras are assumed to be $*$-preserving, and ideals in C^*-algebras are always closed and two-sided. Unless otherwise stated, a representation of a C^*-algebra on a Hilbert space is presumed to be nondegenerate, and our Hilbert spaces are all complex.

I have tried to use fairly standard notation, and I have provided a "Notation and Symbol Index" as well as the usual index. The canonical extension of a representation π of an ideal I in an algebra A to A is usually denoted by $\bar{\pi}$. When using the notation sH for the cosets becomes too cumbersome, I will use $\dot{s}$ instead.

If X is a locally compact Hausdorff space, then $C(X)$, $C^b(X)$, $C_0(X)$, $C_c(X)$ and $C_c^+(X)$ denote, respectively, the algebra of all continuous complex-valued continuous functions on X, the subalgebra of bounded functions in $C(X)$, the subalgebra of functions in $C^b(X)$ which vanish at infinity, the subalgebra of all functions in $C_0(X)$ which have compact support and the cone in $C_c(X)$ of nonnegative functions.

In general, I have not made separability assumptions unless, such as in the case in the proof of the GRS-Theorem, they cannot be avoided. With the possible exception of Section 4.5, this causes little extra effort. Some arguments do use nets in place of sequences, but this should cause no undue difficulty. (I suggest Pedersen's [127, §1.3] as a good reference for nets and subnets.)

Reader's Guide

Chapter 1 provides a very quick overview of the theory of locally compact groups and Haar measure. As we will repeatedly want to integrate continuous compactly supported functions taking values in a Banach space, we review the basics of the sort of vector-valued integration needed. More information on groups can be found in Appendix D, and for those who want a treatment of vector-valued integration complete with measurable functions and the like, I have included Appendix B. Anyone who is comfortable with locally compact groups, and/or the integrals in the text, may want to skip or postpone this chapter.

In Chapter 2, we define dynamical systems and their associated crossed products. We show that the representations of the crossed product are in one-to-one correspondence with covariant representations of the associated dynamical system. We also prove that a crossed product $A \rtimes_\alpha G$ can be defined as the C^*-algebra generated by a universal covariant homomorphism. We talk briefly about examples, although it is difficult to do much at this point. Working out the details of some of the deeper examples will have to wait until we have a bit more technology.

In Chapter 3, partially to give something close to an example and partially to provide motivation, we take some time to work out the structure of the group C^*-algebra of both abelian and compact groups. We also look at some basic tools needed to work with crossed products. In particular, we show that if G is a semidirect product $N \rtimes_\varphi H$, then the crossed product $A \rtimes_\alpha G$ decomposes as an iterated crossed product $(A \rtimes_{\alpha|_N} N) \rtimes_\beta H$. We also prove that G-invariant ideals I in A correspond naturally to ideals $I \rtimes_\alpha G$ in $A \rtimes_\alpha G$.

In Chapter 4, we turn to the guts of the Mackey machine. One of the main results is Raeburn's Symmetric Imprimitivity Theorem, which provides a common generalization of many fundamental Morita equivalences in the subject and implies the imprimitivity theorems we need to define and work with induced representations. Since this finally provides the necessary technology, we give some basic examples of crossed products. In particular, we show that for any group G, $C_0(G) \rtimes_{\mathrm{lt}} G$ is isomorphic to the compact operators on $L^2(G)$. More generally, we also show that $C_0(G/H) \rtimes_{\mathrm{lt}} G$ is isomorphic to $C^*(H) \otimes \mathcal{K}\big(L^2(G/H, \beta)\big)$, where β is any quasi-invariant measure on the homogeneous space G/H. This result is a bit unusual, as it requires we use a suitably measurable cross section for the natural map of G onto G/H. This briefly pulls us away from continuous compactly supported functions and requires some fussing with measure theory.

In Chapter 5, we define induced representations of crossed products and develop the properties we'll need in the sequel. We also introduce the important concept of inducing ideals, and we show that the induction processes are compatible with the decomposition of crossed products with respect to invariant ideals.

In Chapter 6, we expand on our preliminary discussion of orbit spaces in Section 3.5 and add quasi-orbits and the quasi-orbit space to the mix. A particularly important result is the Mackey-Glimm dichotomy for the orbit space for a second countable locally compact group G action on a second countable, not necessarily Hausdorff, locally compact space X. Simply put, the orbit space is either reasonably well-behaved or it is awful. When the orbit space is awful, it is often necessary to pass to quasi-orbits, and we discuss this and its connection with the restriction of representations.

In Chapter 7, we get to the heart of crossed products and prove a number of fundamental results. In Section 7.1, we prove the Takai Duality Theorem, which is the analogue for crossed products by abelian groups of the Pontryagin Duality Theorem. In Section 7.2, we look at the reduced crossed product and show that in analogy with the group C^*-algebra case, the reduced and universal crossed products coincide when G is amenable. (The necessary background on amenable groups is given in Appendix A.) In Section 7.3, we look at the special case where the algebra A is the algebra of compact operators. Since this leads naturally to a discussion of projective and cocycle representations (discussed in Appendix D.3), I couldn't resist including a short aside on twisted crossed products. This also allows us to show that, in analogy with the decomposition result for semidirect products in Section 3.3, we can decompose $A \rtimes_\alpha G$ into an iterated *twisted* crossed product whenever G contains a normal subgroup N. In Section 7.5, we give a very preliminary discussion of when a crossed product is GCR or CCR. This is a ridiculously difficult problem in general, so our results in this direction are very modest. We have more to say in Section 8.3 about the case of a transformation group C^*-algebra $C_0(X) \rtimes_{\mathrm{lt}} G$ with G abelian.

In Chapter 8, we take on the ideal structure of crossed products. In Section 8.1, we see that one can obtain fairly fine information if the action of G on $\operatorname{Prim} A$ is "nice". (In the text, the formal term for nice is "regular". In the literature, the term "smooth" is also used.) In Section 8.2, we face up to the general case. When (A, G, α) is separable and G is amenable, we can use the GRS-Theorem to say quite a bit. However, the proof of this result is very difficult and occupies all of Chapter 9 (and a few appendices). In Chapter 8, we merely concentrate on some of its wide ranging implications. In particular, we devote Section 8.3 to a detailed analysis of the ideal structure of $C_0(X) \rtimes_{\mathrm{lt}} G$ when G is abelian. Although extending the results in Section 8.3 to cases where either G or A is nonabelian appears to be very formidable, this section provides a blueprint from which to start.

The remainder of this book consists of appendices that are meant to be read "as needed" to provide supplements where material is needed which is not part of the prerequisites mentioned above. As a result, I have definitely *not* tried to completely avoid overlap. Appendix A gives a brief overview of the properties of amenable groups which are needed in our treatment of reduced crossed products, and in the proof of the GRS-Theorem. Appendix B is a self-contained treatment of vector-valued integration on locally compact spaces. Although we make passing use of this material in Chapter 9, I have included this appendix primarily to satisfy those who would prefer to think of $A \rtimes_\alpha G$ as a completion of $L^1(G, A)$. This material may be useful to anyone who wishes to extend the approach here to Busby-Smith twisted crossed products.

Appendix C is a self-contained treatment of $C_0(X)$-algebras. These algebras are well known to be a convenient way to view an algebra as being fibred over X. In addition to developing the key properties of these kinds of algebras, we prove an old result of Hofmann's which implies, at least after translating his results onto contemporary terminology, that any $C_0(X)$-algebra is the section algebra of a bona fide bundle which we call an upper semicontinuous-bundle of C^*-algebras over X.

Appendix D contains additional information about groups — particularly Borel structure issues. Appendices E and F give considerable background on representations of C^*-algebras and on direct integrals in particular. The direct integral theory is needed for the proof of the GRS-Theorem. In particular, we need direct integrals to discuss Effros's decomposition result for representations of C^*-algebras, called an ideal center decomposition, which is proved in Appendix G and which is essential to the proof of the GRS-Theorem.

Appendix H is devoted to a discussion of the Fell topology on the closed subsets of a locally compact space and its restriction to the closed subgroups of a locally compact group. This material is essential to our discussion of the topology on the primitive ideal space of $C_0(X) \rtimes_{\mathrm{lt}} G$ with G abelian and to the proof of the GRS-Theorem.

Further Reading

As I mentioned above, no book at this level, or perhaps any level, could adequately cover all there is to talk about when it comes to crossed products. Even a modest expository discussion of the material that I wished I could have included would involve several chapters, or even a short book by itself. Instead, I will just give a short list of topics and references to some of the major omissions. There is a good deal in Pedersen's classic book [126] that won't be found here, and Green's original paper [66] and Echterhoff's *Memoir* [38] are full of ideas and results about the Mackey Machine for crossed products. In fact, both [66] and [38] work with twisted crossed products $A \rtimes_\alpha^\tau G$, which are quotients of $A \rtimes_\alpha G$, and which get only a brief aside in this text. There is even another flavor of twisted crossed products, called Busby-Smith twisted crossed products, which is not dealt with at all [15, 100, 119–121]. One serious research question about crossed products is under what conditions is a crossed product simple. This is particularly important today with the intense study of separable, nuclear, purely infinite nuclear C^*-algebras [89, 90, 128, 152]. One of the tools for this sort of question is the Connes spectrum. Its relationship to crossed products was studied extensively in [63, 114–116]. Many of the key points can be found in [126]. The fundamental obstacle to extending many of the results for transformation group C^*-algebras to the general case is the appearance of Mackey obstructions as touched on in Section 7.3. Understanding the ideal structure in this case requires a subtle analysis involving the symmetrizer subgroup of the stability group. For example, see [42, 83] and [38] in particular. We have nothing here to say about the K-theory of crossed products. This is a difficult subject, and the key results here are the Pimsner-Voculescu six-term exact sequence for $\mathbf{Z}$-actions, the Thom isomorphism for $\mathbf{R}$-actions, and a variety of results about embedding certain transformation group C^*-algebras into AF C^*-algebras. Summaries and references for these results can be found in Blackadar's treatise [8]. As we shall see in Chapter 4, proper actions of groups on spaces are especially important and there is a bit of an industry on extending this notion to actions of groups on noncommutative C^*-algebras [79, 80, 107, 150, 151]. Also, there was no time to discuss Morita equivalence of dynamical systems and its implications [16, 19, 20, 41, 79, 95, 122, 172], nor was there space to talk about imprimitivity theorems for the reduced crossed product [78, 131]. We have not touched on coactions, their crossed products, and the powerful theory of noncommutative duality. An overview of non-abelian duality for reduced crossed products is provided in [40, Appendix A]. The theory for full crossed products is less well-developed, but this is a topic of current interest (see, for example, [85]). There is also a short introduction to the main ideas of the subject in [135].

Acknowledgments

This is survey of a huge subject, so almost all of the results in the book are due to someone besides the author. I have tried to give references for further reading, but I have not always been able to cite original sources. No doubt there are some glaring omissions, and I apologize for all omissions — glaring or not. I am very indebted to the people who taught me the subject. In particular, I am grateful to my advisor, Marc Rieffel, who pointed me to the subject in the first place, and to Phil Green whose work was both groundbreaking and inspirational. I owe a special debt to my friend Gert Pedersen whose inspiration extended far beyond mathematics. Naturally, I could have done nothing without the love and support of my wife and family.

I also want to thank those people who helped with the book. In addition to those Dartmouth students who sat through seminars and courses based on early drafts, I would especially like to thank Lisa Clark, Siegfried Echterhoff, Astrid an Huef, Marius Ionescu, Tobias Kauch, Katharina Neumann, Iain Raeburn, and Mark Tomforde, all of whom read large parts of early drafts and made helpful comments. I would especially like to thank Geoff Goehle, who read the entire draft and suggested countless improvements. Of course, the remaining errors, obscurities, and omissions are *this* author's fault alone.

Chapter 1

Locally Compact Groups

Crossed products are built from locally compact group actions on C^*-algebras. Therefore, we will need to have a bit of expertise in both of these components. Fortunately, there are several nice introductory treatments on C^*-algebras available. As a result, this material is considered as a prerequisite for this book. On the other hand, the background on locally compact groups needed is a bit more difficult to sort out of the literature, and this chapter is devoted to providing a quick overview of the material we need here. Of course, there is nothing here that can't be found in [71] and/or [56]. A more classical reference is [129], and much more than we need on group C^*-algebras can be found in the second part of [28].

The first two sections cover topological considerations. In Section 1.3, we talk a bit about Haar measure and its basic properties (including the modular function). In Section 1.4, we take a short detour to talk about harmonic analysis on locally compact abelian groups. The primary purpose of this excursion is to provide a bit of motivation for the constructions to come. (A secondary purpose is to provide a bit of a break from definitions and preliminary results.) After this, we finish with a long section on integration on groups with particular emphasis on vector-valued integration of continuous compactly supported functions. The material in the last section concerning integration on groups is fairly technical and may be skipped until needed.

While it is the philosophy of this book that group C^*-algebras and crossed products should be thought of as completions of families of continuous (compactly supported) functions, it is possible to work with L^1-algebras as in classical abelian harmonic analysis. For those interested, the details of vector-valued integration of measurable functions are discussed in Appendix B. We have also shunted off to Appendix D some technicalities on groups and measure theory on groups as this material won't be needed until much later.

1.1 Preliminaries on Topological Groups

Definition 1.1. A *topological group* is a group $(G, \cdot)$ together with a topology τ such that

(a) points are closed in (G, τ), and
(b) the map $(s, r) \mapsto sr^{-1}$ is continuous from $G \times G$ to G.

Condition (b) is easily seen to be equivalent to

(b′) the map $(s, r) \mapsto sr$ is continuous, and
(b″) the map $s \mapsto s^{-1}$ is continuous.

It is shown in [71, §4.20] that neither (b′) nor (b″) by itself implies (b).

Remark 1.2. To avoid pathologies, we've insisted that points are closed in a topological group.[1] As we will see in Lemma 1.13 on page 4, this assumption actually guarantees that a topological group is necessarily Hausdorff and regular.

Topological groups are ubiquitous in mathematics. We'll settle for a few basic examples here.

Example 1.3. Any group G equipped with the discrete topology is a topological group.

Example 1.4. The groups $\mathbf{R}^n$, $\mathbf{T}^m$ and $\mathbf{Z}^d$ are all topological groups in their usual topologies.

Example 1.5. If G and H are topological groups, then $G \times H$ is a topological group in the product topology. In particular, if F is any finite group (with the discrete topology), then $G = \mathbf{R}^n \times \mathbf{T}^m \times \mathbf{Z}^d \times F$ is a topological group. If F is abelian, these groups are called elementary abelian groups in the literature.

Examples 1.4–1.5 are of course examples of abelian groups. Our next example is certainly not abelian, nor is the topology discrete.

Example 1.6. Let $\mathcal{H}$ be a complex Hilbert space of dimension at least two and let

$$U(\mathcal{H}) := \{\, U \in B(\mathcal{H}) : U^*U = UU^* = 1_{\mathcal{H}} \,\}$$

with the strong operator topology. Then $U(\mathcal{H})$ is a topological group.

Verifying that $U(\mathcal{H})$ is a topological group requires a bit of work. It is immediate that $U(\mathcal{H})$ is Hausdorff — in fact, $B(\mathcal{H})$ is Hausdorff in the strong operator topology. Suppose that $U_n \to U$ and $V_n \to V$. Since $\|U_n\| = 1$ for all n, it is easy to see that $h_n \to h$ in $\mathcal{H}$ implies that $U_n h_n \to Uh$ in $\mathcal{H}$. Using this, it is immediate that $U_n V_n h \to UVh$ for all $h \in \mathcal{H}$. Thus (b′) is easy. To see that (b″) holds, we have to prove that $U_n^* h \to U^* h$ for any h. But

$$\|U_n^*h - U^*h\|^2 = \big(U_n^*h - U^*h \mid U_n^*h - U^*h\big) = 2\|h\|^2 - 2\operatorname{Re}(U_n^*h \mid U^*h).$$

[1] In terms of the classical separation axioms of point-set topology, a topological space X is called a T_1-space when points in X are closed. We say that X is a T_0-space if distinct points have distinct closures, and a Hausdorff space is said to be T_2. Some authors call a space which is regular and T_1 a T_3-space.

But $(U_n^*h \mid U^*h) = (h \mid U_nU^*h) \to \|h\|^2$, and it follows that $\|U_n^*h - U^*h\| \to 0$ as required.

The case where $\dim \mathcal{H} < \infty$ in Example 1.6 is particularly important. Our assertions below follow from elementary linear algebra.

Example 1.7. If $\mathcal{H} = \mathbf{C}^n$, then we can identify $U(\mathbf{C}^n)$ with the set U_n of invertible $n \times n$-matrices whose inverse is given by the conjugate transpose. Specifically, we identify U with the matrix (u_{ij}) where $u_{ij} := (Ue_j \mid e_i)$. If we give U_n the relative topology coming by viewing U_n as a *closed* subset of $\mathbf{C}^{n^2}$, then $U \mapsto (u_{ij})$ is a homeomorphism of $U(\mathbf{C}^n)$ onto U_n.

Example 1.7 is merely one of many classical matrix groups which are topological groups in their inherited Euclidean topology.

Example 1.8. Let A be a C^*-algebra. Then the collection $\operatorname{Aut} A$ of $*$-automorphisms of A is a group under composition. We give $\operatorname{Aut} A$ the *point-norm topology*; that is, $\alpha_n \to \alpha$ if and only if $\alpha_n(a) \to \alpha(a)$ for all $a \in A$. Then $\operatorname{Aut} A$ is a topological group.

As in Example 1.6, the only real issue involved in showing that $\operatorname{Aut} A$ is a topological group is to show that $\alpha_n \to \alpha$ implies $\alpha_n^{-1} \to \alpha^{-1}$. This is not hard to show using the fact that any automorphism of a C^*-algebra is isometric (cf. [139, Lemma 1.3]).

Lemma 1.9. *If G is a topological group, then $s \mapsto s^{-1}$ is a homeomorphism of G onto itself. Similarly, if $r \in G$, then $s \mapsto sr$ and $s \mapsto rs$ are also homeomorphisms.*

Proof. These statements are easy consequences of Definition 1.1 on the preceding page; for example, $s \mapsto sr^{-1}$ is a continuous inverse for $s \mapsto sr$. □

Although straightforward, Lemma 1.9 has important consequences. In particular, it asserts that the topology of G is determined by the topology near the identity e. The topology near a point is usually described via a neighborhood basis.[2]

Definition 1.10. A family $\mathcal{N}$ of neighborhoods of $x \in X$ is called a *neighborhood basis* at x if given a neighborhood W of x, there is a $N \in \mathcal{N}$ such that $N \subset W$.

Therefore the statement that the topology on G is determined locally amounts to the following corollary of Lemma 1.9.

Corollary 1.11. *Let $\mathcal{N}$ be a neighborhood basis of e in a topological group G. Then $\{Nr\}_{N \in \mathcal{N}}$ and $\{rN\}_{N \in \mathcal{N}}$ are both neighborhood bases of r in G. In particular, if $\mathcal{N}$ consists of open neighborhoods of e, then $\beta = \{Vr : V \in \mathcal{N} \text{ and } r \in G\}$ is a basis for the topology on G.*

Of course, rV simply stands for the elements of G that are products of r and an element of V. More generally, if A and B are subsets of G, $AB := \{ab : a \in A \text{ and } b \in B\}$. Thus A^2 is the set of products ab with both $a, b \in A$; it is not the set of squares of elements in A. Also, $A^{-1} := \{a^{-1} : a \in A\}$.

[2] In these notes, a neighborhood of a point is any set containing an open set containing the point.

Lemma 1.12. *Let V be a neighborhood of e in G. Then $V \subset \overline{V} \subset V^2$.*

Proof. Note that $s \in \overline{V}$ if and only if every neighborhood of s meets V. Since sV^{-1} is a neighborhood of s, it must meet V. Thus there is a $t \in V$ of the form sr^{-1} with $r \in V$. But then $s = tr \in V^2$, and $\overline{V} \subset V^2$. □

Lemma 1.13. *If G is a topological group, then G is Hausdorff and regular.*

Proof. Suppose that $F \subset G$ is closed and that $s \notin F$. Multiplying by s^{-1} allows us to assume that $s = e$. Since F is closed, $W := G \setminus F$ is an open neighborhood of e, and the continuity of multiplication implies there is an open neighborhood V of e such that $V^2 \subset W$. Thus $\overline{V} \subset W$ by Lemma 1.12. Then $U := G \setminus \overline{V}$ is an open neighborhood of F which is disjoint from V. This shows that G is regular. Since points in G are closed by assumption, G is also Hausdorff. □

Comments on Nets and Subnets

Before proceeding, a few comments about nets and subnets are in order. Anyone who is willing to restrict to second countable spaces—a perfectly reasonable decision—can ignore this discussion and continue to replace the word "net" with sequence, and "subnet" with subsequence.

A good reference for nets and subnets is Pedersen's *Analysis Now* [127, §1.3]. For now, recall that a *net* is formally a function x from a directed set I into a space X. A sequence is just a net based on $I = \mathbf{N}$. Just as with sequences, it is standard to write x_i in place of $x(i)$. Except for the fact that convergent nets need not be bounded, nets behave very much like sequences. However, subnets are a bit more subtle than subsequences.

Definition 1.14. A net $\{ y_j \}_{j \in J}$ is a *subnet* of a net $\{ x_i \}_{i \in I}$ if there is a function $N : J \to I$ such that

(a) $y = x \circ N$; that is $y_j = x_{N_j}$ for all $j \in J$, and
(b) for all $i_0 \in I$, there is $j_0 \in J$ such that $j \geq j_0$ implies that $N_j \geq i_0$.

We generally write $\{ x_{N_j} \}_{j \in J}$ in place of $\{ y_j \}_{j \in J}$ to suggest the similarity to a subsequence, and some authors like to replace N_j with i_j to further strengthen the analogy. Generally, in this text, we prefer to dispose of double subscripts if at all possible. When we pass to a subnet, we will usually have no more need of the original net and so we will relabel the subnet and replace the original net, say $\{ x_i \}$, with the subnet and not change the notation.

Proposition 1.15 ([54, Proposition II.13.2]). *Let $p : X \to Y$ be a surjection. Then p is an open map if and only if given a net $\{ y_i \}_{i \in I}$ converging to $p(x)$ in Y, there is a subnet $\{ y_{N_j} \}_{j \in J}$ and a net $\{ x_j \}_{j \in J}$ indexed by the same set which converges to x in X, and which also satisfies $p(x_j) = y_{N_j}$.*

Proof. Suppose that $p : X \to Y$ is an open map and that $\{ y_i \}_{i \in I}$ a net converging to $p(x)$ in Y. Let $\mathcal{N}(x)$ be the collection of neighborhoods of x in X and define

$$J = \{ (i, U) \in I \times \mathcal{N}(x) : U \cap p^{-1}(y_i) \neq \emptyset \}.$$

Define

$$(i, U) \le (i', U') \iff i \le i' \text{ and } U \supset U'.$$

Then given (i', U') and (i'', U'') in J, $p(U' \cap U'')$ is a neighborhood of $p(x)$ and we eventually have y_i in $p(U' \cap U'')$. Therefore we can find i_0 dominating both i' and i'' with $y_{i_0} \in p(U' \cap U'')$. Thus $(i_0, U' \cap U'') \in J$ and dominates both (i', U') and (i'', U''). Thus J is a directed set. Let $N : J \to I$ be the obvious map: $N_{(i,U)} = i$. Thus $\{ z_{(i,U)} \}_{(i,U) \in J} = \{ y_{N_{(i,U)}} \}_{(i,U) \in J}$ is easily seen to be a subnet of $\{ y_i \}_{i \in I}$. For each (i, U), let $x_{(i,U)}$ be an element of U such that $p(x_{(i,U)}) = y_i = y_{N_{(i,U)}}$. Now we clearly have $\{ x_j \}_{j \in J}$ converging to x, and by construction $p(x_j) = z_j$ for all $j \in J$. This proves the "only if" portion of the proposition.

Now assume that p is a surjection with the given lifting property. Suppose that U is an open set in X such that $p(U)$ is not open in Y. Then there is a net $\{ y_i \}_{i \in I}$ such that $y_i \to p(x) \in p(U)$ with $y_i \notin p(U)$ for all $i \in I$. By assumption, there is a subnet $\{ z_j \}_{j \in J}$ and a net $\{ x_j \}_{j \in J}$ converging to x such that $p(x_j) = z_j$. But x_j is eventually in U, and so z_j is eventually in $p(U)$. But the $z_j = y_{N_j}$ (for some $N : J \to I$), so this is nonsense. It follows that p is open. □

1.2 Preliminaries on Locally Compact Groups

Definition 1.16. A topological space is called *locally compact* if every point has a neighborhood basis consisting of compact sets.

If a space is Hausdorff — such as a topological group — then Definition 1.16 can be replaced with a simpler condition that makes no mention of neighborhood bases.

Lemma 1.17. *If X is Hausdorff, then X is locally compact if and only if every point in X has a compact neighborhood.*

Proof. Suppose that K is a compact neighborhood of x in X. We'll show that x has a neighborhood basis of compact sets. (This will suffice as the other implication is clear.) Thus we need to see that any neighborhood U of x contains a compact neighborhood of x. Let V be the interior of $U \cap K$. Then $\overline{V}$ is compact and Hausdorff, and therefore regular. Furthermore, $\overline{V} \setminus V$ is a closed subset of $\overline{V}$ not containing x. Thus there is an open set W in $\overline{V}$ such that $x \in W \subset \overline{W} \subset V$. Thus, W is open in X and $\overline{W}$ is a compact neighborhood of x with $\overline{W} \subset U$. □

Definition 1.18. A *locally compact group* is a topological group for which the underlying topology is locally compact.

Remark 1.19. Since topological groups are Hausdorff, Corollary 1.11 on page 3 implies a topological group G is locally compact if and only if there is a compact neighborhood of e. In fact, G is locally compact if and only if there is a nonempty open set with compact closure.

Now some basic examples.

Example 1.20. Any discrete group G is a locally compact group.

Example 1.21. Any elementary abelian group $G = \mathbf{R}^n \times \mathbf{T}^m \times \mathbf{Z}^d \times F$ is locally compact. More generally, any Lie group[3] is a locally compact group.

Example 1.22. Since U_n is (topologically) a closed subset of $\mathbf{C}^{n^2}$, U_n is a locally compact group. The set $\mathrm{GL}_n(\mathbf{R})$ of invertible $n \times n$ matrices may be viewed as an open subset of $\mathbf{R}^{n^2}$ and is easily seen to be a locally compact group. Similar considerations apply to other classical matrix groups. (Alternatively, these are all Lie groups, but it requires a bit more work to see this.)

Remark 1.23. In the previous example, we made use of the observation that open and closed subsets of a locally compact Hausdorff space are locally compact in the relative topology. It is not hard to see that the same is true in possibly non-Hausdorff locally compact spaces, and that the intersection of an open subset and a closed subset is locally compact. However, we will see shortly that not every subset of a locally compact Hausdorff space will be locally compact in the relative topology.

Definition 1.24. A subset Y of a space X is called *locally closed* if each point in Y has an open neighborhood P in X such that $P \cap Y$ is closed in P.

Lemma 1.25. *If X is a topological space and $Y \subset X$, then the following are equivalent.*

(a) *Y is locally closed in X.*

(b) *Y is open in $\overline{Y}$.*

(c) *$Y = C \cap O$ where C is closed in X and O is open in X.*

Proof. (a) $\Longrightarrow$ (b) Fix $y \in Y$ and P as in Definition 1.24. It will suffice to see that $\overline{Y} \cap P \subset Y$. But if $x \in \overline{Y} \cap P$, then there is a net $\{ x_n \} \subset Y$ such that $x_n \to x$. Since $x \in P$ and P is open, we eventually have $x_n \in P \cap Y$. Since $P \cap Y$ is closed in P, $x \in P \cap Y \subset Y$ as required.

(b) $\Longrightarrow$ (c) This is clear: by definition, there is an open subset O such that $Y = O \cap \overline{Y}$.

(c) $\Longrightarrow$ (a) Let $Y = C \cap O$ as above and fix $y \in Y$. Choose P open such that $y \in P \subset O$. (We could take $P = O$.) Now I claim $P \cap Y$ is closed in P. To see this, suppose that $\{ x_n \} \subset Y \cap P \subset C$ and $x_n \to x \in P$. Since C is closed, $x \in C$. On the other hand, $x \in P \subset O$. Thus $x \in Y \cap P$. □

Lemma 1.26. *Let Y be a subspace of a topological space X. If X is Hausdorff and Y is locally compact, then Y is locally closed in X. If X is a (possibly non-Hausdorff) locally compact space and if Y is locally closed in X, then Y is locally compact.*

Proof. The second assertion follows from Remark 1.23 and Lemma 1.25. To prove the first assertion, fix $y \in Y$. Let U be an open neighborhood of y in X such that $U \cap Y$ has compact closure B in Y. Since B is compact in Y, it is also compact in X, and therefore closed as X is Hausdorff. It suffices to see that $U \cap Y$ is closed

[3] A Lie group is a topological group for which underlying space is a manifold and the structure maps are smooth.

in U. Let $\{x_n\} \subset U \cap Y \subset B$ be such that $x_n \to x \in U$. Since B is closed, $x \in B \subset Y$. Therefore $x \in Y \cap U$ and Y is locally closed. □

Example 1.27 (The $ax + b$ group). Let $G = \{(a, b) \in \mathbf{R}^2 : a > 0\}$. Then G is a group with the operations $(a, b)(c, d) := (ac, ad + b)$ and $(a, b)^{-1} = (\frac{1}{a}, -\frac{b}{a})$. We call G the $ax + b$ group as it is easily identified with the affine transformations $x \mapsto ax + b$ of the real line.

Example 1.28. If $\dim \mathcal{H} = \infty$, then $U(\mathcal{H})$ is not locally compact. If it were, then some basic open neighborhood

$$V = \{U \in U(\mathcal{H}) : |Uh_i - h_i| < \epsilon \text{ for } i = 1, 2, \ldots, n\}$$

would have compact closure. But the span of $\{h_1, \ldots, h_n\}$ is certainly finite dimensional. Therefore there is an infinite orthonormal set $\{e_i\}_{i=1}^{\infty}$ contained in the orthogonal complement of the span of the h_i. Then we can find unitaries U_n such that $U_n h_i = h_i$ for all i and $U_n e_1 = e_n$. But $\{U_n e_1\}$ converges weakly to 0 in $\mathcal{H}$, as does any subnet. Consequently, $\{U_n\} \subset V$ has no subnet converging to a unitary U in the strong operator topology. Thus, $\overline{V}$ can't be compact.

If A is infinite dimensional, then $\operatorname{Aut} A$ is another example of a topological group which is rarely locally compact, but is nonetheless important to our purpose here. We now take a slight detour to give a description of $\operatorname{Aut} A$ in the case A is abelian and therefore isomorphic to $C_0(X)$ for some locally compact Hausdorff space X.

Definition 1.29. Suppose that X and Y are topological spaces and that $C(X, Y)$ is the collection of continuous functions from X to Y. Then the *compact-open topology* on $C(X, Y)$ is the topology with subbasis[4] consisting of the sets

$$\mathcal{U}(K, V) := \{f \in C(X, Y) : f(K) \subset V\}$$

for each compact set $K \subset X$ and open subset $V \subset Y$.

The compact-open topology[5] is usually as nice as Y is; for example, $C(X, Y)$ will be Hausdorff in the compact-open topology if Y is. If X is discrete, then the compact-open topology is the topology of pointwise convergence, and $C(X, Y)$ with the compact-open topology is homeomorphic to the product Y^X with the product topology. If $Y = \mathbf{C}$, then the compact-open topology on $C(X) := C(X, \mathbf{C})$ is the topology of uniform convergence on compacta [168, Theorem 43.7]. The following lemma will prove useful in understanding the compact-open topology in the examples we have in mind.

Lemma 1.30. *Let X and Y be locally compact Hausdorff spaces and give $C(X, Y)$ the compact-open topology. Then $f_n \to f$ in $C(X, Y)$ if and only if whenever $x_n \to x$ in X we also have $f_n(x_n) \to f(x)$ in Y.*

[4] A subbasis for a topology on a set X is any collection of subsets of X. The associated topology is the collection of all unions of finite intersections of elements of the subbasis. The whole space arises as the union over the empty set. Thus the collection of all finite intersections of elements of the subbasis is a basis for the associated topology which may also be viewed as the smallest topology on X containing the subbasis.

[5] The compact-open topology is treated in detail in many topology books. Section 43 of Willard's book [168] is a nice reference.

Proof. Suppose that $f_n \to f$ and $x_n \to x$. Let V be an open neighborhood of $f(x)$ and let K be a compact neighborhood of x such that $f(K) \subset V$. Then $f \in \mathcal{U}(K,V)$. Thus we eventually have both $f_n \in \mathcal{U}(K,V)$ and $x_n \in K$. Therefore we eventually have $f_n(x_n) \in V$. Since V was arbitrary, this proves that $f_n(x_n) \to f(x)$.

Now assume that $x_n \to x$ implies $f_n(x_n) \to f(x)$. To show that $f_n \to f$, it is enough to show that f_n is eventually in any $\mathcal{U}(K,V)$ that contains f. Thus if $f_n \not\to f$, then we can pass to a subnet and assume that there is a $K \subset X$ and $V \subset Y$ such that $f \in \mathcal{U}(K,V)$ and $f_n \notin \mathcal{U}(K,V)$ for all n. Therefore there are $x_n \in K$ such that $f_n(x_n) \notin V$. We can then pass to another subnet and assume that $x_n \to x \in K$. By assumption $f_n(x_n) \to f(x)$. Since $X \setminus V$ is closed, we must have $f(x) \notin V$. This is a contradiction. □

Definition 1.31. Let X be a locally compact Hausdorff space. We give $\operatorname{Homeo}(X)$ the topology with subbasis consisting of the sets

$$\mathcal{U}(K,K',V,V') := \{\, h \in \operatorname{Homeo}(X) : h(K) \subset V \text{ and } h^{-1}(K') \subset V' \,\}$$

with K and K' compact in X and V and V' open in X.

Remark 1.32. Notice that $h_n \to h$ in $\operatorname{Homeo}(X)$ if and only if *both* $h_n \to h$ and $h_n^{-1} \to h^{-1}$ in the compact-open topology. In particular, $h_n \to h$ in $\operatorname{Homeo}(X)$ if and only if both $h_n(x_n) \to h(x)$ and $h_n^{-1}(x_n) \to h^{-1}(x)$ whenever $x_n \to x$.

Now it is clear that $\operatorname{Homeo}(X)$ is a topological group under composition; our subbasis has been constructed so that $h_n \to h$ clearly implies $h_n^{-1} \to h^{-1}$. If X is locally compact Hausdorff and locally connected, then the topology on $\operatorname{Homeo}(X)$ reduces to the compact open topology [12, Chap. X.3, Exercise 17a]. In general it is necessary to add the condition on the inverse (cf. [12, Chap. X.3, Exercise 17b][6] or [23]).

Now we have the technology to consider the special case of Example 1.8 on page 3 where $A = C_0(X)$. This example will be of interest later.

Lemma 1.33. *Let $A = C_0(X)$. If $\alpha \in \operatorname{Aut} C_0(X)$, then there is a $h \in \operatorname{Homeo}(X)$ such that $\alpha(f)(x) = f\big(h(x)\big)$ for all $f \in C_0(X)$. The map $\alpha \mapsto h$ is a homeomorphism of $\operatorname{Aut} C_0(X)$ with $\operatorname{Homeo}(X)$.*

Proof. The theory of commutative Banach algebras implies that $x \mapsto \mathrm{ev}_x$ is a homeomorphism of X onto the set Δ of nonzero complex homomorphisms from $C_0(X)$ to $\mathbf{C}$ when Δ is given the relative topology as a subset of $C_0(X)^*$ equipped

---

[6] Let $X = \{\, x \in \mathbf{R} : x = 0 \text{ or } x = 2^n \text{ for } n \in \mathbf{Z} \,\}$. Then X is closed in $\mathbf{R}$ and therefore locally compact. For each $n \in \mathbf{Z}$, let h_n be defined by

$$h_n(x) = \begin{cases} 0 & \text{if } x = 0, \\ 2^{(k-1)n} & \text{if } x = 2^{kn} \text{ with } k \in \mathbf{Z} \setminus \{0,1\}, \\ 2^{-n} & \text{if } x = 2^n, \\ 1 & \text{if } x = 1, \text{ and} \\ x & \text{otherwise.} \end{cases}$$

Then $h_n^{-1} = h_{-n}$, each $h_n \in \operatorname{Homeo}(X)$ and $h_n \to \mathrm{id}$ as $n \to \infty$. However since $h_{-n}(2^{-n}) = 2^n$, $h_n^{-1} = h_{-n} \not\to \mathrm{id}$ as $n \to \infty$.

with the weak-$*$ topology. (See [127, 4.2.5] or [139, Example A.23].) The transpose of α is a continuous map $\alpha^* : C_0(X)^* \to C_0(X)^*$ defined by $\alpha^*(\varphi) := \varphi \circ \alpha$. Since the restriction of α^* to Δ is clearly a homeomorphism, we get a commutative diagram of homeomorphisms

$$\begin{array}{ccc} \Delta & \xrightarrow{\alpha^*} & \Delta \\ {\scriptstyle \mathrm{ev}}\uparrow & & \uparrow{\scriptstyle \mathrm{ev}} \\ X & \overset{h}{\dashrightarrow} & X \end{array}$$

where h is given by $\mathrm{ev}_{h(x)} = \alpha^* \circ \mathrm{ev}_x$. In particular,

$$\alpha(f)(x) = \mathrm{ev}_x \circ \alpha(f) = \alpha^* \circ \mathrm{ev}_x(f) = \mathrm{ev}_{h(x)}(f) = f\big(h(x)\big).$$

We still have to check that $\alpha \mapsto h$ is a homeomorphism. Let $\alpha_n \in \operatorname{Aut} C_0(X)$ be given by $\alpha_n(f) = f \circ h_n$. Suppose first that $\alpha_n \to \alpha$. If, contrary to what we wish to show, $x_n \to x$ and $h_n(x_n) \not\to h(x)$, then there must be a $f \in C_0(X)$ such that $f\big(h_n(x_n)\big) \not\to f\big(h(x)\big)$. But then $\alpha_n(f) \not\to \alpha(f)$ in $C_0(X)$. This is a contradiction and we must have $h_n \to h$ in the compact-open topology. Since $\alpha_n^{-1} \to \alpha^{-1}$, we also have $h_n^{-1} \to h^{-1}$ in the compact-open topology. Thus $h_n \to h$ in $\mathrm{Homeo}(X)$ by Remark 1.32 on the preceding page.

Now assume that $h_n \to h$ in $\mathrm{Homeo}(X)$. We need to show that given $f \in C_0(X)$, $f \circ h_n \to f \circ h$ in $C_0(X)$. If not, then we can pass to a subnet so that there are $x_n \in X$ such that

$$\big|f\big(h_n(x_n)\big) - f\big(h(x_n)\big)\big| \geq \epsilon_0 > 0 \quad \text{for all } n. \tag{1.1}$$

Now we can pass to a subnet and assume that *either* $\{\, h_n(x_n) \,\}$ or $\{\, h(x_n) \,\}$ belongs to the compact set $K := \{\, x \in X : |f(x)| \geq \frac{\epsilon_0}{2} \,\}$. Suppose first that that $\{\, h_n(x_n) \,\} \subset K$. Then we can pass to a subnet and assume that $h_n(x_n) \to y \in K$. Since $h_n^{-1} \to h^{-1}$ in the compact-open topology, Lemma 1.30 on page 7 implies that $x_n \to h^{-1}(y)$. But the continuity of h implies $h(x_n) \to y$. Now the continuity of f and (1.1) lead to a contradiction. On the other hand, if $\{\, h(x_n) \,\} \subset K$, then we can assume $h(x_n) \to y \in K$. Thus $x_n \to h^{-1}(y)$. Since $h_n \to h$ in the compact-open topology, we also have $h_n(x_n) \to y$. Thus we get a contradiction as above. □

1.2.1 Subgroups of Locally Compact Groups

If G is a topological group, then any subgroup H of G is a topological group in the relative topology. However, if G is locally compact, then H can be locally compact if and only if it is locally closed (Lemma 1.26 on page 6).

Lemma 1.34. *A locally closed subgroup of a topological group is closed.*

Proof. Let H be a locally closed subgroup of a topological group G. Therefore, there is an open neighborhood W of e in G such that $W \cap H$ is closed in W (Definition 1.24 on page 6). Let U and V be neighborhoods of e in G such that

$$V^2 \subset U \subset U^2 \subset W.$$

Now fix $x \in \overline{H}$. Let $\{ x_n \}$ be a net in H converging to x. Notice that $x^{-1} \in \overline{H}$ as well. Since $x^{-1}V$ is a neighborhood of x^{-1}, it must meet H. Let $y \in x^{-1}V \cap H$. Now x_n is eventually in Vx and $x_n y$ is eventually in

$$Vx(x^{-1}V \cap H) \subset V^2 \cap H \subset U \cap H \subset W \cap H.$$

It follows that $xy \in \overline{U} \subset U^2 \subset W$. Since $W \cap H$ is closed in W, we have $xy \in W \cap H$. But then $x = (xy)y^{-1} \in H$. □

Corollary 1.35. *If H is a subgroup of a topological group G and if* either

(a) *H is open, or*

(b) *H is discrete, or*

(c) *H is locally compact,*

then H is closed.

Proof. In all three cases, H is locally closed and therefore closed: in (a) this is trivial and in (c), this follows from Lemma 1.26 on page 6. In (b), there must be an open set W such that $W \cap H = \{ e \}$. Now use Definition 1.24 on page 6. □

Remark 1.36. That open subgroups are closed also follows from Lemma 1.12 on page 4: $H \subset \overline{H} \subset H^2 = H$.

Definition 1.37. A locally compact Hausdorff space X is called *σ-compact* if $X = \bigcup_{i=1}^{\infty} C_n$ where each C_n is a compact subset of X.

Note that σ-compactness is a relatively weak condition. Every second countable locally compact space is σ-compact.[7] Note that any uncountable discrete set fails to be σ-compact; more generally, the topological disjoint union of uncountably many locally compact spaces will fail to be σ-compact.

Lemma 1.38. *Every locally compact group G has an* open *σ-compact subgroup H (which could be all of G). In particular, every locally compact group is the topological disjoint union of σ-compact spaces.*

Proof. Let V be a symmetric open neighborhood of e in G with compact closure.[8] Let $H := \bigcup_{n=1}^{\infty} V^n$. Then H is an open subgroup of G. Since Lemma 1.12 on page 4 implies that $\overline{V}^n \subset V^{2n}$, $H = \bigcup \overline{V}^n$ is σ-compact. Therefore, for each $s \in G$, sH is a clopen σ-compact subset of G, and G is the disjoint union $\bigcup_{s \in G} sH$. □

Corollary 1.39. *Every connected locally compact group is σ-compact.*

---

[7] To see this, let $\{ O_n \}$ be a countable basis for the topology. Since X is locally compact, the subset consisting of those O_n which have compact closure is still a basis.

[8] That is, $V = V^{-1}$. Symmetric neighborhoods of e are easy to find. If W is any neighborhood, let $V = W \cap W^{-1}$.

1.2.2 Remarks on Separability Hypotheses

In a monograph such as this, it would not be profitable to devote enormous amounts of time and space to issues particular to locally compact spaces and groups which are not second-countable or even σ-compact. Nevertheless, we don't want to invoke blanket assumptions when much of the general theory can be presented relatively painlessly without such assumptions. One price that must be paid for this is that we must work with nets; even a compact group — such as an uncountable product — can fail to have a countable neighborhood basis at the identity. However, with some exceptions, nets can be treated like sequences and can certainly be replaced by sequences if the group or space in question is assumed to have a second countable topology.

Definition 1.40. A Hausdorff topological space X is *paracompact* if every open cover of X has a locally finite refinement. (Recall that a cover is locally finite if each point in X has a neighborhood that meets only finitely many elements of the cover.)

A σ-compact locally compact Hausdorff space is always paracompact [127, Proposition 1.7.11]. Thus it follows from Lemma 1.38 on the preceding page that all locally compact *groups* are paracompact and are therefore normal as topological spaces [168, Theorem 20.10]. Thus we can use Urysohn's Lemma [127, Theorem 1.5.6] and the Tietze Extension Theorem [127, Theorem 1.5.8] at will.

The situation for locally compact spaces is less clean. Although a compact Hausdorff space is always normal, a locally compact Hausdorff space need not be. However there are suitable versions of Urysohn and Tietze available provided we work with compact sets and compactly supported functions.

Lemma 1.41 (Urysohn's Lemma). *Suppose that X is a locally compact Hausdorff space and that V is a an open neighborhood of a compact set K in X. Then there is a $f \in C_c(X)$ such that $0 \le f(x) \le 1$ for all x, $f(x) = 1$ for all $x \in K$ and $f(x) = 0$ if $x \notin V$.*

Proof. Since X is locally compact, there is an open neighborhood W of K with compact closure $\overline{W} \subset V$. Since $\overline{W}$ is compact and therefore normal, we can find $h \in C(\overline{W})$ such that $0 \le h(x) \le 1$ for all x, $h(x) = 1$ for all $x \in K$ and $h(x) = 0$ if $x \notin W$. Now define

$$f(x) := \begin{cases} h(x) & \text{if } x \in \overline{W}, \text{ and} \\ 0 & \text{if } x \notin W. \end{cases}$$ □

Lemma 1.42 (Tietze Extension Theorem). *Suppose that X is a locally compact Hausdorff space and that $K \subset X$ is compact. If $g \in C(K)$, then there is a $f \in C_c(X)$ such that $f(x) = g(x)$ for all $x \in K$.*

Proof. Let W be an open neighborhood of K with compact closure. By the usual Tietze Extension Theorem for normal spaces, there is a $k \in C(\overline{W})$ extending g. By Lemma 1.41, there is a $h \in C_c(X)$ such that $h(x) = 1$ for all $x \in K$ and $h(x) = 0$

if $x \notin W$. Now let

$$f(x) := \begin{cases} k(x)h(x) & \text{if } x \in \overline{W}, \text{ and} \\ 0 & \text{if } x \notin W. \end{cases}$$ □

Lemma 1.43 (Partitions of Unity). *Suppose that X is a locally compact space and that $\{U_i\}_{i=1}^n$ is a cover of a compact set $K \subset X$ by open sets with compact closures. Then for $i = 1, \dots, n$ there are $\varphi_i \in C_c(X)$ such that*

(a) $0 \leq \varphi_i(x) \leq 1$ *for all* $x \in X$,

(b) $\operatorname{supp} \varphi_i \subset U_i$,

(c) *if* $x \in K$, *then* $\sum_{i=1}^n \varphi_i(x) = 1$, *and*

(d) *if* $x \notin K$, *then* $\sum_{i=1}^n \varphi_i(x) \leq 1$.

Proof. Let $C = \bigcup_{i=1}^n \overline{U_i}$. Then C is a compact neighborhood of K. Since C is compact, there is a partition of unity $\{\varphi_i\}_{i=0}^n \subset C(C)$ subordinate to the cover $\{C \smallsetminus K, U_1, \dots, U_n\}$ [139, Lemma 4.34]. Since $\operatorname{supp} \varphi_i \subset U_i$ $(i \geq 1)$ and continuous on C, we may view φ_i as a continuous compactly supported function on all of X by setting it equal to zero off U_i. Now the φ_i satisfy the requirements of the lemma. □

1.2.3 Homogeneous Spaces

Suppose that H is a subgroup of a topological group G. The set of left cosets, G/H, inherits a topology — called the *quotient topology* — from G which is the smallest topology making the *quotient map* $q : G \to G/H$ continuous. Since

$$\tau := \{U \subset G/H : q^{-1}(U) \text{ is open in } G\}$$

is a topology on G/H, we have $U \subset G/H$ open in the quotient topology if and only if $q^{-1}(U)$ is open in G.

Lemma 1.44. *If H is a subgroup of a topological group G, then the quotient map $q : G \to G/H$ is open and continuous.*

Proof. Since q is continuous by definition of the topology on G/H, it suffices to see that it is open as well. If V is open in G, then

$$q^{-1}(q(V)) = \bigcup_{h \in H} Vh.$$

Since Vh is open in G for all h, $q^{-1}(q(V))$ is open in G. This shows that $q(V)$ is open in G/H. Hence q is open as claimed. □

Remark 1.45 (Homogeneous Spaces). A topological space X is called *homogeneous* if for all $x, y \in X$, there is a homeomorphism $h \in \operatorname{Homeo}(X)$ such that $h(x) = y$. If H is a subgroup of G and if $s \in G$, then $\varphi_s(rH) := srH$ defines a bijection of G/H onto itself with inverse $\varphi_{s^{-1}}$. If V is open in G, then $U := \{rH : r \in V\}$ is a typical open set in G/H. Furthermore, $\varphi_s(U) = \{srH : r \in V\} = \{aH : a \in sV\}$ is open in G/H. Thus φ_s is an open map. Since $\varphi_s^{-1} = \varphi_{s^{-1}}$ is also open, φ_s is a homeomorphism. It follows that G/H is a homogeneous space.

Lemma 1.46. *Suppose that H is a subgroup of a topological group G and that $\{ s_i H \}_{i \in I}$ is a net in G/H converging to sH. Then there is a subnet $\{ s_{i_j} \}_{j \in J}$ and $h_{i_j} \in H$ such that $s_{i_j} h_{i_j} \to s$.*

Remark 1.47. When applying this result, we are often only interested in the subnet. Then, provided we're careful to keep in mind that we've passed to a subnet, we can relabel everything, and assume that there are $h_i \in H$ such that $s_i h_i \to s$. This allows us to dispense with multiple subscripts.

Proof. Since $q : G \to G/H$ is open, Proposition 1.15 on page 4 implies that there is a subnet $\{ s_{i_j} \}_{j \in J}$ and a net $\{ r_j \}_{j \in J}$ such that $r_j \to s$ and such that $r_j H = s_{i_j} H$. But then there are h_{i_j} such that $r_j = s_{i_j} h_{i_j}$. □

Proposition 1.48. *Let H be a subgroup of a topological group G. The left coset space G/H equipped with the quotient topology is Hausdorff if and only if H is closed in G. If G is locally compact, then G/H is locally compact. If G is second countable, then G/H is second countable.*

Proof. Suppose that H is closed, we'll show that G/H is Hausdorff. For this, it suffices to see that convergent nets have unique limits. So suppose $\{ s_n H \}$ converges to both sH and rH. Since the quotient map $q : G \to G/H$ is open, we can pass to a subnet, relabel, and assume that $s_n \to s$ (see Remark 1.47). Passing to a further subnet, and relabeling, we can also assume that there are $h_n \in H$ such that $s_n h_n \to r$. Since we also have $s_n^{-1} \to s^{-1}$,

$$h_n = s_n^{-1}(s_n h_n) \to s^{-1} r.$$

Since H is closed, $s^{-1} r \in H$, and $sH = rH$. Therefore G/H is Hausdorff as claimed.

On the other hand, if H is not closed, then there is a net $h_n \to s$ with $s \in G \setminus H$. Since q is continuous, it follows that $sH \in \overline{\{ eH \}}$. Thus G/H is certainly not Hausdorff.

Suppose that $\{ V_n \}$ is a countable basis for the topology on G. Let $U_n := q(V_n)$. Let W be an open set in G/H and let $sH \in W$. Since the V_n are a basis, there is a n such that $s \in V_n \subset q^{-1}(W)$. Then $sH \in U_n \subset W$, and the U_n are a basis for the topology on G/H.

Suppose that G is locally compact. Since G/H may not be Hausdorff, we have to verify that Definition 1.16 on page 5 holds. However, the above argument shows that the forward image of any basis in G is a basis for G/H. Since G has a basis of compact sets, and since the continuous image of a compact set is compact, it follows that G/H has a basis of compact sets. That is, G/H is locally compact. □

Corollary 1.49. *If H is a normal closed subgroup of a locally compact group G, then G/H is a locally compact group with respect to the usual multiplication of cosets: $(sH)(rH) = srH$.*

Proof. We only have to see that multiplication and inversion are continuous. Suppose $s_n H \to sH$ and $r_n H \to rH$. Since it suffices to see that every subnet of

$\{ s_n r_n H \}$ has a subnet converging to srH, we can pass to subnets at will. In particular, we can pass to a subnet, relabel, and assume that there are $h_n \in H$ such that $s_n h_n \to s$ in G. Similarly, we can pass to another subnet, relabel again, and assume that there are $h'_n \in H$ such that $r_n h'_n \to r$ in G. But then $s_n h_n r_n h'_n \to sr$ and the continuity of the quotient map q implies that

$$s_n h_n r_n h'_n H = s_n r_n H \to srH.$$

Similarly, $s_n^{-1} H \to s^{-1} H$, and G/H is a topological group. □

1.2.4 Group Extensions

Theorem 1.50. *Suppose that H is a closed subgroup of a topological group G. Then G is locally compact if and only if both H and G/H are locally compact.*

Proof. Suppose that G is locally compact. Since any closed subset of G is locally compact, H is locally compact, and G/H is locally compact by Proposition 1.48 on the previous page.

Now assume that H and G/H are locally compact. Since G is Hausdorff, to show that G is locally compact, it will suffice to produce a compact neighborhood of e in G. Let U_1 be a neighborhood of e in G such that $U_1 \cap H$ has compact closure in H. Since $e^2 = e$, we can find a neighborhood U of e in G such that $U^2 \subset U_1$. Thus $\overline{U} \subset U_1$ (Lemma 1.12 on page 4), and $\overline{U} \cap H$ is a compact neighborhood of e in H. Choose an open neighborhood V of e in G such that $V^4 \subset U$ so that we have $(\overline{V})^2 \subset \overline{U}$. Since $\pi(V)$ is a neighborhood of eH in G/H and since G/H is locally compact, we can find a compact neighborhood C of $\pi(e) = eH$ in G/H with $C \subset \pi(V)$.

Let $K = \pi^{-1}(C) \cap \overline{V}$. Then K is a closed neighborhood of e in G and it will suffice to see that K is compact. Since K is closed, it will suffice to see that any net $\{ s_\alpha \}$ in K has a convergent subnet (in G). Since $\{ \pi(s_\alpha) \} \subset C$ and since $C \subset \pi(V)$ we can pass to subnet, relabel, and assume that $\pi(s_\alpha) \to \pi(s)$ for some $s \in V$. Using Lemma 1.46 on the previous page, we can pass to another subnet, relabel, and assume that there are $t_\alpha \in H$ such that $s_\alpha t_\alpha \to s$ in G. Since V is open and $s \in V$, we eventually have $s_\alpha t_\alpha \in V$. But then

$$t_\alpha \in s_\alpha^{-1}(V \cap H) \subset (\overline{V})^2 \cap H \subset \overline{U} \cap H.$$

Since $\overline{U} \cap H$ is compact, we can pass to a subnet, relabel, and assume that $t_\alpha \to t$ in H. But then

$$s_\alpha = (s_\alpha t_\alpha) t_\alpha^{-1} \to st^{-1}.$$ □

If N, G and E are groups, then E is called an extension of N by G if there is a short exact sequence of groups

$$e \longrightarrow N \xrightarrow{i} E \xrightarrow{j} G \longrightarrow e\,.$$

This allows us to *algebraically* identify N with a normal subgroup of E, and to *algebraically* identify the quotient group E/N with G. If N, G and E are topological groups, then we certainly are going to want i and j to be continuous. This forces $i(N)$ to be closed in E and the natural map $\bar{j} : E/i(N) \to G$ to be a continuous bijection. However a continuous bijection between locally compact groups need not be a homeomorphism. For example, consider the identity map from $\mathbf{R}_d$ — the real line with the discrete topology — to $\mathbf{R}$ with the usual topology. Therefore we make the following definition.

Definition 1.51. If N, G and E are topological groups, then an algebraic short exact sequence

$$e \longrightarrow N \xrightarrow{\ i\ } E \xrightarrow{\ j\ } G \longrightarrow e \tag{1.2}$$

of groups is called a *short exact sequence of topological groups* if i is a homeomorphism onto its range and j is a continuous open surjection.

Note that if (1.2) is a short exact sequence of topological groups, then we can identify N with $i(N)$ and E/N with G as topological groups. In particular, we get as an immediate corollary to Theorem 1.50 on the facing page the following.

Corollary 1.52. *If*

$$e \longrightarrow N \xrightarrow{\ i\ } E \xrightarrow{\ j\ } G \longrightarrow e$$

is a short exact sequence of topological groups, then E is locally compact if and only if N and G are.

The definition of a short exact sequence of topological groups is certainly what we want, but it would seem that a lot of checking is required to show that a particular algebraic short exact sequence is topological. But in the presence of some separability, it suffices merely to check that the connecting maps are continuous. To prove this requires some fairly hard, but very important, results showing that Borel homomorphisms are often necessarily continuous. These "automatic continuity results" are proved in the appendices.

Proposition 1.53. *Suppose that N, G and E are second countable locally compact groups and that*

$$e \longrightarrow N \xrightarrow{\ i\ } E \xrightarrow{\ j\ } G \longrightarrow e \tag{1.3}$$

is an algebraic *short exact sequence with i and j continuous. Then* (1.3) *is a short exact sequence of topological groups.*

Proof. We have to show that i is a homeomorphism onto its range and that j is open. Recall that a second countable locally compact group is Polish and therefore analytic as a Borel space (Lemma D.9 on page 372). Since $i(N)$ is closed in E, it is Polish, and the generalized Souslin Theorem ([2, Corollary 2 of Theorem 3.3.4]) implies that $i : N \to i(N)$ is a Borel isomorphism. Thus i^{-1} is a Borel homomorphism, and therefore continuous (Theorem D.3 on page 370). Thus i is a homeomorphism as required.

On the other hand, $E/i(N)$ is locally compact and therefore an analytic group. Just as above $\bar{j}$ is a Borel isomorphism and $\bar{j}^{-1}$ is a continuous homomorphism. Since $q : E \to E/i(N)$ and $\bar{j}$ are open, so is j. □

Remark 1.54. With just a bit more work (using [2, Chap. 3] or Theorem D.11 on page 372), we can weaken the hypotheses in Proposition 1.53 on the preceding page to allow N, G and E to be Polish, and the connecting maps to be Borel.

1.3 Haar Measure

We concentrate on locally compact groups because they have a uniquely defined measure class which respects the group structure. First some terminology. A measure μ on a locally compact space G is called a *Borel measure* if each open set is measurable. If for each open set $V \subset G$,

$$\mu(V) = \sup\{\, \mu(C) : C \subset V \text{ and } C \text{ is compact}\,\},$$

and for each measurable set A,

$$\mu(A) = \inf\{\, \mu(V) : A \subset V \text{ and } V \text{ is open}\,\},$$

then μ is called *Radon measure*.[9] If G is a group, then we say that μ is left-invariant if

$$\mu(sA) = \mu(A) \quad \text{for all } s \in G \text{ and } A \text{ measurable.}$$

Of course, if $\mu(As) = \mu(A)$ we say that μ is right-invariant. If μ is both left- and right-invariant, we say it is bi-invariant.

Definition 1.55. A left-invariant Radon measure on a locally compact group G is called a left *Haar measure*. A right-invariant Radon measure is called a right Haar measure. The term Haar measure is reserved for left-invariant measures.

Remark 1.56. Our preference for left over right Haar measures is strictly convention. If μ is a left Haar measure, then $\nu(E) := \mu(E^{-1})$ is a right Haar measure. We'll have more to say about the relationship between left and right Haar measures in Lemma 1.67 on page 20.

The fundamental result is the following.

Theorem 1.57. *Every locally compact group G has a Haar measure which is unique up to a strictly positive scalar.*

The proof of Theorem 1.57 need not concern us here. (I suggest [56, Theorems 2.10 and 2.20] as a succinct and handy reference.) What we need to keep in mind is that obtaining a Haar measure is equivalent to constructing a linear functional

$$I : C_c(G) \to \mathbf{C} \tag{1.4}$$

which is *positive* in that $I(f) \geq 0$ if $f(s) \geq 0$ for all $s \in G$, and which satisfies

$$I(\lambda(r)f) = I(f) \quad \text{for all } r \in G \text{ and } f \in C_c(G), \tag{1.5}$$

[9] For more on Radon measures, see the beginning of Appendix B.1. For the moment, be aware that the precise formulation of the regularity conditions (given by the supremum and infimum conditions) varies a bit from reference to reference (if the group is not second countable).

where

$$\lambda(r)f(s) := f(r^{-1}s).$$

Then the Riesz Representation Theorem [156, Theorem 2.14] implies that (1.4) gives us a Radon measure μ such that

$$I(f) = \int_G f(s)\,d\mu(s), \tag{1.6}$$

and (1.5) implies that

$$\int_G f(r^{-1}s)\,d\mu(s) = \int_G f(s)\,d\mu(s),$$

which is equivalent to μ being left-invariant.[10] A positive linear functional (1.4) satisfying (1.5) is called a *Haar functional* on G.

It is not hard to see that every Haar measure assigns finite measure to each compact set, and [56, Proposition 2.19] implies that a Haar measure assigns strictly positive measure to each nonempty open set.

These elementary properties guarantee that

$$\|f\|_1 := \int_G |f(s)|\,d\mu(s)$$

defines a *norm* on $C_c(G)$. The completion with respect to this norm is $L^1(G)$. I'll generally try to avoid working with $L^1(G)$ and measure theory if at all possible. Even so, from time to time, we'll refer to $\|\cdot\|_1$ as the L^1-norm.

In practice, the Haar measure on a given group G is usually the natural choice. This is especially true for the sort of groups we work with in dealing with crossed products.

Example 1.58. If G is discrete, then counting measure is a Haar measure on G. Thus if $G = \mathbf{Z}$, then

$$\int_{\mathbf{Z}} f(n)\,d\mu(n) = \sum_{n\in\mathbf{Z}} f(n).$$

Example 1.59. If G is $\mathbf{R}^n$ or $\mathbf{T}^m$, then Haar measure is Lebesgue measure.

Note that the measures in Examples 1.58 and 1.59 are bi-invariant. This need not always be true. Folland gives a number of additional examples [56, p. 41], and several of these are not bi-invariant. In fact, Haar measure on the $ax + b$ group is not bi-invariant, and we look at this measure in the next example.

Example 1.60. Let $G = \{\,(a, b) \in \mathbf{R}^2 : a > 0\,\}$ be the $ax + b$ group described in Example 1.27 on page 7. Then Haar measure on G is given by the functional $I : C_c(G) \to \mathbf{C}$ defined by

$$I(f) := \int_{-\infty}^{\infty}\int_0^{\infty} f(x, y)\frac{1}{x^2}\,dx\,dy. \tag{1.7}$$

[10] Once we know (1.5) holds for all $f \in C_c(G)$, it then holds for all L^1 functions. Since left-invariance is just (1.5) for characteristic functions, we're in business — at least for all compact sets. Then regularity takes care of the rest.

Since I is clearly a positive linear functional, to verify the above assertion, we only have to check left-invariance. But

$$\begin{aligned}
I\big(\lambda(a,b)f\big) &= \int_{-\infty}^{\infty}\int_{0}^{\infty} f\big((a,b)^{-1}(x,y)\big)\frac{1}{x^2}\,dx\,dy. \\
&= \int_{-\infty}^{\infty}\int_{0}^{\infty} f\Big(\frac{x}{a},\frac{y}{a}-\frac{b}{a}\Big)\Big(\frac{a}{x}\Big)^2\frac{dx}{a}\frac{dy}{a}. \\
&= \int_{-\infty}^{\infty}\int_{0}^{\infty} f\Big(u,\frac{y}{a}-\frac{b}{a}\Big)\frac{1}{u^2}\,du\,\frac{dy}{a} \\
&= \int_{-\infty}^{\infty}\int_{0}^{\infty} f\Big(u,v-\frac{b}{a}\Big)\frac{1}{u^2}\,du\,dv \\
&= I(f).
\end{aligned}$$

Thus I gives us a Haar measure on G. Note that if we have a measurable rectangle

$$R = \{\,(x,y)\in G : a<x<b \text{ and } c<y<d\,\},$$

then

$$\mu(R) = \int_c^d\int_a^b \frac{1}{x^2}\,dx\,dy = \Big(\frac{1}{a}-\frac{1}{b}\Big)(d-c).$$

To see that Haar measure on the $ax+b$ group is not right-invariant, notice that a Haar measure μ is right-invariant exactly when

$$\int_G f(sr)\,d\mu(s) = \int_G f(s)\,d\mu(s) \quad \text{for all } r\in G \text{ and } f\in C_c(G). \tag{1.8}$$

Alternatively, if we define $\rho(r)f(s) := f(sr)$, then a Haar functional I yields a bi-invariant measure when $I\big(\rho(r)f\big) = I(f)$ for all r and f.

Thus to see that the Haar measure on the $ax+b$ group is not bi-invariant we compute

$$\begin{aligned}
I\big(\rho(a,b)f\big) &= \int_{-\infty}^{\infty}\int_{0}^{\infty} f(ax,bx+y)\frac{1}{x^2}\,dx\,dy \\
&= \int_{-\infty}^{\infty}\int_{0}^{\infty} f(u,\frac{b}{a}u+y)\frac{1}{u^2}\,du\,a dy \\
&= a\int_{0}^{\infty}\int_{-\infty}^{\infty} f(u,y+\frac{b}{a}u)\frac{1}{u^2}\,dy\,du \\
&= aI(f).
\end{aligned}$$

Since (1.8) holds only for $r=(a,b)$ if $a=1$, it follows that Haar measure on the $ax+b$ group is not right-invariant.

Lemma 1.61. *Let μ be a Haar measure on a locally compact group G. Then there is a continuous homomorphism $\Delta : G\to \mathbf{R}^+$ such that*

$$\Delta(r)\int_G f(sr)\,d\mu(s) = \int_G f(s)\,d\mu(s) \tag{1.9}$$

for all $f \in C_c(G)$*. The function* Δ *is independent of choice of Haar measure and is called the* modular function *on* G*.*

Proof of all but continuity assertion: Let $J_r : C_c(G) \to \mathbf{C}$ be defined by

$$J_r(f) := \int_G f(sr)\,d\mu(s).$$

Then a straightforward computation shows that $J_r\big(\lambda(t)f\big) = J_r(f)$ for all f and all $t \in G$. The uniqueness of Haar measure then implies that there is a strictly positive scalar $\Delta(r)$ such that (1.9) holds. Furthermore, on the one hand

$$\Delta(tr)\int_G f(str)\,d\mu(s) = \int_G f(s)\,d\mu(s).$$

On the other hand,

$$\begin{aligned}\Delta(t)\Delta(r)\int_G f(str)\,d\mu(s) &= \Delta(r)\int_G f(sr)\,d\mu(s)\\ &= \int_G f(s)\,d\mu(s).\end{aligned}$$

Since we can certainly find a f with nonzero integral, it follows that $\Delta(tr) = \Delta(t)\Delta(r)$, and Δ is a homomorphism.[11] □

To show that Δ is continuous, we need a bit more technology. In particular, we need to know that functions in $C_c(G)$ are uniformly continuous.

Lemma 1.62. *Suppose that* $f \in C_c(G)$ *and that* $\epsilon > 0$*. Then there is a neighborhood* V *of* e *in* G *such that either* $s^{-1}r \in V$ *or* $sr^{-1} \in V$ *implies*

$$|f(s) - f(r)| < \epsilon. \tag{1.10}$$

Proof. The cases $s^{-1}r \in V$ and $sr^{-1} \in V$ can be treated separately; we can take the intersection of the two neighborhoods. Here I'll show only that there is a V so that $sr^{-1} \in V$ implies (1.10).

Let W be a symmetric compact neighborhood of e and let $K = \operatorname{supp} f$. Notice that $sr^{-1} \in V \subset W$ and $|f(s) - f(r)| > 0$ implies that both s and r belong to the compact set WK.[12] Thus if our assertion were false, there is an $\epsilon_0 > 0$ such that for each neighborhood $V \subset W$ there are s_V and r_V in WK such that

$$|f(s_V) - f(r_V)| \geq \epsilon_0. \tag{1.11}$$

Notice that $\{\, s_V \,\}$ and $\{\, r_V \,\}$ are nets, directed by decreasing V, such that $s_V r_V^{-1} \to e$ in G. Since WK is compact, we can pass to a subnet, relabel, and assume that $s_V \to s$ and $r_V \to s$. Now continuity of f and (1.11) lead to a contradiction. □

[11] Many authors prefer to work directly with the measure rather than the associated linear functional. Then one can define Δ by noting that $\nu(E) := \mu(Er)$ is a Haar measure; hence, $\nu = \Delta(r)\mu$ for some scalar $\Delta(r)$.

[12] To see this, notice that we certainly have either s or r in K. If $r \in K$, then $sr^{-1} \in W$ implies $s \in WK$. If $s \in K$, then $r^{-1} \in s^{-1}W \subset K^{-1}W$. Then $r \in WK$.

Proof of continuity in Lemma 1.61: Using Lemma 1.62 on the previous page, it is clear that if $r_i \to r$ in G, then $\rho(r_i)f \to \rho(r)f$ uniformly. Furthermore, the supports are all eventually contained in the same compact subset.[13] From this, it easy to check[14] that

$$\int_G \rho(r_i)f\,d\mu(s) \to \int_G \rho(r)f\,d\mu(s). \tag{1.12}$$

But by the first part of Lemma 1.61 on page 18, this implies

$$\Delta(r_i)\int_G f(s)\,d\mu(s) \to \Delta(r)\int_G f(s)\,d\mu(s). \tag{1.13}$$

Thus if we choose f such that $\int_G f\mu \neq 0$, then $\Delta(r_i) \to \Delta(r)$. This shows that Δ is continuous and finishes the proof. □

Example 1.63. The modular function on the $ax + b$ group is given by $\Delta(a,b) = \frac{1}{a}$.

Remark 1.64. Notice that Haar measure on a group G is bi-invariant if and only if $\Delta \equiv 1$. Such groups are called *unimodular.*

Lemma 1.65. *If G is compact, then G is unimodular.*

Proof. Since G is compact and Δ is continuous, $\Delta(G)$ is a bounded compact subset of $R^+ = (0,\infty)$. But if Δ is not identically one, then there is a $s \in G$ such that $\Delta(s) > 1$. But then $\Delta(s^n) \to \infty$. This is a contradiction and completes the proof. □

Remark 1.66. If μ is a Haar measure on G, then we noted in Remark 1.56 on page 16 that $\nu(E) := \mu(E^{-1})$ is a right Haar measure. Then, working with characteristic functions, it is easy to see that

$$\int_G f(s)\,d\nu(s) = \int_G f(s^{-1})\,d\mu(s).$$

Lemma 1.67. *For all $f \in C_c(G)$,*

$$\int_G f(s^{-1})\Delta(s^{-1})\,d\mu(s) = \int_G f(s)\,d\mu(s). \tag{1.14}$$

It follows that any left-Haar measure and right-Haar measure are mutually absolutely continuous. In particular, if ν is the right-Haar measure $\nu(E) := \mu(E^{-1})$, then

$$\frac{d\nu}{d\mu}(s) = \Delta(s^{-1}),$$

and $d\nu(s) = \Delta(s^{-1})\,d\mu(s)$.

[13] One says that $\rho(r_i)f \to \rho(r)f$ in the "inductive limit topology". We'll have more to say about this unfortunate terminology later (see Remark 1.86 on page 29 and [139, Appendix D]).

[14] Since $\{\rho(r_i)f\}$ is potentially a net (and not necessarily a sequence), we can't apply the Dominated Convergence Theorem. Otherwise, pointwise convergence would suffice and we could have postponed Lemma 1.62 until later.

Remark 1.68. Since we are not assuming that G is σ-compact, there is no reason to believe that Haar measure is σ-finite. Hence, the usual Radon-Nikodym Theorem does not apply.[15] But once we have proved that (1.14) holds, then since the left-hand side of (1.14) is simply

$$\int_G f(s)\Delta(s)\,d\nu(s),$$

it certainly follows that μ and ν have the same null sets, and that we can obtain ν by integration against $\Delta(s^{-1})\,d\mu(s)$, and we have employed the usual notation for the Radon-Nikodym derivative.

Proof. Define $J : C_c(G) \to \mathbf{C}$ by

$$J(f) := \int_G f(s^{-1})\Delta(s^{-1})\,d\mu(s).$$

Then J is clearly a positive linear functional, and

$$\begin{aligned} J\big(\lambda(r)f\big) &= \Delta(r)\int_G f\big((sr)^{-1}\big)\Delta(sr)^{-1}\,d\mu(s) \\ &= \int_G f(s^{-1})\Delta(s^{-1})\,d\mu(s) \\ &= J(f). \end{aligned}$$

Therefore (1.14) holds up to a strictly positive scalar c. But we can find a nonzero $g \in C_c(G)$ with $g(s) \geq 0$ for all s. Then

$$h(s) := g(s) + \Delta(s^{-1})g(s^{-1})$$

defines a function h in $C_c(G)$ such that $h(s) = \Delta(s^{-1})h(s^{-1})$. Plugging h into (1.14) implies that $c = 1$ and we've proved that (1.14) holds.

The other assertions are clear. □

Remark 1.69 (Quasi-invariant measures). Suppose that G is σ-compact. Two measures that are mutually absolutely continuous are said to be equivalent, and equivalence is an equivalence relation on the set of Radon measures on G. An equivalence class is called a *measure class*. Lemma 1.67 on the preceding page implies that Haar measure and right-Haar measure are in the same measure class $\mathcal{C}$. Since all measures $\mu' \in \mathcal{C}$ have the same null sets, it follows that if N is a μ'-null set, then rE and Er are μ'-null for all $r \in G$. A Radon measure on G is called *quasi-invariant* if N null implies that rN is null for all $r \in G$. We will show in Lemma H.14 on page 463 that any two quasi-invariant measures on G must be equivalent. Therefore any quasi-invariant measure on G is in the same measure class as Haar measure.

[15]For a version of the Radon-Nikodym Theorem that does apply to general Radon measures, see [71, Theorem 12.17].

1.4 An Interlude: Abelian Harmonic Analysis

This section is meant to serve as motivation for many of the constructions and techniques in the sequel. In this section, G will always be a locally compact *abelian* group with a fixed Haar measure μ. In order to appeal to the classical theory of commutative Banach algebras, we'll work with $L^1(G)$ rather than $C_c(G)$ in this section.[16]

Remark 1.70. If G is second countable, then its Haar measure is certainly σ-finite, and we can apply the familiar results of measure theory such as Fubini's Theorem without much concern. But if we allow G to be arbitrary, then questions about Fubini's Theorem and other results are bound to come up. However many of these questions can be finessed provided the functions one wants to integrate are supported on σ-finite subsets of the product $G \times G$. (A discussion of these sorts of issues may be found in [56, §2.3].)

A situation that will arise several times in this section is the following. Suppose that f and g are measurable L^1-functions. With a little hard work, it follows that $(s, r) \mapsto f(r)g(s-r)$ is a measurable function with respect to the product measure $\mu \times \mu$.[17] Then one wants to apply Fubini or Tonelli to see that $F(s, r) = f(r)g(s-r)$ is in $L^1(G \times G)$. For this it suffices to note that f and g have σ-finite supports A and B, respectively, and that F is supported in the σ-finite set $A \times (A + B)$. Then we can apply Fubini with a clear conscience. In particular,

$$f * g(s) := \int_G f(r)g(s-r)\, d\mu(r) \tag{1.15}$$

is in $L^1(G)$. The properties of Haar measure guarantee that

$$f^*(s) := \overline{f(-s)} \tag{1.16}$$

is in $L^1(G)$ if f is, and that $\|f^*\|_1 = \|f\|_1$, $f * g = g * f$ and that $(f * g)^* = g^* * f^*$. Now it is a relatively straightforward matter to check that $L^1(G, \mu)$ is a commutative Banach $*$-algebra with respect to the operations (1.15) and (1.16).

[16]A nice reference for both commutative Banach algebras and applications to harmonic analysis is [101]. For example, Proposition 1.76 on page 24 essentially came from [101, Chap. VII]. Unfortunately, [101] is out of print. A modern treatment of commutative Banach algebras can be found in [155], and all the basics (and a good deal more) about abelian harmonic analysis is to be found in [71] (and especially [71, Chap. V]).

[17]The issue is to see that $(s, r) \mapsto g(s-r)$ is measurable. Since $(s, r) \mapsto s - r$ is continuous, there is no problem if g is continuous or even Borel. However, in general, the composition of a measurable function with a continuous function need not be measurable. But since g is the pointwise limit of simple functions in $L^1(G)$, we can assume g is a characteristic function. Thus it suffices to see that

$$\sigma(E) := \{\, (s, r) \in G \times G : s - r \in E \,\}$$

is $\mu \times \mu$-measurable if E is μ-measurable. As above, this is automatic if E is a Borel set. If E is a null set, then there is a Borel null set F with $E \subset F$. Now $\sigma(F)$ is measurable and $\sigma(F)_s := \{\, r \in G : (s, r) \in \sigma(F) \,\} = s - F$ is a null set. It follows from Tonelli that $\sigma(F)$ is a $\mu \times \mu$-null set. Since $\sigma(E) \subset \sigma(F)$, $\sigma(E)$ is $\mu \times \mu$-measurable (and a null set). But if E is any $\mu \times \mu$-measurable set, $E = B \cup N$ for a Borel set B and a μ-null set N. Since $\sigma(E) = \sigma(B) \cup \sigma(N)$, it follows that $\sigma(E)$ is measurable.

Definition 1.71. A net $\{ u_i \}$ of self-adjoint[18] elements of norm at most one in $L^1(G)$ is called an *approximate identity* for $L^1(G)$ if for all $f \in L^1(G)$, $f * u_i = u_i * f$ converges to f in norm.

Lemma 1.72. *$L^1(G)$ has an approximate identity in $C_c(G)$.*

Proof. If V is a neighborhood of e in G, then let u_V be a nonnegative element of $C_c(G)$ with integral 1, $\operatorname{supp} u_V \subset V$ and $u_V = u_V^*$.[19] Now fix $f \in C_c(G)$. If $\epsilon > 0$, then there is a compact neighborhood W of e such that $s - r \in W$ implies $|f(s) - f(r)| < \epsilon$ (Lemma 1.62 on page 19). Now if $V \subset W$ we have $\operatorname{supp} u_V * f \subset \operatorname{supp} f + W$ and

$$\begin{aligned} |u_V * f(s) - f(s)| &= \int_G u_V(r)\big(f(s-r) - f(s)\big)\, d\mu(r) \\ &\leq \int_G u_V(r)\big|\big(f(s-r) - f(s)\big)\big|\, d\mu(r) \\ &\leq \epsilon \int_G u_V(r)\, d\mu(r) \\ &= \epsilon. \end{aligned}$$

It follows that $u_V * f \to f$ uniformly and the support of $u_V * f$ is eventually contained in a fixed compact set. In particular, $u_V * f$ converges to f in $L^1(G)$. This suffices since $C_c(G)$ is dense in $L^1(G)$ and each u_V has norm one. □

Definition 1.73. Let $\Delta(G)$ be the set of nonzero complex homomorphisms of $L^1(G)$ to $\mathbf{C}$ equipped with the relative weak-$*$ topology coming from $\Delta(G) \subset L^1(G)^*$.

Definition 1.74. We let $\widehat{G}$ denote the set of continuous homomorphisms from G to $\mathbf{T}$. Under pointwise multiplication, $\widehat{G}$ is a group called the *character group* of G or the *Pontryagin dual* of G.

Lemma 1.75. *If $\omega \in \widehat{G}$, then $h_\omega \in \Delta(G)$ where*

$$h_\omega(f) := \int_G f(s)\omega(s)\, d\mu(s). \tag{1.17}$$

Proof. Since $\|\omega\|_\infty = 1$, $|h_\omega(f)| \leq \|f\|_1$ and h_ω is certainly continuous. Now using Fubini,

$$\begin{aligned} h_\omega(f * g) &= \int_G \int_G f(r)g(s-r)\omega(s)\, d\mu(r)\, d\mu(s) \\ &= \int_G \int_G f(r)g(s-r)\omega(s)\, d\mu(s)\, d\mu(r) \\ &= \int_G \int_G f(r)g(s)\omega(s+r)\, d\mu(s)\, d\mu(r) \\ &= h_\omega(f)h_\omega(g). \end{aligned}$$

[18] An element a in a Banach $*$-algebra is called self-adjoint if $x^* = x$.

[19] Let u_1 be any nonnegative compactly supported function with integral 1 and support in a symmetric neighborhood contained in V. Then $u_V := \frac{1}{2}(u_1 + u_1^*)$ will do.

The rest is clear.[20] □

Proposition 1.76. *The map $\omega \mapsto h_\omega$ is a bijection of $\widehat{G}$ onto $\Delta(G)$.*

Proof. If $h_\omega = h_{\omega'}$, then

$$\int_G f(s)\bigl(\omega(s) - \omega'(s)\bigr)\,d\mu(s) = 0 \quad \text{for all } f \in L^1(G).$$

Thus ω and ω' must agree almost everywhere and therefore everywhere. This proves that $\omega \mapsto h_\omega$ is one-to-one.

To motivate what follows, note that $h_\omega\bigl(\lambda(r)f\bigr) = \omega(r)h_\omega(f)$ for all $f \in L^1(G)$. Now let $h \in \Delta(G)$. Since Haar measure is translation invariant, it is easy to check that

$$\lambda(r)u * f = u * \lambda(r)f. \tag{1.18}$$

Let $\{\, u_i \,\} \subset C_c(G)$ be an approximate identity for $L^1(G)$. Then (1.18) implies that

$$h\bigl(\lambda(r)u_i * f\bigr) = h\bigl(\lambda(r)u_i\bigr)h(f) \to h\bigl(\lambda(r)f\bigr).$$

If $h(f) \neq 0$, then there is a $\omega_h(r) \in \mathbf{C}$ such that

$$h\bigl(\lambda(r)u_i\bigr) \to \frac{h\bigl(\lambda(r)f\bigr)}{h(f)} := \omega_h(r). \tag{1.19}$$

Since the left-hand side of (1.19) is independent of our choice of f, $\omega_h(r) = h\bigl(\lambda(r)g\bigr)/h(g)$ provided only that $h(g) \neq 0$.

Notice that

$$|\omega_h(r)| = \Bigl|\frac{h\bigl(\lambda(r)f\bigr)}{h(f)}\Bigr| \le \frac{\|f\|_1}{|h(f)|} < \infty,$$

so $\|\omega_h\|_\infty < \infty$.

Replacing g by $\lambda(s)f$ in $\lambda(r)f * g = f * \lambda(r)g$ gives

$$\bigl(\lambda(r)f\bigr) * \bigl(\lambda(s)f\bigr) = f * \lambda(rs)f$$

which, after applying h to both sides and dividing by $h(f)^2$, gives

$$\frac{h\bigl(\lambda(r)f\bigr)}{h(f)}\frac{h\bigl(\lambda(s)f\bigr)}{h(f)} = \frac{h\bigl(\lambda(rs)f\bigr)}{h(f)}$$

and this implies that

$$\omega_h(r)\omega_h(s) = \omega_h(rs).$$

[20]Notice that h_ω is a $*$-homomorphism: that is, $h(f^*) = \overline{h(f)}$. In view of Proposition 1.76, it will follow that every $h \in \Delta(G)$ is a $*$-homomorphism.

This shows that ω_h is a homomorphism of G into $\mathbf{C} \setminus \{0\}$. Since translation is continuous in $L^1(G)$ and

$$\begin{aligned}|\omega_h(s) - \omega_h(r)| &= \frac{1}{|h(f)|}|h\big(\lambda(s)f - \lambda(r)f\big)| \\ &\le \frac{1}{|h(f)|}\|\lambda(s)f - \lambda(r)f\|_1,\end{aligned}$$

it follows that ω_h is continuous. If $|\omega_h(s)| > 1$, then $\omega_h(s^n) = \omega_h(s)^n \to \infty$. Since $\|\omega_h\|_\infty < \infty$, we must have $\omega_h(G) \subset \mathbf{T}$. Therefore ω_h is a character in $\widehat{G}$.

All that remains to show is that $h_{\omega_h} = h$. However, $h \in \Delta(G)$ is a bounded linear functional on $L^1(G)$, so there is an $\alpha \in L^\infty(G)$ such that

$$h(f) = \int_G f(s)\alpha(s)\,d\mu(s).$$

Now we compute that

$$\begin{aligned}\int_G f(s)\omega_h(s)\,d\mu(s) &= \lim_i \int_G f(s)h\big(\lambda(s)u_i\big)\,d\mu(s) \\ &= \lim_i \int_G\int_G f(s)u_i(r-s)\alpha(r)\,d\mu(r)\,d\mu(s) \\ &= \lim_i \int_G f * u_i(r)\alpha(r)\,d\mu(r) \\ &= \lim_i h(f * u_i) \\ &= h(f)\end{aligned}$$

□

Lemma 1.77. *The map $(s,h) \mapsto \omega_h(s)$ is continuous from $G \times \Delta(G)$ to $\mathbf{T}$.*

Proof. Let $(r,h) \in G \times \Delta(G)$. Choose $f \in L^1(G)$ such that $h(f) \ne 0$. Now $|h(f)| \cdot |\omega_k(s) - \omega_h(r)|$ is bounded by

$$\begin{aligned}&|\omega_k(s)\big(h(f) - k(f)\big)| + |\omega_k(s)k(f) - \omega_h(r)h(f)| \\ &\quad\le |h(f) - k(f)| + |k\big(\lambda(s)f\big) - h\big(\lambda(r)f\big)| \\ &\quad\le |h(f) - k(f)| + |k\big(\lambda(s)f\big) - k\big(\lambda(r)f\big)| \\ &\qquad\qquad + |k\big(\lambda(r)f\big) - h\big(\lambda(r)f\big)| \\ &\quad\le |h(f) - k(f)| + \|\lambda(s)f - \lambda(r)f\|_1 + |k\big(\lambda(r)f\big) - h\big(\lambda(r)f\big)|.\end{aligned}$$

The result follows. □

In view of Proposition 1.76 on the facing page, we can identify $\widehat{G}$ with $\Delta(G)$ and give $\widehat{G}$ the weak-$*$ topology coming from $\Delta(G)$.

Lemma 1.78. *The weak-$*$ topology on $\widehat{G}$ coincides with the topology of uniform convergence on compact sets (a.k.a. the compact-open topology).*

Proof. If $\omega_{h_n} \to \omega_h$ uniformly on compacta, then it is straightforward to see that $h_n \to h$ in $\Delta(G)$. (It suffices to check on $f \in C_c(G)$.) On the other hand, if $h_n \to h$ and if there is a compact set C on which $\omega_{h_n} \not\to \omega_h$ uniformly, then we can pass to a subnet so that there are $s_n \in C$ and an $\epsilon > 0$ such that

$$|\omega_{h_n}(s_n) - \omega_h(s_n)| \geq \epsilon \quad \text{for all } n.$$

We can pass to a subnet and assume that $s_n \to s$ in G. This leads to a contradiction in view of Lemma 1.77 on the previous page. □

Corollary 1.79. *Suppose that G is a locally compact abelian group. Then the character group $\widehat{G}$ is a locally compact abelian group in the topology of uniform convergence on compact sets.*

Proof. It is clear that $\widehat{G}$ is a topological group. Furthermore, the weak-$*$ topology is a locally compact topology on the maximal ideal space of any commutative Banach algebra. Since $\widehat{G}$ is homeomorphic to $\Delta(G)$, $\widehat{G}$ is locally compact. □

1.4.1 The Fourier Transform

The Gelfand transform on a commutative Banach algebra A maps $a \in A$ to the function $\hat{a} \in C_0\big(\Delta(A)\big)$ defined by $\hat{a}(h) := h(a)$ (where $\Delta(A)$ is the set of complex homomorphisms of A). In our situation, the Gelfand transform of $f \in L^1(G)$ is given by the function $\hat{f} \in C_0(\widehat{G})$ defined by

$$\hat{f}(\omega) := \int_G f(s)\omega(s)\,d\mu(s). \tag{1.20}$$

Of course, $\hat{f}$ is also known as the *Fourier transform* of f, and there are libraries full of material, both pure and applied, relating the properties of $\hat{f}$ and f.

Example 1.80. If $G = \mathbf{R}$, then it is an exercise to show that every character on $\mathbf{R}$ is of the form $\omega(x) = e^{-ixy}$ for some $y \in \mathbf{R}$. Thus we can identify $\widehat{\mathbf{R}}$ with $\mathbf{R}$, and a few moments thought shows that the topology is the usual one. With this identification, (1.20) is written as

$$\hat{f}(y) = \int_{-\infty}^{\infty} f(x)e^{-ixy}\,dx,$$

which is the classical Fourier transform on the real line. A good deal of the residue calculus taught in undergraduate complex analysis courses is devoted to evaluating these sorts of integrals. For example, if $f(x) := 1/(x^2+1)$, then

$$\hat{f}(x) = \int_{-\infty}^{\infty} \frac{\cos(xy) + i\sin(xy)}{y^2+1}\,dy = \pi e^{-|x|}.$$

Example 1.81. If $G = \mathbf{T}$, then it can be shown that the characters are of the form $\omega(z) = z^n$ for some $n \in \mathbf{Z}$. Thus we can identify $\widehat{\mathbf{T}}$ with $\mathbf{Z}$ by associating n with the character $z \mapsto z^{-n}$. (The introduction of the minus sign is a convention which merely allows us to tie up with the classical theory in (1.21).) Since $\mathbf{T}$ is compact, it is traditional to normalize Haar measure μ so that $\mu(\mathbf{T}) = 1$. Thus

$$\int_{\mathbf{T}} f(z)\, d\mu(z) = \frac{1}{2\pi} \int_{-\pi}^{\pi} f(e^{i\theta})\, d\theta.$$

Then computing (1.20) amounts to computing the classical Fourier coefficients of f:

$$\hat{f}(n) = \frac{1}{2\pi} \int_{-\pi}^{\pi} f(e^{i\theta}) e^{-in\theta}\, d\theta. \tag{1.21}$$

1.4.2 Non Abelian Groups and Beyond

It is both proper and natural to try to generalize the results of abelian harmonic analysis to nonabelian locally compact groups. In the abelian case, $\widehat{G}$ is a locally compact abelian group, and has its own dual group, and the content of the Pontryagin Duality Theorem is that the map sending $s \in G$ to the character on $\widehat{G}$ defined by $\omega \mapsto \omega(s)$ is a topological group isomorphism [56, Theorem 4.31]. However, in the general case, it quickly becomes apparent that there will not be enough characters to yield very much information about G. In fact, there may only be the trivial character $1 : G \to \mathbf{T}$ mapping all elements to 1; this happens exactly when the closure of the commutator subgroup $[G, G]$ is all of G. Therefore, we will have to expand the notion of a character.

Definition 1.82. A *unitary representation* of a locally compact group G is a continuous homomorphism $U : G \to U(\mathcal{H})$ where $U(\mathcal{H})$ is equipped with the strong operator topology. (We are merely insisting that the maps $s \mapsto U_s h$ be continuous from G to $\mathcal{H}$ for all $h \in \mathcal{H}$.) The *dimension* of U is $d_U := \dim \mathcal{H}$. We say that U is *equivalent* to $V : G \to B(\mathcal{H}')$ if there is a unitary $W : \mathcal{H} \to \mathcal{H}'$ such that $V_s = WU_sW^*$ for all $s \in G$.

Example 1.83. Let G be a locally compact group and $\mathcal{H} = L^2(G)$. If $r \in G$, then $\lambda(r)$ and $\rho(r)$ are unitary operators on $L^2(G)$ where

$$\lambda(r)f(s) := f(r^{-1}s) \quad \text{and} \quad \rho(r)f(s) = \Delta(r)^{\frac{1}{2}} f(sr).$$

Since translation is continuous in $L^2(G)$, it follows that $\lambda : G \to U\big(L^2(G)\big)$ and $\rho : G \to U\big(L^2(G)\big)$ are representations of G called, respectively, the *left-regular representation* and the *right-regular representation*.

Definition 1.84. Let $U : G \to U(\mathcal{H})$ be a unitary representation. A subspace $V \subset \mathcal{H}$ is *invariant* for U if $U_s V \subset V$ for all $s \in G$. If the only closed invariant subspaces for U are the trivial ones — $\{0\}$ and $\mathcal{H}$ — then U is called an *irreducible representation*.

Remark 1.85. As we shall see in due course (cf., Section 3.1), if G is abelian, then the irreducible representations of G are one-dimensional and correspond to the characters of G in the obvious way. For compact (not necessarily abelian) groups G, the Peter-Weyl Theorem [56, Theorems 5.2 and 5.12] implies, among other things, that every irreducible representation of G is finite dimensional. However, there are locally compact groups with infinite dimensional irreducible representations. However, even for nice groups — for example abelian groups — we'll still want to work with infinite-dimensional representations in the sequel.

Part of the point of Remark 1.85 on the preceding page is that in the nonabelian case, the natural analogue for characters is going to be unitary representations and irreducible representations in particular. Thus if we want to generalize the Fourier transform to accommodate nonabelian groups, then we are going to have to replace ω in (1.20) with U where U is a potentially infinite-dimensional representation. However, now the integrand in (1.20) is taking values in $B(\mathcal{H})$. There is work to do to make sense of this.

1.5 Integration on Groups

In order to make sense of integrals where the integrand is a function taking values in a C^*-algebra, Hilbert space or Hilbert module, we need a workable theory of what is referred to in the literature as *vector-valued* integration. In general, the properties of vector-valued integration closely parallel those for scalar-valued functions, but the theory is more complicated, and there are a number of subtle differences and pitfalls that make the theory a bit more formidable. Fortunately, the theory simplifies significantly when it is possible to restrict to Haar measure on a locally compact group G, and to integrands which are continuous with compact support on G taking values in a Banach space $\mathcal{D}$. This was the approach taken in [139, §C.2] for C^*-algebra valued functions and we'll modify that treatment here to accommodate Banach space values.[21]

The idea is to assign to a function $f \in C_c(G, \mathcal{D})$ an element $I(f)$ of $\mathcal{D}$ which is meant to be the integral of f:

$$I(f) := \int_G f(s)\, d\mu(s). \tag{1.22}$$

Naturally, we want I to be linear and be bounded in some sense. Note that if $f \in C_c(G, \mathcal{D})$, then $s \mapsto \|f(s)\|$ is in $C_c(G)$ and

$$\|f\|_1 := \int_G \|f(s)\|\, d\mu(s) \leq \|f\|_\infty \cdot \mu(\operatorname{supp} f) < \infty.$$

If $z \in C_c(G)$ and $a \in \mathcal{D}$, then the function $s \mapsto z(s)a$ is called an *elementary tensor* and will be denoted by $z \otimes a$. We also want our integral to have the following properties:

$$\Big\| \int_G f(s)\, d\mu(s) \Big\| \leq \|f\|_1 \quad \text{and} \quad \int_G (z \otimes a)(s)\, d\mu(s) = a \int_G z(s)\, d\mu(s). \tag{1.23}$$

[21] For much of what appears here, we could replace G by a paracompact locally compact space, and Haar measure by any Radon measure on X. Our approach is loosely based on [13, Chap. III, §3].

We call $\|\cdot\|_1$ the L^1-*norm*. We claim that once we have (1.23), we have uniquely specified our integral. To see this, notice that if $\{ f_i \} \subset C_c(G, \mathcal{D})$ is a net such that

$$f_i \to f \text{ uniformly on } G, \text{ and such that} \tag{1.24}$$

$$\text{there is a compact set } K \text{ such that we eventually have } \operatorname{supp} f_i \subset K, \tag{1.25}$$

then $f_i \to f$ in the L^1-norm and

$$\int_G f_i(s)\, d\mu(s) \to \int_G f(s)\, d\mu(s). \tag{1.26}$$

Then our assertion about uniqueness follows once we have verified that given $f \in C_c(G, \mathcal{D})$, there is a net $\{ f_i \}$ of elementary tensors satisfying (1.24) and (1.25) above.

Remark 1.86 (Inductive Limit Topology). It is common parlance in the literature (and in this book) to say that a net $\{ f_i \}$ satisfying (1.24) and (1.25) *converges to f in the inductive limit topology.* However, there is a bona fide topology on $C_c(G, \mathcal{D})$ — called the inductive limit topology — such that nets satisfying (1.24) and (1.25) converge in that topology. This topology is discussed in detail in [139, Appendix D.2] for the case $\mathcal{D} = \mathbf{C}$. The possibility of confusion arises as it need not be the case that every convergent net in the inductive limit topology satisfy both (1.24) and (1.25) (cf. [139, Example D.9]). However, in this book we usually just want to show that a subset, such as the span of elementary tensors in $C_c(G, \mathcal{D})$, is dense in the inductive limit topology, and then it clearly suffices to produce nets with properties (1.24) and (1.25). The use of the term becomes controversial only if we insist on making statements like "if $f_i \to f$ in the inductive limit topology, then (1.26) holds," because we don't know that (1.24) and (1.25) hold. Nevertheless, if $j : C_c(G, \mathcal{D}) \to M$ is a linear map into a locally convex space M, then j is continuous with respect to the inductive limit topology on $C_c(G, \mathcal{D})$ if and only if $j(f_i) \to j(f)$ for all nets satisfying (1.24) and (1.25). This is proved in the case $\mathcal{D} = \mathbf{C}$ in [139, Lemma D.10]. Rather than prove the general assertion here, it will suffice for the purposes of this book to use the term "$f_j \to f$ in the inductive limit topology" to mean that (1.24) and (1.25) are satisfied, and to forget about the technical definition of the inductive limit topology altogether.

Lemma 1.87. *Suppose that $\mathcal{D}_0$ is a dense subset of a Banach space $\mathcal{D}$. Then*

$$C_c(G) \odot \mathcal{D}_0 := \operatorname{span}\{ z \otimes a : z \in C_c(G) \text{ and } a \in \mathcal{D}_0 \} \tag{1.27}$$

is dense in $C_c(G, \mathcal{D})$ in the inductive limit topology, and therefore for the topology on $C_c(G, \mathcal{D})$ induced by the L^1-norm.

We'll need the following assertion about the uniform continuity of functions in $C_c(G, \mathcal{D})$.

Lemma 1.88. *Suppose that $\mathcal{D}$ is a Banach space, that $f \in C_c(G, \mathcal{D})$ and that $\epsilon > 0$. Then there is a neighborhood V of e in G such that either $sr^{-1} \in V$ or $s^{-1}r \in V$ implies*

$$\|f(s) - f(r)\| < \epsilon.$$

Proof. The proof is exactly the same as in the scalar case (Lemma 1.62 on page 19). □

Proof of Lemma 1.87. Fix $f \in C_c(G, \mathcal{D})$ with $K := \operatorname{supp} f$. Let W be a compact neighborhood of e in G. If $\epsilon > 0$, then, using Lemma 1.88 on the preceding page, choose a symmetric neighborhood $V \subset W$ of e such that $sr^{-1} \in V$ implies that

$$\|f(s) - f(r)\| < \frac{\epsilon}{2}.$$

Then there are $r_i \in K$ such that $K \subset \bigcup_{i=1}^{n} Vr_i$. Then $K^c \cup \{ Vr_i \}_{i=1}^{n}$ is an open cover of G. Since all locally compact groups are paracompact (see Definition 1.40 on page 11 and following remarks), there is a partition of unity $\{ z_i \}_{i=0}^{n}$ such that $\operatorname{supp} z_0 \subset K^c$ and $\operatorname{supp} z_i \subset Vr_i$ for $i = 1, 2, \ldots, n$ (Lemma 1.43 on page 12). Let $x_i \in \mathcal{D}_0$ be such that $\|x_i - f(r_i)\| < \epsilon/2$, and define $g := \sum_{i=1}^{n} z_i \otimes x_i$. Then $g \in C_c(G) \odot \mathcal{D}_0$ and $\operatorname{supp} g \subset VK \subset WK$. (Note that WK is a compact set and does not depend on ϵ.) Since $\sum_{i=1}^{n} z_i(s) = 1$ if $s \in K = \operatorname{supp} f$ and $\sum_{i=1}^{n} z_i(s) \le 1$ otherwise,

$$\begin{aligned} \|f(s) - g(s)\| &= \Big\| \sum_{i=1}^{n} z_i(s)\big(f(s) - x_i\big) \Big\| \\ &\le \Big\| \sum_{i=1}^{n} z_i(s)\big(f(s) - f(r_i)\big) \Big\| + \Big\| \sum_{i=1}^{n} z_i(s)\big(f(r_i) - x_i\big) \Big\| \\ &\le \frac{\epsilon}{2} + \frac{\epsilon}{2} = \epsilon. \end{aligned}$$

This suffices. □

We'll need the following result from Banach space theory. Recall that $C \subset \mathcal{D}$ is called *convex* if $x, y \in C$ and $t \in (0,1)$ imply that $tx + (1-t)y \in C$. If $A \subset \mathcal{D}$, then the *convex hull* of A is the smallest convex set $c(A)$ containing A. Equivalently, $c(A)$ is the intersection of all convex sets containing A. The closure of $c(A)$ is called the *closed convex hull* of A. Notice that $a_1, \ldots, a_n \in A$ and if $\lambda_i \ge 0$ for all i, then $\sum_i \lambda_i a_i \in \lambda c(A)$, where $\lambda := \lambda_1 + \cdots + \lambda_n$.

Lemma 1.89. *If A is a compact convex set in a Banach space $\mathcal{D}$, then its closed convex hull is compact.*

Proof. It suffices to see that $c(A)$ is totally bounded. Let $\epsilon > 0$. As usual, let $B_\epsilon(d)$ be the ϵ-ball centered at $d \in \mathcal{D}$. Since A is compact, there are $d_1, \ldots, d_n \in A$ such that

$$A \subset \bigcup_{i=1}^{n} B_{\frac{\epsilon}{2}}(d_i) = \bigcup_{i=1}^{n} d_i + V,$$

where $V = B_{\frac{\epsilon}{2}}(0)$. Let $C := \{ \sum_i \lambda_i d_i : \sum_i \lambda_i = 1 \text{ and each } \lambda_i \ge 0 \}$. Then C is convex, and as it is the continuous image of a compact subset of $\mathbf{R}^n$, it is a compact subset of $\mathcal{D}$. Therefore there are $c_1, \ldots, c_m \in C$ such that

$$C \subset \bigcup_{i=1}^{m} B_{\frac{\epsilon}{2}}(c_i) = \bigcup_{i=1}^{m} c_i + V.$$

But $C + V$ is convex and contains A. Therefore $c(A) \subset C + V$. It follows that

$$c(A) \subset \bigcup_{i=1}^{m} c_i + 2V = \bigcup_{i=1}^{m} B_\epsilon(c_i).$$

Since ϵ was arbitrary, $c(A)$ is totally bounded as required. □

If $f \in C_c(G, \mathcal{D})$ and $\varphi \in \mathcal{D}^*$, then we can define

$$L_f(\varphi) := \int_G \varphi\big(f(s)\big)\, d\mu(s).$$

Since we clearly have

$$\big|L_f(\varphi)\big| \leq \|\varphi\| \|f\|_1, \tag{1.28}$$

L_f is a bounded linear functional in the double dual $\mathcal{D}^{**}$ of norm at most $\|f\|_1$. Let $\iota : \mathcal{D} \to \mathcal{D}^{**}$ be the natural isometric inclusion of $\mathcal{D}$ into its double dual:

$$\iota(a)(\varphi) := \varphi(a).$$

It is straightforward to check that

$$L_{z\otimes a} = \iota(ca) = c\iota(a) \quad \text{where} \quad c = \int_G z(s)\, d\mu(s).$$

Notice that if $f_i \to f$ in the inductive limit topology on $C_c(G, \mathcal{D})$, then $L_{f_i}(\varphi) \to L_f(\varphi)$ for all $\varphi \in \mathcal{D}^*$. In other words, $L_{f_i} \to L_f$ in the weak-$*$ topology on in $\mathcal{D}^{**}$. Since $f \mapsto L_f$ is certainly linear, we can construct our integral by showing that $L_f \in \iota(\mathcal{D})$ for all $f \in C_c(G, \mathcal{D})$, and then defining $I : C_c(G, \mathcal{D}) \to \mathcal{D}$ by

$$I(f) := \iota^{-1}(L_f).$$

Lemma 1.90. *If $f \in C_c(G, \mathcal{D})$, then $L_f \in \iota(\mathcal{D})$.*

Proof. Let W be a compact neighborhood of $\operatorname{supp} f$ in G, and let $K := f(G) \cup \{0\}$. Also let C be the closed convex hull of K in $\mathcal{D}$. Since K is compact, C is compact by Lemma 1.89 on the preceding page.

Fix $\epsilon > 0$. By Lemma 1.88 on page 29, there is neighborhood V of e in G such that $\|f(s) - f(r)\| < \epsilon$ provided $s^{-1}r \in V$. We can shrink V if necessary so that $\operatorname{supp}(f)V \subset W$. Let $s_1, \ldots, s_n \in \operatorname{supp} f$ be such that

$$\operatorname{supp} f \subset \bigcup_{i=1}^{n} s_i V.$$

By Lemma 1.43 on page 12, there are $\varphi_i \in C_c^+(G)$ such that $\operatorname{supp} \varphi_i \subset s_i V$ and such that $\sum_i \varphi_i(s)$ is bounded by 1 for all s, and equal to 1 when $s \in \operatorname{supp} f$. Let $g(s) = \sum_i f(s_i)\varphi_i(s)$. Then $\operatorname{supp} g \subset W$ and $\|g(s) - f(s)\| < \epsilon$ for all $s \in G$. Furthermore,

$$L_g = \sum_{i=1}^{n} \Big(\int_G \varphi_i(s)\, d\mu(s) \Big) \iota\big(f(s_i)\big).$$

Since

$$\sum_{i=1}^{n} \int_G \varphi_i(s)\, d\mu(s) := d \le \mu(W),$$

it follows that $L_g \in d\iota(C)$ which is in $\mu(W)\iota(C)$ since $0 \in C$.

Since ϵ was arbitrary, it follows that there are $g_i \to f$ in the inductive limit topology such that $L_{g_i} \in \mu(W)\iota(C)$. Since $L_{g_i} \to L_f$ in the weak-$*$ topology, it follows that L_f is in the weak-$*$ closure of $\mu(W)\iota(C)$. But C is compact in the norm topology, and therefore in the weak topology. Untangling definitions shows that this immediately implies $\iota(C)$ is compact in the weak-$*$ topology. But then $\iota(C)$ is closed in the weak-$*$ topology, and it follows that $L_f \in \mu(W)\iota(C) \subset \iota(D)$. □

Lemma 1.91. *Suppose that $\mathcal{D}$ is a Banach space and G is a locally compact group with left Haar measure μ. Then there is a linear map*

$$f \mapsto \int_G f(s)\, d\mu(s)$$

from $C_c(G, \mathcal{D})$ to $\mathcal{D}$ which is characterized by

$$\varphi\Big(\int_G f(s)\, d\mu(s)\Big) = \int_G \varphi\big(f(s)\big)\, d\mu(s) \quad \textit{for all } \varphi \in \mathcal{D}^*. \tag{1.29}$$

Both formulas in (1.23) hold, and if $L : \mathcal{D} \to \mathcal{Y}$ is a bounded linear operator and $f \in C_c(G, \mathcal{D})$, then

$$L\Big(\int_G f(s)\, d\mu(s)\Big) = \int_G L\big(f(s)\big)\, d\mu(s). \tag{1.30}$$

Proof. In view of Lemma 1.90 on the preceding page, we can define $\int_G f(s)\, d\mu(s)$ to be $\iota^{-1}(L_f)$. The formulas in (1.23) follow immediately from (1.29). If $L : \mathcal{D} \to \mathcal{Y}$ is bounded and linear, then $L \circ f \in C_c(G, \mathcal{Y})$ and $\Psi \circ L \in \mathcal{D}^*$ for $\Psi \in \mathcal{Y}^*$. Therefore

$$\Psi\Big(L\Big(\int_G f(s)\, d\mu(s)\Big)\Big) = \int_G \Psi\big(L\big(f(s)\big)\big)\, d\mu(s) \quad \text{for all } \Psi \in \mathcal{Y}^*,$$

and (1.30) follows from (1.29). □

We'll need to recall the definition of the multiplier algebra $M(A)$ of a C^*-algebra A. Loosely put, $M(A)$ is the largest unital C^*-algebra containing A as an essential ideal. There are several ways to define $M(A)$, but we will follow and refer to [139, §2.3] for the definition and basic properties. Thus we define $M(A)$ to be the adjointable operators $\mathcal{L}(A_A)$ on A viewed as a right Hilbert module over itself. We view A as an ideal in $M(A)$ by identifying A with the ideal $\mathcal{K}(A)$ of compact operators in $\mathcal{L}(A)$ by sending $a \in A$ to left multiplication by itself.

Lemma 1.92 ([139, Lemma C.3]). *Suppose that A is a C^*-algebra and that G is a locally compact group with Haar measure μ. Then the integral defined in*

Lemma 1.91 has the following properties in addition to (1.23) and (1.30). Suppose that $f \in C_c(G, A)$. If $\pi : A \to B(\mathcal{H}_\pi)$ is a representation and if $h, k \in \mathcal{H}_\pi$, then

$$\Bigl(\pi\Bigl(\int_G f(s)\,d\mu(s)\Bigr)h \mid k\Bigr) = \int_G \bigl(\pi\bigl(f(s)\bigr)h \mid k\bigr)\,d\mu(s). \tag{1.31}$$

Furthermore,

$$\Bigl(\int_G f(s)\,d\mu(s)\Bigr)^* = \int_G f(s)^*\,d\mu(s), \tag{1.32}$$

and if $a, b \in M(A)$, then

$$a\int_G f(s)\,d\mu(s)b = \int_G af(s)b\,d\mu(s). \tag{1.33}$$

Proof. Let $\varphi \in A^*$ be defined by

$$\varphi(a) := \bigl(\pi(a)h \mid k\bigr).$$

Then (1.31) follows from (1.29). The rest follows easily from (1.31). For example, let π be a faithful representation of A with $a, b \in M(A)$. Then for all $h, k \in \mathcal{H}_\pi$,

$$\begin{aligned}\Bigl(\pi\Bigl(a\int_G f(s)\,d\mu(s)b\Bigr)h \mid k\Bigr) &= \Bigl(\pi\Bigl(\int_G f(s)\,d\mu(s)\Bigr)\bar\pi(b)h \mid \bar\pi(a^*)k\Bigr)\\ &= \int_G \bigl(\pi\bigl(f(s)\bigr)\bar\pi(b)h \mid \bar\pi(a^*)k\bigr)\,d\mu(s)\\ &= \int_G \bigl(\pi\bigl(af(s)b\bigr)h \mid k\bigr)\,d\mu(s)\end{aligned}$$

which, since $s \mapsto af(s)b$ belongs to $C_c(G, A)$, is

$$= \Bigl(\pi\Bigl(\int_G af(s)b\,d\mu(s)\Bigr)h \mid k\Bigr).$$

Since π is faithful and h and k are arbitrary, (1.33) follows. □

Now Lemma 1.92 on the preceding page is a fine and impressive as far as it goes, but recall our goal was to make sense out of integrals like

$$\int_G f(s)U_s\,d\mu(s), \tag{1.34}$$

where $f \in C_c(G)$, or even $C_c\bigl(G, B(\mathcal{H})\bigr)$, and $U : G \to U(\mathcal{H})$ is a unitary representation of G. Then the integrand is not necessarily a continuous function into $B(\mathcal{H})$ with the norm topology, and Lemma 1.92 on the facing page can't be applied. However the integrand is continuous in the strong operator topology, and we now want to see that that is sufficient to define a well behaved integral. To do this in sufficient generality to be useful later, we need to talk a bit more about the multiplier algebra.

Definition 1.93. Let A be a C^*-algebra. If $a \in A$, then let $\|\cdot\|_a$ be the seminorm on $M(A)$ defined by $\|b\|_a := \|ba\| + \|ab\|$. The *strict topology* on $M(A)$ is the topology generated by all the seminorms $\{\, \|\cdot\|_a : a \in A \,\}$.

Example 1.94. A net $\{\, b_i \,\}$ in $M(A)$ converges strictly to b if and only if $ab_i \to ab$ and $b_i a \to ba$ for all $a \in A$. Since $B(\mathcal{H}) = M\big(\mathcal{K}(\mathcal{H})\big)$, $B(\mathcal{H})$ has a strict topology, and $T_i \to T$ strictly if and only if $T_iK \to TK$ and $KT_i \to KT$ for all compact operators K.

We aim to show that the integrand in (1.34) is continuous in the strict topology on $B(\mathcal{H})$. At the moment all that seems clear is that the integrand is a continuous compactly supported function into $B(\mathcal{H})$ with the strong operator topology.

Definition 1.95. The *$*$-strong operator topology* on $B(\mathcal{H})$ has subbasic open sets

$$N(T,h,\epsilon) := \{\, S \in B(\mathcal{H}) : \|Sh - Th\| + \|S^*h - T^*h\| < \epsilon \,\}$$

where $T \in B(\mathcal{H})$, $h \in \mathcal{H}$ and $\epsilon > 0$.

Remark 1.96. A net $T_i \to T$ $*$-strongly if and only if both $T_i \to T$ strongly and $T_i^* \to T^*$ strongly.

The connection to the strict topology is given by the following lemma.

Lemma 1.97 ([139, Lemma C.6]). *On norm bounded subsets of $B(\mathcal{H})$, the strict and $*$-strong topologies coincide.*

This lemma is a special case of the corresponding result for Hilbert modules. If X is a Hilbert A-module, then recall that $\mathcal{L}(\mathsf{X}) \cong M\big(\mathcal{K}(\mathsf{X})\big)$ and therefore has a strict topology. It also has a $*$-strong topology defined as in Definition 1.95 in which a net $T_i \to T$ if and only if $T_i(x) \to T(x)$ and $T_i^*(x) \to T^*(x)$ for all $x \in \mathsf{X}$. Of course, we can view a Hilbert space $\mathcal{H}$ as a Hilbert $\mathbf{C}$-module and then we recover the strict and $*$-strong topologies as above. Then Lemma 1.97 follows from

Proposition 1.98 ([139, Proposition C.7]). *If X is a Hilbert A-module, then strict convergence implies $*$-strong convergence, and the strict and $*$-strong topologies coincide on norm bounded subsets of $\mathcal{L}(\mathsf{X})$.*

At the moment our main interest in Proposition 1.98 is the following corollary.

Corollary 1.99 ([139, Corollary C.8]). *Suppose that $u : G \to U\mathcal{L}(\mathsf{X})$ is a homomorphism into the unitary group of $\mathcal{L}(\mathsf{X})$. Then u is strictly continuous if and only if it is strongly continuous.*

Proof. In view of Proposition 1.98, it suffices to see that $s \mapsto u_s(x)$ continuous for all $x \in \mathsf{X}$ implies $s \mapsto u_s^*(x)$ is continuous. But

$$\begin{aligned}\|u_s^*(x) - u_t^*(x)\|_A^2 &= \big\langle u_s^*(x) - u_t^*(x) \,,\, u_s^*(x) - u_t^*(x)\big\rangle_A \\ &= 2\|x\|_A^2 - \big\langle u_s\big(u_t^*(x)\big) \,,\, x\big\rangle_A - \big\langle x \,,\, u_s\big(u_t^*(x)\big)\big\rangle_A.\end{aligned}$$

But $s \to t$, implies $u_s\big(u_t^*(x)\big) \to x$ by assumption. □

Remark 1.100. In [139, Example C.10], we exhibit a sequence $S := \{T_n\} \subset B(\mathcal{H})$ such that 0 belongs to the $*$-strong closure of S but not to the strict closure. Therefore the strict and $*$-strong topologies are not the same. Notice too that the principle of uniform boundedness implies that a strongly convergent *sequence* is bounded. Therefore a $*$-strong convergent sequence is bounded and also strictly convergent by Proposition 1.98 on the preceding page. Therefore there must be a *net* in S converging to 0 $*$-strongly, but no sequence in S can converge to 0 $*$-strongly. Thus the $*$-strong topology fails to be first countable even on a separable Hilbert space.

We'll write $M_s(A)$ to denote $M(A)$ with the strict topology. Notice that if $f \in C_c\big(G, M_s(A)\big)$, then $s \mapsto f(s)a$ is in $C_c(G, A)$ for each $a \in A$. In particular, $\{ f(s)a : s \in G \}$ is bounded and the uniform boundedness principle implies that $\{ \|f(s)\| : s \in G \}$ is bounded.

Lemma 1.101 ([139, Lemma C.11]). *Let A be a C^*-algebra. There is a unique linear map $f \mapsto \int_G f(s)\,d\mu(s)$ from $C_c\big(G, M_s(A)\big)$ to $M(A)$ such that for any nondegenerate representation $\pi : A \to B(\mathcal{H}_\pi)$ and all $h, k \in \mathcal{H}_\pi$*

$$\Big(\bar\pi\Big(\int_G f(s)\,d\mu(s)\Big)h \mid k\Big) = \int_G \big(\bar\pi\big(f(s)\big)h \mid k\big)\,d\mu(s). \tag{1.35}$$

We have

$$\Big\|\int_G f(s)\,d\mu(s)\Big\| \le \|f\|_\infty \cdot \mu(\operatorname{supp} f).$$

Equations (1.32) and (1.33) are valid is this context. If $L : A \to B$ is a nondegenerate homomorphism into a C^*-algebra B, then

$$\bar L\Big(\int_G f(s)\,d\mu(s)\Big) = \int_G \bar L\big(f(s)\big)\,d\mu(s).$$

Proof. Uniqueness is clear from (1.35). Recall that $M(A) = \mathcal{L}(A_A)$. For each $f \in C_c\big(G, M_s(A)\big)$, define $L_f : A \to A$ by

$$L_f(a) := \int_G f(s)a\,d\mu(s).$$

Then using the properties laid out in Lemma 1.92 on page 32,

$$\big\langle L_f(a)\,,\,b\big\rangle_A = L_f(a)^*b = \int_G a^*f(s)^*b\,d\mu(s) = \big\langle a\,,\,L_{f^*}(b)\big\rangle_A.$$

Thus $L_f \in \mathcal{L}(A_A)$ with adjoint L_{f^*}. We define

$$\int_G f(s)\,d\mu(s) := L_f.$$

The analogue of (1.32) follows from $L_f^* = L_{f^*}$. Furthermore (1.31) implies

$$\begin{aligned}\Big(\bar\pi\Big(\int_G f(s)\,d\mu(s)\Big)\pi(a)h \mid k\Big) &= \int_G \big(\pi\big(f(s)a\big)h \mid k\big)\,d\mu(s)\\ &= \int_G \big(\bar\pi\big(f(s)\big)\pi(a)h \mid k\big)\,d\mu(s).\end{aligned}$$

Now (1.35) follows from the nondegeneracy of π. The analogue of (1.33) is straightforward to check and the analogue of (1.30) follows from

$$\begin{aligned}\bar{L}\Big(\int_G f(s)\,d\mu(s)\Big)L(a)b &= L\Big(\int_G f(s)b\,d\mu(s)\Big)\\ &= \int_G \bar{L}\big(f(s)\big)\,d\mu(s)L(a)b. \qquad \square\end{aligned}$$

Shortly, we'll want to interchange the order of integration in our vector-valued integrals. Since our integrands will be continuous with compact support, we can avoid appealing to theorems about general vector-valued integrals, and appeal instead to the scalar-valued version of Fubini's Theorem. First, some preliminary lemmas. Here we are once again working with Banach space-valued functions.

Lemma 1.102. *Suppose that X is a locally compact space and that $F \in C_c(X \times G, \mathcal{D})$. Then the function*

$$x \mapsto \int_G F(x,s)\,d\mu(s)$$

is an element of $C_c(X, \mathcal{D})$.

Proof. It will clearly suffice to prove continuity in x. So suppose that $x_i \to x$ and that $\epsilon > 0$. Let K and C be compact sets such that $\operatorname{supp} F \subset K \times C$. For each $x \in X$, let $\varphi(x)$ be the element of $C_c(G, \mathcal{D})$ defined by $\varphi(x)(s) := F(x,s)$. We claim that $\varphi(x_i) \to \varphi(x)$ uniformly on G. If this were not the case, then after passing to a subnet and relabeling, there would be an $\epsilon_0 > 0$ and $r_i \in G$ such that

$$\|F(x_i, r_i) - F(x, r_i)\| \geq \epsilon_0 \quad \text{for all } i. \tag{1.36}$$

But we certainly have $\{\, r_i \,\} \subset C$, and since C is compact, we can pass to subnet, relabel, and assume that $r_i \to r$. However this leads to a contradiction as (1.36) would fail for large i. This establishes the claim.

Now we can assume that for large i we have

$$\|\varphi(x_i) - \varphi(x)\|_\infty < \frac{\epsilon}{\mu(C)}.$$

Since $\operatorname{supp} \varphi(x_i) \subset C$ for all i,

$$\begin{aligned}\Big\|\int_G F(x_i, s)\,d\mu(s) - \int_G F(x,s)\,d\mu(s)\Big\| &\leq \int_C \|\varphi(x_i)(s) - \varphi(x)(s)\|\,d\mu(s)\\ &\leq \epsilon.\end{aligned}$$

This suffices. $\square$

The following corollary will be very useful in the sequel.

Corollary 1.103. *Let H be a closed subgroup of G. Suppose that $F \in C(G \times H, \mathcal{D})$ is such that there is a compact set $K \subset G$ such that $F(s,t) = 0$ if $s \notin K$. Then the function*

$$\psi(s) \mapsto \int_H F(st, t)\,d\mu_H(t)$$

is a well-defined element of $C(G)$ with support in KH.

Proof. Fix $s_0 \in G$. Then $t \mapsto F(s_0t, t)$ is in $C_c(H)$, and $\psi(s_0)$ is well-defined. If C is a compact neighborhood of s_0, then $F'(s,t) := F(st,t)$ defines an element of $C_c(C \times H, \mathcal{D})$. Since

$$\psi(s) = \int_H F'(s,t)\,d\mu_H(t) \quad \text{for all } s \in C,$$

the continuity of follows from Lemma 1.102 on the preceding page. □

Corollary 1.104. *Suppose that $F \in C_c(G \times G', \mathcal{D})$. Then the functions*

$$s \mapsto \int_{G'} F(s,r)\,d\mu_{G'}(r) \quad \text{and} \quad r \mapsto \int_G F(s,r)\,d\mu_G(s)$$

are in $C_c(G, \mathcal{D})$ and $C_c(G', \mathcal{D})$, respectively.

Proof. This follows immediately from Lemma 1.102 on the facing page. □

Proposition 1.105. *Suppose that $F \in C_c\big(G \times G, M_s(A)\big)$. Then*

$$s \mapsto \int_G F(s,r)\,d\mu(r) \quad \text{and} \quad r \mapsto \int_G F(s,r)\,d\mu(s)$$

are in $C_c\big(G, M_s(A)\big)$. Then the iterated integrals

$$\int_G \int_G F(s,r)\,d\mu(s)\,d\mu(r) \quad \text{and} \quad \int_G \int_G F(s,r)\,d\mu(r)\,d\mu(s)$$

are defined by Lemma 1.101 on page 35, and have a common value.

A similar statement holds for $F \in C_c(G \times G, \mathcal{D})$ with respect to the norm continuous integral defined in Lemma 1.91 on page 32.

Proof. To show that $s \mapsto \int_G F(s,r)\,d\mu(r)$ is in $C_c\big(G, M_s(A)\big)$, it suffices to see that $s \mapsto a\int_G F(s,r)\,d\mu(r)$ and $s \mapsto \int_G F(s,r)\,d\mu(s)a$ are in $C_c(G,A)$ for all $a \in A$. Since $a\int_G F(s,r)\,d\mu(r) = \int_G aF(s,r)\,d\mu(r)$ and $\int_G F(s,r)\,d\mu(r)a = \int_G F(s,r)a\,d\mu(r)$ by Lemma 1.101 on page 35 and both $(s,r) \mapsto aF(s,r)$ and $(s,r) \mapsto F(s,r)a$ are in $C_c(G \times G, A)$, the first assertion follows from Corollary 1.104.

Now it is clear from Lemma 1.101 on page 35 that both iterated integrals are defined and take values in $M(A)$. To see that they have the same value, let π be a faithful representation of A in $B(\mathcal{H})$. Then $\bar{\pi}$ is faithful on $M(A)$. Then (1.35) and the usual scalar-valued Fubini Theorem imply that

$$\Big(\bar{\pi}\Big(\int_G \int_G F(s,r)\,d\mu(s)\,d\mu(r)\Big)h \mid k\Big) = \Big(\bar{\pi}\Big(\int_G \int_G F(s,r)\,d\mu(r)\,d\mu(s)\Big)h \mid k\Big)$$

for all $h, k \in \mathcal{H}$. Now the result follows since $\bar{\pi}$ is faithful.

If $F \in C_c(G \times G, \mathcal{D})$, then we proceed as above and obtain the equality

$$\int_G \int_G \varphi\big(F(s,r)\big)\,d\mu(s)\,d\mu(r) = \int_G \int_G \varphi\big(F(s,r)\big)\,d\mu(r)\,d\mu(s)$$

for all $\varphi \in \mathcal{D}^*$ using the usual scalar-valued form of Fubini's Theorem. The rest follows from (1.29).

□

1.5.1 Completions

In the sequel, we will often work with Banach spaces, and in particular C^*-algebras, which are defined to be the completion of a vector space V with respect to a given norm or seminorm $\|\cdot\|$. If $\|\cdot\|$ is a bona fide norm, then we can realize the completion as the closure of $i(V)$ in V^{**}, where $i : V \to V^{**}$ is the natural inclusion of V in its double dual. In this case, we can identify V with $i(V)$ and view V as a subspace in its completion. If $\|\cdot\|$ is a seminorm, then $N := \{ v \in V : \|v\| = 0 \}$ is a subspace of V, and $\|\cdot\|$ defines a norm on the vector space quotient V/N in the obvious way. Then the completion of V is defined to be the closure of $q(V)$ in $(V/N)^{**}$ where $q : V \to (V/N)^{**}$ is the composition of the quotient map of V onto V/N with the natural map of V/N into its double dual. Often we will suppress the map q to avoid decorating already complicated formulas with distracting notations.

The most important sort of completion we look at in this book will be the completion B of a function space $C_c(P, \mathcal{D})$ with respect to a norm $\|\cdot\|$, where P is a locally compact space. Much like the L^1-norm, this norm will have the property that it is continuous with respect to the inductive limit topology.[22] Recall that this means simply that if $\{ f_i \}$ is a net in $C_c(P, \mathcal{D})$ which converges uniformly to f and if the supports of the f_i are eventually contained in the same compact subset of P, then $f_i \to f$ in norm (see Remark 1.86 on page 29). Unfortunately, evaluation at a point $s \in P$ will not make sense; that is, evaluation at s will not be continuous with respect to $\|\cdot\|$. However, there will be circumstances where certain *a priori* B-valued integrals will take values in $q\big(C_c(P, \mathcal{D})\big)$, and then it at least makes sense to ask if we can pass evaluation at s through the integral. We show that this is the case in many situations in Lemma 1.108 on the next page. First, we need a preliminary observation.

Recall that a subset U in a locally compact space is called *pre-compact* if its closure is compact.[23]

Lemma 1.106. *Suppose that U is a pre-compact open subset in P. If $f \in C_0(U, \mathcal{D})$, then*

$$j(f)(s) := \begin{cases} f(s) & \text{if } s \in U, \text{ and} \\ 0 & \text{otherwise,} \end{cases}$$

defines an element $j(f) \in C_c(P, \mathcal{D})$. In particular, there is a natural linear map $j : C_0(U, \mathcal{D}) \to C_c(P, \mathcal{D})$, called the inclusion map, *which is continuous when $C_c(P, \mathcal{D})$ is equipped with the inductive limit topology. In particular, if B is the completion of $C_c(P, \mathcal{D})$ with a norm that is continuous with respect to the inductive limit topology, then $q \circ j$ is continuous linear map from $C_0(U, \mathcal{D})$ to B.*

Remark 1.107. We are only claiming that j is a linear map. If $C_c(P, \mathcal{D})$ has a $*$-algebra structure, then $j\big(C_0(U, \mathcal{D})\big)$ will, in general, not be closed under multiplication.

[22] The example we have in mind, of course, is the crossed product $A \rtimes_\alpha G$ which will be defined to be the completion of $C_c(G, A)$ with respect to just such a norm.

[23] The term *relatively compact* is also used. We have avoided it here since, in analogy with relatively closed or relatively Hausdorff, it suggests “compact in some relative topology”.

Proof. We only need to see that $j(f)$ is continuous on G (see Remark 1.86 on page 29). Suppose $x_i \to x$. If $x \in U$, x_i is eventually in U and $j(f)(x_i) \to j(f)(x)$. If $x \notin U$, then $j(f)(x) = 0$. Given $\epsilon > 0$, then since $f \in C_0(U, \mathcal{D})$ it follows that $K := \{\, x \in U : \|f(x)\| \geq \epsilon \,\}$ is compact. Thus K is closed in G, $K^c := G \setminus K$ is a neighborhood of x, and x_i is eventually in K^c. Thus $\|j(f)(x_i)\|$ is eventually less than ϵ. Since ϵ is arbitrary, $j(f)(x) = 0$. □

Lemma 1.108. *Let G be a locally compact group, P a locally compact space and $\mathcal{D}$ a Banach space. Let $\|\cdot\|$ be a norm on $C_c(P, \mathcal{D})$ such that convergence in the inductive limit topology in $C_c(P, \mathcal{D})$ implies convergence in $\|\cdot\|$. Define B to be the completion of $C_c(P, \mathcal{D})$ with respect to $\|\cdot\|$, and let $q : C_c(P, \mathcal{D}) \to B$ be the natural map. Suppose that $Q \in C_c(G \times P, \mathcal{D})$ and that $g : G \to C_c(P, \mathcal{D})$ is defined by $g(s)(p) := Q(s, p)$. Then $s \mapsto q\big(g(s)\big)$ is an element in $C_c(G, B)$ and we can form the element*

$$b := \int_G^B q\big(g(s)\big)\, d\mu(s).$$

Then $b \in q\big(C_c(P, \mathcal{D})\big)$, and $b = q(\tilde{b})$ where

$$\tilde{b}(p) := \int_G^{\mathcal{D}} g(s)(p)\, d\mu(s).$$

Remark 1.109. Suppressing the q in the above, the conclusion of the Lemma 1.108 asserts that $b \in C_c(P, \mathcal{D})$ and that

$$\int_G^B g(s)\, d\mu(s)(p) = \int_G^{\mathcal{D}} g(s)(p)\, d\mu(s).$$

In colloquial terms, we are "passing evaluation through the integral".

Proof. If $s_i \to s$ in G, then it is not hard to check that $g(s_i) \to g(s)$ in the inductive limit topology in $C_c(P, \mathcal{D})$. By assumption, this means $s \mapsto q\big(g(s)\big)$ is in $C_c(G, B)$. On the other hand, Lemma 1.102 on page 36 shows that $\tilde{b} \in C_c(P, \mathcal{D})$.

Let $\operatorname{supp} Q \subset C \times K$ with $C \subset G$ and $K \subset P$ compact. Let U be a pre-compact neighborhood of K. Then we can view $s \mapsto g(s)$ as an element of $C_c\big(G, C_0(U, \mathcal{D})\big)$. Since evaluation at $p \in U$ is a bounded homomorphism from $C_0(U, \mathcal{D})$ to $\mathcal{D}$,

$$\int_G^{C_0(U,\mathcal{D})} g(s)|_U\, d\mu(s)(p) = \int_G^{\mathcal{D}} g(s)(p)\, d\mu(s). \tag{1.37}$$

Note that the right-hand side of (1.37) is just the restriction of $\tilde{b}$ to U. But the inclusion $j : C_0(U, \mathcal{D})$ into $C_c(P, \mathcal{D})$ is continuous in the inductive limit topology (Lemma 1.106 on the preceding page), and therefore $q \circ j$ is a continuous linear map of $C_0(U,\mathcal{D})$ into B. Thus

$$\begin{aligned} q(\tilde{b}) &= q\circ j(\tilde{b}|_U) \\ &= q\circ j\Big(\int_G^{C_0(U,\mathcal{D})} g(s)|_U \, d\mu(s)\Big) \\ &= \int_G^B q\big(g(s)\big)\, d\mu(s) = b. \end{aligned}$$

□

Notes and Remarks

The standard reference for the basic properties and structure of locally compact groups is [71]. The sorts of point-set topology and basic functional analysis required for work in operator algebras is beautifully laid out in [127]. (For questions of point-set topology outside the respectable demands of the subject, I recommend [168].) My authorities on Radon measures, Haar measure in particular and integration on locally compact groups are [56, Chap. 2; 71, Chap. 3 & 4; 156, Theorem 2.14]. The approach to vector-valued integration given here is based on [13, Chap. III, §3].

Chapter 2

Dynamical Systems and Crossed Products

A C^*-dynamical system is a locally compact group G acting by automorphisms on a C^*-algebra A. A crossed product is a C^*-algebra built out of a dynamical system. In abelian harmonic analysis (cf., Section 1.4), an important principle is to recover information about a function f from its Fourier coefficients $\hat{f}(\omega)$, and $f \mapsto \hat{f}(\omega)$ is a representation of the group algebra on a one-dimensional Hilbert space. We are going to study dynamical systems and their crossed products via classes of representations called covariant representations. We define dynamical systems, their crossed products and covariant representations in the first two sections. Just as characters are in one-to-one correspondence with complex homomorphisms (that is, one-dimensional representations) of the L^1-algebra of a locally compact abelian group, covariant representations of a dynamical system are in one-to-one correspondence with representations of the associated crossed product. This correspondence is developed in Sections 2.3 and 2.4. This correspondence is crucial to understanding crossed products and suggests that it is very profitable to view a crossed product as the C^*-algebra generated by a universal covariant representation. We validate this approach in Section 2.6.

2.1 Dynamical Systems

The study of dynamical systems is a subject unto itself, and good references abound. A nice summary of the connections to physics and operator algebras can be found in [136]. In this treatment, we'll just concentrate on what we need to get started.

Definition 2.1. A group G *acts* on the left of a set X if there is a map

$$(s, x) \mapsto s \cdot x \tag{2.1}$$

from $G \times X \to X$ such that for all $s, r \in G$ and $x \in X$

$$e \cdot x = x \quad \text{and} \quad s \cdot (r \cdot x) = sr \cdot x.$$

If G is a topological group and X a topological space, then we say the action is continuous if (2.1) is continuous from $G \times X$ to X.[1] In this case, X is called a *left G-space* and the pair (G, X) is called a *transformation group*. If both G and X are locally compact, then (G, X) is called a *locally compact transformation group*, and X is called a locally compact G-space. A right G-space is defined analogously.

Since we are most concerned with left G-spaces, we will assume here that group actions are on the left unless indicated otherwise.[2]

Example 2.2 ([136, §1]). Consider an ordinary autonomous differential equation of the form

$$\begin{cases} \mathbf{x}' = f(\mathbf{x}) \\ \mathbf{x}(0) = \mathbf{x}_0 \end{cases} \tag{2.2}$$

for a function $f : \mathbf{R}^n \to \mathbf{R}^n$. Assuming f satisfies some mild smoothness conditions, there exists a unique solution $\mathbf{x} : \mathbf{R} \to \mathbf{R}^n$ to (2.2) which depends continuously on the initial condition $\mathbf{x}(0) = \mathbf{x}_0$. If $t \in \mathbf{R}$ and $\mathbf{z} \in \mathbf{R}^n$, then we can define

$$t \cdot \mathbf{z} := \mathbf{x}(t), \tag{2.3}$$

where $\mathbf{x}$ is the unique solution with $\mathbf{x}_0 = \mathbf{z}$. So we trivially have $0 \cdot \mathbf{z} = \mathbf{z}$. Now $s \cdot (t \cdot \mathbf{z}) = \mathbf{y}(s)$ where $\mathbf{y}'(s) = f\big(\mathbf{y}(s)\big)$ and $\mathbf{y}(0) = t \cdot \mathbf{z} = \mathbf{x}(t)$. Now it is straightforward to see that $\mathbf{y}(s) = \mathbf{x}(t+s)$. In other words, $s \cdot (t \cdot \mathbf{z}) = (s+t) \cdot \mathbf{z}$. The continuity of the map $(t, \mathbf{z}) \mapsto t \cdot \mathbf{z}$ is an consequence of the continuous dependence of the solution to (2.2) on its initial conditions, and we get a transformation group $(\mathbf{R}, \mathbf{R}^n)$.

Example 2.3. Let $h \in \operatorname{Homeo}(X)$. Then $\mathbf{Z}$ acts on X by $n \cdot x := h^n(x)$ and $(\mathbf{Z}, X)$ is a transformation group.

Example 2.4. Suppose that G is a locally compact group and H a closed subgroup. Then H is locally compact and acts on G by left translation:

$$h \cdot s := hs.$$

Then (H, G) is a locally compact transformation group.

Now let (G, X) be a locally compact transformation group. Then for each $s \in G$, $x \mapsto s \cdot x$ is in $\operatorname{Homeo}(X)$. (In particular, every $\mathbf{Z}$-space arises from a single homeomorphism as in Example 2.3.) Therefore we obtain a homomorphism

$$\alpha : G \to \operatorname{Aut} C_0(X) \tag{2.4}$$

defined by

$$\alpha_s(f)(x) := f(s^{-1} \cdot x). \tag{2.5}$$

[1]Often we say that the action is *jointly continuous* to emphasize the continuity on the product $G \times X$.

[2]Preferring left over right actions is pure prejudice. If X is a left G-space, then we can view X as a right G-space by defining $x \cdot s := s^{-1} \cdot x$. The same formula may be used to convert from right to left actions.

We certainly have $\alpha_s^{-1} = \alpha_{s^{-1}}$ and

$$\begin{aligned}\alpha_{sr}(f)(x) &= f(r^{-1}s^{-1}\cdot x) = \alpha_r(f)(s^{-1}\cdot x)\\ &= \alpha_s\big(\alpha_r(f)\big)(x).\end{aligned} \tag{2.6}$$

Therefore $\alpha_{sr} = \alpha_s \circ \alpha_r$ as required. (The computation in (2.6) explains the inverse in (2.5).)

Lemma 2.5. *Suppose that (G, X) is a locally compact transformation group and that $\operatorname{Aut} C_0(X)$ is given the point-norm topology. Then the associated homomorphism (2.4) of G into $\operatorname{Aut} C_0(X)$ is continuous.*

Proof. It suffices to see that $\|\alpha_s(f) - f\|_\infty \to 0$ as $s \to e$. If this were to fail, then there would be an $\epsilon > 0$, $s_i \to e$ and $x_i \in X$ such that

$$|f(s_i^{-1}\cdot x_i) - f(x_i)| \geq \epsilon \quad \text{for all } i. \tag{2.7}$$

Since f vanishes at infinity, $K := \{\, x \in X : |f(x)| \geq \epsilon/2 \,\}$ is compact. In order for (2.7) to hold, we must have either $x_i \in K$ or $s_i^{-1}\cdot x_i \in K$. Since we must eventually have the s_i in a compact neighborhood V of e, the x_i eventually lie in the compact set $V \cdot K = \{\, s\cdot x : s \in V \text{ and } x \in K \,\}$. Then we can assume that $x_i \to x_0$, where $x_0 \in V\cdot K$. But then $s_i^{-1}\cdot x_i \to x_0$ and we eventually contradict (2.7). □

Definition 2.6. A *C^*-dynamical system* is a triple (A, G, α) consisting of a C^*-algebra A, a locally compact group G and a continuous homomorphism $\alpha : G \to \operatorname{Aut} A$. We say that (A, G, α) is *separable* if A is separable and G is second countable.

We'll usually shorten C^*-dynamical system to just "dynamical system". Notice that the continuity condition on α in Definition 2.6 amounts to the statement that $s \mapsto \alpha_s(a)$ is continuous for all $a \in A$. Lemma 2.5 simply states that a locally compact transformation group (G, X) gives rise to a dynamical system with A commutative. It turns out that all dynamical systems with A commutative arise from locally compact transformation groups.

Proposition 2.7. *Suppose that $\big(C_0(X), G, \alpha\big)$ is a dynamical system (with X locally compact). Then there is a transformation group (G, X) such that*

$$\alpha_s(f)(x) = f(s^{-1}\cdot x). \tag{2.8}$$

Proof. We saw in Lemma 1.33 on page 8 that there is a $h_s \in \operatorname{Homeo}(X)$ such that

$$\alpha_s(f)(x) = f\big(h_s(x)\big), \tag{2.9}$$

and that the map $s \mapsto h_s$ is continuous from G to $\operatorname{Homeo}(X)$ with the topology described in Definition 1.31 on page 8. Clearly $h_e = \mathrm{id}_X$ and $h_{sr} = h_r \circ h_s$. Therefore we get an action of G on X via

$$s\cdot x := h_s^{-1}(x) = h_{s^{-1}}(x).$$

The map $(s, x) \mapsto s\cdot x$ is continuous in view of Remark 1.32 on page 8. Therefore (G, X) is a transformation group such that (2.8) holds. □

Recall that the irreducible representations of $C_0(X)$ correspond exactly to the point-evaluations ev_x where $\mathrm{ev}_x(f) := f(x)$. Furthermore, the map $x \mapsto [\mathrm{ev}_x]$ is a homeomorphism of X onto $C_0(X)^\wedge$. If $\alpha : G \to \operatorname{Aut} C_0(X)$ is a dynamical system, then (2.8) amounts to

$$\mathrm{ev}_x \circ \alpha_s^{-1} = \mathrm{ev}_{s\cdot x},$$

and the G-action on $C_0(X)^\wedge$ is given by

$$s \cdot [\mathrm{ev}_x] = [\mathrm{ev}_x \circ \alpha_s^{-1}].$$

Notice that if $\alpha : G \to \operatorname{Aut} A$ is any dynamical system and $[\pi] \in \hat{A}$, then $[\pi \circ \alpha_s^{-1}] \in \hat{A}$, and depends only on the class of π.

Proposition 2.7 on the preceding page admits a significant generalization which will be used repeatedly in the sequel. A complete proof is given in [139], and we will not repeat that proof here.

Lemma 2.8 ([139, Lemma 7.1]). *Suppose that (A, G, α) is a dynamical system. Then there is a jointly continuous action of G on the spectrum $\hat{A}$ of A given by $s \cdot [\pi] := [\pi \circ \alpha_s^{-1}]$ called the* action induced by α.

Since $[\pi] \mapsto \ker \pi$ is a continuous, open surjection of $\hat{A}$ onto $\operatorname{Prim} A$ and $\ker(\pi \circ \alpha_s^{-1}) = \alpha_s(\ker \pi)$, there is jointly continuous action of G on $\operatorname{Prim} A$ given by

$$s \cdot P = \alpha_s(P) := \{\, \alpha_s(a) : a \in P \,\}. \tag{2.10}$$

Remark 2.9 (Degenerate Examples). It will be helpful to keep in mind that groups and C^*-algebras are by themselves degenerate examples of dynamical systems. Since the only (algebra) automorphism of $\mathbf{C}$ is the identity, every locally compact group G gives rise to a dynamical system $(\mathbf{C}, G, \mathrm{id})$. Similarly, every C^*-algebra A is associated to a dynamical system with G trivial: $(A, \{\, e \,\}, \mathrm{id})$.

2.2 Covariant Representations

From our point of view in these notes, C^*-dynamical systems are a natural algebraic framework in which to view and to generalize classical dynamical systems. The physical significance of these systems and their representations is described in [136]. Here we will limit the motivation to the idea that C^*-algebras and groups are profitably studied via representations on Hilbert space. The next definition gives a reasonable way to represent a dynamical system on a Hilbert space.

Definition 2.10. Let (A, G, α) be a dynamical system. Then a *covariant representation* of (A, G, α) is a pair (π, U) consisting of a representation $\pi : A \to B(\mathcal{H})$ and a unitary representation $U : G \to U(\mathcal{H})$ on the same Hilbert space such that

$$\pi\big(\alpha_s(a)\big) = U_s \pi(a) U_s^*. \tag{2.11}$$

We say that (π, U) is a possibly degenerate covariant representation if π is a possibly degenerate representation.[3]

[3]Recall that by convention, all representations of C^*-algebras are presumed to be nondegenerate.

Example 2.11 (Degenerate examples). Obviously, covariant representations of degenerate dynamical systems such as $(A, \{e\}, \mathrm{id})$ correspond exactly to representations of A. Covariant representations of dynamical systems $(\mathbf{C}, G, \mathrm{id})$ correspond to unitary representations of G.

Example 2.12. Let G act on itself by left translation, and let $\mathrm{lt} : G \to \operatorname{Aut} C_0(G)$ be the associated dynamical system. Let $M : C_0(G) \to B\big(L^2(G)\big)$ be given by pointwise multiplication:

$$M(f)h(s) := f(s)h(s),$$

and let $\lambda : G \to U\big(L^2(G)\big)$ be the left-regular representation. Then (M, λ) is a covariant representation of $\big(C_0(G), G, \mathrm{lt}\big)$.

Example 2.13. Let $h \in \mathrm{Homeo}(\mathbf{T})$ be "rotation by θ": that is,

$$h(z) := e^{2\pi i\theta} z,$$

and let $\big(C(\mathbf{T}), \mathbf{Z}, \alpha\big)$ be the associated dynamical system:

$$\alpha_n(f)(z) = f(e^{-2\pi i n\theta} z).$$

(Although it might seem natural to think of h as "rotation through the angle $2\pi\theta$", the crucial feature is that h is rotation through θ of the circle. This action, as we shall see, has a very different character depending on whether θ is rational or irrational.)

(a) Let $M : C(\mathbf{T}) \to B\big(L^2(\mathbf{T})\big)$ be the representation given by pointwise multiplication:

$$M(f)h(z) := f(z)h(z).$$

Let $U : \mathbf{Z} \to U\big(L^2(\mathbf{T})\big)$ be the unitary representation given by

$$U_n h(z) := h(e^{-2\pi i n\theta} z).$$

Then it is not hard to check that (M, U) is a covariant representation of $\big(C(\mathbf{T}), \mathbf{Z}, \alpha\big)$.

(b) Now fix $w \in \mathbf{T}$ and let λ be the left-regular representation of $\mathbf{Z}$ on $L^2(\mathbf{Z})$. Define $\pi_w : C(\mathbf{T}) \to B\big(L^2(\mathbf{Z})\big)$ to be the representation

$$\pi_w(f)\xi(n) = f(e^{2\pi i n\theta} w)\xi(n).$$

Then (π_w, λ) is a covariant representation for each $w \in \mathbf{T}$.

In general, it is not obvious that there are any covariant representations of a given dynamical system. However, we know from the GNS theory (see [139, Appendix A.1]) that C^*-algebras have lots of representations. Given this, we can produce covariant representations of any system.

Example 2.14. Let $\rho : A \to B(\mathcal{H}_\rho)$ be any (possibly degenerate) representation of A on $\mathcal{H}_\rho$. Then define $\mathrm{Ind}_e^G \rho$ to be the pair $(\tilde{\rho}, U)$ of representations on the Hilbert space $L^2(G, \mathcal{H}_\rho) \cong L^2(G) \otimes \mathcal{H}_\rho$,[4] where

$$\tilde{\rho}(a)h(r) := \rho\big(\alpha_r^{-1}(a)\big)\big(h(r)\big) \quad \text{and} \quad U_s h(r) := h(s^{-1}r).$$

[4] See Appendix I.4.

Now compute:

$$\begin{aligned}
U_s\tilde{\rho}(a)U_s^*h(r) &= \tilde{\rho}(a)U_s^*h(s^{-1}r)\\
&= \rho\big(\alpha_{s^{-1}r}^{-1}(a)\big)\big(U_s^*h(s^{-1}r)\big)\\
&= \rho\big(\alpha_r^{-1}(\alpha_s(a))\big)\big(h(r)\big)\\
&= \tilde{\rho}\big(\alpha_s(a)\big)h(r).
\end{aligned}$$

Thus, $\operatorname{Ind}_e^G \rho := (\tilde{\rho}, U)$ is a (possibly degenerate) covariant representation.

Example 2.15. The representation in part (b) of Example 2.13 on the previous page is $\operatorname{Ind}_e^G \mathrm{ev}_w$.

Remark 2.16. The representations constructed in Example 2.14 on the preceding page are called *regular representations* of (A, G, α). The notation Ind_e^G is meant to suggest that a regular representation is induced from the system $(A, \{e\}, \alpha|_{\{e\}})$. We will discuss a general theory of induced representations in Section 5.1 and make this suggestion formal in Remark 5.7 on page 156.

Lemma 2.17. $\operatorname{Ind}_e^G \rho$ *is nondegenerate if* ρ *is nondegenerate.*

Proof. Let $(\tilde{\rho}, U) = \operatorname{Ind}_e^G \rho$ as above. Let $\{e_i\}$ be an approximate identity in A. It suffices to see that $\tilde{\rho}(e_i)\xi \to \xi$ for all $\xi \in L^2(G, \mathcal{H})$. Since $\tilde{\rho}$ is norm decreasing and ρ is nondegenerate, we can assume that $\xi(s) = (f \otimes \rho(a)h)(s) := f(s)\rho(a)h$ for $f \in C_c(G)$, $a \in A$ and $h \in \mathcal{H}$. Since

$$\tilde{\rho}(e_i)\xi(r) = f(r)\rho\big(\alpha_r^{-1}(e_i)a\big)h,$$

it suffices to see that given $\epsilon > 0$ we eventually have

$$\|\alpha_r^{-1}(e_i)a - a\| < \epsilon \quad \text{for all } r \in \operatorname{supp} f.$$

If this is not true, then there is an $\epsilon_0 > 0$ and a subset of $\{e_i\}$ and $r_i \in \operatorname{supp} f$ such that after relabeling

$$\|\alpha_{r_i}^{-1}(e_i)a - a\| \geq \epsilon_0. \tag{2.12}$$

Since $\operatorname{supp} f$ is compact, we can assume that $r_i \to r$. But then

$$\begin{aligned}
\|\alpha_{r_i}^{-1}(e_i)a - a\| &= \|\alpha_{r_i}^{-1}\big(e_i\alpha_{r_i}(a)\big) - \alpha_{r_i}^{-1}\big(\alpha_{r_i}(a)\big)\|\\
&= \|e_i\alpha_{r_i}(a) - \alpha_{r_i}(a)\|\\
&\leq \|e_i\big(\alpha_{r_i}(a) - \alpha_r(a)\big)\| + \qquad\qquad (2.13)\\
&\qquad \|e_i\alpha_r(a) - \alpha_r(a)\| + \|\alpha_r(a) - \alpha_{r_i}(a)\|.
\end{aligned}$$

But (2.13) goes to 0 since a subnet of $\{e_i\}$ is still an approximate identity. This contradicts (2.12) and finishes the proof. □

Definition 2.18. Suppose that (A, G, α) is a dynamical system and that (π, U) and (ρ, V) are covariant representations on $\mathcal{H}$ and $\mathcal{V}$ respectively. Their *direct sum* $(\pi, U) \oplus (\rho, V)$ is the covariant representation $(\pi \oplus \rho, U \oplus V)$ on $\mathcal{H} \oplus \mathcal{V}$ given by

$(\pi \oplus \rho)(a) := \pi(a) \oplus \rho(a)$ and $(U \oplus V)_s := U_s \oplus V_s$. A subspace $\mathcal{H}' \subset \mathcal{H}$ is invariant for (π, U) if $\pi(a)(\mathcal{H}') \subset \mathcal{H}'$ and $U_s(\mathcal{H}') \subset \mathcal{H}'$ for all $a \in A$ and $s \in G$. If $\mathcal{H}'$ is invariant, then the restrictions π' of π to $\mathcal{H}'$ and U' of U to $\mathcal{H}'$ are representations and the covariant representation (π', U') on $\mathcal{H}'$ is called a *subrepresentation* of (π, U). We call (π, U) irreducible if the only *closed* invariant subspaces are the trivial ones: $\{ 0 \}$ and $\mathcal{H}$. Finally, we say that (π, U) and (ρ, V) are *equivalent* if there is a unitary $W : \mathcal{H} \to \mathcal{V}$ such that

$$\rho(a) = W\pi(a)W^* \quad \text{and} \quad V_s = WU_sW^* \quad \text{for all } a \in A \text{ and } s \in G.$$

Remark 2.19. If $\mathcal{V} \subset \mathcal{H}$ is invariant for (π, U), then it is not hard to check that $\mathcal{V}^\perp$ is also invariant. Thus if (π', U') and (π'', U'') are the subrepresentations corresponding to $\mathcal{V}$ and $\mathcal{V}^\perp$, respectively, then $(\pi, U) = (\pi', U') \oplus (\pi'', U'')$. In particular, (π, U) is irreducible if and only if it is not equivalent to the direct sum of two nontrivial representations.

2.3 The Crossed Product

When G is abelian, we used the $*$-algebra $L^1(G)$ in Section 1.4 to recover the characters — that is, the irreducible representations of G. Here we want to construct a $*$-algebra — a C^*-algebra called the crossed product of A by G — from which we can recover exactly the covariant representations of a given dynamical system (A, G, α). This will include a special case, the group C^*-algebra as described in [139, Appendix C.3].

Example 2.20. Suppose $1_{\mathcal{H}} \in A \subset B(\mathcal{H})$ and that $u \in U(\mathcal{H})$ is such that $uAu^* \subset A$. For example, we could start with any automorphism of A, take a faithful representation of A and proceed as in Example 2.14 on page 45. Then $\alpha(a) = uau^*$ is an automorphism of A and $(A, \mathbf{Z}, \alpha)$ is a dynamical system. (Here $\alpha_n := \alpha^n$.) In this case, it would be natural to associate the C^*-subalgebra $C^*(A, \{ u \})$ of $B(\mathcal{H})$ generated by A and u to $(A, \mathbf{Z}, \alpha)$. I claim that

$$\mathcal{B} := \Bigl\{ \sum_{i \in \mathbf{Z}} a(i)u^i : a \in C_c(\mathbf{Z}, A) \Bigr\}$$

is a $*$-subalgebra of $B(\mathcal{H})$ and that $C^*(A, \{ u \})$ is the closure of $\mathcal{B}$.[5] For example, keeping in mind that all the sums in sight are finite, we can consider the product

$$\begin{aligned} \Bigl(\sum_i a(i)u^i \Bigr) \Bigl(\sum_j b(j)u^j \Bigr) &= \sum_{i,j} a(i)u^i b(j)u^j \\ &= \sum_i \sum_j a(i)\alpha_i\bigl(b(j)\bigr)u^{i+j} \end{aligned}$$

[5] If $1 \notin A$, then $\mathcal{B}$ is the subalgebra generated by *products* au with $a \in A$.

which, letting $k = j + i$, equals

$$= \sum_i \sum_k a(i)\alpha_i\big(b(k-i)\big)u^k$$
$$= \sum_k a * b(k)u^k,$$

where we have defined $a * b \in C_c(\mathbf{Z}, A)$ by

$$a * b(k) := \sum_i a(i)\alpha_i\big(b(k-i)\big). \tag{2.14}$$

This shows that $\mathcal{B}$ is closed under multiplication. Since a similar computation shows that the adjoint of $\sum_i a(i)u^i$ is given by $\sum_k a^*(k)u^k$ where

$$a^*(k) := \alpha_k\big(a(-k)^*\big), \tag{2.15}$$

$\mathcal{B}$ is closed under taking adjoints, and $\mathcal{B}$ is a $*$-algebra as claimed. Since $1_{\mathcal{H}} \in A$, it follows easily that $C^*(A, \{u\})$ is simply the closure of $\mathcal{B}$.

The previous example, and some experience with the group C^*-algebra construction, suggests we might be able to construct a crossed product by starting with a $*$-algebra based on $C_c(G, A)$ with the multiplication and involution compatible with (2.14) and (2.15). If $f, g \in C_c(G, A)$, then $(s, r) \mapsto f(r)\alpha_r\big(g(r^{-1}s)\big)$ is in $C_c(G \times G, A)$, and Corollary 1.104 on page 37 guarantees that

$$f * g(s) := \int_G f(r)\alpha_r\big(g(r^{-1}s)\big)\, d\mu(r) \tag{2.16}$$

defines an element of $C_c(G, A)$ called the *convolution* of f and g. Using the properties of Haar measure, our "Poor Man's" vector-valued Fubini Theorem (Proposition 1.105 on page 37) and Lemma 1.92 on page 32 it is not hard to check that (2.16) is an associative operation: $f * (g * h) = (f * g) * h$. Similarly, a little more computation shows that

$$f^*(s) := \Delta(s^{-1})\alpha_s\big(f(s^{-1})^*\big) \tag{2.17}$$

is an involution on $C_c(G, A)$ making $C_c(G, A)$ a $*$-algebra. For example:

$$\begin{aligned}
f^* * g^*(s) &= \int_G f^*(r)\alpha_r\big(g^*(r^{-1}s)\big)\, d\mu(r) \\
&= \Delta(s^{-1}) \int_G \alpha_r\big(f(r^{-1})^*\big)\alpha_s\big(g(s^{-1}r)^*\big)\, d\mu(r) \\
&= \Delta(s^{-1}) \int_G \alpha_{sr}\big(f(r^{-1}s^{-1})^*\big)\alpha_s\big(g(r)^*\big)\, d\mu(r) \\
&= \Delta(s^{-1})\alpha_s\Big(\int_G g(r)\alpha_r\big(f(r^{-1}s^{-1})\big)\, d\mu(r)\Big)^* \\
&= \Delta(s^{-1})\alpha_s\big(g * f(s^{-1})\big)^* \\
&= (g * f)^*(s).
\end{aligned}$$

Furthermore

$$\|f\|_1 := \int_G \|f(s)\|\,d\mu(s) \tag{2.18}$$

is a norm on $C_c(G,A)$, and the properties of Haar measure guarantee that $\|f^*\|_1 = \|f\|_1$ and $\|f*g\|_1 \le \|f\|_1\|g\|_1$.

Definition 2.21. A $*$-homomorphism $\pi : C_c(G,A) \to B(\mathcal{H})$ is called a *$*$-representation* of $C_c(G,A)$ on $\mathcal{H}$. We say π is nondegenerate if

$$\{\,\pi(f)h : f \in C_c(G,A) \text{ and } h \in \mathcal{H}\,\}$$

spans a dense subset of $\mathcal{H}$. If $\|\pi(f)\| \le \|f\|_1$, then π is called L^1-norm decreasing.

Example 2.22. Since a locally compact group G is a degenerate dynamical system, $C_c(G)$ is a $*$-algebra with operations

$$f*g(s) = \int_G f(r)g(r^{-1}s)\,d\mu(s) \quad \text{and} \quad f^*(s) = \Delta(s^{-1})\overline{f(s^{-1})}.$$

Proposition 2.23. *Suppose that (π,U) is a (possibly degenerate) covariant representation of (A,G,α) on $\mathcal{H}$. Then*

$$\pi \rtimes U(f) := \int_G \pi\big(f(s)\big)U_s\,d\mu(s) \tag{2.19}$$

defines a L^1-norm decreasing $*$-representation of $C_c(G,A)$ on $\mathcal{H}$ called the integrated form *of (π,U). Furthermore, $\pi \rtimes U$ is nondegenerate if π is nondegenerate. We call $\pi \rtimes U$ the* integrated form *of (π,U).*

Proof. Notice that $s \mapsto U_s$ is strictly continuous by Corollary 1.99 on page 34, and $s \mapsto \pi\big(f(s)\big)$ is in $C_c\big(G,B(\mathcal{H})\big)$. Consequently, the integrand in (2.19) is in $C_c\big(G,B_s(\mathcal{H})\big)$, and $\pi \rtimes U(f)$ is defined by Lemma 1.101 on page 35. If h and k are unit vectors in $\mathcal{H}$, then (1.35) implies that $|\big(\pi \rtimes U(f)h \mid k\big)| \le \int_G |\big(\pi\big(f(s)\big)U_sh \mid k\big)|\,d\mu(s)$. The Cauchy-Schwarz inequality implies that $|\big(\pi\big(f(s)\big)U_sh \mid k\big)| \le \|f(s)\|$. Thus $|\big(\pi \rtimes U(f)h \mid k\big)| \le \|f\|_1$. Since h and k are arbitrary, $\|\pi \rtimes U(f)\| \le \|f\|_1$.

To see that $\pi \rtimes U$ is a $*$-homomorphism, we compute using Lemma 1.101 on page 35:

$$\begin{aligned}
\pi \rtimes U(f)^* &= \int_G \big(\pi(f(s))U_s\big)^*\,d\mu(s)\\
&= \int_G U_{s^{-1}}\pi\big(f(s)^*\big)\,d\mu(s)\\
&= \int_G U_s\pi\big(f(s^{-1})^*\big)\Delta(s^{-1})\,d\mu(s)\\
&= \int_G \pi\big(\alpha_s\big(f(s^{-1})^*\Delta(s^{-1})\big)\big)U_s\,d\mu(s)\\
&= \pi \rtimes U(f^*).
\end{aligned}$$

Similarly,

$$\begin{aligned}\pi \rtimes U(f * g) &= \int_G \int_G \pi\big(f(r)\alpha_r\big(g(r^{-1}s)\big)\big)U_s \, d\mu(r)\, d\mu(s) \\ &= \int_G \int_G \pi\big(f(r)\big)U_r\pi\big(g(r^{-1}s)\big)U_{r^{-1}s}\, d\mu(r)\, d\mu(s),\end{aligned}$$

which, by Fubini (Proposition 1.105 on page 37) and left-invariance, is

$$\begin{aligned}&= \int_G \int_G \pi\big(f(r)\big)U_r\pi\big(g(s)\big)U_s \, d\mu(s)\, d\mu(r) \\ &= \pi \rtimes U(f) \circ \pi \rtimes U(g).\end{aligned}$$

Now assume that π is nondegenerate. Let $h \in \mathcal{H}$ and $\epsilon > 0$. Then if $\{ e_i \}$ is an approximate identity for A, we have $\pi(e_i)h \to h$ in $\mathcal{H}$. Thus we can choose an element u of norm one in A such that $\|\pi(u)h - h\| < \epsilon/2$. Let V be a neighborhood of e in G such that $\|U_s h - h\| < \epsilon/2$ if $s \in V$. Let $\varphi \in C_c(G)$ be nonnegative with $\operatorname{supp}\varphi \subset V$ and integral one. Let $f(s) = (\varphi \otimes u)(s) := \varphi(s)u$. Then $f \in C_c(G, A)$ and if k is an element of norm one in $\mathcal{H}$,

$$\begin{aligned}\big|\big(\pi \rtimes U(f)h \mid k\big) - (h \mid k)\big| &= \Big|\int_G \varphi(s)\big(\pi(u)U_s h - h \mid k\big)\, ds\Big| \\ &\le \int_G \varphi(s)\big|\big(\pi(u)(U_s h - h) \mid k\big)\big|\, ds + \\ &\qquad \int_G \varphi(s)\big|\big(\pi(u)h - h \mid k\big)\big|\, ds \\ &\le \frac{\epsilon}{2} + \frac{\epsilon}{2} = \epsilon.\end{aligned}$$

Since k is arbitrary, it follows that $\|\pi \rtimes U(f)h - h\| \le \epsilon$ and therefore $\pi \rtimes U$ is nondegenerate. □

Example 2.24. Notice that the map in Example 2.20 on page 47 sending $a \in C_c(\mathbf{Z}, A)$ to $\sum_i a(i)u^i$ is simply the integrated form of (id, u) where id is the identity representation of A and u is viewed as a representation of $\mathbf{Z}$ in the obvious way: $u_k := u^k$.

Example 2.25. In the case of a locally compact group G and a unitary representation U, the integrated form is:

$$\mathrm{id} \rtimes U(z) := \int_G z(s)U_s \, d\mu(s) \quad \text{for all } z \in C_c(G).$$

Traditionally, $\mathrm{id} \rtimes U$ is shortened to just U. That is, the same letter is used both for a unitary representation and its integrated form.

The following maps will be of considerable importance in Section 2.4 and in Proposition 2.34 on page 54 in particular. For each $r \in G$ let $i_G(r) : C_c(G, A) \to C_c(G, A)$ be defined by

$$i_G(r)f(s) := \alpha_r\big(f(r^{-1}s)\big). \tag{2.20}$$

Note that if (π, U) is a covariant representation of (A, G, α), then

$$\begin{aligned}\pi \rtimes U\big(i_G(r)f\big) &= \int_G \pi\big(i_G(r)f(s)\big)U_s\,d\mu(s) \\ &= \int_G \pi\big(\alpha_r\big(f(r^{-1}s)\big)\big)U_s\,d\mu(s) \\ &= \int_G \pi\big(\alpha_r\big(f(s)\big)\big)U_{rs}\,d\mu(s) \\ &= U_r \circ \pi \rtimes U(f).\end{aligned} \tag{2.21}$$

Lemma 2.26. *Let ρ be a faithful representation of A on $\mathcal{H}$ and let $\operatorname{Ind}_e^G \rho = (\tilde{\rho}, U)$ be the corresponding regular representation. Suppose that $f \in C_c(G, A)$ and $f \neq 0$. Then $\tilde{\rho} \rtimes U(f) \neq 0$. That is, the integrated form of a regular representation corresponding to a faithful representation of A is faithful on $C_c(G, A)$.*

Proof. There is a $r \in G$ such that $f(r) \neq 0$. Since (2.21) implies that $\|\tilde{\rho} \rtimes U(f)\| = \|\tilde{\rho} \rtimes U\big(i_G(r^{-1})f\big)\|$, we can replace f by $i_G(r^{-1})f$ and assume that $r = e$. Since ρ is faithful, there are vectors h and k in $\mathcal{H}$ such that

$$\big(\rho\big(f(e)\big)h \mid k\big) \neq 0.$$

We can find a neighborhood V of e such that $s, r \in V$ implies

$$\big|\big(\rho\big(\alpha_r^{-1}\big(f(s)\big)\big)h \mid k\big) - \big(\rho\big(f(e)\big)h \mid k\big)\big| < \frac{|\big(\rho\big(f(e)\big)h \mid k\big)|}{3}.$$

Choose $\varphi \in C_c^+(G)$ with support contained in symmetric neighborhood W of e such that $W^2 \subset V$, and such that

$$\int_G \int_G \varphi(s^{-1}r)\varphi(r)\,d\mu(r)\,d\mu(s) = 1.$$

Now define ξ and η in $L^2(G, \mathcal{H})$ by

$$\xi(s) := \varphi(s)h \quad \text{and} \quad \eta(s) := \varphi(s)k.$$

Then

$$\begin{aligned}&|\big(\tilde{\rho} \rtimes U(f)\xi \mid \eta\big) - \big(\rho\big(f(e)\big)h \mid k\big)| \\ &\quad = \Big|\int_G \int_G \big(\rho\big(\alpha_r^{-1}\big(f(s)\big)\big)\xi(s^{-1}r) \mid \eta(r)\big)\,d\mu(s)\,d\mu(r) \\ &\qquad\qquad - \big(\rho\big(f(e)\big)h \mid k\big)\Big| \\ &\quad \leq \int_G \int_G \varphi(s^{-1}r)\varphi(r)\Big|\big(\big(\rho\big(\alpha_r^{-1}\big(f(s)\big)\big)h \mid k\big) - \\ &\qquad\qquad \big(\rho\big(f(e)\big)h \mid k\big)\big)\Big|\,d\mu(s)\,d\mu(r) \\ &\quad < \frac{|\big(\rho\big(f(e)\big)h \mid k\big)|}{2}.\end{aligned}$$

Now it follows that $\tilde{\rho} \rtimes U(f) \neq 0$. □

We define the crossed product associated to (A, G, α) as a completion of $C_c(G, A)$ as described in Section 1.5.1.

Lemma 2.27. *Suppose that (A, G, α) is a dynamical system and that for each $f \in C_c(G, A)$ we define*

$$\|f\| := \sup\{\|\pi \rtimes U(f)\| : (\pi, U) \text{ is a (possibly degenerate) covariant representation of } (A, G, \alpha)\}. \quad (2.22)$$

Then $\|\cdot\|$ is a norm on $C_c(G, A)$ called the universal norm. *The universal norm is dominated by the $\|\cdot\|_1$-norm, and the completion of $C_c(G, A)$ with respect to $\|\cdot\|$ is a C^*-algebra called the* crossed product *of A by G and is denoted by $A \rtimes_\alpha G$.*

Remark 2.28. Some might find the supremum in (2.22) suspicious because the collection of covariant representations is not clearly a set. Various finesses for this apparent defect are employed in the literature. For example, it would be possible to consider only covariant representations on a fixed Hilbert space of suitably large dimension and argue that all representations are equivalent to a (possibly degenerate) representation on this space. However, the collection of values in (2.22) is a subclass of the set $\mathbf{R}$ of real numbers and the separation axioms of set theory guarantee that a subclass of a set is a set and that we are taking the supremum of a bounded *set* of real numbers [84, §1.1]. Thus here, and in the sequel, we will not worry about taking such supremums.

Proof. Once we are satisfied that we have made sense of the supremum in (2.22), Proposition 2.23 on page 49 and Lemma 2.26 on the previous page imply that $0 < \|f\| \leq \|f\|_1 < \infty$ (provided $f \neq 0$). Now it easy to see that $\|\cdot\|$ is a norm on $C_c(G, A)$ such that $\|f^* * f\| = \|f\|^2$. Therefore the completion is a C^*-algebra. □

Remark 2.29. Since the universal norm is a norm on $C_c(G, A)$, we can view $C_c(G, A)$ as a $*$-subalgebra of $A \rtimes_\alpha G$. Therefore we will rarely distinguish between an element of $C_c(G, A)$ and its image in $A \rtimes_\alpha G$.

Remark 2.30. Let $\mathcal{B} \subset C_c(G, A)$ be a $*$-subalgebra which is dense in the inductive limit topology. This means that given $f \in C_c(G, A)$ there is a compact set K such that for all $\epsilon > 0$ there exists a $b \in \mathcal{B}$ such that $\operatorname{supp} b \subset K$ and $\|b - f\|_\infty < \epsilon$. This immediately implies that $\mathcal{B}$ is $\|\cdot\|_1$-norm dense in $C_c(G, A)$ and therefore dense in $A \rtimes_\alpha G$ as well.

Lemma 2.31. *Let (π, U) be a (possibly degenerate) covariant representation of (A, G, α) on $\mathcal{H}$. Let*

$$\mathcal{V} := \overline{\operatorname{span}}\{\, \pi(a)h : a \in A \text{ and } h \in \mathcal{H}\,\}$$

be the essential subspace *of π, and let* $\operatorname{ess}\pi$ *be the corresponding subrepresentation. Then $\mathcal{V}$ is also invariant for U, and if U' is the corresponding subrepresentation, then $(\operatorname{ess}\pi, U')$ is a nondegenerate covariant representation on $\mathcal{V}$. For all $f \in C_c(G, A)$,*

$$\|(\operatorname{ess}\pi) \rtimes U'(f)\| = \|\pi \rtimes U(f)\|.$$

In particular,

$$\|f\| = \sup\{\, \|\pi \rtimes U(f)\| : (\pi, U) \text{ is a nondegenerate covariant representation of } (A, G, \alpha).\}$$

Proof. Since $U_s\pi(a)h = \pi\big(\alpha_s(a)\big)U_sh$, it is clear that $\mathcal{V}$ is invariant for U, and that $(\operatorname{ess}\pi, U')$ is a nondegenerate covariant representation. Note that $\pi = \operatorname{ess}\pi \oplus 0$. Thus if $U = U' \oplus U''$, then $\pi \rtimes U = (\operatorname{ess}\pi) \rtimes U' \oplus 0$, and the rest is straightforward. □

It should come as no surprise that crossed products in which the algebra A is commutative are considerably more tractable than the general case. Since there is a one-to-one correspondence between dynamical systems with A commutative and transformations groups, such crossed products are called *transformation group C^*-algebras.* Moreover, it is possible to describe the $*$-algebra structure on $C_c\big(G, C_0(X)\big)$ in terms of functions on $G \times X$. If we agree to identify a function on $G \times X$ with the obvious function from G to functions on X, then it is not hard to prove that we have inclusions

$$C_c(G \times X) \subset C_c\big(G, C_c(X)\big) \subset C_c\big(G, C_0(X)\big). \tag{2.23}$$

If $f \in C_c\big(G, C_0(X)\big)$, then since point evaluation is a homomorphism from $C_0(X)$ to $\mathbf{C}$,

$$\int_G f(s)\, d\mu(s)(x) = \int_G f(s)(x)\, d\mu(s). \tag{2.24}$$

Using (2.24) and the formulas for convolution and involution on $C_c\big(G, C_0(X)\big)$, it is not hard to see that $C_c(G \times X)$ is a $*$-subalgebra. It follows easily (using the first inclusion in (2.23) and Lemma 1.87 on page 29) that $C_c(G \times X)$ is dense in $C_c\big(G, C_0(X)\big)$ and therefore in $C_0(X) \rtimes_\alpha G$. The formulas for convolution and involution on $C_c(G \times X)$ are

$$f * g(s, x) = \int_G f(r, x) g(r^{-1}s, r^{-1} \cdot x)\, d\mu(r), \text{ and} \tag{2.25}$$

$$f^*(s, x) = \Delta(s^{-1})\overline{f(s^{-1}, s^{-1} \cdot x)}. \tag{2.26}$$

(In particular, the formula for convolution is a scalar valued integral.)

Remark 2.32. We mentioned the inclusion of $C_c(G \times X)$ into $C_c\big(G, C_c(X)\big)$ only to make applying Lemma 1.87 on page 29 a bit easier. It would be tempting to guess that this inclusion is always an equality. But this is not the case. Let $G = \mathbf{T}$ and $X = \mathbf{R}$. Let $\varphi : \mathbf{R} \to \mathbf{R}$ be defined by

$$\varphi(x) := \begin{cases} 0 & \text{if } |x| \geq 1, \\ 1 + x & \text{if } -1 \leq x \leq 0, \text{ and} \\ 1 - x & \text{if } 0 \leq x \leq 1. \end{cases}$$

Then it is not hard to check that $f(x+iy)(r) := |y|\varphi(|y|r)$ is in $C_c\big(\mathbf{T}, C_c(\mathbf{R})\big)$. Since

$$\operatorname{supp} f(x+iy) = \begin{cases} \{0\} & \text{if } y = 0, \text{ and} \\ [-1/|y|, 1/|y|] & \text{otherwise,} \end{cases}$$

the support of f as a function on $\mathbf{T} \times \mathbf{R}$ is unbounded and therefore not compact.

Example 2.33. In the degenerate case where our dynamical system reduces to $(A, \{e\}, \mathrm{id})$, then the crossed product gives us back A. In the case $(\mathbf{C}, G, \mathrm{id})$, then covariant representations correspond exactly to unitary representations of G, and the crossed product is the *group C^*-algebra* $C^*(G)$ as defined in [139, C.3]. (We took a slightly different approach in [139, C.3], but [139, Corollary C.18] shows the approaches arrive at the same completion of $C_c(G)$.)

2.4 Representations of the Crossed Product

Except in special cases, the crossed product $A \rtimes_\alpha G$ does not contain a copy of either A or G. However, the multiplier algebra $M(A \rtimes_\alpha G)$ does. Recall that if A is a C^*-algebra, then $M(A)$ is $\mathcal{L}(A_A)$ — that is, the collection of adjointable operators from A to itself [139, §2.3]. When convenient, we will view A as a subalgebra of $M(A)$ — $a \in A$ is identified with the operator $b \mapsto ab$. The unitary group of $M(A)$ is denoted by $UM(A)$.

If we view $C_c(G, A)$ as a $*$-subalgebra of $M(A \rtimes_\alpha G)$, then a $T \in M(A \rtimes_\alpha G)$ may, or may not, map $C_c(G, A)$ into itself. In practice however, a multiplier T is usually defined by first defining it as a map from $C_c(G, A)$ to itself, and then showing it is bounded with respect to the universal norm so that it extends to a map also called T from $A \rtimes_\alpha G$ to itself. It defines a multiplier provided we can find an adjoint T^* which is characterized by

$$T(a)^*b = \big\langle T(a)\,,\, b\big\rangle_{A\rtimes_\alpha G} = \big\langle a\,,\, T^*(b)\big\rangle_{A\rtimes_\alpha G} = a^*T^*(b)$$

for all $a, b \in A$. The next proposition is case in point.

Proposition 2.34. *Suppose that $\alpha : G \to \operatorname{Aut} A$ is a dynamical system. Then there is a nondegenerate faithful homomorphism*

$$i_A : A \to M(A \rtimes_\alpha G)$$

and an injective strictly continuous unitary valued homomorphism

$$i_G : G \to UM(A \rtimes_\alpha G)$$

such that for $f \in C_c(G, A)$, $r, s \in G$ and $a \in A$ we have

$$i_G(r)f(s) = \alpha_r\big(f(r^{-1}s)\big) \quad \textit{and} \quad i_A(a)f(s) = af(s).$$

Moreover (i_A, i_G) is covariant in that

$$i_A\big(\alpha_r(a)\big) = i_G(r)i_A(a)i_G(r)^*.$$

If (π, U) *is nondegenerate, then*

$$(\pi \rtimes U)^{-}\big(i_A(a)\big) = \pi(a) \quad \textit{and} \quad (\pi \rtimes U)^{-}\big(i_G(s)\big) = U_s.$$

Proof. We already considered $i_G(r)$ in Section 2.3 and it follows from (2.21) that for all $r \in G$

$$\|i_G(r)f\| = \|f\|.$$

Thus we can extend $i_G(r)$ to a map of $A \rtimes_\alpha G$ to itself, and

$$\big(i_G(r)f\big)^* * g(t) = \int_G \alpha_{sr}\big(f(r^{-1}s^{-1})^*\big)\Delta(s^{-1})\alpha_s\big(g(s^{-1}t)\big)\,d\mu(s)$$

which, since $\Delta(s^{-1})\,d\mu(s)$ is a right Haar measure on G, is

$$\begin{aligned} &= \int_G \alpha_s\big(f(s^{-1})^*\big)\Delta(s^{-1})\alpha_{sr^{-1}}\big(g(rs^{-1}t)\big)\,d\mu(s) \\ &= \int_G f^*(s)\alpha_s\big(i_G(r^{-1})g(s^{-1}t)\big)\,d\mu(s) \\ &= f^* * i_G(r^{-1})g(t). \end{aligned}$$

Thus each $i_G(r)$ is adjointable with adjoint $i_G(r)^* = i_G(r^{-1})$. Since we certainly have $i_G(rs) = i_G(r)\circ i_G(s)$ and $i_G(r)^{-1} = i_G(r^{-1})$, i_G is a unitary-valued homomorphism into $UM(A \rtimes_\alpha G)$. To show that i_G is strictly continuous, fix $f \in C_c(G, A)$, a compact neighborhood W of e in G, and let $K := \operatorname{supp} f$. Notice that as long as $r \in W$, then

$$\operatorname{supp} i_G(r)f \subset WK.$$

If $\epsilon > 0$, then the uniform continuity of f implies that we can choose $V \subset W$ such that $r \in V$ implies

$$\|f(r^{-1}s) - f(s)\| < \frac{\epsilon}{2\mu(WK)} \quad \text{for all } s \in G.$$

Since f has compact support, we can shrink V if need be so that $r \in V$ also implies

$$\|\alpha_r\big(f(s)\big) - f(s)\| < \frac{\epsilon}{2\mu(WK)} \quad \text{for all } s \in G.$$

Since

$$\|i_G(r)f(s) - f(s)\| \le \|\alpha_r\big(f(r^{-1}s) - f(s)\big)\| + \|\alpha_r\big(f(s)\big) - f(s)\|,$$

it follows that $\|i_G(r)f - f\| \le \|i_G(r)f - f\|_1 < \epsilon$. Therefore $r \mapsto i_G(r)$ is strongly continuous, and i_G is strictly continuous by Corollary 1.99 on page 34.

The argument for i_A is similar. If (π, U) is a covariant representation of (A, G, α), then for $f \in C_c(G, A)$,

$$\pi \rtimes U\big(i_A(a)f\big) = \pi(a) \circ \pi \rtimes U(f), \tag{2.27}$$

and it follows that $\|i_A(a)f\| \le \|a\|\|f\|$. Therefore $i_A(a)$ extends to map from $A \rtimes_\alpha G$ to itself. An easy computation shows that

$$\big(i_A(a)f\big)^* * g = f^* * i_A(a^*)g.$$

Therefore $i_A(a)$ is adjointable with adjoint $i_A(a^*)$, and i_A is a homomorphism of A into $M(A \rtimes_\alpha G)$. The covariance condition can be checked by applying both sides to $f \in C_c(G, A)$.

If (π, U) is nondegenerate, then (2.27) implies that $(\pi \rtimes U)^{-}\big(i_A(a)\big) = \pi(a)$, and the corresponding assertion for $i_G(s)$ follows from (2.21).

To see that i_A and i_G are injective, let $\rho : A \to B(\mathcal{H})$ be a faithful representation of A, and set $(\tilde{\rho}, U) = \operatorname{Ind}_e^G \rho$. Then $\tilde{\rho}$ is faithful and certainly $U_s \neq \mathrm{id}$ if $s \neq e$. Since $\tilde{\rho}(a) = (\tilde{\rho} \rtimes U)^{-} \circ i_A(a)$, it follows that $i_A(a)$ is faithful. Similarly $U_s = (\tilde{\rho} \rtimes U)^{-} \circ i_G(s)$ shows $i_G(s) \neq \mathrm{id}$ if $s \neq e$.

To see that i_A is nondegenerate, note that elementary tensors of the form $\varphi \otimes ab$ span a dense subalgebra of $A \rtimes_\alpha G$ by Lemma 1.87 on page 29, and are also in $i_A(A) \cdot A \rtimes_\alpha G$. □

Lemma 2.35. *There is a homomorphism $\tilde{\imath}_G : C^*(G) \to M(A \rtimes_\alpha G)$ such that*

$$\tilde{\imath}_G(z) = \int_G z(s) i_G(s)\, d\mu(s) \quad \textit{for all } z \in C_c(G).$$

Just as for unitary representations (c.f., Example 2.25 on page 50), we write $i_G(z)$ in place of $\tilde{\imath}_G(z)$.

Proof. If $z \in C_c(G)$, then $\tilde{\imath}_G(z)$ is a well-defined element of $M(A \rtimes_\alpha G)$ by Lemma 1.101 on page 35. Just as in Proposition 2.23 on page 49, $\tilde{\imath}_G$ is a homomorphism on $C_c(G)$. For example,

$$\begin{aligned}\tilde{\imath}(z * w) &= \int_G z * w(s) i_G(s)\, d\mu(s)\\ &= \int_G \int_G z(r) w(r^{-1}s)\, d\mu(r) i_G(s)\, d\mu(s)\end{aligned}$$

which, by our Fubini result Proposition 1.105 and using Lemma 1.101 to move $i_G(s)$ through the integral, is

$$\begin{aligned}&= \int_G \int_G z(r) w(r^{-1}s) i_G(s)\, d\mu(s)\, d\mu(r)\\ &= \int_G \int_G z(r) w(s) i_G(rs)\, d\mu(r)\, d\mu(s)\\ &= \tilde{\imath}(z)\tilde{\imath}(w).\end{aligned}$$

All that remains to show is that $\tilde{\imath}_G$ is bounded with respect to the universal norm on $C^*(G)$. But if π is a nondegenerate faithful representation of $A \rtimes_\alpha G$, then $\bar{\pi}$ is faithful on $M(A \rtimes_\alpha G)$. Let $U_s := \bar{\pi}\big(i_G(s)\big)$. Then $s \mapsto U_s$ is a unitary-valued homomorphism into $U(\mathcal{H}_\pi)$. To see that U is a unitary representation of G, we need to see that $s \mapsto U_s h$ is continuous for all $h \in \mathcal{H}_\pi$. Since π is nondegenerate, it suffices to consider $h = \pi(f)k$ for $f \in C_c(G, A)$. Then $U_s\pi(f)k = \pi\big(i_G(s)f\big)k$. Since $i_G(s)f$ is continuous in s, it follows that U is strongly continuous and therefore a representation. Therefore Lemma 1.101 on page 35 implies that

$$\begin{aligned}\bar{\pi}\big(\tilde{\imath}_G(z)\big) &= \bar{\pi}\Big(\int_G z(s) i_G(s)\, d\mu(s)\Big)\\ &= \int_G z(s) U_s \, d\mu(s)\\ &= U(z).\end{aligned}$$

Thus

$$\|\tilde{\imath}_G(z)\| = \|\bar{\pi}\big(i_G(z)\big)\| = \|U(z)\| \le \|z\|. \qquad \square$$

Corollary 2.36. *Suppose that* $\alpha : G \to \operatorname{Aut} A$ *is a dynamical system. Let* $a \in A$, $z \in C_c(G)$ *and* $g, h \in C_c(G, A)$. *Then* $i_A(a)i_G(z)$, $\int_G i_A\big(g(r)\big)i_G(r)(h)\, d\mu(r)$ *and* $\int_G i_A\big(g(r)\big)i_G(r)\, d\mu(r)$ *are in* $C_c(G, A) \subset A \rtimes_\alpha G \subset M(A \rtimes_\alpha G)$. *In fact,*

$$i_A(a)i_G(z) = z \otimes a, \tag{2.28}$$

$$\int_G i_A\big(g(s)\big)i_G(s)(h)\, d\mu(s) = g * h, \text{ and} \tag{2.29}$$

$$\int_G i_A\big(g(s)\big)i_G(s)\, d\mu(s) = g. \tag{2.30}$$

Proof. Notice that if $T \in M(A \rtimes_\alpha G)$ and $(\pi \rtimes U)^-(T) = 0$ for all nondegenerate covariant pairs, then $T = 0$. (For example, $Tf = 0$ for all $f \in A \rtimes_\alpha G$ by Lemma 2.31 on page 52.) But if (π, U) is nondegenerate,

$$(\pi \rtimes U)^-\Big(\int_G i_A\big(g(s)\big)i_G(s)\, d\mu(s)\Big) = \int_G \pi\big(g(s)\big)U_s\, d\mu(s) = \pi \rtimes U(g).$$

This proves (2.30), and (2.28) is a special case. Finally,

$$g * h = \int_G i_A\big(g(s)\big)i_G(s)\, d\mu(s)h = \int_G i_A\big(g(s)\big)i_G(s)(h)\, d\mu(s)$$

by Lemma 1.101 on page 35. $\square$

Definition 2.37. Suppose that $\alpha : G \to \operatorname{Aut} A$ is a dynamical system and that X is a Hilbert B-module. Then a *covariant homomorphism* of (A, G, α) into $\mathcal{L}(\mathsf{X})$ is a pair (π, u) consisting of a homomorphism $\pi : A \to \mathcal{L}(\mathsf{X})$ and a strongly continuous unitary-valued homomorphism $u : G \to U\mathcal{L}(\mathsf{X})$ such that

$$\pi\big(\alpha_s(a)\big) = u_s \pi(a) u_s^*. \tag{2.31}$$

We say that (π, u) is nondegenerate if π is nondegenerate.

Remark 2.38. If X is a Hilbert space $\mathcal{H}$, then a (nondegenerate) covariant homomorphism into $\mathcal{L}(\mathcal{H}) = B(\mathcal{H})$ is just a (nondegenerate) covariant representation. Corollary 1.99 on page 34 implies that the strong continuity of u in Definition 2.37 on the previous page is equivalent to strict continuity as a map into $\mathcal{L}(\mathsf{X}) = M(\mathcal{K}(\mathsf{X}))$. In fact, a covariant homomorphism can equally well be thought of as maps into $M(B)$ for a C^*-algebra B. Since $M(B) = \mathcal{L}(B_B)$ and $\mathcal{L}(\mathsf{X}) = M(\mathcal{K}(\mathsf{X}))$ there is no real difference.

Proposition 2.39. *Suppose that $\alpha : G \to \operatorname{Aut} A$ is a dynamical system and that X is a Hilbert B-module. If (π, u) is a covariant homomorphism of (A, G, α) into $\mathcal{L}(\mathsf{X})$, then Lemma 1.101 on page 35 implies that the* integrated form

$$\pi \rtimes u(f) := \int_G \pi\big(f(s)\big)u_s \, d\mu(s)$$

is a well-defined operator in $\mathcal{L}(\mathsf{X})$, and $\pi \rtimes u$ extends to a homomorphism of $A \rtimes_\alpha G$ into $\mathcal{L}(\mathsf{X})$ which is nondegenerate whenever π is nondegenerate. In this case, $(\pi \rtimes u)^-\big(i_A(a)\big) = \pi(a)$ and $(\pi \rtimes u)^-\big(i_G(s)\big) = u_s$.

Conversely, if $L : A \rtimes_\alpha G \to \mathcal{L}(\mathsf{X})$ is a nondegenerate homomorphism, then there is a nondegenerate covariant homomorphism (π, u) of (A, G, α) into $\mathcal{L}(\mathsf{X})$ such that $L = \pi \rtimes u$. In fact, if $\bar{L}$ is the canonical extension of L to $M(A \rtimes_\alpha G)$, then

$$u_s := \bar{L}\big(i_G(s)\big) \quad \textit{and} \quad \pi(a) := \bar{L}\big(i_A(a)\big). \tag{2.32}$$

Proof. The map $s \mapsto u_s$ is strictly continuous into $\mathcal{L}(\mathsf{X}) = M(\mathcal{K}(\mathsf{X}))$ by Corollary 1.99 on page 34, and so $s \mapsto \pi\big(f(s)\big)u_s$ is a strictly continuous map for each $f \in C_c(G, A)$. Thus $\pi \rtimes u(f)$ is an operator in $\mathcal{L}(\mathsf{X})$ by Lemma 1.101 on page 35.

Let $\rho : \mathcal{K}(\mathsf{X}) \to B(\mathcal{H}_\rho)$ be a faithful nondegenerate representation of $\mathcal{K}(\mathsf{X})$, and let $\bar{\rho}$ be the canonical (and faithful) extension to $\mathcal{L}(\mathsf{X})$. Let $\Pi(a) := \bar{\rho}\big(\pi(a)\big)$ and $U_s := \bar{\rho}(u_s)$. We claim that $s \mapsto U_s$ is strongly continuous from G into $U(\mathcal{H}_\rho)$. Since ρ is nondegenerate, it suffices to show that $s \mapsto U_s\big(\rho(T)h\big)$ is continuous when $h \in \mathcal{H}_\rho$ and $T \in \mathcal{K}(\mathsf{X})$). But $s \mapsto u_sT$ is continuous since u is strictly continuous, and $U_s\big(\rho(T)h\big) = \rho(u_sT)h$ is continuous in s. It now follows easily that (Π, U) is a covariant representation and Lemma 1.101 on page 35 implies

$$\begin{aligned}
\|\bar{\rho}(\pi \rtimes u(f))\| &= \Big\|\bar{\rho}\Big(\int_G \pi\big(f(s)\big)u_s \, d\mu(s)\Big)\Big\| \\
&= \Big\|\int_G \Pi\big(f(s)\big)U_s \, d\mu(s)\Big\| \\
&= \|\Pi \rtimes U(f)\| \\
&\le \|f\|.
\end{aligned}$$

It follows that $\pi \rtimes u = \bar{\rho}^{-1} \circ (\Pi \rtimes U)$ is a homomorphism of $C_c(G, A)$ into $\mathcal{L}(\mathsf{X})$ which is bounded with respect to the universal norm. Therefore $\pi \rtimes u$ extends to a homomorphism of $A \rtimes_\alpha G$ into $\mathcal{L}(\mathsf{X})$ as claimed. The proof that $\pi \rtimes u$ is nondegenerate when π is proceeds exactly as in Proposition 2.23 on page 49, and the statements about $(\pi \rtimes u)^-$ follow as in Proposition 2.34 on page 54.

To prove the converse, suppose that $L : A \rtimes_\alpha G \to \mathcal{L}(\mathsf{X})$ is a nondegenerate homomorphism, and let π and u be defined as in (2.32). Since i_G is strictly continuous, it is straightforward to check that u is strongly continuous. Since i_A and L are nondegenerate it follows that if $\{ e_i \}$ is an approximate identity in A, then $i_A(e_i)$ converges to strictly to 1 in $M(A \rtimes_\alpha G)$. Therefore $\pi(e_i) = \bar{L}\bigl(i_A(e_i)\bigr)$ converges strictly to 1_X in $\mathcal{L}(\mathsf{X})$, and π is nondegenerate. Since the covariance condition is straightforward to check, it follows that (π, u) is a nondegenerate covariant homomorphism, and the first part of this proof shows the integrated form $\pi \rtimes u$ is a nondegenerate homomorphism into $\mathcal{L}(\mathsf{X})$. Since both $\pi \rtimes u$ and L are nondegenerate, we can apply Lemma 1.101 to conclude that for each $a \in A$ and $z \in C_c(G)$ that

$$\begin{aligned}
\pi \rtimes u\bigl(i_A(a)i_G(z)\bigr) &= \pi(a) \circ (\pi \rtimes u)^{-}\bigl(i_G(z)\bigr) \\
&= \bar{L}\bigl(i_A(a)\bigr) \int_G z(s)u_s \, d\mu(s) \\
&= \bar{L}\bigl(i_A(a)\bigr) \int_G z(s)\bar{L}\bigl(i_G(s)\bigr) \, d\mu(s) \\
&= \bar{L}\bigl(i_A(a)\bigr)\bar{L}\bigl(i_G(z)\bigr) \\
&= L\bigl(i_A(a)i_G(z)\bigr).
\end{aligned}$$

Now Lemma 1.87 on page 29 and Corollary 2.36 on page 57 imply that the span of elements of the form $i_A(a)i_G(z)$ are dense in $A \rtimes_\alpha G$, and it follows that L and $\pi \rtimes u$ are equal. □

Proposition 2.40. *If $\alpha : G \to \operatorname{Aut} A$ is a dynamical system, then the map sending a covariant pair (π, U) to its integrated form $\pi \rtimes U$ is a one-to-one correspondence between nondegenerate covariant representations of (A, G, α) and nondegenerate representations of $A \rtimes_\alpha G$. This correspondence preserves direct sums, irreducibility and equivalence.*

Proof. Proposition 2.39 on the facing page shows that the map $(\pi, U) \mapsto \pi \rtimes U$ is a surjection. It's one-to-one in view of Equations (2.21) and (2.27).

The statement about equivalence is straightforward. Let (π, U), (ρ, V) and W be as in Definition 2.18 on page 46. Then

$$\begin{aligned}
\bigl(W(\pi \rtimes U)(f)h \mid k\bigr) &= \bigl(\pi \rtimes U(f)h \mid W^*k\bigr) \\
&= \int_G \bigl(\pi\bigl(f(s)\bigr)U_s h \mid W^*k\bigr) \, d\mu(s) \\
&= \int_G \bigl(W\pi\bigl(f(s)\bigr)U_s h \mid k\bigr) \, d\mu(s) \\
&= \int_G \bigl(\rho\bigl(f(s)\bigr)V_s W h \mid k\bigr) \, d\mu(s) \\
&= \bigl(\rho \rtimes V(f)Wh \mid k\bigr).
\end{aligned}$$

And it follows that $\pi \rtimes U$ and $\rho \rtimes V$ are equivalent. Conversely, if W intertwines $\pi \rtimes U$ and $\rho \rtimes V$, then

$$\begin{aligned} WU_s\pi \rtimes U(f)h &= W\pi \rtimes U\big(i_G(s)f)\big)h \\ &= \rho \rtimes V\big(i_G(s)f\big)Wh \\ &= V_s\rho \rtimes V(f)Wh \\ &= V_sW\pi \rtimes U(f)h. \end{aligned}$$

Since $\pi \rtimes U$ is nondegenerate, it follows that $WU_s = V_sW$ for all $s \in G$. A similar argument shows that $W\pi(a) = \rho(a)W$ for all $a \in A$.

The other assertions will follow once we show that a closed subspace $\mathcal{V}$ is invariant for a nondegenerate (π, U) if and only if $\mathcal{V}$ is invariant for $\pi \rtimes U$. Suppose first that $\mathcal{V}$ is invariant for (π, U). Let $h \in \mathcal{V}$ and $k \in \mathcal{V}^\perp$. Then

$$\big(\pi \rtimes U(f)h \mid k\big) = \int_G \big(\pi\big(f(s)\big)U_sh \mid k\big),$$

and since $\pi\big(f(s)\big)U_sh \in \mathcal{V}$ for all $s \in G$,

$$\big(\pi \rtimes U(f)h \mid k\big) = 0.$$

It follows that $\mathcal{V}$ is invariant for $\pi \rtimes U$.

Now assume $\mathcal{V}$ is invariant for $\pi \rtimes U$. Since $\pi \rtimes U$ is nondegenerate, $\pi \rtimes U(e_i) \to 1_{\mathcal{H}}$ strongly for any approximate identity $\{ e_i \}$ for $A \rtimes_\alpha G$. If $h \in \mathcal{V}$ and $k \in \mathcal{V}^\perp$, then $\pi \rtimes U\big(i_G(s)e_i\big) \in \mathcal{V}$ and

$$\begin{aligned} (U_sh \mid k) &= \lim\big(U_s\big(\pi \rtimes U(e_i)h\big) \mid k\big) \\ &= \lim\big(\pi \rtimes U\big(i_G(s)e_i\big)h \mid k\big) \\ &= 0. \end{aligned}$$

It follows that $\mathcal{V}$ is invariant for U_s and a similar argument works for $\pi(a)$. □

Remark 2.41. We can now identify the spectrum of $C^*(G)$ with the set $\widehat{G}$ of irreducible unitary representations of G. We give $\widehat{G}$ the topology coming from the spectrum of $C^*(G)$ and refer to $\widehat{G}$ as the *spectrum of* G. We should check to see that this usage is consistent with that for abelian groups. Since $C^*(G)$ is commutative if G is abelian, its irreducible representations are one-dimensional complex homomorphisms [139, Example A.16], and the characters of G correspond exactly the irreducible representations of G. Since a pure state on $C^*(G)$ must be of the form $f \mapsto \big(\omega(f)h \mid h\big) = \omega(f)$ for the integrated form of a character ω and a unimodular scalar h, the pure states also correspond exactly to the characters. [139, Theorem A.38] implies that the weak-$*$ topology on $\widehat{G}$, viewed as the set of pure states, is the same as the topology obtained by viewing $\widehat{G}$ as the spectrum of $C^*(G)$. The proof of Lemma 1.78 on page 25 implies that this topology is the compact-open topology.

Remark 2.42. If $u : G \to U\mathcal{L}(\mathsf{X})$ is a strongly continuous homomorphism, then we'll use the same letter u for the integrated form of u viewed either as a homomorphism of $C_c(G)$ into $\mathcal{L}(\mathsf{X})$ or $C^*(G)$ into $\mathcal{L}(\mathsf{X})$. If $L = (\pi, U)$ is a covariant representation, we'll often write $L(f)$ in place of $\pi \rtimes U(f)$.

Definition 2.43. A $*$-homomorphism $\pi : C_c(G, A) \to B(\mathcal{H})$ is *continuous in the inductive limit topology* if whenever we're given $h, k \in \mathcal{H}$ and a net $f_i \to f$ in the inductive limit topology on $C_c(G, A)$, then we also have $\bigl(\pi(f_i)h \mid k\bigr) \to \bigl(\pi(f)h \mid k\bigr)$.

Example 2.44. If π is L^1-norm decreasing, then π is continuous in the inductive limit topology.

The following clever argument is due to Iain Raeburn.

Lemma 2.45. *Suppose that $\pi : C_c(G, A) \to B(\mathcal{H})$ is a $*$-homomorphism which is continuous in the inductive limit topology. Then π is bounded with respect to the universal norm on $C_c(G, A) \subset A \rtimes_\alpha G$. That is, $\|\pi(f)\| \le \|f\|$ for all $f \in C_c(G, A)$.*

Proof. By reducing to the essential subspace of π, we may assume that π is nondegenerate. The equation

$$(f \otimes h \mid g \otimes k) := \bigl(\pi(g^* * f)h \mid k\bigr)$$

defines a sesquilinear form on the algebraic tensor product $C_c(G, A) \odot \mathcal{H}$. This form is positive since

$$\begin{aligned}
\Bigl(\sum_i f_i \otimes h_i \mid \sum_i f_i \otimes h_i\Bigr) &= \sum_{ij} \bigl(\pi(f_j^* * f_i)h_i \mid h_j\bigr) \\
&= \sum_{ij} \bigl(\pi(f_i)h_i \mid \pi(f_j)h_j\bigr) \\
&= \Bigl(\sum_i \pi(f_i)h_i \mid \sum_i \pi(f_i)h_i\Bigr) \\
&\ge 0.
\end{aligned}$$

Thus we can complete $C_c(G, A) \odot \mathcal{H}$ to get a Hilbert space $\mathcal{V}$. The map $(f, h) \mapsto \pi(f)h$ is bilinear and therefore extends to a map $U : C_c(G, A) \odot \mathcal{H} \to \mathcal{H}$ which has dense range since π is nondegenerate. Moreover

$$\begin{aligned}
\bigl(U(f \otimes h) \mid U(g \otimes k)\bigr) &= \bigl(\pi(f)h \mid \pi(g)k\bigr) \\
&= \bigl(\pi(g^* * f)h \mid k\bigr) \\
&= (f \otimes h \mid g \otimes k).
\end{aligned}$$

It follows that U extends to a unitary operator $U : \mathcal{V} \to \mathcal{H}$. Now for each $b \in M(A)$, we define an operator on $C_c(G, A) \odot \mathcal{H}$ by

$$M(b)(f \otimes h) = \bar{\iota}_A(b) f \otimes h.$$

Since an easy calculation shows that

$$\bigl(M(b)(f \otimes h) \mid g \otimes k\bigr) = \bigl(f \otimes h \mid M(b^*)(g \otimes k)\bigr),$$

we can let $a_0 := (\|a\|^2 1 - a^*a)^{\frac{1}{2}}$ and compute that

$$\|a\|^2(f \otimes h \mid f \otimes h) - \big(M(a)(f \otimes h) \mid M(a)(f \otimes h)\big) = \big(M(a_0)(f \otimes h) \mid M(a_0)(f \otimes h)\big) \geq 0.$$

It follows that $\|M(a)\| \leq \|a\|$ and that M extends to a representation of A on $\mathcal{V}$. Similarly, we define a map from G into operators on $C_c(G, A) \odot \mathcal{H}$ by

$$V_s(f \otimes h) = i_G(s)(f) \otimes h.$$

Easy calculations show that $V_{sr} = V_s \circ V_r$. Since we can also check that $i_G(g)^* * i_G(f) = g^* * f$, it follows that

$$\big(V_s(f \otimes h) \mid V_s(g \otimes k)\big) = (f \otimes h \mid g \otimes k).$$

Therefore it follows that V is a unitary-valued homomorphism from G into $U(\mathcal{V})$. To see that V is strongly continuous, notice that

$$\|V_s(f \otimes h) - f \otimes h\|^2 = 2\big(\pi(f^* * f)h \mid h\big) - 2\operatorname{Re}\big(\pi\big(f^* * i_G(s)(f)\big)h \mid h\big). \tag{2.33}$$

Since $i_G(s)(f) \to f$ in the inductive limit topology as $s \to e$ and since π is continuous in the inductive limit topology, it follows that (2.33) goes to zero if $s \to e$ in G. Thus V is a unitary representation of G, and it is easy to see that (M, V) is covariant. We will prove that π and $M \rtimes V$ are equivalent representations.

Now let $f, g \in C_c(G, A)$. Then $(s, r) \mapsto f(s)i_G(s)(g)(r)$ has support in $(\operatorname{supp} f) \times (\operatorname{supp} f)(\operatorname{supp} g)$. Therefore if U is a pre-compact open neighborhood of the compact set $(\operatorname{supp} f)(\operatorname{supp} g)$, then we can view $q(s) := f(s)i_G(s)(g)$ as defining a function in $C_c\big(G, C_0(U, A)\big)$. Then we can form the $C_0(U, A)$-valued integral

$$\int_G^{C_0(U,A)} f(s)i_G(s)(g)\, d\mu(s). \tag{2.34}$$

Since evaluation at $r \in U$ is a continuous homomorphism from $C_0(U, A)$ to A, we have

$$\int_G^{C_0(U,A)} f(s)i_G(s)(g)\, d\mu(s)(r) = \int_G^{A} f(s)i_G(s)(g)(r)\, d\mu(s) = f * g(r).$$

Thus (2.34) is the restriction of $f * g$ to U. If $h, k \in \mathcal{H}$ and j is the inclusion of $C_0(U, A)$ into $C_c(G, A)$, then the continuity of π and j (Lemma 1.106 on page 38) allows us to define a continuous linear functional

$$L : C_0(U, A) \to \mathbf{C}$$

by

$$L(f) := \big(\pi\big(j(f)\big)h \mid k\big).$$

Now on the one hand,

$$L\Big(\int_G^{C_0(U,A)} f(s)i_G(s)(g)\, d\mu(s)\Big) = \big(\pi(f * g)h \mid k\big) = \big(\pi(f)U(g \otimes h) \mid k\big). \tag{2.35}$$

On the other hand, since L is a continuous linear functional, the left-hand side of (2.35) equals

$$\begin{aligned}\int_G L\big(f(s)i_G(s)(g)\big)\,d\mu(s) &= \int_G \big(\pi\big(f(s)i_G(s)(g)\big)h \mid k\big)\,d\mu(s)\\ &= \int_G \big(U\big(f(s)i_G(s)(g)\otimes h\big) \mid k\big)\,d\mu(s)\\ &= \int_G \big(f(s)i_G(s)(g)\otimes h \mid U^{-1}k\big)\,d\mu(s)\\ &= \int_G \big(M\big(f(s)\big)V_s(g\otimes h) \mid U^{-1}k\big)\,d\mu(s)\\ &= \big(M\rtimes V(f)(g\otimes h) \mid U^{-1}k\big)\\ &= \big(U\circ M\rtimes V(f)(g\otimes h) \mid k\big).\end{aligned}$$

This proves that $\pi(f)\circ U = U\circ M\rtimes V(f)$, and that $\|\pi(f)\| = \|M\rtimes V(f)\| \le \|f\|$ as desired. □

It follows that every $*$-homomorphism $L : C_c(G,A)\to B(\mathcal{H})$ which is L^1-norm bounded or just continuous in the inductive limit topology is bounded with respect to the universal norm and extends to a (possibly degenerate) representation of $A\rtimes_\alpha G$. Thus every such representation is the integrated form of a covariant pair.

Corollary 2.46. *Suppose that (A,G,α) is a dynamical system and that $f\in C_c(G,A)$. Then*

$$\begin{aligned}\|f\| &= \sup\{\,\|L(f)\| : L \textit{ is a } L^1\textit{-norm decreasing representation.}\}\\ &= \sup\{\,\|L(f)\| : L \textit{ is continuous in the inductive limit topology.}\}\end{aligned}$$

Corollary 2.47. *Suppose that (A,G,α) and (B,H,β) are dynamical systems and that $\Phi : C_c(G,A)\to C_c(H,B)$ is a $*$-homomorphism which is continuous in the inductive limit topology. Then Φ is norm-decreasing with respect to the universal norms and extends to an homomorphism of $A\rtimes_\alpha G$ into $B\rtimes_\beta H$.*

Proof. Suppose that L is a representation of $B\rtimes_\beta H$. Then $L\circ\Phi$ is a representation of $C_c(G,A)$ which is continuous in the inductive limit topology. Thus Lemma 2.45 on page 61 implies that $L\circ\Phi$ is bounded for the universal norm, and thus for each $f\in C_c(G,A)$, $\|L(\Phi(f)\|\le\|f\|$. Since L is arbitrary, it follows that $\|\Phi(f)\|\le \|f\|$. □

Corollary 2.48. *Suppose that (A,G,α) and (B,G,β) are dynamical systems and that $\varphi : A\to B$ is an equivariant homomorphism. Then there is a homomorphism $\varphi\rtimes\mathrm{id} : A\rtimes_\alpha G\to B\rtimes_\beta G$ mapping $C_c(G,A)$ into $C_c(G,B)$ such that $\varphi\rtimes\mathrm{id}(f)(s) = \varphi\big(f(s)\big)$.*

Proof. Define $\Phi : C_c(G,A)\to C_c(G,B)$ by $\Phi(f)(s) = \varphi\big(f(s)\big)$. Then Φ is norm decreasing for the L^1-norm and therefore continuous in the inductive limit topology. Now we can apply Corollary 2.47 on the previous page. □

2.5 Comments on Examples

Having worked hard just to define a crossed product it would be natural to want, and to provide, a few illustrative examples other than the degenerate examples in Example 2.33 on page 54. It is a bit frustrating to admit that we still need more technology to do this in a systematic way. (We'll deal with the group C^*-algebras of abelian and compact groups in Section 3.1 and Section 3.2, respectively.)

However, we can work out some simple and provocative examples when we assume that (A, G, α) is a dynamical system with G *finite*. To begin with, suppose that $G = \mathbf{Z}_2 := \mathbf{Z}/2\mathbf{Z}$. This means we are given $\alpha \in \operatorname{Aut} A$ with $\alpha^2 = \mathrm{id}$. Then $C_c(\mathbf{Z}_2, A) = C(\mathbf{Z}_2, A)$ and elements of $C(\mathbf{Z}_2, A)$ are simply functions from $\mathbf{Z}_2 = \{0, 1\}$ to A. Let

$$D := \left\{ \begin{pmatrix} a & b \\ \alpha(b) & \alpha(a) \end{pmatrix} \in M_2(A) : a, b \in A \right\}.$$

It is easy to check that D is a C^*-subalgebra of $M_2(A)$ and that

$$\Phi(f) := \begin{pmatrix} f(0) & f(1) \\ \alpha(f(1)) & \alpha(f(0)) \end{pmatrix}$$

defines an injective $*$-homomorphism of $C(\mathbf{Z}_2, A)$ (with the $*$-algebra structure coming from $(A, \mathbf{Z}_2, \alpha)$) into D. In fact, Φ is clearly surjective. Since we can faithfully represent D on Hilbert space, it follows that the universal norm of $f \in C(\mathbf{Z}_2, A)$ satisfies

$$\|f\| \geq \|\Phi(f)\|.$$

On the other hand, if L is any representation of $A \rtimes_\alpha \mathbf{Z}_2$, then $L \circ \Phi^{-1}$ is a $*$-homomorphism of D into $B(\mathcal{H}_L)$. Since $*$-homomorphisms of C^*-algebras are norm reducing,

$$\|L(f)\| = \|L(\Phi^{-1}(\Phi(f)))\| \leq \|\Phi(f)\|.$$

Therefore,

$$\|f\| = \|\Phi(f)\|.$$

In particular, $C(\mathbf{Z}_2, A)$ is already complete in its universal norm, and Φ is an isomorphism of $A \rtimes_\alpha \mathbf{Z}_2$ onto D. Although this situation is rather atypical — normally $A \rtimes_\alpha G$ is going to be a genuine completion of $C_c(G, A)$ — it will be useful to record the general principle we used above so that we can use it in some other examples of finite group actions.

Lemma 2.49. *Suppose that (A, G, α) is a dynamical system with G* finite *and that $L : C(G, A) \to D$ is a $*$-isomorphism of $C(G, A)$ onto a C^*-algebra D. Then $A \rtimes_\alpha G \cong D$.*

It is an interesting exercise to use Lemma 2.49 to study $A \rtimes_\alpha G$ for various finite groups G — for example, compare $G = \mathbf{Z}_4$ with $G = \mathbf{Z}_2 \times \mathbf{Z}_2$. Here we specialize to transformation groups.

Lemma 2.50. *Suppose that G is a finite group with $|G| = n$. Then G acts on itself by left translation and*

$$C(G) \rtimes_{\mathrm{lt}} G \cong M_n,$$

where M_n denotes the C^*-algebra of $n \times n$-matrices (with complex entries).

Proof. Of course, we give G the discrete topology and use counting measure for Haar measure on G. Let $L = M \rtimes \lambda$ be the natural representation as in Example 2.12 on page 45. Let $G = \{ s_i \}_{i=1}^n$ with $s_1 := e$. Then $L^2(G)$ is a n-dimensional Hilbert space with orthonormal basis $\{ e_s \}_{s \in G}$ where e_s is the function δ_s which is 1 at s and zero elsewhere. We view operators on $L^2(G)$ as $n \times n$ matrices calculated with respect to $\{ e_s \}$. If $f \in C(G \times G)$ and $h \in L^2(G)$, then

$$L(f)h(s) := \sum_{r \in G} f(r,s)h(r^{-1}s) = \sum_{r \in G} f(sr^{-1}, s)h(r).$$

It follows that $L(f)$ is given by the matrix M^f with $(s,r)^{\text{th}}$ entry $M^f_{s,r} = f(sr^{-1}, s)$. If $M = (m_{s,r})_{s,r \in G}$ is any $n \times n$ matrix, then

$$f(s,r) := m_{r,s^{-1}r}$$

satisfies

$$M^f_{s,r} = f(sr^{-1}, s) = m_{s,r}.$$

Thus L is a surjective $*$-isomorphism of $C(G \times G)$ onto M_n. The result now follows from Lemma 2.49 on the facing page. □

Lemma 2.50 can be extended to arbitrary groups.

Example 2.51. Suppose that G is a locally compact group acting on itself by left translation. Then $C_0(G) \rtimes_{\mathrm{lt}} G$ is isomorphic to the compact operators on $L^2(G)$.

Verifying Example 2.51 is a good exercise if G is discrete. In general, some work is required and there are a number of ways to proceed. In this book, we'll eventually prove this in Theorem 4.24 on page 133 using the Imprimitivity Theorems developed in Section 4.3.

Now we suppose that $A = C(X)$ for a *compact* Hausdorff space X.[6] To get a $\mathbf{Z}_2$-action we need a homeomorphism $\sigma \in \mathrm{Homeo}(X)$ such that $\sigma^2 = \mathrm{id}$. For convenience, we'll assume that the $\mathbf{Z}_2$-action is free so that $\sigma(x) \neq x$ for all $x \in X$. We'll let $\big(C(X), \mathbf{Z}_2, \alpha\big)$ be the associated dynamical system (so that $\alpha_1(f)(x) = f\big(\sigma(x)\big)$). For example, we could let X be the n-sphere $S^n = \{ \mathbf{x} \in \mathbf{R}^{n+1} : \|\mathbf{x}\| = 1 \}$, and σ the antipodal map $\sigma(\mathbf{x}) := -\mathbf{x}$.

However, for the moment, we let σ be any period two homeomorphism of a compact space X (without fixed points). We'll view operators on $L^2(\mathbf{Z}_2)$ as 2×2-matrices with respect to the orthonormal basis $\{ e_0, e_1 \}$ where e_i is the function δ_i as above. Let $L^x = \pi \rtimes \lambda = \mathrm{Ind}_e^G \mathrm{ev}_x$ be the regular representation on $L^2(\mathbf{Z}_2)$ coming from evaluation at x. Thus λ is the left-regular representation and

$$\pi(\varphi)h(s) = \mathrm{ev}_x\big(\alpha_s^{-1}(\varphi)\big)h(s).$$

[6] It isn't really necessary to take X compact, but it makes the examples a little easier to digest.

Therefore $\pi(\varphi)$ is given by the matrix

$$\begin{pmatrix} \varphi(x) & 0 \\ 0 & \varphi\big(\sigma(x)\big) \end{pmatrix}$$

If $f \in C(\mathbf{Z}_2 \times X)$, then

$$\begin{aligned} \pi \rtimes \lambda(f) &= \sum_{s=0,1} \pi\big(f(s,\cdot)\big)\lambda(s) \\ &= \begin{pmatrix} f(0,x) & 0 \\ 0 & f\big(0,\sigma(x)\big) \end{pmatrix} + \begin{pmatrix} f(1,x) & 0 \\ 0 & f\big(1,\sigma(x)\big) \end{pmatrix} \begin{pmatrix} 0 & 1 \\ 1 & 0 \end{pmatrix} \\ &= \begin{pmatrix} f(0,x) & f(1,x) \\ f\big(1,\sigma(x)\big) & f\big(0,\sigma(x)\big) \end{pmatrix}. \end{aligned}$$

In this way, we can view L^x as a $*$-homomorphism of $C(\mathbf{Z}_2 \times X)$ into M_2. Since $\sigma(x) \neq x$, L^x is surjective. Let

$$W := \begin{pmatrix} 0 & 1 \\ 1 & 0 \end{pmatrix},$$

and let

$$A := \big\{\, f \in C(X, M_2) : f\big(\sigma(x)\big) = W f(x) W^* \,\big\}.$$

Then A is a C^*-subalgebra of $C(X, M_2)$. Since each L^x is onto M_2, it is not hard to see that the irreducible representations of A correspond to the point evaluations π_x. Since A is a $C(\mathbf{Z}_2\backslash X)$-algebra, we have π_x equivalent to π_y if and only if $y = \sigma(x)$. Thus the spectrum of A is naturally identified with $\mathbf{Z}_2\backslash X$.[7] Since all the irreducible representations of A are 2-dimensional, A is called a 2-homogeneous C^*-algebra.

Now define $\Phi : C(\mathbf{Z}_2 \times X) \to C(X, M_2)$ by

$$\Phi(f)(x) = L^x(f) = \begin{pmatrix} f(0,x) & f(1,x) \\ f\big(1,\sigma(x)\big) & f\big(0,\sigma(x)\big). \end{pmatrix}$$

Since each L^x is a $*$-homomorphism, so is Φ. Furthermore, Φ is injective and

$$\Phi(f)\big(\sigma(x)\big) = \begin{pmatrix} f\big(0,\sigma(x)\big) & f\big(1,\sigma(x)\big) \\ f(1,x) & f(0,x) \end{pmatrix} = W\Phi(f)(x)W^*.$$

Thus Φ maps into A. On the other hand, if $a \in A$, say

$$a(x) = \begin{pmatrix} a_{11}(x) & a_{12}(x) \\ a_{21}(x) & a_{22}(x) \end{pmatrix},$$

and if we define $f \in C(\mathbf{Z}_2 \times X)$ by $f(i,x) := a_{1,1+i}(x)$, then we have $\Phi(f) = a$. After applying Lemma 2.49 on page 64, we have proved the following.

[7] It is not hard to see that $\mathbf{Z}_2\backslash X$ is compact and Hausdorff. This also follows from Corollary 3.43 on page 100.

Proposition 2.52. *Suppose that X is a compact free $\mathbf{Z}_2$-space determined by a period two homeomorphism σ. Then $C(\mathbf{Z}_2 \times X)$ is complete in its universal norm and*

$$\Phi(f)(x) := \begin{pmatrix} f(0,x) & f(1,x) \\ f\big(1,\sigma(x)\big) & f\big(0,\sigma(x)\big) \end{pmatrix}$$

defines an isomorphism of $C(X) \rtimes_\alpha \mathbf{Z}_2 = C(\mathbf{Z}_2 \times X)$ with the 2-homogeneous C^*-algebra with spectrum $\mathbf{Z}_2 \backslash X$

$$A := \{\, f \in C(X, M_2) : f\big(\sigma(x)\big) = Wf(x)W^* \,\},$$

where $W = \left(\begin{smallmatrix} 0 & 1 \\ 1 & 0 \end{smallmatrix}\right)$.

We will have more to say about examples provided by Proposition 2.52 later (cf. Proposition 4.15 on page 127).

As sort of a preview of coming attractions, I also want to mention a few non-trivial examples that we'll take up in due course. An obvious question to ask is what happens if the G-action on A is trivial. It turns out that this isn't too different than allowing the G-action to be implemented by a homomorphism into the unitary group of the multiplier algebra of A.

Example 2.53 (Trivial Dynamical Systems). Suppose that (A, G, α) is a dynamical system and that there is a strictly continuous homomorphism $u : G \to UM(A)$ such that that $\alpha_s(a) = u_s a u_s^*$ for all $s \in G$ and $a \in A$. Then it is a straightforward corollary to Lemma 2.73 on page 76 that $A \rtimes_\alpha G$ is isomorphic to $A \otimes_{\max} C^*(G)$ (see Lemma 2.68 on page 74 and Remark 2.71 on page 75).

Another natural example is to let a closed subgroup H of a locally compact group G act by left translation on the coset space G/H. The imprimitivity theorem (Theorem 4.22 on page 132) implies that $C_0(G/H) \rtimes_{\mathrm{lt}} G$ is Morita equivalent to $C^*(H)$. With considerably more work (Theorem 4.30 on page 138), we can sharpen this to an isomorphism result.

Example 2.54. $C_0(G/H) \rtimes_{\mathrm{lt}} G$ is isomorphic to the tensor product of $C^*(H)$ with the compact operators on $L^2(G/H, \mu)$ where μ is any quasi-invariant measure on G/H.

Example 2.54 is more generally applicable than one might think at first. We'll show later that in many cases — namely when the action of G is "nice" in a sense to made precise later — $C_0(X) \rtimes G$ is fibred over the orbits in X with fibres of the form $C_0(G/H_\omega) \rtimes_{\mathrm{lt}} G$ where H_ω depends on the orbit ω (see Section 8.1).

Example 2.55. Let $\mathbf{Z}$ act on $\mathbf{T}$ by rotation through θ as in Example 2.13 on page 45. The resulting crossed product $A_\theta := C_0(\mathbf{T}) \rtimes_\tau \mathbf{Z}$ is called a *rational or irrational rotation algebra* depending on whether θ is rational or irrational.

The structure of the rotation algebras, both irrational and rational, is surprisingly intricate and was a hot research topic in the early 1980's. These algebras continue to be of interest to this day. For example, just seeing that A_θ and $A_{\theta'}$ are not isomorphic for different irrational θ and θ' in $[0, \frac{1}{2}]$ involves more than we're prepared to undertake in this book.[8]

[8] A proof is given in Davidson's text [21, §VI.5].

Proposition 2.56. *If θ is irrational then the irrational rotation algebra A_θ is simple and is generated by two unitaries u and v such that $uv = \rho vu$ where $\rho = e^{2\pi i\theta}$. Furthermore, if $\mathcal{H}$ is a Hilbert space and if U and V are unitaries in $B(\mathcal{H})$ such that $UV = \rho VU$, then there is a representation $L : A_\theta \to B(\mathcal{H})$ such that $L(u) = U$, $L(v) = V$ and L is an isomorphism of A_θ onto the C^*-algebra $C^*(U,V)$ generated by U and V.*

Remark 2.57. One often summarizes the final assertion in Proposition 2.56 by saying that A_θ is the "universal C^*-algebra generated by two unitaries U and V satisfying $UV = \rho VU$".

Remark 2.58. In the proof, we will need to know that $\{\, \rho^n z \,\}_{n\in\mathbf{Z}}$ is dense in $\mathbf{T}$ for all $z \in \mathbf{T}$. This will be proved in Lemma 3.29 on page 96.

Proof. Recall that $A_\theta = C(\mathbf{T}) \rtimes_\tau \mathbf{Z}$ is the completion of the $*$-algebra $C_c(\mathbf{Z} \times \mathbf{T})$ where the convolution product is given by the *finite* sum

$$f * g(n,z) := \sum_{m=-\infty}^{\infty} f(m,z)g(n-m, \rho^{-m}z),$$

and the involution is given by

$$f^*(n,z) = \overline{f(-n, \rho^{-n}z)}.$$

If $\varphi \in C(\mathbf{T})$ and $h \in C_c(\mathbf{Z})$, then $\varphi \otimes h$ is the element of $C_c(\mathbf{Z} \times \mathbf{T})$ given by $\varphi \otimes h(n,z) = \varphi(z)h(n)$. As usual, let δ_n be the function on $\mathbf{Z}$ which is equal to 1 at n and zero elsewhere. In particular, A_θ has an identity in $C_c(\mathbf{Z} \times \mathbf{T})$ given by $1 \otimes \delta_0$. Furthermore, $u = 1 \otimes \delta_1$ is a unitary in $C_c(\mathbf{Z} \times \mathbf{T})$, as is $v = \iota_{\mathbf{T}} \otimes \delta_0$ where $\iota_{\mathbf{T}}(z) := z$ for all $z \in \mathbf{T}$. If $\varphi \in C(\mathbf{T})$, then let $i_{C(\mathbf{T})}(\varphi) := \varphi \otimes \delta_0$. Then $i_{C(\mathbf{T})}$ is a homomorphism of $C(T)$ into $C_c(\mathbf{Z} \times \mathbf{T}) \subset A_\theta$ (and is therefore bounded). Notice that $i_{C(\mathbf{T})}(\varphi) * u^n = \varphi \otimes \delta_n$ for all $n \in \mathbf{Z}$, and that $u * i_{C(\mathbf{T})}(\varphi) * u^* = i_{C(\mathbf{T})}\big(\tau_1(\varphi)\big)$. It follows from Lemma 1.87 on page 29 that

$$\{\, i_{C(\mathbf{T})}(\varphi) * u^n : \varphi \in C(\mathbf{T}) \text{ and } n \in \mathbf{Z} \,\}$$

spans a dense subalgebra of $C_c(\mathbf{Z} \times \mathbf{T})$.

The Stone-Weierstrass Theorem implies that $\iota_{\mathbf{T}}$ generates $C(\mathbf{T})$ as a C^*-algebra. In particular,

$$\{\, v^n : n \in \mathbf{Z} \,\}$$

spans a dense subalgebra of $i_{C(\mathbf{T})}\big(C(\mathbf{T})\big)$. It follows that u and v generate A_θ, and it is easy to verify that $uv = \rho vu$.

Now suppose that $U, V \in B(\mathcal{H})$ are as in the statement of the proposition. Since U and V are unitaries, their spectrums $\sigma(U)$ and $\sigma(V)$ are subsets of $\mathbf{T}$. I claim that $\sigma(V) = \mathbf{T}$.[9] Note that

$$\begin{aligned}\lambda \in \sigma(V) &\iff V - \lambda I \text{ is not invertible}\\ &\iff U^n(V - \lambda I) \text{ is not invertible}\\ &\iff (\rho^n V - \lambda I)U^n \text{ is not invertible}\\ &\iff V - \rho^{-n}\lambda I \text{ is not invertible}\\ &\iff \rho^{-n}\lambda \in \sigma(V).\end{aligned}$$

Since $\sigma(V)$ must be nonempty, $\sigma(V)$ contains a dense subset of $\mathbf{T}$ by Remark 2.58 on the preceding page. Since $\sigma(V)$ is closed, the claim follows.

Thus the Abstract Spectral Theorem [110, Theorem 2.1.13] implies that there is an isomorphism $\pi : C(\mathbf{T}) \to C^*(V) \subset C^*(U, V)$ taking $\iota_{\mathbf{T}}$ to V. Let $W : \mathbf{Z} \to U(\mathcal{H})$ be given by $W_n := U^n$. Since $UV = \rho VU$, it follows that

$$W_n \pi(\iota_{\mathbf{T}}) W_n^* = \pi\big(\tau_n(\iota_{\mathbf{T}})\big).$$

Since $\iota_{\mathbf{T}}$ generates $C(\mathbf{T})$, (π, W) is a covariant representation of $\big(C(\mathbf{T}), \mathbf{Z}, \tau\big)$. Furthermore,

$$\pi \rtimes W(u) = \sum_{m=-\infty}^{\infty} \pi\big(u(m,\cdot)\big) W_m = \pi(1_{C(\mathbf{T})}) W_1 = U, \text{ and}$$

$$\pi \rtimes W(v) = \sum_{m=-\infty}^{\infty} \pi\big(v(m,\cdot)\big) W_m = \pi(\iota_{\mathbf{T}}) = V.$$

Since u and v generate A_θ and since U and V generate $C^*(U, V)$, it follows that $\pi \rtimes W(A_\theta) = C^*(U, V)$ and $L := \pi \rtimes W$ is an homomorphism of A_θ onto $C^*(U, V)$ taking u to U and v to V.

Now it will suffice to see that A_θ is simple. However, proving that is going to require more work.

For each $\omega \in \mathbf{T}$, let $\hat{\tau}_\omega : C_c(\mathbf{Z} \times \mathbf{T}) \to C_c(\mathbf{Z} \times \mathbf{T})$ be defined by $\hat{\tau}_\omega(f)(n, z) = \omega^n f(n, z)$. It is not hard to check that $\hat{\tau}_\omega$ is a $*$-isomorphism which is continuous in the inductive limit topology. Therefore $\hat{\tau}_\omega$ extends to an automorphism of A_θ (with inverse $\hat{\tau}_{\bar{\omega}}$). Since $\omega \mapsto \hat{\tau}_\omega(f)$ is continuous from $\mathbf{T}$ to $C_c(\mathbf{Z} \times \mathbf{T})$ in the inductive limit topology for each $f \in C_c(\mathbf{Z} \times \mathbf{T})$, it is not hard to see that $\omega \mapsto \hat{\tau}_\omega(a)$ is continuous from $\mathbf{T}$ to A_θ for all $a \in A_\theta$.[10] Therefore we can define $\Phi : A_\theta \to A_\theta$ by

$$\Phi(a) = \int_{\mathbf{T}} \hat{\tau}_\omega(a)\, d\omega.$$

It is easy to see that Φ is linear and $\|\Phi\| \leq 1$. Since for each $\omega \in \mathbf{T}$, $\hat{\tau}_\omega(v) = v$ and $\hat{\tau}_\omega(u) = \omega u$, we have for each $k, m \in \mathbf{Z}$,

$$\Phi(v^k * u^m) = \int_{\mathbf{T}} \hat{\tau}_\omega(v^k * u^m)\, d\omega = v^k * \Big(\int_{\mathbf{T}} \omega^m\, d\omega\Big) u^m.$$

[9] Of course, we also have $\sigma(U) = \mathbf{T}$.

[10] The automorphism group $\hat{\tau} : \mathbf{T} \to \operatorname{Aut} A_\theta$ is an example of the dual-action which will reappear in Section 7.1.

Therefore,

$$\Phi(v^k * u^m) = \begin{cases} v^k & \text{if } m = 0, \text{ and} \\ 0 & \text{otherwise.} \end{cases} \tag{2.36}$$

Now let $E_n : A_\theta \to A_\theta$ be defined by

$$E_n(a) := \frac{1}{2n+1} \sum_{j=-n}^{n} v^j * a * v^{-j}.$$

Since v is a unitary, E_n is linear with $\|E_n\| \le 1$. Furthermore,

$$E_n(v^k * u^m) = \sum_{j=-n}^{n} v^j * v^k * u^m * v^{-j} = v^k * u^m \sum_{j=-n}^{n} \rho^{jm}. \tag{2.37}$$

Thus if $m = 0$, (2.37) is equal to v^k. Otherwise, notice that the usual sort of manipulations with geometric series gives

$$\begin{aligned} \sum_{j=-n}^{n} \rho^{jm} &= \frac{\rho^{m(n+1)} - \rho^{-mn}}{\rho^m - 1} \\ &= \frac{\rho^{\frac{m}{2}(2n+1)} - \rho^{-\frac{m}{2}(2n+1)}}{\rho^{\frac{m}{2}} - \rho^{-\frac{m}{2}}} \\ &= \frac{e^{i\pi\theta m(2n+1)} - e^{-i\pi\theta m(2n+1)}}{e^{i\pi m\theta} - e^{-i\pi n m\theta}} \\ &= \frac{\sin\big((2n+1)\pi m\theta\big)}{\sin(\pi m\theta)}. \end{aligned}$$

Thus if $m \neq 0$,

$$\lim_{n\to\infty} E_n(v^k * u^m) = v^k * u^m \lim_{n\to\infty} \sum_{j=-n}^{n} \rho^{jm} = 0.$$

It follows that

$$\Phi(a) = \lim_{n\to\infty} E_n(a) \tag{2.38}$$

for all a in the dense subalgebra $A_0 := \operatorname{span}\{\, u^k * v^m : k, m \in \mathbf{Z} \,\}$.

We want to see that (2.38) holds for all $a \in A_\theta$. Fix $\epsilon > 0$ and $a \in A_\theta$. Then there is a $b \in A_0$ such that $\|a - b\| < \epsilon/3$, and a $N \in \mathbf{Z}^+$ such that $n \ge N$ implies $\|\Phi(b) - E_n(b)\| < \epsilon/3$. Since Φ and E_n each have norm 1, it follows that if $n \ge N$,

$$\begin{aligned} \|\Phi(a) - E_n(a)\| &\le \|\Phi(a-b)\| + \|\Phi(b) - E_n(b)\| + \|E_n(b-a)\| \\ &\le \frac{\epsilon}{3} + \frac{\epsilon}{3} + \frac{\epsilon}{3} = \epsilon. \end{aligned}$$

Therefore (2.38) holds for all $a \in A_\theta$.

Now suppose that I is a nonzero ideal in A_θ. Let a be a nonzero positive element in I. Let ρ be a state on A_θ such that $\rho(a) > 0$. Then $\omega \mapsto \rho\big(\hat{\tau}_\omega(a)\big)$ is a nonzero continuous nonnegative function on $\mathbf{T}$. Thus

$$\rho\big(\Phi(a)\big) = \int_{\mathbf{T}} \rho\big(\hat{\tau}_\omega(a)\big)\,d\omega > 0,$$

and $\Phi(a) > 0$.

It follows from (2.36) that $\Phi(a) \in i_{C(\mathbf{T})}\big(C(\mathbf{T})\big)$ and from (2.38) that $\Phi(a) \in I$. Thus there is a nonzero nonnegative function $h \in C(\mathbf{T})$ such that $i_{C(\mathbf{T})}(h) \in I$. Let O be a neighborhood of $z_0 \in \mathbf{T}$ such that $h(z) > 0$ for all $z \in O$. Since $\{\,\rho^k z_0 : k \in \mathbf{Z}\,\}$ is dense in $\mathbf{T}$ (Remark 2.58 on page 68), there is a n such that $\bigcup_{k=-n}^{n} \rho^k V$ covers $\mathbf{T}$. Then

$$g := \sum_{k=-n}^{n} \tau_k(h)$$

is positive and nonzero on $\mathbf{T}$. Hence g is invertible in $C(\mathbf{T})$ and $i_{C(\mathbf{T})}(g)$ is invertible in A_θ (with inverse $i_{C(\mathbf{T})}(g^{-1})$). However,

$$i_{C(\mathbf{T})}(g) = \sum_{k=-n}^{n} i_{C(\mathbf{T})}\big(\tau_k(h)\big) = \sum_{k=-n}^{n} u^k * i_{C(\mathbf{T})}(h) * u^{-k}$$

clearly belongs to I. Thus I contains an invertible element and must be all of A_θ. Thus A_θ is simple. □

Example 2.59. To obtain a more concrete realization of A_θ for irrational θ we can let $\mathcal{H} = L^2(\mathbf{T})$. Define U and V as follows:

$$U(f)(z) = zf(z) \quad \text{and} \quad V(f)(z) = f(\bar{\rho}z).$$

Then its not hard to see that U and V are unitaries on $\mathcal{H}$. Furthermore, $VU(f)(z) = U(f)(\bar{\rho}z) = \bar{\rho}zf(\bar{\rho}z)$ and $UV(f)(z) = zV(f)(z) = zf(\bar{\rho}z)$. Thus, $UV = \rho VU$ and $C^*(U,V) \cong A_\theta$.

Remark 2.60 (The Whole Story). Since $C^*(U,V) = C^*(V,U)$, it is not so hard to see that A_θ is isomorphic to $A_{\theta'}$ whenever *either* $\theta - \theta'$ or $\theta + \theta'$ is an integer; that is, whenever $\theta' = \theta \mod 1$. In fact, $A_\theta \cong A_{\theta'}$ if and only if $\theta' = \theta \mod 1$. This was proved for θ irrational by Rieffel in [147]. The same result for rational rotation C^*-algebras was proved later by Høegh-Krohn and Skjelbred [74] (see also [149]). Rieffel also showed that the rational rotation algebras are all Morita equivalent to $C_0(\mathbf{T}^2)$ [149],[11] and that two irrational rotation C^*-algebras A_θ and $A_{\theta'}$ are Morita equivalent if and only if θ and θ' are in the same $\mathrm{GL}_2(\mathbf{Z})$ orbit [147].[12]

[11] See also footnote 11 on page 255.

[12] If α is an irrational number and $\left(\begin{smallmatrix} a & b \\ c & d \end{smallmatrix}\right) \in \mathrm{GL}_2(\mathbf{Z})$, then $\left(\begin{smallmatrix} a & b \\ c & d \end{smallmatrix}\right)(\alpha) = \frac{a\alpha+b}{c\alpha+d}$.

2.6 Universal Property

The notion of a universal object can be a powerful one. Good examples are the direct limit (see [139, Appendix D.1]), the maximal tensor product (see [139, Theorem B.27]) and graph C^*-algebras (see [6]); it is often easiest to exhibit such objects by verifying a particular representation has the required universal property, rather than working directly with the definition. Starting with [133] it has become apparent that the crossed product constructed in Section 2.3 can often profitably be thought of as a universal object for covariant representations of the dynamical system.

Theorem 2.61 (Raeburn). *Let (A, G, α) be a dynamical system. Suppose that B is a C^*-algebra such that*

(a) *there is a covariant homomorphism (j_A, j_G) of (A, G, α) into $M(B)$,*

(b) *given a nondegenerate covariant representation (π, U) of (A, G, α), there is a (nondegenerate) representation $L = L_{(\pi,U)}$ of B such that $\bar{L} \circ j_A = \pi$ and $\bar{L} \circ j_G = U$, and*

(c) $B = \overline{\operatorname{span}}\{\, j_A(a) j_G(z) : a \in A$ *and* $z \in C_c(G)\,\}$.

Then there is an isomorphism

$$j : B \to A \rtimes_\alpha G$$

such that

$$\bar{\jmath} \circ j_A = i_A \quad \textit{and} \quad \bar{\jmath} \circ j_G = i_G, \tag{2.39}$$

where (i_A, i_G) is the canonical covariant homomorphism of (A, G, α) into $M(A \rtimes_\alpha G)$ defined in Proposition 2.34 on page 54.

Remark 2.62. Notice that the crossed product $A \rtimes_\alpha G$ is an example of a C^*-algebra B satisfying (a), (b) and (c) above where $(j_A, j_G) = (i_A, i_G)$ and $L_{(\pi,U)} := \pi \rtimes U$. Property (a) follows from Propositions 2.34 on page 54, and Property (b) from Proposition 2.39 on page 58. Property (c) follows from Lemma 1.87 on page 29 together with Equation (2.28) of Corollary 2.36 on page 57.

Lemma 2.63. *Suppose that (B, j_A, j_G) satisfies* (a), (b) *and* (c) *as in Theorem 2.61. Then $j_A : A \to M(B)$ is nondegenerate, and if $\{\, e_i \,\}$ is an approximate identity in A, $j_A(e_i) \to 1$ strictly in $M(B)$.*

Proof. It is easy to see that A^2 is dense in A.[13] Therefore, $\{\, j_A(ab) j_G(z) : a, b \in A$ and $z \in C_c(G)\,\}$ is dense in B by property (c). Thus j_A is nondegenerate. The last assertion follows easily. □

Proof of Theorem 2.61. Suppose that (B, j_A, j_G) satisfies (a), (b) and (c) as above. Let $\rho : A \rtimes_\alpha G \to B(\mathcal{H}_\rho)$ be a faithful representation of $A \rtimes_\alpha G$. Let

$$\pi := \bar{\rho} \circ i_A \quad \text{and} \quad u := \bar{\rho} \circ i_G$$

[13] In fact, $A^2 = A$ by the Cohen Factorization Theorem [139, Proposition 2.33].

Then (π, u) is a nondegenerate covariant representation of (A, G, α) (Proposition 2.39 on page 58). Property (b) implies that there is a nondegenerate representation $L : B \to B(\mathcal{H}_\rho)$ such that $\bar{L} \circ j_A = \pi$ and $\bar{L} \circ j_G = u$. Note that if $a \in A$ and $z \in C_c(G)$ then

$$\begin{aligned} L\big(j_A(a)j_G(z)\big) &= \pi(a) \int_G z(s)u_s \, d\mu(s) \\ &= \bar{\rho}\big(i_A(a)\big) \int_G z(s)\bar{\rho}\big(i_G(s)\big) \, d\mu(s) \\ &= \bar{\rho}\big(i_A(a)\big)\bar{\rho}\big(i_G(z)\big) \\ &= \rho\big(i_A(a)i_G(z)\big) \end{aligned}$$

It follows that $j := \rho^{-1} \circ L$ is a homomorphism from B to $A \rtimes_\alpha G$ mapping generators to generators; thus j is surjective and clearly satisfies (2.39).

To finish the proof, we provide an inverse for j. We could do this by reversing the roles of B and $A \rtimes_\alpha G$ above, and noticing we used nothing about $(A \rtimes_\alpha G, i_A, i_G)$ except that it satisfies (a), (b) and (c). Alternatively, we can invoke Proposition 2.40 on page 59 to conclude that there is a nondegenerate homomorphism $j_A \rtimes j_G : A \rtimes_\alpha G \to M(B) = \mathcal{L}(B_B)$ such that

$$j_A \rtimes j_G\big(i_A(a)i_G(z)\big) = j_A(a)j_G(z).$$

Then it follows that $j_A \rtimes j_G(A \rtimes_\alpha G) \subset B$. To see that $j_A \rtimes j_G$ is the required inverse, just note that $j_A \rtimes j_G \circ j$ and $j \circ j_A \rtimes j_G$ are the identity on generators. □

Definition 2.64. Two dynamical systems (A, G, α) and (D, G, δ) are *equivariantly isomorphic* if there is an isomorphism $\varphi : A \to D$ such that $\varphi\big(\alpha_s(a)\big) = \delta_s\big(\varphi(a)\big)$ for all $s \in G$ and $a \in A$. We call φ an *equivariant isomorphism.*

Lemma 2.65. *Suppose that φ is an equivariant isomorphism of (A, G, α) onto (D, G, δ). Then the map $\varphi \rtimes \mathrm{id} : C_c(G, A) \to C_c(G, D)$ defined by*

$$\varphi \rtimes \mathrm{id}(f)(s) := \varphi\big(f(s)\big)$$

extends to an isomorphism of $A \rtimes_\alpha G$ onto $D \rtimes_\delta G$.

The result follows immediately from Corollary 2.48 on page 63. However, it might also be instructive to see a proof using Theorem 2.61 on the facing page.

Proof. We'll produce an isomorphism $j : A \rtimes_\alpha G \to D \rtimes_\delta G$ using Theorem 2.61 (with $A \rtimes_\alpha G$ playing the role of B). Let (i_A, i_G) and (k_D, k_G) be the canonical covariant homomorphisms for (A, G, α) and (D, G, δ), respectively. Then define (j_D, j_G) from (D, G, δ) into $M(A \rtimes_\alpha G)$ by letting $j_D(d) := i_A\big(\varphi^{-1}(d)\big)$ and $j_G = i_G$. It is easy to see that (j_D, j_G) is covariant. If (π, U) is a covariant representation of (D, G, δ), then $(\pi \circ \varphi, U)$ is a covariant representation of (A, G, α) and we can define $L = L_{(\pi, U)}$ to be $(\pi \circ \varphi) \rtimes U$. Clearly

$$\big((\pi \circ \varphi) \rtimes U\big)^{-} \circ j_D(d) = \pi(d) \quad \text{and} \quad \big((\pi \circ \varphi) \rtimes U\big)^{-} \circ j_G = U_s.$$

Thus conditions (a) and (b) of Theorem 2.61 on page 72 are satisfied. Since $j_D(d)j_G(z) = i_A(\varphi^{-1}(d))i_G(z)$, it is clear that condition (c) is also satisfied. Therefore there is an isomorphism $j : A \rtimes_\alpha G \to D \rtimes_\delta G$ such that

$$j\big(i_A(a)i_G(z)\big) = j\big(j_D(\varphi(a))j_G(z)\big) = k_D\big(\varphi(a)\big)k_G(z).$$

It follows that $j = \varphi \rtimes \mathrm{id}$. □

Definition 2.66. Two dynamical systems (A, G, α) and (A, G, β) are called *exterior equivalent* if there is a strictly continuous unitary-valued function $u : G \to UM(A)$ such that

(a) $\alpha_s(a) = u_s\beta_s(a)u_s^*$ for all $s \in G$ and $a \in A$, and

(b) $u_{st} = u_s\bar{\beta}_s(u_t)$ for all $s, t \in G$.

The map u is called a *unitary* 1*-cocycle.*

Remark 2.67. It should be noted that Definition 2.66 is symmetric in α and β: $v_s := u_s^*$ is a unitary 1-cocycle implementing an equivalence between β and α.

Lemma 2.68. *Suppose that (A, G, α) and (A, G, β) are exterior equivalent via a* 1*-cocycle u as in Definition 2.66. Then the map sending $f \in C_c(G, A)$ to $\varphi(f)$, where*

$$\varphi(f)(s) := f(s)u_s$$

extends to an isomorphism between $A \rtimes_\alpha G$ and $A \rtimes_\beta G$.

Remark 2.69. As with Lemma 2.65 on the preceding page, this result could be proved using either Theorem 2.61 or Corollary 2.47. Instead, we give a proof using only the properties of covariant homomorphisms.

Proof. Let (i_A, i_G) be the canonical covariant homomorphism of (A, G, α) into $M(A \rtimes_\alpha G)$, and let (k_A, k_G) be the canonical covariant homomorphism of (A, G, β) into $M(A \rtimes_\beta G)$. Let

$$j_A(a) := k_A(a) \quad \text{and} \quad j_G(s) := \bar{k}_A(u_s)k_G(s).$$

Since

$$\begin{aligned} j_G(st) &= \bar{k}_A(u_{st})k_G(st) \\ &= \bar{k}_A(u_s)\bar{k}_A\big(\bar{\beta}_s(u_t)\big)k_G(s)k_G(t) \\ &= \bar{k}_A(u_s)k_G(s)\bar{k}_A(u_t)k_G(t) \\ &= j_G(s)j_G(t), \end{aligned}$$

j_G is a strictly continuous unitary-valued homomorphism. Furthermore

$$\begin{aligned} j_G(s)j_A(a)j_G(s)^* &= \bar{k}_A(u_s)k_G(s)k_A(a)k_G(s)^*\bar{k}_A(u_s^*) \\ &= \bar{k}_A(u_s)k_A\big(\beta_s(a)\big)\bar{k}_A(u_s^*) \\ &= k_A\big(\alpha_s(a)\big). \end{aligned}$$

and (j_A, j_G) is covariant. Therefore $j_A \rtimes j_G$ is a nondegenerate homomorphism of $A \rtimes_\alpha G$ into $M(A \rtimes_\beta G)$. Furthermore if $f \in C_c(G, A)$, then

$$\begin{aligned} j_A \rtimes j_G(f) &= \int_G j_A\bigl(f(s)\bigr) j_G(s)\, d\mu(s) \\ &= \int_G k_A\bigl(f(s)u_s\bigr) k_G(s)\, d\mu(s) \\ &= k_A \rtimes k_G\bigl(\varphi(f)\bigr). \end{aligned}$$

This proves that φ extends to a homomorphism of $A \rtimes_\alpha G$ into $A \rtimes_\beta G$. Reversing the roles of α and β shows that $\varphi^{-1}(f)(s) := f(s)u_s^*$ is an inverse for φ, and φ is an isomorphism as claimed. □

Definition 2.70. A dynamical system $\alpha : G \to \operatorname{Aut} A$ is *unitarily implemented* or just *unitary* if there is a strictly continuous homomorphism $u : G \to UM(A)$ such that $\alpha_s(a) = u_s a u_s^*$ for all $a \in A$ and $s \in G$.

Remark 2.71. A dynamical system $\alpha : G \to \operatorname{Aut} A$ is unitary if and only if it is exterior equivalent to the trivial system $\iota : G \to \operatorname{Aut} A$ where $\iota_s := \mathrm{id}_A$ for all $s \in G$.

Before continuing further, we need some remarks on tensor products and the tensor product of two dynamical systems.

Remark 2.72 (Tensor Products of C^*-algebras). The theory of the tensor product of C^*-algebras has some subtleties which impact the study of crossed products. All that we need, and more, can be found in [139, Appendix B]. For convenience, we mention some properties we'll need here. Usually, the algebraic tensor product $A \odot B$ of two C^*-algebras is not a C^*-algebra. Instead, one looks for a norm $\|\cdot\|_\alpha$ (called a C^*-norm) on $A \odot B$ so that the completion $A \otimes_\alpha B$ is a C^*-algebra. It is a fact of life that there can be more than one C^*-norm on $A \odot B$. The *spatial norm* $\|\cdot\|_\sigma$ has the property that given two representation $\rho_A : A \to B(\mathcal{H}_A)$ and $\rho_B : B \to B(\mathcal{H}_B)$ there is a representation $\rho_A \otimes \rho_B : A \otimes_\sigma B \to B(\mathcal{H}_A \otimes \mathcal{H}_B)$ such that $\rho_A \otimes \rho_B(a \otimes b) = \rho_A(a) \otimes \rho_B(b)$. If ρ_A and ρ_B are faithful, then

$$\Bigl\|\sum a_i \otimes b_i\Bigr\|_\sigma = \Bigl\|\sum \rho_A(a_i) \otimes \rho_B(b_i)\Bigr\|.$$

On the other hand, the *maximal norm* $\|\cdot\|_{\max}$ has the property that if $\pi_A : A \to B(\mathcal{H})$ and $\pi_B : B \to B(\mathcal{H})$ are representations with *commuting ranges* — that is, $\pi_A(a)\pi_B(b) = \pi_B(b)\pi_A(a)$ for all $a \in A$ and $b \in B$ — then there is a representation $\pi_A \otimes_{\max} \pi_B : A \otimes_{\max} B \to B(\mathcal{H})$ such that $\pi_A \otimes_{\max} \pi_B(a \otimes b) = \pi_A(a)\pi_B(b)$. In fact,[14]

$$\Big\|\sum a_i \otimes b_i\Big\|_{\max} = \sup\Big\{ \Big\|\sum \pi_A(a_i)\pi_B(b_i)\Big\| : \\ \pi_A \text{ and } \pi_B \text{ have commuting ranges} \Big\}.$$

It is nontrivial to show that if $\|\cdot\|_\alpha$ is any C^*-norm on $A \odot B$, then $\|\cdot\|_\sigma \leq \|\cdot\|_\alpha \leq \|\cdot\|_{\max}$. For this reason, the spatial norm is often called the *minimal norm.* For a very large class of C^*-algebras A the spatial norm coincides with maximal norm on $A \odot B$ for *every* C^*-algebra B and there is a unique C^*-norm on $A \odot B$. By definition, A is called *nuclear* when $A \odot B$ has a unique C^*-norm for all B. The class of nuclear C^*-algebras includes all GCR C^*-algebras.

As illustrated by our next result and Lemma 2.75 on page 78, it is the maximal tensor product which seems to play the key role in the study of crossed products.[15] In fact, the next two results motivate the assertion that a crossed product can be though of as a twisted maximal tensor product of A and $C^*(G)$.

Lemma 2.73. *If $\iota : G \to \operatorname{Aut} A$ is trivial, then*

$$A \rtimes_\iota G \cong A \otimes_{\max} C^*(G).$$

Proof. We'll use Theorem 2.61 on page 72 to produce an isomorphism

$$j : A \otimes_{\max} C^*(G) \to A \rtimes_\iota G.$$

Let $j_A : A \to M(A \otimes_{\max} C^*(G))$ and $j_G : C^*(G) \to M(A \otimes_{\max} C^*(G))$ be the natural commuting homomorphisms defined, for example, in [139, Theorem B.27]. Since j_G is certainly nondegenerate, it is the integrated form of the strictly continuous unitary-valued homomorphism $j_G : G \to UM(A \otimes_{\max} C^*(G))$, where $j_G(s) := \bar{j}_G\big(i_G(s)\big)$. Note that

$$\begin{aligned} j_G(s)j_A(a)(b \otimes z) &= j_G(s)j_G(z)j_A(ab) \\ &= j_G\big(i_G(s)z\big)j_A(a)j_A(b) \\ &= j_A(a)j_G(s)(b \otimes z). \end{aligned}$$

Thus $j_G(s)j_A(a)j_G(s)^* = j_A(a)$ and (j_A, j_G) is a covariant homomorphism of (A, G, ι) into $M(A \otimes_{\max} C^*(G))$.

If (π, U) is a nondegenerate covariant representation of (A, G, ι), then $U_s \pi(a) U_s^* = \pi(a)$ for all $s \in G$ and $a \in A$. But if $z \in C_c(G)$, we also have

$$\pi(a)U(z) = \int_G z(s)\pi(a)U_s \, d\mu(s) = U(z)\pi(a).$$

[14] As with the definition of the universal norm in Lemma 2.27 on page 52, we have to be a bit careful to see that we are taking the supremum over a set. There are ways to define the maximal norm which avoid this subtly (cf. [139, Proposition B.25]), but the supremum definition gives a better flavor of the properties of the maximal norm, and can be justified with a bit of set theory: see Remark 2.28 on page 52.

[15] When we get around to defining reduced crossed products in Section 7.2, it will turn out that the appropriate norm for reduced crossed products is the spatial norm.

Therefore π and U are commuting representations of A and $C^*(G)$, respectively, and we can define a representation of $A \otimes_{\max} C^*(G)$ by

$$L = L_{(\pi,U)} = \pi \otimes_{\max} U$$

as in [139, Theorem B.27(b)]. We also have $\bar{L} \circ j_A(a) = \pi(a)$ and $\bar{L} \circ j_G(z) = U(z)$ for $a \in A$ and $z \in C_c(G)$. Then

$$\begin{aligned}\bar{L}\big(j_G(s)\big)L(a \otimes z) &= L\big(j_G(s)j_G(z)j_A(a)\big)\\ &= L\big(j_G\big(i_G(s)z\big)j_A(a)\big)\\ &= U\big(i_G(s)z\big)\pi(a)\\ &= U_sU(z)\pi(a)\\ &= U_sL(a \otimes z).\end{aligned}$$

Thus $\bar{L} \circ j_G(s) = U_s$, and we've established conditions (a) and (b) of Theorem 2.61 on page 72. But the elementary tensors $a \otimes z = j_A(a)j_G(z)$ are certainly dense in $A \otimes_{\max} C^*(G)$, so we're done. □

Remark 2.74 (Tensor product systems). If A and B are C^*-algebras with $\alpha \in \operatorname{Aut} A$ and $\beta \in \operatorname{Aut} B$, then [139, Lemma B.31] implies that there is a $\alpha \otimes_{\max} \beta \in \operatorname{Aut}(A \otimes_{\max} B)$ such that $\alpha \otimes_{\max} \beta(a \otimes b) = \alpha(a) \otimes \beta(b)$. Similarly, [139, Proposition B.13] implies there is a $\alpha \otimes \beta \in \operatorname{Aut}(A \otimes_\sigma B)$. If (A, G, α) and (B, G, β) are dynamical systems, then we clearly obtain homomorphisms

$$\alpha \otimes_{\max} \beta : G \to \operatorname{Aut}(A \otimes_{\max} B) \quad \text{and} \quad \alpha \otimes \beta : G \to \operatorname{Aut}(A \otimes_\sigma B).$$

To see that these are also dynamical systems, we still have to verify that the actions are strongly continuous. However, since automorphisms are isometric and since $A \odot B$ is dense in any completion, this follows from estimates such as

$$\begin{aligned}&\big\|(\alpha \otimes \beta)_s(t) - (\alpha \otimes \beta)_r(t)\big\|_\sigma\\ &\quad\le \Big\|(\alpha \otimes \beta)_s\Big(t - \sum a_i \otimes b_i\Big)\Big\|_\sigma\\ &\qquad\quad + \Big\|\sum \alpha_s(a_i) \otimes \beta_s(b_i) - \alpha_r(a_i) \otimes \beta_r(b_i)\Big\|_\sigma\\ &\qquad\qquad + \Big\|(\alpha \otimes \beta)_r\Big(\sum a_i \otimes b_i - t\Big)\Big\|_\sigma\\ &\quad\le 2\Big\|t - \sum a_i \otimes b_i\Big\|_\sigma + \sum \big\|\alpha_s(a_i) \otimes \beta_s(b_i) - \alpha_r(a_i) \otimes \beta_r(b_i)\big\|_\sigma.\end{aligned}$$

Furthermore, if ρ_A and ρ_B are faithful representations of A and B, respectively, then the same is true of $\rho_A \circ \alpha$ and $\rho_B \circ \beta$. As in Remark 2.72 on page 75, if $\kappa : A \otimes_{\max} B \to A \otimes_\sigma B$ is the natural map,

$$\begin{aligned}\ker \kappa &= \ker\big((\rho_A \otimes 1) \otimes_{\max} (1 \otimes \rho_B)\big)\\ &= \ker\big((\rho_A \otimes 1) \otimes_{\max} (1 \otimes \rho_B) \circ (\alpha \otimes_{\max} \beta)\big).\end{aligned}$$

Consequently, $\ker \kappa$ is $\alpha \otimes_{\max} \beta$-invariant and $\alpha \otimes \beta$ is the induced action, $(\alpha \otimes_{\max} \beta)^{\ker \kappa}$, on $A \otimes_\sigma B \cong A \otimes_{\max} B / \ker \kappa$ (see Section 3.4). Except when confusion is likely, it is common practice to simply write $\alpha \otimes \beta$ in place of $\alpha \otimes_{\max} \beta$.

At this point, we only want to consider the product of the trivial action with an arbitrary dynamical system. Although an apparently straightforward example, it has a number of interesting applications such as Corollary 7.18 on page 203.

Lemma 2.75. *Suppose that (C, G, γ) is a dynamical system and that D is a C^*-algebra. Then there is an isomorphism*

$$(C \otimes_{\max} D) \rtimes_{\gamma\otimes \mathrm{id}} G \cong C \rtimes_\gamma G \otimes_{\max} D$$

which carries $(c\otimes_{\max} d)\otimes f \mapsto (c\otimes f)\otimes_{\max} d$, and which intertwines the representation $(\pi_C \rtimes V)\otimes_{\max} \pi_D$ of $C \rtimes_\gamma G \otimes_{\max} D$ with the representation $(\pi_C \otimes_{\max} \pi_D) \rtimes V$ of $(C \otimes_{\max} D) \rtimes_{\gamma\otimes\mathrm{id}} G$.

Proof. We want to apply Theorem 2.61 on page 72 to the system $(C\otimes_{\max} D, G, \gamma\otimes \mathrm{id})$ and the C^*-algebra $C \rtimes_\gamma G \otimes_{\max} D$. For any C^*-algebras A and B, we'll use the letter 'k' for the natural nondegenerate maps $k_A : A \to M(A \otimes_{\max} B)$ and $k_B : B \to M(A \otimes_{\max} B)$ as in [139, Theorem B27]. First we need a covariant homomorphism $(j_{C\otimes_{\max} D}, j_G)$ of $(C \otimes_{\max} D, G, \gamma \otimes \mathrm{id})$ into $M\bigl((C \rtimes_\gamma G) \otimes_{\max} B\bigr)$. Let (i_C, i_G) be the canonical covariant homomorphism of (A, G, γ) into $M(C \rtimes_\gamma G)$, and define j_C to be the composition

$$C \xrightarrow{i_C} M(C \rtimes_\gamma G) \xrightarrow{\bar{k}_{C\rtimes_\gamma G}} M(C \rtimes_\gamma G \otimes_{\max} B).$$

Let $j_D := k_D : D \to M\bigl((C \rtimes_\gamma G) \otimes_{\max} D\bigr)$. Then j_C and j_D have commuting ranges and we can let $j_{C\otimes_{\max} D} := j_C \otimes_{\max} j_D$. We let j_G be the composition

$$G \xrightarrow{i_G} M(C \rtimes_\gamma G) \xrightarrow{\bar{k}_{C\rtimes_\gamma G}} M\bigl((C \rtimes_\gamma G) \otimes_{\max} D\bigr).$$

Since (i_C, i_G) is covariant, it is not hard to check that $(j_{C\otimes_{\max} D}, j_G)$ is covariant and we've established (a) of Theorem 2.61 on page 72.

Now suppose that (π, U) is a nondegenerate covariant representation of $(C\otimes_{\max} D, G, \gamma \otimes \mathrm{id})$. Then $\pi = \pi_C \otimes_{\max} \pi_D$ (where π_C and π_D are characterized by $\pi(c\otimes_{\max} d) = \pi_C(c)\pi_D(d)$). Then it is not hard to verify that (π_C, U) is a covariant representation of (C, G, γ):

$$\begin{aligned} U_s\pi_C(c)\pi(c' \otimes_{\max} d) &= U_s\pi(cc' \otimes_{\max} d) \\ &= \pi\bigl(\gamma_s(cc') \otimes_{\max} d\bigr)U_s \\ &= \pi_C\bigl(\gamma_s(c)\bigr)\pi\bigl(\gamma_s(c') \otimes_{\max} d\bigr)U_s \\ &\pi_C\bigl(\gamma_s(c)\bigr)U_s\pi(c' \otimes_{\max} d). \end{aligned}$$

We also want to check that $\pi_C \rtimes U$ and π_D have commuting ranges:

$$\begin{aligned}
\pi_C \rtimes U(c\otimes z)\pi_D(d)\pi(c'\otimes_{\max} d') &= \pi_C(c)U(z)\pi(c'\otimes_{\max} dd') \\
&= \pi_C(c)\int_G z(s)U_s\pi(c'\otimes_{\max} dd')\,d\mu(s) \\
&= \pi_C(c)\int_G z(s)\pi\bigl(\gamma_s(c')\otimes_{\max} dd'\bigr)U_s\,d\mu(s) \\
&= \pi_C(c)\pi_D(d)\int_G z(s)\pi\bigl(\gamma_s(c')\otimes_{\max} d'\bigr)U_s\,d\mu(s) \\
&= \pi_D(d)\pi_C(c)\int_G z(s)U_s\pi(c'\otimes_{\max} d')\,d\mu(s) \\
&= \pi_D(d)\pi_C \rtimes U(c\otimes z)\pi(c'\otimes_{\max} d').
\end{aligned}$$

Therefore we can form $L = L_{(\pi,U)} := \pi_C \rtimes U \otimes_{\max} \pi_D$ and then verify that $L \circ j_{C\otimes_{\max} D} = \pi$ and $L\circ j_G = U$. This verifies part (b).

For part (c), consider

$$\begin{aligned}
j_{C\otimes_{\max} D}(c\otimes_{\max} d)j_G(z) &= j_C(c)j_D(d)j_G(z) \\
&= j_C(c)j_G(z)j_D(d) \\
&= i_C(c)i_G(z)\otimes_{\max} d \\
&= c\otimes z\otimes_{\max} d.
\end{aligned}$$

It follows from Theorem 2.61 on page 72 that $j := j_{C\otimes_{\max} D} \rtimes j_G$ is the required isomorphism, and that $j\circ L_{((\pi_C\otimes_{\max}\pi_D)\rtimes V)} = (\pi_C\otimes_{\max}\pi_D)\rtimes V$. Since $L_{((\pi_C\otimes_{\max}\pi_D)\rtimes V)} = (\pi_C\rtimes V)\otimes_{\max}\pi_D$, the assertion about intertwining representations follows. □

Notes and Remarks

The notion of a crossed product — at least as a purely algebraic object — dates to the beginning of the twentieth century. For example, the basic construction appears in [22] and [166] and a nice survey of this early work can be found in [14]. (These references were pointed out by an anonymous reviewer.) The idea of associating an operator algebra to a group of automorphisms of another operator algebra probably originates with the pioneering work of Murray and von Neumann [111]. A (reduced) C^*-crossed product construction appears in [163], and the connections to physics and covariant representations begins in earnest in [30]. A systematic study of transformation group C^*-algebras can be found in [49] and full formed crossed products appear in [100, 162, 173]. The current notation of $A\rtimes_\alpha G$ and the terminology "C^*-dynamical system" was popularized in the work of Olesen and Pedersen [114–116]. Although viewing the crossed product as a universal object for covariant representations is fundamental even to the early work on C^*-crossed products, the concept was formalized by Raeburn in [133].

Chapter 3

Special Cases and Basic Constructions

It is a bit frustrating not to work out some nontrivial examples of crossed products now that we have the definition in hand. However, rather than pausing to grind out some *ad hoc* examples here, we'll wait until Chapter 4 when we will have proved the Imprimitivity Theorem. Before taking on the formidable technology required for imprimitivity theorems, we'll concentrate on some basic properties of crossed products. Since group C^*-algebras can be viewed as degenerate crossed products and since crossed products can be thought of as analogues of group C^*-algebras in which scalar valued functions are replaced by C^*-algebra-valued functions, group C^*-algebras serve as excellent sources of motivation. In this chapter, we look at the two classes where we can say something significant in general: namely we examine the C^*-algebras of abelian and compact groups, respectively. After that, in Section 3.3, we see that when G is a semidirect product, the crossed product $A \rtimes_\alpha G$ can be decomposed as an iterated crossed product. (Generalizing this result to the case where G contains a normal subgroup is discussed in Section 7.4 — although this requires a brief introduction to twisted crossed products.) In Section 3.4 we note that G-invariant ideals in A correspond to ideals in $A \rtimes_\alpha G$ is a natural way. In Section 3.5, we start the study of orbit spaces, the properties of which will be increasingly more important as we get deeper into the book. In the final section, we look at proper group actions and induced C^*-algebras — both of which are going to be important in Chapter 4.

3.1 Another Interlude: The C^*-Algebra of an Abelian group

In this section, we satisfy our desire for an example by considering the group C^*-algebra of an abelian group and examining what this example says about C^*-

algebras and representation theory.[1]

Let G be a locally compact *abelian* group. If $f \in C_c(G)$, then Theorem A.14 of [139] implies that there is an irreducible representation π of $C^*(G)$ such that $\|f\| = \|\pi(f)\|$. Recall that π is necessarily the integrated form of an irreducible unitary representation of G (which we denote with the same letter). Since $C^*(G)$ is abelian, it follows from [139, Example A.16] that every irreducible representation is one-dimensional and therefore corresponds to a character $\omega \in \widehat{G}$. Thus the universal norm of $f \in C_c(G)$ is given by

$$\|f\| = \|\hat{f}\|_\infty$$

where $\hat{f}$ is the Fourier transform

$$\hat{f}(\omega) := \int_G f(s)\omega(s)\,d\mu(s) \quad (\omega \in \widehat{G})$$

of f (see (1.20) in Section 1.4.1). Even without invoking the Gelfand Theory, it is easy to see that $\hat{f}$ is continuous on $\widehat{G}$ (using the dominated convergence theorem for example). Since we can identify $\widehat{G}$ with the spectrum of $C^*(G)$ (Remark 2.41 on page 60), we can conclude that $\hat{f}$ vanishes at infinity on $\widehat{G}$ since $[\pi] \mapsto \|\pi(f)\|$ always vanishes at infinity on the spectrum by [139, Lemma A.30]. Thus, $\hat{f} \in C_0(\widehat{G})$, and $\mathfrak{A} := \{\, \hat{f} : f \in C_c(G)\,\}$ is a subalgebra of $C_0(\widehat{G})$ which is closed under complex conjugation since $\widehat{f^*}(\omega) = \omega(f^*) = \overline{\omega(f)} = \overline{\hat{f}(\omega)}$, and which separates points since the irreducible representations of $C^*(G)$ must separate points. Thus the Stone-Weierstrass Theorem implies that $\mathfrak{A}$ is dense in $C_0(\widehat{G})$ in the sup-norm. This gives us the following proposition.

Proposition 3.1. *If G is a locally compact abelian group, then the Fourier transform extends to an isomorphism of $C^*(G)$ with $C_0(\widehat{G})$.*

If you're only interested in Harmonic analysis or the unitary representations of G, it might be reasonable to ask "why bother with Proposition 3.1?" For example, suppose your goal is to describe all the unitary representations of G. Then it turns out that Proposition 3.1 does allow us to describe the representations of G, provided we can invoke some standard results about representations of C^*-algebras. In fact, much of the early work on abstract C^*-algebras was motivated by the fact that C^*-algebras have a nice representation theory which should shed some light on the representation theory of groups via the group C^*-algebra construction. The classic books [28, 29] are beautiful illustrations of this point of view. Another very readable treatment, focusing on the representation theory of Type I algebras, is Arveson's book [2] where he works out the following.

Example 3.2 ([2, pp. 54–56]). Let A be a *separable* commutative C^*-algebra. Then we may assume that $A = C_0(X)$ for a second countable locally compact Hausdorff space X. Then every representation of $C_0(X)$ on a separable Hilbert space is

[1] A useful and extensive survey of the "state of the art" (as of 1976) of classes of locally compact groups and properties of their C^*-algebras is given in [123].

equivalent to one constructed as follows. Let $\{X_\infty, X_1, X_2, \dots\}$ be a partition of X into Borel sets. For each $n = \infty, 1, 2, \dots$, let μ_n be a finite Borel measure on X_n and let $\mathcal{H}_n$ be a Hilbert space of dimension n. (If $n = \infty$, then any separable infinite-dimensional Hilbert space will do.) Let

$$\mathcal{H} := \bigoplus_{n=1}^{n=\infty} L^2(X_n, \mu_n) \otimes \mathcal{H}_n.$$

Now define $\pi : C_0(X) \to B(\mathcal{H})$ by $\pi := \bigoplus_{n=1}^{n=\infty} \pi_{\mu_n} \otimes 1_{\mathcal{H}_n}$, where $\pi_{\mu_n} : C_0(X) \to B\big(L^2(X_n)\big)$ is given by pointwise multiplication:

$$\pi_{\mu_n}(f)h(x) := f(x)h(x).$$

Combining Example 3.2 with Proposition 3.1 on the preceding page, we see that if G is a second countable locally compact abelian group, then every representation of $C^*(G)$ is given on $f \in C_c(G)$ by

$$\pi(f) = \bigoplus_{n=1}^{n=\infty} \pi_{\mu_n}(\hat{f}) \otimes 1_{\mathcal{H}_n}$$

for a partition $\{X_n\}$ of $\widehat{G}$ and measures μ_n as above. Furthermore, it is not hard to see that π is the integrated form of

$$U = \bigoplus_{n=1}^{n=\infty} U^n \otimes 1_{\mathcal{H}_n}, \tag{3.1}$$

where $U^n : G \to U\big(L^2(X_n)\big)$ is given by

$$U^n_s h(\omega) := \omega(s)h(\omega).$$

Notice that (3.1) is a description, up to equivalence, of all unitary representations of G on a separable Hilbert space. This answers our original question and illustrates the power of the group C^*-algebra construction.

Our characterization of the representations of a locally compact abelian group has a curious corollary. Recall that if H is a closed subgroup of a locally compact abelian group G, then H is also a locally compact abelian group. It is a consequence of the Pontryagin Duality Theorem that the restriction map from $\widehat{G}$ to $\widehat{H}$ induces an isomorphism of locally compact abelian groups:

$$\widehat{G}/H^\perp \cong \widehat{H},$$

where $H^\perp := \{\omega \in \widehat{G} : \omega(h) = 1 \text{ for all } h \in H\}$ [56, Theorem 4.39]. If we assume that G is second countable, then there is a Borel cross-section

$$c : \widehat{H} \to \widehat{G}$$

such that $c(\sigma)(h) = \sigma(h)$ for all $\sigma \in \widehat{H}$ and $h \in H$ ([2, Theorem 3.4.1] will do). Now if V is a unitary representation of H of the form

$$V = \bigoplus_{n=1}^{\infty} V^n \otimes 1_{\mathcal{H}_n},$$

then we can define unitary representations $U^n : G \to U\big(L^2(X_n)\big)$ by

$$U^n_s h(\sigma) = c(\sigma)(s)h(\sigma).$$

Then U^n extends V^n and $U := \bigoplus_{n=1}^{n=\infty} U^n \otimes 1_{\mathcal{H}_n}$ extends V. We have proved the following.

Proposition 3.3. *Suppose that H is a closed subgroup of a second countable locally compact* abelian *group G. Then every unitary representation of H can be extended to a unitary representation of G.*

3.2 Third Interlude: The C^*-Algebra of a Compact group

Now suppose that G is compact, but not necessarily abelian. The Peter-Weyl Theorem [56, §5.2] tells us that if $[\pi] \in \widehat{G}$, then π is a finite dimensional representation and that π is equivalent to a subrepresentation of the left-regular representation $\lambda : G \to U\big(L^2(G)\big)$. For the sake of definiteness, if $c \in \widehat{G}$, let π_c be a specific subrepresentation of λ such that $[\pi_c] = c$. Note that if $d \neq c$ in $\widehat{G}$, then π_c and π_d act on orthogonal subspaces by [56, Proposition 5.3]. Since $d_c := \dim \mathcal{H}_{\pi_c} < \infty$, we'll identify $B(\mathcal{H}_{\pi_c}) = \mathcal{K}(\mathcal{H}_{\pi_c})$ with the C^*-algebra of $d_c \times d_c$-matrices M_{d_c}. Note that $\pi_c\big(C^*(G)\big) = \mathcal{K}(\mathcal{H}_{\pi_c}) \cong M_{d_c}$ since an irreducible representation that contains one nonzero compact operator in its range must contain all compact operators [110, Theorem 2.4.9].

Proposition 3.4. *Suppose that G is a compact group. Then*

$$C^*(G) \cong \bigoplus_{c \in \widehat{G}} M_{d_c}.$$

The isomorphism sends f to $\big(\pi_c(f)\big)_{c \in \widehat{G}}$.

To prove the proposition, we'll need a couple of straightforward lemmas.

Lemma 3.5. *If G is a compact group and $f \in C^*(G)$, then $\lambda(f)$ is a compact operator on $L^2(G)$.*

Proof. Suppose first that $f \in C_c(G) \subset C^*(G)$ and $h, k \in C_c(G) \subset L^2(G)$. Then

$$\begin{aligned}(\lambda(f)h \mid k) &= \int_G (f(s)\lambda(s)h \mid k)\,d\mu(s)\\ &= \int_G\int_G f(s)h(s^{-1}r)\overline{k(r)}\,d\mu(r)\,d\mu(s)\\ &= \int_G f * h(r)\overline{k(r)}\,d\mu(r).\end{aligned}$$

Thus $\lambda(f)h = f * h$ in $L^2(G)$. But using the unimodularity of G,

$$\begin{aligned}f * h(r) &= \int_G f(s)h(s^{-1}r)\,d\mu(s)\\ &= \int_G f(rs^{-1})h(s)\,d\mu(s).\end{aligned}$$

Thus $\lambda(f)$ is a Hilbert-Schmidt operator on $L^2(G)$ with kernel $K(s,r) = f(s^{-1}r)$, and therefore compact. Since $C_c(G)$ is dense in $C^*(G)$, it follows that $\lambda(f)$ is compact for all $f \in C^*(G)$. □

Lemma 3.6. *Let $f \in C^*(G)$ and $\epsilon > 0$. Then the collection F of $c \in \widehat{G}$ such that $\|\pi_c(f)\| > \epsilon$ is finite. In particular, the spectrum $\widehat{G}$ of $C^*(G)$ is a discrete topological space.*

Proof. Let $\mathcal{H}_c \subset L^2(G)$ be the subspace corresponding to the subrepresentation π_c. Then the $\{\mathcal{H}_c\}$ are pairwise orthogonal. Furthermore, we can choose a unit vector $h_c \in \mathcal{H}_c$ such that $\|\lambda(f)h_c\| = \|\pi_c(f)h_c\| > \epsilon$. But $\lambda(f)h_c \in \mathcal{H}_c$. Thus if F were infinite, $\{\lambda(f)h_c\}_{c\in F}$ would have no convergent subnet. This would contradict the compactness of $\lambda(f)$.

However, F is an open subset of $\widehat{G}$ by [139, Lemma A.30]. Since M_n has, up to equivalence, a unique irreducible representation, we can identify $\widehat{G}$ with $\operatorname{Prim} C^*(G)$. Since $\ker\pi$ is clearly a maximal ideal if $[\pi] \in \widehat{G}$, $\{c\}$ is closed for all $c \in \widehat{G}$. However, given $c = [\pi] \in \widehat{G}$, there is a $f \in C^*(G)$ such that $\|\pi(f)\| > 1$. The first part of this proof implies there is a finite neighborhood of c in $\widehat{G}$. Since points are closed, $\{c\}$ is open. Since c was arbitrary, $\widehat{G}$ is discrete as claimed. □

Lemma 3.7. *Suppose that G is a compact group, $c \in \widehat{G}$, and $T \in M_{d_c}$. Then there is a $f \in C^*(G)$ such that*

$$\rho(f) = \begin{cases} T & \text{if } \rho = \pi_c, \text{ and}\\ 0 & \text{if } [\rho] \neq c.\end{cases}$$

Proof. Since $\widehat{G}$ is discrete, $\{c\}$ is open and is not in the closure of $\{c' \in \widehat{G} : c' \neq c\}$. Therefore

$$J := \bigcap_{c'\neq c} \ker\pi_{c'} \not\subset \ker\pi_c.$$

In particular $J \neq \{0\}$ and $\pi_c(J)$ is a nonzero ideal in $M_{d_{[\pi_c]}}$. Since matrix algebras are simple, $\pi_c(J) = M_{d_{[\pi_c]}}$, and there is a $f \in J$ such that $\pi_c(f) = T$. Since $f \in J$, we're done. □

Proof of Proposition 3.4 on page 84. Define Φ from $C^*(G)$ to the product $\prod_{c\in\widehat{G}} M_{d_c}$ by $\Phi(f) = \big(\pi_c(f)\big)_{c\in\widehat{G}}$. Since the irreducible representations of any C^*-algebra separate points, Φ is an isomorphism onto its range. Lemma 3.6 on the previous page implies that the range of Φ lies in the direct sum. Lemma 3.7 on the preceding page implies that the range of Φ is dense in the direct sum. Since Φ necessarily has closed range, we're done. □

Since any representation of the compacts is equivalent to a multiple of the identity representation and since we've shown that $C^*(G)$ is a direct sum of matrix algebras if G is compact, we get following result essentially for free.

Corollary 3.8. *Every representation of a compact group is the direct sum of (necessarily finite dimensional) irreducible representations.*

3.3 Semidirect Products

Suppose that H and N are locally compact groups and that $\varphi : H \to \operatorname{Aut} N$ is a homomorphism such that $(h, n) \mapsto \varphi(h)(n)$ is continuous from $H \times N$ to N. (In other words, H acts continuously on N via automorphisms.) The semidirect product $N \rtimes_\varphi H$ is the group with underlying set $N \times H$ and group operations

$$(n,h)(m,k) := \big(n\varphi(h)(m), hk\big) \quad \text{and} \quad (n,h)^{-1} := \big(\varphi(h^{-1})(n^{-1}), h^{-1}\big).$$

If $N \times H$ has the product topology, then $N \rtimes_\varphi H$ is a locally compact group.

Example 3.9. Let $H := (\mathbf{R}^+, \cdot)$, $N := (\mathbf{R}, +)$ and define $\varphi : \mathbf{R}^+ \to \operatorname{Aut} \mathbf{R}$ by $\varphi(a)(x) := ax$. Then $\mathbf{R} \rtimes_\varphi \mathbf{R}^+$ is isomorphic to the $ax + b$-group of Example 1.27 on page 7.

Example 3.10. Let G be a locally compact group with a closed normal subgroup N and a closed subgroup H such that $N \cap H = \{e\}$ and $G = NH$. Then we can define $\varphi : H \to \operatorname{Aut} N$ by $\varphi(h)(n) = hnh^{-1}$, and $(n, h) \mapsto nh$ is an isomorphism of locally compact groups between $N \rtimes_\varphi H$ and G. Conversely, if $G = N \rtimes_\varphi H$, then $N' := \{(n, e_H) \in G : n \in N\}$ is a closed normal subgroup of G and $H' := \{(e_N, h) \in G : h \in H\}$ is a closed subgroup of G such that $G = N'H'$ and $N' \cap H' = \{e\}$.

In this section, we'll assume that $G = NH$ as in Example 3.10 so that $\varphi(h)(n) = hnh^{-1}$. Let μ_N be a Haar measure on N and μ_H a Haar measure on H. Considerations similar to those used to prove that the modular function is a continuous homomorphism (Lemma 1.61 on page 18) show that there is a continuous homomorphism

$$\sigma : H \to \mathbf{R}^+$$

such that

$$\sigma(h)\int_N f(hnh^{-1})\,d\mu_N(n)=\int_N f(n)\,d\mu_N(n). \tag{3.2}$$

Then a Haar measure on G is given by

$$I(f):=\int_H\int_N f(nh)\sigma(h)^{-1}\,d\mu_N(n)\,d\mu_H(h)=\int_H\int_N f(hn)\,d\mu_N(n)\,d\mu_H(h).$$

Proposition 3.11. *Suppose that G is the semidirect product $N\rtimes_\varphi H$ of two locally compact groups as above, and that (A,G,α) is a dynamical system. Then there is a dynamical system*

$$\beta: H\to \operatorname{Aut}(A\rtimes_{\alpha|_N}N)$$

such that for all $g\in C_c(N,A)$

$$\beta_h(g)(n)=\sigma(h)^{-1}\alpha_h\big(g(h^{-1}nh)\big).$$

Furthermore, if ι is the natural map from $C_c\big(H,C_c(N,A)\big)\subset C_c(H,A\rtimes_{\alpha|_N}N)$ into $C_c(N\times H,A)\subset C_c(G,A)$, then ι extends to an isomorphism of $(A\rtimes_{\alpha|_N}N)\rtimes_\beta H$ with $A\rtimes_\alpha G$.

Proof. For convenience, we'll assume $G=NH$ as in Example 3.10 on the facing page. Let (j_A,j_N) be the canonical covariant homomorphisms for $A\rtimes_{\alpha|_N}N$. To define an action of H on $A\rtimes_{\alpha|_N}N$, fix $h\in H$ and define $\pi^h: A\to M(A\rtimes_{\alpha|_N}N)$ and $U^h:N\to UM(A\rtimes_{\alpha|_N}N)$ by

$$\pi^h(a):=j_A\big(\alpha_h(a)\big)\quad\text{and}\quad U^h_n:=j_N(hnh^{-1}).$$

Since j_A is nondegenerate and j_N is strictly continuous, it follows that (π^h,U^h) is a nondegenerate covariant homomorphism of $(A,N,\alpha|_N)$ into $M(A\rtimes_{\alpha|_N}N)$. Thus $\beta_h:=\pi^h\rtimes U^h$ is a nondegenerate homomorphism of $A\rtimes_{\alpha|_N}N$ into $M(A\rtimes_{\alpha|_N}N)$. If $f\in C_c(N,A)$, then

$$\begin{aligned}\beta_h(f)&:=\int_N \pi^h\big(f(n)\big)U^h_n\,d\mu_N(n)\\&=\int_N j_A\big(\alpha_h\big(f(n)\big)\big)j_N\big(hnh^{-1}\big)\,d\mu_N(n)\\&=\int_N j_A\big(\sigma(h^{-1})\alpha_h\big(f(h^{-1}nh)\big)\big)j_N(n)\,d\mu_N(n)\\&=j_A\rtimes j_N({}^hf),\end{aligned}$$

where ${}^hf(n):=\sigma(h^{-1})\alpha_h\big(f(h^{-1}nh)\big)$. In particular, $\beta_h(f)={}^hf\in C_c(N,A)\subset A\rtimes_{\alpha|_N}N$. Therefore β_h is a homomorphism of $A\rtimes_{\alpha|_N}N$ into itself. Since routine computations show that $\beta_e=\mathrm{id}$ and $\beta_{hk}=\beta_h\circ\beta_k$, it follows that each β_h is a automorphism with inverse $\beta_{h^{-1}}$, and

$$\beta: H\to\operatorname{Aut}(A\rtimes_{\alpha|_N}N) \tag{3.3}$$

is a homomorphism.

Lemma 3.12. *If H, N, G and β are as above, then* (3.3) *is a dynamical system.*

Proof. At this point, we only have to see that β is strongly continuous. For this, it suffices to show that given $f \in C_c(N, A)$ and $\delta > 0$, then there is a neighborhood V of e in H such that $h \in V$ implies that $\|\beta_h(f) - f\|_1 < \delta$. Let $K := \operatorname{supp} f$ and let W be a compact neighborhood of e in H. Then if $h \in W$, $\operatorname{supp} \beta_h(f) \subset WKW$. Thus it suffices to show that given $\epsilon > 0$ there is a neighborhood V of e in H such that $h \in V$ implies $\|\beta_h(f) - f\|_\infty \leq \epsilon$. To this end, choose V' such that $h \in V'$ implies $|\sigma(h^{-1}) - 1| < \epsilon/(3\|f\|_\infty)$. Since f is uniformly continuous, the triangle inequality implies there is a $V'' \subset V'$ such that $h \in V''$ implies

$$\|f(h^{-1}nh) - f(n)\| < \frac{\epsilon}{3} \quad \text{for all } n \in N.$$

A compactness argument implies there is a $V \subset V''$ such that $h \in V$ implies

$$\|\alpha_h\big(f(n)\big) - f(n)\| < \frac{\epsilon}{3} \quad \text{for all } n \in N.$$

Therefore $h \in V$ implies

$$\begin{aligned} \|\sigma(h)^{-1}\alpha_h\big(f(h^{-1}nh)\big) - f(n)\| &\leq |\sigma(h)^{-1} - 1|\|\alpha_h\big(f(h^{-1}nh)\big)\| + \\ &\qquad \|\alpha_h\big(f(h^{-1}nh) - f(n)\big)\| + \|\alpha_h\big(f(n)\big) - f(n)\| \\ &\leq |\sigma(h)^{-1} - 1|\|f\|_\infty + \|f(h^{-1}nh) - f(n)\| + \frac{\epsilon}{3} \\ &< \epsilon. \end{aligned}$$

□

Lemma 3.13. *Let X be a Hilbert B-module and $G = NH$ a semidirect product as above. If $U : N \to U\mathcal{L}(\mathsf{X})$ and $V : H \to U\mathcal{L}(\mathsf{X})$ are strictly continuous unitary-valued homomorphisms such that*

$$V_h U_n V_h^* = U_{hnh^{-1}}, \tag{3.4}$$

then $W_{nh} := U_n V_h$ defines a strictly continuous unitary-valued homomorphism $W : G \to U\mathcal{L}(\mathsf{X})$.

Conversely, if $W : G \to U\mathcal{L}(\mathsf{X})$ is a strictly continuous unitary-valued homomorphism, then $U_n := W_n$ and $V_h := W_h$ define strictly continuous unitary-valued homomorphisms satisfying (3.4).

Proof. Straightforward. □

Lemma 3.14. *Suppose that (π, U) is a nondegenerate covariant homomorphism of (A, G, α) in $\mathcal{L}(\mathsf{X})$, and that $T \in M(A)$. Then*

$$U_s \bar{\pi}(T) U_s^* = \bar{\pi}\big(\bar{\alpha}_s(T)\big), \tag{3.5}$$

and if (i_A, i_G) is the canonical covariant homomorphism of (A, G, α) into $M(A \rtimes_\alpha G)$, then

$$(\pi \rtimes U)^{-}\big(\bar{\imath}_A(T)\big) = \bar{\pi}(T). \tag{3.6}$$

Proof. Let $x \in \mathsf{X}$ and $a \in A$. Then

$$\begin{aligned} U_s\bar{\pi}(T)\big(\pi(a)(x)\big) &= U_s\pi(Ta)(x) \\ &= \pi\big(\alpha_s(Ta)\big)U_s(x) \\ &= \bar{\pi}\big(\bar{\alpha}_s(T)\big)\pi\big(\alpha_s(a)\big)U_s(x) \\ &= \bar{\pi}\big(\bar{\alpha}_s(T)\big)U_s\big(\pi(a)(x)\big) \end{aligned}$$

Since π is nondegenerate, $\{\,\pi(a)(x) : a \in A \text{ and } x \in \mathsf{X}\,\}$ spans a dense subspace of X, and (3.5) follows.

Similarly,

$$\begin{aligned} (\pi \rtimes U)^{-}\big(\bar{\imath}_A(T)\big)\pi \rtimes U\big(i_A(a)f\big) &= \pi \rtimes U\big(\bar{\imath}_A(T)i_A(a)f\big) \\ &= \pi \rtimes U\big(i_A(Ta)f\big) \\ &= \pi(Ta)\pi \rtimes U(f) \\ &= \bar{\pi}(T)\pi(a)\pi \rtimes U(f) \\ &= \bar{\pi}(T)\pi \rtimes U\big(i_A(a)f\big). \end{aligned}$$

Now (3.6) follows since both i_A and $\pi \rtimes U$ are nondegenerate. □

Now we want to apply Theorem 2.61 on page 72 with $B = (A \rtimes_{\alpha|_N} N) \rtimes_\beta H$ with the object of showing that B is isomorphic to $A \rtimes_\alpha G$. Let $k_{A\rtimes_{\alpha|_N}N} : A \rtimes_{\alpha|_N} N \to M((A \rtimes_{\alpha|_N} N) \rtimes_\beta H)$ and $k_H : H \to M((A \rtimes_{\alpha|_N} N) \rtimes_\beta H)$ be the canonical covariant homomorphism for $(A \rtimes_{\alpha|_N} N, H, \beta)$. Define maps

$$l_A : A \to M((A \rtimes_{\alpha|_N} N) \rtimes_\beta H) \quad \text{and} \quad l_G : G \to M((A \rtimes_{\alpha|_N} N) \rtimes_\beta H)$$

as follows. Let $l_A(a) := (k_{A\rtimes_{\alpha|_N}N})^{-}\big(j_A(a)\big)$ and $l_G(nh) = W_nV_h$, where $W_n = (k_{A\rtimes_{\alpha|_N}N})^{-}\big(j_N(n)\big)$ and $V_h = k_H$. Of course V is strictly continuous since k_H is. The strict continuity of W follows from the strict continuity of j_N and the nondegeneracy of $k_{A\rtimes_{\alpha|_N}N}$: note that

$$(k_{A\rtimes_{\alpha|_N}N})^{-}\big(j_N(n)\big)k_{A\rtimes_{\alpha|_N}N}(f)k_H(z) = k_{A\rtimes_{\alpha|_N}N}\big(j_N(n)f\big)k_H(z).$$

is continuous in n and the span of elements of the form $k_{A\rtimes_{\alpha|_N}N}(f)k_H(z)$ is dense in $(A \rtimes_{\alpha|_N} N) \rtimes_\beta H$.

Now using Lemma 3.14 on the facing page we have

$$\begin{aligned} V_hW_nV_{h^{-1}} &= k_H(h)(k_{A\rtimes_{\alpha|_N}N})^{-}\big(j_N(n)\big)k_H(h)^* \\ &= (k_{A\rtimes_{\alpha|_N}N})^{-}\big(\bar{\beta}_h\big(j_N(n)\big)\big) \end{aligned}$$

which, since $\beta_h = \pi^h \rtimes U^h$, equals

$$\begin{aligned} &= (k_{A\rtimes_{\alpha|_N}N})^{-}\big(j_N(hnh^{-1})\big) \\ &= W_{hnh^{-1}}. \end{aligned}$$

Therefore l_G is a strictly continuous unitary-valued homomorphism (Lemma 3.13 on page 88). To check that (l_A, l_G) is covariant, consider

$$\begin{aligned} l_G(nh)l_A(a)l_G(nh)^* &= W_n V_h l_A(a) V_h^* U_n^* \\ &= W_n k_H(h) (k_{A\rtimes_{\alpha|_N} N})^{-}\big(j_A(a)\big) k_H(h)^* W_n^* \end{aligned}$$

which, by Lemma 3.14, is

$$= W_n (k_{A\rtimes_{\alpha|_N} N})^{-}\big(\bar{\beta}_h(j_A(a))\big) W_n^*$$

which, since $\beta_h = \pi^h \rtimes U^h$, is

$$\begin{aligned} &= W_n (k_{A\rtimes_{\alpha|_N} N})^{-}\big(j_A(\alpha_h(a))\big) W_n^* \\ &= (k_{A\rtimes_{\alpha|_N} N})^{-}\big(j_N(n) j_A(\alpha_h(a)) j_N(n)^*\big) \\ &= (k_{A\rtimes_{\alpha|_N} N})^{-}\big(j_A(\alpha_{nh}(a))\big) \\ &= l_A(\alpha_{nh}(a)). \end{aligned}$$

All this establishes condition (a) of Theorem 2.61 on page 72.

Now suppose that (π, T) is a nondegenerate covariant representation of (A, G, α). Define W and V as in Lemma 3.13 on page 88: $T_{nh} = W_n V_h$. Then (π, W) is a covariant representation of $(A, N, \alpha|_N)$. Moreover the computation

$$\begin{aligned} V_h \pi \rtimes W(f) V_h^* &= V_h \int_N \pi\big(f(n)\big) W_n \, d\mu_N(n) V_h^* \\ &= \int_N T_h \pi\big(f(n)\big) T_{nh^{-1}} \, d\mu_N(n) \\ &= \int_N \pi\big(\beta_h(f)(n)\big) W_n \, d\mu_N(n) \\ &= \pi \rtimes W\big(\beta_h(f)\big) \end{aligned}$$

shows that $(\pi \rtimes W, V)$ is covariant. Thus we define $L = L_{(\pi,T)}$ to be $(\pi \rtimes W) \rtimes V$. Using Lemma 3.14 on page 88, we have

$$\begin{aligned} \bar{L}\big(l_A(a)\big) &= ((\pi \rtimes W) \rtimes V)^{-}\big((k_{A\rtimes_{\alpha|_N} N})^{-}\big(j_A(a)\big)\big) \\ &= (\pi \rtimes W)^{-}\big(j_A(a)\big) \\ &= \pi(a). \end{aligned}$$

Similarly,

$$\begin{aligned} \bar{L}\big(l_G(nh)\big) &= ((\pi \rtimes W) \rtimes V)^{-}\big((k_{A\rtimes_{\alpha|_N} N})^{-}\big(j_N(n)\big)\big)((\pi \rtimes W) \rtimes V)^{-}\big(k_H(h)\big) \\ &= (\pi \rtimes W)^{-}\big(j_N(n)\big) V_h \\ &= W_n V_h \\ &= T_{nh}. \end{aligned}$$

Now we've established condition (b) of Theorem 2.61.

To verify condition (c), let $z \in C_c(G)$ be defined by

$$z(nh) = \sigma(h)x(n)y(h)$$

for functions $x \in C_c(N)$ and $y \in C_c(H)$. Then

$$l_G(z) = \int_H \int_N x(n)y(h)l_G(nh)\,d\mu_N(n)\,d\mu_H(h) = W(x)V(y).$$

Therefore

$$l_A(a)l_G(z) = l_A(a)W(x)V(y) = k_{A\rtimes_{\alpha|_N}N}\big(j_A(a)j_N(x)\big)k_H(y).$$

Now condition (c) of Theorem 2.61 follows as the span of elements of the form $j_A(a)j_N(x)$ is dense in $A \rtimes_{\alpha|_N} N$.

Now Theorem 2.61 on page 72 implies that there is an isomorphism

$$j : (A \rtimes_{\alpha|_N} N) \rtimes_\beta H \to A \rtimes_\alpha G$$

such that $j \circ l_G = i_G$ and $j \circ l_A = i_A$, where (i_A, i_G) are the canonical maps associated to (A, G, α).

Now suppose $F \in C_c\big(H, C_c(N, A)\big)$ (viewed as a subspace of $C_c(H, A \rtimes_{\alpha|_N} N)$) is of the form $F(h)(n) = y(h)x(n)a$. Then $F = k_{A\rtimes_{\alpha|_N}N} \rtimes k_H(F) = k_{A\rtimes_{\alpha|_N}N}\big(j_A(a)j_N(x)\big)k_H(y)$, and $j(F) = i_A(a)i_G(x \otimes y)$, where $x \otimes y$ denotes the function on G given by $nh \mapsto x(n)y(h)$. In other words, $j(F) = \iota(F)$. This completes the proof of Proposition 3.11 on page 87. □

Example 3.15 (The $ax+b$-group). Let $G = \mathbf{R} \rtimes_\varphi \mathbf{R}^+$ be (isomorphic to) the $ax+b$-group; thus $\varphi(a)(r) := ar$. Then $\sigma : \mathbf{R}^+ \to \mathbf{R}$ is defined by

$$\sigma(a) \int_{-\infty}^{\infty} f(ax)\,dx = \int_{-\infty}^{\infty} f(x)\,dx,$$

and $\sigma(a) = a$. Then Proposition 3.11 on page 87 implies that $C^*(G) \cong \mathbf{C} \rtimes_\iota G$ is isomorphic to $C^*(\mathbf{R}) \rtimes_\beta \mathbf{R}^+$ where $\beta_a(f)(r) = \frac{1}{a} f(\frac{r}{a})$. The Fourier transform is an isomorphism $\mathcal{F} : C^*(\mathbf{R}) \to C_0(\mathbf{R})$ such that $\mathcal{F}\big(\beta_a(f)\big)(y) = \hat{f}(ay)$. Thus we get an equivariant isomorphism of $\big(C^*(\mathbf{R}), \mathbf{R}^+, \beta\big)$ and $\big(C_0(\mathbf{R}), \mathbf{R}^+, \tau\big)$, where $\mathbf{R}^+$ acts on $\mathbf{R}$ via $a \cdot r := \frac{r}{a}$. Thus $C^*(G)$ is isomorphic to the transformation group C^*-algebra $C_0(\mathbf{R}) \rtimes_\tau \mathbf{R}^+$.

Example 3.16 (Semidirect Products of Abelian Groups). Assume that N and K are *abelian* locally compact groups and that $G = N \rtimes_\alpha H$. As usual, we assume that $G = NH$ with N normal in G and $N \cap H = \{e\}$ as in Example 3.10 on page 86. Since $C^*(G) = \mathbf{C} \rtimes_\iota G$, Proposition 3.11 on page 87 implies there is an isomorphism $\Phi : C^*(G) \to C^*(N) \rtimes_\beta H$ where

$$\beta_h(f)(n) = \sigma(h)^{-1} f(h^{-1}nh) \quad \text{for } f \in C_c(N),$$

and $\sigma(h)$ is determined by (3.2). The Fourier transform

$$\hat{f}(\omega) := \int_N f(n)\omega(n)\,d\mu_N(n)$$

is an isomorphism of $C^*(N)$ onto $C_0(\widehat{N})$ (Proposition 3.1 on page 82), and

$$\begin{aligned}(\beta_h(f))^\wedge(\omega) &= \int_N \sigma(h)^{-1} f(h^{-1}nh)\omega(n)\,d\mu_N(n)\\ &= \int_N f(n)\omega(hnh^{-1})\,d\mu_N(n)\\ &= \hat{f}(h^{-1}\cdot\omega),\end{aligned}$$

where, by definition, $h \cdot \omega(n) = \omega(h^{-1}nh)$. Therefore, as in the previous example, we get an isomorphism of $C^*(N) \rtimes_\beta H$ with the transformation group C^*-algebra $C_0(\widehat{N}) \rtimes_\tau H$, where

$$\tau_h \hat{f}(\omega) := \hat{f}(h^{-1}\cdot\omega).$$

Therefore $C^*(N \rtimes_\varphi H) \cong C_0(\widehat{N}) \rtimes_\tau H$.

3.4 Invariant Ideals

If (A, G, α) is a dynamical system, then $\mathcal{I}_G(A)$ will denote the α-invariant (closed two-sided) ideals in A. If $I \in \mathcal{I}_G(A)$, then each α_s restricts to an automorphism of I and we obtain a dynamical system (I, G, α) as well as a quotient system $(A/I, G, \alpha^I)$ defined in the obvious way:

$$\alpha^I_s(a + I) := \alpha_s(a) + I.$$

The inclusion map $\iota : I \to A$ and the quotient map $q : A \to A/I$ are equivariant homomorphisms and therefore define homomorphisms $\iota \rtimes \mathrm{id} : I \rtimes_\alpha G \to A \rtimes_\alpha G$ and $q \rtimes \mathrm{id} : A \rtimes_\alpha G \to A/I \rtimes_{\alpha^I} G$ (Corollary 2.48 on page 63).

Note that $C_c(G, I)$ sits in $C_c(G, A)$ as a $*$-closed two-sided ideal. Therefore the closure is an ideal which is usually denoted by $\operatorname{Ex} I$. Of course, $\operatorname{Ex} I$ is exactly the image of $\iota \rtimes \mathrm{id}$. The next lemma will allow us to identify $\operatorname{Ex} I$ and $I \rtimes_\alpha G$ from now on.

Lemma 3.17. *If (A, G, α) is a dynamical system and if I is a G-invariant ideal in A, then $\iota \rtimes \mathrm{id}$ is an isomorphism of $I \rtimes_\alpha G$ onto $\operatorname{Ex} I$.*

Proof. At this point, it suffices to see that $\iota \rtimes \mathrm{id}$ is isometric on $C_c(G, I)$. Let $\|\cdot\|$ be the universal norm on $C_c(G, A)$ and $\|\cdot\|_I$ the universal norm on $C_c(G, I)$. Let $\pi \rtimes V$ be a faithful representation of $A \rtimes_\alpha G$. Then (π, V) restricts to a (possibly degenerate) covariant representation of (I, G, α) denoted (π_I, V). Then for all $f \in C_c(G, I)$,

$$\|f\| = \|\pi \rtimes V(f)\| = \|\pi_I \rtimes V(f)\| \le \|f\|_I.$$

On the other hand, if $\eta \rtimes V$ is a faithful representation of $I \rtimes_\alpha G$, then since η in nondegenerate, η has a canonical extension $\bar{\eta}$ to A characterized by $\bar{\eta}(a)\eta(b)h := \eta(ab)h$ for all $a \in A$, $b \in I$ and $h \in \mathcal{H}_\eta$. A quick computation verifies that $(\bar{\eta}, V)$ is a covariant representation of (A, G, α). Thus

$$\|f\|_I = \|\eta \rtimes V(f)\| = \|\bar{\eta} \rtimes V(f)\| \leq \|f\|.$$

Thus $\|f\| = \|f\|_I$ for all $f \in C_c(G, I)$. □

Lemma 3.18. *Suppose that X is locally compact, that I is an ideal in A and that $q : A \to A/I$ is the quotient map. If $g \in C_0(X, A/I)$, then there is a $f \in C_0(X, A)$ such that $g = q \circ f$. If g has compact support, then we can find f with compact support. In particular, we have a short exact sequence*

$$0 \longrightarrow C_0(X, I) \xrightarrow{\ i\ } C_0(X, A) \xrightarrow{\ q\circ\cdot\ } C_0(X, A/I) \longrightarrow 0$$

where i is the inclusion map and $q \circ \cdot$ sends f to $q \circ f$.

Proof. Suppose that $g \in C_c(X, A/I)$ with $\operatorname{supp} g = K$. Let W be a pre-compact open neighborhood of K. Fix $\epsilon > 0$. Let $\{ V_i \}_{i=1}^n$ be a cover of K by pre-compact open sets $V_i \subset W$ so that there is a $r_i \in V_i \cap K$ such that

$$\|g(r) - g(r_i)\| < \epsilon \quad \text{for all } r \in V_i.$$

Since $\overline{W}$ is compact, there is a partition of unity $\{ \varphi_i \}_{i=0}^n$ in $C(\overline{W})$ such that $\operatorname{supp} \varphi_0 \subset \overline{W} \setminus K$ and $\operatorname{supp} \varphi_i \subset V_i$ for $1 \leq i \leq n$. Let $a_i \in A$ be such that $q(a_i) = g(r_i)$, and let $f = \sum_{i=1}^n \varphi_i \otimes a_i$. Then $f \in C_c(X, A)$ and

$$\|q \circ f - g\|_\infty < \epsilon.$$

It follows that $f \mapsto q \circ f$ has dense range in $C_0(X, A/I)$ and must therefore be surjective since C^*-homomorphisms have closed range. Replacing X by W, the assertion about compact supports follows as well. The rest is straightforward. □

Proposition 3.19. *Suppose that (A, G, α) is a dynamical system and that I is an α-invariant ideal in A. Then $\iota \rtimes \mathrm{id}$ is an isomorphism identifying $I \rtimes_\alpha G$ with $\operatorname{Ex} I = \ker q \rtimes \mathrm{id}$, and we have a short exact sequence*

$$0 \longrightarrow I \rtimes_\alpha G \xrightarrow{\iota \rtimes \mathrm{id}} A \rtimes_\alpha G \xrightarrow{q \rtimes \mathrm{id}} A/I \rtimes_{\alpha^I} G \longrightarrow 0$$

of C^*-algebras.

Proof. Note that $\iota \rtimes \mathrm{id}$ is an isomorphism by Lemma 3.17 on the facing page. Lemma 3.18 implies that $q \rtimes \mathrm{id}$ maps $C_c(G, A)$ onto $C_c(G, A/I)$ so that $q \rtimes \mathrm{id}$ is surjective. Thus we only have to show that $\ker(q \rtimes \mathrm{id}) = \operatorname{Ex} I$. Since $C_c(G, I)$ is dense in $\operatorname{Ex} I$, we certainly have

$$\operatorname{Ex} I \subset \ker(q \rtimes \mathrm{id}). \tag{3.7}$$

Now suppose that $\pi \rtimes V$ is any representation of $A \rtimes_\alpha G$. If $I \subset \ker \pi$, then $C_c(G, I) \subset \ker(\pi \rtimes V)$ and $\operatorname{Ex} I \subset \ker(\pi \rtimes V)$. But $\pi = \pi' \circ q$ for some representation π' of A/I and (π', V) is a covariant representation of $(A/I, G, \alpha^I)$ such that

$$(\pi' \rtimes V) \circ (q \rtimes \mathrm{id}) = \pi \rtimes V.$$

Therefore, $\ker(q \rtimes \mathrm{id}) \subset \ker(\pi \rtimes V)$ as well.

On the other hand, if $I \not\subset \ker \pi$, then there is a $a \in I$ such that $\pi(a) \neq 0$. If $\{ u_i \} \subset C_c(G)$ is an approximate identity for $C^*(G)$, then for all i, $a \otimes u_i \in C_c(G, I)$ and so belongs to both $\operatorname{Ex} I$ and $\ker(q \rtimes \mathrm{id})$. But

$$\pi \rtimes V(a \otimes u_i) = \pi(a)V(u_i)$$

which converges strongly to $\pi(a)$. Thus neither $\operatorname{Ex} I$ nor $\ker(q \rtimes \mathrm{id})$ is contained in $\ker(\pi \rtimes V)$ in this case. It follows that we have equality in (3.7) as required. □

It is a useful exercise to consider the special case of Proposition 3.19 on the preceding page in which A is commutative. In particular, let $(C_0(X), G, \mathrm{lt})$ be a dynamical system. Then the G-invariant ideals I of A correspond to G-invariant open sets $U \subset X$: $I = C_0(U)$, were $C_0(U)$ is viewed as an ideal in $C_0(X)$ in the obvious way. Let $F := X \smallsetminus U$. Then F is a G-invariant closed set and we can identify $C_0(F)$ with the quotient $C_0(X)/C_0(U)$. Notice that is identification intertwines the quotient action lt^I with the automorphism group on $C_0(F)$ induced by left-translation. Therefore we have derived the following corollary to Proposition 3.19 on the previous page.

Corollary 3.20. *Suppose that (G, X) is a transformation group, and that U is an open G-invariant subset of X. Then we have a short exact sequence*

$$0 \longrightarrow C_0(U) \rtimes_{\mathrm{lt}} G \xrightarrow{\iota \rtimes \mathrm{id}} C_0(X) \rtimes_{\mathrm{lt}} G \xrightarrow{q \rtimes \mathrm{id}} C_0(X \smallsetminus U) \rtimes_{\mathrm{lt}} G \longrightarrow 0$$

of C^*-algebras.

3.5 The Orbit Space

Every dynamical system (A, G, α) gives rise to a G-action on $\operatorname{Prim} A$ (Lemma 2.8 on page 44). In the case of transformation groups, where $A = C_0(X)$, the G-action on X is the salient feature. In this section, we record some basic observations about G-spaces and their orbit spaces.

Definition 3.21. Suppose that X is a (left) G-space and that $x \in X$. The *orbit* through x is the set $G\cdot x := \{ s\cdot x \in X : s \in G \text{ and } x \in X \}$. The *stability group* at x is $G_x := \{ s \in G : s \cdot x = x \}$. The G-action is called free if $G_x = \{ e \}$ for all $x \in X$. The set of orbits is denoted by $G\backslash X$. The natural map $p : X \to G\backslash X$ is called the *orbit map*, and $G\backslash X$ is called the *orbit space* when equipped with the *quotient topology* (which is the largest topology making the orbit map p continuous).

Remark 3.22. If X is a Hausdorff G-space, then it is straightforward to check that each stability group G_x is a closed subgroup of G. But if (A, G, α) is a dynamical system, then $\operatorname{Prim} A$ is a G-space (Lemma 2.8 on page 44), and the stability groups G_P play an important role in the theory. Since $\operatorname{Prim} A$ is always a T_0-topological space, so that distinct points have distinct closures, the next lemma implies that each G_P is a closed subgroup of G

Lemma 3.23. *Suppose that (G, X) is a topological transformation group and that X is a T_0-topological space. Then each stability group G_x is closed.*

Proof. Let $\{ s_i \}$ be a net in G_x converging to s in G. We want to see that $s \in G_x$. But $x = s_i \cdot x \to s \cdot x$. Thus, $s \cdot x \in \overline{\{ x \}}$. On the other hand, $s \cdot x = ss_i^{-1} \cdot x \to x$. Thus $x \in \overline{\{ s \cdot x \}}$. Since X is T_0, this forces $s \cdot x = x$. That is, $x \in G_x$. □

Remark 3.24. If X is a right G-space, then the orbit space is denoted by X/G and the orbits by $x \cdot G$.

It follows easily from the definition that a subset $U \subset G\backslash X$ is open if and only if $p^{-1}(U)$ is open in X. However, we can say a bit more.

Lemma 3.25. *If X is a G-space, then the orbit map is continuous and open.*

Proof. The continuity of $p : X \to G\backslash X$ holds by definition. If W is open in X, then

$$p^{-1}\big(p(W)\big) = G \cdot W := \{ s \cdot x : s \in G \text{ and } x \in W \} = \bigcup_{s \in G} s \cdot W,$$

which is a union of open sets. Therefore $p(W)$ is open. □

Any topological space Y can arise as an orbit space. For example, let $X = Y \times G$ and let G act by left translation on the second factor. Then $G\backslash X$ is naturally identified with Y. Of course, in this example, any pathological behavior in $G\backslash X$ is already exhibited in X. However, even if we insist that X be a nice topological space, examples show that the orbit space can vary from nice to pathological.

Example 3.26 (cf. Example 2.13 on page 45). Let $X = \mathbf{T}$ and $G = \mathbf{Z}$ with $\mathbf{Z}$ acting by "irrational rotation":

$$n \cdot z := e^{2\pi i n\theta} z,$$

where θ is an irrational number in $(0, 1)$. As we will show in Lemma 3.29 on the next page, each orbit is dense in $\mathbf{T}$. Consequently, $G\backslash X$ is an uncountable set with the trivial topology: $\tau = \{ \emptyset, G\backslash X \}$. Furthermore, even though for each $z \in \mathbf{T}$, $n \mapsto n \cdot z$ is a bijection of $\mathbf{Z}$ onto the orbit through z, each orbit is topologically quite different from $\mathbf{Z}$. In fact, the orbits are not even locally compact subsets of $\mathbf{T}$.

We still have to prove the claim in the above example. For this, we'll want the following lemma.

Lemma 3.27. *If H is a proper closed subgroup of the circle $\mathbf{T}$, then H is finite.*

Remark 3.28. It is possible to give a very short proof of this result using the duality theory for locally compact abelian groups. It is a direct consequence of the Pontryagin duality that the character group $\widehat{H}$ of H is isomorphic to the quotient $\widehat{T}/H^{\perp}$ where $H^{\perp} := \{\, \gamma \in \widehat{T} : \gamma(H) = \{\,1\,\}\,\}$ [56, Theorem 4.39]. Since $\widehat{T} \cong \mathbf{Z}$, $\widehat{H}$ is isomorphic to a quotient of $\mathbf{Z}$ by a nontrivial subgroup. Thus $\widehat{H}$ is finite. The Pontryagin Duality Theorem [56, Theorem 4.31] implies that the dual of $\widehat{H}$ is isomorphic to H, and therefore H itself must be finite. However, we will give an elementary proof using only the topology of $\mathbf{T}$.

Proof. We'll assume that H is not finite and derive a contradiction. Let U be an open set such that $U \cap H = \emptyset$. Since H is infinite and compact, there is a sequence of *distinct* elements $\{\, h_n \,\}$ converging to $h \in H$. Multiplying by h^{-1} we can assume we have $h_n \to 1$ with $h_n \neq 1$ for all n. Let $h_n = e^{2\pi i \theta_n}$ with $\theta_n \in (0,1)$. Since we also have $h_n^{-1} \to 1$, we can replace h_n by h_n^{-1} whenever $\theta_n > \frac{1}{2}$ and assume that $\theta_n \to 0$. Choose $z = e^{2\pi i \theta} \in U$ with $\theta \in (0,1)$. Since U is open, there is $\epsilon > 0$ such that $e^{2\pi i \psi} \in U$ provided $\psi \in (\theta, \theta + \epsilon)$. But if $\theta_n < \epsilon$, then there is a $m \in \mathbf{Z}$ such that $m\theta_n \in (\theta, \theta + \epsilon)$. Therefore $h_n^m \in U$, which contradicts our choice of U. □

Lemma 3.29. *Suppose that* $\mathbf{Z}$ *acts on* $\mathbf{T}$ *by an irrational rotation (as in Example 3.26 on the preceding page). Then for every* $z \in \mathbf{T}$*, the orbit* $\mathbf{Z} \cdot z$ *is dense in* $\mathbf{T}$.

Proof. The orbit of 1 is an infinite subgroup $H := \mathbf{Z} \cdot 1$. Since the closure of a subgroup is a subgroup, Lemma 3.27 on the previous page implies that H is dense. But if $z \in \mathbf{T}$, then $\omega \mapsto \omega z$ is a homeomorphism of $\mathbf{T}$ taking $\mathbf{Z} \cdot 1$ to $\mathbf{Z} \cdot z$. Thus every orbit is dense as claimed. □

Example 3.30. Let $X = \mathbf{C}$, and let $G = \mathbf{T}$ act by multiplication: $z \cdot \omega := z\omega$. Then the origin is a fixed point and the remaining orbits are circles $\{\, \omega \in C : |\omega| = r \,\}$ for some $r > 0$. The orbit space is easily identified with the non-negative real axis with its natural (Hausdorff) topology. In particular, each orbit is closed and all the nondegenerate orbits are homeomorphic to $G = \mathbf{T}$.

Example 3.31. Let $X = \mathbf{C}$ and let $G = \mathbf{R}$ act via $r \cdot \omega := e^{2\pi i r}\omega$. Then the orbits and orbit space are exactly as in Example 3.30. However, now the action is no longer free at points on the nondegenerate orbits. Note that each nondegenerate orbit is homeomorphic to the quotient of $\mathbf{R}$ by the stability group $\mathbf{Z}$.

Example 3.32. Let $X = \mathbf{R}^2$ and let G be the positive reals under multiplication. Let G act on X by $r \cdot (x, y) := (rx, r^{-1}y)$. The orbits consist of the origin, the remaining four bits of the coordinate axes and the hyperbolas $\{\, (x, y) : xy = b \,\}$ with $b \in \mathbf{R} \setminus \{\, 0 \,\}$. (Note that each b corresponds to two orbits.) Then the orbit space can be identified with the union of the two lines $y = \pm x$ (denoted by A) and the four points $a_1 = (1, 0)$, $a_2 = (0, 1)$, $a_3 = (-1, 0)$ and $a_4 = (0, -1)$. Let Q_i be the right angle formed from A having a_i in its interior. Given A and the Q_i's the relative topology from $\mathbf{R}^2$. Using the openness of the orbit map, it is easy to describe the topology on $G\backslash X = A \cup \{\, a_1, a_2, a_3, a_4 \,\}$. Note that every neighborhood of $(0, 0)$ contains all the a_i. (In particular, points in $G\backslash X$ need not be closed.) Then a basis for the open sets in $G\backslash X$ consists of the following.

(a) $\{ U \subset A : (0,0) \notin U \text{ and } U \text{ is open in } A \}$,
(b) $\{ W \subset G\backslash X : W \cap A \text{ is open in } A, \text{ and } (0,0), a_1, \dots, a_4 \in W \}$,
(c) $\{ V \subset Q_i \cup \{ a_i \} : a_i \in V,\ (0,0) \notin V \text{ and } \{ (0,0) \} \cup (V \cap Q_i) \text{ is a neighborhood of } (0,0) \text{ in } Q_i \}$.

Note that each orbit is locally compact and homeomorphic to $\mathbf{R}^+$ (or a point). All orbits except those corresponding to the a_i are closed and that $G\backslash X \smallsetminus \{ a_1, a_2, a_3, a_4 \}$ is actually Hausdorff.

Example 3.33. Let G be the real line equipped with the discrete topology and let X be the real line with its usual topology. Let G act on X by translation: $x \cdot y := x + y$. Then there is only one orbit — one says G acts *transitively*. However the natural bijections of G onto X are not homeomorphisms.

Example 3.34. Suppose that H is a subgroup of a topological group G. Then G is a *right* H-space if we let H act on the right of G by multiplication: $s \cdot h := sh$. The H-orbits $s \cdot H$ are exactly the left cosets sH, and the orbit space is the set G/H of left cosets equipped with the largest topology making the quotient map continuous.

Even in view of examples like 3.26 and 3.32, we still want to investigate orbit spaces as they will play a significant role in the structure theory of crossed products and especially of transformation group C^*-algebras. As we shall see later (Theorem 6.2 on page 173), under mild hypotheses on G and X, there is a sharp dichotomy in orbit space structure. Orbit spaces are either reasonably well behaved with Example 3.32 on the facing page about as bad as it gets, or they exhibit rather extreme pathological behavior similar to Example 3.26 on page 95.

Lemma 3.35. *Let X be a G-space. If X is second countable, then so is $G\backslash X$. If X is a (*not necessarily Hausdorff*) locally compact space, then $G\backslash X$ is a (*not necessarily Hausdorff*) locally compact space.*[2]

Remark 3.36. In this work, when we speak of a locally compact space, it is tacitly understood that the space is Hausdorff. It just seems too cumbersome to constantly be adding Hausdorff to the statement of every other result. As a penalty, we now have to add an annoying "not necessarily Hausdorff" when we want to look at spaces, such as orbit spaces, which may naturally have non Hausdorff topologies. Of course, there are difficulties to keep in mind when working with potentially non Hausdorff spaces. We must appeal to Definition 1.16 on page 5 rather than Lemma 1.17 on page 5 when we want to show a possibly non-Hausdorff space is locally compact. We also have to keep in mind that compact subsets need not be closed, and that convergent nets need not have unique limits.

Proof. Let $p : X \to G\backslash X$ be the orbit map. Suppose that $\{ U_n \}$ is a countable basis of open sets for the topology on X. Let $V_n := p(U_n)$. Let V be an open set in $G\backslash X$ with $q \in V$. Choose $x \in p^{-1}(q)$. Since p is continuous $p^{-1}(V)$ is a neighborhood of x, and there exists a U_n such that $x \in U_n \subset p^{-1}(V)$. Thus $q \in V_n \subset V$, and $\{ V_n \}$ is a countable basis for the topology on $G\backslash X$.

[2]Notice that $G\backslash X$ can fail to be Hausdorff, even if X is locally compact Hausdorff to begin with.

Now suppose that X is locally compact. Since $G\backslash X$ may not be Hausdorff, we have to verify that Definition 1.16 on page 5 holds. However, the above argument shows that the forward image of any basis in X is a basis in $G\backslash X$. Since X has a basis of compact sets, and since the continuous image of a compact set is compact, it follows that $G\backslash X$ has a basis of compact sets. Therefore $G\backslash X$ is locally compact. □

Lemma 3.37. *Suppose that X is a not necessarily Hausdorff locally compact G-space, and let $p : X \to G\backslash X$ be the quotient map. If $T \subset G\backslash X$ is compact, then there is a compact set $D \subset X$ such that $p(D) \supset T$. If $G\backslash X$ is Hausdorff, then $T \subset G\backslash X$ is compact if and only if there is a compact set $D \subset X$ such that $p(D) = T$.*

Proof. Suppose that T is compact in $G\backslash X$, and that $G \cdot x \in T$. If V is a compact neighborhood of x in X, then since p is open and continuous, $p(V)$ is a compact neighborhood of $G \cdot x$. Therefore there are finitely many compact sets $\{ V_i \}$ with $p(\bigcup V_i) \supset T$. Since $\bigcup V_i$ is compact, this proves the first assertion. If $G\backslash X$ is Hausdorff, then T is closed and

$$D := p^{-1}(T) \cap \bigcup V_i$$

is a closed subset of a compact set and therefore compact. Clearly, $p(D) = T$ as required. The other implication is an easy consequence of the continuity of p. □

As should be evident from the above examples and the proof of Lemma 3.35 on the previous page, the openness of the orbit map is a useful tool. Recall that if $p : X \to Y$ is open, then convergent nets in Y have subnets witch lift to convergent nets in X — see Proposition 1.15 on page 4. As a consequence we have the following useful observation.

Lemma 3.38. *Suppose that X is a G-space and that $p : X \to G\backslash X$ is the orbit map. If $\{ x_i \}_{i\in I}$ is a net in X such that $p(x_i) \to p(x)$ in $G\backslash X$, then there is a subnet $\{ x_{N_j} \}_{j\in J}$ and elements $s_{N_j} \in G$ such that $s_{N_j} \cdot x_{N_j} \to x$.*

Remark 3.39. It is often the case, as in Remark 1.47 on page 13, that when we apply this result, we are only interested in the subnet. Then, provided we're careful to keep in mind that we've passed to a subnet, we can relabel everything and assume that there are $s_i \in G$ such that $s_i \cdot x_i \to x$. The added readability resulting from the dropping the double subscripts more than makes up for any confusion that might result.

Proof. Proposition 1.15 on page 4 implies that there is a subnet $\{ p(x_{N_j}) \}_{j\in J}$ and $z_{N_j} \in X$ such that $z_{N_j} \to x$ and $p(z_{N_j}) = p(x_{N_j})$ for all j. But then there must be a $s_{N_j} \in G$ such that $z_{N_j} = s_{N_j} \cdot x_{N_j}$ for all j. □

3.6 Proper Actions and Induced Algebras

Although group actions and their orbit spaces can be very complicated, there is nevertheless a class of actions which are both extremely well-behaved and important to the development of the subject. If X and Y are locally compact Hausdorff spaces, then a continuous map $f : X \to Y$ induces a map $f^* : C^b(Y) \to C^b(X)$ via $f^*(\varphi)(y) := \varphi\big(f(y)\big)$. But it need not be the case that

$$f^*\big(C_0(Y)\big) \subset C_0(X). \tag{3.8}$$

In order that (3.8) hold, we need $f^{-1}(K)$ to be compact in X whenever K is compact in Y. Such maps f are called *proper maps*. Proper maps are the natural morphisms in the category of locally compact spaces when the latter are identified with abelian C^*-algebras. Naturally, proper maps play an important role.

Definition 3.40. A locally compact G-space P is called *proper* if the map $(s, x) \mapsto (s \cdot x, x)$ is a proper map from $G \times P$ to $P \times P$.

If P is a proper G-space, then we say that G *acts properly on* P. Of course, there is an analogous definition for right G-spaces. Proper actions, and their connection with principal G-bundles in particular, were discussed at the end of §4.2 of [139].

Example 3.41. Let H be a closed subgroup of a locally compact group G. Then H acts properly on the left of G. To prove this, we need to see that the map

$$(s, t) \mapsto (st, s) \tag{3.9}$$

is proper from $H \times G$ to $G \times G$. But, viewed as a map from $G \times G \to G \times G$, (3.9) has a continuous inverse $(s, r) \mapsto (r, r^{-1}s)$. Thus (3.9) defines a proper map from $G \times G \to G \times G$. Therefore the restriction to $H \times G$ is also proper.

It is useful to keep in mind that the action of any compact group is always proper. More generally, we have the following criteria in terms of nets.

Lemma 3.42. *Let P be a locally compact G-space. Then G acts properly on P if and only if whenever we are given convergent nets $x_i \to x$ and $s_i \cdot x_i \to y$, the net $\{ s_i \}$ has a convergent subnet.*

Proof. Let $\Phi : G \times P \to P \times P$ be given by $\Phi(s, x) = (s \cdot x, x)$. Suppose that P is a proper G-space and that $\{ x_i \}$ and $\{ s_i \cdot x_i \}$ are as in the lemma. Let K be a compact neighborhood of x and y. Then we eventually have x_i and $s_i \cdot x_i$ in K. Thus we eventually have $(s_i, x_i) \in \Phi^{-1}(K \times K)$. Since $\Phi^{-1}(K \times K)$ is compact by assumption, $\{ s_i \}$ must eventually lie in a compact subset of G, and the assertion follows

For the converse, suppose that C is compact in $P \times P$. Let $\{ (s_i, x_i) \}$ be a net in $\Phi^{-1}(C)$. It suffices to see that $\{ (s_i, x_i) \}$ has a convergent subnet. But $(s_i \cdot x_i, x_i) \in C$. Since C is compact, we can pass to a subnet, relabel, and assume that $x_i \to x$ and $s_i \cdot x_i \to y$. By assumption, we can pass to another subnet, relabel, and assume that $s_i \to s$ in G. But then $(s_i, x_i) \to (s, x)$ in $G \times P$. □

Corollary 3.43. *If P is a proper locally compact G-space, then $G\backslash P$ is a locally compact Hausdorff space.*

Proof. Since $G\backslash P$ is always locally compact (Lemma 3.35 on page 97), we just need to see that $G\backslash P$ is Hausdorff. Suppose that $\{ w_i \}$ is a net $G\backslash P$ converging to $G \cdot x$ and $G \cdot y$. Passing to a subnet and relabeling, Proposition 1.15 on page 4 implies that there are $x_i \in P$ such that $w_i = G \cdot x_i$ and such that $x_i \to x$. Since $G \cdot x_i \to G \cdot y$, Lemma 3.38 on page 98 implies that we can pass to subnet, relabel, and assume that there are $s_i \in G$ such that $s_i \cdot x_i \to y$. By Lemma 3.42 on the preceding page, we can pass to a subnet, relabel, and assume that $s_i \to s$ in G. But then $s_i \cdot x_i \to s \cdot x$. Since P is Hausdorff, $s \cdot x = y$ and $G \cdot x = G \cdot y$. Thus, $G\backslash P$ is Hausdorff as claimed. □

Remark 3.44. Examples such as Example 3.31 on page 96 show that we can have $G\backslash P$ Hausdorff even if the action in not proper. It is also possible, but not so easy, to find examples of free actions with Hausdorff orbit spaces which are nevertheless not proper actions (cf., [65, p. 95] and [151, Example 1.18]).

Corollary 3.45. *Suppose that P is a proper G-space. Then for each $x \in P$, $sG_x \mapsto s \cdot x$ is a homeomorphism of G/G_x onto $G \cdot x$.*

Remark 3.46. When G and P are second countable, it turns out that the maps $sG_x \mapsto G \cdot x$ are homeomorphisms if and only if the orbit space $G\backslash X$ is a T_0-topological space (Theorem 6.2 on page 173).

Proof. Since $s \mapsto s \cdot x$ is continuous and since the quotient map of G to G/G_x is open, it suffices to see that $N \cdot x$ is open in $G \cdot x$ whenever N is open in G. If $N \cdot x$ is not open in $G \cdot x$, then there is a net $\{ s_i \} \subset G$ and $n \in N$ such that $s_i \cdot x \to n \cdot x$ with $s_i \cdot x \notin N \cdot x$ for all i. Using Lemma 3.42 on the previous page, we can pass to a subnet, relabel, and assume that $s_i \to s$ in G. Since P is Hausdorff, $s \cdot x = n \cdot x$ and $s \in n \cdot G_x \subset N \cdot G_x$. Since $N \cdot G_x$ is open in G, we must eventually have $s_i \in N \cdot G_x$, and $s_i \cdot x$ is eventually in $N \cdot x$. This contradicts our assumptions. Therefore $N \cdot x$ is open. □

Suppose that P is a G-space and that (A, G, α) is a dynamical system. If $f : P \to A$ is a continuous function such that

$$f(s \cdot x) = \alpha_s\big(f(x)\big) \quad \text{for all } x \in P \text{ and } s \in G, \tag{3.10}$$

then $x \mapsto \|f(x)\|$ is constant on G-orbits and gives a well-defined function on $G\backslash P$. By definition, the *induced algebra* is

$$\begin{aligned} \operatorname{Ind}_G^P(A, \alpha) := \{\, f \in C^b(P, A) : f &\text{ satisfies (3.10) and} \\ &G \cdot x \mapsto \|f(x)\| \text{ is in } C_0(G\backslash P). \,\} \end{aligned}$$

Since $\operatorname{Ind}_G^P(A, \alpha)$ is a closed $*$-subalgebra of $C^b(P, A)$, it is a C^*-algebra with respect to the supremum norm. Induced algebras have been studied extensively (see [139, §6.15]), and they will play a fundamental role in the imprimitivity theorems for crossed products in Section 4.1. When the context is clear, we will shorten $\operatorname{Ind}_G^P(A,\alpha)$ to $\operatorname{Ind}_G^P \alpha$ or even $\operatorname{Ind}\alpha$. If P is a right G-space, then we replace (3.10) with

$$f(x\cdot s) = \alpha_s^{-1}\big(f(x)\big). \tag{3.10$'$}$$

Then $x \mapsto \|f(x)\|$ is continuous on P/G and we make the obvious modifications to the definition of $\operatorname{Ind}_G^P(A,\alpha)$.

One rather important example, and an example in which a right action is more convenient than the usually preferred left action, is the following.

Example 3.47. Let H be a closed subgroup of a locally compact group G and let (D,H,β) be a dynamical system. Then G is a right H-space with orbit space the set of left cosets G/H (Example 3.34 on page 97). Then

$$\begin{aligned}\operatorname{Ind}_H^G(D,\beta) = \{\, f \in C^b(G,D) : f(sh) = \beta_h^{-1}\big(f(s)\big)&\\ \text{for } s\in G \text{ and } h \in H, \text{ and } sH \mapsto \|f(s)\| \text{ is in } C_0(G/H).\,\}&\end{aligned}$$

It should be observed that $\operatorname{Ind}_H^G\beta$ can be nontrivial only if β can't be lifted to an action of G. More precisely, if $\beta = \alpha|_H$ for some dynamical system (A,G,α), then $\varphi : C^b(G,D) \to C^b(G,D)$ given by $\varphi(f)(s) := \alpha_s\big(f(s)\big)$ defines an isomorphism of $\operatorname{Ind}_H^G(D,\beta)$ onto $C_0(G/H,D)$ (viewed as functions on G which are constant on orbits).

The algebra $\operatorname{Ind}_H^G\beta$ above is rather special to the imprimitivity theory and will reoccur several times below. One of its attractions is that it admits a nice G-action.

Lemma 3.48. *Suppose that H is a closed subgroup of a locally compact group and that (D,H,β) is a dynamical system. Let $\operatorname{Ind}_H^G\beta$ be as in Example 3.47 above. Then there is a dynamical system $\big(\operatorname{Ind}_H^G\beta, G, \mathrm{lt}\big)$ where*

$$\mathrm{lt}_r(f)(s) := f(r^{-1}s). \tag{3.11}$$

Proof. It is straightforward to check that (3.11) defines a homomorphism $\alpha : G \to \operatorname{Aut}(\operatorname{Ind}\beta)$. The only issue is to verify that $s \mapsto \alpha_s(f)$ is continuous for each $f \in \operatorname{Ind}\beta$. As this will follow from Lemma 3.54 on page 106, we'll settle for a forward reference and not repeat the argument here. □

In general, our definition of $\operatorname{Ind}_G^P(A,\alpha)$ does not guarantee that it contains any functions other than the zero function. For example, if $G\backslash P$ is not Hausdorff, then $C_0(G\backslash P)$ could consist of only the zero function. However, if the action of G on P is proper, as in the case of Example 3.47, then we can show that $\operatorname{Ind}_G^P(A,\alpha)$ may be thought of as the section algebra of an (upper semicontinuous) C^*-bundle over $G\backslash P$ with fibres all isomorphic to A. More precisely, we show that $\operatorname{Ind}_G^P(A,\alpha)$ is a $C_0(G\backslash P)$-algebra. The basic properties of such algebras and their associated C^*-bundles are detailed in Appendix C and Theorem C.26 on page 367 in particular.

Proposition 3.49. *If P is a free and proper G-space and if (A,G,α) a dynamical system, then $\operatorname{Ind}_G^P\alpha$ is a $C_0(G\backslash P)$-algebra such that*

$$\varphi\cdot f(x) = \varphi(G\cdot x)f(x) \quad \textit{for all } \varphi\in C_0(G\backslash P),\ f\in \operatorname{Ind}\alpha \textit{ and } x \in P, \tag{3.12}$$

and such that the map $f \mapsto f(x)$ induces an isomorphism of the fibre over $G \cdot x$ onto A.

It is not hard to see that (3.12) gives a nondegenerate homomorphism of $C_0(G\backslash P)$ into the center of the multiplier algebra of $\operatorname{Ind}\alpha$, so the only real issue is to see that $\Phi(f) := f(x)$ defines a surjective homomorphism of $\operatorname{Ind}\alpha$ onto A with kernel

$$I_{G\cdot x} = \overline{\operatorname{span}}\{\, \varphi \cdot f : \varphi(G \cdot x) = 0 \text{ and } f \in \operatorname{Ind}\alpha.\,\} \tag{3.13}$$

In order to prove this, we need a few preliminary results. We state and prove these in considerable more generality than needed for the proof of Proposition 3.49 on the previous page; the extra generality will be useful in the proof of the Symmetric Imprimitivity Theorem (Theorem 4.1 on page 110).

Let $\operatorname{Ind}_c \alpha$ be the subalgebra of $\operatorname{Ind}\alpha$ consisting of those f for which $G \cdot x \mapsto \|f(x)\|$ is in $C_c(G\backslash P)$. Since $\operatorname{Ind}\alpha$ is a $C_0(G\backslash P)$-module, it is easy to see that $\operatorname{Ind}_c \alpha$ is dense in $\operatorname{Ind}\alpha$. Although in the following, we could work directly with $C_c\bigl(Y, \operatorname{Ind}_c \alpha\bigr)$, it will be more convenient to work with a smaller dense subalgebra of compactly supported functions in analogy with $C_c(G \times X) \subset C_c\bigl(G, C_0(X)\bigr)$ in the case of transformation group C^*-algebras (see page 53).

Lemma 3.50. *Suppose that Y is a locally compact space, that P is a locally compact G-space and that (A, G, α) is a dynamical system. Let $C_{cc}\bigl(Y, \operatorname{Ind}_c \alpha\bigr)$ be the collection of functions f in $C(Y \times P, A)$ such that*

(a) *$f(y, s \cdot x) = \alpha_s\bigl(f(y, x)\bigr)$ for $y \in Y$, $x \in P$ and $s \in G$, and*
(b) *there are compact sets $C \subset Y$ and $T \subset G\backslash P$ such that $f(y, x) = 0$ if $(y, G{\cdot}x) \notin C \times T$.*

Then $y \mapsto f(y, \cdot)$ is in $C_c(Y, \operatorname{Ind}\alpha)$ and we can identify $C_{cc}\bigl(Y, \operatorname{Ind}_c \alpha\bigr)$ with a subspace of $C_c\bigl(Y, \operatorname{Ind}\alpha\bigr)$ which is dense in the inductive limit topology.

Proof. Suppose that $f \in C_{cc}\bigl(Y, \operatorname{Ind}_c \alpha\bigr)$ with C and T as above. We show that $y \mapsto f(y, \cdot)$ is continuous from Y into $\operatorname{Ind}\alpha$. If not, then there is a sequence $y_i \to y$ in Y and an $\epsilon > 0$ such that $\|f(y_i, \cdot) - f(y, \cdot)\| \geq \epsilon$ for all i. Then there must be $x_i \in P$ such that

$$\|f(y_i, x_i) - f(y, x_i)\| \geq \epsilon \quad \text{for all } i. \tag{3.14}$$

Clearly, we must have $\{\, G \cdot x_i \,\} \subset T$, and $\{\, G \cdot x_i \,\}$ must have a convergent subnet. Using Lemma 3.38 on page 98, we can pass to a subnet, relabel, and assume that there are $s_i \in G$ such that $x_i \cdot s_i \to x$ for some $x \in P$. But then (3.14) implies

$$\|f(y_i, s_i \cdot x_i) - f(y, s_i \cdot x_i)\| = \bigl\|\alpha_{s_i}\bigl(f(y_i, x_i) - f(y, x_i)\bigr)\bigr\| \geq \epsilon.$$

Since the left-hand side must go to zero, we have a contradiction.

Lemma 1.87 on page 29 implies that elementary tensors $z \otimes f$, with $z \in C_c(Y)$ and $f \in \operatorname{Ind}_c \alpha$ span a dense subspace of $C_c\bigl(Y, \operatorname{Ind}\alpha\bigr)$. Since each $z \otimes f$ is clearly in $C_{cc}\bigl(Y, \operatorname{Ind}_c \alpha\bigr)$, the final assertion follows. $\square$

Remark 3.51. Note that $C_{cc}(Y, \operatorname{Ind}_c \alpha)$ has an inductive limit topology as a subspace of $C_c(Y, \operatorname{Ind} \alpha)$. If $\{ f_i \}$ is a net in $C_{cc}(Y, \operatorname{Ind}_c \alpha)$ converging uniformly to f on $Y \times P$, and if there are compact sets $K \subset Y$ and $T \subset G \backslash P$ such that $\operatorname{supp} f_i \subset K \times p^{-1}(T)$ for all i, then $f_i \to f$ in the inductive limit topology in $C_{cc}(Y, \operatorname{Ind}_c \alpha) \subset C_c(Y, \operatorname{Ind} \alpha)$.

Lemma 3.52. *Suppose that Y is a locally compact space, that P is a free and proper locally compact G-space and that (A, G, α) is a dynamical system. If $F \in C_c(Y \times P, A)$, then*

$$\psi(y, x) := \int_G \alpha_s\big(F(y, s^{-1} \cdot x)\big)\, d\mu(s)$$

defines an element $\psi \in C_{cc}(Y, \operatorname{Ind}_c \alpha) \subset C_c(Y, \operatorname{Ind} \alpha)$.

Proof. Let $\operatorname{supp} F \subset K \times C$ with $K \subset Y$ and $C \subset P$ compact. Then

$$\alpha_s\big(F(y, s^{-1} \cdot x)\big) \neq 0 \quad \text{implies that } y \in K \text{ and } s^{-1} \cdot x \in C.$$

Since the action is free and proper, $s \mapsto s \cdot x$ is a homeomorphism of G onto $G \cdot x$ by Corollary 3.45 on page 100. Therefore $\{ s \in G : s^{-1} \cdot x \in C \}$ is compact, and ψ is defined for all (y, x). Furthermore,

$$\begin{aligned} \psi(y, r \cdot x) &= \int_G \alpha_s\big(F(y, s^{-1} r \cdot x)\big)\, d\mu(s) \\ &= \alpha_r\Big(\int_G \alpha_s\big(F(y, s^{-1} \cdot x)\big)\, d\mu(s)\Big) \\ &= \alpha_r\big(\psi(y, x)\big). \end{aligned}$$

Since $\operatorname{supp} \psi \subset K \times G \cdot C$, it will, in view of Lemma 3.50 on the facing page, suffice to see that ψ is continuous. So suppose that $(y_i, x_i) \to (y, x)$. We can enlarge C a bit if necessary, and assume that $x_i \in C$ for all i. Since the action is proper,

$$\{ (s, x) \in G \times P : (x, s^{-1} \cdot x) \in C \times C \}$$

is compact. Therefore, there is a compact set C' such that

$$\{ s \in G : s^{-1} \cdot x_i \in C \text{ for some } i \} \subset C'.$$

We claim that given $\epsilon > 0$ there is an i_0 such that $i \geq i_0$ implies that

$$\| F(y_i, s^{-1} \cdot x_i) - F(y, s^{-1} \cdot x) \| < \epsilon \quad \text{for all } s \in G. \tag{3.15}$$

If the claim were not true, then after passing to a subnet and relabeling, there would be $s_i \in C'$ such that

$$\| F(y_i, s_i^{-1} \cdot x_i) - F(y, s_i^{-1} \cdot x) \| \geq \epsilon.$$

But after passing to a subnet and relabeling, we can assume that $s_i \to s$, and this leads to a contradiction as F was assumed to be continuous. But if (3.15) holds, then

$$\begin{aligned}\|\psi(y_i, x_i) - \psi(y, x)\| &\leq \int_G \|F(y_i, s^{-1} \cdot x_i) - F(y, s^{-1} \cdot x)\| \, d\mu(s) \\ &\leq \epsilon\mu(C').\end{aligned}$$

Since $\mu(C') < \infty$ and ϵ was arbitrary, it follows that ψ is continuous. □

Proof of Proposition 3.49 on page 101. We still have to show that $\ker \Phi = I_{G \cdot x}$ and $\Phi(\operatorname{Ind} \alpha) = A$. If $f \in \ker \Phi$, then $f(x) = 0$ and therefore $f(s \cdot x) = 0$ for all $s \in G$. Then given $\epsilon > 0$,

$$T := \{\, G \cdot y : \|f(y)\| \geq \epsilon \,\}$$

is a compact set in $G \backslash P$, and since $G \backslash P$ is Hausdorff, there is a $\varphi \in C_0(G \backslash P)$ such that $\varphi(G \cdot x) = 0$, $\varphi(G \cdot y) = 1$ if $G \cdot y \in T$ and $0 \leq \varphi(G \cdot y) \leq 1$ for all $y \in P$. But then $\varphi \cdot f \in I_{G \cdot x}$ and $\|f - \varphi \cdot f\| < \epsilon$. It follows that $\ker \Phi \subset I_{G \cdot x}$. Since the other containment is obvious, we have $\ker \Phi = I_{G \cdot x}$.

Now if $a \in A$ and $\epsilon > 0$, there is a neighborhood V of e in G such that $\|\alpha_s(a) - a\| < \epsilon$ if $s \in V$. Since $s \mapsto s \cdot x$ is a homeomorphism of G onto $G \cdot x$, $V \cdot x$ is an open set in $G \cdot x$ (Corollary 3.45 on page 100). Thus there is an open set $U \subset P$ such that $U \cap G \cdot x = V \cdot x$. Let $z \in C_c^+(P)$ be such that $\operatorname{supp} z \subset U$ and

$$\int_G z(s^{-1} \cdot x) \, d\mu(s) = 1.$$

Then

$$f(x) = \int_G z(s^{-1} \cdot x) \alpha_s(a) \, d\mu(s)$$

is in $\operatorname{Ind} \alpha$ by Lemma 3.52 on the previous page, and

$$\|f(x) - a\| = \Big\| \int_G z(s^{-1} \cdot x) \big(\alpha_s(a) - a\big) \, d\mu(s) \Big\| < \epsilon.$$

It follows that Φ has dense range. Since Φ is a homomorphism, it must have closed range, so $\Phi(\operatorname{Ind} \alpha) = A$ as required. □

Since it has an important role to play in the imprimitivity theory, we'll look a bit more closely at the algebra $\operatorname{Ind}_H^G \beta$ of Example 3.47 on page 101. In [139, Proposition 6.16], we showed that $M(s, \pi)(f) := \pi\big(f(s)\big)$ is an irreducible representation of $\operatorname{Ind} \beta$ for all $s \in G$ and $\pi \in \hat{D}$, and that the map $(s, \pi) \mapsto \ker M(s, \pi)$ induces a homeomorphism of $\operatorname{Prim}(\operatorname{Ind} \beta)$ with the orbit space $(G \times \operatorname{Prim} D)/H$ where $(s, P) \cdot t = (st, t^{-1} \cdot P)$. Furthermore, it is not hard to check that the induced G-action on $\operatorname{Prim}(\operatorname{Ind} \beta)$ is given on the class $[s, P]$ of (s, P) by $r \cdot [s, P] = [rs, P]$. Therefore $[s, P] \mapsto sH$ is a continuous G-equivariant map of $\operatorname{Prim}(\operatorname{Ind} \beta)$ onto G/H. Notice that if (A, G, α) is any dynamical system, then any continuous map of $\operatorname{Prim} A$ onto G/H makes A into a $C_0(G/H)$-algebra (Proposition C.5 on page 355). However, if this map is also G-equivariant, then it follows from the next result that we recover the situation of Example 3.47 on page 101 and Lemma 3.48 on page 101. The statement and proof are taken from [36, 37].

Proposition 3.53. *Suppose that (A, G, α) is a dynamical system, that H is a closed subgroup of G and that $\sigma : \operatorname{Prim} A \to G/H$ is a G-equivariant continuous map. Let*

$$I := \bigcap\{\, P \in \operatorname{Prim} A : \sigma(P) = eH \,\},$$

and let $D := A/I$. Then I is H-invariant and (D, H, α^I) is a dynamical system. Furthermore there is a G-equivariant and $C_0(G/H)$-linear isomorphism $\Phi : A \to \operatorname{Ind}_H^G(D, \alpha^I)$ given by

$$\Phi(a)(s) := q\big(\alpha_s^{-1}(a)\big),$$

where $q : A \to A/I$ is the quotient map.

Proof. Since α is strongly continuous, we certainly have $\Phi(a) \in C^b(G, D)$. Also

$$\begin{aligned} \Phi(a)(sh) &= q\big(\alpha_h^{-1}\big(\alpha_s^{-1}(a)\big)\big) \\ &= \alpha^I_{h^{-1}}\big(q\big(\alpha_s^{-1}(a)\big)\big) \\ &= \alpha^I_{h^{-1}}\big(\Phi(a)(s)\big). \end{aligned}$$

Therefore to see that $\Phi(a) \in \operatorname{Ind}_H^G(D, \alpha^I)$, we just have to check that $sH \mapsto \|\Phi(a)(s)\|$ vanishes at infinity on G/H. However, Proposition C.5 on page 355 implies that A is a $C_0(G/H)$-algebra. Let $\bar\sigma : \hat A \to G/H$ be the lift of σ to $\hat A$: $\bar\sigma\big([\pi]\big) = \sigma(\ker \pi)$. Then $\bar\sigma$ is continuous and G-equivariant. Furthermore,

$$\begin{aligned} \|\Phi(a)(s)\| &= \big\|q\big(\alpha_s^{-1}(a)\big)\big\| \\ &= \sup_{\substack{\pi \in \hat A \\ \ker \pi \supset I}} \big\|\pi\big(\alpha_s^{-1}(a)\big)\big\| \\ &= \sup_{\bar\sigma(\pi) = eH} \big\|\pi\big(\alpha_s^{-1}(a)\big)\big\| \\ &= \sup_{\bar\sigma(\pi) = eH} \|s \cdot \pi(a)\| \\ &= \sup_{\bar\sigma(\pi) = sH} \|\pi(a)\| \\ &= \|a(sH)\|. \end{aligned}$$

But $sH \mapsto \|a(sH)\|$ vanishes at infinity on G/H by Proposition C.10 on page 357. Thus we have shown that $\Phi(a) \in \operatorname{Ind}_H^G(D, \alpha^I)$.[3]

If $\Phi(a) = 0$, then the previous computation shows that $a(sH) = 0$ for all $s \in G$. Therefore $a = 0$ by Proposition C.10 on page 357. W still need to show that Φ is surjective and $C_0(G/H)$-linear. Since $\{\, \Phi(a)(s) : a \in A \,\} = A/I = D$, and in view of Proposition 3.49 on page 101, Proposition C.24 on page 363 and Theorem C.26

[3] Since the above computation shows that $sH \mapsto \|\Phi(a)(s)\| = \|a(sH)\|$ is continuous, it follows from part (b) of Proposition C.10 on page 357 that σ must be an open map.

on page 367, it will suffice to see that Φ is $C_0(G/H)$-linear. But if $\pi \in \hat{A}$ is such that $\ker \pi \supset I$ and if $\varphi \in C_0(G/H)$, then $\bar{\sigma}(\pi) = eH$ and

$$\begin{aligned}\pi\big(\alpha_s^{-1}(\varphi \cdot a)\big) &= s \cdot \pi(\varphi \cdot a) \\ &= \varphi\big(\bar{\sigma}(s \cdot \pi)\big) s \cdot \pi(a) \\ &= \varphi(sH)\pi\big(\alpha_s^{-1}(a)\big).\end{aligned}$$

It follows that

$$q\big(\alpha_s^{-1}(\varphi \cdot a)\big) = \varphi(sH) q\big(\alpha_s^{-1}(a)\big),$$

or equivalently,

$$\Phi(\varphi \cdot a)(s) = \varphi(sH)\Phi(a)(s) = (\varphi \cdot \Phi)(a)(s).$$

Thus the range of Φ is a $C_0(G/H)$-module and it follows that Φ is surjective.
Since it is immediate that $\Phi\big(\alpha_r(a)\big)(s) = \Phi(a)(r^{-1}s)$, Φ is equivariant and we're done. $\square$

The next result will be of use in Section 4.1. It will also complete the proof of Lemma 3.48 on page 101.

Lemma 3.54. *Suppose that we have free and proper actions of locally compact groups K and H on the left and right, respectively, of a locally compact space P such that*

$$t \cdot (p \cdot s) = (t \cdot p) \cdot s \quad \textit{for all } t \in K,\ p \in P \textit{ and } s \in H. \tag{3.16}$$

Let α and β be commuting strongly continuous actions of K and H, respectively, on a C^*-algebra A. Then there are strongly continuous actions

$$\sigma : K \to \operatorname{Aut}\big(\operatorname{Ind}_H^P(A,\beta)\big) \quad \textit{and} \quad \tau : H \to \operatorname{Aut}\big(\operatorname{Ind}_K^P(A,\alpha)\big)$$

given by

$$\sigma_t(f)(p) := \alpha_t\big(f(t^{-1} \cdot p)\big) \quad \textit{and} \quad \tau_s(f)(p) := \beta_s\big(f(p \cdot s)\big). \tag{3.17}$$

Remark 3.55. The actions σ and τ above are often called *diagonal actions* because, for example, σ is the restriction to $\operatorname{Ind}\beta$ of the canonical extension of $\mathrm{lt} \otimes \alpha$ on $C_0(P) \otimes A$ to $C^b(P, A) \subset M\big(C_0(P) \otimes A\big)$.

Proof. By symmetry, it suffices to prove that σ is strongly continuous. Suppose that $t_i \to t$ in K. Since $\operatorname{Ind}_c \beta$ is dense in $\operatorname{Ind}\beta$, it suffices to show that $\sigma_{t_i}(f) \to \sigma_t(f)$ for $f \in \operatorname{Ind}\beta$ with $\operatorname{supp} f \subset D \cdot H$ for D compact in P (Lemma 3.37 on page 98). If N is a compact neighborhood of t, then there is an i_0 such that $i \geq i_0$ implies $t_i \in N$. Then the functions $p \mapsto \|\sigma_{t_i}(f)(p)\|$ are constant on H-orbits and vanish outside $N \cdot D \cdot H$ for large i. Therefore if $i \geq i_0$, then

$$\|\sigma_{t_i}(f) - \sigma_t(f)\| = \sup_{p \in N \cdot D} \|\sigma_{t_i}(f)(p) - \sigma_t(f)(p)\|.$$

Since $N \cdot D$ is compact in P and $f(t^{-1} \cdot N \cdot D)$ is compact in A, there is an $i_1 \geq i_0$ such that $i \geq i_1$ implies that

$$\begin{aligned} \|f(t_i^{-1} \cdot p) - f(t^{-1} \cdot p)\| &< \epsilon/2 \quad \text{for all } p \in N \cdot D, \text{ and} \\ \|\alpha_{t_i}(a) - \alpha_t(a)\| &< \epsilon/2 \quad \text{for all } a \in f(t^{-1} \cdot N \cdot D). \end{aligned}$$

Then $i \geq i_1$ implies $\|\sigma_{t_i}(f) - \sigma_t(f)\| < \epsilon$ as required. □

The next two technical results will be of use in the next section.

Lemma 3.56. *Let K be a locally compact group, P a locally compact G-space and (A, G, α) a dynamical system. If $f \in C_{cc}\big(K, \operatorname{Ind}_c \alpha\big)$ and $\epsilon > 0$, then there is a neighborhood V of e in K such that $kt^{-1} \in V$ implies*

$$\|f(k, x) - f(t, x)\| < \epsilon \quad \textit{for all } x \in P.$$

Proof. Suppose no such V exists. Let N be a symmetric compact neighborhood of e in K. By assumption, for each neighborhood $V \subset N$, there are $x_V \in P$, $k_V, t_V \in K$ such that $k_V t_V^{-1} \in V$ and

$$\|f(k_V, x_V) - f(t_V, x_V)\| \geq \epsilon. \tag{3.18}$$

But there are compact sets $C \subset K$ and $T \subset G\backslash P$ such that $\operatorname{supp} f \subset C \times p^{-1}(T)$. Clearly, at least one of k_V and t_V must be in C. Since $k_V t_V^{-1} \in V \subset N$, both t_V and k_V are in the compact set NC. Thus we can pass to a subnet, relabel, and assume that there is a $t \in K$ such that $t_V \to t$ and $k_V \to t$. Since $\{\, p(x_V) \,\} \subset T$ and since we can replace x_V by $s^{-1} \cdot x_V$ in (3.18), we can use Lemma 3.38 on page 98 to pass to a subnet, relabel, and assume that $x_V \to x$ in P. Now the continuity of f and (3.18) lead to a contradiction. □

Lemma 3.57. *Suppose that P is a free and proper G-space, and that $f \in C_c^+(P)$. Then given $\epsilon > 0$ there is a $g \in C_c^+(P)$ such that $\operatorname{supp} g \subset \operatorname{supp} f$ and*

$$\Big| f(x) - g(x) \int_G g(s^{-1} \cdot x)\, d\mu(s) \Big| < \epsilon.$$

Proof. Define

$$F(G \cdot x) := \int_G f(s^{-1} \cdot x)\, d\mu(s).$$

A simple variation on the proof of Lemma 3.52 on page 103, shows that $F \in C_c(G\backslash P)$. Let

$$C = \{\, x \in P : f(x) \geq \epsilon \,\}.$$

Since $F(G \cdot x) > 0$ if $x \in C$ and since $p(C)$ is compact in $G\backslash P$, $m := \inf\{\, F(G \cdot x) : x \in C \,\} > 0$. Let

$$U = \{\, G \cdot x \in G\backslash P : F(G \cdot x) > m/2 \,\}.$$

Then U is a neighborhood of $p(C)$ and there is a $Q \in C_c(G\backslash P)$ such that $Q(G\cdot x) = 1$ if $x \in C$, $Q(G \cdot x) = 0$ if $G \cdot x \notin U$ and $Q(G \cdot x) \leq 1$ otherwise. Then $H(G \cdot x) = Q(G \cdot x)F(G \cdot x)^{-\frac{1}{2}}$ is in $C_c(G\backslash P)$. Then if $g(x) := f(x)Q(G \cdot x)F(G \cdot x)^{-\frac{1}{2}}$,

$$g(x) \int_G \Delta(s)^{-\frac{1}{2}} g(s^{-1} \cdot x)\, d\mu(s) = \begin{cases} f(x) & \text{if } G \cdot x \in G \cdot C, \text{ and} \\ f(x)Q(G \cdot x)^2 & \text{otherwise.} \end{cases}$$

But if $G \cdot x \notin G \cdot C$, then $|f(x)| < \epsilon$ and

$$\begin{aligned} |f(x) - f(x)Q(G \cdot x)^2| &\leq |f(x)||1 - Q(G \cdot x)^2| \\ &\leq |f(x)| < \epsilon. \end{aligned}$$

Thus g suffices. □

Notes and Remarks

Propositions 3.1 and 3.4 are relatively straightforward observations and have been well-known for a long time. Proposition 3.3 has also been around for a long time. The proof given here is based on a comment in [17]. A compilation of the properties of an extraordinary varied collection of group C^*-algebras can be found in [123]. Proposition 3.11 is a relatively straightforward consequence of the axioms and is certainly very well known in the case of group C^*-algebras. Proposition 3.19 and Corollary 3.20 are due to Green [65, 67]. The importance of proper actions in the theory was first emphasized by Green in [65]. Induced algebras derive their name from their similarity to the space for Mackey's induced representation. As examples of C^*-algebras, they have been around since at least [117, 137, 138].

Chapter 4

Imprimitivity Theorems

Originally, imprimitivity theorems, such as Mackey's Imprimitivity Theorem, were meant to tell us which representations of a locally compact group G are induced from a closed subgroup H. Rieffel's theory of Morita equivalence allows us to recast Mackey's theorem as the statement that $C^*(H)$ and the transformation group C^*-algebra $C_0(G/H) \rtimes_{\mathrm{lt}} G$ are Morita equivalent via an imprimitivity bimodule X. As an added bonus, this new approach allows us to define induced representations using the calculus of imprimitivity bimodules and Morita theory. In this chapter, we want to see that Rieffel's approach extends to dynamical systems in a deep and highly nontrivial way. The key step is proving a Morita equivalence result called the Symmetric Imprimitivity Theorem which is due to Raeburn. We then investigate its consequences. In particular, this will allow us to give some nontrivial examples of crossed products. In Chapter 5 we will use this machinery to develop a theory of induced representations for crossed products that will play a crucial role in our program to understand the ideal structure of crossed products in Chapters 8 and 9.

The Symmetric Imprimitivity Theorem was proved by Raeburn as a common generalization of most of the significant Morita equivalence results for crossed products in the literature. We state and then take on the proof of the Symmetric Imprimitivity Theorem in the first section. In Sections 4.2 and 4.3 we look at some special cases of the Symmetric Imprimitivity Theorem, and these so-called special cases constitute the centerpiece of this chapter.

In Sections 4.4 and 4.5, we tackle some interesting and nontrivial examples. In particular, we show that $C_0(G) \rtimes_{\mathrm{lt}} G$ is isomorphic to the compact operators on $L^2(G)$ — which should be viewed as a modern version of the von Neumann Uniqueness Theorem — and we show that $C_0(G/H) \rtimes_{\mathrm{lt}} G$ is isomorphic to $C^*(H) \otimes \mathcal{K}\big(L^2(G/H, \beta^H)\big)$, where β^H is any quasi-invariant measure on G/H. Proving this last assertion involves some technical gymnastics using a suitable *measurable* cross section for the quotient map of G onto G/H.

In Chapter 5, we will use the material in this chapter to define and study induced representations of crossed products.

4.1 The Symmetric Imprimitivity Theorem

There are a number of Morita equivalences that play a fundamental role in the study of the representation theory of crossed products. These equivalences go by the name of imprimitivity theorems as the original motivation and statements can be traced back through Rieffel's work [145] and from there to Mackey's systems of imprimitivity [102]. Most of these are subsumed by Raeburn's Symmetric Imprimitivity Theorem [132] which we reproduce here. The set-up requires two commuting free and proper actions of locally compact groups K and H on a locally compact space P. To reduce confusion, it is convenient to have one group, K in this case, act on the left and the other, H, on the right. Then the fact that the actions commute simply amounts to the condition

$$t \cdot (p \cdot h) = (t \cdot p) \cdot s \quad \text{for all } t \in K,\ p \in P \text{ and } s \in H. \tag{4.1}$$

In addition we suppose that there are commuting strongly continuous actions α and β of K and H, respectively, on a C^*-algebra A. Then Lemma 3.54 on page 106 implies there are dynamical systems

$$\big(\operatorname{Ind}_H^P(A,\beta), K, \sigma\big) \qquad \big(\operatorname{Ind}_K^P(A,\alpha), H, \tau\big)$$

as defined in (3.17). The Symmetric Imprimitivity Theorem states that the crossed products $\operatorname{Ind}\beta \rtimes_\sigma K$ and $\operatorname{Ind}\alpha \rtimes_\tau H$ are Morita equivalent. As is usually the case, it is convenient to work with dense $*$-subalgebras. We'll invoke Lemma 3.50 on page 102 and define

$$E_0 = C_{cc}\big(K, \operatorname{Ind}_c \beta\big) \subset \operatorname{Ind}\beta \rtimes_\sigma K \quad \text{and} \quad B_0 = C_{cc}\big(H, \operatorname{Ind}_c \alpha\big) \subset \operatorname{Ind}\alpha \rtimes_\tau H.$$

Therefore E_0 consists of A-valued functions on $K \times P$, and B_0 of A-valued functions on $H \times P$. Both B_0 and E_0 are are viewed as dense $*$-subalgebras of the corresponding crossed products as indicated above.

Theorem 4.1 (Raeburn's Symmetric Imprimitivity Theorem). *Suppose that we have commuting free and proper actions of locally compact groups K and H on the left and right, respectively, of a locally compact space P, and commuting strongly continuous actions α and β of K and H, respectively, on a C^*-algebra A. As described above, let E_0 and B_0 be viewed as dense $*$-subalgebras of* $\operatorname{Ind}\beta \rtimes_\sigma K$ *and* $\operatorname{Ind}\alpha \rtimes_\tau H$*, respectively, and $\mathsf{Z}_0 = C_c(P, A)$. If $c \in E_0$, $b \in B_0$ and $f, g \in \mathsf{Z}_0$, then define*

$$c \cdot f(p) = \int_K c(t,p)\alpha_t\big(f(t^{-1} \cdot p)\big)\Delta_K(t)^{\frac{1}{2}}\, d\mu_K(t) \tag{4.2}$$

$$f \cdot b(p) = \int_H \beta_s\big(f(p \cdot s)b(s^{-1}, p \cdot s)\big)\Delta_H(s)^{-\frac{1}{2}}\, d\mu_H(s) \tag{4.3}$$

$${}_{E_0}\langle f\,,\, g\rangle(t,p) = \Delta_K(t)^{-\frac{1}{2}} \int_H \beta_s\big(f(p \cdot s)\alpha_t\big(g(t^{-1} \cdot p \cdot s)^*\big)\big)\, d\mu_H(s) \tag{4.4}$$

$$\langle f\,,\, g\rangle_{B_0}(s,p) = \Delta_H(s)^{-\frac{1}{2}} \int_K \alpha_t\big(f(t^{-1} \cdot p)^* \beta_s\big(g(t^{-1} \cdot p \cdot s)\big)\big)\, d\mu_K(t). \tag{4.5}$$

Then Z_0 *is a* $E_0 - B_0$*-pre-imprimitivity bimodule. The completion* $\mathsf{Z} = \mathsf{Z}_H^K$ *is a* $\operatorname{Ind}\beta \rtimes_\sigma K - \operatorname{Ind}\alpha \rtimes_\tau H$*-imprimitivity bimodule, and* $\operatorname{Ind}\beta \rtimes_\sigma K$ *is Morita equivalent to* $\operatorname{Ind}\alpha \rtimes_\tau H$.

Before we begin the proof of Theorem 4.1, we need some technical preliminaries. The most formidable of these will involve finding approximate identities of a specific form. Fortunately, we can save ourselves a bit of work by exploiting symmetry.

Remark 4.2. At first glance, it might not seem that the statement of Theorem 4.1 on the preceding page is completely symmetric in K and H. But it is important to keep in mind that the right action of H is purely a convention to keep the formulas easier to digest. We could easily have H act on the left via

$$s : p := p \cdot s^{-1} \quad \text{for } s \in H \text{ and } p \in P.$$

Similarly, we can let K act on the right via

$$p : t := t^{-1} \cdot p \quad \text{for } t \in K \text{ and } p \in P.$$

Then we can swap the roles of K and H in (4.2)–(4.5), and obtain formulas for a left $\operatorname{Ind}\alpha \rtimes_\tau H$-action and a right $\operatorname{Ind}\beta \rtimes_\sigma K$-action as follows:

$$b : f(p) = \int_H b(s,p)\beta_s\big(f(s^{-1} : p)\big)\Delta_H(s)^{\frac{1}{2}}\, d\mu_H(s) \tag{4.6}$$

$$f : c(p) = \int_K \alpha_t\big(f(p : t)c(t^{-1}, p : t)\big)\Delta_K(t)^{-\frac{1}{2}}\, d\mu_K(t) \tag{4.7}$$

$${}_{B_0}\langle f\, , g\rangle(s,p) = \Delta_H(s)^{-\frac{1}{2}} \int_K \alpha_t\big(f(p : t)\beta_s\big(g(s^{-1} : p : t)^*\big)\big)\, d\mu_K(t) \tag{4.8}$$

$$\langle f\, , g\rangle_{E_0}(t,p) = \Delta_K(t)^{-\frac{1}{2}} \int_H \beta_s\big(f(s^{-1} : p)^* \alpha_t\big(g(s^{-1} : p : t)\big)\big)\, d\mu_H(s). \tag{4.9}$$

Therefore if Z_0 is a $E_0 - B_0$-pre-imprimitivity bimodule with respect to (4.2)–(4.5), then it is a $B_0 - E_0$-pre-imprimitivity bimodule with respect to (4.6)–(4.9). However, the symmetry runs even deeper.

Recall that if $c \in E_0 \subset \operatorname{Ind}\beta \rtimes_\sigma K$, then

$$c^*(t,p) := \alpha_t\big(c(t^{-1}, t^{-1} \cdot p)^*\big)\Delta_K(t^{-1}).$$

If $\Phi : \mathsf{Z}_0 \to \mathsf{Z}_0$ is defined by $\Phi(f)(p) = f(p)^*$, then we can compute as follows:

$$\begin{aligned}
\Phi(c \cdot f)(p) &= \int_K \alpha_t\big(f(t^{-1} \cdot p)^*\big)c(t,p)^*\Delta_K(t)^{\frac{1}{2}}\, d\mu_K(t)\\
&= \int_K \alpha_t\big(f(t^{-1} \cdot p)^* \alpha_t^{-1}\big(c(t,p)^*\big)\Delta_K(t)\big)\Delta_K(t)^{-\frac{1}{2}}\, d\mu_K(t)\\
&= \int_K \alpha_t\big(f(t^{-1} \cdot p)^* c^*(t^{-1}, t^{-1} \cdot p)\big)\Delta_K(t)^{-\frac{1}{2}}\, d\mu_K(t)\\
&= \int_K \alpha_t\big(f(p : t)^* c^*(t^{-1}, p : t)\big)\Delta_K(t)^{-\frac{1}{2}}\, d\mu_K(t)\\
&= \Phi(f) : c^*(p).
\end{aligned}$$

Similar sorts of computations result in the following symmetry lemma.

Lemma 4.3. *Define* $\Phi : \mathsf{Z}_0 \to \mathsf{Z}_0$ *by* $\Phi(f)(p) = f(p)^*$. *Then for all* $c \in E_0$, $b \in B_0$ *and* $f, g \in \mathsf{Z}_0$*:*

$$\Phi(c \cdot f) = \Phi(f) : c^* \qquad \Phi(f \cdot b) = b^* : \Phi(f)$$
$${}_{E_0}\langle \Phi(f)\,,\,\Phi(g)\rangle = \langle f\,,\,g\rangle_{E_0} \qquad \langle \Phi(f)\,,\,\Phi(g)\rangle_{B_0} = {}_{B_0}\langle f\,,\,g\rangle.$$

Remark 4.2 on the preceding page and Lemma 4.3 will allow us some shortcuts in the following. For example, if we show that for all $f \in \mathsf{Z}_0$, $\langle f\,,\,f\rangle_{B_0} \geq 0$ in $\operatorname{Ind}\alpha \rtimes_\tau H$, then just by exchanging the roles of K and H, we know that $\langle f\,,\,f\rangle_{E_0} \geq 0$ in $\operatorname{Ind}\beta \rtimes_\sigma K$. In particular, given $f \in \mathsf{Z}_0$, $\langle \Phi(f)\,,\,\Phi(f)\rangle_{E_0} \geq 0$, then Lemma 4.3 then implies ${}_{E_0}\langle f\,,\,f\rangle \geq 0$. We'll often use this sort of reasoning in our proof of the Symmetric Imprimitivity Theorem.

Lemma 4.4. *If* $c \in E_0$, $f, g \in \mathsf{Z}_0$ *and* $b \in B_0$, *then* $c \cdot f$ *and* $f \cdot b$ *are in* Z_0, $\langle f\,,\,g\rangle_{B_0} \in B_0$ *and* ${}_{E_0}\langle f\,,\,g\rangle \in E_0$. *If* $f_i \to f$ *and* $g_i \to g$ *in the inductive limit topology on* Z_0, *then* ${}_{E_0}\langle f_i\,,\,g_i\rangle \to {}_{E_0}\langle f\,,\,g\rangle$ *in the inductive limit topology on* E_0 *while* $\langle f_i\,,\,g_i\rangle_{B_0} \to \langle f\,,\,g\rangle_{B_0}$ *in the inductive limit topology on* B_0.

Proof. Since $(t,p) \mapsto c(t,p)\alpha_t\big(f(t^{-1}\cdot p)\big)$ is in $C_c(K \times P, A)$, we have $c \cdot f \in \mathsf{Z}_0$ by Lemma 1.102 on page 36. A similar observation shows that $f \cdot b \in \mathsf{Z}_0$.

To see that ${}_{E_0}\langle f\,,\,g\rangle \in E_0$, let $D_f := \operatorname{supp} f$ and let

$$F(t,p) := f(p)\alpha_t\big(g(t^{-1}\cdot p)^*\big).$$

Since the K-action is proper,

$$\{\, t \in K : D_f \cap t \cdot D_g \neq \emptyset \,\} = \{\, t \in K : (p, t^{-1}\cdot p) \in D_f \times D_g \text{ for some } p \,\}$$

is contained in a compact set $C_K \subset K$, and $\operatorname{supp} F \subset C_K \times D_f$. Thus Lemma 3.52 on page 103 implies that ${}_{E_0}\langle f\,,\,g\rangle \in E_0$.

Notice that

$${}_{E_0}\langle f\,,\,g\rangle(t,p) = 0 \quad \text{if } (t,p) \notin C_K \times D_f \cdot H,$$

and that

$$\|{}_{E_0}\langle f\,,\,g\rangle\|_\infty = \sup_{(t,p)\in C_K \times D_f} \|{}_{E_0}\langle f\,,\,g\rangle(t,p)\| \tag{4.10}$$

Since the H-action is proper, then there is a compact set $C_H \subset H$ such that

$$\{\, s \in H : D_f \cap D_f \cdot s^{-1} \neq \emptyset \,\} \subset C_H.$$

It follows from (4.10) that

$$\|{}_{E_0}\langle f\,,\,g\rangle\|_\infty \leq \Big(\sup_{t\in C_K} \Delta_K(t)^{-\frac{1}{2}}\Big)\|f\|_\infty\|g\|_\infty\mu_H(C_H). \tag{4.11}$$

Since

$${}_{E_0}\langle f_i\,,\,g_i\rangle - {}_{E_0}\langle f\,,\,g\rangle = {}_{E_0}\langle f_i - f\,,\,g_i\rangle - {}_{E_0}\langle f\,,\,g_i - g\rangle,$$

and since the compact set C_K in (4.11) depends only on D_f and D_g, it is now easy to see that ${}_{E_0}\langle f_i \, , g_i\rangle \to {}_{E_0}\langle f \, , g\rangle$ in the inductive limit topology.

The corresponding statements for $\langle \cdot \, , \cdot \rangle_{B_0}$ can be proved similarly, or we can invoke symmetry. For example, if $f_i \to f$ and $g_i \to g$ in the inductive limit topology, then $\Phi(f_i) \to \Phi(f)$ and $\Phi(g_i) \to \Phi(g)$. By switching the roles of K and H, we know that ${}_{B_0}\big\langle \Phi(f_i) \, , \Phi(g_i)\big\rangle \to {}_{B_0}\big\langle \Phi(f) \, , \Phi(g)\big\rangle$, and the result follows from Lemma 4.3 on the preceding page. □

The key technical tool for the proof of the Symmetric Imprimitivity Theorem is the following proposition which says that we can construct approximate identities of a very special form. The use of these sorts of constructions is now very common in the literature. The original idea goes back at least to [66, Lemma 2].

Proposition 4.5. *There is a net* $\{ e_m \}_{m\in M}$ *in* E_0 *such that*

(a) *for all* $c \in E_0$*,* $e_m * c \to c$ *in the inductive limit topology on* E_0*,*

(b) *for all* $f \in \mathsf{Z}_0$*,* $e_m \cdot f \to f$ *in the inductive limit topology on* Z_0 *and*

(c) *for each* $m \in M$ *there are* $f_i^m \in \mathsf{Z}_0$ *such that*

$$e_m = \sum_{i=1}^{n_m} {}_{E_0}\langle f_i^m \, , f_i^m\rangle.$$

The proof of the above proposition is a bit involved. In order not to obscure the basic ideas of the Symmetric Imprimitivity Theorem, the proof of this proposition has been exiled to Section 4.1.1 on page 117.

Proof of the Symmetric Imprimitivity Theorem. As observed in Lemma 4.4 on the facing page, (4.2)–(4.5) take values in the appropriate spaces. We need to show that Z_0 is a $E_0 - B_0$-pre-imprimitivity bimodule. (The rest follows from [139, Proposition 3.12].) First, we check that Z_0 is a $E_0 - B_0$-bimodule. This amounts to checking that for $f \in \mathsf{Z}_0$, $b, b' \in B_0$ and $c, c' \in E_0$,

$$f \cdot (b * b') = (f \cdot b) \cdot b', \tag{4.12}$$
$$(c' * c) \cdot f = c' \cdot (c \cdot f) \text{ and} \tag{4.13}$$
$$(c \cdot f) \cdot b = c \cdot (f \cdot b). \tag{4.14}$$

Fortunately, these are all fairly routine computations. For example, to verify (4.14) we recall that α and β commute with each other by assumption, and with integrals by Lemma 1.92 on page 32. Then we compute that

$$\begin{aligned}(c \cdot f) \cdot b(p) &= \int_H \beta_s\big(c \cdot f(p \cdot s) b(s, p \cdot s)\big) \Delta_H(s)^{-\frac{1}{2}} \, d\mu_H(s) \\ &= \int_H \int_K \beta_s\big(c(t, p \cdot s) \alpha_t\big(f(t^{-1} \cdot p \cdot s)\big) b(s, p \cdot s)\big) \\ &\qquad\qquad \Delta_K(t)^{\frac{1}{2}} \Delta_H(s)^{-\frac{1}{2}} \, d\mu_K(t) \, d\mu_H(s)\end{aligned}$$

which, utilizing Fubini's Theorem (Proposition 1.105 on page 37) and that $c(t,\cdot) \in \operatorname{Ind}\beta$ and $b(s,\cdot) \in \operatorname{Ind}\alpha$,

$$\begin{aligned}
&= \int_K \int_H c(t,p)\alpha_t\big(\beta_s\big(f(t^{-1}\cdot p\cdot s)b(s,t^{-1}\cdot p\cdot s)\big)\big) \\
&\qquad\qquad\qquad\qquad \Delta_K(t)^{\frac12}\Delta_H(s)^{-\frac12}\,d\mu_H(s)\,d\mu_K(t) \\
&= \int_K c(t,p)\alpha_t\Big(\int_H \beta_s\big(f(t^{-1}\cdot p\cdot s) \\
&\qquad\qquad b(s,t^{-1}\cdot p\cdot s)\big)\Delta_H(s)^{-\frac12}\,d\mu_H(s)\Big)\Delta_K(t)^{\frac12}\,d\mu_K(t) \\
&= \int_K c(t,p)\alpha_t\big(f\cdot b(t^{-1}\cdot p)\big)\Delta_K(t)^{\frac12}\,d\mu_K(t) \\
&= c\cdot(f\cdot b)(p).
\end{aligned}$$

Equations (4.12) and (4.13) can be verified via equally exciting computations.

Now we proceed to verify that Z_0 satisfies the axioms of a pre-imprimitivity bimodule as laid out in [139, Definition 3.9]. Specifically, we have to verify the following:

(IB1) The bimodule Z_0 is both a left pre-inner product E_0-module and a right pre-inner product B_0-module.

(IB2) The range ${}_{E_0}\langle \mathsf{Z}_0\,,\,\mathsf{Z}_0\rangle$ and $\langle \mathsf{Z}_0\,,\,\mathsf{Z}_0\rangle_{B_0}$ of the inner products span dense ideals.

(IB3) The E_0- and B_0-actions are bounded with respect to the norms induced by $\langle\cdot\,,\cdot\rangle_{B_0}$ and ${}_{E_0}\langle\cdot\,,\cdot\rangle$, respectively.

(IB4) For all $f,g,h \in \mathsf{Z}_0$, ${}_{E_0}\langle f\,,g\rangle\cdot h = f\cdot\langle g\,,h\rangle_{B_0}$.

Verifying (IB4) amounts to a computation:

$$\begin{aligned}
{}_{E_0}\langle f\,,g\rangle\cdot h(p) &= \int_K {}_{E_0}\langle f\,,g\rangle(t,p)\alpha_t\big(h(t^{-1}\cdot p)\big)\Delta_K(t)^{\frac12}\,d\mu_K(t) \\
&= \int_K\int_H \beta_s\big(f(p\cdot s)\alpha_t\big(g(t^{-1}\cdot p\cdot s)^*\big)\big)\alpha_t\big(h(t^{-1}\cdot p)\big)\,d\mu_H(s)\,d\mu_K(t).
\end{aligned}$$

Since the actions of K and H are free and proper, the above integrand is a continuous compactly supported function on $K\times H$. Therefore Fubini's Theorem (Proposition 1.105) applies and the above

$$\begin{aligned}
&= \int_H \beta_s\Big(f(p\cdot s)\int_K \alpha_t\big(g(t^{-1}\cdot p\cdot s)^*\beta_s^{-1}\big(h(t^{-1}\cdot p)\big)\big)\,d\mu_K(t)\Big)\,d\mu_H(s) \\
&= \int_H \beta_s\big(f(p\cdot s)\langle g\,,h\rangle_{B_0}(s^{-1},p\cdot s)\big)\Delta_H(s)^{-\frac12}\,d\mu_H(s) \\
&= f\cdot\langle g\,,h\rangle_{B_0}(p).
\end{aligned}$$

Thus (IB4) holds.

To verify (IB1), we have to check properties (a)–(d) of [139, Definition 2.1]. For the E_0 action, this means we must show

(a) ${}_{E_0}\langle \cdot \,, \cdot \rangle$ is linear in its first variable,
(b) ${}_{E_0}\langle c \cdot f \,, g \rangle = c * {}_{E_0}\langle f \,, g \rangle$,
(c) ${}_{E_0}\langle f \,, g \rangle^* = {}_{E_0}\langle g \,, f \rangle$, and
(d) ${}_{E_0}\langle f \,, f \rangle \geq 0$ as an element of $\operatorname{Ind} \beta \rtimes_\sigma K$.

Property (a) is easy to check. Properties (b) and (c) follow from straightforward computations. For example,

$$\begin{aligned}
{}_{E_0}\langle f \,, g \rangle^*(t,p) &= \Delta_K(t^{-1})\alpha_t\big({}_{E_0}\langle f \,, g \rangle(t^{-1}, t^{-1} \cdot p)\big)^* \\
&= \Delta_K(t)^{-\frac{1}{2}}\alpha_t\Big(\int_H \beta_s\big(f(t^{-1} \cdot p \cdot s)\alpha_t^{-1}\big(g(p \cdot s)^*\big)\big)\, d\mu_H(s)\Big)^* \\
&= \Delta_K(t)^{-\frac{1}{2}} \int_H \beta_s\big(g(p \cdot s)\alpha_t\big(f(t^{-1} \cdot p \cdot s)^*\big)\big)\, d\mu_H(s) \\
&= {}_{E_0}\langle g \,, f \rangle(t,p).
\end{aligned}$$

However, Property (d) is nontrivial, and we have prepared our approximate identity $\{ e_m \}_{m \in M}$ in Proposition 4.5 on page 113 with this in mind. If $f \in E_0$, then $e_m \cdot f \to f$ in the inductive limit topology in Z_0. Lemma 4.4 on page 112 implies that $\langle e_m \cdot f \,, f \rangle_{B_0} \to \langle f \,, f \rangle_{B_0}$ in the inductive limit topology in B_0 and therefore in the C^*-norm topology as well. Thus, in the C^*-norm, we have

$$\begin{aligned}
\langle f \,, f \rangle_{B_0} &= \lim_m \langle e_m \cdot f \,, f \rangle_{B_0} \\
&= \lim_m \sum_{i=1}^{n_m} \big\langle {}_{E_0}\langle f_i^m \,, f_i^m \rangle \cdot f \,, f \big\rangle_{B_0} \\
&= \lim_m \sum_{i=1}^{n_m} \big\langle f_i^m \cdot \langle f_i^m \,, f \rangle_{B_0} \,, f \big\rangle_{B_0} \\
&= \lim_m \sum_{i=1}^{n_m} \langle f_i^m \,, f \rangle_{B_0}^* * \langle f_i^m \,, f \rangle_{B_0}.
\end{aligned}$$

Since elements of the form $b^* * b$ are positive, and since $\langle f \,, f \rangle_{B_0}$ is the norm limit of positive elements, it follows that $\langle f \,, f \rangle_{B_0} \geq 0$. The corresponding statements for the E_0-valued inner product follow by symmetry as we remarked following the proof of Lemma 4.3 on page 112. Thus (IB1) is satisfied.

Our approximate identity also gives an easy proof of (IB2). If $c \in E_0$ and if $\{ e_m \}_{m \in M} \subset E_0$ is as above, then $e_m * c \to c$ in the inductive limit topology. Since convergence in the inductive limit topology implies convergence in the C^*-norm, it suffices to see that $e_m * c$ belongs to the span of ${}_{E_0}\langle \mathsf{Z}_0 \,, \mathsf{Z}_0 \rangle$. But

$$e_m * c = \sum_{i=1}^{n_m} {}_{E_0}\langle f_i^m \,, f_i^m \rangle * c = \sum_{i=1}^{n_m} {}_{E_0}\langle f_i^m \,, c^* \cdot f_i^m \rangle.$$

Thus ${}_{E_0}\langle \mathsf{Z}_0 \, , \, \mathsf{Z}_0 \rangle$ is dense in E_0. Exchanging the roles of K and H implies that ${}_{B_0}\langle \mathsf{Z}_0 \, , \, \mathsf{Z}_0 \rangle$ is dense in B_0. Since $\Phi(\mathsf{Z}_0) = \mathsf{Z}_0$, it follows that $\langle \mathsf{Z}_0 \, , \, \mathsf{Z}_0 \rangle_{B_0}$ is dense in B_0 by Lemma 4.3 on page 112.

To verify (IB3), we have to show that

$$\langle c \cdot f \, , c \cdot f \rangle_{B_0} \leq \|c\|^2 \langle f \, , f \rangle_{B_0}, \quad \text{and} \tag{4.15}$$

$$ {}_{E_0}\langle f \cdot b \, , f \cdot b \rangle \leq \|b\|^2 \, {}_{E_0}\langle f \, , f \rangle. \tag{4.16}$$

Let Z_1 be the completion of Z_0 as a Hilbert $\operatorname{Ind} \alpha \rtimes_\tau H$-module as in [139, Definition 2.16]. (Most of the time, we'll suppress the map $q : \mathsf{Z}_0 \to \mathsf{Z}_1$.) For each $t \in K$, define $v_t : \mathsf{Z}_0 \to \mathsf{Z}_0$ by

$$v_t(f)(p) := \Delta_K(t)^{\frac{1}{2}} \alpha_t\big(f(t^{-1} \cdot p)\big). \tag{4.17}$$

Let $E = \{ \varphi \in C^b(P, M(A)) : \varphi(p \cdot s) = \bar{\beta}_s^{-1}(\varphi(p)) \}$. Then E is a closed $*$-subalgebra of $C^b(P, M(A))$ and is a C^*-algebra containing $\operatorname{Ind}\beta$ with identity given by the function $1_E(p) = 1_A$ for all $p \in P$. If $\varphi \in E$, then we can define $M(\varphi) : \mathsf{Z}_0 \to \mathsf{Z}_0$ by

$$M(\varphi)(f)(p) = \varphi(p) f(p). \tag{4.18}$$

Then a straightforward computation shows that

$$\langle M(\varphi)(f) \, , g \rangle_{B_0} = \langle f \, , M(\varphi^*)(g) \rangle_{B_0}. \tag{4.19}$$

If $\varphi \in \operatorname{Ind}\beta$, then $\|\varphi\|^2 1_E - \varphi^*\varphi$ is positive in E, and therefore is of the form $\psi^*\psi$ for some $\psi \in E$. The positivity of the inner product together with (4.19) implies that

$$\begin{aligned} \|\varphi\|^2 \langle f \, , f \rangle_{B_0} - \big\langle M(\varphi)(f) \, , M(\varphi)(f) \big\rangle_{B_0} &= \big\langle M(\|\varphi\|^2 1_E - \varphi^*\varphi)(f) \, , f \big\rangle_{B_0} \\ &= \big\langle M(\psi)(f) \, , M(\psi)(f) \big\rangle_{B_0} \geq 0. \end{aligned}$$

Thus $M(\varphi)$ is bounded and extends to an operator on Z_1 with adjoint $M(\varphi^*)$. It is now easy to see that $M : \operatorname{Ind}\beta \to \mathcal{L}(\mathsf{Z}_1)$ is a homomorphism.

Another computation shows that for all $k \in K$

$$\big\langle v_k(f) \, , v_k(g) \big\rangle_{B_0} = \langle f \, , g \rangle_{B_0}.$$

Thus v_r is isometric and extends to an operator on Z_1. Then it is straightforward to check that $v : K \to U\mathcal{L}(\mathsf{Z}_1)$ is a unitary valued homomorphism. Computations such as in Proposition 2.34 on page 54 show that $k \mapsto v_k(f)$ is continuous from K into Z_0 in the inductive limit topology. Thus

$$\begin{aligned} \|v_k(f) - f\|_{B_0}^2 &= \|\big\langle v_k(f) - f \, , v_k(f) - f \big\rangle_{B_0}\| \\ &= \|2\langle f \, , f \rangle_{B_0} - \big\langle v_k(f) \, , f \big\rangle_{B_0} - \big\langle f \, , v_k(f) \big\rangle_{B_0}\| \end{aligned} \tag{4.20}$$

tends to zero as $k \to e$ in K. Therefore v is strongly continuous — and therefore strictly continuous. Proposition 4.5 on page 113, applied in the case $K = \{ e \}$, implies that given $f \in \mathsf{Z}_0$ there is a net $\{\varphi_i\} \subset \operatorname{Ind}_c \beta$ such that $\varphi_i \cdot f \to f$ in the inductive limit topology. Since $\varphi_i \cdot f = M(\varphi_i)(f)$ (when $K = \{e\}$), it follows that $M(\operatorname{Ind}\beta)(\mathsf{Z}_0)$ is dense in Z_0 in the inductive limit topology. Therefore M is a nondegenerate homomorphism. Since

$$\begin{aligned} v_k\bigl(M(\varphi)(f)\bigr)(p) &= \Delta_K(k)^{\frac{1}{2}}\alpha_k\bigl(M(\varphi)(f)(k^{-1}\cdot p)\bigr) \\ &= \Delta_K(k)^{\frac{1}{2}}\alpha_k\bigl(\varphi(k^{-1}\cdot p)f(k^{-1}\cdot p)\bigr) \\ &= \sigma_k(\varphi)(p)v_k(f)(p) \\ &= M\bigl(\sigma_k(\varphi)\bigr)\bigl(v_k(f)\bigr)(p), \end{aligned}$$

(M, v) is a nondegenerate covariant homomorphism of $(\operatorname{Ind}\beta, K, \sigma)$ into $\mathcal{L}(\mathsf{Z}_1)$. If $\langle\cdot\,,\cdot\rangle_B$ is the inner product on Z_1, then the integrated form $M \rtimes v$ maps E_0 into $\mathcal{L}(\mathsf{Z}_1)$, and [139, Corollary 2.22] implies that

$$\begin{aligned} \bigl\langle M \rtimes v(c)(f)\,, M \rtimes v(c)(f)\bigr\rangle_B &\le \|M \rtimes v(c)\|^2 \langle f\,, f\rangle_B \\ &\le \|c\|^2 \langle f\,, f\rangle_B. \end{aligned}$$

Thus to prove (4.15), we just need to see that $M \rtimes v(c)(f) = c \cdot f$. (More precisely, we have to see that $M \rtimes v(c)\bigl(q(f)\bigr) = q(c \cdot f)$.) However,

$$M \rtimes v(c)(f) = \int_K M\bigl(c(t,\cdot)\bigr)v_t(f)\,d\mu_K(t).$$

Thus if we set $Q(t,p) := \Delta_K(t)^{\frac{1}{2}} c(t,p)\alpha_t\bigl(f(t^{-1}\cdot p)\bigr)$, then $Q \in C_c(K \times P, A)$ and $g(t) := Q(t,\cdot) = M\bigl(c(t,\cdot)\bigr)v_t(f)$. Suppressing the map $q : \mathsf{Z}_0 \to \mathsf{Z}_1$, Lemma 1.108 on page 39 implies that $M \rtimes v(c)(f) \in \mathsf{Z}_0$ and that

$$M \rtimes v(c)(f)(p) = \int_K c(t,p)\alpha_t\bigl(f(t^{-1}\cdot p)\bigr)\Delta_K(t)^{\frac{1}{2}}\,d\mu_K(t) = c \cdot f(p).$$

This establishes (4.15). To get (4.16), we use symmetry; that is, we reverse the roles of K and H and apply Lemma 4.3 on page 112:

$$\begin{aligned} {}_{E_0}\langle f\cdot b\,, f\cdot b\rangle &= \bigl\langle \Phi(f\cdot b)\,, \Phi(f\cdot b)\bigr\rangle_{E_0} \\ &= \bigl\langle b^* : \Phi(f)\,, b^* : \Phi(f)\bigr\rangle_{E_0} \\ &\le \|b\|^2 \bigl\langle \Phi(f)\,, \Phi(f)\bigr\rangle_{E_0} \\ &= \|b\|^2\,{}_{E_0}\langle f\,, f\rangle. \end{aligned}$$

This completes the proof. □

4.1.1 Proof of Proposition 4.5

Although the real work in proving Proposition 4.5 is to construct e_m which are sums of inner products, it is still necessary for $\{e_m\}$ to form an approximate identity for both the action on Z_0 and on E_0. For this, the next result is useful.

Proposition 4.6. *Let* $\{ b_l \}_{l\in L}$ *be an approximate identity for* $\operatorname{Ind}_H^P(A,\beta)$*. Suppose that for each 4-tuple* (T,U,l,ϵ) *consisting of a compact set* T *in* P/H*, a pre-compact open neighborhood* U *of* e *in* K*, an* $l\in L$ *and an* $\epsilon>0$*, there is an*

$$e = e_{(T,U,l,\epsilon)} \in E_0$$

such that

(a) $e(t,p)=0$ *if* $t\notin U$*,*

(b) $\displaystyle\int_K \|e(t,p)\|\,d\mu_K(t)\le 4$ *if* $p\cdot H\in T$ *and*

(c) $\displaystyle\Big\|\int_K e(t,p)\,d\mu_K(t) - b_l(p)\Big\| < \epsilon$ *if* $p\cdot H\in T$.

Then the net $\{ e_{(T,U,l,\epsilon)} \}$*, directed by increasing* T *and* l*, and decreasing* U *and* ϵ*, satisfies* (a) *and* (b) *of Proposition 4.5 on page 113.*

Before proceeding with the proof of the proposition, we need a couple of preliminary observations.

Lemma 4.7. *Suppose that* $z\in \mathsf{Z}_0 = C_c(P,A)$*, that* $\{ a_j \}_{j\in J}$ *is an approximate identity for* A*, that* C_H *is a compact subset of* H *and that* $\epsilon>0$*. Then there is a* $j_0\in J$ *such that* $j\ge j_0$ *implies that for all* $s\in C_H$ *we have*

$$\big\|\beta_s(a_j)z(p)-z(p)\big\| < \epsilon \quad \textit{for all } p\in P.$$

Proof. If the lemma were false, then we could pass to a subnet, relabel, and find $s_j\in C_H$ and $p_j\in \operatorname{supp} z$ such that

$$\|\beta_{s_j}(a_j)z(p_j)-z(p_j)\|\ge\epsilon \quad \text{for all } j. \tag{4.21}$$

Since C_H and $\operatorname{supp} z$ are compact, we can pass to another subnet, relabel, and assume that $s_j\to s\in H$ and $p_j\to p$ in P. Since each β_{s_j} is isometric, the left-hand side of (4.21) is equal to

$$\|a_j\beta_{s_j}^{-1}\big(z(p_j)\big)-\beta_{s_j}^{-1}\big(z(p_j)\big)\|.$$

This is bounded by

$$\|a_j\big(\beta_{s_j}^{-1}\big(z(p_j)\big)-\beta_s^{-1}\big(z(p)\big)\big)\| \\ + \|a_j\beta_s^{-1}\big(z(p)\big)-\beta_s^{-1}\big(z(p)\big)\| + \|\beta_s^{-1}\big(z(p)\big)-\beta_{s_j}^{-1}\big(z(p_j)\big)\|$$

The middle term tends to zero since the a_j form an approximate identity. Notice that

$$\|\beta_{s_j}^{-1}\big(z(p_j)\big)-\beta_s^{-1}\big(z(p)\big)\| \le \|\beta_{s_j}^{-1}\big(z(p_j)-z(p)\big)\| + \|\beta_{s_j}^{-1}\big(z(p)\big)-\beta_s\big(z(p)\big)\|.$$

Since z is continuous, since β is strongly continuous and each β_s is isometric, and since $\|a_j\|\le 1$, it follows that the first and third terms tend to zero as well. This eventually contradicts (4.21). □

Lemma 4.8. *Suppose that* $\{ b_l \}$ *is an approximate identity for* $\operatorname{Ind}_H^P \beta$. *For* $z \in \mathsf{Z}_0$ *and* $c \in E_0$, *define* $b_l \cdot z(p) = b_l(p)z(p)$ *and* $b_l \cdot c(t,p) = b_l(p)c(t,p)$.

(a) *For each* $z \in \mathsf{Z}_0$, $b_l \cdot z \to z$ *in the inductive limit topology on* Z_0.

(b) *If* $c \in E_0$, *then* $b_l \cdot c \to c$ *in the inductive limit topology on* E_0.

Proof. Because $\operatorname{supp} b_l \cdot c \subset \operatorname{supp} c$, to prove part (b), it will suffice to see that given $\epsilon > 0$ there is a $l_0 \in L$ such that $l \geq l_0$ implies

$$\|b_l \cdot c(t,\cdot) - c(t,\cdot)\| < \epsilon \quad \text{for all } t \in K. \tag{4.22}$$

If (4.22) fails, then we can pass to a subnet and relabel so that the left-hand side of (4.22) is always greater than or equal to ϵ. Then there is a net $\{ t_l \} \subset \operatorname{supp}(t \mapsto c(t,\cdot))$ such that

$$\|b_l \cdot c(t_l,\cdot) - c(t_l,\cdot)\| \geq \epsilon. \tag{4.23}$$

Passing to a subnet and relabeling, we may assume that $t_l \to t$ in K. But the left-hand side of (4.23) is bounded by

$$\|b_l \cdot \bigl(c(t_l,\cdot) - c(t,\cdot)\bigr)\| + \|b_l \cdot c(t,\cdot) - c(t,\cdot)\| + \|c(t,\cdot) - c(t_l,\cdot)\|.$$

However, since $\|b_l\| \leq 1$ and since $t \mapsto c(t,\cdot)$ is a compactly supported continuous function of K into $\operatorname{Ind}_H^P \beta$ by Lemma 3.50 on page 102, we eventually contradict (4.23). This completes the proof of part (b).

To prove part (a), it suffices to show that given $\epsilon > 0$ there is a l_0 such that $l \geq l_0$ implies that

$$\|b_l(p)z(p) - z(p)\| < \epsilon \quad \text{for all } p \in P.$$

Since the H action on P is free and proper, it follows as in the proof of Lemma 3.57 on page 107, that there is a $w \in C_c^+(P)$ such that

$$\int_H w(p \cdot s)\, d\mu_H(s) = 1 \quad \text{for all } p \in \operatorname{supp} z.$$

Furthermore, there is a compact set C_H such that

$$\{ s \in H : p \in \operatorname{supp} z \text{ and } p \cdot s \in \operatorname{supp} w \} \subset C_H.$$

Lemma 4.7 on the facing page implies that there is an $a \in A$ such that for all $s \in C_H$ we have

$$\|\beta_s(a)z(p) - z(p)\| < \delta,$$

where $\delta = \min(\epsilon/3, \epsilon/(3\|z\|_\infty + 1))$. Define $b \in \operatorname{Ind}_H^P \beta$ by

$$b(p) = \int_H w(p \cdot s)\beta_s(a)\, d\mu_H(s).$$

Then

$$\|b(p)z(p) - z(p)\| = \Bigl\| \int_H w(p \cdot s)\bigl(\beta_s(a)z(p) - z(p)\bigr)\, d\mu_H(s) \Bigr\| \leq \delta.$$

Now if l_0 is such that $\|b_l b - b\| < \delta$ for $l \geq l_0$, then $l \geq l_0$ implies that

$$\begin{aligned}\|b_l(p)z(p) - z(p)\| &\leq \|b_l(p)\big(z(p) - b(p)z(p)\big)\| + \|\big(b_l(p)b(p) - b(p)\big)z(p)\| \\ &\qquad\qquad + \|b(p)z(p) - z(p)\| \\ &\leq \delta\|z\|_\infty + 2\delta.\end{aligned}$$

Our choice of δ implies this is less than ϵ. This completes the proof of part (a). □

Proof of Proposition 4.6 on page 118. We start with the left-action of E_0 on Z_0. Let $z \in \mathsf{Z}_0$.

Let U_0 be a pre-compact neighborhood of e in K. Notice that $U_0 \cdot \operatorname{supp} z$ is pre-compact in P. Let T_1 be a compact set containing the image of $U_0 \cdot \operatorname{supp} z$ in P/H.

Claim. There is a neighborhood U_1 of e in K such that $U_1 \subset U_0$ and such that $t \in U_1$ implies that

$$\|\alpha_t\big(z(t^{-1} \cdot p)\big) - z(p)\| < \epsilon \quad \text{for all } p \in P. \tag{4.24}$$

Proof of Claim. If the claim were false, then we could produce a net $\{\, t_i \,\} \subset K$ with $t_i \to e$ and $\{\, p_i \,\} \subset P$ such that

$$\|\alpha_{t_i}\big(z(t_i^{-1} \cdot p_i)\big) - z(p_i)\| \geq \epsilon. \tag{4.25}$$

But we must have $p_i \in U_0 \cdot \operatorname{supp} z$. Since the latter is pre-compact, we can pass to a subnet, relabel, and assume that $p_i \to p$, However, this would eventually contradict (4.25). This establishes the claim. □

By Lemma 4.8 on the preceding page, we can choose l_1 such that $l \geq l_1$ implies that

$$\|b_l(p)z(p) - z(p)\| < \epsilon \quad \text{for all } p \in P. \tag{4.26}$$

Now suppose that $(T, U, l, \epsilon) \geq (T_1, U_1, l_1, \epsilon)$, and let $e = e_{(T,U,l,\epsilon)}$. Then $e \cdot z(p) - z(p) = 0$ if $p \notin U \cdot \operatorname{supp} z$. Since $U \subset U_1 \subset U_0$, we must have $p \cdot H \in T_1 \subset T$ whenever $e \cdot z(p) - z(p) \neq 0$. Thus, if $e \cdot z(p) - z(p) \neq 0$, then we have

$$\begin{aligned}\|e \cdot z(p) - z(p)\| &\leq \Big\| \int_K e(t,p)\big(\alpha_t\big(z(t^{-1} \cdot p)\big) - z(p)\big)\, d\mu_K(t) \Big\| &(4.27)\\ &\qquad + \Big\| \Big(\int_K e(t,p)\, d\mu_K(t) - b_l(p) \Big) z(p) \Big\| &(4.28)\\ &\qquad + \|b_l(p)z(p) - z(p)\| &(4.29)\\ &\leq 4\epsilon + \|z\|_\infty \epsilon + \epsilon, &(4.30)\end{aligned}$$

where we used (b) and (4.24) to bound the right-hand side of (4.27), (c) to bound (4.28) and (4.26) to bound (4.29). Since $\operatorname{supp}(e \cdot z) \subset U_1 \cdot \operatorname{supp} z$, this completes the proof for the action on Z_0.

The proof for the left-action on E_0 proceeds similarly. Let $c \in E_0$. Then there are compact sets $C_0 \subset K$ and $T_0 \subset P/H$ such that

$$c(v, p) = 0 \quad \text{if } v \notin C_0 \text{ or } p \cdot H \notin T_0.$$

Let U_0 be a pre-compact neighborhood of e in K. Then $U_0 \cdot T_0$ is pre-compact in P/H and we let $T_1 := \overline{U_0 \cdot T_0}$. Similarly, let C_1 be the compact set $\overline{U_0 C_0}$ in K. Since the orbit map of P onto P/H is open, there is a compact set $D_1 \subset P$ such that T_1 is the image of D_1 — and of $D_1 H$ as well (Lemma 3.37 on page 98).

Claim. There is a neighborhood $U_1 \subset U_0$ such that $t \in U_1$ implies that

$$\|\alpha_t\big(c(t^{-1}v, t^{-1} \cdot p)\big) - c(v,p)\| < \epsilon \quad \text{for all } (v,p) \in K \times P.$$

Proof of Claim. If the claim were false, then we can find $\{ t_i \}$ converging to $e \in K$, $\{ p_i \} \subset D_1$ and $\{ v_i \} \subset C_1$ such that

$$\|\alpha_{t_i}\big(c(t_i^{-1}v_i, t_i^{-1} \cdot p_i)\big) - c(v_i,p_i)\| \geq \epsilon. \tag{4.31}$$

Since D_1 and C_1 are compact, we can pass to a subnet, and relabel, so that $v_i \to v$ and $p_i \to p$. This eventually contradicts (4.31). The claim follows. □

By Lemma 4.8 on page 119, there is a l_1 such that $l \geq l_1$ implies that

$$\|b_l(p)c(v,p) - c(v,p)\| < \epsilon \quad \text{for all } (v,p) \in K \times P.$$

Now if $(T,U,l,\epsilon) \geq (T_1,U_1,l_1,\epsilon)$ and if $e = e_{(T,U,l,\epsilon)}$, then $e * c(v,p) - c(v,p) \neq 0$ implies that $p \cdot H \in T_1 \subset T$. Therefore, if $e * c(v,p) - c(v,p) \neq 0$, then we have

$$\begin{aligned} \|e * c(v,p) - c(v,p)\| \leq \Big\| \int_K & e(t,p)\big(\alpha_t\big(c(t^{-1}v, t^{-1}\cdot p)\big) - c(v,p)\big)\, d\mu_K(t) \Big\| \\ &+ \Big\| \Big(\int_K e(t,p)\, d\mu_K(t) - b_l(p)\Big) c(v,p) \Big\| \\ &+ \|b_l(p)c(v,p) - c(v,p)\| \\ \leq 4\epsilon &+ \|c\|_\infty \epsilon + \epsilon. \end{aligned}$$

Since $\operatorname{supp}(e * c) \subset C_1 \times D_1 H$, this suffices. □

To clean up some calculations below, which would otherwise be marred by an annoying modular function, it will be convenient to have a slight variation on Proposition 4.6 on page 118. This approach was suggested by [148, p. 307].

Corollary 4.9. *Let $\{ \hat e_m \}_{m \in M}$ be a net as constructed in Proposition 4.6 on page 118. For each $m = (T,U,l,\epsilon)$, define*

$$e_m(t,p) := \Delta_K(t)^{-\frac12} \hat e(t,p).$$

Then $\{ e_m \}_{m \in M}$ satisfies (a) *and* (b) *of Proposition 4.5 on page 113.*

Proof. Let $\check e_m(t,p) := \hat e_m(t,p) - e_m(t,p) = \big(1 - \Delta_K(t)^{-\frac12}\big)\hat e_m(t,p)$. Let U_0 be a neighborhood of e in K such that $|1 - \Delta_K(t)^{-\frac12}| < \epsilon$ for all $t \in U_0$. Then if $U \subset U_0$ and if $m = (T,U,l,\epsilon)$, then using (b) of Proposition 4.6 on page 118, we have

$$\|\check e_m * c\|_\infty < 4\epsilon \|c\|_\infty \quad \text{and} \quad \|\check e_m \cdot z\|_\infty \leq 4\epsilon\|z\|_\infty$$

for all $c \in E_0$ and $z \in \mathsf{Z}_0$. Now the result follows easily from Proposition 4.6 on page 118. □

Lemma 4.10. *If K acts freely and properly on a locally compact space P and if N is a neighborhood of e in K, then each p in P has a neighborhood U such that*

$$\{\, t \in K : t \cdot U \cap U \neq \emptyset \,\} \subset N.$$

Proof. Let W be the interior of N. If the lemma is false, then there is a $p \in P$ such that for each neighborhood U of p there is a $t_U \in K \setminus W$ and a $p_U \in U$ such that $t_U \cdot p_U \in U$. Then $p_U \to p$ and $t_U \cdot p_U \to p$. Lemma 3.42 on page 99 implies that we can pass to a subnet, relabel, and assume that $t_U \to t \in K \setminus W$. In particular, $t \neq e$ and, since P is Hausdorff, $p = t \cdot p$. This contradicts the freeness of the action. □

With these preliminaries in hand, we can turn to the construction of the approximate identity we need. The basic ideas for these constructions goes back to [148].

Proof of Proposition 4.5 on page 113. Let $\{\, b_l \,\}_{l \in L}$ be an approximate identity for $\operatorname{Ind}_H^P(A, \beta)$. Fix a compact set $T \subset P/H$, a pre-compact open neighborhood U of e in K, $l \in L$ and $\epsilon > 0$. In view of Corollary 4.9 on the preceding page, it will suffice to produce a function $e = e_{(T,U,l,\epsilon)} \in E_0$ which is a sum of inner products, and such that $\tilde{e}(t,p) := \Delta_K(t)^{\frac{1}{2}} e(t,p)$ satisfies conditions (a), (b) and (c) of Proposition 4.6 on page 118.

Let $\delta < \min(\epsilon/3, 1/2)$. Since the orbit map of P onto P/H is open, there is a compact set $D \subset P$ such that $D \cdot H$ is the pre-image of T in P. Fix a compact neighborhood C of D. Let $\varphi \in C_c^+(P)$ be such that $\varphi(p) = 1$ for all $p \in C$. Then $z(p) := \varphi(p) b_l(p)^{\frac{1}{2}}$ defines an element of Z_0 such that $z(p) = b_l(p)^{\frac{1}{2}}$ for all $p \in C$.[1]

Claim. There is a neighborhood W of e in K such that $W \subset U$ and such that $t \in W$ implies that

$$\|z(p)\alpha_t\big(z(t^{-1} \cdot p)\big) - b_l(p)\| < \delta \quad \text{for all } p \in C. \tag{4.32}$$

Proof of Claim. If the claim were false, then there would a net $\{\, t_i \,\}$ converging to e in K and $\{\, p_i \,\} \subset C$ such that

$$\|z(p_i)\alpha_{t_i}\big(z(t_i^{-1} \cdot p_i)\big) - b_l(p_i)\| \geq \delta. \tag{4.33}$$

Since C is compact, we can pass to a subnet, relabel, and assume that $p_i \to p \in C$. But then $z(p_i)\alpha_{t_i}\big(z(t_i^{-1} \cdot p_i)\big)$ converges to $z(p)^2$, which equals $b_l(p)$ since $p \in C$. This eventually contradicts (4.33), and completes the proof of the claim. □

By Lemma 4.10, each point in D has an open neighborhood $V \subset C$ such that

$$\{\, t \in K : t \cdot V \cap V \neq \emptyset \,\} \subset W.$$

[1] To see that z is continuous, it suffices to see that $a \mapsto a^{\frac{1}{2}}$ is continuous on A^+. But this follows from the functional calculus as $t \mapsto \sqrt{t}$ can be uniformly approximated by polynomials on bounded subsets of $\mathbf{R}^+$.

Let $\{ V_i \}_{i=1}^n$ be a cover of D by such sets. By Lemma 1.43 on page 12, there are $h_i \in C_c^+(P)$ such that $\operatorname{supp} h_i \subset V_i$ and such that

$$\sum_i h_i(p) = 1 \quad \text{if } p \in D \text{, and} \quad \sum_i h_i(p) \le 1 \quad \text{otherwise.}$$

Define

$$H(p) := \sum_{i=1}^n \int_H h_i(p \cdot s) \, d\mu_H(s),$$

and notice that $H(p) > 0$ if $p \in D \cdot H$. Let

$$m := \inf_{p \in D \cdot H} H(p) = \inf_{p \in D} H(p).$$

Since D is compact, $m > 0$, and

$$G(p) := \max(H(p), m/2)$$

is a continuous nowhere vanishing function on P. Let

$$k_i(p) := h_i(p) G(p)^{-1}.$$

Then $k_i \in C_c^+(P)$, $\operatorname{supp} k_i \subset V_i$ and

$$\sum_{i=1}^n \int_H k_i(p \cdot s) \, d\mu_H(s) \begin{cases} = 1 & \text{if } p \in D \cdot H \text{, and} \\ \le 1 & \text{otherwise.} \end{cases}$$

Since the H-action is proper, there is compact set $C_H \subset H$ such that

$$\{ s \in H : C \cdot s \cap C \ne \emptyset \} \subset C_H.$$

By Lemma 3.57 on page 107, there are $g_i \in C_c^+(P)$ such that $\operatorname{supp} g_i \subset V_i$ and such that for all $p \in P$,

$$\Bigl| k_i(p) - g_i(p) \int_K g_i(t^{-1} \cdot p) \, d\mu_K(t) \Bigr| < \frac{\delta}{n \mu_H(C_H)}.$$

Since $\bigcup_{i=1}^n V_i \subset C$, we have for all $p \in C$,

$$\Bigl| \int_H \int_K g_i(p \cdot s) g_i(t^{-1} \cdot p \cdot s) \, d\mu_K(t) \, d\mu_H(s) - \int_H k_i(p \cdot s) \, d\mu_H(s) \Bigr| < \frac{\delta}{n}. \tag{4.34}$$

By left-invariance, (4.34) must hold for all $p \in C \cdot H$. But if $p \notin C \cdot H$, then the left-hand side of (4.34) is zero. Hence (4.34) holds for all p. Thus if we define

$$F(t, p) := \sum_{i=1}^n g_i(p) g_i(t^{-1} \cdot p), \tag{4.35}$$

then our choice of the V_i guarantees that

$$F(t, p) = 0 \quad \text{if } t \notin W \text{ or } p \notin C. \tag{4.36}$$

Our choice of the g_i implies that

$$\Bigl|\int_H\int_K F(t,p\cdot s)\,d\mu_K(t)\,d\mu_H(s)-\sum_{i=1}^n\int_H k_i(p\cdot s)\,d\mu_H(s)\Bigr|<\delta.$$

As a consequence, we have

$$\Bigl|\int_K\int_H F(t,p\cdot s)\,d\mu_H(s)\,d\mu_K(t)-1\Bigr|<\delta\quad\text{if } p\in D\cdot H,\tag{4.37}$$

and since $\delta<\frac{1}{2}$,

$$0\le\int_K\int_H F(t,p\cdot s)\,d\mu_H(s)\,d\mu_K(t)\le 1+\delta\le 2\quad\text{for all } p.\tag{4.38}$$

Define

$$f_i(p):=g_i(p)z(p).$$

Then compute

$$\begin{aligned}
e(t,p)&:=\sum_{i=1}^n {}_{E_0}\langle f_i\,,\,f_i\rangle(t,p)\\
&=\sum_{i=1}^n\Delta_K(t)^{-\frac12}\int_H\beta_s\bigl(f_i(p\cdot s)\alpha_t\bigl(f_i(t^{-1}\cdot p\cdot s)^*\bigr)\bigr)\,d\mu_H(s)\\
&=\Delta_K(t)^{-\frac12}\sum_{i=1}^n\int_H g_i(p\cdot s)g_i(t^{-1}\cdot p\cdot s)\\
&\qquad\qquad\beta_s\bigl(z(p\cdot s)\alpha_t\bigl(z(t^{-1}\cdot p\cdot s)\bigr)\bigr)\,d\mu_H(s)\\
&=\Delta_K(t)^{-\frac12}\int_H F(t,p\cdot s)\beta_s\bigl(\theta(t,p\cdot s)\bigr)\,d\mu_H(s),
\end{aligned}$$

where F is defined in (4.35) and we have let

$$\theta(t,p):=z(p)\alpha_t\bigl(z(t^{-1}\cdot p)\bigr)$$

to make some of the formulas below a bit easier to write. Notice that

$$\theta(e,p)=b_l(p)\quad\text{if } p\in C,\tag{4.39}$$

and our choice of W implies that

$$\|\theta(t,p)-\theta(e,p)\|=\|\theta(t,p)-b_l(p)\|<\delta\quad\text{if } t\in W\text{ and } p\in C.$$

Since $\|b_l(p)\|\le 1$ for all p,

$$\|\theta(t,p)\|\le 1+\delta\le 2\quad\text{if } t\in W\text{ and } p\in C.\tag{4.40}$$

To finish the proof, we need only check that $\tilde e$ satisfies properties (a), (b) and (c) of Proposition 4.6 on page 118, where $\tilde e(t,p):=\Delta_K(t)^{\frac12}e(t,p)$. But (4.36) implies that $\tilde e(t,p)=0$ if $t\notin W$, and since $W\subset U$, property (a) is satisfied.

If $p \in D \cdot H$, then

$$\int_K \|\tilde{e}(t,p)\| \, d\mu_H(s) \leq \int_K \int_H F(t, p\cdot s) \|\beta_s\big(\theta(t, p\cdot s)\big)\| \, d\mu_H(s) \, d\mu_K(t)$$

which, since $F(t, p \cdot s) = 0$ off $W \times C$, is

$$\leq 2 \int_K \int_H F(t, p\cdot s) \, d\mu_H(s) \, d\mu_K(t)$$
$$\leq 4,$$

where we used (4.38) for the final inequality. Thus property (b) is satisfied.

Similarly, if $p \in D \cdot H$, then

$$\Big\| \int_K \tilde{e}(t,p) \, d\mu_K(t) - b_l(p) \Big\|$$
$$= \Big\| \int_K \int_H F(t, p\cdot s)\beta_s\big(\theta(t, p\cdot s)\big) \, d\mu_H(s) \, d\mu_K(t) - b_l(p) \Big\|$$

which, since for any $s \in H$, we have $b_l(p) = \beta_s\big(b_l(p \cdot s)\big)$, is

$$\leq \Big\| \int_K \int_H F(t, p\cdot s)\beta_s\big(\theta(t, p\cdot s) - b_l(p \cdot s)\big) \, d\mu_H(s) \, d\mu_K(t) \Big\|$$
$$+ \Big| \int_K \int_H F(t, p\cdot s) \, d\mu_H(s) \, d\mu_K(t) - 1 \Big| \|b_l(p)\|.$$

Since β_s is isometric and $\|\theta(t, p\cdot s) - b_l(p\cdot s)\| < \delta$ if $F(t, p\cdot s) \neq 0$, the first of the integrals in the last expression is bounded by

$$\delta \int_K \int_H F(t, p\cdot s) \, d\mu_H(s) \, d\mu_K(t) \leq 2\delta.$$

Since $\|b_l(p)\| \leq 1$ and $p \in D \cdot H$, (4.37) implies that the second integral is bounded by δ. Since we've chosen δ so that $3\delta < \epsilon$, we've shown that property (c) holds. This completes the proof. □

4.2 Some Special Cases

There are a number of special cases of the Symmetric Imprimitivity Theorem that deserve special mention. All of these pre-date Theorem 4.1 [132]. The easiest to describe is the case where A is the one-dimensional algebra $\mathbf{C}$ of complex numbers. This version is due to Green and is proved in [148, Situation 10]. It follows immediately from Theorem 4.1 on page 110 together with the observation that $\operatorname{Ind}_H^P(\mathbf{C}, \mathrm{id})$ and $\operatorname{Ind}_K^P(\mathbf{C}, \mathrm{id})$ are easily identified with $C_0(P/H)$ and $C_0(K\backslash P)$, respectively. The K-action on $C_0(P/H)$ will be denoted by lt and the H-action on $C_0(K\backslash P)$ by rt.[2]

[2]When working with the actions on induced algebras, and in particular when working with situations derived from the Symmetric Imprimitivity Theorem, we often have to sort out a confusing

Corollary 4.11 (Green's Symmetric Imprimitivity Theorem). *Suppose that H and K are locally compact groups acting freely and properly on the right and left, respectively, of a locally compact space P. If the actions commute*[3]*, then*

$$C_0(P/H) \rtimes_{\mathrm{lt}} K \quad \textit{and} \quad C_0(K\backslash P) \rtimes_{\mathrm{rt}} H$$

are Morita equivalent via an imprimitivity bimodule Z which is the completion of $\mathsf{Z}_0 = C_c(P)$ equipped with actions and inner products given by

$$c \cdot f(p) = \int_K c(t, p \cdot H) f(t^{-1} \cdot p) \Delta_K(t)^{\frac{1}{2}} \, d\mu_K(t) \tag{4.41}$$

$$f \cdot b(p) = \int_H f(p \cdot s) b(s^{-1}, K \cdot p \cdot s) \Delta_H(s)^{-\frac{1}{2}} \, d\mu_H(s) \tag{4.42}$$

$${}_{E_0}\langle f\,, g\rangle(t, p \cdot H) = \Delta_K(t)^{-\frac{1}{2}} \int_H f(p \cdot s) \overline{g(t^{-1} \cdot p \cdot s)} \, d\mu_H(s) \tag{4.43}$$

$$\langle f\,, g\rangle_{B_0}(s, K \cdot p) = \Delta_H(s)^{-\frac{1}{2}} \int_K \overline{f(t^{-1} \cdot p)} g(t^{-1} \cdot p \cdot s) \, d\mu_K(t) \tag{4.44}$$

for all $c \in C_c(K \times P/H)$, $b \in C_c(H \times K\backslash P)$ and $f, g \in C_c(P)$.

Remark 4.12. By taking $H = \{ e \}$ in Corollary 4.11, we see that if K acts freely and properly on P then $C_0(P) \rtimes_{\mathrm{lt}} K$ is Morita equivalent to the abelian C^*-algebra $C_0(K\backslash P)$. In particular, $C_0(P) \rtimes_{\mathrm{lt}} K$ has continuous trace [139, Proposition 5.15] with trivial Dixmier-Douady class [139, Proposition 5.32]. In [65], Green showed that if K and P are second countable and if K acts freely on P, then $C_0(P) \rtimes_{\mathrm{lt}} K$ has continuous trace if and only if K acts properly. The non-free case is a bit more complicated and is treated in [170]. In particular, it is possible for a transformation group C^*-algebra to have continuous trace with a nontrivial Dixmier-Douady class [137, Example 4.6].

Example 4.13. There are a number of impressive applications of Green's Symmetric Imprimitivity Theorem. A nice source for many of these is Rieffel's expository article in [148]. There isn't space to recount all of his examples here, but one example deserves special mention. Namely, suppose that K and H are closed subgroups of a locally compact group G. Then we can let K act by left multiplication on G and let H act by right multiplication. Then we get immediately from Corollary 4.11 that $C_0(K\backslash G) \rtimes_{\mathrm{rt}} H$ is Morita equivalent to $C_0(G/H) \rtimes_{\mathrm{lt}} K$.

Example 4.14 (Rotation algebras revisited). Example 4.13 has an interesting application to rotation algebras. Let $G = \mathbf{R}$ and $K = \mathbf{Z}$. If $\alpha \in \mathbf{R}$, then let $H = \alpha\mathbf{Z}$. Since $t \mapsto e^{2\pi i t}$ induces the usual identification of $\mathbf{Z}\backslash\mathbf{R}$ with $\mathbf{T}$, it is not hard to see that $C_0(\mathbf{Z}\backslash\mathbf{R}) \rtimes_{\mathrm{rt}} \alpha\mathbf{Z}$ is isomorphic to the rotation algebra A_α corresponding to rotation by $e^{2\pi i\alpha}$. Similarly, since $t \mapsto e^{2\pi i \frac{t}{\alpha}}$ also induces an identification of

array of actions on spaces and actions on C^*-algebras. Thus we have adopted the convention that the dynamical systems associated to left group actions will be denoted by lt (for "left-translation") and right actions by rt (for "right translation"). I hope that the meaning will be clear from context, and that the total confusion will be minimized.

[3]That is, equation (4.1) holds.

$\mathbf{R}/\alpha\mathbf{Z}$ with $\mathbf{T}$, $C_0(\mathbf{R}/\alpha\mathbf{Z}) \rtimes_{\mathrm{lt}} \mathbf{Z}$ is isomorphic to $C_0(\mathbf{R}/\mathbf{Z}) \rtimes_{\mathrm{lt}} \frac{1}{\alpha}\mathbf{Z} \cong A_{\alpha^{-1}}$. Thus A_α is always Morita equivalent to $A_{\alpha^{-1}}$. This Morita equivalence was a bit of a mystery when it was first discovered by Rieffel in [146, 147], and plays a significant role in determining the Morita equivalence classes of the irrational rotation algebras described in Remark 2.60 on page 71.

Let X be a compact free $\mathbf{Z}_2$-space. Then the action of $\mathbf{Z}_2$ is certainly proper, and Corollary 4.11 on the preceding page and Example 4.13 on the facing page implies that $C(X) \rtimes_\alpha \mathbf{Z}_2$ is Morita equivalent to $C(\mathbf{Z}_2\backslash X)$. Moreover, Proposition 2.52 on page 67 shows that $C(X) \rtimes_\alpha \mathbf{Z}_2$ is a 2-homogeneous C^*-algebra, and it would be natural to guess that $C(X) \rtimes_\alpha \mathbf{Z}_2$ is isomorphic to $C(\mathbf{Z}_2\backslash X, M_2)$. However, this need not be the case. For example, let X be the n-sphere

$$S^n := \{ (x_0, \dots, x_n) \in \mathbf{R}^{n=1} : x_0^2 + x_1^2 + \cdots + x_n^2 = 1 \},$$

and let $\sigma : S^n \to S^n$ be the antipodal map $\sigma(\mathbf{x}) = -\mathbf{x}$. The orbit space $\mathbf{R}P^n := \mathbf{Z}_2\backslash S^n$ is known as *real projective n-space.*

Proposition 4.15 ([25, p. 25]). *Let $\alpha : \mathbf{Z}_2 \to \operatorname{Aut} C(S^n)$ be the dynamical system induced by the antipodal map on S^n (for $n \geq 2$). Then $C(S^n) \rtimes \mathbf{Z}_2$ is a 2-homogeneous C^*-algebra which is Morita equivalent to $C(\mathbf{R}P^n)$ but which is not isomorphic to $C(\mathbf{R}P^n, M_2)$.*

Proof. By Proposition 2.52 on page 67, $C(S^2) \rtimes_\alpha \mathbf{Z}_2$ is isomorphic to

$$A = \{ f \in C(S^2, M_2) : f(-\mathbf{x}) = W f(\mathbf{x}) W^* \},$$

where $W = \left(\begin{smallmatrix} 0 & 1 \\ 1 & 0 \end{smallmatrix}\right)$. Suppose to the contrary that there is an isomorphism $\Phi : A \to C(\mathbf{R}P^n, M_2)$. Since both these algebras have spectrum identified with $\mathbf{R}P^n$, and since Φ must induce a homeomorphism of the spectra, we can adjust Φ so that the induced map on $\mathbf{R}P^n$ is the identity. In view of the Dauns-Hofmann Theorem [139, Theorem A.34], this implies that Φ is a $C(\mathbf{R}P^n)$-linear map with respect to the natural actions of $C(\mathbf{R}P^n)$ on both algebras. Thus if $f, g \in A$ and if $f(\mathbf{x}) = g(\mathbf{x})$, then it follows that $\Phi(f)(\mathbf{Z}_2 \cdot \mathbf{x}) = \Phi(g)(\mathbf{Z}_2 \cdot \mathbf{x})$. Therefore we can define $\alpha : S^n \to \operatorname{Aut} M_2$ by

$$\alpha(\mathbf{x})\big(f(\mathbf{x})\big) := \Phi(f)(\mathbf{Z}_2 \cdot \mathbf{x}).$$

Claim. $\mathbf{x} \mapsto \alpha(\mathbf{x})$ is continuous from S^n to $\operatorname{Aut} M_2$, where $\operatorname{Aut} M_2$ has the point-norm topology.

Proof of Claim. We need to see that $\mathbf{x} \mapsto \alpha(\mathbf{x})(T)$ is continuous for each $T \in M_2$. Fix $\mathbf{y} \in S^n$ and $f \in A$ such that $f(\mathbf{y}) = T$. Then

$$\begin{aligned} \|\alpha(\mathbf{x})(T) - \alpha(\mathbf{y})(T)\| &= \|\alpha(\mathbf{x})(T) - \alpha(\mathbf{x})\big(f(\mathbf{x})\big)\| \\ &\qquad + \|\alpha(\mathbf{x})\big(f(\mathbf{x})\big) - \alpha(\mathbf{y})\big(f(\mathbf{y})\big)\| \\ &\leq \|T - f(\mathbf{x})\| + \|\Phi(f)(\mathbf{Z}_2 \cdot \mathbf{x}) - \Phi(f)(\mathbf{Z}_2 \cdot \mathbf{y})\|, \end{aligned}$$

which tends to 0 as $\mathbf{x} \to \mathbf{y}$. This proves the claim. □

Since $\mathbf{x}$ and $-\mathbf{x}$ are in the same $\mathbf{Z}_2$-orbit,

$$\alpha(\mathbf{x})\big(f(\mathbf{x})\big) = \alpha(-\mathbf{x})\big(f(-\mathbf{x})\big) = \alpha(-\mathbf{x})\big(Wf(\mathbf{x})W^*\big).$$

Thus $\alpha(\mathbf{x}) = \alpha(-\mathbf{x}) \circ \operatorname{Ad} W$.

As in [139, Chap. 1], we can identify $\operatorname{Aut} M_2$ with the projective unitary group PU_2 which is the quotient of the unitary group U_2 of unitary matrices by the scalar unitaries $\mathbf{T}I_2$. Let $\{\, \mathbf{e}_i \,\}_{i=0}^n$ be the standard orthonormal basis for $\mathbf{R}^{n+1}$. Since $n \geq 2$, we can define a map[4]

$$\psi : S^1 \to PU_2 \quad \text{by} \quad \psi(e^{i\theta}) = \alpha(\cos(\theta)\mathbf{e}_0 + \sin(\theta)\mathbf{e}_1).$$

Then

$$H(e^{i\theta}, t) := \alpha(t\cos(\theta)\mathbf{e}_0 + t\sin(\theta)\mathbf{e}_1 + (\sqrt{1-t^2})\mathbf{e}_2)$$

is a homotopy of ψ to a constant path. If

$$SU_2 := \{\, U \in U_2 : \det U = 1 \,\}$$

is the special unitary group, then the natural map $\operatorname{Ad} : SU_2 \to PU_2$ is a 2-sheeted covering map. Since ψ is null homotopic, the homotopy lifting theorem implies that there is a lift of ψ to a map

$$V : S^1 \to SU_2$$

such that $\psi(z) = \operatorname{Ad} V(z)$. Furthermore,

$$\begin{aligned}\psi(-e^{i\theta}) &= \alpha(-\cos(\theta)\mathbf{e}_0 - \sin(\theta)\mathbf{e}_1) \\ &= \alpha(\cos(\theta)\mathbf{e}_0 + \sin(\theta)\mathbf{e}_1) \circ \operatorname{Ad} W \\ &= \psi(e^{i\theta}) \circ \operatorname{Ad} W.\end{aligned}$$

Therefore

$$\begin{aligned}\operatorname{Ad} V(-z) &= \psi(-z) \\ &= \psi(z) \circ \operatorname{Ad} W \\ &= \operatorname{Ad} V(z)W.\end{aligned}$$

Thus there is a continuous function $\lambda : S^1 \to S^1$ such that

$$V(-z) = \lambda(z)V(z)W. \tag{4.45}$$

Since $\det W = -1$, taking determinants of both sides of (4.45) implies that $\lambda(z)^2 = -1$ for all z. Thus $z \mapsto \lambda(z)$ is constant — say, $\lambda(z) = \lambda$ for all z. But then,

$$V(1) = \lambda V(-1)W = \lambda\big(\lambda V(1)W\big)W = \lambda^2 V(1),$$

which is a contradiction. This completes the proof. □

---

[4]If $n = 2$, then we are simply restricting α to the equator. In general, we simply restricting α to a great circle passing through $\mathbf{e}_0$ and $-\mathbf{e}_0$.

Remark 4.16 ([65, Corollary 15]). It is interesting to note that the lack of triviality in Proposition 4.15 on page 127 is *less likely* to occur in examples where the fibres are infinite dimensional. That is, if G is an infinite second countable locally compact group acting freely and properly on a second countable locally compact space X, then there are fairly general conditions which force $C_0(X) \rtimes_{\mathrm{lt}} G$ to be isomorphic to $C_0\big(G\backslash X, \mathcal{K}\big(L^2(G)\big)\big)$. We don't have the technology at hand to give precise examples. Instead, we'll give a sketch. Since G acts freely and properly, $C_0(X) \rtimes_{\mathrm{lt}} G$ is Morita equivalent to $C_0(G\backslash X)$. Therefore $C_0(X) \rtimes_{\mathrm{lt}} G$ is isomorphic to the generalized compact operators $\mathcal{K}(\mathsf{X})$ on a suitable Hilbert $C_0(X)$-module [139, Proposition 3.8]. In fact, X consists of the sections of a fibre bundle $\mathcal{B}$ over $G\backslash X$ with fibre $\mathcal{K}\big(L^2(G)\big)$ (and structure group $PU\big(L^2(G)\big)$). If G is infinite, then $L^2(G)$ is infinite-dimensional, and the unitary group of $L^2(G)$ is contractible in the strong operator topology [139, Lemma 4.72]. It follows that if $\mathcal{B}$ is even locally trivial, then it must be trivial [139, Corollary 4.79]. It turns out that there are many cases where topological considerations imply that $\mathcal{B}$ must be locally trivial. For example, if $G\backslash X$ has finite covering dimension, then $\mathcal{B}$ is automatically locally trivial, and hence trivial, whenever $L^2(G)$ is infinite-dimensional [28, Proposition 10.8.7] (see also [65, Corollary 15]).

We turn now to another important corollary of the Symmetric Imprimitivity Theorem.

Corollary 4.17. *Suppose that H is a closed subgroup of a locally compact group G, and that (D, H, β) is dynamical system. Let $\mathrm{lt} : G \to \mathrm{Aut}\big(\mathrm{Ind}_H^G(D,\beta)\big)$ be defined by $\mathrm{lt}_r(f)(t) = f(r^{-1}t)$ as in Lemma 3.48 on page 101. View $E_0 := C_{cc}(G, \mathrm{Ind}_c\,\beta)$ and $B_0 := C_c(H,D)$ as dense subalgebras of $\mathrm{Ind}_H^G(D,\beta) \rtimes_{\mathrm{lt}} G$ and $D \rtimes_\beta H$, respectively. Let $\mathsf{Z}_0 = C_c(G,D)$. If $c \in E_0$, $f,g \in \mathsf{Z}_0$ and $b \in B_0$, then define*

$$c \cdot f(r) = \int_G c(t,r) f(t^{-1}r) \Delta_G(t)^{\frac{1}{2}}\, d\mu_G(t) \tag{4.46}$$

$$f \cdot b(r) = \int_H \beta_s^{-1}\big(f(rs^{-1})b(s)\big) \Delta_H(s)^{-\frac{1}{2}}\, d\mu_H(s) \tag{4.47}$$

$${}_{E_0}\langle f\,,\, g\rangle(t,r) = \Delta_G(t)^{-\frac{1}{2}} \int_H \beta_s\big(f(rs) g(t^{-1}rs)^*\big)\, d\mu_H(s) \tag{4.48}$$

$$\langle f\,,\, g\rangle_{B_0}(s) = \Delta_H(s)^{-\frac{1}{2}} \int_G f(t^{-1})^* \beta_s\big(g(t^{-1}s)\big)\, d\mu_G(t). \tag{4.49}$$

Then Z_0 is a $E_0 - B_0$-pre-imprimitivity bimodule. The completion Z is a $\mathrm{Ind}_H^G(D,\beta) \rtimes_{\mathrm{lt}} G - D \rtimes_\beta H$-imprimitivity bimodule, and $\mathrm{Ind}_H^G(D,\beta) \rtimes_{\mathrm{lt}} G$ is Morita equivalent to $D \rtimes_\beta H$.

Proof. To see that this situation is covered by the Symmetric Imprimitivity Theorem, let $\alpha : G \to \mathrm{Aut}\, D$ be the trivial system; that is, let $\alpha_t = \mathrm{id}_D$ for all t. Then $\mathrm{Ind}_G^G(D,\alpha)$ consists of constant functions on G and is easily identified with D via $f \mapsto f(e)$. This identification intertwines $\tau : H \to \mathrm{Aut}\big(\mathrm{Ind}_G^G(D,\alpha)\big)$ — defined by $\tau_s(f)(t) := \beta_s\big(f(ts)\big)$ — with β. Therefore we can identify $\mathrm{Ind}_G^G(D,\alpha) \rtimes_\tau H$ with $D \rtimes_\beta H$. The rest follows almost immediately from Theorem 4.1 on page 110 and the left-invariance of Haar measure. □

Combining the above corollary with Proposition 3.53 on page 105 we get another imprimitivity result due to Green.

Corollary 4.18 ([66, Theorem 17]). *Suppose that H is a closed subgroup of a locally compact group G and that (A, G, α) is a dynamical system. Assume that there is a continuous G-equivariant map $\varphi : \operatorname{Prim} A \to G/H$. Let*

$$I := \bigcap \{\, P \in \operatorname{Prim} A : \varphi(P) = eH \,\}.$$

If $D := A/I$ and if $\alpha^I : H \to \operatorname{Aut} D$ is the quotient system, then $A \rtimes_\alpha G$ is Morita equivalent to $D \rtimes_{\alpha^I} H$.

To illustrate some of the power of the above corollary, it is instructive to consider what it says for transformation group C^*-algebras.[5]

Corollary 4.19. *Let P be a locally compact G-space and H a closed subgroup of G. Suppose that $\sigma : P \to G/H$ is a G-equivariant continuous map. Let $Y := \sigma^{-1}(eH)$. Then Y is a H-space, and $C_0(P) \rtimes_{\mathrm{lt}} G$ is Morita equivalent to $C_0(Y) \rtimes_{\mathrm{lt}} H$.*

Example 4.20. Suppose that H is a closed subgroup of G, and that Y is a H-space. Let $P := G \times_H Y$ be the orbit space for $G \times Y$ with respect to the right H-action given by $(s, p) \cdot h := (sh, h^{-1} \cdot p)$. Then P is a left G-space, the action of r on the class of (s, p) is given by the class of (rs, p), and $(s, p) \mapsto sH$ induces a G-equivariant map of P onto G/H. It follows from Corollary 4.19 that $C_0(G \times_H Y) \rtimes_{\mathrm{lt}} G$ is Morita equivalent to $C_0(Y) \rtimes_{\mathrm{lt}} H$.

Example 4.21. In the previous example, we can let $Y = \mathbf{T}$, and let $\mathbf{Z}$ act by an irrational rotation θ. In this case, $(r, e^{2\pi i \alpha}) \mapsto (e^{2\pi i r}, e^{2\pi i(\alpha - \theta r)})$ is a homeomorphism of $\mathbf{R} \times_{\mathbf{Z}} \mathbf{T}$ onto $\mathbf{T}^2$ which intertwines the left $\mathbf{R}$ action on $\mathbf{R} \times_{\mathbf{Z}} \mathbf{T}$ with a flow on $\mathbf{T}^2$ along a line of irrational slope. Therefore the irrational rotation algebra $A_\theta = C(\mathbf{T}) \rtimes \mathbf{Z}$ is Morita equivalent to $C(\mathbf{T}^2) \rtimes \mathbf{R}$. Notice that A_θ is a simple *unital* NGCR algebra, while $C(\mathbf{T}^2) \rtimes \mathbf{R}$ is a simple *non-unital* NGCR algebra.

4.3 Green's Imprimitivity Theorem

Although the Symmetric Imprimitivity Theorem is an important tool in its own right, in this book we are most interested in the special case where H is a subgroup of K, $\beta : H \to \operatorname{Aut} A$ is trivial and $P = K$ with $t \cdot k = tk$ and $k \cdot h = kh$ for $t, k \in K$ and $h \in H$. Then we can identify $\operatorname{Ind}_H^K(A, \beta)$ with $C_0(K/H, A)$ by viewing elements of $C_0(K/H, A)$ as functions on K which are constant on orbits. Then $\sigma : K \to \operatorname{Aut} C_0(K/H, A)$ is given by

$$\sigma_t(\varphi)(kH) = \alpha_t\bigl(\varphi(t^{-1}kH)\bigr) \quad \text{for } t, k \in K \text{ and } \varphi \in C_0(K/H, A). \tag{4.50}$$

On the other hand, $f \mapsto f(e)$ is an isomorphism of $\operatorname{Ind}_K^K(A, \alpha)$ onto A which intertwines τ with $\alpha|_H$:

$$\tau_h(f)(e) = \beta_h\bigl(f(h)\bigr) = f(h) = \alpha_h\bigl(f(e)\bigr).$$

[5]Corollary 4.19 can also be derived directly from Green's Symmetric Imprimitivity Theorem [148, Situation 7]. Examples 4.20 and 4.21 also come from [148].

Therefore the Symmetric Imprimitivity Theorem implies that $\mathsf{Z}_0 = C_c(K, A)$ completes to a $C_0(K/H, A) \rtimes_\sigma K - A \rtimes_{\alpha|_H} H$-imprimitivity bimodule. The dense $*$-subalgebras E_0 and B_0 are given by

$$C_c(K \times K/H, A) \quad \text{and} \quad C_c(H, A),$$

respectively. The module actions and inner products on Z_0 are given by

$$F \cdot f(s) := \int_G F(r, sH)\alpha_r\big(f(r^{-1}s)\big)\Delta_G(r)^{\frac{1}{2}}\, d\mu_G(r) \tag{4.51}$$

$$f \cdot b(s) := \int_H f(sh)\alpha_{sh}\big(b(h^{-1})\big)\Delta_H(h)^{-\frac{1}{2}}\, d\mu_H(h) \tag{4.52}$$

$$\langle f\,,\, g\rangle_{B_0}(h) := \Delta_H(h)^{-\frac{1}{2}} \int_G \alpha_s\big(f(s^{-1})^* g(s^{-1}h)\big)\, d\mu_G(s) \tag{4.53}$$

$${}_{E_0}\langle f\,,\, g\rangle(s, rH) := \Delta_G(s)^{-\frac{1}{2}} \int_H f(rh)\alpha_s\big(g(s^{-1}rh)^*\big)\, d\mu_H(h). \tag{4.54}$$

However, the modular functions decorating (4.51)–(4.54) do not quite agree with those in treatments coming from Rieffel's original formulas (in the case $A = \mathbf{C}$) (cf., [145] or [139, Theorem C.23]). This is a minor irritation, and it not so surprising as there are lots of bimodules giving a Morita equivalence between $C_0(K/H, A) \rtimes_\sigma K$ and $A \rtimes_{\alpha|_H} H$. In fact, there will even be several pre-imprimitivity bimodule structures on $C_c(K, A)$. To explain the different formulas in the literature, we want to show how to derive additional structures on $C_c(K, A)$, and in particular, we want to derive the formulas we prefer to work with in Section 5.1. We start with some general observations.

Let X_0 be a $E_0 - B_0$-pre-imprimitivity bimodule. Let u be any vector space isomorphism of the vector space X_0 onto itself. Then we can equip X_0 with an new $E_0 - B_0$-pre-imprimitivity bimodule structure as follows. For $f, g \in \mathsf{X}_0$, $F \in E_0$ and $b \in B_0$, let

$$\begin{aligned} F : f &:= u^{-1}\big(F \cdot u(f)\big) \\ f : b &:= u^{-1}\big(u(f) \cdot b\big) \\ \langle\!\langle f\,,\, g\rangle\!\rangle_{B_0} &:= \big\langle u(f)\,,\, u(g)\big\rangle_{B_0} \\ {}_{E_0}\langle\!\langle f\,,\, g\rangle\!\rangle &:= {}_{E_0}\big\langle u(f)\,,\, u(g)\big\rangle. \end{aligned}$$

It is routine to see that X_0 is an $E_0 - B_0$-bimodule with these new operations. For example,

$$\begin{aligned} (F * G) : f &= u^{-1}\big(F * G \cdot u(f)\big) \\ &= u^{-1}\big(F \cdot \big(G \cdot u(f)\big)\big) \\ &= u^{-1}\big(F \cdot u\big(u^{-1}\big(G \cdot u(f)\big)\big)\big) \\ &= F : u^{-1}\big(G \cdot u(f)\big) \\ &= F : (G : f). \end{aligned}$$

Proving that $\langle\!\langle \cdot\,,\,\cdot \rangle\!\rangle_{B_0}$ and ${}_{E_0}\langle\!\langle \cdot\,,\,\cdot \rangle\!\rangle$ are pre-inner products is also straightforward. For example,

$$\begin{aligned}\langle\!\langle x\,,\,y : b \rangle\!\rangle_{B_0} &= \big\langle u(x)\,,\,u(y:b)\big\rangle_{B_0}\\ &= \big\langle u(x)\,,\,u(y)\cdot b\big\rangle_{B_0}\\ &= \langle\!\langle x\,,\,y \rangle\!\rangle_{B_0} b.\end{aligned}$$

We claim that X_0 is an $E_0 - B_0$-pre-imprimitivity bimodule. As above, this follows from easy computations. For example, to see that the E_0-action is bounded:

$$\begin{aligned}\langle\!\langle a:x\,,\,a:x \rangle\!\rangle_{B_0} &= \big\langle u(a:x)\,,\,u(a:x)\big\rangle_{B_0}\\ &= \big\langle a\cdot u(x)\,,\,a\cdot u(x)\big\rangle_{B_0}\\ &\leq \|a\|^2 \langle\!\langle x\,,\,x \rangle\!\rangle_{B_0}.\end{aligned}$$

Using such calculations, the claim follows.

To obtain the particular version we want — called Green's Imprimitivity Theorem — we define $u : \mathsf{Z}_0 \to \mathsf{Z}_0$ by $u(f)(k) := \Delta_K(k)^{-\frac{1}{2}} f(k)$. Then we obtain formulas compatible with those in [38, 145, 170], rather than (4.51)–(4.54) which are compatible with the formulas in [39, 79, 132, 148]. To emphasize that we've altered the formulas in Theorem 4.1 on page 110, we replace Z_0 by X_0. To make the formulas easier to use in Section 5.1, we also replace the group K by G. Then the next result is a corollary of the Symmetric Imprimitivity Theorem and the above discussion.

Theorem 4.22 (Green's Imprimitivity Theorem). *Suppose that (A, G, α) is a dynamical system, that H is a closed subgroup of G and that $\sigma = \mathrm{lt}\otimes\alpha$ is the diagonal action of G on $C_0(G/H, A)$ defined in (4.50). For notational convenience, let $\gamma_H(s) := \Delta_G(s)^{\frac{1}{2}}\Delta_H(s)^{-\frac{1}{2}}$. View $E_0 := C_c(G \times G/H, A)$ and $B_0 := C_c(H, A)$ as $*$-subalgebras of $C_0(G/H, A) \rtimes_\sigma G$ and $A \rtimes_{\alpha|_H} H$, respectively. Let $\mathsf{X}_0 := C_c(G, A)$. If $c \in E_0$, $f, g \in \mathsf{X}_0$ and $b \in B_0$, then define*

$$c\cdot f(s) := \int_G c(r, sH)\alpha_r\big(f(r^{-1}s)\big)\,d\mu_G(r) \tag{4.55}$$

$$f\cdot b(s) := \int_H f(sh)\alpha_{sh}\big(b(h^{-1})\big)\gamma_H(h)\,d\mu_H(h) \tag{4.56}$$

$$\langle f\,,\,g\rangle_{B_0}(h) := \gamma_H(h)\int_G \alpha_s^{-1}\big(f(s)^*g(sh)\big)\,d\mu_G(s) \tag{4.57}$$

$${}_{E_0}\langle f\,,\,g\rangle(s, rH) := \int_H f(rh)\alpha_s\big(g(s^{-1}rh)^*\big)\Delta(s^{-1}rh)\,d\mu_H(h). \tag{4.58}$$

Then X_0 is an $E_0 - B_0$-pre-imprimitivity bimodule. The completion $\mathsf{X} := \mathsf{X}_H^G$ is a $C_0(G/H, A) \rtimes_\sigma G - A \rtimes_{\alpha|_H} H$-imprimitivity bimodule, and $C_0(G/H, A) \rtimes_\sigma G$ is Morita equivalent to $A \rtimes_{\alpha|_H} H$.

Furthermore, in the proof of Theorem 4.1 on page 110, we introduced a covariant homomorphism (M, v) of $\big(C_0(G/H, A), G, \sigma\big)$ into $\mathcal{L}(\mathsf{Z}_H^G)$. The formulas for M and v on Z_0 are given by (4.18) and (4.17), respectively. However, if we're working with X_0, and therefore with the formulas (4.55)–(4.58), the formula for v loses the modular function. In particular, we have the following Corollary of the proof of Theorems 4.1 on page 110 and 4.22 on the facing page. Note that as usual, we are suppressing the map $q : \mathsf{X}_0 \to \mathsf{X}$.

Corollary 4.23. *Let* $\mathsf{X} = \mathsf{X}_H^G$ *be Green's* $C_0(G/H, A) \rtimes_\sigma G - A \rtimes_{\alpha|_H} H$*-imprimitivity bimodule defined in Theorem 4.22 on the preceding page, and let* $\mathsf{X}_0 = C_c(G, A) \subset \mathsf{X}_H^G$. *Then there is a nondegenerate covariant homomorphism* (M, v) *of* $\big(C_0(G/H, A), G, \mathrm{lt} \otimes \alpha\big)$ *into* $\mathcal{L}(\mathsf{X})$ *such that for each* $c \in C_c(G \times G/H, A) \subset C_0(G/H, A) \rtimes_\sigma G$ *and* $f \in \mathsf{X}_0$ *we have*

$$v_s(f)(r) = \alpha_s\big(f(s^{-1}r)\big) \qquad M(\varphi)f(r) = \varphi(rH)f(r)$$
$$M \rtimes v(c)(f) = c \cdot f.$$

Furthermore, if $k_A : A \to M\big(C_0(G/H, A)\big)$ *is the natural map sending* a *to the corresponding constant function and if* $N := \bar{M} \circ k_A$, *then* (N, v) *is a nondegenerate covariant homomorphism of* (A, G, α) *into* $\mathcal{L}(\mathsf{X})$ *such that if* $g \in C_c(G, A) \subset A \rtimes_\alpha G$ *and* $f \in \mathsf{X}_0$, *then*

$$N(a)f(r) = af(r)$$
$$N \rtimes v(g)(f) = g * f.$$

Proof. The assertions about $M \rtimes v$ were detailed in the proof of Theorem 4.1 on page 110. It is straightforward to check that (N, v) is covariant, and the proofs that $N \rtimes v(g)(f) = g * f$ and that N is nondegenerate are nearly identical to those for $M \rtimes v$ and M. □

4.4 The Stone-von Neumann Theorem

With Theorem 4.22 on the preceding page in hand, we can give a fairly short proof of a classical result which, in the language of dynamical systems, describes the crossed product $C_0(G) \rtimes_{\mathrm{lt}} G$ where G acts on itself by left-translation. We'll give a more classical formulation as well. (This material is taken from [139, §C.6].)

Theorem 4.24 (Stone-von Neumann). *Suppose that* G *is a locally compact group. Then*

$$C_0(G) \rtimes_{\mathrm{lt}} G \cong \mathcal{K}\big(L^2(G)\big).$$

In particular, if (M, λ) *is the natural covariant representation of* $\big(C_0(G), G, \mathrm{lt}\big)$ *on* $L^2(G)$ *given in Example 2.12 on page 45, then* $M \rtimes \lambda$ *is a faithful irreducible representation of* $C_0(G) \rtimes_{\mathrm{lt}} G$ *onto* $\mathcal{K}\big(L^2(G)\big)$.

Proof. The Imprimitivity Theorem implies that $C_0(G) \rtimes_{\mathrm{lt}} G$ is Morita equivalent to $\mathbf{C}$. This implies $C_0(G) \rtimes_{\mathrm{lt}} G$ is simple, and that $M \rtimes \lambda$ is faithful. Since $\mathcal{K}\big(L^2(G)\big)$ is an irreducible subalgebra of $B(\mathcal{H})$, it will suffice to see that $M \rtimes \lambda$ maps $C_0(G) \rtimes_{\mathrm{lt}} G$ onto $\mathcal{K}\big(L^2(G)\big)$.

If $K \in C_c(G \times G)$, then

$$f_K(r,s) := \Delta(r^{-1}s)K(s, r^{-1}s) \tag{4.59}$$

defines an element in $C_c(G \times G)$, and if $h, k \in C_c(G) \subset L^2(G)$, then

$$\begin{aligned}
\big(M \rtimes \lambda(f_K)h \mid k\big) &= \int_G \big(M\big(f_K(r,\cdot)\big)\lambda(r)h \mid k\big)_{L^2(G)}\, d\mu(r) \\
&= \int_G \int_G f_K(r,s)h(r^{-1}s)\overline{k(s)}\, d\mu(s)\, d\mu(r) \\
&= \int_G \int_G \Delta(r^{-1}s)K(s,r^{-1}s)h(r^{-1}s)\overline{k(s)}\, d\mu(r)\, d\mu(s) \\
&= \int_G \int_G \Delta(r^{-1})K(s,r^{-1})h(r^{-1})\overline{k(s)}\, d\mu(r)\, d\mu(s) \\
&= \int_G \int_G K(s,r)h(r)\, d\mu(r)\overline{k(s)}\, d\mu(s).
\end{aligned}$$

Thus

$$M \rtimes \lambda(f_K)h(s) = \int_G K(s,r)h(r)\, d\mu(r),$$

and $M \rtimes \lambda(f_K)$ is a Hilbert-Schmidt operator with kernel $K \in C_c(G \times G) \subset L^2(G \times G)$ [33, Chap. XI §6]. In particular, $M \rtimes \lambda(f_K)$ is a compact operator. If $g, h \in C_c(G) \subset L^2(G)$ and $K(r,s) = g(r)\overline{h(s)}$, then $M \rtimes \lambda(f_K)$ is the rank-one operator $\theta_{g,h}$. Since $C_c(G)$ is dense in $L^2(G)$, it follows that $\mathcal{K}\big(L^2(G)\big) \subset M \rtimes \lambda\big(C_0(G) \rtimes_{\mathrm{lt}} G\big)$.

If $f \in C_c(G \times G)$, then we can set

$$K(r,s) := \Delta(s^{-1})f(rs^{-1}, r).$$

Then $f_K = f$, and it follows that $M \rtimes \lambda\big(C_c(G \times G)\big) \subset \mathcal{K}\big(L^2(G)\big)$. Since $C_c(G \times G)$ is dense in $C_0(G) \rtimes_{\mathrm{lt}} G$, we're done. □

Example 4.25. Let $[-\infty, \infty]$ be the extended reals which are obtained from $\mathbf{R} = (\infty, \infty)$ by adjoining points at $+\infty$ and $-\infty$. We let $\mathbf{R}$ act on $[-\infty, \infty]$ by fixing the endpoints and by translation in the interior. Then $U := (-\infty, \infty)$ is a $\mathbf{R}$-invariant open subset, and the complement F consists of two fixed-points. Theorem 4.24 on the previous page implies that $C_0(U) \rtimes_{\mathrm{lt}} \mathbf{R}$ is isomorphic to the compacts $\mathcal{K}\big(L^2(\mathbf{R})\big)$. Of course, $C(F) \rtimes_{\mathrm{lt}} \mathbf{R} \cong C^*(\mathbf{R}) \oplus C^*(\mathbf{R})$ which is isomorphic to $C_0(\mathbf{R}) \oplus C_0(\mathbf{R})$ by Proposition 3.1 on page 82. Therefore Corollary 3.20 on page 94 implies that we have an exact sequence

$$0 \longrightarrow \mathcal{K}\big(L^2(\mathbf{R})\big) \longrightarrow C_0([-\infty,\infty]) \rtimes_{\mathrm{lt}} \mathbf{R} \longrightarrow C_0(\mathbf{R}) \oplus C_0(\mathbf{R}) \longrightarrow 0.$$

That is, $C_0([-\infty,\infty]) \rtimes_{\mathrm{lt}} \mathbf{R}$ is an extension of the compacts by a commutative C^*-algebra. In general, the collection of (equivalence classes) of extensions of the compacts by a commutative C^*-algebra $C_0(X)$ is an abelian group called $\operatorname{Ext}(X)$. A nice starter reference for $\operatorname{Ext}(X)$ is [21, Chap. IX]. Considerably more detail can be found in [8, Chap. VII]. In [65, §1 & §2], Green studies a variety of transformation group C^*-algebras which, as in this example, turn out to be extensions of the compact by commutative C^*-algebras and calculates their Ext classes.

To give a classical reformulation of Theorem 4.24 on page 133, we assume that G is abelian. Let $\widehat{G}$ be the dual of G, and recall that the Pontryagin Duality Theorem [56, Theorem 4.31] allows us to identify G with the dual of $\widehat{G}$. If $\hat{\mu}$ is a Haar measure on $\widehat{G}$, then the Fourier transform

$$F : C_c(\widehat{G}) \to C_0(G)$$

defined by

$$F(\varphi)(s) = \int_{\widehat{G}} \varphi(\gamma)\gamma(s)\, d\hat{\mu}(\gamma)$$

extends to an isomorphism of $C^*(\widehat{G})$ onto $C_0(G)$. Let $C^b_s(G) = M_s\big(C_0(G)\big)$ be $C^b(G)$ equipped with the strict topology. Then if $\varphi \in C_c(\widehat{G})$, $\gamma \mapsto \varphi(\gamma)\gamma$ is in $C_c\big(G, C^b_s(G)\big)$ and we can form the $C^b(G)$-valued integral

$$\int_{\widehat{G}} \varphi(\gamma)\gamma\, d\hat{\mu}(\gamma). \tag{4.60}$$

Since evaluation at $s \in G$ is a continuous functional on $C^b(G)$, (4.60) is equal to $F(\varphi)$. If π is a representation of $C_0(G)$, then we can define a unitary representation S^π of $\widehat{G}$ on $\mathcal{H}_\pi$ by

$$S^\pi_\gamma := \bar{\pi}(\gamma),$$

and then

$$\pi\big(F(\varphi)\big) = \int_{\widehat{G}} \varphi(\gamma) S^\pi_\gamma\, d\hat{\mu}(\gamma). \tag{4.61}$$

It follows that there is a one-to-one correspondence between representations π of $C_0(G)$ and unitary representations S^π of $\widehat{G}$. Since it is straightforward to see that $\bar{F}\big(i_{\widehat{G}}(\gamma)\big) = \gamma$, this correspondence is the usual one between representations of $C^*(\widehat{G})$ and unitary representations of $\widehat{G}$ combined with the isomorphism F.

Definition 4.26. A pair (R, S) consisting of unitary representations $R : G \to U(\mathcal{H})$ and $S : \widehat{G} \to U(\mathcal{H})$ is called a *Heisenberg representation* if

$$S_\gamma R_s = \gamma(s) R_s S_\gamma \quad \text{for all } s \in G \text{ and } \gamma \in \widehat{G}.$$

Example 4.27. Let $\mathcal{H} = L^2(G)$. Define $V : \widehat{G} \to U\big(L^2(G)\big)$ by $V_\gamma h(r) = \gamma(r)h(r)$. Then if $\lambda : G \to U\big(L^2(G)\big)$ is the left-regular representation, then (λ, V) is a Heisenberg representation of G called the *Schrödinger representation.*

Let $\varphi \in C_c(\widehat{G})$ and $s \in G$. Then we can define $s \cdot \varphi(\gamma) \in C_c(G)$ by $s \cdot \varphi(\gamma) := \overline{\gamma(s)}\varphi(\gamma)$. Then

$$\begin{aligned}\operatorname{lt}_r\big(F(\varphi)\big)(s) &= F(\varphi)(r^{-1}s)\\ &= \int_{\widehat{G}} \varphi(\gamma)\gamma(r^{-1}s)\,d\hat{\mu}(\gamma)\\ &= \int_{\widehat{G}} (r\cdot\varphi)(\gamma)\gamma(s)\,d\hat{\mu}(\gamma);\end{aligned}$$

thus

$$\operatorname{lt}_r\big(F(\varphi)\big) = F(r\cdot\varphi). \tag{4.62}$$

Lemma 4.28. *Suppose R and S are unitary representations of G and $\widehat{G}$, respectively, on $\mathcal{H}$. Let π be the representation of $C_0(G)$ corresponding to S via (4.61). Then (R, π) is a nondegenerate covariant representation of $\big(C_0(G), G, \operatorname{lt}\big)$ if and only if (R, S) is a Heisenberg representation of G.*

Proof. Suppose (R, S) is a Heisenberg representation of G. Then

$$\begin{aligned}R_s\pi\big(F(\varphi)\big) &= R_s\int_{\widehat{G}} \varphi(\gamma)S_\gamma\,d\hat{\mu}(\gamma)\\ &= \int_{\widehat{G}} \varphi(\gamma)\overline{\gamma(s)}S_\gamma\,d\hat{\mu}(\gamma)R_s\\ &= \int_{\widehat{G}} s\cdot\varphi(\gamma)S_\gamma\,d\hat{\mu}(\gamma)R_s\end{aligned}$$

which, using (4.62), is

$$= \pi\big(\operatorname{lt}_s\big(F(\varphi)\big)\big)R_s.$$

Thus, (π, R) is covariant.

Now suppose (π, R) is a nondegenerate covariant representation. Observe that

$$\begin{aligned}F\big(i_{\widehat{G}}(\gamma)(s\cdot\varphi)\big)(r) &= \int_{\widehat{G}} (s\cdot\varphi)(\bar{\gamma}\sigma)\sigma(r)\,d\hat{\mu}(\sigma)\\ &= \int_{\widehat{G}} \gamma(s)\overline{\sigma(s)}\varphi(\bar{\gamma}\sigma)\sigma(r)\,d\hat{\mu}(\sigma)\\ &= \gamma(s)\int_{\widehat{G}} \varphi(\bar{\gamma}\sigma)\sigma(s^{-1}r)\,d\hat{\mu}(\sigma)\\ &= \gamma(s)F\big(i_{\widehat{G}}(\gamma)\varphi\big)(s^{-1}r)\\ &= \gamma(s)\operatorname{lt}_s\big(F\big(i_{\widehat{G}}(\gamma)\varphi\big)\big)(r).\end{aligned}$$

Therefore

$$\begin{aligned} S_\gamma R_s \pi\bigl(F(\varphi)\bigr) &= S_\gamma \pi\bigl(\operatorname{lt}_s\bigl(F(\varphi)\bigr)\bigr) R_s \\ &= S_\gamma \pi\bigl(F(s\cdot\varphi)\bigr) R_s \\ &= \pi\bigl(F\bigl(i_{\widehat{G}}(\gamma)(s\cdot\varphi)\bigr)\bigr) R_s \\ &= \gamma(s)\pi\bigl(\operatorname{lt}_s\bigl(F(i_{\widehat{G}}(\gamma)\varphi)\bigr)\bigr) R_s \\ &= \gamma(s) R_s \pi\bigl(F\bigl(i_{\widehat{G}}(\gamma)\varphi\bigr)\bigr) \\ &= \gamma(s) R_s S_\gamma \pi\bigl(F(\varphi)\bigr). \end{aligned}$$

Since π is nondegenerate, $S_\gamma R_s = \gamma(s) R_s S_\gamma$ and (R, S) is a Heisenberg representation. □

Theorem 4.29 (von Neumann Uniqueness Theorem). *Suppose that G is a locally compact abelian group. Then every Heisenberg representation of G is equivalent to a direct sum of copies of the Schrödinger representation of G on $L^2(G)$.*

Proof. Let (R, S) be a Heisenberg representation of G on $\mathcal{H}$. Let (π, R) be the corresponding nondegenerate covariant representation. Then any invariant subspace for $\pi \rtimes R$ is invariant for both π and R, and therefore for (R, S) (Proposition 2.40 on page 59). However, $\pi \rtimes R$ is a nondegenerate representation of $C_0(G) \rtimes_{\mathrm{lt}} G \cong \mathcal{K}\bigl(L^2(G)\bigr)$, and is equivalent to a direct sum of copies of $M \rtimes \lambda$. Since M clearly corresponds to V, this completes the proof. □

4.5 Transitive Transformation Groups

Let H be a closed subgroup of a locally compact group G. Green's Imprimitivity Theorem on page 132 implies that $\mathsf{X}_0 := C_c(G)$ is a $B_0 - E_0$-pre-imprimitivity bimodule where $E_0 := C_c(G \times G/H)$ is viewed a subalgebra of $C_0(G/H) \rtimes_{\mathrm{lt}} G$ and $B_0 = C_c(H)$ is viewed as a $*$-subalgebra of $C^*(H)$. In particular, we can complete X_0 to a $C_0(G/H) \rtimes_{\mathrm{lt}} G - C^*(H)$-imprimitivity bimodule X, and $C_0(G/H) \rtimes_{\mathrm{lt}} G$ is Morita equivalent to $C^*(H)$. If, say, G is second countable, then the Brown-Green-Rieffel Theorem [139, Theorem 5.55] applies, and $C_0(G/H) \rtimes_{\mathrm{lt}} G$ is stably isomorphic to $C^*(H)$. Our object here is to give a considerable sharpening of this result due to Green [67]. Green's argument is a bit surprising as it relies on factoring G as a product of G/H and H which is usually impossible to do topologically. Therefore, we can do this only at the expense of introducing measurable functions in place of the continuous functions with which we are considerably more comfortable. This will force us to deal with some issues about measurability below.

As our result involves the compact operators on $L^2(G/H)$, and our first bit of measure theory will concern the existence of appropriate measures on G/H. Note that G/H is always a locally compact Hausdorff space in the quotient topology. If H isn't normal, then G/H is not a group and need not possess a measure which is invariant for the left action of G on G/H. If σ is a Borel measure on G/H, then for each $r \in G$, we get a new measure $r \cdot \sigma$ on G/H defined by $r \cdot \sigma(E) := \sigma(r^{-1} \cdot E)$. If $r \cdot \sigma$ is mutually absolutely continuous with σ for all $r \in G$, then

σ is called *quasi-invariant*. In general, G/H always has a quasi-invariant measure (for example, see [139, Lemma C.2] or [56, Theorem 2.56 and Proposition 2.54]).[6] As in [56, 139], we will work with quasi-invariant measures associated to continuous functions $\rho : G \to (0, \infty)$ such that

$$\rho(sh) = \gamma_H(h)^{-2}\rho(s) = \frac{\Delta_H(h)}{\Delta_G(h)}\rho(s) \quad \text{for all } s \in G \text{ and } h \in H.$$

Such functions are constructed in Appendix H.2. Given any such ρ, there is a quasi-invariant Borel measure $\mu_{G/H}$ on G/H such that for all $f \in C_c(G)$ we have

$$\int_G f(s)\rho(s)\,d\mu_G(s) = \int_{G/H}\int_H f(sh)\,d\mu_H(h)\,d\mu_{G/H}(sH). \tag{4.63}$$

Theorem 4.30 (Green). *Suppose that H is a closed subgroup of a locally compact group G. Then the imprimitivity algebra $C_0(G/H) \rtimes_{\mathrm{lt}} G$ is isomorphic to $C^*(H) \otimes \mathcal{K}(L^2(G/H, \mu_{G/H}))$ where $\mu_{G/H}$ is a quasi-invariant measure satisfying* (4.63).

The proof is considerably simpler if G is second countable. A key ingredient in the proof is a suitable *cross section* s for the natural map $q : G \to G/H$. Even in elementary examples — such as $G = \mathbf{R}$ and $H = \mathbf{Z}$ — the best we can hope for is a Borel map $c : G/H \to G$. But if G is not second countable, even a Borel map will be difficult to deal with. As a result, we'll need a short detour if we want the theorem for general groups. Anyone smart enough to settle for a proof of Theorem 4.30 in the case G is second countable can skim this material and skip ahead to Section 4.5.2 on page 143.

Remark 4.31. If (G, X) is a second countable locally compact transformation group and if G acts transitively on X — that is, if given $x, y \in X$ there is a $s \in G$ such that $y = s \cdot x$, then it will follow from the Mackey-Glimm Dichotomy (Theorem 6.2 on page 173) that X is equivariantly homeomorphic to G/H for some closed subgroup H and Theorem 4.30 applies. If G is not second countable, then the situation can be more complicated. For example if G is a connected second countable *abelian* group such as $\mathbf{R}$, and if we let G_d be the group G equipped with the discrete topology, then it can be shown that $C_0(G) \rtimes_{\mathrm{lt}} G_d$ is a simple NGCR algebra.[7]

4.5.1 Baire Sets

If G is second countable, then it is well known (e.g., [104, Lemma 1.1] or [2, Theorem 3.4.1]) that there are Borel cross sections $c : G/H \to G$ such that $p\big(c(rH)\big) = rH$ for all $r \in G$ and which are *locally bounded* in that $c(K)$ is pre-compact whenever K is compact in G/H.[8] However, for a number of reasons, there seem to be

[6] In fact, any two are mutually absolutely continuous [56, Theorem 2.59] (see also Lemma H.14 on page 463).

[7] This is an unpublished observation due to Dorte Olesen, Gert Pedersen and the author.

[8] Theorem 3.4.1 in [2] provides a Borel cross section, but does not deal with locally bounded sections. But if G is second countable, then it is the countable union of compact sets K_n such that K_n is contained in the interior of K_{n+1}. Then the restriction of the quotient map to K_n has a Borel cross section, and we get a regular section by piecing these sections together.

problems with Borel sections on general locally compact groups. Part of the reason is that in non-second countable groups, there are more Borel sets than we really want to pay attention to. The solution is to restrict attention to only those Borel sets needed to ensure that continuous compactly supported functions are measurable. Recall that a subset of a topological space is called a G_δ set if it is the intersection of countably many open sets.

Definition 4.32. Let X be a locally compact space. The *Borel sets* $\mathscr{B}(X)$ in X are the elements of the σ-algebra generated by the open subsets of X. The *Baire sets* $\mathscr{B}_\delta(X)$ in X are the elements of the σ-algebra generated by the compact G_δ subsets of X. Suppose that Y is locally compact, and that $f : X \to Y$ is a function. We say that f is *locally bounded* if $f(C)$ is pre-compact for each compact set C in X. We say that f is *Baire measurable* (*Borel measurable*) if $f^{-1}(E)$ is Baire (Borel) in X for every Baire (Borel) set $E \subset Y$.

Remark 4.33. If $f : X \to Y$ is a function and if X and Y are locally compact, then f is Borel[9] if and only if $f^{-1}(V)$ is Borel for every open set V in Y. Similarly, f is Baire if and only if $f^{-1}(C)$ is Baire for each compact G_δ subset of Y. Since not every open subset of Y need be Baire, a Baire measurable function $f : X \to Y$ need not be Borel. As far as I can see, there is no reason why a continuous function $f : X \to Y$ need be Baire. (However, see Lemma 4.41 on the next page.)

The following observations are routine.

Lemma 4.34. *Suppose X is locally compact Hausdorff.*

(a) *Every Baire set is Borel. If X is second countable, every Borel set is Baire.*
(b) *The Baire sets form the smallest σ-algebra for which each $f \in C_c(X)$ is measurable.*
(c) *A Baire measurable function $f : X \to \mathbf{C}$ is Borel.*
(d) *If $q : Y \to X$ is Baire measurable, and $f : X \to \mathbf{C}$ is Baire measurable, then $f \circ q$ is Baire measurable.*

We'll use the following result of Kehlet without proof.

Theorem 4.35 ([88]). *If H is a closed subgroup of a locally compact group G, then there is a locally bounded Baire cross-section $c : G/H \to G$ for the natural map $q : G \to G/H$. That is, c is Baire measurable map such that $p\big(c(rH)\big) = rH$ for all $r \in G$, and $c(K)$ is pre-compact if K is compact in G/H.*

Some care must be taken in citing results about Baire sets as not all sources agree on the definition. Kehlet uses Halmos [70] as an authority on measure theory, and Halmos defines the Baire sets to be the σ*-ring* generated by the compact G_δ's. That we can use σ-algebras instead is guaranteed by the following.

Definition 4.36. If X is a locally compact space, then we'll write $\mathscr{M}_\delta(X)$ for the σ-ring generated by the compact G_δ's.

[9] It is common practice to shorten "Borel measurable" and "Baire measurable" to simply "Borel" and "Baire", respectively.

Lemma 4.37 ([154, Chap. 13 §1]). *If X is a locally compact Hausdorff space, then a Baire set $E \subset X$ is in $\mathscr{M}_\delta(X)$ if and only if E is contained in a σ-compact set. (Such sets are called* σ-bounded.*) If E is a Baire set, then either $E \in \mathscr{M}_\delta(X)$ or $E^c := X \setminus E \in \mathscr{M}_\delta(X)$. If both E and E^c are σ-bounded, then X is σ-compact.*

Corollary 4.38. *Suppose X and Y are locally compact Hausdorff spaces and that $f : X \to Y$ is a function such that $f^{-1}(B) \in \mathscr{M}_\delta(X)$ for all $B \in \mathscr{M}_\delta(Y)$. Then f is a Baire function.*

That is, a "Baire function" with respect to σ-rings as defined in Halmos or Kehlet is a Baire measurable function as in Definition 4.32 on the preceding page.

Definition 4.39. A function $f : X \to Y$ is *locally Baire* if $f^{-1}(V) \cap W$ is a Baire set for each $V \in \mathscr{B}_\delta(Y)$ and each σ-bounded Baire set $W \in \mathscr{M}_\delta(X)$.

Lemma 4.40. *A function $f : X \to Y$ is locally Baire if and only if $f^{-1}(C) \cap K$ is Baire for each compact G_δ set $C \subset Y$ and compact G_δ set $K \subset X$.*

Proof. The assertion follows easily from the observations that for each $W \in \mathscr{M}_\delta(X)$, the set

$$\{\, V \subset Y : f^{-1}(V) \cap W \in \mathscr{B}_\delta(X) \,\}$$

is a σ-algebra in Y, and that for all $V \subset Y$ the set

$$\{\, W \subset X : f^{-1}(V) \cap W \in \mathscr{B}_\delta(X) \,\}$$

is a σ-ring. □

Lemma 4.41. *Suppose that $f : X \to Y$ is continuous. Then f is Borel and locally Baire.*

Proof. The first assertion is easy. For the second, let $C = \bigcap O_n$ be a compact G_δ in Y and $K = \bigcap U_m$ a compact G_δ in X. Then

$$f^{-1}(C) = \bigcap f^{-1}(O_n)$$

is a closed G_δ in X. Thus

$$f^{-1}(C) \cap K = \bigcap_{n,m} f^{-1}(O_n) \cap U_m$$

is a compact G_δ in X. Thus f is locally Baire by Lemma 4.40. □

Remark 4.42. If C is a compact G_δ subset of X, then it is not hard to verify that

$$\mathscr{B}_\delta(C) = \{\, B \cap C : B \in \mathscr{B}_\delta(X) \,\}.$$

On the other hand, if $f : X \to Y$ is a function, then $f^{-1}(V) \cap C = (f|_C)^{-1}(V)$. Hence f is locally Baire if and only if $f|_C$ is Baire for all compact G_δ sets $C \subset X$.

Lemma 4.43. *Suppose that X is locally compact Hausdorff, and that $f : X \to \mathbf{C}$ is a complex-valued function which vanishes off a compact G_δ-set $C \subset X$. If $f|_C$ is Baire, then f is Baire. In particular, any locally Baire function $f : X \to \mathbf{C}$ that vanishes off a compact set is Baire.*

Proof. Suppose that $f|_C$ is Baire. It suffices to verify that $f^{-1}(V) \in \mathscr{B}_\delta(X)$ for V open in $\mathbf{C}$. But by assumption, $f^{-1}(V) \cap C \in \mathscr{B}_\delta(C) \subset \mathscr{B}_\delta(X)$. But if $0 \notin V$, then $f^{-1}(V) = f^{-1}(V) \cap C$. If $0 \in V$, then $f^{-1}(V) = f^{-1}(V) \cap C \cup X \setminus C$. In either case, $f^{-1}(V) \in \mathscr{B}_\delta(X)$, and f is a Baire function.

The last assertion follows from the above and the observation that Urysohn's Lemma implies that every compact subset of X is contained in a compact G_δ subset. □

Now we must think just a bit about product spaces. If $\mathscr{A}$ and $\mathscr{B}$ are σ-algebras in X and Y respectively, then we'll write $\mathscr{A} \times \mathscr{B}$ for the σ-algebra in $X \times Y$ generated by the rectangles $A \times B$ with $A \in \mathscr{A}$ and $B \in \mathscr{B}$.

Lemma 4.44. *Let X be a locally compact Hausdorff space. Then*

$$\mathscr{B}_\delta(X \times X) \subset \mathscr{B}_\delta(X) \times \mathscr{B}_\delta(X), \text{ and} \tag{4.64}$$

$$\mathscr{B}(X) \times \mathscr{B}(X) \subset \mathscr{B}(X \times X). \tag{4.65}$$

If X is compact, then equality holds in (4.64). *If X second countable, then equality holds in* (4.65).

Proof. Suppose that K is a compact G_δ subset of $X \times X$. Then there is a $f \in C_c(X \times X)$ such that $K = f^{-1}(1)$ [31, Chap. VII Corollary 4.2]. The Stone-Weierstrass Theorem implies there is sequence $\{ f_i \} \subset C_c(X \times X)$ converging to f uniformly such that each f_i has the form $f_i(x, y) = \sum_{k=1}^{n_i} \varphi_k^i(x)\psi_k^i(y)$ for functions $\varphi_k^i, \psi_k^i \in C_c(X)$. In particular, each such f_i is measurable with respect to the product σ-algebra $\mathscr{B}_\delta(X) \times \mathscr{B}_\delta(X)$. Thus f is too, and $K \in \mathscr{B}_\delta(X) \times \mathscr{B}_\delta(X)$. This suffices to prove (4.64).

To prove (4.65) we need to see that $A \times B \in \mathscr{B}(X \times X)$ if $A, B \in \mathscr{B}(X)$. However,

$$\{ C \subset X : C \times X \in \mathscr{B}(X \times X) \}$$

is a σ-algebra in X containing all open sets. Thus $A \times X$ is Borel as is $X \times B$. Thus $A \times B$ is Borel as required.

If X is compact and C is a compact G_δ in X, then $C \times X$ is a compact G_δ in $X \times X$. Arguing as above, $A \times X \in \mathscr{B}_\delta(X) \times \mathscr{B}_\delta(X)$ for all $A \in \mathscr{B}_\delta(X)$. Thus $A \times B \in \mathscr{B}_\delta(X \times X)$ for all $A, B \in \mathscr{B}_\delta(X)$. Thus $\mathscr{B}_\delta(X) \times \mathscr{B}_\delta(X) \subset \mathscr{B}_\delta(X \times X)$.

If X is second countable and $O \subset X \times X$ is open, then O is a countable union of open rectangles. Therefore $O \in \mathscr{B}(X) \times \mathscr{B}(X)$. This suffices. □

Remark 4.45. The equality $\mathscr{B}(X) \times \mathscr{B}(X) = \mathscr{B}(X \times X)$ can fail even if X is a compact group [7, §2].

Lemma 4.46. *Suppose that X, Y and Z are locally compact Hausdorff spaces, and that $f : X \to Y$ is continuous. If $g : Y \to Z$ is Borel, then $p := g \circ f$ is Borel. If $g : Y \to Z$ is Baire, then $p := g \circ f$ is locally Baire.*

Proof. The assertion about Borel functions is standard. Let $C \subset Z$ and $K \subset X$ be compact G_δ's. Then

$$p^{-1}(C) \cap K = f^{-1}\big(g^{-1}(C)\big) \cap K.$$

The result now follows since $g^{-1}(C)$ is Baire and f is locally Baire (Lemma 4.41 on page 140). □

Proposition 4.47. *Suppose that G is a locally compact group, that $g, h : G \to G$ are functions and that $f(r) := g(r)h(r)$. If g and h are locally bounded, locally Baire functions, then f is locally Baire. If G is second countable and g and h are Borel, then f is Borel.*

Proof. Suppose that g and h are locally bounded, locally Baire functions. Let K be a compact G_δ subset of G. By Remark 4.42 on page 140, it suffices to show that $f|_K$ is Baire. Since g and h are locally bounded, there is a compact G_δ set K' containing $g(K)$ and $h(K)$.[10] But $f|_K$ is the composition

$$K \xrightarrow{\quad i \quad} K' \times K' \xrightarrow{\ m|_{K'\times K'}\ } G,$$

where $i(r) := \big(g(r), h(r)\big)$ and m is multiplication. Since $g|_K$ and $h|_K$ are Baire, i is measurable from $\big(K, \mathscr{B}_\delta(K)\big)$ to $\big(K' \times K', \mathscr{B}_\delta(K') \times \mathscr{B}_\delta(K')\big)$. By Lemma 4.44 on the preceding page, $\mathscr{B}_\delta(K') \times \mathscr{B}_\delta(K') = \mathscr{B}_\delta(K' \times K')$. On the other hand, $m : G \times G \to G$ is locally Baire (Lemma 4.41 on page 140), so $m|_{K'\times K'}$ is Baire. Thus $f|_K$ is Baire.

If G is second countable and both g and h are Borel, then $\mathscr{B}(G \times G) = \mathscr{B}(G) \times \mathscr{B}(G)$. Now the proof proceeds as above. □

Definition 4.48. Suppose $c : G/H \to G$ is a cross section for the natural map $q : G \to G/H$. Then we define $d := d_c : G \to H$ by $d(r) := c\big(q(r)\big)^{-1} r$. Thus for all $r \in G$,

$$r = c(rH)d(r).$$

Corollary 4.49. *If $c : G/H \to G$ is a locally bounded Baire cross section, then d_c is a locally bounded, locally Baire function. If G is second countable, and c is a locally bounded Borel section, then d_c is a locally bounded Borel function.*

Proof. If K is compact in G, then $d(K) \subset c\big(q(K)\big)^{-1} K \subset \overline{c\big(q(K)\big)}^{-1} K$. Thus d is locally bounded if c is. Since $r \mapsto r^{-1}$ is a homeomorphism, $c'(rH) := c(rH)^{-1}$ defines a Baire (Borel) function if c is Baire (Borel). Suppose c is locally bounded and Baire. Lemma 4.46 on the preceding page implies $g(r) = c'\big(q(r)\big)$ is locally Baire. Since g is locally bounded, the first assertion follows from Proposition 4.47 with $h(r) = r$.

If G is second countable and c Borel, then f is Borel and the result follows again from Proposition 4.47. □

[10]By Urysohn's Lemma, every compact set is contained in a compact G_δ set.

4.5.2 The Proof

Our starting point is the Green $C_0(G/H)\rtimes_{\mathrm{lt}} G - C^*(H)$-pre-imprimitivity bimodule $\mathsf{X}_0 := C_c(G)$ obtained via Theorem 4.22 on page 132 with $A = \mathbf{C}$. In order to make use of our locally bounded Baire cross section $c : G/H \to G$, we'll need to enlarge X_0 a bit.

Let $\mathcal{B}^b_{\delta c}(G)$ be the collection of bounded Baire functions on G which vanish off a compact set.[11] We want to turn $\mathsf{X}_{00} := \mathcal{B}^b_{\delta c}(G)$ into a pre-inner product $C^*(H)$-module which contains $\mathsf{X}_0 := C_c(G)$ as a dense sub-module. If $f \in \mathsf{X}_{00}$ has support contained in a compact set K and $b \in C_c(G)$ has support C, then

$$\varphi(s,h) := f(sh)b(h^{-1})\gamma(h).$$

defines a bounded function on $G \times H$ with support contained in $KC^{-1} \times C^{-1}$. (Recall that $\gamma(h) := \Delta_G(h)^{\frac{1}{2}}\Delta_H(h)^{-\frac{1}{2}}$.) Let K' and C' be compact G_δ's such that $KC^{-1} \subset K'$ and $C^{-1} \subset C'$. However, $(s,h) \mapsto f(sh)$, is locally Baire by Lemma 4.46 on page 141. Since b and γ are continuous, it follows that $\varphi|_{K'\times C'}$ is Baire. We can mimic (4.56) and define

$$\begin{aligned} f\cdot b(s) &:= \int_H f(sh)b(h^{-1})\gamma(h)\,d\mu_H(h) \qquad (4.66)\\ &= \int_H \varphi(s,h)\,d\mu_H(h)\\ &= \int_{C'} \varphi(s,h)\,d\mu_H(h). \end{aligned}$$

Then $f\cdot b$ has support in K' and is bounded by $\|\varphi\|_\infty \cdot \mu_H(C)$. Therefore to see that $f\cdot b \in \mathsf{X}_{00}$, we need only see that $f \cdot b$ is Baire. By Lemma 4.43 on page 141, it suffices to see that the restriction of $f\cdot b$ to K' is a Baire function. But $\varphi|_{K'\times C'}$ is measurable with respect to $\mathscr{B}_\delta(K'\times C')$, and $\mathscr{B}_\delta(K'\times C') = \mathscr{B}_\delta(K')\times\mathscr{B}_\delta(C')$ by Lemma 4.44 on page 141. Since $(K'\times C', \mathscr{B}_\delta(K')\times\mathscr{B}_\delta(C'), \mu_G\times\mu_H)$ is a finite measure space, we can apply Fubini's Theorem (as in [156, Theorem 8.8]) to conclude that $f\cdot b$ is a Baire function on K'. Now it is easy to see that X_{00} is a right B_0-module. If $f \in \mathsf{X}_0$, then $f\cdot b$ coincides with (4.56) and X_0 is a sub-module of X_{00}.

Next we define a sesquilinear form on X_{00} using (4.57) from Theorem 4.22 on page 132:

$$\langle f\,,g\rangle_{B_0}(h) := \gamma(h)\int_G \overline{f(s)}g(sh)\,d\mu_G(s). \qquad (4.67)$$

It is easy to see that (4.67) defines a function on H with support in $(\operatorname{supp} f)^{-1}(\operatorname{supp} g)$. It may come as a bit of a surprise that (4.67) is continuous in

[11] If G is second countable, the Baire and Borel sets coincide and we could work with Borel functions throughout.

h. To see this, let $g^h(s) := g(sh)$ and compute

$$\begin{aligned}\bigl|\langle f\,,g\rangle_{B_0}(h) - \langle f\,,g\rangle_{B_0}(t)\bigr| &\le \bigl|\gamma(h)-\gamma(t)\bigr|\Bigl|\int_G \overline{f(s)}g(sh)\,d\mu_G(s)\Bigr| + \\ &\qquad |\gamma(t)|\Bigl|\int_G \overline{f(s)}\bigl(g(sh)-g(st)\bigr)\,d\mu_G(s)\Bigr| \\ &\le \bigl|\gamma(h)-\gamma(t)\bigr|\|f\|_1\|g\|_\infty + |\gamma(t)|\|f\|_\infty\|g^h - g^t\|_1.\end{aligned}$$

Since right translation is continuous in $L^1(G)$, the claim about continuity follows.

To show that X_{00} is a pre-inner product B_0-module with the extended definitions of the module action (4.66) and inner product (4.67), we proceed as follows. To show that

$$\langle f\,,g\cdot b\rangle_{B_0} = \langle f\,,g\rangle_{B_0} * b,$$

we compute using Fubini after restricting to compact subsets as above. To show positivity, notice that

$$\langle f\,,g\rangle_{B_0}(h) - \gamma(h)(g^h \mid f)_{L^2(G)}, \tag{4.68}$$

where $g^h(s) = g(sh)$ with $\operatorname{supp}\langle f\,,g\rangle_{B_0} \subset \operatorname{supp}(f)^{-1}\operatorname{supp}(g)\cap H$. Furthermore, using the Cauchy-Schwarz inequality we have

$$\begin{aligned}\bigl|\langle f\,,f\rangle_{B_0}(h) - \langle g\,,g\rangle_{B_0}(h)\bigr| &= \bigl|\langle f\,,f-g\rangle_{B_0}(h) + \langle f-g\,,g\rangle_{B_0}(h)\bigr| \\ &\le \gamma(h)\|(f-g)^h\|_2\|f\|_2 + \|g^h\|_2\|f-g\|_2 \\ &= \gamma(h)\Delta_G(h)^{-\frac12}\|f-g\|_2(\|f\|_2+\|g\|_2).\end{aligned} \tag{4.69}$$

Now if $f\in\mathsf{X}_{00}$ with $\operatorname{supp} f\subset K$, then there are $g_i\in\mathsf{X}_0$ with $\operatorname{supp} g_i\subset K$ such that $g_i\to f$ in $L^2(G)$. Since $\gamma(h)\Delta_G(h)^{-\frac12}$ is bounded on $K^{-1}K\cap H$, it follows that $\langle g_i\,,g_i\rangle_{B_0}\to\langle f\,,f\rangle_{B_0}$ in the inductive limit topology and therefore in the C^*-norm. Since each $\langle g_i\,,g_i\rangle_{B_0}\ge 0$, it follows that $\langle f\,,f\rangle_{B_0}\ge 0$.

We'll write $\|f\|_B$ for the seminorm $\|\langle f\,,f\rangle_{B_0}\|^{\frac12}$. The following lemma will prove useful.

Lemma 4.50. *Let $f\in\mathsf{X}_{00}$. Suppose that K is a compact set in G and that $\{g_i\}$ is a sequence in X_{00} such that g_i vanishes off K for all i and such that $g_i\to f$ in $L^2(G)$. Then $\|g_i - f\|_B\to 0$.*

Proof. We can assume that f vanishes off K. If $M := \max|\Delta_H(h)^{-\frac12}|$ for $h\in K^{-1}K\cap H$, then we have

$$\begin{aligned}\|g_i - f\|_B^2 &= \|\langle g_i - f\,,g_i - f\rangle_{B_0}\| \\ &\le \int_H \bigl|\langle g_i - f\,,g_i - f\rangle_{B_0}(h)\bigr|\,d\mu_H(h) \\ &\le \int_{K^{-1}K\cap H} \|(g_i - f)^h\|_2\|g_i - f\|_2\gamma(h)\,d\mu_H(h)\end{aligned}$$

which, since $\|g^h\|_2 = \Delta_G(h)^{-\frac{1}{2}}\|g\|_2$, is

$$= \|g_i - f\|_2^2 \int_{K^{-1}K\cap H} \Delta_H(h)^{-\frac{1}{2}}\, d\mu_H(h)$$
$$\leq \|g_i - f\|_2^2 \cdot M \cdot \mu_H(K^{-1}K \cap H).$$

Thus if $\{\, g_i \,\}$ is as stated, $\|g_i - f\|_B \to 0$. □

As an immediate corollary, we obtain the following.

Corollary 4.51. *The submodule X_0 is $\|\cdot\|_B$-seminorm dense in X_{00}. In particular, the completion X of X_{00} is the same as the completion of X_0. Thus $\mathcal{K}(\mathsf{X})$ is isomorphic to $C_0(G/H) \rtimes_{\mathrm{lt}} G$.*

We can also prove the following corollary which will be needed in the last step of the proof.

Corollary 4.52. *Let Y be the subspace of X_{00} spanned by the characteristic functions of compact G_δ subsets of G. Then Y is dense in X_{00} in the $\|\cdot\|_B$-seminorm.*

Proof. If $f \in \mathsf{X}_{00}$, then there is a sequence $\{\, f_i \,\}$ of Baire simple functions converging uniformly to f such that $|f_i(s)| \leq |f(s)|$ for all s [57, Theorem 2.10(b)]. Thus the subspace Y' spanned by characteristic functions of Baire sets which are contained in a compact set is dense in X_{00} by Lemma 4.50 on the facing page. Thus it will suffice to see that we can approximate $\mathbb{1}_F$ where F is a Baire set contained in a compact set K. Since G is locally compact, we can assume that F is contained in the interior V of K. If $\epsilon > 0$, then by regularity of Haar measure, there is a compact set C and an open set O such that

$$C \subset F \subset O \subset V,$$

and such that $\mu_G(O \setminus C) < \epsilon^2$. Using Urysohn's Lemma, there is a compact G_δ set E such that $C \subset E \subset O$. Then

$$\|\mathbb{1}_F - \mathbb{1}_E\|_2 < \epsilon.$$

Therefore there are $g_i \in \mathsf{Y}$ such that g_i vanishes off K and such that $g_i \to \mathbb{1}_F$ in $L^2(G)$. This suffices by Lemma 4.50 on the preceding page. □

Now we'll switch gears and build a new Hilbert $C^*(H)$-module Z. Let $\mathcal{B}^b_{\delta c}(H)$ be the bounded Baire functions on H which vanish off a compact set. If $b' \in \mathcal{B}^b_{\delta c}(H)$ and $b \in B_0$, then the convolution

$$b' * b(h) := \int_H b'(t)b(t^{-1}h)\, d\mu_H(t)$$

is continuous in h and compactly supported. Thus $b' * b$ is certainly in $\mathcal{B}^b_{\delta c}(H)$. Now it is not hard to see that $\mathcal{B}^b_{\delta c}(H)$ is a pre-Hilbert $C^*(H)$-module containing $C_c(H)$ as a dense submodule. (This is a special case of Corollary 4.51 with both G and H equal to H.) In particular, the completion of $\mathcal{B}^b_{\delta c}(H)$ is $C^*(H)$.

Let Z_0 be the algebraic tensor product

$$\mathsf{Z}_0 := \mathcal{B}^b_{\delta c}(H) \odot \mathcal{B}^b_{\delta c}(G/H).$$

We can define a $C_c(H)$-action on Z_0 by

$$(b' \otimes n) \cdot b := b' * b \otimes n. \tag{4.70}$$

We define a sesquilinear form on Z_0 by

$$\begin{aligned} \langle\!\langle a \otimes n \,, b' \otimes m \rangle\!\rangle_{B_0}(h) &:= (m \mid n)_{L^2(G/H)} a^* * b'(h) \qquad (4.71)\\ &= \int_{G/H} m(sH)\overline{n(sH)}\, d\mu_{G/H}(sH) \int_H \overline{c(t)} b'(th)\, d\mu_H(t). \end{aligned}$$

Note that (4.71) is just the product of the usual inner product on the Hilbert space $L^2(G/H)$ and the inner product on the pre-Hilbert $C^*(H)$-module $\mathcal{B}^b_{\delta c}(H)$. Now the next lemma follows easily.

Lemma 4.53. *With the module action* (4.70) *and inner product* (4.71), Z_0 *is a pre-inner product* B_0*-module. The completion* Z *is (isomorphic to) the external tensor product* $C^*(H) \otimes L^2(G/H)$.

Corollary 4.54. $\mathcal{K}(\mathsf{Z}) \cong \mathcal{K}\big(L^2(G/H)\big) \otimes C^*(H)$.

Proof. By [139, Corollary 3.38], $\mathcal{K}(\mathsf{Z}) \cong \mathcal{K}\big(L^2(G/H)\big) \otimes \mathcal{K}\big(C^*(H)\big)$, and $\mathcal{K}\big(C^*(H)\big) = C^*(H)$ by [139, Example 2.26]. □

Lemma 4.55. *Suppose that* X *and* Z *are Hilbert* B*-modules and that* $\theta : \mathsf{X} \to \mathsf{Z}$ *is a Hilbert module isomorphism. Then* $T \mapsto \theta T \theta^{-1}$ *is an isomorphism* Φ *of* $\mathcal{K}(\mathsf{X})$ *onto* $\mathcal{K}(\mathsf{Z})$ *such that*

$$\Phi\big({}_{\mathcal{K}(\mathsf{X})}\langle x \,, y\rangle\big) = {}_{\mathcal{K}(\mathsf{Z})}\big\langle \theta(x) \,, \theta(y)\big\rangle. \tag{4.72}$$

Proof. Let $\Phi(T) := \theta T \theta^{-1}$. Clearly Φ maps $\mathcal{L}(\mathsf{X})$ into $\mathcal{L}(\mathsf{Z})$. If $x, y \in \mathsf{X}$ and $z \in \mathsf{Z}$, then

$$\begin{aligned} \theta\, {}_{\mathcal{K}(\mathsf{X})}\langle x \,, y\rangle \theta^{-1}(z) &= \theta\big(x \cdot \langle y \,, \theta^{-1}(z)\rangle_B\big) \\ &= \theta\big(x \cdot \langle \theta(y) \,, z\rangle_B\big) \\ &= \theta(x)\big\langle \theta(y) \,, z\big\rangle_B \\ &= {}_{\mathcal{K}(\mathsf{Z})}\big\langle \theta(x) \,, \theta(y)\big\rangle(z). \end{aligned}$$

This establishes (4.72). It also follows that $\Phi\big(\mathcal{K}(\mathsf{X})\big) \subset \mathcal{K}(\mathsf{Z})$ and has dense range. Thus $\Phi\big(\mathcal{K}(\mathsf{X})\big) = \mathcal{K}(\mathsf{Z})$. This suffices as Φ is clearly injective. □

Now it suffices by Corollary 4.54 and Lemma 4.55 to show that X and Z are isomorphic Hilbert $C^*(H)$-modules. Recall that $q : G \to G/H$ is the quotient map and that d is defined in Definition 4.48 on page 142. If $a \otimes n \in \mathsf{Z}_0$, then we define $\alpha(a \otimes n) : G \to \mathbf{C}$ by

$$\alpha(a \otimes n)(s) := \rho(s)^{\frac{1}{2}} a\big(d(s)\big) n\big(q(s)\big),$$

where ρ is as in (4.63). We want to see that $\alpha(a \otimes n) \in \mathsf{X}_{00}$. Let $\operatorname{supp} a \subset K$ and $\operatorname{supp} n \subset C$. Then if $\alpha(a \otimes n)(r) \neq 0$, then $rH \in C$ and $d(r) = c(rH)^{-1} r \in K$. Thus $r \in \overline{c(C)} \cdot K$. Thus there is a compact G_δ set K_0 such that $\operatorname{supp} \alpha(a \otimes n) \subset K_0$. To see that $\alpha(a \otimes n)$ is Baire, it suffices to show that

$$f(s) := a\big(d(s)\big) n\big(q(s)\big)$$

defines a Baire function on K_0 (Lemma 4.43 on page 141). Note that f is a composition

$$K_0 \xrightarrow{\ i\ } \mathbf{C} \times \mathbf{C} \xrightarrow{\ m\ } \mathbf{C},$$

where m is multiplication and $i(s) := \big(a\big(d(s)\big), n\big(q(s)\big)\big)$. Then if V is open in $\mathbf{C}$,

$$f^{-1}(V) = i^{-1}\big(m^{-1}(V)\big).$$

Since $\mathbf{C}$ is second countable, $m^{-1}(V)$ is a countable union of open rectangles. Thus it suffices to see that

$$i^{-1}(V \times U) = K_0 \cap i^{-1}(V \times U) = K_0 \cap d^{-1}\big(a^{-1}(U)\big) \cap (n \circ q)^{-1}(V)$$

is Baire. Since d is locally Baire and $a^{-1}(U)$ is Baire, $K_0 \cap d^{-1}\big(a^{-1}(U)\big)$ is a σ-bounded Baire set. Since Lemma 4.46 on page 141 implies $n \circ q$ is locally Baire, $i^{-1}(V \times U)$ is Baire.

Now we've proved that we have a map

$$\alpha : \mathsf{Z}_0 \to \mathsf{X}_{00}.$$

To complete the proof of Theorem 4.30 on page 138 we just need to see that α preserves the $C^*(H)$-valued inner products and has dense range. Then it follows that α extends to a unitary — which is automatically $C^*(H)$-linear by [139, Lemma 2.18].

To see that α preserves inner products, we calculate

$$\big\langle \alpha(a \otimes n)\, ,\, \alpha(b \otimes m) \big\rangle_{B_0}(h) = \gamma(h) \int_G \overline{\alpha(a \otimes n)(s)} \alpha(b \otimes m)(sh)\, d\mu_G(s)$$

and, noting that $\gamma(h)\rho(s)^{\frac{1}{2}}\rho(sh)^{\frac{1}{2}} = \rho(s)$, this

$$= \int_G \overline{a\big(d(s)\big) n(sH)} b\big(d(sh)\big) m(sH) \rho(s)\, d\mu_G(s)$$

and using (4.63) and $d(st) = d(s)t$ for $s \in G$ and $t \in H$,

$$\begin{aligned}
&= \int_{G/H} \int_H \overline{a\big(d(s)t\big)n(sH)} b\big(d(s)th\big) m(sH)\, d\mu_H(t)\, d\mu_{G/H}(sH) \\
&= \int_{G/H} \overline{n(sH)} m(sH)\, d\mu_{G/H}(sH) \int_H \overline{a(t)} b(th)\, d\mu_H(t) \\
&= (m \mid n)_{L^2(G/H)} \int_H \overline{a(t^{-1})} \Delta_H(t^{-1}) b(t^{-1}h)\, d\mu_H(t) \\
&= (m \mid n)_{L^2(G/H)} a^* * b(h) \\
&= \langle\!\langle a \otimes n \,,\, b \otimes m \rangle\!\rangle_{B_0}(h).
\end{aligned}$$

Now all that remains to be shown is that $\alpha(\mathsf{Z}_0)$ is dense in X_{00}. We'll show that we can approximate those $h \in \mathsf{X}_{00}$ of the form $\mathbb{1}_E$ for a compact G_δ subset E of G. This will suffice in view of Corollary 4.52 on page 145.

Let C be a compact neighborhood of E. Given $\epsilon > 0$, we'll find $h \in \alpha(\mathsf{Z}_0)$ such that $\operatorname{supp} h \subset C$ and $\|\mathbb{1}_E - h\|_2^2 < \epsilon$. This will suffice by Lemma 4.50 on page 144.

Let O be an open set such that $E \subset O \subset C$ and $\mu(O - E) < \epsilon$. Using normality (of C), there is an open set P such that $E \subset P \subset \overline{P} \subset O$.

Lemma 4.56. *There is a neighborhood V of e in H such that $c(sH)Vd(s) \subset P$ for all $s \in E$*

Proof. It suffices to produce a neighborhood in G with the same property. Suppose no such neighborhood exists. Let $\{ V_i \}$ be the family of neighborhoods of e ordered by inclusion. For all i there exits an $s_i \in E$ such that $c(s_iH)V_id(s_i) \not\subset P$. By compactness we may extract a subnet, relabel, and assume that $c(s_iH) \to a$ and $d(s_i) \to b$. But $c(s_iH)d(s_i) \in E$ for all i. Thus, $ab \in E$, and by continuity there exist neighborhoods U, Q, V of a, e, and b respectively such that $UQV \subset P$. But the $c(s_iH)$ and the $d(s_i)$ are eventually in U and V, respectively. Thus there is a i_0 such that $i \geq i_0$ implies that $c(s_iH)Qd(s_i) \subset P$. Now take i so large that $V_i \subseteq Q$. This provides a contradiction. □

We'll complete the proof by producing an h such that $\alpha(h)$ is a characteristic function $\mathbb{1}_F$ such that $E \subset F \subset O$.

Choose V as above, and let W be a Baire neighborhood of e in H such that $W = W^{-1}$, $W^3 \subseteq V$, and $\overline{W}$ is compact. Since E is compact, $d(E)$ has compact closure. Thus, there are elements $s_i \in G$ such that $Ws_1, \dots, Ws_m$ cover $d(E)$. Let $B_1, \dots, B_m$ be disjoint Baire sets covering $d(E)$ with the property that $B_i \subset Ws_i$. Since $\overline{W}$ is compact, it follows that $\overline{B}_i$ is compact.

Let $T_i = \{ s \in G : s\overline{B}_i \subset \overline{P} \}$, and $U_i = \{ s \in G : s\overline{B}_i \subset O \}$. Notice that $T_i \subset U_i$, each T_i is compact and that each U_i is open with compact closure. In particular, there is a Baire set V_i such that $T_i \subseteq V_i \subseteq U_i$.[12] Let $F_i = c^{-1}(V_i) = q(V_i)$; notice that F_i is a Baire set with compact closure. Furthermore, $sH \in F_i$ implies

[12] Simply choose $f \in C_c^+(G)$ which is strictly positive on T_i and which vanishes off U_i. Let $V_i = \{ f \neq 0 \}$.

that $c(sH)\overline{B}_i \subset O$. Also $c(sH)\overline{B}_i \subset P$ implies that $c(sH) \in T_i$, and hence that $sH \in F_i$. Set $a_k(t) = \rho(t)^{-\frac{1}{2}}\mathbb{1}_{B_k}(t)$, and let $n_k\big(q(s)\big) = \rho\big(c(q(s))\big)^{-\frac{1}{2}}\mathbb{1}_{F_k}\big(q(s)\big)$. Note that each $a_k \otimes n_k \in \mathsf{Z}_0$, and that since $\rho\big(d(s)\big)\rho\big(c(q(s))\big) = \rho(s)$, we have $\alpha(a_k \otimes n_k)(s) = \mathbb{1}_{B_k}\big(d(s)\big)\mathbb{1}_{F_k}\big(q(s)\big)$. Moreover,

$$a_i\big(d(s)\big) = \begin{cases} 1 & \text{if } s \in c(sH)B_i \iff d(s) \in B_i \\ 0 & \text{otherwise.} \end{cases}$$

In particular, $\alpha(a_k \otimes n_k)$ is a characteristic function. Now we observe the following:

(a) If there is a $s \in G$ such that $\alpha(a_i \otimes n_i)(s) = 1 = \alpha(a_j \otimes n_j)(s)$, then $d(s) \in B_i \cap B_j$. Therefore $i = j$.

(b) If $s \in E$, then $d(s) \in B_i$ for some i. Say $B_i \subseteq Ws_i$, so that $d(s) = w_0 s_i$ with $w_0 \in W$. Thus, $c(sH)\overline{B}_i \subset c(sH)\overline{W}s_i \subset c(sH)W^2 s_i = c(sH)W^2 w_0^{-1} w_0 s_i \subset c(sH)W^3 d(s) \subset c(sH)Vd(s) \subseteq P$. Therefore, $sH \in F_i$, which implies that $\alpha(a_i \otimes n_i)(s) = 1$.

(c) If $\alpha(a_i \otimes n_i)(s) = 1$, then $d(s) \in B_i$ and $c(sH)\overline{B_i} \subset O$. Thus, $c(sH)d(s) \in O$.

The point is that $\alpha(\sum_1^n a_i \otimes n_i)$ is the characteristic function of a Baire set F with the property that $E \subset F \subset O$. Hence, $\|\mathbb{1}_E - \alpha(\sum_1^n a_i \otimes n_i)\|_2^2 = \mu(F - E) \leq \epsilon$. This completes the proof.

Notes and Remarks

Originally, imprimitivity theorems were developed by Mackey and others to determine when a given group representation was induced from a subgroup [102]. Thinking of a representation which is not induced as "primitive", it was natural to describe these results (telling us certain representations were not primitive) as "imprimitivity" theorems. In the modern theory, these theorems are now formulated as Morita equivalences. Although we consider the question of determining when a representation is induced from a subsystem in the next chapter (where we look at induced representations), in this chapter the focus is that imprimitivity theorems describe important Morita equivalences exposing the structure and representation theory of crossed products.

The Symmetric Imprimitivity Theorem is due to Raeburn [132]. Corollary 4.11 is due to Green, but it appeared in [148]. Corollary 4.18 is also due to Green [66], as is Theorem 4.22. Our proof of the the Stone-von Neumann Theorem and its variants in Sections 4.4 and 4.5 use the imprimitivity theorems in this chapter. Historically, these results pre-date the machinery developed in this chapter and have proofs which do not depend on the imprimitivity theorems given here. A nice history of the subject is given in [153]. Theorem 4.30 is due to Green and is a special case of the results in [67].

Chapter 5

Induced Representations and Induced Ideals

The Morita equivalences that parade under the name of imprimitivity theorems in the previous chapter are fundamental to the theory for (at least) two reasons. The first we explored in the previous chapter: they provide deep insight into the structure of crossed products. The second reason, which we explore in this chapter, is that they form the basis for a theory of inducing representations from $A \rtimes_{\alpha|_H} H$ up to $A \rtimes_\alpha G$ which lies at the heart of the Mackey machine and our analysis of the ideal structure of crossed products. The machinery also allows us to induce ideals, and we look carefully at this process in Section 5.3. In Section 5.4, we show that the inducing process is compatible with the decomposition of $A \rtimes_\alpha G$ with respect to an invariant ideal I in A as described in Section 3.4.

5.1 Induced Representations of Crossed Products

Suppose that (A, G, α) is a dynamical system and that H is a closed subgroup of G. We want to induce representations $L = \pi \rtimes u$ of $A \rtimes_{\alpha|_H} H$ on $\mathcal{H}_L$ up to $A \rtimes_\alpha G$. We'll follow Rieffel's formalism as outlined in §2.4 of [139]. Therefore we require a right Hilbert $A \rtimes_{\alpha|_H} H$-module X and a homomorphism $\varphi_{A\rtimes_\alpha G}$ of $A \rtimes_\alpha G$ into $\mathcal{L}(\mathsf{X})$. The induced representation $\mathsf{X}\text{–}\mathrm{Ind}_{A\rtimes H}^{A\rtimes_\alpha G}(L)$ acts on the Hilbert space $\mathsf{X} \otimes_{A\rtimes H} \mathcal{H}_L$ which is the completion of the algebraic tensor product $\mathsf{X} \odot \mathcal{H}_L$ with respect to the pre-inner product

$$(x \otimes h \mid y \otimes k) := \big(L(\langle y\,,x\rangle_{A\rtimes H})h \mid k\big).$$

Then $\mathsf{X}\text{–}\mathrm{Ind}_{A\rtimes H}^{A\rtimes_\alpha G}(L)$ is given on $\mathsf{X} \odot \mathcal{H}_L$ by the formula

$$\mathsf{X}\text{–}\mathrm{Ind}_{A\rtimes H}^{A\rtimes_\alpha G}(L)(a)(x \otimes h) = \varphi_{A\rtimes_\alpha G}(a)(x) \otimes h.$$

Note that $\mathsf{X} \otimes_{A \rtimes H} \mathcal{H}_L$ is the internal tensor product of the right Hilbert $A \rtimes_{\alpha|_H} H$-module X with the Hilbert space $\mathcal{H}_L$ (viewed as a right Hilbert $\mathbf{C}$-module) together with the homomorphism L of $A \rtimes_{\alpha|_H} H$ into $\mathcal{L}(\mathcal{H}_L) \cong B(\mathcal{H}_L)$.[1]

For dynamical systems (A, G, α), the natural choice for X is Green's $C_0(G/H, A) \rtimes_{\mathrm{lt} \otimes \alpha} G - A \rtimes_{\alpha|_H} H$-imprimitivity bimodule X_H^G. Recall that X_H^G was constructed in Section 4.3, and that it is the completion of $\mathsf{X}_0 = C_c(G, A)$. The homomorphism $\varphi_{A \rtimes_\alpha G}$ is given by $N \rtimes v$ in Corollary 4.23 on page 133.

Definition 5.1. Suppose that (A, G, α) is a dynamical system and that L is a representation of $A \rtimes_{\alpha|_H} H$. Then $\operatorname{Ind}_H^G L$ will denote the representation of $A \rtimes_\alpha G$ induced from L via Green's imprimitivity bimodule $\mathsf{X} = \mathsf{X}_H^G$ and the homomorphism $N \rtimes v : A \rtimes_\alpha G \to \mathcal{L}(\mathsf{X})$.

Remark 5.2. Since the action $N \rtimes v$ of $A \rtimes_\alpha G$ on X_H^G is nondegenerate (Corollary 4.23 on page 133), it follows easily (see [139, Proposition 2.66]) that $\operatorname{Ind}_H^G L$ is nondegenerate.

Remark 5.3. If $f, g \in C_c(G, A)$, then $N \rtimes v(g)(f) = g * f$ (Corollary 4.23 on page 133). Therefore, on $\mathsf{X}_0 \odot \mathcal{H}_L$,

$$\operatorname{Ind}_H^G L(g)(f \otimes h) = g * f \otimes h.$$

Furthermore, it is not hard to check that the canonical extension of $\operatorname{Ind}_H^G L$ to $M(A \rtimes_\alpha G)$ is given by the canonical extension of $N \rtimes v$:

$$(\operatorname{Ind}_H^G L)^{\bar{}}(m)(f \otimes h) = (N \rtimes v)^{\bar{}}(m)(f) \otimes h.$$

Since $(N \rtimes v)^{\bar{}}\big(i_A(a)\big) = N(a)$ and $(N \rtimes v)^{\bar{}}\big(i_G(s)\big) = v_s$, it follows that $\operatorname{Ind}_H^G L = (N \otimes 1) \rtimes (v \otimes 1)$, where

$$\begin{aligned} (N \otimes 1)(a)(f \otimes h) &:= N(a)(f) \otimes h \quad \text{and} \\ (v \otimes 1)_s(f \otimes h) &= v_s(f) \otimes h. \end{aligned}$$

Even though the description of induced representations via module actions as described above is very convenient algebraically, it still helpful to have a concrete description of $\operatorname{Ind}_H^G L$ and the space on which it acts which is independent of the bimodule X. This will also bring us into contact with the original treatments of Mackey and Takesaki as generalized to arbitrary groups by Blattner. (Details and references can be found in Chapter XI of [55].[2])

[1] A little more formally, $\mathsf{X} \otimes_{A \rtimes H} \mathcal{H}_L$ is isomorphic to $\mathsf{X} \otimes_L \widetilde{H}_L$, where $\widetilde{H}_L$ is the dual Hilbert space with $\mathbf{C}$ acting on the right. Thus, if $\flat : \mathcal{H}_L \to \widehat{\mathcal{H}}_L$ is the identity map, then $\flat(h)\lambda = \flat(\bar{\lambda} h)$, $L(f)\flat(h) = \flat\big(L(f^*)h\big)$ and $\big(\flat(h) \mid \flat(k)\big) = (h \mid k)$.

[2] Blattner and Fell work with functions which transform slightly differently than ours:

$$\xi(rt) = \Delta_H(t)^{\frac{1}{2}} \Delta_G(t)^{-\frac{1}{2}} u_t^{-1}\big(f(r)\big).$$

The map sending the function $(s \mapsto f(s))$ to $(s \mapsto \rho(s)^{\frac{1}{2}} f(s))$ induces a unitary isomorphism of our $L^2_u(G, \mu_{G/H}, \mathcal{H}_L)$ onto the Blattner/Fell space which implements an equivalence between the induced representations.)

We suppose that $L = \pi \rtimes u$ is a representation of $A \rtimes_{\alpha|_H} H$ on $\mathcal{H}_L$. Define

$$\mathcal{V}_1 := \{\, \xi \in C_b(G, \mathcal{H}_L) : \xi(rt) = u_t^{-1}\big(\xi(r)\big) \text{ for all } r \in G \text{ and } t \in H \,\}.$$

Notice that if $\xi \in \mathcal{V}_1$, then $\|\xi(rt)\| = \|\xi(r)\|$ for $t \in H$, and hence $rH \mapsto \|\xi(r)\|$ is a well-defined function on G/H. We set

$$\mathcal{V}_c := \{\, \xi \in \mathcal{V}_1 : rH \mapsto \|\xi(r)\| \text{ belongs to } C_c(G/H) \,\}.$$

As shown in Appendix H.2 or [139, Lemma C.2], there always exists a continuous function $\rho : G \to (0, \infty)$ such that $\rho(rt) = \big(\Delta_H(t)/\Delta_G(t)\big)\rho(r)$ for all $r \in G$ and $t \in H$. Then there is a quasi-invariant measure $\mu_{G/H}$ on G/H satisfying (4.63) on page 138 ([139, Lemma C.2]). Given $\mu_{G/H}$, it follows that

$$(\xi \mid \eta) := \int_{G/H} \big(\xi(r) \mid \eta(r)\big)\, d\mu_{G/H}(rH) \tag{5.1}$$

is a well-defined positive-definite sesquilinear form on $\mathcal{V}_c$. The completion $\mathcal{V} := L^2_u(G, \mu_{G/H}, \mathcal{H}_L)$ is a Hilbert space.[3]

The proof of the next result closely follows that of [139, Theorem C.33].

Proposition 5.4. *Suppose that (A, G, α) is a dynamical system, that H is a closed subgroup of a locally compact group G, and that $L = \pi \rtimes u$ is a representation of $A \rtimes_{\alpha|_H} H$ on $\mathcal{H}_L$. Let $\rho : G \to (0, \infty)$ be a continuous function such that $\rho(rt) = \big(\Delta_H(t)/\Delta_G(t)\big)\rho(r) = \gamma_H(t)^{-2}\rho(r)$ for all $r \in G$ and $t \in H$. Let $\mu_{G/H}$ be the quasi-invariant measure on G/H constructed from ρ satisfying (4.63), and let $\mathcal{V} = L^2_u(G, \mu_{G/H}, \mathcal{H}_L)$ be the Hilbert space constructed above. Then* $\operatorname{Ind}_H^G L$ *is unitarily equivalent to the representation $\Pi \rtimes U$ on $\mathcal{V}$ where*

$$\Pi(a)\xi(r) = \pi\big(\alpha_r^{-1}(a)\big)\xi(r), \text{ and} \tag{5.2}$$

$$U_s\xi(r) = \Big(\frac{\rho(s^{-1}r)}{\rho(r)}\Big)^{\frac{1}{2}} \xi(s^{-1}r) \tag{5.3}$$

for $s, r \in G$, $a \in A$, and $\xi \in \mathcal{V}_c$.

Proof. We want to define a unitary W from $\mathsf{X} \otimes_{A \rtimes H} \mathcal{H}_L$ onto $\mathcal{V}$. For an elementary tensor $f \otimes h \in \mathsf{X}_0 \odot \mathcal{H}_L$, we define $W(f \otimes h)(r)$ to be the element of $\mathcal{H}$ given by

$$W(f \otimes h)(r) = \int_H \rho(rt)^{-\frac{1}{2}} \pi\big(\alpha_r^{-1}\big(f(rt)\big)\big) u_t h \, d\mu_H(t),$$

for each $r \in G$. That $W(f \otimes h) \in \mathcal{V}_c$ follows from Corollary 1.103 on page 36, left-invariance of μ_H and the covariance of (π, u). To see that W is isometric, and to see

[3] The notation $\mathcal{V} := L^2_u(G, \mu_{G/H}, \mathcal{H}_L)$ is fairly standard because $\mathcal{V}$ can be realized as a space of (equivalence classes) of functions on G. Precisely, each vector in $\mathcal{V}$ is a μ_G-almost everywhere equivalence class of functions $\xi : G \to \mathcal{H}_L$ such that $\xi(rt) = u_t^{-1}\big(\xi(r)\big)$ everywhere and such that $rH \mapsto \|\xi(r)\|$ belongs to $L^2(G/H, \mu_{G/H})$. The details are worked out in a more general setting in Section 9.3.

where the formula for W comes from, recall that the inner product in $\mathsf{X} \otimes_{A \rtimes H} \mathcal{H}_L$ is given on elementary tensors by

$$\begin{aligned}(f \otimes h \mid g \otimes k) &= \big(L(\langle g\,,\, f\rangle_{A\rtimes H})h \mid k\big)\\ &= \int_H \big(\pi(\langle g\,,\, f\rangle_{A\rtimes H}(t))u_t h \mid k\big)\, d\mu_H(t)\\ &= \int_H \int_G \gamma_H(t)\big(\pi\big(\alpha_r^{-1}\big(g(r)^* f(rt)\big)\big)u_t h \mid k\big)\, d\mu_G(r)\, d\mu_H(t)\end{aligned}$$

which, after decomposing μ_G as in (4.63) and abbreviating our notation for measures in an obvious way, is

$$\begin{aligned}&= \int_H \int_{G/H} \int_H \gamma_H(t)\rho(rv)^{-1}\big(\pi\big(\alpha_{rv}^{-1}\big(g(rv)^* f(rvt)\big)\big)u_t h \mid k\big)\, dv\, d\dot{r}\, dt\\ &= \int\limits_{G/H} \int\limits_H \int\limits_H \gamma_H(t)\rho(rv)^{-1}\big(\pi\big(\alpha_{rv}^{-1}\big(f(rvt)\big)\big)u_t h \mid \pi\big(\alpha_{rv}^{-1}\big(g(rv)\big)\big)k\big)\, dv\, dt\, d\dot{r}\end{aligned}$$

which, using left-invariance and covariance, is

$$= \int\limits_{G/H} \int\limits_H \int\limits_H \gamma_H(v^{-1}t)\rho(rv)^{-1}\big(\pi\big(\alpha_r^{-1}\big(f(rt)\big)\big)u_t h \mid \pi\big(\alpha_r^{-1}\big(g(rv)\big)\big)u_v k\big)\, dv\, dt\, d\dot{r}$$

which, since $\rho(rt) = \gamma_H(t)^{-2}\rho(r)$ implies $\rho(rv)^{-1}\gamma_H(v^{-1}t) = \rho(rt)^{-\frac{1}{2}}\rho(rv)^{-\frac{1}{2}}$, is

$$= \int_{G/H} \big(W(f \otimes h)(r) \mid W(g \otimes k)(r)\big)\, d\mu_{G/H}(rH).$$

Therefore it follows that W extends to an isometry from $\mathsf{X} \otimes_{A \rtimes H} \mathcal{H}_L$ into $\mathcal{V}$. To show that W is surjective, and therefore a unitary, it will suffice to show that given $\xi \in \mathcal{V}_c$ and $\epsilon > 0$ there is a $\eta \in \operatorname{im} W$ such that $\|\xi - \eta\| < \epsilon$. Notice that $\operatorname{im} W$ is closed under multiplication by elements of $C_c(G/H)$. Let D be a compact subset of G such that $\operatorname{supp} \xi \subset DH$, and let C be a compact neighborhood of D. We claim it will suffice to produce for each $r \in D$ an element $\xi_r \in \operatorname{im} W$ such that

$$\|\xi_r(r) - \xi(r)\| < \frac{\epsilon}{\mu_{G/H}\big(q(C)\big)^{\frac{1}{2}}}, \tag{5.4}$$

where $q : G \to G/H$ is the quotient map. Then if (5.4) holds, there is a neighborhood N_r of r in G such that $N_r \subset C$ and

$$\|\xi_r(s) - \xi(s)\| < \frac{\epsilon}{\mu_{G/H}\big(q(C)\big)^{\frac{1}{2}}} \quad \text{for all } s \in N_r; \tag{5.5}$$

since the norms of functions in $\mathcal{V}_c$ are constant on H-cosets, (5.5) holds for all $s \in N_r H$. By compactness there are $r_1, \dots, r_n$ in D such that $D \subset \bigcup_{i=1}^n N_{r_i}$, and a partition of unity argument gives functions $\psi_i \in C_c^+(G/H)$ such that $\operatorname{supp}\psi_i \subseteq q(N_{r_i})$, $0 \le \psi_i \le 1$, $\sum_i \psi_i(\dot s) = 1$ for $s \in DH$ and $\sum_i \psi_i(\dot s) \le 1$ for all s. Then $\sum_i \psi_i \xi_{r_i}$ belongs to $\operatorname{im} W$, and

$$\Big\|\sum_{i=1}^n \psi_i(\dot s)\xi_{r_i}(s) - \xi(s)\Big\| < \frac{\epsilon}{\mu_{G/H}\big(q(C)\big)^{\frac12}} \quad \text{for all } s \in G. \tag{5.6}$$

Since (5.6) vanishes off CH, we have $\|\sum_i \psi_i\xi_{r_i} - \xi\| < \epsilon$. This proves the claim.

For convenience, let $\delta := \epsilon \cdot \mu_{G/H}\big(q(C)\big)^{-\frac12}$. Fix $r \in D$. Since π is nondegenerate, $\{\pi(e_i)\}$ converges strongly to $1_{\mathcal{H}_L}$ whenever $\{e_i\}$ is an approximate identity for A. In particular, there is a positive element a of norm at most 1 such that $\|\pi\big(\alpha_r^{-1}(a)\big)\xi(r) - \xi(r)\| < \delta/2$. Furthermore, since $t \mapsto u_t h$ is continuous for each $h \in \mathcal{H}_L$, there is a neighborhood N of e in G such that

$$\|u_t\xi(r) - \xi(r)\| < \frac{\delta}{2} \quad \text{provided } t \in N \cap H.$$

Let $z \in C_c^+(G)$ be such that $\operatorname{supp} z \subset rN$ and such that

$$\int_H \rho(rt)^{-\frac12} z(rt)\, d\mu_H(t) = 1.$$

Define $f \in \mathsf{X}_0$ by $f(r) := z(r)a$. Then if $k \in \mathcal{H}_L$ with $\|k\| \le 1$,

$$\begin{aligned}
&\big|\big(W(f\otimes\xi(r))(r) \mid k\big) - \big(\xi(r) \mid k\big)\big| \\
&\quad < \big|\big(W(f\otimes\xi(r))(r) \mid k\big) - \big(\pi\big(\alpha_r^{-1}(a)\big)\xi(r) \mid k\big)\big| + \frac{\delta}{2} \\
&\quad = \Big|\int_H \rho(rt)^{-\frac12} z(rt)\big(\pi\big(\alpha_r^{-1}(a)\big)\big(u_t\xi(r) - \xi(r)\big) \mid k\big)\, d\mu_H(t)\Big| + \frac{\delta}{2} \\
&\quad \le \frac{\delta}{2} + \frac{\delta}{2} = \delta := \frac{\epsilon}{\mu_{G/H}\big(q(C)\big)^{\frac12}}.
\end{aligned}$$

Since k is arbitrary, we've shown that (5.4) holds with $\xi_r := W(f \otimes \xi(r))$. This completes the proof that W is a unitary.

As described in Remark 5.3 on page 152, $\operatorname{Ind}_H^G L$ is the integrated form of $(N \otimes 1, v \otimes 1)$, where

$$N\otimes 1(a)(f\otimes h) := N(a)f\otimes h \quad\text{and}\quad (v\otimes 1)_r(f\otimes h) := v_r(f)\otimes h.$$

Therefore

$$\begin{aligned}
\big(W\big(N\otimes 1(a)(f\otimes h)(r) \mid k\big) &= \big(W\big(N(a)f\otimes h\big)(r) \mid k\big) \\
&= \int_H \rho(rt)^{-\frac12}\big(\pi\big(\alpha_r^{-1}\big(af(rt)\big)\big)u_t h \mid k\big)\, d\mu_H(t) \\
&= \int_H \rho(rt)^{-\frac12}\big(\pi\big(\alpha_r^{-1}\big(f(rt)\big)\big)u_t h \mid \pi\big(\alpha_r^{-1}(a^*)\big)k\big)\, d\mu_H(t) \\
&= \big(\pi\big(\alpha_r^{-1}(a)\big)W(f\otimes h)(r) \mid k\big) \\
&= \big(\Pi(a)\big(W(f\otimes h)\big)(r) \mid k\big).
\end{aligned}$$

Thus $N \otimes 1$ is equivalent to Π as claimed.

Similarly, we compute that

$$\begin{aligned}
\big(W(1\otimes v)_s(f\otimes h)(r)\mid k\big) &= \big(W\big(v_s(f)\otimes h\big)(r)\mid k\big)\\
&= \int_H \rho(rt)^{-\frac12}\big(\pi\big(\alpha_r^{-1}\big(v_s(f)(rt)\big)\big)u_t h\mid k\big)\,d\mu_H(t)\\
&= \int_H \rho(rt)^{-\frac12}\big(\pi\big(\alpha_{s^{-1}r}^{-1}\big(f(s^{-1}rt)\big)\big)u_t h\mid k\big)\,d\mu_H(t)\\
&= \frac{\rho(s^{-1}r)^{\frac12}}{\rho(r)^{\frac12}}\int_H \rho(s^{-1}rt)^{-\frac12}\big(\pi\big(\alpha_{s^{-1}r}^{-1}\big(f(s^{-1}rt)\big)\big)u_t h\mid k\big)\,d\mu_H(t)\\
&= \frac{\rho(s^{-1}r)^{\frac12}}{\rho(r)^{\frac12}}\big(W(f\otimes h)(s^{-1}r)\mid k\big)\\
&= \big(U_sW(f\otimes h)(r)\mid k\big).
\end{aligned}$$

Thus $(1\otimes v)$ is equivalent to U, and this completes the proof. □

It is important to notice that $L^2_u(G,\mu_{G/H},\mathcal{H}_L)$ is independent of A and α and depends only on G, H and the unitary representation u. The same is true for the representation U defined by (5.3).

Definition 5.5. Suppose that G is a locally compact group with closed subgroup H. Then a unitary representation $u : H \to U(\mathcal{H})$ can be viewed as representation of $\mathbf{C}\rtimes_{\mathrm{id}} H$ in the usual way. Since $(\mathbf{C}, G, \mathrm{id})$ is a dynamical system, we can form the induced representation $\operatorname{Ind}_H^G(\mathrm{id}\rtimes u)$. This representation is denoted $\operatorname{Ind}_H^G u$, and is called the representation of G induced from u on H.

Notice that $\operatorname{Ind}_H^G u$ is equivalent to the representation U of Proposition 5.4 on page 153. Thus we have the following interesting corollary.

Corollary 5.6. *Let (A,G,α) be a dynamical system, and suppose that $L=\pi\rtimes u$ is a nondegenerate representation of $A\rtimes_{\alpha|_H} H$. Then the unitary part of $\operatorname{Ind}_H^G L$ is (equivalent to) $\operatorname{Ind}_H^G u$.*

Remark 5.7. If $H=\{e\}$, then Proposition 5.4 on page 153 shows that $\operatorname{Ind}_e^G \pi\rtimes \mathrm{id}$ equivalent to the regular representation $\operatorname{Ind}_e^G \pi$ of (A,G,α) associated to π as defined in Remark 2.16 on page 46. This explains the notation used for regular representations.

One of the properties of induction is that it respects the subgroup structure of the ambient group. This is exemplified by the following two results. The first states that induction respects the natural G-action on its subgroups. The second result — called *induction in stages* — says we can proceed via an intermediate subgroup.

For each $s\in G$, let $s\cdot H := sHs^{-1}$. If we have fixed Haar measures μ_H and $\mu_{s\cdot H}$ on H and $s\cdot H$, respectively, then the uniqueness of Haar measure implies there is a positive number $\omega(s,H)$ such that

$$\int_H g(shs^{-1})\,d\mu_H(h) = \omega(s,H)\int_{s\cdot H} g(h)\,d\mu_{s\cdot H}(h)$$

for all $g \in C_c(G)$.[4] We will use this observation and notation in the next lemma.

Lemma 5.8. *Suppose that (A, G, α) is a dynamical system, that H is a closed subgroup of G, and that $L = (\pi, u)$ is a covariant representation of $(A, H, \alpha|_H)$ on $\mathcal{H}$. Let $s \cdot H := sHs^{-1}$, $s \cdot \pi := \pi \circ \alpha_s^{-1}$ and $s \cdot u(h) := u(s^{-1}hs)$. Then $s \cdot L := (s \cdot \pi, s \cdot u)$ is a covariant representation of $(A, s \cdot H, \alpha|_{s \cdot H})$. Furthermore,*

$$\operatorname{Ind}_H^G L \quad \textit{is equivalent to} \quad \operatorname{Ind}_{s\cdot H}^G s \cdot L.$$

Proof. It's not hard to see that $(s \cdot \pi, s \cdot u)$ is covariant. Recall that $\operatorname{Ind}_H^G L$ acts on the completion of $C_c(G, A) \otimes \mathcal{H}$ with respect to the inner product

$$\big(f \otimes h \mid g \otimes k\big) = \big(L(\langle g \, , f \rangle_{A \rtimes H})h \mid k\big).$$

On the other hand, $\operatorname{Ind}_{s\cdot H}^G s \cdot L$ acts on the completion of $C_c(G, A) \otimes \mathcal{H}$ with respect to the inner product given by the same formula, but with L replaced by $s \cdot L$. (The formulas for the $A \rtimes_{\alpha|_H} H$-valued inner products are given in Theorem 4.22 on page 132.) If we define $V : C_c(G, A) \to C_c(G, A)$ by

$$Vf(r) := \omega(s, H)^{\frac{1}{2}} \Delta_G(s)^{\frac{1}{2}} f(rs)$$

and recall that $\gamma_H(h) = \Delta_G(h)^{\frac{1}{2}} \Delta_H(h)^{-\frac{1}{2}}$, then we compute that

$$\begin{aligned}
L(\langle g \, , f \rangle_{A \rtimes H}) &= \int_H \pi\big(\langle g \, , f \rangle_{A \rtimes H}(h)\big) u(h) \, d\mu_H(h) \\
&= \int_H \gamma_H(h) \int_G \pi\big(\alpha_r^{-1}\big(g(r)^* f(rh)\big)\big) u(h) \, d\mu_G(r) \, d\mu_H(h) \\
&= \int_H \int_G \gamma_H(h) \Delta_G(s) \pi\big(\alpha_{rs}^{-1}\big(g(rs)^* f(rsh)\big)\big) u(h) \, d\mu_G(r) \, d\mu_H(h) \\
&= \int_{s\cdot H} \int_G \gamma_{s\cdot H}(h) \Delta_G(s) \omega(s, H) \\
&\qquad\qquad s \cdot \pi\big(\alpha_r^{-1}\big(g(rs)^* f(rhs)\big)\big) s \cdot u(h) \, d\mu_G(r) \, d\mu_H(h) \\
&= s \cdot L\big(\langle V(g) \, , V(f) \rangle_{A \rtimes s\cdot H}\big).
\end{aligned}$$

Since V is clearly onto, $f \otimes h \mapsto V(f) \otimes h$ extends to a unitary from the space of $\operatorname{Ind}_H^G L$ to the space of $\operatorname{Ind}_{s\cdot H}^G s \cdot L$. Since $V(f * g) = f * V(g)$, this unitary provides the desired equivalence. □

Theorem 5.9 (Induction in Stages). *Suppose that (A, G, α) is a dynamical system and that K and H are closed subgroups of G with $H \subset K$. If L is a representation of $A \rtimes_{\alpha|_H} H$, then the representations*

$$\operatorname{Ind}_H^G L \quad \textit{and} \quad \operatorname{Ind}_K^G \operatorname{Ind}_H^K L$$

are equivalent.

[4] We have more to say about $\omega(s, H)$ in Lemma H.10 on page 461.

For the proof we'll have to look at the imprimitivity bimodules X_H^G, X_H^K and X_K^G from Theorem 4.22 on page 132. The space of $\operatorname{Ind}_H^G L$ is $\mathsf{X}_H^G \otimes_{A\rtimes H} \mathcal{H}_L$, and the space of $\operatorname{Ind}_K^G \operatorname{Ind}_H^K L$ is

$$\mathsf{X}_K^G \otimes_{A\rtimes K} (\mathsf{X}_H^K \otimes_{A\rtimes H} \mathcal{H}_L). \tag{5.7}$$

Since we can view (5.7) as an iterated internal tensor product, Lemma I.4 on page 481 implies that (5.7) is isomorphic to

$$(\mathsf{X}_K^G \otimes_{A\rtimes K} \mathsf{X}_H^K) \otimes_{A\rtimes H} \mathcal{H}_L$$

where the map of $A \rtimes_{\alpha|_K} K$ into $\mathcal{L}(\mathsf{X}_H^K)$ is given by $N \rtimes v$ defined in Corollary 4.23 on page 133 (with K in place of G). Thus the technical part of the proof of Theorem 5.9 on the previous page is provided by the following lemma.

Lemma 5.10. *Let H, K, G and α be as in Theorem 5.9. Then there is an isomorphism Φ of Hilbert $A \rtimes_{\alpha|_H} H$-modules $\mathsf{X}_K^G \otimes_{A\rtimes K} \mathsf{X}_H^K$ and X_H^G defined by the map sending the elementary tensor $f \otimes b \in C_c(G,A) \odot C_c(K,A) \subset \mathsf{X}_K^G \otimes_{A\rtimes K} \mathsf{X}_H^K$ to $f \cdot b \in \mathsf{X}_H^G$ where $C_c(K,A)$ acts on the right of $C_c(G,A)$ via the action of $A \rtimes_{\alpha|_K} K$ on X_K^G given by* (4.56) *(with K in place of H).*

Proof. We'll write $\gamma_H^K(t)$ for $(\Delta_K(t)/\Delta_H(t))^{\frac{1}{2}}$. To see that the indicated map is isometric, we compute as follows:

$$\begin{aligned}
\langle f\otimes b\,,g\otimes c\rangle_{A\rtimes H}(t) &= \bigl\langle b\,,\langle f\,,g\rangle_{A\rtimes K} * c\bigr\rangle_{A\rtimes H}(t)\\
&= \gamma_H^K(t)\int_K \alpha_z^{-1}\Bigl(b(z)^*\langle f\,,g\rangle_{A\rtimes K} * c(zt)\Bigr)\,d\mu_K(z)\\
&= \gamma_H^K(t)\int_K\int_K \alpha_z^{-1}\Bigl(b(z)^*\langle f\,,g\rangle_{A\rtimes K}(w)\\
&\qquad\qquad \alpha_w\bigl(c(w^{-1}zt)\bigr)\Bigr)\,d\mu_K(w)\,d\mu_K(z)\\
&= \int_K\int_K\int_G \gamma_H^K(t)\gamma_K^G(w)\alpha_z^{-1}\Bigl[b(z)^*\alpha_s^{-1}\bigl(f(s)^*g(sw)\bigr)\\
&\qquad\qquad \alpha_w\bigl(c(w^{-1}zt)\bigr)\Bigr]\,d\mu_G(s)\,d\mu_K(w)\,d\mu_K(z)
\end{aligned}$$

which, after sending $s \mapsto sz^{-1}$, is

$$\begin{aligned}
&= \int_G\int_K\int_K \gamma_H^K(t)\gamma_K^G(w)\Delta_G(z^{-1})\alpha_z^{-1}\bigl(b(z)^*\bigr)\alpha_s^{-1}\bigl(f(sz^{-1})^*g(sz^{-1}w)\bigr)\\
&\qquad\qquad \alpha_{z^{-1}w}\bigl(c(w^{-1}zt)\bigr)\,d\mu_K(w)\,d\mu_K(z)\,d\mu_G(s)
\end{aligned}$$

which, after sending $z \mapsto z^{-1}$, is

$$\begin{aligned}
&= \int_G\int_K\int_K \gamma_H^K(t)\gamma_K^G(w)\Delta_G(z)\Delta_K(z^{-1})\alpha_s^{-1}\Bigl[\alpha_{sz}\bigl(b(z^{-1})^*\bigr)f(sz)^*\\
&\qquad\qquad g(szw)\alpha_{szw}\bigl(c(w^{-1}z^{-1}t)\bigr)\Bigr]\,d\mu_K(w)\,d\mu_K(z)\,d\mu_G(s)
\end{aligned}$$

which, after sending $w \mapsto z^{-1}tw$, is

$$= \int_G \int_K \int_K \gamma_H^K(t)\gamma_K^G(z^{-1}tw)\gamma_K^G(z)^2 \alpha_s^{-1}\Big[\alpha_{sz}\big(b(z^{-1})^*\big)f(sz)^*$$
$$g(stw)\alpha_{stw}\big(c(w^{-1})\big)\Big]\,d\mu_K(w)\,d\mu_K(z)\,d\mu_G(s)$$
$$= \gamma_H^G(t)\int_G \alpha_s^{-1}\Big[\Big(\int_K f(sz)\alpha_{sz}\big(b(z^{-1})\big)\gamma_K^G(z)\,d\mu_K(z)\Big)^*$$
$$\Big(\int_K g(stw)\alpha_{stw}\big(c(w^{-1})\big)\gamma_K^G(w)\,d\mu_K(w)\Big)\Big]\,d\mu_G(s)$$
$$= \gamma_H^G(t)\int_G \alpha_s^{-1}\big(f\cdot b(s)^* g\cdot c(st)\big)\,d\mu_G(s)$$
$$= \langle f\cdot b\,,\, g\cdot c\rangle_{A\rtimes H}(t).$$

To see that the map has dense range, it suffices to check that

$$\{\, f\cdot b : f\in C_c(G,A) \text{ and } b\in C_c(K,A)\,\} \tag{5.8}$$

is dense in $C_c(G,A)$ in the inductive limit topology. To see this, recall that we can view $C_c(G,A)$ as a $C_c(G\times G/K,A)-C_c(K,A)$ pre-imprimitivity bimodule Z_0 coming from the Symmetric Imprimitivity Theorem with respect to the module actions and inner products given in (4.51)–(4.54). If we write $f : b$ for the right action of $b\in C_c(K,A)$ on f viewed as an element in Z_0, then it follows from symmetry and Proposition 4.5 on page 113 that there is an approximate identity in $C_c(K,A)$, $\{\, b_i\,\}$, such that $f : b_i \to f$ in the inductive limit topology on $C_c(G,A)$. However if $u(f)(s) := \Delta_G(s)^{-\frac12} f(s)$, then $u(f:b) = u(f)\cdot b$. Therefore, $f\cdot b_i\to f$ in the inductive limit topology, and (5.8) is dense as required. □

Proof of Theorem 5.9. We define a unitary U to be the composition $U_2\circ U_1$

$$\begin{array}{ccc} \mathsf{X}_K^G \otimes_{A\rtimes K} (\mathsf{X}_H^K \otimes_{A\rtimes H} \mathcal{H}_L) & \xrightarrow{U_1} & (\mathsf{X}_K^G \otimes_{A\rtimes K} \mathsf{X}_H^K)\otimes_{A\rtimes H}\mathcal{H}_L \\ & \overset{U}{\searrow} & \downarrow U_2 \\ & & \mathsf{X}_H^G \otimes_{A\rtimes H} \mathcal{H}_L, \end{array}$$

where U_2 is defined using Lemma 5.10 on the preceding page, and U_1 is defined using Lemma I.4 on page 481. Thus on elementary tensors

$$U\big(f\otimes(b\otimes h)\big) = f\cdot b\otimes h.$$

Since

$$U\circ \operatorname{Ind}_K^G \operatorname{Ind}_H^K L(g)\big(f\otimes(b\otimes h)\big) = U\big(g*f\otimes(b\otimes h)\big)$$
$$= (g*f)\cdot b\otimes h,$$

and since

$$\begin{aligned}(g*f)\cdot b(s) &= \int_K g*f(sz)\alpha_{sz}\big(b(z^{-1})\big)\gamma_K^G(z)\,d\mu_K(z)\\ &= \int_G\int_K g(r)\alpha_r\big(f(r^{-1}sz)\big)\alpha_{sz}\big(b(z^{-1})\big)\gamma_K^G(z)\,d\mu_K(z)\,d\mu_G(r)\\ &= \int_G g(r)\alpha_r\Big(\int_K f(r^{-1}sz)\alpha_{r^{-1}sz}\big(b(z^{-1})\big)\gamma_K^H(z)\,d\mu_K(z)\Big)\,d\mu_G(r)\\ &= \int_G g(r)\alpha_r\big(f\cdot b(r^{-1}s)\big)\,d\mu_G(r)\\ &= g*(f\cdot b)(s),\end{aligned}$$

we have

$$U\circ \operatorname{Ind}_K^G \operatorname{Ind}_H^K L(g) = \operatorname{Ind}_H^G L(g)\circ U.$$

This completes the proof. □

Having defined induced representations of $A\rtimes_\alpha G$, it is natural to ask which representations of $A\rtimes_\alpha G$ are induced from $A\rtimes_{\alpha|_H} H$ for a given closed subgroup H of G. Such results are usually called imprimitivity theorems, and the Symmetric Imprimitivity Theorem and Green's Imprimitivity Theorem derive their names from the fact that they provide the key ingredient in answering the above question. The following remarks will be helpful.

Remark 5.11. By identifying an elementary tensor $\varphi\otimes a$ with a function on G/H in the obvious way, we obtain a map of $C_0(G/H)\odot A$ into $C_0(G/H,A)$. This extends to an isomorphism of $C_0(G/H)\otimes A$ with $C_0(G/H,A)$ [139, Proposition B.16]. Using [139, Corollary B.22], we note that every representation R of $C_0(G/H,A)$ is determined by a pair (η,π) of commuting representations $\eta: C_0(G/H)\to B(\mathcal{H})$ and $\pi: A\to B(\mathcal{H})$ such that

$$R(\varphi\otimes a)=\eta(\varphi)\pi(a). \tag{5.9}$$

In particular, if $k_{C_0(G/H)}$ and k_A are the natural maps of $C_0(G/H)$ and A, respectively, into $M\big(C_0(G/H,A)\big)$, then $\eta=\bar{R}\circ k_{C_0(G/H)}$ and $\pi=\bar{R}\circ k_A$. Since commutative C^*-algebras are nuclear ([139, Theorem B.43]), $C_0(G/H)\otimes A = C_0(G/H)\otimes_{\max} A$, and every pair (η,π) of commuting representations as above gives rise to a representation $R=\eta\otimes_{\max}\pi$ satisfying (5.9) ([139, Theorem B.27]).

To cut down on distracting notation in the sequel, let $E_H^G(A)$ be the imprimitivity algebra $C_0(G/H,A)\rtimes_{\mathrm{lt}\otimes\alpha} G$. Suppose that L is a representation of $A\rtimes_{\alpha|_H} H$. Let X–$\operatorname{Ind} L$ be the representation of $E_H^G(A)$ coming from Green's $E_H^G(A)$ – $A\rtimes_{\alpha|_H} H$-imprimitivity bimodule $\mathsf{X}=\mathsf{X}_H^G$. Then by definition, X–$\operatorname{Ind} L$ acts on the space of $\operatorname{Ind}_H^G L$, and

$$\mathsf{X}\text{–}\operatorname{Ind} L(F)(g\otimes h) := M\rtimes v(F)(g)\otimes h,$$

where (M,v) is the covariant homomorphism giving the natural action of $E_H^G(A)$ on X (Corollary 4.23 on page 133). As in Remark 5.3 on page 152, it follows that X–$\operatorname{Ind} L=(M\otimes 1)\rtimes(v\otimes 1)$, where

$$(M\otimes 1)(z)(g\otimes h)=M(z)(g)\otimes h.$$

In view of Remark 5.11 on the preceding page, we have $M \otimes 1 = \eta \otimes_{\max} \pi$ for commuting representations η of $C_0(G/H)$ and π of A. On the other hand, by Remark 5.3 on page 152, $\operatorname{Ind}_H^G L = (N \otimes 1) \rtimes (v \otimes 1)$. Since $N = M^{-} \circ k_A$, it follows that $N \otimes 1 = (M \otimes 1)^{-} \circ k_A$. Therefore, $N \otimes 1 = \pi$. We have proved the first part of our imprimitivity theorem.

Theorem 5.12 (Imprimitivity Theorem for Crossed Products). *Suppose that (A, G, α) is a dynamical system and that H is a closed subgroup of G. If L is a representation of $A \rtimes_{\alpha|_H} H$ and if $\mathsf{X}\text{–}\operatorname{Ind} L = (\eta \otimes_{\max} \pi) \rtimes U$ is the corresponding representation of Green's imprimitivity algebra $E_H^G(A) = C_0(G/H, A) \rtimes_{\mathrm{lt} \otimes \alpha} G$, then $\operatorname{Ind}_H^G L = \pi \rtimes U$.*

In particular, suppose that $R = \pi \rtimes U$ is a nondegenerate representation of $A \rtimes_\alpha G$ on $\mathcal{H}$. Then R is unitarily equivalent to a representation induced from $A \rtimes_{\alpha|_H} H$ if and only if there is a nondegenerate representation $\eta : C_0(G/H) \to B(\mathcal{H})$ such that η and π are commuting representations, and such that (η, U) is a covariant representation of $\big(C_0(G/H), G, \mathrm{lt}\big)$.

Proof. The first assertion follows from the above discussion. Let $R = \pi \rtimes U$ and η be as in the statement of the theorem. Then it is straightforward to see that $(\eta \otimes_{\max} \pi, U)$ is a nondegenerate covariant representation of $\big(C_0(G/H, A), G, \mathrm{lt} \otimes \alpha\big)$. The integrated form is a representation of $E_H^G(A)$. Since $\mathsf{X} = \mathsf{X}_H^G$ is a $E_H^G(A) - A \rtimes_{\alpha|_H} H$-imprimitivity bimodule, R is equivalent to a representation of the form $\mathsf{X}\text{–}\operatorname{Ind} L = (\eta' \otimes_{\max} \pi') \rtimes U'$ for some representation L of $A \rtimes_{\alpha|_H} H$ ([139, Theorem 3.29]). By the above, $\operatorname{Ind}_H^G L = \pi' \rtimes U'$. Using Proposition 2.40 on page 59, we must have $\eta \otimes_{\max} \pi$ equivalent to $\eta' \otimes_{\max} \pi'$, and U equivalent to U'. This implies that $\pi = (\eta \otimes_{\max} \pi)^{-} \circ k_A$ is equivalent to $\pi' = (\eta' \otimes_{\max} \pi')^{-} \circ k_A$. Thus $R = \pi \rtimes U$ is equivalent to $\operatorname{Ind}_H^G L = \pi' \rtimes U'$, which is what we wanted.

On the other hand, if $\pi \rtimes U = \operatorname{Ind}_H^G L$, then $\mathsf{X}\text{–}\operatorname{Ind} L = (\eta \otimes_{\max} \pi) \rtimes U$ for some representation η of $C_0(G/H)$. It is easy to see that η has the required properties. □

Remark 5.13. The relationship between $\operatorname{Ind}_H^G L$ and $\mathsf{X}\text{–}\operatorname{Ind} L$ is made explicit by the Imprimitivity theorem on this page. Another take is as follows. Let $(j_{C_0(G/H,A)}, j_G)$ be the canonical covariant homomorphism of $\big(C_0(G/H, A), G, \mathrm{lt} \otimes \alpha\big)$ into $M\big(E_H^G(A)\big)$. Then $j'_A := \bar{j}_{C_0(G/H,A)} \circ k_A$ is an nondegenerate homomorphism of A into $M\big(E_H^G(A)\big)$ such that (j'_A, j_G) is a covariant homomorphism of (A, G, α) into $M\big(E_H^G(A)\big)$. By checking on generators, it follows that

$$(\mathsf{X}\text{–}\operatorname{Ind} L)^{-} \circ (j'_A \rtimes j_G) = \operatorname{Ind}_H^G L. \tag{5.10}$$

It is also helpful to keep in mind that the natural identification of $E_H^G(A)$ with $\mathcal{K}(\mathsf{X}_H^G)$ intertwines $j'_A \rtimes j_G$ with $N \rtimes v$.

5.2 An Example

Although Proposition 5.4 on page 153 gives a fairly concrete realization of an induced representation, there is still plenty of work to do to make sense of specific examples. A case where we can give a fairly complete description is that of inducing a character of an abelian group. We will use this material in Section 8.3.

Suppose now that G is abelian. Then the Fourier transform $\varphi \mapsto \hat{\varphi}$ extends to an isomorphism of $C^*(G)$ onto $C_0(\widehat{G})$ (Proposition 3.1 on page 82), where $\widehat{G}$ is the Pontryagin dual and

$$\hat{\varphi}(\sigma) := \int_G \varphi(s)\sigma(s)\,d\mu_G(s).$$

If τ is a character on G, then since we can identify $C_c(G) \odot \mathbf{C}$ with $C_c(G)$, the induced representation $L := \operatorname{Ind}_H^G(\tau|_H)$ acts on the completion of $C_c(G)$ with respect to the inner product

$$(\varphi \mid \psi)_L = \int_H \int_G \overline{\psi(r)}\varphi(rt)\tau(t)\,d\mu_G(r)\,d\mu_H(t). \tag{5.11}$$

Of course, $L(s)[\varphi] = [\mathrm{lt}_s\,\varphi]$.

The left-regular representation of G on $C_c(G) \subset L^2(G)$ is given by $\lambda(r)\varphi(s) = \varphi(r^{-1}s)$, and as a representation of $C_c(G)$, λ is given by convolution: $\lambda(\psi)(\varphi) = \psi * \varphi$. Since $(\varphi * \psi)\hat{\ } = \hat{\varphi}\hat{\psi}$, it follows that as a representation of $C_0(\widehat{G})$, λ is equivalent to the multiplication representation M of $C_0(\widehat{G})$ on $L^2(\widehat{G})$ given by

$$M(g)f = gf.$$

Since $\lambda = \operatorname{Ind}_{\{e\}}^G \iota$, a modest generalization of the above discussion is the following. Recall that if H is a closed subgroup of an abelian group G, then $H^\perp := \{\sigma \in \widehat{G} : \sigma(t) = 1 \text{ for all } t \in H\}$, and that we can identify $H^\perp$ with the dual of G/H [56, Theorem 4.39].

Proposition 5.14. *Suppose that G is a locally compact abelian group and that $\tau \in \widehat{G}$. If H is a closed subgroup, then $\operatorname{Ind}_H^G(\tau|_H)$ is equivalent to the representation M^τ of $C_0(\widehat{G})$ on $L^2(H^\perp)$ given by*

$$M^\tau(f)\xi(\sigma) = f(\tau\sigma)\xi(\sigma).$$

In particular,

$$\ker\bigl(\operatorname{Ind}_H^G(\tau|_H)\bigr) = \{\, f \in C_0(\widehat{G}) : f \text{ vanishes on } \tau H^\perp \,\}.$$

Since $\tau \in \widehat{G}$ is, as a representation of $C_0(\widehat{G})$, simply evaluation at τ, we get the following result as an immediate corollary.

Corollary 5.15. *If $\tau \in \widehat{G}$ and if H is a closed subgroup of G, then*

$$\ker\bigl(\operatorname{Ind}_H^G(\tau|_H)\bigr) \subset \ker\tau.$$

Proof of Proposition 5.14. We realize $L := \operatorname{Ind}_H^G(\tau|_H)$ on the completion of $C_c(G)$ as in (5.11). We can choose a Haar measure $\mu_{G/H}$ on G/H such that for all $f \in C_c(G)$ we have

$$\int_G f(s)\,d\mu_G(s) = \int_{G/H} \int_H f(st)\,d\mu_H(t)\,d\mu_{G/H}(\dot{s}).$$

Then we define $U : C_c(G) \to C_c(G/H)$ by

$$U(f)(\dot{r}) := \int_H f(rt)\tau(rt)\,d\mu_H(t).$$

Using a Bruhat approximate cross section (Proposition H.17 on page 466), we see that U is surjective. Furthermore, interchanging the order of integration as needed, we have

$$\begin{aligned}
\big(U(\varphi) \mid U(\psi)\big)_{L^2(G/H)} &= \int_{G/H} U(\varphi)(\dot{s})\overline{U(\psi)(\dot{s})}\,d\mu_{G/H}(\dot{s}) \\
&= \int_{G/H}\int_H\int_H \varphi(st)\overline{\psi(sv)}\tau(tv^{-1})\,d\mu_H(t)\,d\mu_H(v)\,d\mu_{G/H}(\dot{s}) \\
&= \int_H\int_{G/H}\int_H \overline{\psi(sv)}\varphi(svt)\tau(t)\,d\mu_H(v)\,d\mu_{G/H}(\dot{s})\,d\mu_H(t) \\
&= \int_H\int_G \overline{\psi(s)}\varphi(st)\tau(t)\,d\mu_G(s)\,d\mu_H(t) \\
&= (\varphi \mid \psi)_L.
\end{aligned}$$

Therefore U extends to a unitary of $\mathcal{H}_L$ onto $L^2(G/H)$. Also,

$$\begin{aligned}
U\big(L(s)(\varphi)\big)(\dot{r}) &= \int_H L(s)\varphi(rt)\tau(rt)\,d\mu_H(t) \\
&= \int_H \varphi(s^{-1}rt)\tau(rt)\,d\mu_H(t) \\
&= \tau(s)U(\varphi)(s^{-1}\cdot\dot{r}).
\end{aligned}$$

Therefore U intertwines $L = \operatorname{Ind}_H^G(\tau|_H)$ with the representation R on $L^2(G/H)$ given by

$$R(s)F(\dot{r}) = \tau(s)F(s^{-1}\cdot\dot{r}).$$

Since we can identify $H^\perp$ with $(G/H)^\wedge$, the Plancherel Theorem [56, 4.25] implies that the Fourier transform induces a unitary $V : L^2(G/H) \to L^2(H^\perp)$ given on $F \in C_c(G/H)$ by

$$V(F)(\sigma) = \int_{G/H} F(\dot{r})\sigma(r)\,d\mu_{G/H}(\dot{r}) := \hat{F}(\sigma).$$

If $\varphi \in C_c(G)$, then

$$R(\varphi)F(\dot{r}) = \int_G \varphi(s)\tau(s)F(s^{-1}\cdot\dot{r})\,d\mu_G(s),$$

and

$$\begin{aligned}
V\big(R(\varphi)F\big)(\sigma) &= \int_{G/H} R(\varphi)F(\dot{r})\sigma(r)\,d\mu_{G/H}(\dot{r}) \\
&= \int_{G/H}\int_G \varphi(s)\tau(s)F(s^{-1}\cdot\dot{r})\sigma(r)\,d\mu_G(s)\,d\mu_{G/H}(\dot{r})
\end{aligned}$$

which, after interchanging the integrals and using the invariance of the Haar measure $\mu_{G/H}$ to replace $s^{-1}\cdot \dot r = s^{-1}rH = s^{-1}HrH$ with $\dot r = rH$, is

$$\begin{aligned}&= \int_G\int_{G/H} \varphi(s)\tau(s)F(\dot r)\sigma(sr)\,d\mu_{G/H}(\dot r)\,d\mu_G(s)\\ &= \hat\varphi(\tau\sigma)V(F)(\sigma).\end{aligned}$$

Thus VU intertwines $L = \operatorname{Ind}_H^G(\tau|_H)$ with the representation M^τ. The rest follows easily. □

5.3 Inducing Ideals

If A is a C^*-algebra, then $\mathcal{I}(A)$ denotes the set of (closed two-sided) ideals in A. The set $\mathcal{I}(A)$ is partially ordered by inclusion: that is, $I \le J$ means $I \subset J$. Then $\mathcal{I}(A)$ is a *lattice* with $I\wedge J := I\cap J$ and $I \vee J$ is the ideal generated by I and J. It is sometimes convenient to equip $\mathcal{I}(A)$ with the topology with subbasic open sets indexed by $J \in \mathcal{I}(A)$ given by

$$\mathcal{O}_J = \{\, I \in \mathcal{I}(A) : J \not\subset I \,\}.$$

(Thus the open sets are the unions of finite intersections $\bigcap\{\mathcal{O}_J : J \in F\}$; the whole space arises as the intersection over the empty set F.) It is important to keep in mind that when restricted to $\operatorname{Prim} A$, the $\mathcal{O}_J$ form a *basis* for a topology — the hull-kernel or Jacobson topology. In particular, the relative topology on $\operatorname{Prim} A$ as a subset of $\mathcal{I}(A)$ is the natural one.

We need the following result which is essentially contained in [139, §3.3].

Lemma 5.16. *If H is a closed subgroup of G, then there is a containment preserving continuous map*

$$\operatorname{Ind}_H^G : \mathcal{I}(A \rtimes_{\alpha|_H} H) \to \mathcal{I}(A\rtimes_\alpha G) \tag{5.12}$$

such that

$$\operatorname{Ind}_H^G \ker L = \ker \operatorname{Ind}_H^G L \tag{5.13}$$

for all nondegenerate representations of $A \rtimes_{\alpha|_H} H$. Moreover

$$\operatorname{Ind}_H^G J = \{\, f\in A\rtimes_\alpha G : N \rtimes v(f)(\mathsf{X}) \subset \overline{\mathsf{X}\cdot J}\,\}, \tag{5.14}$$

where $\mathsf{X} = \mathsf{X}_H^G$ is the $C_0(G/H, A) \rtimes_{\mathrm{lt}\otimes\alpha} G$ – $A\rtimes_{\alpha|_H} H$-imprimitivity bimodule of Theorem 4.22 on page 132, and (N, v) is the natural covariant homomorphism of (A, G, α) into $\mathcal{L}(\mathsf{X})$ given in Corollary 4.23 on page 133.

Remark 5.17. The Cohen Factorization Theorem [139, Proposition 2.33] implies that

$$\overline{\mathsf{X}\cdot J} := \overline{\operatorname{span}}\{\, x\cdot a : x \in \mathsf{X} \text{ and } a \in J\,\} = \{\, x\cdot a : x\in \mathsf{X} \text{ and } a \in J\,\}.$$

In particular, we could replace $\overline{\mathsf{X}\cdot J}$ with $\mathsf{X}\cdot J$ in formulas such as (5.14). However, while this observation can be useful in certain cases, I have chosen to retain the overlines for emphasis.

Proof. The map Ind_H^G is defined on page 61 of [139] to be the composition of the restriction map $\operatorname{Res}_{N\rtimes v} : \mathcal{I}(\mathcal{K}(\mathsf{X})) \to \mathcal{I}(A \rtimes_\alpha G)$ and the Rieffel homeomorphism $\mathsf{X}\text{–Ind} : \mathcal{I}(A \rtimes_{\alpha|_H} H) \to \mathcal{I}(\mathcal{K}(\mathsf{X}))$ where

$$\operatorname{Res}_{N\rtimes v} I := \{\, f \in A \rtimes_\alpha G : N \rtimes v(f)\big(\mathcal{K}(\mathsf{X})\big) \subset I \,\}.$$

Thus

$$\operatorname{Ind}_H^G J = \{\, f \in A \rtimes_\alpha G : N \rtimes v(f)\big(\mathcal{K}(\mathsf{X})\big) \subset \mathsf{X}\text{–Ind}\, J \,\}. \tag{5.15}$$

If $J := \ker L$, then the right-hand side of (5.15) is the kernel of $\operatorname{Ind}_H^G L$ by [139, Proposition 3.34].

We have $\overline{(\mathsf{X}\text{–Ind}\, J)\cdot \mathsf{X}} = \overline{\mathsf{X}\cdot J}$ by [139, Proposition 3.24]. Since $\mathcal{K}(\mathsf{X})$ acts nondegenerately on X, it follows that whenever $f \in \operatorname{Ind}_H^G J$, we have

$$N \rtimes v(f)(\mathsf{X}) \subset \overline{\mathsf{X}\cdot J}.$$

On the other hand, if $f \in A \rtimes_\alpha G$ and $x \in \mathsf{X}$ is such that $N \rtimes v(f)(x) \notin \overline{\mathsf{X}\cdot J}$, then we can write $x = Ty$ for $y \in \mathsf{X}$ and $T \in \mathcal{K}(\mathsf{X})$ by [139, Proposition 2.31], and conclude that $(N \rtimes v(f)T)y \notin \overline{\mathsf{X}\cdot J}$. Thus

$$N \rtimes v(f)\big(\mathcal{K}(\mathsf{X})\big) \not\subset \mathsf{X}\text{–Ind}\, J,$$

and we have proved (5.14).

The continuity of Ind_H^G is proved in [139, Corollary 3.35]. It will also follow from the next two lemmas. □

We take a slight detour to show that induction of ideals preserves intersections. This will be useful in Section 8.1, and together with the following observation, it also takes care of the continuity assertion in Lemma 5.16 on the facing page.

Lemma 5.18. *Let A and B be C^*-algebras with ideal spaces $\mathcal{I}(A)$ and $\mathcal{I}(B)$, respectively. If $\varphi : \mathcal{I}(A) \to \mathcal{I}(B)$ preserves arbitrary intersections, then φ is continuous.*

Proof. Notice that if $I \in \overline{\{ I_k \}}$ in $\mathcal{I}(A)$, then $\bigcap_k I_k \subset I$. (If not, then $\mathcal{O}_{\bigcap_k I_k}$ is a neighborhood of I which is disjoint from $\{ I_k \}$.) If φ preserves intersections and $I \subset J$, then $\varphi(I) = \varphi(I \cap J) = \varphi(I) \cap \varphi(J) \subset \varphi(J)$. Therefore φ also preserves containment. If $\{ I_k \}$ is a net in $\mathcal{I}(A)$ converging to I, it will suffice to see that $\{ \varphi(I_k) \}$ is eventually in any basic neighborhood $\mathcal{O}_J$ of $\varphi(I)$. If this fails for an ideal J, then we can pass to a subnet, relabel, and assume that $\varphi(I_k) \notin \mathcal{O}_J$ for all k. Since $\varphi(I_k) \supset J$ for all k and since $I_k \to I$ implies that $\bigcap I_k \subset I$, we have

$$\begin{aligned} \varphi(I) &\supset \varphi\Big(\bigcap I_k\Big) \quad \text{(since } \varphi \text{ preserves containment)} \\ &= \bigcap \varphi(I_k) \quad \text{(since } \varphi \text{ preserves intersections)} \\ &\supset J. \end{aligned}$$

But this contradicts our choice of J. □

Lemma 5.19. *Suppose that (A, G, α) is a dynamical system, that H is a closed subgroup of G and that $\{ I_k \}_{k \in K}$ is a (possibly infinite) family of ideals in $\mathcal{I}(A \rtimes_{\alpha|_H} H)$. Then*

$$\operatorname{Ind}_H^G \Big(\bigcap_{k \in K} I_k \Big) = \bigcap_{k \in K} \operatorname{Ind}_H^G I_k.$$

Proof. Let $\mathsf{X} = \mathsf{X}_H^G$ be the $C_0(G/H, A) \rtimes_{\mathrm{lt} \otimes \alpha} G - A \rtimes_{\alpha|_H} H$-imprimitivity bimodule of Theorem 4.22 on page 132. Then the result will follow immediately from (5.14) once we verify that

$$\overline{\mathsf{X} \cdot \bigcap I_k} = \bigcap \overline{\mathsf{X} \cdot I_k}. \tag{5.16}$$

Since the right-hand side of (5.16) is a closed sub-bimodule of X and since the Rieffel correspondence [139, Theorem 3.22] establishes a bijection between closed submodules of X and ideals in $A \rtimes_{\alpha|_H} H$,

$$\bigcap \overline{\mathsf{X} \cdot I_k} = \overline{\mathsf{X} \cdot J},$$

where

$$J = \{ \langle x \, , y \rangle_{A \rtimes H} : x \in \mathsf{X} \text{ and } y \in \bigcap \overline{\mathsf{X} \cdot I_k} \}.$$

But

$$\langle \mathsf{X} \, , \bigcap \overline{\mathsf{X} \cdot I_k} \rangle_{A \rtimes H} \subset \bigcap I_k$$

and therefore

$$\bigcap \overline{\mathsf{X} \cdot I_k} = \overline{\mathsf{X} \cdot J} \subset \overline{\mathsf{X} \cdot \bigcap I_k}.$$

Since the other containment is clear, we're done. □

Remark 5.20. One can also prove Lemma 5.19 by showing that Ind_H^G preserves direct sums. Then the result follows from (5.13).

5.4 Invariant Ideals and the Induction Process

If I is an α-invariant ideal in A and if H is a closed subgroup of G, then we'd like to see that the decomposition of $A \rtimes_\alpha G$ given in Proposition 3.19 on page 93 is compatible with the inducing process. We continue to write $E_H^G(A)$ for the imprimitivity algebra $C_0(G/H, A) \rtimes_{\mathrm{lt} \otimes \alpha} G$, and similarly with $E_H^G(I) := C_0(G/H, I) \rtimes_{\mathrm{lt} \otimes \alpha} G$ and $E_H^G(A/I) := C_0(G/H, A/I) \rtimes_{\mathrm{lt} \otimes \alpha^I} G$. Let X_H^G be the $E_H^G(A) - A \rtimes_{\alpha|_H} H$-imprimitivity bimodule of Theorem 4.22 on page 132. We also have a $E_H^G(I) - I \rtimes_{\alpha|_H} H$-imprimitivity bimodule Y_H^G and a $E_H^G(A/I) - A/I \rtimes_{\alpha^I|_H} H$-imprimitivity bimodule Z_H^G. As usual, we'll write X_0, Y_0 and Z_0 for the dense subspaces $C_c(G, A)$, $C_c(G, I)$ and $C_c(G, A/I)$ of X_H^G, Y_H^G and Z_H^G, respectively.

Now suppose that L is a nondegenerate representation of $I \rtimes_{\alpha|_H} H$. Let $\mathsf{Y}\text{–}\operatorname{Ind}_H^G L$ be the representation of $I \rtimes_\alpha G$ induced via Y_H^G. Since $\mathsf{Y}\text{–}\operatorname{Ind}_H^G L$ is nondegenerate, it extends to a representation $\mathsf{Y}\text{–}\overline{\operatorname{Ind}}_H^G L$ of $A \rtimes_\alpha G$. On the other hand, L has a canonical extension to a representation $\bar{L}$ of A, and we can form

$\mathsf{X}\text{–}\mathrm{Ind}_H^G \bar{L}$. Naturally, we expect that $\mathsf{Y}\text{–}\overline{\mathrm{Ind}}_H^G L$ and $\mathsf{X}\text{–}\mathrm{Ind}_H^G \bar{L}$ are equivalent. Proving this requires untangling a bit of the machinery of the Rieffel correspondence as described in [139, §3.3]. Recall that if X is any $E-A$-imprimitivity bimodule and $I \in \mathcal{I}(A)$, then $\overline{\mathsf{X}\cdot I}$ is a closed $E-A$-sub-bimodule of X. The Rieffel map $\mathsf{X}\text{–}\mathrm{Ind} : \mathcal{I}(A) \to \mathcal{I}(E)$ sending I to

$$ {}_E\langle \mathsf{X}\cdot I\,, \mathsf{X}\cdot I\rangle = \overline{\operatorname{span}}\{\, {}_E\langle x\,, y\rangle : x, y \in \mathsf{X}\cdot I\,\} $$

is a lattice bijection ([139, Proposition 3.24]), and $\overline{\mathsf{X}\cdot I}$ is a $\mathsf{X}\text{–}\mathrm{Ind}\, I - I$-imprimitivity bimodule with respect to the inherited actions and inner products. ([139, Proposition 3.25]).

Proposition 5.21. *Suppose that (A,G,α) is a dynamical system and that I is an α-invariant ideal in A. If we identify $E_H^G(I)$ with the ideal $\mathrm{Ex}\big(C_0(G/H,I)\big)$ in $E_H^G(A)$, then*

$$ E_H^G(I) = \mathsf{X}\text{–}\mathrm{Ind}(I \rtimes_{\alpha|_H} H), $$

and $\mathsf{X}_{I\rtimes H} := \overline{\mathsf{X}_H^G \cdot (I \rtimes_{\alpha|_H} H)}$ is a $E_H^G(I) - I \rtimes_{\alpha|_H} H$-imprimitivity bimodule. The inclusion of $C_c(G,I)$ into $C_c(G,A)$ induces an imprimitivity bimodule isomorphism of Y_H^G onto $\mathsf{X}_{I\rtimes H}$. If L is a nondegenerate representation of $I \rtimes_{\alpha|_H} H$, then

$$ \mathsf{Y}\text{–}\overline{\mathrm{Ind}}_H^G L \quad \textit{and} \quad \mathsf{X}\text{–}\mathrm{Ind}_H^G \bar{L} $$

are equivalent representations of $A \rtimes_\alpha G$.

We need a simple observation.

Lemma 5.22. *The submodule $\mathsf{X}_{I\rtimes H}$ is the closure of $C_c(G,I)$ in X_H^G.*

Proof. Using Equation (4.56) of Theorem 4.22 on page 132, it is easy to see that $f\cdot b$ belongs to $C_c(G,I)$ when $f \in C_c(G,A)$ and $b \in C_c(H,I)$. Thus we have $\mathsf{X}\cdot(I \rtimes_{\alpha|_H} H) \subset \overline{C_c(G,I)}$. Let $\{\, u_i \,\} \subset C_c(H,I)$ be an approximate identity for $I \rtimes_{\alpha|_H} H$. Thus $b * u_i \to b$ and $u_i * b \to b$ in $I \rtimes_{\alpha|_H} H$, and therefore in $A \rtimes_{\alpha|_H} H$, for all $b \in C_c(H,I)$. If $f \in C_c(G,I)$, then $\langle f\,, f\rangle_{A\rtimes H} \in C_c(H,I)$ and

$$ \begin{aligned} &\|f\cdot u_i - f\|^2_{\mathsf{X}_H^G} = \\ &\qquad \|u_i * \langle f\,, f\rangle_{A\rtimes H} * u_i - \langle f\,, f\rangle_{A\rtimes H} * u_i - u_i * \langle f\,, f\rangle_{A\rtimes H} + \langle f\,, f\rangle_{A\rtimes H}\| \end{aligned} $$

tends to zero as i increases. Since each $f\cdot u_i \in \mathsf{X}\cdot(I \rtimes_{\alpha|_H} H)$, we have $C_c(G,I) \subset \mathsf{X}_{I\rtimes H}$. This completes the proof. □

Proof of Proposition 5.21. Since we have identified $E_H^G(I)$ with the corresponding ideal in $E_H^G(A)$, it follows from (4.58) in Theorem 4.22 on page 132 that

$$ {}_{E_H^G(I)}\langle f\,, g\rangle = {}_{E_H^G(A)}\langle f\,, g\rangle \quad \text{for all } f, g \in C_c(G,I). \tag{5.17} $$

Since

$$ \mathsf{X}\text{–}\mathrm{Ind}_H^G(I \rtimes_{\alpha|_H} H) = \overline{{}_{E_H^G(A)}\langle \mathsf{X}_{I\rtimes H}\,, \mathsf{X}_{I\rtimes H}\rangle} $$

and since $C_c(G,I)$ is dense in $\mathsf{X}_{I\rtimes H}$, it follows from (5.17) that

$$\mathsf{X}\text{–}\mathrm{Ind}_H^G(I\rtimes_{\alpha|_H} H) = E_H^G(I) = C_0(G/H,I)\rtimes_{\mathrm{lt}\otimes\alpha} G,$$

and $\mathsf{X}_{I\rtimes H}$ is a $E_H^G(I)$ – $I\rtimes_{\alpha|_H} H$-imprimitivity bimodule as claimed.

It now follows from Lemma 5.22 on the previous page and (5.17) that the inclusion map induces an isomorphism Φ_I of Y_H^G onto $\mathsf{X}_{I\rtimes H}$.

Notice that each $T\in\mathcal{L}(\mathsf{X}_H^G)$ is $I\rtimes_{\alpha|_H} H$-linear so that $\mathsf{X}_{I\rtimes H}$ is an invariant subspace for T. Thus restriction gives us a map $r:\mathcal{L}(\mathsf{X}_H^G)\to\mathcal{L}(\mathsf{X}_{I\rtimes H})$. If (N^A,v^A) is the covariant homomorphism of (A,G,α) into $\mathcal{L}(\mathsf{X})$ given in Corollary 4.23 on page 133 and if $f\in I\rtimes_\alpha G$ (which we view as an ideal in $A\rtimes_\alpha G$), then

$$N^I\rtimes v^I(f) = \Phi_I^{-1}\circ r\big(N^A\rtimes v^A(f)\big)\circ\Phi_I.$$

Thus $\mathsf{Y}\text{–}\overline{\mathrm{Ind}_H^G}\, L$ is equivalent to the representation on $\mathsf{X}_{I\rtimes H}\otimes_{I\rtimes H}\mathcal{H}_L$ given by the left-action of $A\rtimes_\alpha G$ on $\mathsf{X}_{I\rtimes H}$.

On the other hand, $\mathsf{X}\text{–}\mathrm{Ind}_H^G\,\bar{L}$ acts on $\mathsf{X}_H^G\otimes_{A\rtimes H}\mathcal{H}_L$ by the left-action of $A\rtimes_\alpha G$ on X_H^G. If $f_1,\dots,f_n,g_1,\dots,g_n\in C_c(G,A)\subset\mathsf{X}_H^G$ and if $b_1,\dots,b_n,c_1,\dots,c_n\in I\rtimes_{\alpha|_H} H$, then

$$\begin{aligned}\Big(\sum_i f_i\otimes L(b_i)h_i \mid \sum_j g_j\otimes L(c_j)k_j\Big) &= \sum_{ij}\big(L(\langle g_j\cdot c_j\,,\, f_i\cdot b_i\rangle_{A\rtimes H})h_i\mid k_j\big)\\ &= \sum_{ij}\big(L(\langle g_j\cdot c_j\,,\, f_i\cdot b_i\rangle_{I\rtimes H})h_i\mid k_j\big)\\ &= \Big(\sum_i f_i\cdot b_i\otimes h_i\mid\sum_j g_j\cdot c_j\otimes k_j\Big).\end{aligned}$$

Since L is nondegenerate, it follows that

$$\sum_i f_i\otimes L(b_i)h_i\mapsto\sum_i f_i\cdot b_i\otimes h_i$$

extends to a unitary isomorphism of $\mathsf{X}\otimes_{A\rtimes H}\mathcal{H}_L$ onto $\mathsf{X}_{I\rtimes H}\otimes_{I\rtimes H}\mathcal{H}_L$ intertwining the left $A\rtimes_\alpha G$-actions. This proves that $\mathsf{Y}\text{–}\overline{\mathrm{Ind}_H^G}\, L$ and $\mathsf{X}\text{–}\mathrm{Ind}_H^G\,\bar{L}$ are equivalent. □

If R is a representation of $A/I\rtimes_{\alpha^I|_H} H$ and if $j^H: A\rtimes_{\alpha|_H} H\to A/I\rtimes_{\alpha^I|_H} H$ is the quotient map as in Proposition 3.19 on page 93, then $R\circ j^H$ is a nondegenerate representation of $A\rtimes_{\alpha|_H} H$. On the other hand, $\mathsf{Z}\text{–}\mathrm{Ind}_H^G R$ is a representation of $A/I\rtimes_{\alpha^I} G$, and we want to compare

$$\mathsf{X}\text{–}\mathrm{Ind}_H^G(R\circ j^H)\quad\text{and}\quad(\mathsf{Z}\text{–}\mathrm{Ind}_H^G R)\circ j^G,$$

where $j^G: A\rtimes_\alpha G\to A/I\rtimes_{\alpha^I} G$ is the quotient map. To do this, we'll want to work with the quotient module $\mathsf{X}^{I\rtimes H} := \mathsf{X}_H^G/\mathsf{X}_{I\rtimes H}$ which is a $E_H^G(A)/E_H^G(I)$ – $A\rtimes_{\alpha|_H} H/I\rtimes_{\alpha|_H} H$-imprimitivity bimodule by [139, Proposition 3.25]. Proposition 3.19 on page 93 allows us to identify $A\rtimes_{\alpha|_H} H/I\rtimes_{\alpha|_H} H$ with $A/I\rtimes_{\alpha^I|_H} H$, and $E_H^G(A)/E_H^G(I)$ with $C_0(G/H,A)/C_0(G/H,I)\rtimes_{(\mathrm{lt}\otimes\alpha)^{E_H^G(I)}} G$. The latter crossed product can then be identified with $E_H^G(A/I)$ in view Lemma 3.18 on page 93.

Proposition 5.23. *Suppose that (A, G, α) is a dynamical system and that I is an α-invariant ideal in A. If we identify $E^G_H(A)/E^G_H(I)$ with $E^G_H(A/I)$ as above, then the natural map q' of $C_c(G, A)$ onto $C_c(G, A/I)$ given in Lemma 3.18 on page 93 induces an imprimitivity bimodule isomorphism of $\mathsf{X}^{I\rtimes H}$ onto Z^G_H. If R is a nondegenerate representation of $A/I \rtimes_{\alpha^I|_H} H$ and $j^H : A \rtimes_{\alpha|_H} H \to A/I \rtimes_{\alpha^I|_H} H$ is the quotient map of Proposition 3.19 on page 93, then*

$$\mathsf{X}\text{-}\mathrm{Ind}^G_H(R \circ j^H) \quad \textit{and} \quad (\mathsf{Z}\text{-}\mathrm{Ind}^G_H R) \circ j^G$$

are equivalent representations of $A \rtimes_\alpha G$. Furthermore, we always have

$$I \rtimes_\alpha G \subset \mathrm{Ind}^G_H(I \rtimes_{\alpha|_H} H)$$

with equality if and only if the canonical map $N^{A/I} \rtimes v^{A/I}$ of $A/I \rtimes_{\alpha^I} G$ into $\mathcal{L}(\mathsf{Z}^G_H)$ is faithful.

Remark 5.24. In Section 7.2, when we define the reduced crossed product $A \rtimes_{\alpha,r} G$, we will prove in Lemma 7.12 on page 199 that $A/I \rtimes_{\alpha^I,r} G = A/I \rtimes_{\alpha^I} G$ if and only if $N^{A/I} \rtimes v^{A/I}$ is faithful, and therefore if and only if $\mathrm{Ind}^G_H(I \rtimes_{\alpha|_H} H) = I \rtimes_\alpha G$. Thus it will follow from Theorem 7.13 on page 199, that $\mathrm{Ind}^G_H(I \rtimes_{\alpha|_H} H) = I \rtimes_\alpha G$ whenever G is amenable.

Proof. As described above, we view $\mathsf{X}^{I\rtimes H}$ as a $E^G_H(A/I) - A/I \rtimes_{\alpha^I|_H} H$-imprimitivity bimodule using [139, Proposition 3.25]. If $q : \mathsf{X} \to \mathsf{X}^{I\rtimes H}$ is the quotient map, then the inner products on $\mathsf{X}^{I\rtimes H}$ are given by

$$\big\langle q(f)\,,\, q(g)\big\rangle_{A/I\rtimes H} := j^H\big(\langle f\,,\, g\rangle_{A\rtimes H}\big), \text{ and}$$
$${}_{E^G_H(A/I)}\big\langle q(f)\,,\, q(g)\big\rangle := j'\big({}_{E^G_H(A)}\langle f\,,\, g\rangle\big),$$

where j' is the composition of the maps

$$\underbrace{E^G_H(A) \longrightarrow \frac{E^G_H(A)}{E^G_H(I)} \longrightarrow \frac{C_0(G/H, A)}{C_0(G/H, I)} \rtimes_{(\mathrm{lt}\otimes\alpha)^{E^G_H(I)}} G \longrightarrow E^G_H(A/I)}_{j'}.$$

Since, for example,

$$j^H\big(\langle f\,,\, g\rangle_{A\rtimes H}\big) = \big\langle q'(f)\,,\, q'(g)\big\rangle_{A/I\rtimes_{\alpha^I} G},$$

q' is isometric and induces an isomorphism of $\Phi^I : \mathsf{X}^{I\rtimes H} \to \mathsf{Z}^G_H$ as claimed.

Let $(N^{A/I}, v^{A/I})$ be the covariant homomorphism of $(A/I, G, \alpha^I)$ into $\mathcal{L}(\mathsf{Z}^G_H)$ given by Corollary 4.23 on page 133. Since each $T \in \mathcal{L}(\mathsf{X}^G_H)$ is $I \rtimes_{\alpha|_H} H$-linear, we get a map $p : \mathcal{L}(\mathsf{X}) \to \mathcal{L}(\mathsf{X}^{I\rtimes H})$ given by $p(T)\big(q(f)\big) := q(Tf)$. Then for all $f \in A \rtimes_\alpha G$,

$$(\Phi^I)^{-1} \circ (N^{A/I} \rtimes v^{A/I})\big(j^G(f)\big) \circ \Phi^I = p \circ (N^A \rtimes v^A)(f). \tag{5.18}$$

Since (5.14) of Lemma 5.16 on page 164 implies that

$$\operatorname{Ind}_H^G(I \rtimes_{\alpha|_H} H) = \ker\bigl(p \circ (N^A \rtimes v^A)\bigr),$$

we have

$$\operatorname{Ind}_H^G(I \rtimes_{\alpha|_H} H) = \ker\bigl((N^{A/I} \rtimes v^{A/I}) \circ j^G\bigr),$$

and it follows that

$$\ker j^G = \operatorname{Ex} I = I \rtimes_\alpha G \subset \operatorname{Ind}_H^G(I \rtimes_{\alpha|_H} H),$$

with equality if and only if $N^{A/I} \rtimes v^{A/I}$ is faithful.

Furthermore, (5.18) implies that $(\mathsf{Z}\text{–}\operatorname{Ind}_H^G R) \circ j^G$ is equivalent to the representation L of $A \rtimes_\alpha G$ on $\mathsf{X}^{I \rtimes H} \otimes_{A/I \rtimes H} \mathcal{H}_R$ given by left multiplication: that is, $L(f)(q(g) \otimes h) = q(f * g) \otimes h$.

On the other hand, $\mathsf{X}\text{–}\operatorname{Ind}_H^G(R \circ j^H)$ is the representation of $A \rtimes_\alpha G$ acting on $\mathsf{X}_H^G \otimes_{A \rtimes H} \mathcal{H}_R$ by left multiplication. Since

$$\begin{aligned}(f \otimes h \mid g \otimes k) &= \bigl(R \circ j^H(\langle g\,,f\rangle_{A \rtimes H})h \mid k\bigr) \\ &= \bigl(R(\langle q(g)\,,q(f)\rangle_{A/I \rtimes H})h \mid k\bigr) \\ &= \bigl(q(f) \otimes h \mid q(g) \otimes k\bigr),\end{aligned}$$

the map $f \otimes h \mapsto q(f) \otimes h$ extends to a unitary isomorphism of $\mathsf{X}_H^G \otimes_{A \rtimes H} \mathcal{H}_R$ onto $\mathsf{X}^{I \rtimes H} \otimes_{A/I \rtimes H} \mathcal{H}_R$ intertwining the $A \rtimes_\alpha G$-actions. This shows that

$$\mathsf{X}\text{–}\operatorname{Ind}_H^G(R \circ j^H) \quad \text{and} \quad (\mathsf{Z}\text{–}\operatorname{Ind}_H^G R) \circ j^G$$

are equivalent as claimed. □

Notes and Remarks

The notion of an induced representation goes back to Frobenius [58]. Our jumping off point is Mackey's theory for locally compact groups [102–104]. Blattner extended the theory to general locally compact groups in [10]. Rieffel recast that theory in terms of Hilbert modules and Morita equivalence beginning with [145]. There is an extensive treatment of the whole subject with complete references in [55, Chap. XI & XII]. Using a "Mackey-like" approach, Takesaki extended the theory to crossed products in [162]. The treatment in this chapter is based on Green's "Rieffel-like" theory of induced representations and ideals from [66].

Chapter 6

Orbits and Quasi-orbits

As we continue to uncover the general structure of crossed-products, we will find that a great deal of information is encoded in the stability groups for the action of G on $\operatorname{Prim} A$, and in the topology of the orbit space $G\backslash \operatorname{Prim} A$. This is a particularly powerful paradigm in the case of transformation group C^*-algebras. Although the theory is more difficult in the general case, due to the appearance of projective representations of the stability groups arising from Mackey obstructions, it is still a central feature of the Effros-Hahn conjecture that much of the ideal structure can be understood from the stability groups and orbit structure. Since we want to consider general dynamical systems, we will have to pay attention to G-spaces such as $\operatorname{Prim} A$ which fail to be Hausdorff.

6.1 The Mackey-Glimm Dichotomy

Earlier, we remarked that the orbit space $G\backslash X$ of a locally compact G-space X was either pathological as in Example 3.26 on page 95, or enjoyed a number of reasonable regularity conditions. It is time to address this vague assertion. There are a number of conditions one might want to impose on a G-action. On the one hand, we can ask that the orbit space $G\backslash X$ have a well-behaved topology. However, examples show that requiring $G\backslash X$ to be Hausdorff is too restrictive. As a minimum requirement, recall that a topological space is said to be T_0 (or, if you're being formal, a topological space is said to satisfy the T_0 axiom of separability) if given distinct points p and q, there *either* exists an open set containing p and not q or there exists an open set containing q and not p. (Alternatively, distinct points must have distinct closures.) The textbook example of a T_0 topological space is the real line equipped with the topology whose nontrivial open sets are (a, ∞) for $a \in \mathbf{R}$. The orbit space for Example 3.26 on page 95 is not T_0, while that for Example 3.32 on page 96 is.

It should be kept in mind that the primitive ideal space $\operatorname{Prim} A$ of a C^*-algebra A is locally compact [28, Corollary 3.3.8], and although usually not Hausdorff, it is always at least a T_0-topological space [110, Theorem 5.4.7]. The primitive ideal space of GCR or postliminary C^*-algebras (see the discussion on page 221 of [139]) always have a dense open subspace which is Hausdorff ([28, Theorem 4.4.5] or [126, Theorem 6.2.11]).[1] Since quotients of GCR algebras are GCR, it will follow from Lemma 6.3 on the next page and the correspondence between closed subsets of $\operatorname{Prim} A$ with the primitive ideal spaces of the corresponding quotient, that the primitive ideal space of GCR C^*-algebras are "almost Hausdorff" in the sense defined below.

Definition 6.1. A not necessarily Hausdorff locally compact space X is said to be *almost Hausdorff* if each locally compact subspace V contains a relatively open nonempty Hausdorff subset.

Note that with the exception of Example 3.26 on page 95, the orbit spaces in our examples have all been almost Hausdorff. We'll have more to say about almost Hausdorff spaces in Lemma 6.3 on the next page. (In particular, an almost Hausdorff space has a dense open Hausdorff subset as in Example 3.32 on page 96.)

Instead of focusing on the topology of $G\backslash X$, we might want the orbits themselves to have nice properties. For example, if X is any G-space and if $x \in X$, then $sG_x \mapsto s \cdot x$ is a continuous bijection of G/G_x onto the orbit $G \cdot x$, and it would be reasonable to insist that these maps always be homeomorphisms. As Example 3.26 on page 95 shows, this need not always be the case. Another property we can ask of the orbits is that they at least be "nice" subspaces of X (in the relative topology). For example, we might want them to be locally compact which, at least when X is Hausdorff, is equivalent to insisting that they be locally closed subsets of X (Lemma 1.26 on page 6).[2] (Or, more simply, we want each $G \cdot x$ to be open in its closure $\overline{G \cdot x}$.) A generally weaker condition is that each orbit be a Baire space.[3] Any locally compact Hausdorff space is a Baire space [168, Corollary 25.4], and any G_δ subset of a locally compact Hausdorff space is again Baire [168, Theorem 25.3].[4] It is easy to see that if a Baire space is the countable union of closed subsets, then at least one of these sets has nonempty interior.[5]

As our next theorem implies, all the above properties on our wish list for the orbits and orbit space are guaranteed provided the orbit space is tolerably nice and that the spaces involved are second countable.[6] Our result is a simplified version of results due to Glimm [59] and extended to a wider class of G-spaces by Effros [46, 48]. However, as examples such as Example 3.33 on page 97 show, some sort of separability is crucial. Although the full results extend the dichotomy to the

[1] When we assert that U is a Hausdorff subspace of a space X, we are asserting that distinct points in U have disjoint neighborhoods *in* U. It does not follow that distinct points in U have disjoint neighborhoods *in* X.

[2] It is not hard to modify Lemma 1.26 to show that locally compact subsets of an almost Hausdorff locally compact spaces are necessarily locally closed.

[3] Recall that Y is a *Baire space* if the countable intersection of open dense subsets of Y is again dense.

[4] The result is proved for almost Hausdorff locally compact spaces in Lemma 6.4 on page 175.

[5] A subset A of a topological space Y is said to be of *first category* if can be written as a countable union $\bigcup F_n$ with each $\overline{F_n}$ having empty interior. Otherwise a set is said to be of *second category*. A space is Baire if and only if every open subset is of second category.

[6] Classically, G-spaces with these regularity properties were called *smooth*.

underlying Borel structure (cf. Appendix D.2 on page 374) and trace their origins back to Mackey's pioneering work [105], we'll restrict to the topological issues here.

Theorem 6.2 (Mackey-Glimm Dichotomy). *Suppose that X is a second countable almost Hausdorff locally compact G-space with G locally compact and second countable. Then the following statements are equivalent.*

(a) *The orbit space $G\backslash X$ is a T_0 topological space.*
(b) *Each orbit $G \cdot x$ is locally closed in X.*
(c) *Each orbit is a Baire space in its relative topology in X.*
(d) *For each $x \in X$, the map $sG_x \mapsto s \cdot x$ is a homeomorphism of G/G_x onto $G \cdot x$.*
(e) *The orbit space $G\backslash X$ is almost Hausdorff.*

Before beginning the proof we need a couple of lemmas. The first is of some interest in its own right. It says, in particular, that almost Hausdorff spaces really are very close to being Hausdorff. Most importantly, it says that we can always find a dense open relatively Hausdorff subset. However, the complement has a dense open relatively Hausdorff subset, and so on. Unfortunately, this process, at least in theory, can continue infinitely, and, even if the number of sets involved is countable, the structure can be too complicated to allow us to naturally index the resulting sets by positive integers. Instead, the natural indexing is provided by a set $\Gamma = \{\alpha \le \gamma\}$ of *ordinals*. This means that Γ is a totally ordered set in which each nonempty subset $S \subset \Gamma$ has a smallest element $\min S \in S$. We let $0 := \min \Gamma$. Note that each element $\alpha \in \Gamma \setminus \{\gamma\}$ has an *immediate successor* $\alpha + 1 := \min\{\lambda \in \Gamma : \lambda > \alpha\}$. An element $\alpha > 0$ has an *immediate predecessor* if there is an element α' such that $\alpha' + 1 = \alpha$. An element α without an immediate predecessor is called a *limit ordinal*. One says γ is finite if $\Gamma_0 := \{\alpha : 0 \le \alpha < \gamma\}$ is finite in which case γ can be identified with the subset of integers $0 \le k \le n$ for a positive integer n. We say that γ is countable if Γ_0 is countable. Notice that Γ_0 can still contain limit ordinals even if γ is countable.

Lemma 6.3. *Suppose that X is a not necessarily Hausdorff locally compact space. Then the following are equivalent.*

(a) *X is almost Hausdorff.*
(b) *Every nonempty* closed *subspace of X has a relatively open nonempty Hausdorff subspace.*
(c) *Every closed subspace of X has a dense relatively open Hausdorff subspace.*
(d) *There is an ordinal γ and open sets $\{U_\alpha : \alpha \le \gamma\}$ such that*
 (i) *$\alpha < \beta \le \gamma$ implies that $U_\alpha \subsetneq U_\beta$,*
 (ii) *$\alpha < \gamma$ implies that $U_{\alpha+1} \setminus U_\alpha$ is a dense Hausdorff subspace of $X \setminus U_\alpha$.*
 (iii) *if δ is limit ordinal, then*

$$U_\delta = \bigcup_{\alpha < \delta} U_\alpha,$$

(iv) *$U_0 = \emptyset$ and $U_\gamma = X$.*

(e) *Every subspace of X has a relatively open dense Hausdorff subspace.*

Furthermore, if X is a second countable almost Hausdorff locally compact space, then we can take γ to be countable in part (c).

Proof. Clearly (a) $\implies$ (b). Suppose (b) holds and let F be a nonempty closed subset of X. By assumption there is an open subset $V \subset X$ such that $V \cap F$ is nonempty and Hausdorff in F. A simple Zorn's Lemma argument implies there is a maximal such V. We claim that $V \cap F$ is dense in F. Suppose not. Then $F \setminus V$ contains an open subset of F. Thus there is an open set $W \subset X$ such that $W \cap F$ is a nonempty and contained in $F \setminus V$. By assumption, $\overline{W \cap F}$ contains a nonempty relatively open Hausdorff subset which must be of the form $W' \cap (\overline{W \cap F})$ for some open subset $W' \subset X$. If $W' \cap W \cap F = \emptyset$, then $X \setminus W'$ is a closed set containing $W \cap F$. This is nonsense as it implies $W' \cap (\overline{W \cap F}) = \emptyset$. Therefore $W' \cap W \cap F$ is nonempty open Hausdorff subset of F. Let $V' := V \cup (W' \cap W)$. Since $W' \cap W \cap F$ and $V \cap F$ are disjoint, $V' \cap F$ is Hausdorff. Since $V \subsetneq V'$, this contradicts the maximality of V. Thus V is dense and we've proved that (b) $\implies$ (c).

(c) $\implies$ (d): We proceed by transfinite induction. Let $U_0 := \emptyset$, and suppose that U_α has been defined for all $\alpha < \beta$. Suppose that β is not a limit ordinal and has an immediate predecessor β_-. If $U_{\beta_-} = X$, then we're done. Otherwise, $X \setminus U_{\beta_-}$ is a nonempty closed subset and by assumption there is an open set $V \subset X$ such that $(X \setminus U_{\beta_-}) \cap V$ is dense and relatively Hausdorff in $X \setminus U_{\beta_-}$. Therefore we can set $U_\beta := V \cup U_{\beta_-}$.

If β is a limit ordinal, then we can set

$$U_\beta := \bigcup_{\alpha<\beta} U_\alpha.$$

This completes the induction and the proof that (c) $\implies$ (d).

(d) $\implies$ (e): Let $\{U_\alpha\}_{\alpha\le\gamma}$ be as specified in the lemma. Let $S \subset X$ be any nonempty set. Let β be the least ordinal such that $S \cap U_\beta \neq \emptyset$. If β were a limit ordinal, then $S \cap \bigcup_{\alpha<\beta} U_\alpha \neq \emptyset$ implies that $S \cap U_\alpha \neq \emptyset$ for some $\alpha < \beta$ which contradicts our choice of β. Thus, β has an immediate predecessor β_-, and $S \cap U_{\beta_-} = \emptyset$. Then $\emptyset \neq S \cap U_\beta = S \cap (U_\beta \setminus U_{\beta_-})$ is an nonempty open Hausdorff subset of S. It follows that S has a maximal open Hausdorff subset V. If V is not dense in S, then S has a nonempty open subset S' disjoint from V. By the above, S' has a nonempty open Hausdorff subset V'. Since V and V' are disjoint open subsets of S, $V \cup V'$ is Hausdorff. This contradicts the maximality of V. We have shown that (d) $\implies$ (e), and (e) $\implies$ (a) is trivial.

Now assume that X is a second countable almost Hausdorff locally compact space. Let $\{U_\alpha\}_{\alpha\le\gamma}$ be as in (c), and let $\{P_n\}_{n=1}^\infty$ be a countable basis for the topology on X. Since each U_α is a union of P_n's, there is at least one n such that $P_n \subset U_{\alpha+1}$ while $P_n \cap (U_{\alpha+1} \setminus U_\alpha) \neq \emptyset$. Let $n(\alpha)$ be the least such integer. If $\alpha < \beta$, then $P_{n(\alpha)} \subset U_{\alpha+1} \subset U_\beta$ so that $P_{n(\alpha)} \cap (U_{\beta+1} \setminus U_\beta) = \emptyset$. Thus $n(\alpha) \neq n(\beta)$ and $\alpha \mapsto n(\alpha)$ is a one-to-one map into $\mathbf{N}$ and $\{\alpha \le \gamma\}$ must be countable. □

Now we record some topological technicalities for use in the proof of the theorem.

Lemma 6.4. *Every G_δ subset of an almost Hausdorff locally compact space X is a Baire space in its relative topology.*

Proof. We first observe that X itself is Baire. Let G_n be a dense open subspace of X for $n = 1, 2, 3, \ldots$. By Lemma 6.3 on page 173, X has an open dense Hausdorff subspace U. Then U is a locally compact Hausdorff space, and therefore a Baire space. Therefore

$$\bigcap_{n=1}^{\infty} G_n \cap U$$

is dense in U, and it follows that $\bigcap G_n$ is dense in X. Thus X is a Baire space.

Now assume that U_n is open in X for $n = 1, 2, 3, \ldots$, and that $A = \bigcap U_n$ is a G_δ subspace of X. Since $\overline{A}$ is an almost Hausdorff locally compact space, we can replace X by $\overline{A}$ and U_n by $U_n \cap \overline{A}$. Thus we can assume that A is dense in X, and that each U_n is open and dense in X. Now assume that G_n is open and dense in A for $n = 1, 2, 3, \ldots$. Then there are open dense subsets J_n of X such that $G_n = J_n \cap A$. Since X is a Baire space, $\bigcap (U_n \cap J_n)$ is dense in X. But

$$\begin{aligned}\bigcap (U_n \cap J_n) &= \bigcap U_n \cap \bigcap J_n \\ &= A \cap \bigcap J_n \\ &= \bigcap G_n.\end{aligned}$$

It follows that $\bigcap G_n$ is dense in A. This completes the proof. □

Lemma 6.5. *A second countable locally compact Hausdorff space X is completely metrizable; that is there is a complete metric on X inducing the given locally compact topology on X.*

Remark 6.6. A completely metrizable second countable space is called a *Polish space.* Thus the lemma simply says that second countable locally compact spaces are Polish.

Proof. The Urysohn Metrization Theorem [168, Theorem 23.1] certainly implies that X is metrizable. To get completeness requires a bit more.

Let X^+ be the one point compactification of X, and let $\tau = \{ U_n \}$ be a countable basis for the topology on X. We can also find compact sets K_n such that K_n belongs to the interior of K_{n+1}, and such that

$$X = \bigcup_{n=1}^{\infty} K_n.$$

Note that the requirement on the interiors implies that any compact set in X is contained in some K_n. By definition, $\beta = \{ X^+ \setminus K_n \}_{n=1}^{\infty}$ are open neighborhoods of ∞ in X^+. In fact, any neighborhood V of ∞ must contain a neighborhood of the form $X^+ \setminus K$ for $K \subset X$ compact. But then there is an n such that

$$X^+ \setminus K_n \subset X^+ \setminus K \subset V.$$

It follows that $\tau \cup \beta$ is a countable basis for X^+. Therefore X^+ is a second countable compact Hausdorff space. It follows that X^+ is regular, and then Urysohn's Metrization theorem implies that X^+ is metrizable. Since any metric on a compact space is complete, X^+ is completely metrizable. Since X is open in X^+, it too is completely metrizable by [168, Theorem 24.12]. To see this directly, let d be the metric on X^+ and define $f(s) = 1/d(s,\infty)$. Then f is continuous on X, and $s \mapsto (s, f(s))$ is a homeomorphism of X onto its image in the complete metric space $X^+ \times \mathbf{R}$. It is not hard to see that this image is closed, and a closed subset of a complete metric space is still complete. $\square$

Lemma 6.7. *Suppose that X is a second countable almost Hausdorff locally compact G-space, that N is a compact neighborhood of e in G and that P and U are open Hausdorff subsets in X such that $N \cdot P \subset U$. If there is a $x \in P$ such that*

$$N \cdot x \cap P \subsetneq G \cdot x \cap P,$$

then there is a neighborhood V of x such that $N \cdot V \cap P \not\supset G \cdot x \cap P$.

Proof. Let $\{ V_k \}$ be a countable neighborhood basis at x of open sets with $V_{k+1} \subset V_k$. If $G \cdot x \cap P \subset N \cdot V_k \cap P$ for all k, then given $s \cdot x \in G \cdot x \cap P$, there exist $n_k \in N$, $v_k \in V_k$ such that $n_k \cdot v_k = s \cdot x$. Since N is compact, there is a subsequence n_{k_j} converging to $n \in N$. Furthermore, $v_k \to x$. Since $s \cdot x$ and $n \cdot x$ are contained in the Hausdorff set U, it follows that $n \cdot x = s \cdot x$. This implies $N \cdot x \cap P = G \cdot x \cap P$. The result follows. $\square$

Lemma 6.8. *Let $q : X \to G\backslash X$ be the orbit map. Then*

$$q^{-1}\bigl(\overline{\{ q(x) \}}\bigr) = \overline{G \cdot x}.$$

Proof. Since $q^{-1}\bigl(\overline{\{ q(x) \}}\bigr)$ is a closed set containing $G \cdot x$, we clearly have $q^{-1}\bigl(\overline{\{ q(x) \}}\bigr) \supset \overline{G \cdot x}$. However, $X \smallsetminus \overline{G \cdot x}$ is a G-invariant open set and $q(X \smallsetminus \overline{G \cdot x})$ is an open set disjoint from $q(x)$. Thus $\overline{\{ q(x) \}} \subset q(\overline{G \cdot x})$ which gives the other inclusion. $\square$

Proof of Theorem 6.2 on page 173. We'll show that

$$(a) \implies (c) \implies (d) \implies (e) \implies (b) \implies (a).$$

We'll start with $(a) \implies (c)$. Fix $u = q(x)$ in $G\backslash X$. By Lemma 3.35 on page 97, there is a countable basis $\{ U_n \}$ of open sets for the topology on $G\backslash X$. Let

$$B_n := \begin{cases} U_n & \text{if } u \in U_n, \text{ and} \\ G\backslash X \smallsetminus U_n & \text{if } u \notin U_n. \end{cases}$$

Clearly, $u \in \bigcap B_n$. Suppose $v \neq u$. If there is an open set V containing u and not v, then there is a U_k such that $u \in U_k \subset V$. It follows that $v \notin \bigcap B_n$. Similarly, if there is a W containing v but not u, then there is a U_k such that

$v \in U_k \subset W$. Again, it follows that $v \notin \bigcap B_n$. Since $G\backslash X$ is a T_0 space, we always have $\bigcap B_n = \{ u \}$. Consequently,

$$G \cdot x = q^{-1}(u) = \bigcap q^{-1}(B_n).$$

Since each $q^{-1}(B_n)$ is either open or closed, it follows easily that $G \cdot x$ is the intersection of a G_δ subset of X with a closed subset of X. Since any closed subset of X is an almost Hausdorff locally compact space, $G \cdot x$ must be a Baire space by Lemma 6.4 on page 175.

$(c) \implies (d)$: Fix $x \in X$. Since the natural map of $G \to G/G_x$ is open and continuous, it will suffice to show that $V \cdot x$ is open in $G \cdot x$ whenever V is open in G. We claim it suffices to see that $N \cdot x$ has nonempty interior in $G \cdot x$ whenever N is a compact neighborhood of e in G. To prove the claim, let $s \in V$. Choose compact neighborhoods P and N of e in G such that $sP \subset V$ and $N^{-1}N \subset P$. By assumption, there is a $r \in N$ such that $r \cdot x \in \operatorname{int}(N \cdot x)$. (Here, $\operatorname{int}(N \cdot x)$ will always denote the interior of $N \cdot x$ in $G \cdot x$.) Since multiplication by r and r^{-1} continuous, it follows that multiplication by r^{-1} is a homeomorphism of $G \cdot x$ onto itself, and $x \in \operatorname{int}(r^{-1}N \cdot x) \subset \operatorname{int}(P \cdot x)$. Similarly, this implies $s \cdot x \in \operatorname{int}(sP \cdot x) \subset V \cdot x$. Since $s \in V$ was arbitrary, $V \cdot x$ is open in $G \cdot x$. This establishes the claim.

Let N be a compact neighborhood of e in G. Since X is almost Hausdorff, $G \cdot x$ has an open dense Hausdorff subspace V (Lemma 6.3 on page 173). Since multiplication by elements of G induces a homeomorphism of $G \cdot x$ to itself, we can assume that $x \in V$. Since V is a dense open subspace of a Baire space, V is a Baire space. As $s \mapsto s \cdot x$ is continuous from G to $G \cdot x$, there is an open neighborhood U of e in G such that $V = U \cdot x$. Since G is second countable, there are compact neighborhoods $\{ V_n \}_{n=1}^{\infty}$ of e in G and $\{ s_n \}_{n=1}^{\infty}$ in G such that each $V_n \subset N$ and $U = \bigcup_{n=1}^{\infty} s_n V_n$.[7] Thus

$$V = U \cdot x = \bigcup_{n=1}^{\infty} s_n V_n \cdot x. \tag{6.1}$$

Since each V_n is compact, each $s_n V_n \cdot x$ is compact in the Hausdorff space V. Therefore (6.1) is a countable union of closed subspaces of V, and since V is Baire, at least one of these sets, say $s_n V_n \cdot x$, has interior in V — and therefore in $G \cdot x$. Multiplying by s_n^{-1}, we see that $V_n \cdot x \subset N \cdot x$ has interior as required.

$(d) \implies (e)$: In view of Lemma 6.3 on page 173, it suffices to see that every nonempty closed subset of $G\backslash X$ has a nonempty relatively open Hausdorff subset. But if $F \subset G\backslash X$ is closed, then $q^{-1}(F)$ is a second countable almost Hausdorff locally compact G-space in which each orbit is homeomorphic to G/G_x via the natural map. Therefore, it will suffice to see that $G\backslash X$ has a nonempty open Hausdorff subspace.

Let U be a nonempty Hausdorff open subspace of X. If $x \in U$, then there is a compact neighborhood N of e in G and an open neighborhood P_0 of x such that

[7] Let $\{ V_n \}$ be a countable neighborhood basis for e consisting of compact neighborhoods of e such that $V_n \subset N$. If $\{ s_k \}$ is dense in G, then given any open set W and $s \in W$, there is a n such that $sV_n^{-1}V_n \subset W$ and a k such that $s_k \in sV_n^{-1}$. But then $s \in s_k V_n$ and $s_k V_n \subset sV_n^{-1}V_n \subset W$. This suffices as the collection $\{ s_k V_n \}$ is countable.

$N \cdot P_0 \subset U$. We make the following claim: if (d) holds and if N and P_0 are as above, then there is a nonempty open set $P \subset P_0$ such that

$$N \cdot x \cap P = G \cdot x \cap P \quad \text{for all } x \in P. \tag{6.2}$$

We prove the claim. Suppose no such set P exists. Since U is a second countable locally compact Hausdorff space, there is a metric d on U compatible with the topology on U (Lemma 6.5 on page 175). If $A \subset U$, then $\operatorname{dia} A := \sup\{\, d(x,y) : x,y \in A \,\}$. Following Effros's argument in [46, p. 48], we will inductively define a sequence $\{\, P_n \,\}_{n=1}^{\infty}$ of open subsets of P_0 and elements $\{\, s_n \,\}_{n=1}^{\infty} \subset G$ such that

(a) the closure of P_{n+1} in U is compact and is contained in P_n,
(b) $\operatorname{dia}(P_n) < \frac{1}{n}$ for all $n \geq 1$, and
(c) $s_{n+1}P_{n+1} \subset P_n \setminus N \cdot P_{n+1}$.

Let P_0 be as above, and assume that we have defined $\{\, P_k \,\}_{k=0}^{n}$ and $\{\, s_k \,\}_{k=1}^{n}$. By assumption, there is at least one $x \in P_n$ such that

$$N \cdot x \cap P_n \subsetneq G \cdot x \cap P_n.$$

It follows from Lemma 6.7 on page 176, that there is a compact neighborhood M of x in P_n such that $G \cdot x \cap P_n \not\subset N \cdot M \cap P_n$. Thus there is a $s_{n+1} \in G$ such that

$$s_{n+1} \cdot x \in P_n \setminus N \cdot M. \tag{6.3}$$

Since N and M are compact, $N \cdot M$ is a compact subset of U, and therefore closed in U. Thus the right-hand side of (6.3) is open, and since G acts continuously on X, there is a open neighborhood P_{n+1} of x such that

$$s_{n+1} \cdot P_{n+1} \subset P_n \setminus N \cdot M.$$

Since U is a locally compact Hausdorff space, we can shrink P_{n+1} a bit, if necessary, and arrange that the closure of P_{n+1} in U is contained in $P_n \cap M$ and that $\operatorname{dia} P_{n+1} < \frac{1}{n+1}$. Then we also have

$$s_{n+1} \cdot P_{n+1} \subset P_n \setminus N \cdot M \subset P_n \setminus N \cdot P_{n+1}.$$

Continuing in this way, we get our sequences $\{\, P_n \,\}$ and $\{\, s_n \,\}$ as required. Since the closure of P_{n+1} in U is compact and contained in P_n, and since $\operatorname{dia} P_n \to 0$, it easy to see that there is a $x_0 \in X$ such that

$$\{\, x_0 \,\} = \bigcap_{n=0}^{\infty} P_n.$$

Furthermore, $s_{n+1} \cdot x_0 \in P_n$ for all n, and it follows that $s_n \cdot x_0 \to x_0$. However, condition (c) implies that for all n, $s_n \cdot x_0 \notin N \cdot x_0$. This means $N \cdot x_0$ is not a neighborhood of x_0 which contradicts the openness of the map $sG_{x_0} \mapsto s \cdot x_0$. Thus we have established the claim.

Now let N and P be as in (6.2). It suffices to see that $q(P)$ is Hausdorff in $G \backslash X$. To prove this, we'll show that a convergent sequence in $q(P)$ has a unique limit in $q(P)$. Let $\{ p_i \}$ be a sequence in $q(P)$ converging to both $q(x)$ and $q(y)$ for $x, y \in P$. Since $q : X \to G\backslash X$ is open we can use Proposition 1.15 on page 4, pass to a subsequence, relabel, and assume that there are $x_i \in X$ with $q(x_i) = p_i$ and $x_i \to x$. Since $q(x_i) \to q(y)$, we can apply Lemma 3.38 on page 98, pass to subsequence, relabel, and assume that there are $s_i \in G$ such that $s_i \cdot x_i \to y$. Since P is a neighborhood of x, we can assume the x_i are eventually in P. Therefore we eventually have

$$N \cdot x_i \cap P = G \cdot x_i \cap P.$$

Thus we can assume that the s_i are eventually in the compact neighborhood N. Thus we can pass to a subsequence, relabel, and assume that $s_i \to s$ in $N \subset G$. Thus $s_i \cdot x_i$ converges to $s \cdot x$ as well as y. Since $N \cdot P \subset N \cdot P_0 \subset U$, both y and $s \cdot x$ are in the open Hausdorff set U. Therefore $y = s \cdot x$ and $q(x) = q(y)$. This completes the proof that $(d) \implies (e)$.

$(e) \implies (b)$: In view of Lemma 6.8 on page 176, it suffices to see that for each $v \in G\backslash X$, $\{ v \}$ is open in $\overline{\{ v \}}$. But $G\backslash X$ almost Hausdorff implies $\overline{\{ v \}}$ has a dense Hausdorff open subset C. Since $C \subset \overline{\{ v \}}$, every neighborhood of every point in C must contain v. Therefore $v \in C$. Suppose that u is a point of C distinct from v. Since C is Hausdorff in its relative topology in $\overline{\{ v \}}$, there must be an open set U in $\overline{\{ v \}}$ which contains u and not v. But then $u \notin \overline{\{ v \}}$ which is nonsense. Thus $C = \{ v \}$, and since C is open, we're done.

$(b) \implies (a)$: Suppose that $q(x)$ and $q(y)$ are distinct points in $G\backslash X$. Then $G \cdot x$ and $G \cdot y$ are distinct orbits in X. Suppose that $\overline{G \cdot x} \cap G \cdot y = \emptyset$. Then $X \smallsetminus \overline{G \cdot x}$ is an open G-invariant set disjoint from $G \cdot x$. Therefore $q(X \smallsetminus \overline{G \cdot x})$ is an open set in $G\backslash X$ containing $q(y)$ and not containing $q(x)$.

If, on the other hand, $\overline{G \cdot x} \cap G \cdot y \neq \emptyset$, then, as both sets are G-invariant, we must have $G \cdot y \subset \overline{G \cdot x}$. By assumption, there is an open set $U \subset X$ such that $U \cap \overline{G \cdot x} = G \cdot x$. Thus $U \cap G \cdot y = \emptyset$. Let

$$V = \bigcup_{s \in G} s \cdot U = G \cdot U.$$

Since $G \cdot x$, $G \cdot y$ and $\overline{G \cdot x}$ are G-invariant, we still have

$$V \cap \overline{G \cdot x} = G \cdot x \quad \text{and} \quad V \cap G \cdot y = \emptyset.$$

Since V is open and G-invariant, $q(V)$ is an open set in $G\backslash X$ which contains $q(x)$ but not $q(y)$. This shows that $G\backslash X$ is a T_0 topological space.

This completes the proof of the theorem. □

6.2 The Res Map and Quasi-Orbits

If (A, G, α) is a dynamical system, then the structure of the G-space $\operatorname{Prim} A$ (cf. Lemma 2.8 on page 44) is an important tool in understanding the ideal structure. In general, $\operatorname{Prim} A$ need not be almost Hausdorff so Theorem 6.2 on page 173 will not apply without some additional hypotheses. Nevertheless, we will need to look carefully at the orbit space $G\backslash \operatorname{Prim} A$. Since the primitive ideal space of $A \rtimes_\alpha G$, or of any C^*-algebra, is always a T_0 topological space, the sort of information about $\operatorname{Prim} A \rtimes_\alpha G$ we expect to glean from $G \backslash \operatorname{Prim} A$ will come from a topological T_0-quotient of the orbit space which we call the T_0-ization (for lack of a better term).

Definition 6.9. If X is a topological space, then the *T_0-ization* of X is the quotient space $(X)^\sim = X/\sim$ where $\sim$ is the equivalence relation on X defined by $x \sim y$ if $\overline{\{x\}} = \overline{\{y\}}$. We give $(X)^\sim$ the quotient topology, which is the largest topology making the quotient map $q : X \to (X)^\sim$ continuous. A subset of X is said to be *saturated* if it is a union of equivalence classes.

Naturally, we want the T_0-ization to be T_0. In fact, the $(X)^\sim$ is the largest quotient of X which is T_0.

Lemma 6.10. *If X is a topological space, then $(X)^\sim$ is a T_0 topological space. If Y is any T_0 topological space and if $f : X \to Y$ is continuous, then there is a continuous map $f' : (X)^\sim \to Y$ such that*

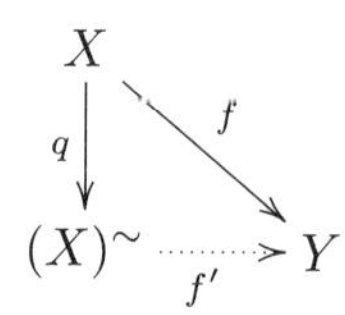

commutes.

Proof. Suppose that $q(x)$ and $q(y)$ are distinct elements in $(X)^\sim$. If $\overline{\{x\}} \subset \overline{\{y\}}$, then $U = X \setminus \overline{\{x\}}$ is a saturated open set containing y but not x. Since $q^{-1}\big(q(U)\big) = U$, $q(U)$ is an open set in $(X)^\sim$ containing $q(y)$ but not $q(x)$. If $\overline{\{x\}} \not\subset \overline{\{y\}}$, then $V = X \setminus \overline{\{y\}}$ is a saturated open set containing x but not y. Thus $q(V)$ is an open set containing $q(x)$ but not $q(y)$. Thus $(X)^\sim$ is T_0.

Now suppose that Y is T_0, and that $f : X \to Y$ is continuous. If $f(x) \neq f(y)$, then, interchanging x and y if necessary, there is an open set U in Y such that $f(x) \in U$ and $f(y) \notin U$. Thus $f^{-1}(U)$ is an open set in X containing x but not containing y. Thus $x \notin \overline{\{y\}}$ and $q(x) \neq q(y)$. It follows that there is a well-defined function $f' : (X)^\sim \to Y$ given by $f'\big(q(x)\big) = f(x)$. If U is open in Y, then $q^{-1}\big((f')^{-1}(U)\big) = f^{-1}(U)$. Thus $(f')^{-1}(U)$ is open and f' is continuous. $\square$

Definition 6.11. If (A, G, α) is a dynamical system, then the *quasi-orbit space* is the T_0-ization $\mathcal{Q}$ of $G \backslash \operatorname{Prim} A$. Each class in $\mathcal{Q}$ is called a *quasi-orbit.*

If $G \backslash \operatorname{Prim} A$ is T_0, then $\mathcal{Q} = G \backslash \operatorname{Prim} A$ and quasi-orbits are just orbits. In general, P and Q in $\operatorname{Prim} A$ determine the same quasi-orbit if and only if $\overline{G \cdot P} = \overline{G \cdot Q}$ in $\operatorname{Prim} A$ (using Lemma 6.8 on page 176). In view of the definition of the topology on $\operatorname{Prim} A$, P and Q determine the same quasi-orbit exactly when $\bigcap_{s \in G} s \cdot P = \bigcap_{s \in G} s \cdot Q$.

Lemma 6.12. *If (A, G, α) is a dynamical system, then the natural map $k : \operatorname{Prim} A \to \mathcal{Q}$ is continuous and open. In particular, if A is separable, then $\mathcal{Q}$ is second countable. The map k is called the* quasi-orbit map.

Proof. Note that k is the composition of the orbit map from $\operatorname{Prim} A$ to $G\backslash \operatorname{Prim} A$ and the quotient map from $G\backslash \operatorname{Prim} A$ to $\mathcal{Q}$. Hence k is continuous.

To show that k is open, it will suffice to see that $k^{-1}\big(k(\mathcal{O}_I)\big)$ is open in $\operatorname{Prim} A$, where $\mathcal{O}_I := \{\, P \in \operatorname{Prim} A : P \not\supset I \,\}$. In fact, we will show that

$$k^{-1}\big(k(\mathcal{O}_I)\big) = \bigcup_{s\in G} \mathcal{O}_{s\cdot I}, \tag{6.4}$$

where $s \cdot I = \{\, \alpha_s(a) : a \in I \,\}$. This will suffice. Let J be the α-invariant ideal generated by $\bigcup_{s\in G} s\cdot I$. Note that if $P \in \mathcal{O}_I$, then $\bigcap_{s\in G} s\cdot P \not\supset J$. On the other hand, if $\bigcap_{s\in G} s\cdot Q \not\supset J$, then there must be a $s \in G$ such that $s\cdot Q \not\supset I$, and hence $k(Q) = k(s\cdot Q) \in k(\mathcal{O}_I)$. Thus,

$$\begin{aligned} k^{-1}\big(k(\mathcal{O}_I)\big) &= \{\, Q \in \operatorname{Prim} A : \bigcap_{s\in G} s\cdot Q \not\supset J \,\} \\ &= \{\, Q \in \operatorname{Prim} A : \text{there is a } s \in G \text{ such that } s\cdot Q \not\supset I \,\} \\ &= \bigcup_{s\in G} \mathcal{O}_{s\cdot I} \end{aligned}$$

which establishes (6.4).

If A is separable, then $\operatorname{Prim} A$ is second countable. (This follows from [139, Theorem A.38] and [139, Proposition A.46].) Since the continuous open image of a second countable space is second countable (as in Lemma 3.35 on page 97), the final assertion follows. □

To see the importance of quasi-orbits, it will be helpful to pause and introduce the *restriction map.* Let (A, G, α) be a dynamical system and suppose that (π, U) is a covariant representation of (A, G, α) on $\mathcal{H}$. If H is a closed subgroup of G, then $(\pi, U|_H)$ is a covariant representation of $(A, H, \alpha|_H)$. Since (π, U) is, by definition, nondegenerate whenever π is, $(\pi, U|_H)$ is nondegenerate whenever (π, U) is. The representation $\pi \rtimes U|_H$ is called the *restriction of* $\pi \rtimes U$, and is denoted $\operatorname{Res}_H^G(\pi \rtimes U)$. Since every representation of $A \rtimes_\alpha G$ is the integrated form of a unique nondegenerate covariant representation (Proposition 2.40 on page 59), it make perfect sense to apply Res_H^G to any representation of $A \rtimes_\alpha G$. More generally, we make the following definition.

Definition 6.13. Suppose that X is Hilbert B-module. If (π, u) is a covariant homomorphism of (A, G, α) into $\mathcal{L}(\mathsf{X})$ and if H is a closed subgroup of G, then the *restriction of* $\pi \rtimes u$ *to* $A \rtimes_{\alpha|_H} H$ is the representation $\operatorname{Res}_H^G(\pi \rtimes u) := \pi \rtimes u|_H$. If $H = \{\, e \,\}$, then we write Res in place of $\operatorname{Res}_{\{e\}}^G$.

Lemma 6.14. *Suppose that (A, G, α) is a dynamical system and that L is a representation of $A \rtimes_\alpha G$. If H is a closed subgroup of G and if $z \in A \rtimes_{\alpha|_H} H$, then*

$$\operatorname{Res}_H^G L(z) = \bar{L}\big(\operatorname{Res}_H^G(i_A \rtimes i_G)(z)\big),$$

where (i_A, i_G) is the canonical covariant homomorphism of (A, G, α) into $M(A \rtimes_\alpha G)$ defined in Proposition 2.34 on page 54.

Proof. Suppose that $L = \pi \rtimes U$. It suffices to consider $z \in C_c(H, A)$. Then using Lemma 1.101 on page 35 and $\bar{L} \circ i_A = \pi$ and $\bar{L} \circ i_G = U$,

$$\begin{aligned}\bar{L}\big(\operatorname{Res}_H^G(i_A \rtimes i_G)(z)\big) &= \bar{L}\Big(\int_H i_A\big(z(t)\big) i_G(t)\, d\mu_H(t)\Big) \\ &= \int_H \pi\big(z(t)\big) U_t \, d\mu_H(t) \\ &= \operatorname{Res}_H^G L(z). \end{aligned}$$

□

Corollary 6.15. *If (A, G, α) is a dynamical system and if H is a closed subgroup of G, then there is a continuous map*

$$\operatorname{Res}_H^G : \mathcal{I}(A \rtimes_\alpha G) \to \mathcal{I}(A \rtimes_{\alpha|_H} H)$$

such that for all representations L of $A \rtimes_\alpha G$

$$\operatorname{Res}_H^G(\ker L) = \ker(\operatorname{Res}_H^G L). \tag{6.5}$$

Proof. Let J be an ideal in $A \rtimes_\alpha G$. Define

$$\operatorname{Res}_H^G J := \{\, z \in A \rtimes_{\alpha|_H} H : \operatorname{Res}_H^G(i_A \rtimes i_G)(z)(A \rtimes_\alpha G) \subset J \,\}. \tag{6.6}$$

Let $z \in A \rtimes_{\alpha|_H} H$, $f \in A \rtimes_\alpha G$ and $h \in \mathcal{H}$. Then Lemma 6.14 on the preceding page implies that

$$\operatorname{Res}_H^G(L)(z)\big(L(f)h\big) = L\big(\operatorname{Res}_H^G(i_A \rtimes i_G)(z)(f)\big)h. \tag{6.7}$$

Thus if $z \in \operatorname{Res}_H^G(\ker L)$, then $\operatorname{Res}_H^G(i_A \rtimes i_G)(z)(f) \in \ker L$, and the right-hand side of (6.7) is zero. Since L is nondegenerate, this implies $\operatorname{Res}_H^G(\ker L) \subset \ker \operatorname{Res}_H^G L$.

On the other hand, if $z \in \ker \operatorname{Res}_H^G L$, then (6.7) implies that $\operatorname{Res}_H^G(i_A \rtimes i_G)(z)(f) \in \ker L$ for all $f \in A \rtimes_\alpha G$. Thus $z \in \operatorname{Res}_H^G(\ker L)$ and we've proved that (6.5) holds. Since the right-hand side of (6.5) is always an ideal, it follows that (6.6) is always an ideal and that Res_H^G maps $\mathcal{I}(A \rtimes_\alpha G)$ into $\mathcal{I}(A \rtimes_{\alpha|_H} H)$.[8]

We want to see that Res_H^G is continuous.[9] It suffices to see that $(\operatorname{Res}_H^G)^{-1}(\mathcal{O}_I)$ is open in $\mathcal{I}(A \rtimes_\alpha G)$, where I is an ideal in $A \rtimes_{\alpha|_H} H$, and by definition,

$$\mathcal{O}_I := \{\, I' \in \mathcal{I}(A \rtimes_{\alpha|_H} H) : I' \not\supset I \,\}.$$

Let K be the ideal in $A \rtimes_\alpha G$ generated by $\operatorname{Res}_H^G(i_A \rtimes i_G)(I)(A \rtimes_\alpha G)$. Since $J \supset K$ if and only if $\operatorname{Res}_H^G J \supset I$, it follows easily that $J \in (\operatorname{Res}_H^G)^{-1}(\mathcal{O}_I)$ if and only if $J \in \mathcal{O}_K \subset \mathcal{I}(A \rtimes_\alpha G)$. This suffices. □

Lemma 6.16. *Suppose that (A, G, α) is a dynamical system. Recall that* $\operatorname{Res} := \operatorname{Res}_{\{e\}}^G$. *For all $J \in \mathcal{I}(A \rtimes_\alpha G)$,* $\operatorname{Res} J$ *is an α-invariant ideal. In particular, if $\pi \rtimes U$ is a covariant representation of $A \rtimes_\alpha G$, then $\ker \pi$ is α-invariant, and $\pi \rtimes U$ factors through $A/\ker \pi \rtimes_{\alpha^{\ker \pi}} G$. If I is an α-invariant ideal, then* $\operatorname{Res}(I \rtimes_\alpha G) = I$.

[8]Every ideal in a C^*-algebra is the kernel of some representation.

[9]The continuity can also be derived from Lemma 8.35 on page 249.

Proof. Suppose that $a \in \operatorname{Res} J$. Then by definition, $i_A(a)(A \rtimes_\alpha G) \subset J$. Thus if $f \in A \rtimes_\alpha G$, then since J is an ideal in $M(A \rtimes_\alpha G)$,

$$i_A\big(\alpha_s(a)\big)(f) = i_G(s) i_A(a) i_G(s^{-1})(f) \in J.$$

Since $f \in A \rtimes_\alpha G$ was arbitrary, we have shown that $\alpha_s(a) \in \operatorname{Res} J$ whenever a is. Therefore $\operatorname{Res} J$ is α-invariant.

Since $\ker \pi = \operatorname{Res}\big(\ker(\pi \rtimes U)\big)$ (Corollary 6.15 on the facing page), it follows that $\ker \pi$ is α-invariant. If I is any α-invariant ideal in A, then Proposition 3.19 on page 93 allows us to identify $I \rtimes_\alpha G$ with an ideal in $A \rtimes_\alpha G$, and implies that the quotient map $q : A \to A/I$ induces an isomorphism of $(A \rtimes_\alpha G)/(I \rtimes_\alpha G)$ with $A/I \rtimes_{\alpha^I} G$. If $I = \ker \pi$, then $\pi = \pi' \circ q$ for a representation π' of A/I and (π', U) is the covariant representation of $(A/I, G, \alpha^I)$ corresponding to (π, U). This shows that $\pi \rtimes V$ factors through $A/I \rtimes_{\alpha^I} G$. On the other hand, if $\rho \rtimes V$ is a faithful representation of $A/I \rtimes_{\alpha^I} G$, then Corollary 6.15 on the facing page implies that ρ must be a faithful representation of A/I. Then $(\rho \circ q, V)$ is a covariant representation of (A, G, α) and $\ker(\rho \circ q \rtimes V) = I \rtimes_\alpha G$. Thus

$$\operatorname{Res}(I \rtimes_\alpha G) = \operatorname{Res}\big(\ker(\rho \circ q \rtimes V)\big) = \ker(\rho \circ q) = I. \qquad \square$$

The connection between quasi-orbits and restriction begins to come into focus with the next definition and remark.

Definition 6.17. A representation $L = \pi \rtimes U$ of $A \rtimes_\alpha G$ *lives on a quasi-orbit* if there is a $P \in \operatorname{Prim} A$ such that

$$\operatorname{Res}(\ker L) = \ker \pi = \bigcap_{s \in G} s \cdot P.$$

A dynamical system (A, G, α) is called *quasi-regular* if every irreducible representation lives on a quasi-orbit.

Remark 6.18. If L lives on the quasi-orbit associated to $P \in \operatorname{Prim} A$, then Lemma 6.16 on the preceding page implies L factors through $A/I \rtimes_{\alpha^I} G$ where $I := \bigcap_{s \in G} s \cdot P$. Note that A/I is the quotient of A associated to the closed set $\overline{G \cdot P}$ in $\operatorname{Prim} A$ — this is supposed to justify the "lives on" terminology. The concept of quasi-regularity will play a role in Sections 7.5 and 8.3.

Induced representations are examples of representations that often live on quasi-orbits.

Lemma 6.19. *Suppose that (A, G, α) is a dynamical system, that H is closed subgroup of G and that L is a representation of $A \rtimes_{\alpha|_H} H$. Then*

$$\operatorname{Res}\big(\ker(\operatorname{Ind}_H^G L)\big) = \bigcap_{s \in G} s \cdot \big(\operatorname{Res}(\ker L)\big).$$

In particular, if $\operatorname{Res}(\ker L)$ *is a primitive ideal* $P \in \operatorname{Prim} A$, *then* $\operatorname{Ind}_H^G L$ *lives on the quasi-orbit corresponding to* P.

Proof. Let $L = \pi \rtimes u$. Using Proposition 5.4 on page 153, we can realize $\operatorname{Ind}_H^G L$ as the representation $\Pi \rtimes U$ on $L^2_u(G, \mu_{G/H}, \mathcal{H}_L) = \overline{\mathcal{V}_c}$ (adopting the notations of Proposition 5.4). Let $P = \operatorname{Res}(\ker L) = \ker \pi$ (Corollary 6.15 on page 182), and define

$$J := \bigcap s \cdot P.$$

If $a \in J$, then $\pi\bigl(\alpha_s^{-1}(a)\bigr) = 0$ for all s. Using Equation (5.2), it follows that $\Pi(a) = 0$. Therefore $J \subset \ker \Pi = \operatorname{Res}\bigl(\ker\bigl(\operatorname{Ind}_H^G L\bigr)\bigr)$.

If $a \notin J$, then there is a $r \in G$ such that $\pi\bigl(\alpha_r^{-1}(a)\bigr) \neq 0$. We want to see that $\Pi(a) \neq 0$. Since the definition of $L^2_u(G, \mu_{G/H}, \mathcal{H}_L)$ is complicated, this requires a bit of fussing. Let h be a unit vector in $\mathcal{H}_L$ and $\epsilon > 0$ be such that

$$\|\pi\bigl(\alpha_r^{-1}(a)\bigr)h\| \geq 2\epsilon > 0.$$

Choose $g \in C_c^+(G)$ such that $\int_H g(t)\,d\mu_H(t) = 1$ and such that $t \in \operatorname{supp}(g) \cap H$ implies that $\|u_t(h) - h\| < \epsilon$. Let $f(s) := g(r^{-1}s)$ and define $\xi \in \mathcal{V}_c \subset L^2_u(G, \mu_{G/H}, \mathcal{H}_L)$ by

$$\xi(s) := \int_H f(st)u_t(h)\,d\mu_H(t).$$

Then

$$\|\xi(r) - h\| = \Bigl\|\int_H f(rt)\bigl(u_t(h) - h\bigr)\,d\mu_H(t)\Bigr\| \leq \epsilon/\|a\|.$$

Therefore $\Pi(a)\xi(r) = \pi\bigl(\alpha_r^{-1}(a)\bigr)\xi(r) \neq 0$. Since $\Pi(a)\xi \in \mathcal{V}_c \smallsetminus \{0\}$, $\Pi(a) \neq 0$. Thus $\operatorname{Res}\bigl(\ker\bigl(\operatorname{Ind}_H^G L\bigr)\bigr) = \ker \Pi = J$. The last statement is an easy consequence. □

Naturally, having bothered to define the term quasi-regular, we want to see that there are conditions that force a dynamical system to be quasi-regular. As we shall see, most systems we're interested in will be quasi-regular and it may even be the case that all are.[10] Let $L = \pi \rtimes U$ be an irreducible representation of $A \rtimes_\alpha G$. Let $I := \operatorname{Res}(\ker L) = \ker \pi$, and $D := \{\, P \in \operatorname{Prim} A : P \supset I \,\}$. Then D is closed, and since I is α-invariant by Lemma 6.16 on page 182, D is G-invariant. If $P \in D$ and $Q \in \operatorname{Prim} A$ is such that $\bigcap s \cdot P = \bigcap s \cdot Q$, then $Q \in \overline{G \cdot P}$ and $Q \in D$. Thus D is saturated with respect to the equivalence relation induced by the map $k : \operatorname{Prim} A \to \mathcal{Q}$. Therefore, $C := k(D)$ is closed in $\mathcal{Q}$. We claim that C is *irreducible* in that it is not possible to decompose it as the union of two proper closed subsets. To prove the claim, we suppose to the contrary that

$$C = C_1 \cup C_2 \tag{6.8}$$

with both C_j proper closed subsets of C. Let

$$I_j := \bigcap_{k(P) \in C_j} P.$$

[10]This question is closely related to the question of whether prime ideals in a C^*-algebra must be primitive. This question was answered negatively in [165], so there may be some pathological nonseparable systems out there which fail to be quasi-regular.

Since $k^{-1}(C_j)$ is closed, $P \supset I_j$ if and only if $k(P) \in C_j$. Since every ideal in a C^*-algebra is the intersection of the primitive ideals which contain it, $I = I_1 \cap I_2$. Thus

$$(I_1 \rtimes_\alpha G)(I_2 \rtimes_\alpha G) \subset I \rtimes_\alpha G \subset \ker L.$$

Since L is irreducible, $\ker L$ is primitive and therefore a prime ideal [139, Proposition A.17]. Thus for some j,

$$I_j \rtimes_\alpha G \subset \ker L.$$

Therefore Lemma 6.16 on page 182 and Corollary 6.15 on page 182 imply that

$$I_j = \operatorname{Res}(I_j \rtimes_\alpha G) \subset \operatorname{Res}(\ker L) = I.$$

But then

$$C_j = \{\, k(P) : P \supset I_j \,\} \supset C = \{\, k(P) : P \supset I \,\},$$

which contradicts our assumption that C_j is a proper subset of C. Therefore C is irreducible as claimed.

Note that a point closure is always an example of an irreducible closed set.

Lemma 6.20. *Suppose that L is an irreducible representation of $A \rtimes_\alpha G$ and let*

$$C := \{\, k(P) \in \mathcal{Q} : P \in \operatorname{Prim} A \text{ and } P \supset \operatorname{Res}(\ker L) \,\}$$

be the irreducible closed set considered above. If $C = \overline{\{\, k(P) \,\}}$ for some $P \in \operatorname{Prim} A$, then $k^{-1}(C) = \overline{G \cdot P}$ and L lives on the quasi-orbit associated to P.

Proof. Let $Q \in k^{-1}(C)$. If $Q \notin \overline{G \cdot P}$, then $\operatorname{Prim} A \setminus \overline{G \cdot P}$ is a saturated open subset of $\operatorname{Prim} A$ containing Q but not P. Thus $k(\operatorname{Prim} A \setminus \overline{G \cdot P})$ is a neighborhood of Q in $\mathcal{Q}$ containing $k(Q)$ but not $k(P)$. This implies $k(Q) \notin \overline{\{\, k(P) \,\}}$ which is a contradiction.

On the other hand, if $Q \in \overline{G \cdot P}$, then there are $s_i \in G$ and a net $s_i \cdot P \to Q$. Since k is continuous and $k(s_i \cdot P) = k(P)$, $k(Q) \in \overline{\{\, k(P) \,\}}$. □

In view of Lemma 6.20, quasi-regularity can be established by showing that all irreducible subsets of a quasi-orbit space are point closures. There are certainly nasty topological spaces where this fails; for example, Weaver [165] has recently shown that there are (necessarily nonseparable) C^*-algebras which contain prime ideals which are not primitive. As in the proof of [139, Theorem A.50], if I is prime, then $C = \{\, P \in \operatorname{Prim} A : P \supset I \,\}$ is irreducible, and then I is primitive if and only if C is a point closure.

To exhibit quasi-regular systems, we will want to take advantage of the fact that the primitive ideal space of a C^*-algebra is a *totally Baire space.* (A space is totally Baire if each closed subset is a Baire space.) Since any closed subspace of a primitive ideal space of a C^*-algebra is the primitive ideal space of a quotient, and since the continuous open image of a totally Baire space is totally Baire, it follows from [139, Corollary A.47] that the primitive ideal spaces of separable C^*-algebras are totally Baire. The general case is proved in [28, Corollary 3.4.13]. Since $\mathcal{Q}$ is the continuous open image of $\operatorname{Prim} A$ (Lemma 6.12 on page 180), $\mathcal{Q}$ is totally Baire. (Since we only want to use this fact when $\mathcal{Q}$ is second countable, the proof in [139] suffices.)

Proposition 6.21. *Let $\mathcal{Q}$ be the quasi-orbit space of (A, G, α). If $\mathcal{Q}$ is either second countable or almost Hausdorff, then (A, G, α) is quasi-regular. In particular, if A is separable, then (A, G, α) is quasi-regular.*

Proof. In view of Lemma 6.20 on the previous page, it suffices to see that every irreducible closed subset of $\mathcal{Q}$ is a point closure. First suppose that $\mathcal{Q}$ is second countable. If C is an irreducible closed subset, then C is second countable and has a basis $\{ U_n \}$ of open sets. But any nonempty open subset of C is necessarily dense, and since $\mathcal{Q}$ is totally Baire, we have $\bigcap U_n \neq \emptyset$. However, any point in the intersection meets every open set in C and is therefore dense in C. Therefore, C contains a dense point as required.

Now suppose that $\mathcal{Q}$ is almost Hausdorff. Then C contains a dense open Hausdorff subset W. But if W contains two distinct points p and q, then there are disjoint nonempty open sets U and V such that $p \in U$ and $q \in V$. This gives a contradiction as $C = (C \smallsetminus U) \cup (C \smallsetminus V)$ is a decomposition of C into proper closed subsets. Therefore W must be a single point and C is a point closure.

The final assertion follows since $\mathcal{Q}$ is second countable if A is separable (Lemma 6.12 on page 180). □

Definition 6.22. Let (A, G, α) be a dynamical system. An orbit $G \cdot P$ in $\operatorname{Prim} A$ is called *regular* if it is locally closed in $\operatorname{Prim} A$ and if $sG_P \mapsto s \cdot P$ is a homeomorphism of G/G_P onto $G \cdot P$. We say that (A, G, α) is *regular* if (A, G, α) is quasi-regular and every orbit is regular.

Remark 6.23. The word regular is over used in mathematics. I took the term and definition from Green [66]. Unfortunately, the use described in Definition 6.22 is not universally accepted. For example, Quigg and Speilberg have used "regular" to describe systems in which the reduced norm coincides with the universal norm [131]. Nevertheless, I like the term and will use it here. The next result can be seen as partial justification for this. It says that the Mackey-Glimm Dichotomy implies that regular systems are more common than one might suspect and have many pleasing properties justifying the terminology. To see this, we have to appeal to the result that the primitive ideal spaces of GCR algebras are almost Hausdorff.

Proposition 6.24. *Suppose that A is a separable* GCR *algebra and that G is a second countable locally compact group. Then (A, G, α) is regular if and only if $G \backslash \operatorname{Prim} A$ is a T_0 space.*

Proof. Since $\operatorname{Prim} A$ must be second countable if A is separable, $\mathcal{Q}$ is second countable, and (A, G, α) is quasi-regular by Proposition 6.21. Since A is GCR, $\operatorname{Prim} A$ is locally Hausdorff by either [28, Theorem 4.4.5] or [126, Theorem 6.2.11]. Thus $(G, \operatorname{Prim} A)$ satisfies the hypotheses for Theorem 6.2 on page 173, and the result follows. □

Notes and Remarks

Theorem 6.2 and its proof are a blend of Effros's main result from [46] and Glimm's main result in [59]. Effros and Glimm's results include important tie ins to Borel structures, based on Mackey's work in [105], which weren't necessary for the modest goals here. The material on Res and quasi-orbits is based primarily on Green's treatment in [66].

Chapter 7

Properties of Crossed Products

Now that we've built up a fair amount of technology for crossed products, it's definitely time to prove some general structure results. In Section 7.1 we generalize the Pontryagin duality theorem for abelian groups to crossed products by abelian groups. This result is due to Takai [161], although the proof given here is strongly influenced by both Raeburn [133] and Echterhoff. In Section 7.2, we develop the crossed product analogue of the reduced group C^*-algebra. (The reduced group C^*-algebra is defined to be the closure in $B\bigl(L^2(G)\bigr)$ of the image of the left-regular representation.) If G is amenable, then this is a faithful representation of $C^*(G)$.[1] We give the corresponding construction of the reduced crossed product, and prove an analogous result for actions of amenable groups.

In Section 7.3, we consider crossed products determined by projective representations $\alpha : G \to \operatorname{Aut}\bigl(\mathcal{K}(\mathcal{H})\bigr)$. This leads naturally to discussion of cocycle representations in Section 7.3.[2] We use this discussion as an excuse to give a brief aside on (Green) twisted crossed products in Section 3.3. In particular, we show that if G has a normal subgroup N, then $A \rtimes_\alpha G$ is an iterated (twisted) crossed product in the spirit of our results for semidirect products.

We finish this chapter with a preliminary look at the question of when a crossed product is CCR or GCR. These sorts of questions are much easier for transformation group C^*-algebras, and we come back to the subject for transformation group C^*-algebras at the end of Section 8.3.

7.1 The Takai Duality Theorem

In these notes, the Takai Duality Theorem is a theorem about abelian groups. There are versions for nonabelian groups involving coactions, but we'll have little

[1] Amenable groups are discussed in detail in Appendix A.

[2] Projective representations and cocycle representations are discussed in Appendix D.3.

to say about that here (cf., [130, 134]). So in this section, G will be an abelian group unless specifically stated otherwise. Let (A, G, α) be a dynamical system. For each $\gamma \in \widehat{G}$, define a $*$-isomorphism $\hat{\alpha}_\gamma : C_c(G, A) \to C_c(G, A)$ by $\hat{\alpha}_\gamma(f)(s) := \overline{\gamma(s)} f(s)$. Then $\hat{\alpha}_\gamma$ is continuous with respect to the inductive limit topology, and therefore bounded with respect to the universal norm (Corollary 2.47). Thus $\hat{\alpha}_\gamma$ extends to an element of $\operatorname{Aut}(A \rtimes_\alpha G)$, and we get a homomorphism

$$\hat{\alpha} : \widehat{G} \to \operatorname{Aut}(A \rtimes_\alpha G).$$

If $\gamma_i \to \gamma$ in $\widehat{G}$ and if $f \in C_c(G, A)$, then $\hat{\alpha}_{\gamma_i}(f) \to \hat{\alpha}_\gamma(f)$ in the inductive limit topology. It follows that $\gamma \mapsto \hat{\alpha}_\gamma(f)$ is continuous on $A \rtimes_\alpha G$. Therefore $(A \rtimes_\alpha G, \widehat{G}, \hat{\alpha})$ is a dynamical system called the *dual system*, and $\hat{\alpha}$ is called the *dual action*.

The Duality Theorem implies that the iterated crossed product $(A \rtimes_\alpha G) \rtimes_{\hat{\alpha}} \widehat{G}$, is isomorphic to $A \otimes \mathcal{K}\big(L^2(G)\big)$.[3]

The Pontryagin Duality Theorem allows us to identify G with the dual of $\widehat{G}$ where $s \in G$ is associated to the character $\gamma \mapsto \gamma(s)$ on $\widehat{G}$. Therefore there is a *double dual action* $\hat{\hat{\alpha}}$ of G on the iterated crossed product, and it is of interest to see what the induced action on $A \otimes \mathcal{K}\big(L^2(G)\big)$ is. For this we need the following. Let $\rho : G \to U\big(L^2(G)\big)$ be the right-regular representation and form the dynamical system $\operatorname{Ad} \rho : G \to \operatorname{Aut} \mathcal{K}\big(L^2(G)\big)$ where $(\operatorname{Ad} \rho)_s(T) := \rho(s) T \rho(s^{-1})$. Then we can form the tensor product dynamical system $\alpha \otimes \operatorname{Ad} \rho : G \to \operatorname{Aut}\big(A \otimes \mathcal{K}\big(L^2(G)\big)\big)$.

Theorem 7.1 (Takai Duality). *Suppose that G is an abelian group and that (A, G, α) is a dynamical system. Then there is an isomorphism Ψ from the iterated crossed product $(A \rtimes_\alpha G) \rtimes_{\hat{\alpha}} \widehat{G}$ onto $A \otimes \mathcal{K}\big(L^2(G)\big)$ which is equivariant for the double dual action $\hat{\hat{\alpha}}$ of G on $(A \rtimes_\alpha G) \rtimes_{\hat{\alpha}} \widehat{G}$, and the action $\alpha \otimes \operatorname{Ad} \rho$ of G on $A \otimes \mathcal{K}\big(L^2(G)\big)$.*

The proof given here is a variation of Raeburn's proof in [133] and was shown to me by Siegfried Echterhoff. We will produce the isomorphism Ψ as a composition:

$$(A \rtimes_\alpha G) \rtimes_{\hat{\alpha}} \widehat{G} \xrightarrow{\Phi_1} (A \rtimes_{\mathrm{id}} \widehat{G}) \rtimes_{\widehat{\mathrm{id}}^{-1} \otimes \alpha} G \xrightarrow{\Phi_2} C_0(G, A) \rtimes_{\mathrm{lt} \otimes \alpha} G$$
$$\xrightarrow{\Phi_3} C_0(G, A) \rtimes_{\mathrm{lt} \otimes \mathrm{id}} G \longrightarrow C_0(G) \rtimes_{\mathrm{lt}} G \otimes A \longrightarrow A \otimes \mathcal{K}\big(L^2(G)\big).$$

As usual, we want to work with dense subalgebras, and this will pose some technical problems in the case of Φ_1. In general, if we have an iterated crossed product

$$(A \rtimes_\beta K) \rtimes_\delta H, \tag{7.1}$$

[3]When working with crossed products, the tensor product of choice is usually the maximal tensor product $\otimes_{\max}$. Here since one of the factors is the compacts, all tensor product norms coincide [139, Corollary B.44], and we are free to use whatever definition we like for the tensor product. In this situation, it is standard to use an undecorated symbol $\otimes$.

then it is straightforward to check that $C_c(H \times K, A)$ can be viewed as a dense *subspace*. For example, if $F \in C_c(H \times K, A)$, then

$$\lambda_F(h)(k) := F(h,k)$$

defines an element $\lambda_F \in C_c(H, A \rtimes_\beta K)$ and we can appeal to Lemma 1.87 on page 29. In many cases, such as that for transformation group C^*-algebras, $C_c(H \times K, A)$ will be a $*$-sub*algebra*. To ensure that this is the case here, we want to add the assumption that $C_c(K, A) \subset A \rtimes_\beta K$ is invariant for δ, and that

$$(h, h', k) \mapsto \delta_h\big(\lambda_F(h')\big)(k) \tag{7.2}$$

is continuous with compact support in h' and k. In this event, we say that the action δ is *compatible* with β. Then, if F_1 and F_2 are in $C_c(H \times K, A)$, (7.2) implies that

$$(h, h', k', k) \mapsto \lambda_{F_1}(h)(k)\beta_k\big(\delta_h\big(\lambda_{F_2}(h^{-1}h')\big)(k^{-1}k')\big)$$

is in $C_c(H \times H \times K \times K, A)$. Therefore Corollary 1.104 implies that

$$\begin{aligned}(h, h', k') \mapsto \int_K &\lambda_{F_1}(h)(k)\beta_k\big(\delta_h\big(\lambda_{F_2}(h^{-1}h')\big)(k^{-1}k')\big)\,d\mu_K(k)\\ &= \lambda_{F_1}(h) * \delta_h\big(\lambda_{F_2}(h^{-1}h')\big)(k')\end{aligned}$$

is in $C_c(H \times H \times K, A)$, and that

$$(h', k') \mapsto \int_H \lambda_{F_1}(h) * \delta_h\big(\lambda_{F_2}(h^{-1}h')\big)(k')\,d\mu_H(h) \tag{7.3}$$

is in $C_c(H \times K, A)$. To see that (7.3) really represents $\lambda_{F_1} * \lambda_{F_2}$, recall that the product of the λ_{F_i} in $(A \rtimes_\beta K) \rtimes_\delta H$ is formally given by the $A \rtimes_\beta K$-valued integral

$$\lambda_{F_1} * \lambda_{F_2}(h') = \int_H^{A \rtimes_\beta K} \lambda_{F_1}(h) * \delta_h\big(\lambda_{F_2}(h^{-1}h')\big)\,d\mu_H(h). \tag{7.4}$$

However, we can apply Lemma 1.108 on page 39 — with $Q(h,k) = \lambda_{F_1}(h) * \delta_h\big(\lambda_{F_2}(h^{-1}h')\big)(k)$ — to conclude that the (7.4) takes its value in $C_c(K, A) \subset A \rtimes_\beta K$, and that

$$\begin{aligned}\lambda_{F_1} * \lambda_{F_2}(h')(k') &= \int_H \lambda_{F_1}(h) * \delta_h\big(\lambda_{F_2}(h^{-1}h')\big)(k')\,d\mu_H(h)\\ &= \int_H \int_K \lambda_{F_1}(h)(k)\beta_k\big(\delta_h\big(\lambda_{F_2}(h^{-1}h')\big)(k^{-1}k')\big)\,d\mu_K(k)\,d\mu_H(h),\end{aligned} \tag{7.5}$$

as you would expect. Therefore we can view $C_c(H \times K, A)$ as a $*$-subalgebra of $(A \rtimes_\beta K) \rtimes_\delta H$ with the convolution product given by (7.5) (the involution presents no problem).

Now we want to apply (7.5) when $K = G$, $\beta = \alpha$, $H = \widehat{G}$ and $\delta = \hat{\alpha}$. Then it is easy to see that $\hat{\alpha}$ is compatible with α, and we can plug into (7.5) to get

$$\lambda_{F_1} * \lambda_{F_2}(\gamma)(s) = \int_{\widehat{G}} \int_G \lambda_{F_1}(\sigma)(r)\overline{\sigma(r^{-1}s)}\alpha_r\bigl(\lambda_{F_2}(\bar{\sigma}\gamma)(r^{-1}s)\bigr)\,d\mu(r)\,d\hat{\mu}(\sigma). \quad (7.6)$$

To be a little less pedantic, we can rewrite (7.6) as

$$F_1 * F_2(\gamma, s) = \int_{\widehat{G}} \int_G F_1(\sigma, r)\overline{\sigma(r^{-1}s)}\alpha_r\bigl(F_2(\bar{\sigma}\gamma, r^{-1}s)\bigr)\,d\mu(r)\,d\hat{\mu}(\sigma),$$

which is what you would expect.

Now let $K = \widehat{G}$ and $\beta = \mathrm{id}$. In this case, $A \rtimes_\beta K = A \rtimes_{\mathrm{id}} \widehat{G} \cong C^*(\widehat{G}) \otimes A$ by Lemma 2.73.[4] Let $\widehat{\mathrm{id}}$ be the dual action on $C^*(\widehat{G})$ which is given on $\varphi \in C_c(\widehat{G})$ by $\widehat{\mathrm{id}}_s(\varphi)(\gamma) = \overline{\gamma(s)}\varphi(\gamma)$. Since $\widehat{G}$ is abelian, $\widehat{\mathrm{id}}^{-1}$ is an action, and we can form the tensor product action $\widehat{\mathrm{id}}^{-1} \otimes \alpha$ on $C^*(\widehat{G}) \otimes A$. We'll use the same notation for the action on $A \rtimes_{\mathrm{id}} \widehat{G}$, and note that it is given on $C_c(\widehat{G}, A) \subset A \rtimes_\beta K$ by $(\widehat{\mathrm{id}}^{-1} \otimes \alpha)_s(f)(\gamma) = \gamma(s)\alpha_s\bigl(f(\gamma)\bigr)$. So, if we set $H = G$ and $\delta = \widehat{\mathrm{id}}^{-1} \otimes \alpha$, then $\widehat{\mathrm{id}}^{-1} \otimes \alpha$ is compatible with id. Therefore if $\tilde{F}_1$ and $\tilde{F}_2$ are in $C_c(G \times \widehat{G}, A) \subset (A \rtimes_{\mathrm{id}} \widehat{G}) \rtimes_{\widehat{\mathrm{id}}^{-1} \otimes \alpha} G$, then (7.5) implies

$$\begin{aligned}
\tilde{F}_1 * \tilde{F}_2(s, \gamma) &= \lambda_{\tilde{F}_1} * \lambda_{\tilde{F}_2}(s)(\gamma) \\
&= \int_G \int_{\widehat{G}} \lambda_{\tilde{F}_1}(r)(\sigma)(\widehat{\mathrm{id}}^{-1} \otimes \alpha)_r\bigl(\lambda_{\tilde{F}_2}(r^{-1}s)\bigr)(\bar{\sigma}\gamma)\,d\hat{\mu}(\sigma)\,d\mu(r) \\
&= \int_G \int_{\widehat{G}} \lambda_{\tilde{F}_1}(r)(\sigma)\bar{\sigma}(r)\gamma(r)\alpha_r\bigl(\lambda_{\tilde{F}_2}(r^{-1}s)(\bar{\sigma}\gamma)\bigr)\,d\hat{\mu}(\sigma)\,d\mu(r) \\
&= \int_G \int_{\widehat{G}} \tilde{F}_1(r, \sigma)\bar{\sigma}(r)\gamma(r)\alpha_r\bigl(\tilde{F}_2(r^{-1}s, \bar{\sigma}\gamma)\bigr)\,d\hat{\mu}(\sigma)\,d\mu(r).
\end{aligned} \quad (7.7)$$

Lemma 7.2. *Suppose that G is an abelian group and that (A, G, α) is a dynamical system. Then there is an isomorphism*

$$\Phi_1 : (A \rtimes_\alpha G) \rtimes_{\hat{\alpha}} \widehat{G} \to (A \rtimes_{\mathrm{id}} \widehat{G}) \rtimes_{\widehat{\mathrm{id}}^{-1} \otimes \alpha} G$$

which maps the dense subalgebra $C_c(\widehat{G} \times G, A)$ onto the dense subalgebra $C_c(G \times \widehat{G}, A)$ and satisfies

$$\Phi_1(F)(s, \gamma) = \gamma(s)F(\gamma, s) \quad F \in C_c(\widehat{G} \times G, A).$$

Proof. It follows easily from (7.6) and (7.7), together with an application of Fubini's Theorem (as in Proposition 1.105), that Φ_1 defines a homomorphism from $C_c(\widehat{G} \times$

[4] Note that $C^*(\widehat{G})$ is commutative and so nuclear [139, Proposition B.43].

$G, A)$ onto $C_c(G\times\widehat{G}, A)$. To see that Φ_1 is a $*$-homomorphism, we need to see that $\lambda^*_{\Phi_1(F)} = \Phi_1(\lambda^*_F)$. But

$$\begin{aligned}\lambda^*_{\Phi_1(F)}(s)(\gamma) &= (\widehat{\mathrm{id}}^{-1}\otimes\alpha)_s\big(\lambda_{\Phi_1(F)}(s^{-1})^*\big)(\gamma)\\ &= \gamma(s)\alpha_s\big(\lambda_{\Phi_1(F)}(s^{-1})^*(\gamma)\big)\\ &= \gamma(s)\alpha_s\big(\lambda_{\Phi_1(F)}(s^{-1})(\bar\gamma)\big)^*\\ &= \gamma(s)\alpha_s\big(\Phi_1(F)(s^{-1},\bar\gamma)\big)^*\\ &= \alpha_s\big(F(\bar\gamma,s^{-1})\big)^*\\ &= \alpha_s\big(\lambda_F(\bar\gamma)(s^{-1})\big)^*\\ &= \lambda_F(\bar\gamma)^*(s)\\ &= \gamma(s)\lambda^*_F(\gamma)(s)\\ &= \Phi_1(\lambda^*_F)(s)(\gamma).\end{aligned}$$

Since Φ_1 maps a dense subalgebra onto a dense subalgebra, we only need to see that Φ_1 is isometric for the universal norms. We could do this using the fact that Φ_1 is continuous with the inductive limit topology and proving a slightly fussier version of Lemma 2.45. Instead, we'll follow the more traditional route of examining covariant representations.

Let L be a representation of $(A\rtimes_{\mathrm{id}}\widehat{G})\rtimes_{\widehat{\mathrm{id}}^{-1}\otimes\alpha} G$. Then $L = R\rtimes U$ for a covariant representation (R,U) of $(A\rtimes_{\mathrm{id}}\widehat{G}, G, \widehat{\mathrm{id}}^{-1}\otimes\alpha)$, and $R=\pi\rtimes V$ for a covariant representation (π, V) of $(A,\widehat{G},\mathrm{id})$. Let $j_{\widehat{G}}$ be the canonical map of G into $M\big(C^*(G)\big)$. Since $j_{\widehat{G}}(\gamma)\varphi(\sigma) := \varphi(\bar\gamma\sigma)$, an easy computation shows that

$$\widehat{\mathrm{id}}_s^{-1}\circ j_{\widehat{G}}(\gamma) = \gamma(s) j_{\widehat{G}}(\gamma)\circ\widehat{\mathrm{id}}_s^{-1}.$$

Now if $a\in A$, $\varphi\in C_c(\widehat{G})$ and $f\in C_c(G)$, then functions which are elementary tensors of the form $\varphi\otimes f\otimes a$ span a dense subset of $C_c(\widehat{G}\times G, A)$ and

$$\begin{aligned}U_sV_\gamma L(\varphi\otimes f\otimes a) &= U_sV_\gamma\pi(a)V(\varphi)U(f)\\ &= U_s\pi(a)V_\gamma V(\varphi)U(f)\\ &= U_s\pi(a)V\big(j_{\widehat{G}}(\gamma)\varphi\big)U(f)\\ &= U_s\pi\rtimes V\big(j_{\widehat{G}}(\gamma)(\varphi)\otimes a\big)U(f)\\ &= \pi\big(\alpha_s(a)\big)V\big(\widehat{\mathrm{id}}_s^{-1}\circ j_{\widehat{G}}(\gamma)\varphi\big)U_sU(f)\\ &= \gamma(s)V_\gamma\pi\big(\alpha_s(a)\big)V\big(\widehat{\mathrm{id}}_s^{-1}\varphi\big)U_sU(f)\\ &= \gamma(s)V_\gamma U_s\pi(a)V(\varphi)U(f)\\ &= \gamma(s)V_\gamma U_s L(\varphi\otimes f\otimes a).\end{aligned}$$

Since L is nondegenerate, it follows that

$$U_sV_\gamma = \gamma(s)V_\gamma U_s. \tag{7.8}$$

A similar computation shows that

$$\begin{aligned} U_s\pi(b)\pi(a)V(\varphi)U(f) &= U_s\pi(ba)V(\varphi)U(f) \\ &= U_s\pi \rtimes V\big(\varphi \otimes ba\big)U(f) \\ &= \pi\big(\alpha_s(ba)\big)V\big(\widehat{\mathrm{id}}_s^{-1}(\varphi)\big)U_sU(f) \\ &= \pi\big(\alpha_s(b)\big)U_s\pi(a)V(\varphi)U(f). \end{aligned}$$

Thus (π, U) is a covariant representation of (A, G, α). We also claim that $(\pi \rtimes U, V)$ is a covariant representation of $(A \rtimes_\alpha G, \widehat{G}, \hat{\alpha})$. This is straightforward using (7.8):

$$\begin{aligned} V_\gamma \pi \rtimes U(a \otimes f) &= \pi(a)V_\gamma \int_G f(s)U_s\,d\mu(s) \\ &= \pi(a) \int_G f(s)\overline{\gamma(s)}U_s\,d\mu(s)V_\gamma \\ &= \pi \rtimes U\big(\hat{\alpha}_\gamma(a \otimes f)\big)V_\gamma. \end{aligned}$$

From the above, $L' = (\pi \rtimes U) \rtimes V$ is a representation of $(A \rtimes_\alpha G) \rtimes_{\hat{\alpha}} \widehat{G}$ and

$$\begin{aligned} L\big(\Phi_1(F)\big) &= \int_G \pi \rtimes V\big(\lambda_{\Phi_1(F)}(s)\big)U_s\,d\mu(s) \\ &= \int_G \int_{\widehat{G}} \pi\big(\Phi_1(F)(s,\gamma)\big)V_\gamma U_s\,d\hat{\mu}(\gamma)\,d\mu(s) \\ &= \int_G \int_{\widehat{G}} \pi\big(F(\gamma, s)\big)\gamma(s)V_\gamma U_s\,d\hat{\mu}(\gamma)\,d\mu(s) \end{aligned}$$

which, by Fubini (Proposition 1.105) and (7.8), is

$$\begin{aligned} &= \int_{\widehat{G}} \int_G \pi\big(F(\gamma, s)\big)U_sV_\gamma\,d\mu(s)\,d\hat{\mu}(\gamma) \\ &= L'(F). \end{aligned}$$

It follows that $\|\Phi_1(F)\| \le \|F\|$. Reversing the above argument gives $\|F\| \le \|\Phi_1(F)\|$, and this completes the proof. □

Now $C_0(G, A) \cong C_0(G) \otimes A$ (e.g., [139, Propositions B.16 and B.43]), and we can form the tensor product action $\mathrm{lt} \otimes \alpha$ on $C_0(G, A)$:

$$(\mathrm{lt} \otimes \alpha)_s(f)(r) = \alpha_s\big(f(s^{-1}r)\big).$$

Furthermore, since G is the dual of $\widehat{G}$, Proposition 3.1 tells us that $C^*(\widehat{G})$ is isomorphic to $C_0(G)$ via the Fourier transform. However if we use the formula for the Fourier transform given in section 3.1 and the identification of G with the dual of $\widehat{G}$ given above, we don't get quite the formula we want for the next proof: we're off by a complex conjugate. This is often the case when working with the Fourier transform because the formula depends on how we identify elements in the dual with actual functions on the group. For example, we think of $\widehat{R}$ as being equal to $\mathbf{R}$, but this is because we make the identification of $x \in \mathbf{R}$ with the character $y \mapsto e^{-ixy}$. Thus the Fourier transform becomes

$$\hat{f}(x) = \int_{-\infty}^{\infty} f(y)e^{-ixy}\,dy.$$

Of course, if we associate x to $y \mapsto e^{ixy}$ then the formula above changes by a conjugate. Similarly, if we identify G with the dual of $\widehat{G}$ by associating $s \in G$ to the character $\gamma \mapsto \overline{\gamma(s)}$, the isomorphism of $C^*(\widehat{G})$ with $C_0(G)$ is given on functions $\varphi \in C_c(\widehat{G})$ by

$$\hat{\varphi}(s) = \int_{\widehat{G}} \varphi(\gamma)\overline{\gamma(s)}\,d\hat{\mu}(\gamma).$$

Lemma 7.3. *Suppose that G is abelian and that (A, G, α) is a dynamical system. Then there is an isomorphism*

$$\Phi_2 : (A \rtimes_{\mathrm{id}} \widehat{G}) \rtimes_{\widehat{\mathrm{id}}^{-1} \otimes \alpha} G \to C_0(G, A) \rtimes_{\mathrm{lt} \otimes \alpha} G$$

which maps $F \in C_c(G \times \widehat{G}, A)$ into $C_c\big(G, C_0(G, A)\big)$ by the formula

$$\Phi_2(F)(s, r) = \int_{\widehat{G}} F(s, \gamma)\overline{\gamma(r)}\,d\hat{\mu}(\gamma).$$

Proof. The map $a \otimes \varphi \mapsto a \otimes \hat{\varphi}$ defines a $*$-homomorphism from the span of elementary tensors in $A \rtimes_{\mathrm{id}} \widehat{G} \cong A \otimes C^*(\widehat{G})$ into $C_0(G, A) \cong C_0(G) \otimes A$ which extends to an isomorphism $\varphi_2 : A \rtimes_{\mathrm{id}} \widehat{G} \to C_0(G, A)$ satisfying

$$\varphi_2(\psi)(r) := \int_{\widehat{G}} \psi(\gamma)\overline{\gamma(r)}\,d\hat{\mu}(\gamma) \quad \text{for } \psi \in C_c(\widehat{G}, A).$$

Since

$$\begin{aligned} \varphi_2\big((\widehat{\mathrm{id}}^{-1} \otimes \alpha)_s(\psi)\big)(r) &= \int_{\widehat{G}} (\widehat{\mathrm{id}}^{-1} \otimes \alpha)_s(\psi)(\gamma)\overline{\gamma(r)}\,d\hat{\mu}(\gamma) \\ &= \int_{\widehat{G}} \gamma(s)\alpha_s\big(\psi(\gamma)\big)\overline{\gamma(r)}\,d\hat{\mu}(\gamma) \\ &= \alpha_s\Big(\int_{\widehat{G}} \psi(\gamma)\overline{\gamma(s^{-1}r)}\,d\hat{\mu}(\gamma)\Big) \\ &= (\mathrm{lt} \otimes \alpha)_s\big(\varphi_2(\psi)\big)(r), \end{aligned}$$

φ_2 is equivariant, and $\Phi_2 := \varphi_2 \rtimes \mathrm{id}$ is the required map (cf. Corollary 2.48 on page 63). □

Lemma 7.4. *Suppose that G is an abelian group and that (A, G, α) is a dynamical system. Then there is an isomorphism*

$$\Phi_3 : C_0(G, A) \rtimes_{\mathrm{lt} \otimes \alpha} G \to C_0(G, A) \rtimes_{\mathrm{lt} \otimes \mathrm{id}} G$$

which satisfies

$$\Phi_3(F)(s,r) = \alpha_r^{-1}\big(F(s,r)\big)$$

for $F \in C_c\big(G, C_0(G,A)\big)$.

Proof. We can define an isomorphism $\varphi_3 : C_0(G,A) \to C_0(G,A)$ by

$$\varphi_3(f)(r) := \alpha_r^{-1}\big(f(r)\big).$$

Then

$$\begin{aligned}\varphi_3\big((\mathrm{lt}\otimes\alpha)_s(f)\big)(r) &= \alpha_r^{-1}\big(\alpha_s\big(f(s^{-1}r)\big)\big)\\ &= \alpha_{s^{-1}r}^{-1}\big(f(s^{-1}r)\big)\\ &= (\mathrm{lt}\otimes\mathrm{id})_s\big(\varphi_3(f)\big)(r).\end{aligned}$$

Therefore φ_3 is equivariant and $\Phi_3 := \varphi_3 \rtimes \mathrm{id}$ is the required isomorphism. □

Let rt denote right-translation on $C_0(G)$: $\mathrm{rt}_r(f)(s) := f(sr)$. Since right and left translation commute, rt_r is equivariant on $\big(C_0(G), G, \mathrm{lt}\big)$ for each $r \in G$, and it is not hard to see that we get a dynamical system $\mathrm{rt}\otimes\mathrm{id} : G \to \mathrm{Aut}\big(C_0(G) \rtimes_{\mathrm{lt}} G\big)$.

Lemma 7.5. *If G is a (possibly nonabelian) locally compact group, then there is an equivariant isomorphism φ_4 from*

$$\big(C_0(G) \rtimes_{\mathrm{lt}} G, G, \mathrm{rt}\otimes\mathrm{id}\big) \quad \textit{onto} \quad \big(\mathcal{K}\big(L^2(G)\big), G, \mathrm{Ad}\,\rho\big)$$

defined on $f \in C_c(G\times G) \subset C_0(G)\rtimes_{\mathrm{lt}} G$ and $h \in C_c(G) \subset L^2(G)$ by

$$\varphi_4(f)(h)(r) = \int_G f(s,r)h(s^{-1}r)\,d\mu(s).$$

Proof. We proved that φ_4 is an isomorphism in Theorem 4.24 on page 133. But

$$\begin{aligned}\rho(v)\varphi_4(f)(h)(r) &= \Delta(v)^{\frac12}\varphi_4(f)(h)(rv)\\ &= \Delta(v)^{\frac12}\int_G f(s,rv)h(s^{-1}rv)\,d\mu(s)\\ &= \int_G (\mathrm{rt}\otimes\mathrm{id})_v(f)(s,r)\rho(v)(h)(s^{-1}r)\,d\mu(s)\\ &= \int_G \varphi_4\big((\mathrm{rt}\otimes\mathrm{id})_v(f)\big)\rho(v)(h)(r).\end{aligned}$$

It follows that $\rho(v)\varphi_4(f)\rho(v^{-1}) = \varphi_4\big((\mathrm{rt}\otimes\mathrm{id})_v(f)\big)$ as required. □

Combining Lemma 7.5 with Lemma 2.75, we obtain the following.

Lemma 7.6. *If G is an locally compact group and (A,G,α) is a dynamical system, then there is an equivariant isomorphism $\Phi_4 := \varphi_4 \otimes \mathrm{id}$ from*

$$\big(C_0(G,A) \rtimes_{\mathrm{lt}\otimes\mathrm{id}} G, G, (\mathrm{rt}\otimes\alpha)\otimes\mathrm{id}\big) \quad \textit{onto} \quad \big(A\otimes\mathcal{K}\big(L^2(G)\big), G, \alpha\otimes\mathrm{Ad}\,\rho\big).$$

Proof of Theorem 7.1. In view of Lemmas 7.2, 7.3 and 7.4, there is an isomorphism $\Phi := \Phi_3 \circ \Phi_2 \circ \Phi_1$ from the iterated crossed product $(A \rtimes_\alpha G) \rtimes_{\hat{\alpha}} \widehat{G}$ onto $C_0(G, A) \rtimes_{\mathrm{lt} \otimes \mathrm{id}} G$ mapping $F \in C_c(\widehat{G} \times G, A)$ to the element in $C_c\big(G, C_0(G, A)\big)$ given by

$$\begin{aligned}
\Phi(F)(s,r) &= \Phi_3 \circ \Phi_2 \circ \Phi_1(F)(s,r) \\
&= \alpha_r^{-1}\big(\Phi_2 \circ \Phi_1(F)(s,r)\big) \\
&= \alpha_r^{-1}\Big(\int_{\widehat{G}} \Phi_1(F)(s,\gamma)\overline{\gamma(r)}\, d\hat{\mu}(\gamma)\Big) \\
&= \int_{\widehat{G}} \alpha_r^{-1}\big(F(\gamma,s)\big)\overline{\gamma(s^{-1}r)}\, d\hat{\mu}(\gamma).
\end{aligned}$$

Now $\hat{\hat{\alpha}}$ is given on $C_c(\widehat{G} \times G, A) \subset (A \rtimes_\alpha G) \rtimes_{\hat{\alpha}} \widehat{G}$ by

$$\hat{\hat{\alpha}}_v(F)(\gamma, s) = \overline{\gamma(v)}F(\gamma, s).$$

Therefore

$$\begin{aligned}
\Phi\big(\hat{\hat{\alpha}}_v(F)\big)(s,r) &= \int_{\widehat{G}} \alpha_r^{-1}\big(\hat{\hat{\alpha}}_v(F)(\gamma,s)\big)\overline{\gamma(s^{-1}r)}\, d\hat{\mu}(\gamma) \\
&= \alpha_v\Big(\int_{\widehat{G}} \alpha_{rv}^{-1}\big(F(\gamma,s)\big)\overline{\gamma(s^{-1}rv)}\, d\hat{\mu}(\gamma)\Big) \\
&= \alpha_v\big(\Phi(F)(s,rv)\big).
\end{aligned}$$

Thus $\Phi \circ \hat{\hat{\alpha}}_v = (\mathrm{rt} \otimes \alpha)_v \otimes \mathrm{id} \circ \Phi$, and the Theorem follows from Lemma 7.6. □

7.2 The Reduced Crossed Product

Since it can be very difficult to pin down the universal norm, it can be useful to exhibit an concrete realization of $A \rtimes_\alpha G$. This is done via regular representations which give rise to the reduced norm on $C_c(G, A)$. The completion of $C_c(G, A)$ in the reduced norm is called the reduced crossed product. The reduced norm is dominated by the universal norm, but agrees with the universal norm at least when G is amenable. We formalize this discussion in this section. Amenable groups are discussed in Appendix A.

An important consequence of Lemma 5.16 on page 164 is that the kernel of $\mathrm{Ind}_H^G L$ depends only on the kernel of L. Thus if $J := \ker L$ and $f \in A \rtimes_\alpha G$, then

$$\| \mathrm{Ind}_H^G L(f) \| = \| f + \ker \mathrm{Ind}_H^G L \| = \| f + \mathrm{Ind}_H^G J \|$$

in the quotient $A \rtimes_\alpha G / \mathrm{Ind}_H^G J$. Therefore if $\ker L = \ker L'$, then

$$\| \mathrm{Ind}_H^G L(f) \| = \| \mathrm{Ind}_H^G L'(f) \|.$$

In particular, if ρ and ρ' are both *faithful* representations of A, then the associated regular representations $\mathrm{Ind}_e^G \rho$ and $\mathrm{Ind}_e^G \rho'$ have the same kernel, and

$$\| \mathrm{Ind}_e^G \rho(f) \| = \| \mathrm{Ind}_e^G \rho'(f) \| \quad \text{for all } f \in A \rtimes_\alpha G. \tag{7.9}$$

This insures that the following definition is independent of the choice of a faithful representation ρ.

Definition 7.7. If (A, G, α) is a dynamical system, then the *reduced norm* on $C_c(G, A)$ is given by

$$\|f\|_r := \|\operatorname{Ind}_e^G \rho(f)\|,$$

where ρ is any faithful representation of A. The completion $A \rtimes_{\alpha,r} G$ of $C_c(G,A)$ with respect to $\|\cdot\|_r$ is called the *reduced crossed product.*

If π is any representation of A, then we certainly have $\operatorname{Ind}_e^G(\ker\pi) \supset \operatorname{Ind}_e^G\{0\}$. Thus for any f and π,

$$\|\operatorname{Ind}_e^G \pi(f)\| \le \|f\|_r.$$

Recall that a representation π of A is said to "factor through" a quotient A/I when $I \subset \ker\pi$. Such a representation determines a representation $\tilde{\pi}$ of A/I by $\tilde{\pi}(a+I) := \pi(a)$. We can summarize some of the above discussion as follows.

Lemma 7.8. *Suppose that (A, G, α) is a dynamical system. Then the reduced cross product $A \rtimes_{\alpha,r} G$ is (isomorphic to) the quotient of $A \rtimes_\alpha G$ by the kernel of $\operatorname{Ind}_e^G \pi$ for any faithful representation π of A. Every regular representation $\operatorname{Ind}_e^G \pi$ factors through $A \rtimes_{\alpha,r} G$, and for each $f \in C_c(G,A)$,*

$$\|f\|_r = \sup\{\, \|\operatorname{Ind}_e^G \pi(f)\| : \pi \text{ is a representation of } A\,\}.$$

Example 7.9. As always, the group C^*-algebra is an important example. If $A = \mathbf{C}$, then the reduced crossed product $\mathbf{C} \rtimes_{\mathrm{id},r} G$ is denoted $C_r^*(G)$ and is called the *reduced group C^*-algebra.* The only regular representation of $C_c(G)$ is the left-regular representation, and $C_r^*(G)$ is (isomorphic to) the closure in $B\big(L^2(G)\big)$ of

$$\{\, \lambda(f) : f \in C_c(G) \subset C^*(G)\,\},$$

where

$$\lambda(f)(g) = f * g \quad \text{for all } g \in C_c(G) \subset L^2(G).$$

In particular, $\|f\|_r = \|\lambda(f)\|$.

Example 7.10. If G is compact, then the Peter-Weyl Theorem [56, Theorem 5.12] implies that each $u \in \widehat{G}$ is equivalent to a subrepresentation of the left-regular representation λ. In particular, $\|u(f)\| \le \|\lambda(f)\|$ for all $f \in C_c(G)$. Since [139, Theorem A.14] implies that

$$\|f\| = \sup_{u \in \widehat{G}} \|u(f)\|,$$

it follows that $\|f\| \le \|\lambda(f)\|$. Since we always have $\|\lambda(f)\| \le \|f\|$, it follows that $\|\cdot\|_r = \|\cdot\|$, and $C_r^*(G) = C^*(G)$ for compact groups.

Example 7.11. If G is an abelian group, then the Plancherel Theorem [56, Theorem 4.25] implies that for each $f \in C_c(G)$, $\|f\|_2 = \|\hat{f}\|_2$, where $\hat{f}$ is the Fourier Transform defined in (1.20). But it is straightforward to check that

$$\lambda(f)(g)^\wedge = \hat{f}\hat{g},$$

and it follows that

$$\|f\|_r = \|\lambda(f)\| = \|\hat{f}\|_\infty = \|f\|,$$

where the last equality follows from Proposition 3.1 on page 82. Thus we again have $C_r^*(G) = C^*(G)$.

There are groups, and therefore crossed products, for which the reduced C^*-algebra is a proper quotient of the universal one. However, there is a complete description of those groups for which the universal norm and reduced norm coincide: these are the amenable groups discussed in Section A.2. The main result is Theorem A.18 on page 326 which verifies that the left-regular representation of G is faithful on $C^*(G)$ if and only if G is amenable. Although abelian, compact, nilpotent and solvable groups are amenable, there are many groups which are not. An important example is the free group on two generators $\mathbb{F}_2$; therefore the reduced norm on $C_c(\mathbb{F}_2)$ does not agree with the universal norm.

If (A, G, α) is a dynamical system, then the equality of the reduced and universal norms can be decided by looking at the natural map $N \rtimes v$ of $A \rtimes_\alpha G$ into $\mathcal{L}(\mathsf{X})$, where $\mathsf{X} = \mathsf{X}_e^G$ is the $C_0(G, A) \rtimes_{\mathrm{lt}\otimes\alpha} G$ – A-imprimitivity bimodule of Green's Imprimitivity Theorem on page 132. Let ρ be a faithful representation of A. Since X is an imprimitivity bimodule, the Rieffel correspondence implies that X–$\operatorname{Ind}\rho$ is a faithful representation of the imprimitivity algebra $E_e^G(A) = C_0(G, A) \rtimes_{\mathrm{lt}\otimes\alpha} G$. Therefore the canonical extension X–$\overline{\operatorname{Ind}}\,\rho$ of X–$\operatorname{Ind}\rho$ to $\mathcal{L}(\mathsf{X})$ is also faithful. It follows easily that $\operatorname{Ind}_e^G \rho = \mathsf{X}\text{–}\overline{\operatorname{Ind}}\,\rho \circ N \rtimes v$. Hence it follows that $\operatorname{Ind}_e^G \rho$ is faithful if and only if $N \rtimes v$ is. We summarize this observation below.

Lemma 7.12. *Suppose that (A, G, α) is a dynamical system. Then the reduced norm on $C_c(G, A)$ coincides with the universal norm if and only if the canonical embedding $N \rtimes v$ of $A \rtimes_\alpha G$ into $\mathcal{L}(\mathsf{X}_e^G)$ is faithful.*

Unlike the situation for group C^*-algebras, for crossed products, there is no definitive characterization of the dynamical systems (A, G, α) for which $N \rtimes v$ is faithful. It is clear that any such characterization will have to involve more than just properties of the group. For example, for any locally compact group G, amenable or not, $C_0(G) \rtimes_{\mathrm{lt}} G$ is simple (Theorem 4.24 on page 133), and any nonzero regular representation is necessarily faithful. However, the remainder of this section will be devoted to proving $A \rtimes_\alpha G = A \rtimes_{\alpha,r} G$ when G is amenable. The proof given here is taken from [66]. (We give a variation of this result, using EH-regularity, in Theorem 8.19 on page 240).

Theorem 7.13. *If G is amenable, then the reduced norm $\|\cdot\|_r$ coincides with the universal norm on $C_c(G, A)$ and $A \rtimes_{\alpha,r} G = A \rtimes_\alpha G$. In particular, if π is a faithful representation of A, then $\tilde{\pi} \rtimes \lambda = \operatorname{Ind}_e^G \pi$ is a faithful representation of $A \rtimes_\alpha G$.*

Proof. Let $\pi \rtimes U$ be a faithful representation of $A \rtimes_\alpha G$ on $\mathcal{H}$. Since it suffices to produce a faithful regular representation (Lemma 7.8 on the facing page), it will suffice to produce a regular representation L of $A \rtimes_\alpha G$ containing $\pi \rtimes U$ as a subrepresentation.

Let $\lambda : G \to U\big(L^2(G)\big)$ be the left-regular representation, and let $\omega : C_0(G) \to B\big(L^2(G)\big)$ be given by pointwise multiplication: $\omega(g)\xi(s) := g(s)\xi(s)$ for all $g \in$

$C_0(G)$ and $\xi \in L^2(G)$. Let W be the unitary representation of G on the Hilbert space tensor product $L^2(G) \otimes \mathcal{H}$ given by $W_s(\xi \otimes h) := \lambda_s \xi \otimes U_s h$. ($W$ is called the internal tensor product of λ and U, and is often denoted by $\lambda \otimes U$.) Let ρ be the representation of $C_0(G, A) \cong C_0(G) \otimes A$ on $L^2(G) \otimes \mathcal{H}$ given by $\omega \otimes \pi$. That is, if $g \otimes a$ is an elementary tensor in $C_0(G, A)$, then $\rho(g \otimes a)(\xi \otimes h) := \omega(g)\xi \otimes \pi(a)h$. Since

$$W_s \rho(g \otimes a)(\xi \otimes h) = \lambda_s \omega(g)\xi \otimes U_s \pi(a)h = \omega(\mathrm{lt}_s\, g)\lambda_s \xi \otimes \pi\big(\alpha_s(a)\big)U_s h,$$

it follows that (ρ, W) is a covariant representation of $\big(C_0(G, A), G, \mathrm{lt} \otimes \alpha\big)$. Since $E_e^G(A) = C_0(G, A) \rtimes_{\mathrm{lt} \otimes \alpha} G$ is Morita equivalent to A via X, every representation of $E_e^G(A)$ is of the form $\mathsf{X}\text{–}\mathrm{Ind}\, \eta$ for a representation η of A. In particular,

$$\rho \rtimes W = (\omega \otimes \pi) \rtimes W = \mathsf{X}\text{–}\mathrm{Ind}\, \eta$$

for some η. Since $\omega \otimes \pi = (\omega \otimes 1) \otimes_{\max} (1 \otimes \pi)$, it follows from the Imprimitivity Theorem on page 161 that $L := \pi' \rtimes W = \mathrm{Ind}_e^G \eta$ is a regular representation, where $\pi'(a) = 1 \otimes \pi(a)$. We can complete the proof by showing that L is faithful.

Let D be the concrete spatial tensor product

$$\lambda\big(C^*(G)\big) \otimes \pi \rtimes U(A \rtimes_\alpha G) \subset B\big(L^2(G) \otimes \mathcal{H}\big);$$

that is,

$$D := \overline{\mathrm{span}}\{\, \lambda(z) \otimes \pi \rtimes U(f) \in B\big(L^2(G) \otimes \mathcal{H}\big) : \ z \in C_c(G) \text{ and } f \in C_c(G, A)\,\}.$$

By [139, Proposition 2.53], we can identify $M(D)$ with the operators

$$\{\, T \in B\big(L^2(G) \otimes \mathcal{H}\big) : TS \in D \text{ and } ST \in D \text{ for all } S \in D\,\}. \tag{7.10}$$

Since $\pi'(a) = 1 \otimes \pi(a) = 1 \otimes (\pi \rtimes U)\bar{}\,\big(i_A(a)\big)$ and $W_s = \lambda_s \otimes U_s = \lambda_s \otimes (\pi \rtimes U)\bar{}\,\big(i_G(s)\big)$ clearly belong to (7.10), we can view (π', W) as a nondegenerate covariant homomorphism of (A, G, α) into $M(D)$. Thus $L = \pi' \rtimes W$ is a nondegenerate homomorphism of $A \rtimes_\alpha G$ into $M(D)$ (Proposition 2.39 on page 58 and Remark 2.38 on page 58).

Let $\iota : G \to \mathbf{C}$ be the trivial representation of G. Since G is amenable, λ is faithful on $C^*(G)$ (Theorem A.18 on page 326), and the integrated form of ι must factor through $\lambda\big(C^*(G)\big)$. Therefore we can define a representation S' of D on $(L^2(G) \oplus \mathbf{C}) \otimes \mathcal{H}$ by $S' = (\mathrm{id}_1 \oplus \iota') \otimes \mathrm{id}_2$, where id_1 is the identity representation of $\lambda\big(C^*(G)\big)$ and id_2 denotes the identity representation of $\pi \rtimes U(A \rtimes_\alpha G)$, and ι' is the representation of $\lambda\big(C^*(G)\big)$ induced by ι: $\lambda(z) \mapsto \iota(z)$. Identifying $(L^2(G) \oplus \mathbf{C}) \otimes \mathcal{H}$ with $(L^2(G) \otimes \mathcal{H}) \oplus \mathcal{H}$ in the obvious way, we see that S' is equivalent to a representation S defined on elementary tensors by

$$S\big(\lambda(z) \otimes \pi \rtimes U(f)\big) = \big(\lambda(z) \otimes \pi \rtimes U(f)\big) \oplus \big(\iota(z) \cdot \pi \rtimes U(f)\big).$$

Since S is nondegenerate, we can extend S to $\bar{S}$ on all of $M(D)$. Notice that

$$\bar{S}(\lambda_s \otimes U_s) = (\lambda_s \otimes U_s) \oplus U_s \quad \text{and} \quad \bar{S}(1 \otimes \pi(a)) = (1 \otimes \pi(a)) \oplus \pi(a).$$

Thus if $R := \bar{S} \circ L$, then R is a representation of $A \rtimes_\alpha G$ having a subrepresentation (on the invariant subspace $0 \oplus \mathcal{H}$) equivalent to $\pi \rtimes U$. Since $\pi \rtimes U$ was assumed to be faithful, it follows that R and hence L must be faithful. This completes the proof. □

Suppose that (A, G, α) is a dynamical system with I an α-invariant ideal in A. Let $\rho : A \to B(\mathcal{H})$ be a faithful representation. Then $\rho|_I$ is a faithful representation of I — although it may be degenerate. If $\operatorname{ess}(\rho|_I)$ is the essential part, then realizing $\operatorname{Ind}_e^G \rho$ on $L^2(G, \mathcal{H}_\rho)$, we easily see that $\operatorname{Ind}_e^G\big(\operatorname{ess}(\rho|_I)\big) = \operatorname{ess}\big((\operatorname{Ind}_e^G \rho)|_{I \rtimes_\alpha G}\big)$. Thus for any $f \in C_c(G, I)$,

$$\big\|\operatorname{Ind}_e^G\big(\operatorname{ess}(\rho|_I)\big)(f)\big\| = \big\|\operatorname{Ind}_e^G(\rho)(f)\big\|.$$

It follows that we have an injection $\iota \rtimes_r \mathrm{id}$ of $I \rtimes_{\alpha,r} G$ into $A \rtimes_{\alpha,r} G$.

Remark 7.14. If (A, G, α) is a dynamical system, if I is an α-invariant ideal of A, if $q : A \to A/I$ is the natural map and if ρ' is a representation of A/I, then it is not hard to check that $\operatorname{Ind}_e^G(\rho' \circ q) = (\operatorname{Ind}_e^G \rho') \circ (q \rtimes \mathrm{id})$.

If ρ' is a faithful representation of A/I and if $f \in C_c(G, A)$, then using Remark 7.14,

$$\begin{aligned} \|q \circ f\|_{A/I \rtimes_{\alpha^I,r} G} &= \big\|\operatorname{Ind}_e^G(\rho')(q \circ f)\big\| \\ &= \big\|\operatorname{Ind}_e^G(\rho' \circ q)(f)\big\| \\ &\le \|f\|_{A \rtimes_{\alpha,r} G}. \end{aligned}$$

Thus $q \rtimes \mathrm{id}$ induces a surjection $q \rtimes_r \mathrm{id}$ of $A \rtimes_{\alpha,r} G$ onto $A/I \rtimes_{\alpha^I,r} G$. In view of Proposition 3.19 on page 93, it is reasonable to suspect that $\ker(q \rtimes_r \mathrm{id}) = I \rtimes_{\alpha,r} G$ (viewed as an ideal in $A \rtimes_{\alpha,r} G$) so that we would get an exact sequence

$$0 \longrightarrow I \rtimes_{\alpha,r} G \xrightarrow{\iota \rtimes_r \mathrm{id}} A \rtimes_{\alpha,r} G \xrightarrow{q \rtimes_r \mathrm{id}} A/I \rtimes_{\alpha^I,r} G \longrightarrow 0. \tag{7.11}$$

However, the exactness of (7.11) is a subtle question. Groups for which (7.11) is exact for all dynamical systems (A, G, α) with I α-invariant in A are called *exact*. Of course amenable groups are exact, but Gromov has announced the existence of finitely generated discrete groups which fail to be exact. However, no other examples of non-exact groups are known as this is written. See [92] for further discussion and references.

In Section 2.6, we observed the important role of the maximal tensor product in the theory of crossed products. Here we want to observe that the spatial tensor product plays a similar role for reduced crossed products.

Remark 7.15. If A and B are potentially non-nuclear C^*-algebras, then, since the maximal norm dominates the minimal norm, we can identify the spatial tensor product $A \otimes_\sigma B$ as a quotient of the maximal tensor product $A \otimes_{\max} B$. We do this as follows. Let $\rho_A : A \to B(\mathcal{H}_A)$ and $\rho_B : B \to B(\mathcal{H}_B)$ be faithful representations so that $\rho_A \otimes \rho_B$ is a faithful representation of $A \otimes_\sigma B$ on $\mathcal{H} := \mathcal{H}_A \otimes \mathcal{H}_B$. Then $\pi_A := \rho_A \otimes 1$ and $\pi_B := 1 \otimes \rho_B$ are commuting faithful representations and we get a representation $\pi_A \otimes_{\max} \pi_B$ whose image is isomorphic to $A \otimes_\sigma B$. In fact, if $\kappa := \kappa_{A,B} : A \otimes_{\max} B \to A \otimes_\sigma B$ is the natural map, then

$$\pi_A \otimes_{\max} \pi_B = (\rho_A \otimes \rho_B) \circ \kappa_{A,B}.$$

Suppose that (A, G, α) is a dynamical system and that B is a C^*-algebra. Let $\rho_A : A \to B(\mathcal{H}_A)$ and $\rho_B : B \to B(\mathcal{H}_B)$ be faithful representations, and let π_A and π_B be the corresponding commuting representations on $\mathcal{H}_A \otimes \mathcal{H}_B$ as above. As in Example 2.14 on page 45, let $(\tilde{\pi}_A, U)$ be the covariant pair such that

$$\tilde{\pi}_A \rtimes U = \operatorname{Ind}_e^G(\rho_A \otimes 1_{\mathcal{H}_B})$$

on the Hilbert space $L^2(G, \mathcal{H}_A \otimes \mathcal{H}_B)$. If we identify $L^2(G, \mathcal{H}_A \otimes \mathcal{H}_B)$ with $L^2(G, \mathcal{H}_A) \otimes \mathcal{H}_B$, then its straightforward to check that

$$\operatorname{Ind}_e^G(\rho_A \otimes 1_{\mathcal{H}_B}) = \operatorname{Ind}_e^G(\rho_A) \otimes 1_{\mathcal{H}_B}.$$

If $\tilde{\pi}_B := 1_{L^2(G,\mathcal{H}_A)} \otimes \rho_B$, then as in Remark 7.15 on the preceding page, we have

$$(\tilde{\pi}_A \rtimes U) \otimes_{\max} \tilde{\pi}_B = \big((\operatorname{Ind}_e^G \rho_A) \otimes \rho_B\big) \circ \kappa,$$

where $\kappa := \kappa_{A \rtimes_\alpha G, B} : A \rtimes_\alpha G \otimes_{\max} B \to A \rtimes_\alpha G \otimes_\sigma B$ is the quotient map. Thus we can identify the range of $(\tilde{\pi}_A \rtimes U) \otimes_{\max} \tilde{\pi}_B$ with $(A \rtimes_{\alpha,r} G) \otimes_\sigma B$.

On the other hand, the kernel of $\kappa' := \kappa_{A,B} : A \otimes_{\max} B \to A \otimes_\sigma B$ is $\alpha \otimes \mathrm{id}$-invariant (Remark 2.74 on page 77). Thus in view of Remark 7.14 on the preceding page, we have

$$\begin{aligned} \operatorname{Ind}_e^G(\rho_A \otimes \rho_B) \circ (\kappa' \rtimes \mathrm{id}) &= \operatorname{Ind}_e^G\big((\rho_A \otimes \rho_B) \circ \kappa'\big) \\ &= \operatorname{Ind}_e^G(\pi_A \otimes_{\max} \pi_B) \\ &= (\pi_A \otimes_{\max} \pi_B)^\sim \rtimes U \\ &= (\tilde{\pi}_A \otimes_{\max} \tilde{\pi}_B) \rtimes U. \end{aligned} \tag{7.12}$$

Thus we can identify the range of $(\tilde{\pi}_A \otimes_{\max} \tilde{\pi}_B) \rtimes U$ with $(A \otimes_\sigma B) \rtimes_{\alpha \otimes \mathrm{id}, r} G$. Lemma 2.75 on page 78 implies there is an isomorphism $L : (A \otimes_{\max} B) \rtimes_{\alpha\otimes\mathrm{id}} G$ onto $(A \rtimes_\alpha G) \otimes_{\max} B$ which intertwines $(\tilde{\pi}_A \rtimes U) \otimes_{\max} \tilde{\pi}_B$ and $(\tilde{\pi}_A \otimes_{\max} \tilde{\pi}_B) \rtimes U$. Thus, with our identifications, we obtain a commutative diagram

$$\begin{array}{ccc} (A \otimes_{\max} B) \rtimes_{\alpha\otimes\mathrm{id}} G & \xrightarrow{\cong} & (A \rtimes_\alpha G) \otimes_{\max} B \\ \Big\downarrow {\scriptstyle (\tilde{\pi}_A \otimes_{\max} \tilde{\pi}_B) \rtimes U} & & \Big\downarrow {\scriptstyle (\tilde{\pi}_A \rtimes U) \otimes_{\max} \tilde{\pi}_B} \\ (A \otimes_\sigma B) \rtimes_{\alpha\otimes\mathrm{id},r} G & \xrightarrow{\cong} & (A \rtimes_{\alpha,r} G) \otimes_\sigma B. \end{array} \tag{7.13}$$

We can summarize this discussion in the following result which is an analogue for reduced crossed products to Lemma 2.75 on page 78.

Lemma 7.16. *Suppose that (A, G, α) is a dynamical system and that B is a C^*-algebra. Then the isomorphism of Lemma 2.75 on page 78 between $(A \otimes_{\max} B) \rtimes_{\alpha\otimes\mathrm{id}} G$ onto $(A \rtimes_\alpha G) \otimes_{\max} B$ induces an isomorphism of*

$$(A \otimes_\sigma B) \rtimes_{\alpha\otimes\mathrm{id},r} G \quad \textit{onto} \quad (A \rtimes_{\alpha,r} G) \otimes_\sigma B$$

taking $(a \otimes_\sigma b) \otimes f$ to $(a \otimes f) \otimes_\sigma b$ as in (7.13).

Corollary 7.17. *If ι is the trivial action of G on A, then $A \rtimes_{\iota,r} G \cong A \otimes_\sigma C_r^*(G)$.*

As with the question of when the reduced crossed product coincides with the universal one, there is no known characterization of which crossed products are nuclear C^*-algebras. Even for group C^*-algebras the situation is nontrivial and not completely understood. Guichardet has noticed that if G is amenable, then $C^*(G)$ is nuclear [69] (this also follows from Corollary 7.18). Lance has proved that for a discrete group G, $C_r^*(G)$ is nuclear if and only if G is amenable [97]. Since quotients of nuclear C^*-algebras are nuclear, $C^*(G)$ nuclear implies that $C_r^*(G)$ is too. Thus Lance's result also holds for the universal group algebra. Naturally, the situation is muddier for crossed products. As an example, notice for any group G, $C_0(G) \rtimes_{\mathrm{lt}} G$ is isomorphic to the compacts (Theorem 4.24 on page 133) and hence nuclear. Nevertheless, the following result is very useful. This result goes back to at least [66, Proposition 14], and the proof given here is based on an idea of Echterhoff's.

Corollary 7.18. *If (A, G, α) is a dynamical system with A nuclear and G amenable, then $A \rtimes_\alpha G$ is nuclear.*

Proof. If A is nuclear, then, by definition, $\kappa' := \kappa_{A,B}$ is an isomorphism as is $\kappa' \rtimes \mathrm{id}$. Since G is amenable, $\mathrm{Ind}_e^G(\rho_A \otimes \rho_B)$ is faithful (Theorem 7.13 on page 199). This forces $(\pi_A \rtimes U) \otimes_{\max} \pi_B$ to be an isomorphism in (7.13). In particular, $\kappa := \kappa_{A \rtimes_\alpha G, B}$ is an isomorphism. Since this holds for any C^*-algebra B, $A \rtimes_\alpha G$ is nuclear. □

Remark 7.19. Note that in proving Corollary 7.18, it would not suffice to use Lemma 2.75 on page 78 and Lemma 7.16 on the preceding page to show that $(A \rtimes_\alpha G) \otimes_{\max} B$ and $(A \rtimes_\alpha G) \otimes_\sigma B$ are merely isomorphic. We have to see that the minimal norm and the maximal norm coincide — which happens exactly when κ is injective.

Remark 7.20 (Exact Groups). It follows from Corollary 7.17 that if G is exact, then $C_r^*(G)$ is an exact C^*-algebra (cf. [164]).[5] A great number of properties of exact groups can be found in [91, 92].

7.3 Crossed Products Involving the Compacts

In this section, we want to consider dynamical systems $(\mathcal{K}(\mathcal{H}), G, \alpha)$ where $\mathcal{K}(\mathcal{H})$, as usual, denotes the compact operators on a complex Hilbert space $\mathcal{H}$. A strongly continuous homomorphism $\alpha : G \to \operatorname{Aut} \mathcal{K}(\mathcal{H})$ is also called a projective representation of G, and as we shall see, the material in this section is closely related to projective and multiplier representations. These notions are discussed briefly in Appendix D.3, and we will make some use of that material here.

[5] A C^*-algebra A is called exact if

$$0 \longrightarrow A \otimes_\sigma I \longrightarrow A \otimes_\sigma B \longrightarrow A \otimes_\sigma B/I \longrightarrow 0$$

is exact for all C^*-algebras B and every ideal $I \subset B$. A C^*-algebra is exact if and only if it can be embedded in a nuclear C^*-algebra [89].

It is well known (cf. [139, Chap. 1]) that any automorphism α of $\mathcal{K}(\mathcal{H})$ must be of the form $\operatorname{Ad} U$ for some unitary $U \in U(\mathcal{H})$; that is, each such α must be inner. Furthermore, if $\operatorname{Ad} U = \operatorname{Ad} V$ on $\mathcal{K}(\mathcal{H})$, then there is a unimodular scalar $z \in \mathbf{T}$ such that $V = zU$. Let G' be the group

$$G' := \{ (s, U) \in G \times U(\mathcal{H}) : \alpha_s = \operatorname{Ad} U \}$$

equipped with the relative topology from $G \times U(\mathcal{H})$ (where $U(\mathcal{H})$ has the strong operator topology). Then we get an algebraic short exact sequence

$$1 \longrightarrow \mathbf{T} \xrightarrow{\ i\ } G' \xrightarrow{\ j\ } G \longrightarrow e, \tag{7.14}$$

with $i(z) := (e, z1_{\mathcal{H}})$ and $j(s, U) = s$. Clearly, i and j are continuous. To see that (7.14) is a short exact sequence of topological groups, we also need to verify that j is open and that i is a homeomorphism onto its range. To see that j is open, we'll use Proposition 1.15 on page 4. Suppose that $s_n \to s_0$ in G, and that $(s_0, U_0) \in G'$. By Lemma D.24 on page 380, there is a neighborhood N of α_{s_0} in $\operatorname{Aut} \mathcal{K}(\mathcal{H})$ and a continuous map $c : N \to U(\mathcal{H})$ such that $\operatorname{Ad} c(\alpha_s) = \alpha_s$ for all $\alpha_s \in N$ and such that $c(\alpha_{s_0}) = U_0$. Since $\alpha : G \to \operatorname{Aut} \mathcal{K}(\mathcal{H})$ is continuous, we eventually have $\alpha_{s_n} \in N$, and for large n, $\big(s_n, c(\alpha_{s_n})\big) \in G'$. Thus j is an open map by Proposition 1.15 on page 4. A much more straightforward argument shows that i is a homeomorphism onto its range. Thus (7.14) is a short exact sequence of topological groups, and since both $\mathbf{T}$ and G are locally compact, Corollary 1.52 on page 15 implies that G' is too.

To state the main theorem in this section, we need to work with a natural quotient $C^*(G', \tau)$ of $C^*(G')$. The construction is a special case of the one described on pages 383–385 in Appendix D.3, and we'll only sketch the particulars here.

Notice that if $f \in C_c(G')$ and if $(s, U) \in G'$, then the integral

$$\int_{\mathbf{T}} f(s, zU)\, dz$$

depends only on s and that we get a Haar integral on G' by

$$\int_{G'} f(s, U)\, d\mu_{G'}(s, U) := \int_G \int_{\mathbf{T}} f(s, zU)\, dz\, d\mu(s).$$

Let $\tau : \mathbf{T} \to \mathbf{T}$ be given by $\tau(z) = \bar{z}$. Then a representation π of G' is said to *preserve* τ if

$$\pi(e, z1_{\mathcal{H}}) = \bar{z}1_{\mathcal{H}_\pi} \quad \text{for all } z \in \mathbf{T}.$$

We let I be the ideal of $M\big(C^*(G')\big)$ generated by $\{ i_{G'}\big((e, z1_{\mathcal{H}})\big) - \bar{z}1_{G'} : z \in \mathbf{T} \}$, and define

$$I_\tau := I \cap C^*(G').$$

Then it is not hard to see that π preserves τ if and only if $I_\tau \subset \ker \pi$. Thus the representations of $C^*(G')$ which factor through the quotient

$$C^*(G', \tau) := C^*(G')/I_\tau$$

are exactly those which preserve τ. Proposition D.30 on page 385 implies that we can realize $C^*(G',\tau)$ as the completion of the $*$-subalgebra

$$C_c(G',\tau) := \{\, f \in C_c(G') : f(s,zU) = zf(s,U) \quad \text{for all } z \in \mathbf{T}\,\}$$

of $C_c(G') \subset C^*(G')$ with respect to the norm

$$\|f\|_\tau := \sup\{\, \|\pi(f)\| : \pi \text{ is a representation of } G' \text{ which preserves } \tau\,\}.$$

Note that if f and g are in $C_c(G',\tau)$ and if (r,V) and (s,U) are in G', then $(r,V) \mapsto f(r,V)g(r^{-1}s,V^*U)$ is constant on $\mathbf{T}$-cosets. It follows that

$$f * g(s,U) = \int_G f(r,V)g(r^{-1}s,V^*U)\,d\mu(r).$$

Similarly, if π preserves τ, then $f(s,U)\pi(s,U)$ depends only on s and

$$\pi(f) = \int_G f(s,U)\pi(s,U)\,d\mu(s).$$

We also have a surjective $*$-homomorphism $Q_\tau : C_c(G') \to C_c(G',\tau)$, given by

$$Q_\tau(f)(s,U) = \int_{\mathbf{T}} f(s,zU)\bar{z}\,dz,$$

which satisfies

$$\pi\big(Q_\tau(f)\big) = \pi(f)$$

provided π is a representation of G' preserving τ.

Theorem 7.21. *Suppose that $\big(\mathcal{K}(\mathcal{H}),G,\alpha\big)$ is a dynamical system, that G' is the associated extension (7.14) of $\mathbf{T}$ by G and that $\tau(z) = \bar{z}$ for all $z \in \mathbf{T}$. If $f \in C_c(G',\tau)$ and $T \in \mathcal{K}(\mathcal{H})$, then $(s,U) \mapsto f(s,U)TU^*$ is constant on $\mathbf{T}$-cosets. Identifying G with $G'/\mathbf{T}$, the map*

$$\psi : C_c(G',\tau) \odot \mathcal{K}(\mathcal{H}) \to C_c\big(G,\mathcal{K}(\mathcal{H})\big)$$

given by $\psi(f \otimes T)(s) = f(s,U)TU^*$ extends to an isomorphism of $C^*(G',\tau) \otimes \mathcal{K}(\mathcal{H})$ with $\mathcal{K}(\mathcal{H}) \rtimes_\alpha G$.

Remark 7.22. If G is second countable and if $\mathcal{H}$ is separable, then $C^*(G',\tau)$ is isomorphic to $C^*(G,\bar{\omega})$ where $[\omega]$ is the Mackey obstruction for α. Thus the representations of $C^*(G',\tau)$ on separable Hilbert spaces are in one-to-one correspondence with $\bar{\omega}$-representations of G. See Appendix D.3 for further details and definitions.

Proof. It is not too hard to check that if $U(\mathcal{H})$ has the strong operator topology, then the map $U \mapsto UT$ is continuous from $U(\mathcal{H})$ into $\mathcal{K}(\mathcal{H})$ for each compact operator T.[6] Since unitaries have norm one, it follows that $(U,T) \mapsto TU$ is continuous

[6]It's enough to check for T finite rank. Alternatively, we could observe that the strong topology on $U(\mathcal{H})$ is the strict topology [139, Corollary C.8].

from $U(\mathcal{H}) \times \mathcal{K}(\mathcal{H})$ into $\mathcal{K}(\mathcal{H})$. Hence, if $f \in C_c(G', \tau)$, then $(s, U) \mapsto f(s, U)TU^*$ is continuous on G' and constant on $\mathbf{T}$-cosets. Thus $\psi(f \otimes T)$ is a well-defined element of $C_c\big(G, \mathcal{K}(\mathcal{H})\big)$. We want to see that the image of ψ is dense in $C_c\big(G, \mathcal{K}(\mathcal{H})\big)$ in the inductive limit topology. We begin by defining $\psi' : C_c(G') \odot \mathcal{K}(\mathcal{H}) \to C_c\big(G', \mathcal{K}(\mathcal{H})\big)$ by

$$\psi'(h \otimes T)(s, U) := h(s, U)TU^*.$$

Then a quick calculation shows that the diagram

$$\begin{array}{ccc} C_c(G') \odot \mathcal{K}(\mathcal{H}) & \xrightarrow{\psi'} & C_c\big(G', \mathcal{K}(\mathcal{H})\big) \\ {\scriptstyle Q_\tau \otimes \mathrm{id}} \downarrow & & \downarrow {\scriptstyle \beta} \\ C_c(G', \tau) \odot \mathcal{K}(\mathcal{H}) & \xrightarrow{\psi} & C_c\big(G, \mathcal{K}(\mathcal{H})\big) \end{array}$$

commutes when β is defined by

$$\beta(g)(s) := \int_{\mathbf{T}} g(s, zU)\, dz.$$

Since $Q_\tau \otimes \mathrm{id}$ is surjective, we can establish that ψ has dense range by showing that $\beta \circ \psi'$ has dense range. In view of Lemma 1.87 on page 29, we can view $C_c(G') \odot \mathcal{K}(\mathcal{H})$ as a dense subspace of $C_c\big(G', \mathcal{K}(\mathcal{H})\big)$. We can easily extend ψ' to all of $C_c\big(G', \mathcal{K}(\mathcal{H})\big)$ by

$$\bar{\psi}'(f)(s, U) = f(s, U)U^*.$$

Since $\operatorname{supp} \bar{\psi}'(f) = \operatorname{supp} f$ and $\|\bar{\psi}'(f)\|_\infty = \|f\|_\infty$, it follows that $\bar{\psi}'$ is continuous with respect to the inductive limit topology. Since $\varphi(g)(s, U) = g(s, U)U$ is clearly an inverse for $\bar{\psi}'$, $\bar{\psi}'$ is surjective. Therefore $\psi'\big(C_c(G') \odot \mathcal{K}(\mathcal{H})\big)$ is dense in $C_c\big(G', \mathcal{K}(\mathcal{H})\big)$. Since β is onto and easily seen to be continuous in the inductive limit topology, $\beta \circ \psi'$ has dense range. Thus ψ has dense range.

To see that ψ is multiplicative, consider

$$\begin{aligned} \psi(f \otimes T) * \psi(g \otimes T')(s) &= \int_G \psi(f \otimes T)(r) \alpha_r\big(\psi(f \otimes T')(r^{-1}s)\big)\, d\mu(r) \\ &= \int_G f(r, V)TV^* \alpha_r\big(g(r^{-1}s, V^*U)T'U^*V\big)\, d\mu(r) \\ &= \int_G f(r, V) g(r^{-1}s, V^*U)\, d\mu(r) TT'U^* \\ &= \psi(f * g \otimes TT')(s). \end{aligned}$$

On the other hand,

$$\begin{aligned} \psi(f \otimes T)^*(s) &= \Delta(s^{-1})\alpha_s\big(\psi(f \otimes T)(s^{-1})\big)^* \\ &= \Delta(s^{-1})\alpha_s\big(f(s^{-1}, U^*)TU\big)^* \\ &= \Delta(s^{-1})\overline{f(s^{-1}, U^*)}T^*U^* \\ &= \psi(f^* \otimes T^*)(s). \end{aligned}$$

Thus ψ is a $*$-homomorphism.

Therefore it only remains to show that ψ is isometric. Let π be a representation of G' which preserves τ and let ρ be a representation of $\mathcal{K}(\mathcal{H})$ that commutes with (the integrated form of) π. Thus $\pi(s,U)$ and $\rho(T)$ commute for each $(s,U)\in G'$ and each $T\in\mathcal{K}(\mathcal{H})$. Since π preserves τ, $\pi(s,U)\bar\rho(U)$ depends only on s, and

$$\pi'(s) := \pi(s,U)\bar\rho(U)$$

defines a representation of $G\cong G'/\mathbf{T}$. Furthermore,

$$\begin{aligned}\pi'(s)\rho(T) &= \pi(s,U)\rho(UT)\\ &= \rho(UT)\pi(s,U)\\ &= \rho(UTU^*)\bar\rho(U)\pi(s,U)\\ &= \rho\big(\alpha_s(T)\big)\pi(s,U)\bar\rho(U)\\ &= \rho\big(\alpha_s(T)\big)\pi'(s).\end{aligned}$$

Thus (ρ,π') is a covariant representation of $\big(\mathcal{K}(\mathcal{H}),G,\alpha\big)$, and

$$\rho\rtimes\pi'\big(\psi(\textstyle\sum f_i\otimes T_i)\big)=\sum\pi(f_i)\rho(T_i)=\pi\otimes\rho\big(\sum f_i\otimes T_i\big).$$

It follows that

$$\Big\|\psi\big(\textstyle\sum f_i\otimes T_i\big)\Big\|\geq\Big\|\sum f_i\otimes T_i\Big\|. \tag{7.15}$$

On the other hand, if (ρ,π') is a covariant representation of $\big(\mathcal{K}(\mathcal{H}),G,\alpha\big)$, then

$$\pi'(s)\bar\rho(U)=\bar\rho\big(\bar\alpha_s(U)\big)\pi'(s)\quad\text{for all } s\in G \text{ and } U\in U(\mathcal{H}).$$

It follows that

$$\pi(s,U):=\pi'(s)\bar\rho(U^*)$$

is a representation of G':

$$\begin{aligned}\pi(s,U)\pi(r,V) &= \pi'(s)\bar\rho(U^*)\pi'(r)\bar\rho(V^*)\\ &= \pi'(sr)\bar\rho(V^*U^*V)\bar\rho(V^*)\\ &= \pi'(sr)\bar\rho(V^*U^*)\\ &= \pi(sr,UV).\end{aligned}$$

Furthermore,

$$\begin{aligned}\pi(s,U)\rho(T) &= \pi'(s)\rho(U^*T)\\ &= \rho(TU^*)\pi'(s)\\ &= \rho(T)\bar\rho(U^*)\pi'(s)\\ &= \rho(T)\pi'(s)\bar\rho(U^*)=\rho(T)\pi(s,U).\end{aligned}$$

Therefore π and ρ are commuting representations of G' and $\mathcal{K}(\mathcal{H})$, respectively, satisfying

$$\pi'(s)=\pi(s,U)\bar\rho(U).$$

Since π clearly preserves τ, we must have equality in (7.15), and ψ is isometric. This completes the proof. □

7.4 Aside on Twisted Crossed Products

Suppose that (A, G, α) is a dynamical system with a normal subgroup M. In Section 3.3 we saw that if G is a semidirect product, say $G = M \rtimes_{\varphi} H$, then we can decompose $A \rtimes_{\alpha} G$ as an iterated crossed product $(A \rtimes_{\alpha} M) \rtimes_{\beta} H$ (Proposition 3.11 on page 87). In Section 7.3, and in more detail in Appendix D.3, we introduced a C^*-algebra $C^*(E, \tau)$ associated to a locally compact group E having $\mathbf{T}$ as a central normal subgroup. We want to think of $C^*(E, \tau)$ as a "twisted group C^*-algebra" for $G := E/\mathbf{T}$ with respect to a character $\tau : \mathbf{T} \to \mathbf{T}$ (cf. Remark 7.22 on page 205). In this section, we want to generalize this twisting construction to exhibit $A \rtimes_{\alpha} G$ as a twisted crossed product of $A \rtimes_{\alpha} M$ by G/M. However, as in our treatment of $C^*(E, \tau)$, G/M does not appear explicitly. Instead, we exhibit the iterated crossed product as a quotient of a crossed product of $A \rtimes_{\alpha} M$ by G. This approach is due to Green [66] and Dang Ngoc [112]. It is also possible to build a iterated crossed product using G/M explicitly — in analogy with $C^*(G, \omega)$ as defined in Appendix D.3 — using cocycles taking values the the unitary group of the multiplier algebra of $A \rtimes_{\alpha} M$. Although this is done elegantly in [119–121] (based on work in [15] and [100]), we will not discuss this approach here.[7]

7.4.1 Twisted Systems

If (A, G, α) is a dynamical system, then a *twisting map*, or a *Green twisting map*, is a strictly continuous homomorphism $\tau : N_{\tau} \to UM(A)$ from a normal subgroup N_{τ} of G into the unitary group $UM(A)$ of the multiplier algebra of A such that for all $n \in N_{\tau}$, $a \in A$ and $s \in G$

$$\tau(n) a \tau(n)^* = \alpha_n(a) \tag{7.16}$$

$$\bar{\alpha}_s\big(\tau(n)\big) = \tau(sns^{-1}). \tag{7.17}$$

Then (A, G, α, τ) is called a *(Green) twisted dynamical system*. A covariant representation (π, U) of (A, G, α) is said to *preserve* τ if

$$\bar{\pi}\big(\tau(n)\big) = U(n) \quad \text{for all } n \in N_{\tau}. \tag{7.18}$$

The goal is to define the twisted crossed product, $A \rtimes_{\alpha}^{\tau} G$, to be the quotient of $A \rtimes_{\alpha} G$ whose representations are exactly those which preserve τ. To see that there is such a quotient, we proceed as follows. Let I be the ideal of $M(A \rtimes_{\alpha} G)$ generated by

$$\{\, i_G(n) - \bar{\imath}_A\big(\tau(n)\big) : n \in N_{\tau} \,\},$$

and let $I_{\tau} := I \cap A \rtimes_{\alpha} G$.

Proposition 7.23. *Let (A, G, α, τ) be a twisted dynamical system, and let I_{τ} the ideal of $A \rtimes_{\alpha} G$ defined above. Then a representation $\pi \rtimes U$ preserves τ if and only if*

[7] Because there are two approaches to twisted crossed products, the term "twisted crossed product" is slightly ambiguous. In the literature, the type we are interested in here are often called "Green" twisted crossed products. Twisted crossed products defined on G/M using cocycles are called "Busby-Smith" twisted crossed products.

$I_\tau \subset \ker(\pi \rtimes U)$. *In particular, the set of irreducible representations which preserve* τ *is the closed subset of the spectrum* $(A \rtimes_\alpha G)^\wedge$ *corresponding to the quotient*

$$A \rtimes_\alpha^\tau G := A \rtimes_\alpha G/I_\tau.$$

We call $A \rtimes_\alpha^\tau G$ *the (Green) twisted crossed product associated to* (A, G, α, τ).

Proof. It suffices to prove the first assertion. Suppose that $\pi \rtimes U$ is a representation of $A \rtimes_\alpha G$. Then

$$(\pi \rtimes U)^{-}\big(i_G(n)\big) = U(n) \quad \text{and} \quad (\pi \rtimes U)^{-}\big(\bar{\imath}_A(\tau(n))\big) = \bar{\pi}\big(\tau(n)\big).$$

Therefore, if (π, U) preserves τ, we must have $(\pi \rtimes U)^{-}(I) = \{0\}$. Thus

$$I_\tau \subset \ker(\pi \rtimes U). \tag{7.19}$$

On the other hand, if (7.19) holds, then since $\pi \rtimes U$ is nondegenerate, it follows that $(\pi \rtimes U)^{-}(I) = \{0\}$, and $\pi \rtimes U$ preserves τ. □

Instead of working with the quotient $A \rtimes_\alpha G/I_\tau$, it is often preferable to realize $A \rtimes_\alpha^\tau G$ as the completion of a $*$-algebra of functions on G. Let $C_c(G, A, \tau)$ be the set of continuous A-valued functions on G such that

$$f(ns) = f(s)\tau(n)^* \quad \text{for all } s \in G \text{ and } n \in N_\tau, \tag{7.20}$$

and such that there is a compact set $K \subset G$ such that $\operatorname{supp} f \subset KN_\tau$. Since any Haar measure is unique up to a positive scalar, we can assume that we have fixed Haar measures on G, N_τ and G/N_τ such that

$$\int_G g(s)\,d\mu(s) = \int_{G/N_\tau} \int_{N_\tau} g(sn)\,d\mu_{N_\tau}(n)\,d\mu_{G/N_\tau}(\dot{s}) \quad \text{for all } g \in C_c(G).$$

If $f, g \in C_c(G, A, \tau)$, then for each $s \in G$,

$$r \mapsto f(r)\alpha_r\big(g(r^{-1}s)\big)$$

is constant on N_τ-cosets: using (7.20)

$$\begin{aligned} f(rn)\alpha_{rn}\big(g(n^{-1}r^{-1}s)\big) &= f(r)\tau(rnr^{-1})^*\alpha_{rn}\big(g(r^{-1}s)\tau(n^{-1})^*\big) \\ &= f(r)\tau(rn^{-1}r^{-1})\alpha_{rn}\big(g(r^{-1}s)\big)\tau(rn^{-1}r^{-1})^* \end{aligned}$$

which, using (7.16), is

$$= f(r)\alpha_r\big(g(r^{-1}s)\big).$$

Therefore $(\dot{r}, s) \mapsto f(r)\alpha_r\big(g(r^{-1}s)\big)$ is continuous on $G/N_\tau \times G$, and has compact support in the first variable. Using Lemma 1.102 on page 36, we can define $f * g \in C_c(G, A, \tau)$ by

$$f * g(s) := \int_{G/N_\tau} f(r)\alpha_r\big(g(r^{-1}s)\big)\,d\mu_{G/N_\tau}(\dot{r}). \tag{7.21}$$

We can also define $f^* \in C_c(G, A, \tau)$ by

$$f^*(s) := \Delta_{G/N_\tau}(\dot{s}^{-1})\alpha_s\big(f(s^{-1})^*\big). \tag{7.22}$$

Rather than verify directly that these operations turn $C_c(G, A, \tau)$ into a $*$-algebra, we proceed as follows. If $f \in C_c(G, A)$, then $(n, s) \mapsto f(sn)\tau(sns^{-1})$ is continuous on $N_\tau \times G$, and if we restrict the second variable to a compact neighborhood of s, it has compact support. Thus Lemma 1.102 on page 36 implies that

$$\Phi(f)(s) := \int_{N_\tau} f(sn)\tau(sns^{-1})\, d\mu_{N_\tau}(n) \tag{7.23}$$

is continuous in s and defines an element of $C_c(G, A, \tau)$:

$$\Phi(f)(ms) = \int_{N_\tau} f(msn)\tau(msns^{-1})\, d\mu_{N_\tau}(n)\tau(m)^*$$

which, after replacing n by $s^{-1}m^{-1}sn$, is

$$\begin{aligned} &= \int_{N_\tau} f(sn)\tau(sns^{-1})\, d\mu_{N_\tau}(n)\tau(m)^* \\ &= \Phi(f)(s)\tau(m)^*. \end{aligned}$$

If $g \in C_c(G, A, \tau)$ and if b is a Bruhat approximate cross section for G over N_τ (as defined in Appendix H.3), then $f(s) := b(s)g(s)$ is in $C_c(G, A)$ and

$$\begin{aligned} \Phi(f)(s) &= \int_{N_\tau} b(sn)g(sn)\tau(sns^{-1})\, d\mu_{N_\tau}(n) \\ &= g(s) \int_{N_\tau} b(sn)\, d\mu_{N_\tau}(n) \\ &= g(s). \end{aligned}$$

Thus $\Phi : C_c(G, A) \to C_c(G, A, \tau)$ is surjective.

We'll need the following observation concerning how the Haar measure of a normal subgroup behaves under the action of G. If M is a normal subgroup of G and $s \in G$, then the uniqueness of Haar measure implies that there is a positive scalar $\Delta_{G,M}(s)$ such that

$$\int_M g(sms^{-1})\, d\mu_M(m) = \Delta_{G,M}(s) \int_M g(m)\, d\mu_M(m) \tag{7.24}$$

for all $g \in C_c(M)$. If $f \in C_c(G)$, then for an appropriate choice of Haar measure on G/M,

$$\begin{aligned}
\Delta_G(s)\int_G f(r)\,d\mu(r) &= \int_G f(srs^{-1})\,d\mu(r) \\
&= \int_{G/M}\int_M f(srms^{-1})\,d\mu_M(m)\,d\mu_{G/M}(\dot r) \\
&= \Delta_{G,M}(s)\int_{G/M}\int_M f(srs^{-1}m)\,d\mu_M(m)\,d\mu_{G/M}(\dot r) \\
&= \Delta_{G,M}(s)\Delta_{G/M}(\dot s)\int_{G/M}\int_M f(rm)\,d\mu_M(m)\,d\mu_{G/M}(\dot r) \\
&= \Delta_{G,M}(s)\Delta_{G/M}(\dot s)\int_G f(r)\,d\mu(r).
\end{aligned}$$

Therefore, we must have

$$\Delta_{G,M}(s) = \Delta_G(s)\Delta_{G/M}(\dot s^{-1}). \tag{7.25}$$

It follows that $\Delta_{G,M}$ is a continuous homomorphism into the multiplicative positive reals.

Lemma 7.24. *With the operations defined by* (7.21) *and* (7.22), $C_c(G, A, \tau)$ *is a* $*$*-algebra, and* (7.23) *defines a surjective* $*$*-homomorphism*

$$\Phi : C_c(G, A) \to C_c(G, A, \tau).$$

Proof. Since $C_c(G, A)$ is a $*$-algebra and since Φ is surjective, it suffices to show that $\Phi(f * g) = \Phi(f) * \Phi(g)$ and $\Phi(f^*) = \Phi(f)^*$. So compute that

$$\begin{aligned}
\Phi(f * g)(s) &= \int_{N_\tau} f * g(sn)\tau(sns^{-1})\,d\mu_{N_\tau}(n) \\
&= \int_{N_\tau}\int_G f(r)\alpha_r\big(g(r^{-1}sn)\big)\tau(sns^{-1})\,d\mu(s)\,d\mu_{N_\tau}(n) \\
&= \int_{N_\tau}\int_{G/N_\tau}\int_{N_\tau} f(rm)\alpha_{rm}\big(g(m^{-1}r^{-1}sn)\big)\tau(sns^{-1}) \\
&\qquad\qquad d\mu_{N_\tau}(m)\,d\mu_{G/N_\tau}(\dot r)\,d\mu_{N_\tau}(n)
\end{aligned}$$

which, after using Fubini's Theorem and replacing n by $s^{-1}rmr^{-1}sn$, is

$$\begin{aligned}
&= \int_{G/N_\tau}\int_{N_\tau}\int_{N_\tau} f(rm)\alpha_{rm}\big(g(r^{-1}sn)\big)\tau(rmr^{-1}sns^{-1}) \\
&\qquad\qquad d\mu_{N_\tau}(n)\,d\mu_{N_\tau}(m)\,d\mu_{G/N_\tau}(\dot r) \\
&= \int_{G/N_\tau} \Phi(f)(r)\alpha_r\Big(\int_{N_\tau} g(r^{-1}sn)\tau(r^{-1}sns^{-1}r)\,d\mu_{N_\tau}(n)\Big) \\
&\qquad\qquad d\mu_{G/N_\tau}(\dot r) \\
&= \int_{G/N_\tau} \Phi(f)(r)\alpha_r\big(\Phi(g)(r^{-1}s)\big)\,d\mu_{G/N_\tau}(\dot r) \\
&= \Phi(f) * \Phi(g)(s).
\end{aligned}$$

Similarly, we compute that

$$\begin{aligned}\Phi(f)^*(s) &= \Delta_{G/N_\tau}(\dot{s}^{-1})\alpha_s\big(\Phi(f)(s^{-1})^*\big)\\ &= \Delta_{G/N_\tau}(\dot{s}^{-1})\alpha_s\Big(\int_{N_\tau} f(s^{-1}n)\tau(s^{-1}ns)\,d\mu_{N_\tau}(n)\Big)^*\\ &= \Delta_{G/N_\tau}(\dot{s}^{-1})\int_{N_\tau} \alpha_s\big(\tau(s^{-1}n^{-1}s)f(s^{-1}n)^*\big)\,d\mu_{N_\tau}(n)\end{aligned}$$

which, by (7.17), is

$$= \Delta_{G/N_\tau}(\dot{s}^{-1})\int_{N_\tau} \tau(n)^*\alpha_s\big(f(s^{-1}n)\big)^*\,d\mu_{N_\tau}(n)$$

which, by (7.16), is

$$= \Delta_{G/N_\tau}(\dot{s}^{-1})\int_{N_\tau} \alpha_{n^{-1}s}\big(f(s^{-1}n)\big)^*\tau(n)^*\,d\mu_{N_\tau}(n)$$

which, using (7.24), is

$$= \Delta_{G/N_\tau}(\dot{s}^{-1})\Delta_{G,N_\tau}(s^{-1})\int_{N_\tau} \alpha_{sn^{-1}}\big(f(ns^{-1})\big)^*\tau(sn^{-1}s^{-1})\,d\mu_{N_\tau}(n)$$

which, using (7.25), is

$$\begin{aligned}&= \Delta_G(s^{-1})\int_{N_\tau} \alpha_{sn}\big(f(n^{-1}s^{-1})\big)^*\tau(sns^{-1})\Delta_{N_\tau}(n^{-1})\,d\mu_{N_\tau}(n)\\ &= \int_{N_\tau} f^*(sn)\tau(sns^{-1})\,d\mu_{N_\tau}(n)\\ &= \Phi(f^*)(s).\end{aligned}$$

This completes the proof. □

Now suppose that (π, U) is a covariant representation of (A, G, α) on $\mathcal{H}$ which preserves τ. If $g \in C_c(G, A, \tau)$, then

$$s \mapsto \pi\big(g(s)\big)U(s)$$

is constant on N_τ-cosets. Thus we can define an operator on $\mathcal{H}$ by

$$\pi \rtimes^\tau U(g) = \int_{G/N_\tau} \pi\big(g(s)\big)U(s)\,d\mu_{G/N_\tau}(\dot{s}).$$

In fact if $f \in C_c(G, A)$, then

$$\begin{aligned}\pi \rtimes^\tau U\big(\Phi(f)\big) &= \int_{G/N_\tau} \pi\big(\Phi(f)(s)\big)U(s)\,d\mu_{G/N_\tau}(\dot{s})\\ &= \int_{G/N_\tau}\int_{N_\tau} \pi\big(f(sn)\tau(sns^{-1})\big)\,d\mu_{N_\tau}(n)U(s)\,d\mu_{G/N_\tau}(\dot{s})\end{aligned}$$

which, since (π, U) preserves τ, is

$$= \int_{G/N_\tau} \int_{N_\tau} \pi\big(f(sn)\big)U(sn)\, d\mu_{N_\tau}(n)\, d\mu_{G/N_\tau}(\dot{s})$$
$$= \pi \rtimes U(f).$$

It follows from this computation that if (π, U) preserves τ, then $\pi \rtimes^\tau U$ defines a $*$-representation of $C_c(G, A, \tau)$, and that

$$\|g\| := \sup\{\, \|\pi \rtimes^\tau U(g)\| : (\pi, U) \text{ preserves } \tau \,\} \tag{7.26}$$

is a C^*-norm on $C_c(G, A, \tau)$ such that $\|\Phi(f)\|$ is the norm of the image of f in the quotient $A \rtimes_\alpha^\tau G$. We can summarize the discussion to this point in the following lemma.

Lemma 7.25. *Suppose that (A, G, α, τ) is a twisted dynamical system. Then we can identify the completion $\overline{C_c(G, A, \tau)}$ of $C_c(G, A, \tau)$, with respect to the norm $\|\cdot\|$ defined by* (7.26)*, with $A \rtimes_\alpha^\tau G$. Then, after this identification, the $*$-homomorphism Φ of Lemma 7.24 on page 211 extends to the quotient map from $A \rtimes_\alpha G$ onto $A \rtimes_\alpha^\tau G$.*

The following analogue of Corollary 2.46 on page 63 will be useful in the next section.

Lemma 7.26. *Suppose that (A, G, α, τ) is a twisted dynamical system, and that $R : C_c(G, A, \tau)$ is a nondegenerate $\|\cdot\|_1$-norm decreasing $*$-representation. Then R is norm decreasing for the universal norm on $A \rtimes_\alpha^\tau G$ and extends to a representation of $A \rtimes_\alpha^\tau G$.*

Proof. It is not hard to check that $\|\Phi(f)\|_1 \le \|f\|_1$ (with respect to the L^1-norms on $C_c(G, A)$ and $C_c(G, A, \tau)$, respectively). Therefore $R' := R \circ \Phi$ is a $\|\cdot\|_1$-norm decreasing representation of $C_c(G, A)$ which must be bounded with respect to the universal norm on $A \rtimes_\alpha G$ by Corollary 2.46. Since $\varphi\big(\bar{\imath}_A\big(\tau(n)\big)f\big) = \Phi\big(i_G(n)f\big)$ for all $f \in C_c(G, A)$, it follows that $\bar{R}'\big(\bar{\imath}\big(\tau(n)\big)\big) = \bar{R}'\big(i_G(n)\big)$ for all $n \in N_\tau$. Therefore $R' = \pi \rtimes U$ for a covariant pair (π, U) which preserves τ. Thus $R = \pi \rtimes^\tau U$ is norm decreasing with respect to the universal norm as required. □

7.4.2 Decomposition with Respect to a Normal Subgroup

Suppose that (A, G, α, τ) is a twisted dynamical system[8] and that M is a normal subgroup of G such that $N_\tau \subset M$. We want to build a strongly continuous G-action

$$\gamma : G \to \operatorname{Aut}\big(A \rtimes_\alpha^\tau M\big).$$

[8]The decomposition result is most germane to our discussion of ordinary crossed products when τ is trivial and $N_\tau = \{e\}$ so that $A \rtimes_\alpha^\tau G = A \rtimes_\alpha G$. However, even in this case, the conclusion involves a twisted crossed product. Starting with a possibly nontrivial τ adds a little extra generality and doesn't significantly complicate the proof, so we include it for reference.

We'll need the modular function $\Delta_{G/N_\tau, M/N_\tau}$ defined in (7.24). Since the notation is getting a bit unwieldy, let

$$\delta(s) := \Delta_{G/N_\tau, M/N_\tau}(\dot{s}) = \Delta_{G/N_\tau}(\dot{s})\Delta_{G/M}(\dot{s}^{-1}),$$

where the last equality comes from (7.25). (We've identified $(G/N_\tau)/(M/N_\tau)$ with G/M.) If $f \in C_c(M, A, \tau)$, then define

$$\gamma_s(f)(m) := \delta(s)\alpha_s\big(f(s^{-1}ms)\big).$$

Note that if $n \in N_\tau$ and $m \in M$, then

$$\begin{aligned}
\gamma_s(f)(nm) &= \delta(s)\alpha_s\big(f(s^{-1}nms)\big) \\
&= \delta(s)\alpha_s\big(f(s^{-1}nss^{-1}ms)\big) \\
&= \delta(s)\alpha_s\big(f(s^{-1}ms)\tau(s^{-1}ns)^*\big) \\
&= \delta(s)\alpha_s\big(f(s^{-1}ms)\big)\tau(n)^*;
\end{aligned}$$

thus, $\gamma_s(f) \in C_c(M, A, \tau)$. Furthermore, if (ρ, V) is a covariant representation of (A, M, α) which preserves τ, and if $s \in G$, then a straightforward computation shows that we get another covariant τ-preserving representation $(s \cdot \rho, s \cdot V)$ where

$$s \cdot \rho(a) := \rho\big(\alpha_s^{-1}(a)\big) \quad \text{and} \quad s \cdot V(m) := V(s^{-1}ms).$$

Moreover, if $f \in C_c(M, A, \tau)$, then

$$\begin{aligned}
\rho \rtimes^\tau V\big(\gamma_s(f)\big) &= \delta(s) \int_{M/N_\tau} \rho\big(\alpha_s\big(f(s^{-1}ms)\big)\big)V(m)\, d\mu_{M/N_\tau}(\dot{m}) \\
&= \int_{M/N_\tau} \rho\big(\alpha_s\big(f(m)\big)\big)V(sms^{-1})\, d\mu_{M/N_\tau}(\dot{m}) \\
&= s^{-1} \cdot \rho \rtimes^\tau s^{-1} \cdot V(f).
\end{aligned}$$

Since (ρ, V) is arbitrary, the above computation shows that $\|\gamma_s(f)\| \le \|f\|$ for all $f \in C_c(M, A, \tau)$.

Now we compute that

$$\begin{aligned}
\gamma_s(f * g)(m) &= \delta(s)\alpha_s\big(f * g(s^{-1}ms)\big) \\
&= \int_{M/N_\tau} \delta(s)\alpha_s\big(f(n)\alpha_n\big(g(n^{-1}s^{-1}ms)\big)\big)\, d\mu_{M/N_\tau}(\dot{n}) \\
&= \int_{M/N_\tau} \delta(s)^2\alpha_s\big(f(s^{-1}ns)\big)\alpha_{ns}\big(g(s^{-1}n^{-1}ms)\big)\big)\, d\mu_{M/N_\tau}(\dot{n}) \\
&= \gamma_s(f) * \gamma_s(g)(m).
\end{aligned}$$

To check that γ_s is $*$-preserving, we need to know that if M is normal in G, then for all $s \in G$, $\Delta_M(sms^{-1}) = \Delta_M(m)$. But

$$\begin{aligned}\Delta_M(sns^{-1}) \int_M g(msns^{-1})\,d\mu_M(m) &= \int_M g(m)\,d\mu_M(m)\\ &= \Delta_{G,M}(s^{-1}) \int_M g(sms^{-1})\,d\mu_M(m)\\ &= \Delta_{G,M}(s^{-1})\Delta_M(n) \int_M g(smns^{-1})\,d\mu_M(m)\\ &= \Delta_M(n) \int_G g(msns^{-1})\,d\mu_M(m).\end{aligned}$$

Now

$$\begin{aligned}\gamma_s(f^*)(m) &= \delta(s)\alpha_s\big(f^*(s^{-1}ms)\big)\\ &= \delta(s)\Delta_{M/N_\tau}(s^{-1}m^{-1}s)\alpha_{ms}\big(f(s^{-1}m^{-1}s)^*\big)\end{aligned}$$

which, since M/N_τ is normal in G/N_τ, is

$$\begin{aligned}&= \delta(s)\Delta_{M/N_\tau}(m^{-1})\alpha_m\big(\alpha_s\big(f(s^{-1}m^{-1}s)^*\big)\big)\\ &= \Delta_{M/N_\tau}(m^{-1})\alpha_m\big(\gamma_s(f)(m^{-1})^*\big)\\ &= \gamma_s(f)^*(m).\end{aligned}$$

Thus γ_s defines an endomorphism of $A \rtimes_\alpha^\tau M$. Since we certainly have $\gamma_{sr} = \gamma_s \circ \gamma_r$ and $\gamma_e = \mathrm{id}$, $s \mapsto \gamma_s$ is a homomorphism of G into $\mathrm{Aut}\big(A \rtimes_\alpha^\tau M\big)$.

We've made a good start on the following result.

Lemma 7.27. *If M is a normal subgroup of G and if (A, G, α, τ) is a twisted dynamical system, then there is a dynamical system*

$$\gamma : G \to \mathrm{Aut}\big(A \rtimes_\alpha^\tau M\big)$$

such that $\gamma_s(g)(m) = \delta(s)\alpha_s\big(g(s^{-1}ms)\big)$ for $g \in C_c(M, A, \tau)$. Let

$$\tau^M : M \to UM\big(A \rtimes_\alpha^\tau M\big)$$

be the composition of the natural map $i_M : M \to UM(A \rtimes_\alpha M)$ with the quotient map Φ of $A \rtimes_\alpha M$ onto $A \rtimes_\alpha^\tau M$; thus if $g \in C_c(M, A, \tau)$, then

$$\tau^M(n)(g)(m) = \alpha_n\big(g(n^{-1}m)\big). \tag{7.27}$$

Furthermore, τ^M is a twisting map for γ and $(A \rtimes_\alpha^\tau M, G, \gamma, \tau^M)$ is a twisted dynamical system.

Proof. Given the above discussion, in order to show that $(A \rtimes_\alpha^\tau M, G, \gamma)$ is a dynamical system all that remains is to show that γ is strongly continuous. If $g \in C_c(M, A, \tau)$ and if $s_i \to s$ in G, then it will suffice to see that

$$\|\gamma_{s_i}(g) - \gamma_s(g)\| \to 0.$$

Let $K \subset M$ be a compact set such that $\operatorname{supp} g \subset KN_\tau$. Let C be a compact symmetric neighborhood of both K and s. We can assume that each $s_i \in C$. Since N_τ is normal in G, $(CKN_\tau C^{-1}) \cap M = (CKC^{-1} \cap M)N_\tau$, and

$$m \mapsto \gamma_{s_i}(g)(m) - \gamma_s(g)(m)$$

vanishes off the saturation of the compact set $K' := CKC^{-1} \cap M$ in M. Since

$$\|g\| \le \|g\|_1 := \int_{M/N_\tau} \|g(m)\| \, d\mu_{M/N_\tau}(\dot{m}),$$

it suffices to see that $\gamma_{s_i}(g) \to \gamma_s(g)$ uniformly. If this were not true, then we could pass to a subnet, relabel, and assume that there is an $\epsilon > 0$ and $m_i \in M$ such that

$$\|\gamma_{s_i}(g)(m_i) - \gamma_s(g)(m_i)\| \ge \epsilon \quad \text{for all } i. \tag{7.28}$$

Since $\|\gamma_r(g)(m)\| = \|\gamma_r(g)(mn)\|$ for all $m \in M$ and $n \in N_\tau$, we can assume that each $m_i \in K'$. Since K' is compact, we may as well assume that $m_i \to m$ in M. But then the left-hand side of (7.28) tends to zero. This is a contradiction and completes the proof that γ is strongly continuous.

Now let $\tau^M = \bar{\Phi} \circ i_M$. Since i_M is strictly continuous, τ^M is a strictly continuous homomorphism of M into $M(A \rtimes_\alpha^\tau M)$. To verify (7.27), notice that if $f \in C_c(G, A)$, then by Lemma 7.25 on page 213 and by definition,

$$\begin{aligned}
\tau^M(n)\Phi(f)(m) &= \Phi(i_M(n)f)(m) \\
&= \int_{N_\tau} i_M(n)(f)(mv)\tau(mvm^{-1}) \, d\mu_{N_\tau}(v) \\
&= \int_N \alpha_n\big(f(n^{-1}mv)\big)\tau(mvm^{-1}) \, d\mu_{N_\tau}(v) \\
&= \int_{N_\tau} \alpha_n\big(f(n^{-1}mv)\tau(n^{-1}mvm^{-1}n)\big) \, d\mu_{N_\tau}(v) \\
&= \alpha_n\big(\Phi(f)(n^{-1}m)\big).
\end{aligned}$$

Since Φ is surjective, this suffices. To verify (7.16) we use (7.27) to compute that for $n \in N_\tau$ and $f, g \in C_c(M, A, \tau)$ we have

$$\begin{aligned}
\tau^M(n)f\tau^M(n)^* * g(m') &= \tau^M(n)f * \tau^M(n^{-1})g(m') \\
&= \int_{M/N_\tau} \alpha_n\big(f(n^{-1}m)\big)\alpha_{mn^{-1}}\big(g(nm^{-1}m')\big) \, d\mu_{M/N_\tau}(\dot{m})
\end{aligned}$$

which, since the integrand is constant on N_τ-cosets and since δ is identically one on N_τ, is

$$\begin{aligned}
&= \int_{M/N_\tau} \alpha_n\big(f(n^{-1}mn)\big)\alpha_m\big(g(m^{-1}m')\big) \, d\mu_{M/N_\tau}(\dot{m}) \\
&= \gamma_n(f) * g(m').
\end{aligned}$$

To check (7.17) we use (7.27) to compare the values of $\gamma_s\big(\tau^M(n)f\big)(m)$ and $\tau^M(sns^{-1})\big(\gamma_s(f)\big)(m)$. □

Proposition 7.28. *Suppose that (A, G, α, τ) is a twisted dynamical system, and that M is a normal subgroup of G with $N_\tau \subset M$. Then there is an isomorphism*

$$\Psi : A \rtimes_\alpha^\tau G \to \big(A \rtimes_\alpha^\tau M\big) \rtimes_\gamma^{\tau^M} G$$

such that if $f \in C_c(G, A, \tau)$, then for each $s \in G$, $\Psi(f)(s) \in C_c(M, A, \tau)$ and

$$\Psi(f)(s)(m) = \delta(s) f(ms). \tag{7.29}$$

Furthermore, Ψ intertwines a covariant representation (π, U) of (A, G, α, τ) with the covariant representation $(\pi \rtimes U|_M, U)$ of $(A \rtimes_\alpha^\tau M, G, \gamma, \tau^M)$.

We will need the following technical lemma for the proof.

Lemma 7.29. *Let (A, G, α, τ) be a twisted dynamical system. Then for each $s \in G$,*

$$\{\, g(s) : g \in C_c(G, A, \tau) \,\}$$

is dense in A. Conversely, if B is a subspace of $C_c(G, A, \tau)$ such that $\{\, g(s) : g \in B \,\}$ is dense in A for all $s \in G$ and such that B is closed under pointwise multiplication by elements of $C_c(G/N_\tau)$, then B is dense in $A \rtimes_\alpha^\tau G$.

Proof. To prove the first statement, consider $a \in A$, $s \in G$ and $\epsilon > 0$. Since $n \mapsto \tau(n)$ is strictly continuous, there is a neighborhood V of e in G such that $n \in V$ implies $\|a\tau(sns^{-1}) - a\| < \epsilon$. Choose $\varphi \in C_c^+(G)$ such that $\operatorname{supp}\varphi \subset sV$ and such that

$$\int_{N_\tau} \varphi(sn)\, d\mu_{N_\tau}(n) = 1.$$

Let $f(s) = \varphi(s)a$ and $g := \Phi(f)$. Then

$$\begin{aligned}\|g(s) - a\| &= \Big\| \int_{N_\tau} \varphi(sn)\big(a\tau(sns^{-1}) - a\big)\, d\mu_{N_\tau} \Big\| \\ &\le \epsilon \int_{N_\tau} \varphi(sn)\, d\mu_{N_\tau}(n) = \epsilon.\end{aligned}$$

This proves the first assertion.

Now assume that B is as in the statement of the lemma and that $g \in C_c(G, A, \tau)$. Let $K \subset G$ be a compact set such that $\operatorname{supp} g \subset KN_\tau$. Let K' be a compact neighborhood of K. Fix $\epsilon > 0$. By assumption, for each $s \in K$, there is a $g_s \in B$ such that

$$\|g(s) - g_s(s)\| < \epsilon$$

By continuity, there is a neighborhood V_s of s such that $V_s \subset K'$ and $r \in V_s$ implies

$$\|g(r) - g_s(r)\| < \epsilon. \tag{7.30}$$

Since g and g_s are in $C_c(G, A, \tau)$, (7.30) holds for all $r \in V_s N_\tau$. Since K is compact, there are $s_1, \dots, s_n \in K$ such that $V_{s_1}, \dots, V_{s_n}$ cover K. Let $\varphi_1, \dots, \varphi_n$ be elements of $C_c^+(G/N_\tau)$ such that $\operatorname{supp}\varphi_i \subset V_{s_i}N_\tau$ and

$$\sum_{i=1}^n \varphi_i(\dot s) = 1 \quad \text{for all } s \in K.$$

Then we let $g_\epsilon := \sum \varphi_i \cdot g_{s_i}$. By assumption $g_\epsilon \in B$ and

$$\|g - g_\epsilon\|_\infty < \epsilon.$$

Therefore

$$\|g - g_\epsilon\|_1 \le \epsilon \mu_{G/N_\tau}(K').$$

Since ϵ is arbitrary, this suffices. □

Proof of Proposition 7.28 on the preceding page. If $f \in C_c(G, A, \tau)$ and Ψ is defined as in the statement of the proposition, then it is straightforward to check that $\Psi(f)(s) \in C_c(M, A, \tau)$. A similar argument to that in the beginning of the proof of Lemma 7.27 on page 215 shows that $s \mapsto \Psi(f)(s)$ is continuous from G into $A \rtimes_\alpha^\tau M$. If $g \in C_c(M, A, \tau)$, then by expanding $g * \tau^M(n)^* f$, we see that

$$g\tau^M(n)^*(m) = \Delta_{M/N_\tau}(\dot{n}) g(mn).$$

Since M/N_τ is normal in G/N_τ, $\Delta_{M/N_\tau}(\dot{n}) = \Delta_{G/N_\tau}(\dot{n})$, and if $n, m \in M$ and $s \in G$, we have

$$\begin{aligned}
\Psi(f)(ns)(m) &= \delta(ns) f(mns) \\
&= \Delta_{G/N_\tau}(\dot{n}) \delta(s) f(mns) \\
&= \Delta_{M/N_\tau}(\dot{n}) \Psi(f)(s)(mn) \\
&= \Psi(f)(s)\tau^M(n)^*(m).
\end{aligned}$$

Therefore, $\Psi(f) \in C_c(G, A \rtimes_\alpha^\tau M, \tau^M)$, and we get a map

$$\Psi : C_c(G, A, \tau) \to C_c(G, A \rtimes_\alpha^\tau M, \tau^M).$$

To see that Ψ is a $*$-homomorphism, we calculate as follows.

$$\begin{aligned}
\Psi(f * g)(s)(m) &= \delta(s) f * g(ms) \\
&= \delta(s) \int_{G/N_\tau} f(r) \alpha_r\bigl(g(r^{-1}ms)\bigr)\, d\mu_{G/N_\tau}(\dot{r})
\end{aligned}$$

which, after identifying $(G/N_\tau)/(M/N_\tau)$ with G/M, and choosing Haar measures consistently, is

$$\begin{aligned}
&= \delta(s) \int_{G/M} \int_{M/N_\tau} f(rn) \alpha_{rn}\bigl(g(n^{-1}r^{-1}ms)\bigr) \\
&\qquad\qquad\qquad d\mu_{M/N_\tau}(\dot{n})\, d\mu_{G/M}(\dot{r}) \\
&= \delta(s) \int_{G/M} \delta(r) \int_{M/N_\tau} f(nr) \alpha_{nr}\bigl(g(r^{-1}n^{-1}ms)\bigr) \\
&\qquad\qquad\qquad d\mu_{M/N_\tau}(\dot{n})\, d\mu_{G/M}(\dot{r}) \\
&= \int_{G/M} \int_{M/N_\tau} \Psi(f)(r)(n) \alpha_n\Bigl(\delta(r) \alpha_r\bigl(\Psi(g)(r^{-1}s)(r^{-1}n^{-1}mr)\bigr)\Bigr)
\end{aligned}$$

$$d\mu_{M/N_\tau}(\dot n)\,d\mu_{G/M}(\dot r)$$

$$= \int_{G/M}\int_{M/N_\tau} \Psi(f)(r)(n)\alpha_n\big(\gamma_r\big(\Psi(g)(r^{-1}s)\big)(n^{-1}m)\big)$$

$$d\mu_{M/N_\tau}(\dot n)\,d\mu_{G/M}(\dot r)$$

$$= \int_{G/M} \Psi(f)(r) * \gamma_r\big(\Psi(g)(r^{-1}s)\big)(m)\,d\mu_{G/M}(\dot r). \tag{7.31}$$

Now, as in the discussion on pages 190–192, we can apply Lemma 1.108 on page 39 to conclude that $\Psi(f) * \Psi(g)(s) \in C_c(M, A, \tau)$, and that (7.31) is

$$= \Psi(f) * \Psi(g)(s)(m).$$

To see that Ψ is $*$-preserving, we first compute

$$\begin{aligned}\Psi(f)^*(s)(m) &= \Delta_{G/M}(\dot s^{-1})\gamma_s\big(\Psi(f)(s^{-1})\big)^*(m)\\ &= \Delta_{G/M}(\dot s^{-1})\alpha_m\big(\gamma_s\big(\Psi(f)(s^{-1})\big)(m^{-1})^*\big)\Delta_{M/N_\tau}(\dot m^{-1})\\ &= \Delta_{G/M}(\dot s^{-1})\alpha_m\big(\delta(s)\alpha_s\big(\Psi(f)(s^{-1})(s^{-1}m^{-1}s)\big)^*\big)\Delta_{M/N_\tau}(\dot m^{-1})\\ &= \Delta_{G/M}(\dot s^{-1})\Delta_{M/N_\tau}(\dot m^{-1})\alpha_{ms}\big(f(s^{-1}m^{-1})^*\big).\end{aligned}$$

On the other hand,

$$\begin{aligned}\Psi(f^*)(s)(m) &= \delta(s)f^*(ms)\\ &= \delta(s)\alpha_{ms}\big(f(s^{-1}m^{-1})^*\big)\Delta_{G/N_\tau}(\dot m\dot s)^{-1}\\ &= \Delta_{G/N_\tau}(\dot s)\Delta_{G/M}(\dot s^{-1})\alpha_{ms}\big(f(s^{-1}m^{-1})^*\big)\Delta_{G/N_\tau}(\dot m\dot s)^{-1}\end{aligned}$$

which, since Δ_{G/N_τ} and Δ_{M/N_τ} agree on the normal subgroup M/N_τ, is

$$= \Delta_{G/M}(\dot s^{-1})\Delta_{M/N_\tau}(\dot m^{-1})\alpha_{ms}\big(f(s^{-1}m^{-1})^*\big).$$

Therefore, Ψ is $*$-preserving.

Also,

$$\begin{aligned}\|\Psi(f)\|_1 &:= \int_{G/M} \|\Psi(f)(s)\|\,d\mu_{G/M}(\dot s)\\ &\le \int_{G/M} \|\Psi(f)(s)\|_1\,d\mu_{G/M}(\dot s)\\ &= \int_{G/M}\int_{M/N_\tau} \|\Psi(f)(s)(m)\|\,d\mu_{M/N_\tau}(\dot m)\,d\mu_{G/M}(\dot s)\\ &= \int_{G/M}\int_{M/N_\tau} \delta(s)\|f(ms)\|\,d\mu_{M/N_\tau}(\dot m)\,d\mu_{G/M}(\dot s)\\ &= \int_{G/M}\int_{M/N_\tau} \|f(sm)\|\,d\mu_{M/N_\tau}(\dot m)\,d\mu_{G/M}(\dot s)\\ &= \|f\|_1.\end{aligned}$$

It follows that if R is any representation of $\bigl(A \rtimes_\alpha^\tau M\bigr) \rtimes_\gamma^{\tau^M} G$, then $R \circ \Psi$ is a representation of $A \rtimes_\alpha^\tau G$ (Lemma 7.26 on page 213). Thus, $\|\Psi(f)\| \le \|f\|$.

On the other hand, if (π, U) is a covariant representation of (A, G, α, τ), then we shall see that $(\pi \rtimes^\tau U|_M, U)$ is a covariant representation of $(A \rtimes_\alpha^\tau M, G, \gamma, \tau^M)$. Let $g \in C_c(M, A, \tau)$. Then

$$\begin{aligned} U(s)(\pi \rtimes^\tau U|_M)(g) &= \int_{M/N_\tau} U(s)\pi\bigl(g(m)\bigr)U(m)\,d\mu_{M/N_\tau}(\dot m) \\ &= \int_{M/N_\tau} \pi\bigl(\alpha_s\bigl(g(m)\bigr)\bigr)U(sm)\,d\mu_{M/N_\tau}(\dot m) \\ &= \int_{M/N_\tau} \delta(s)\pi\bigl(\alpha_s\bigl(g(s^{-1}ms)\bigr)\bigr)U(m)\,d\mu_{M/N_\tau}(\dot m)U(s) \\ &= \pi \rtimes^\tau U|_M\bigl(\gamma_s(g)\bigr)U(s). \end{aligned}$$

Therefore, $(\pi \rtimes^\tau U|_M, U)$ is a covariant representation of $(A \rtimes_\alpha^\tau M, G, \gamma)$. To see that it preserves τ^M as well, we need to check that

$$(\pi \rtimes^\tau U|_M)^{\bar{}}\bigl(\tau^M(m)\bigr) = U(m);$$

but this follows immediately as $\tau^M = \bar\Phi \circ i_M(m)$.

Now we compute that

$$\begin{aligned} (\pi \rtimes^\tau U|_M) \rtimes U\bigl(\Psi(f)\bigr) &= \int_{G/M} \pi \rtimes U|_M\bigl(\Psi(f)(s)\bigr)U(s)\,d\mu_{G/M}(\dot s) \\ &= \int_{G/M}\int_{M/N_\tau} \pi\bigl(\Psi(f)(s)(m)\bigr)U(ms)\,d\mu_{M/N_\tau}(\dot m)\,d\mu_{G/M}(\dot s) \\ &= \int_{G/M}\int_{M/N_\tau} \delta(s)\pi\bigl(f(ms)\bigr)U(ms)\,d\mu_{M/N_\tau}(\dot m)\,d\mu_{G/M}(\dot s) \\ &= \int_{G/M}\int_{M/N_\tau} \pi\bigl(f(sm)\bigr)U(sm)\,d\mu_{M/N_\tau}(\dot m)\,d\mu_{G/M}(\dot s) \\ &= \pi \rtimes^\tau U(f). \end{aligned}$$

In particular,

$$\|\pi \rtimes^\tau U(f)\| = \|(\pi \rtimes^\tau U|_M) \rtimes U\bigl(\Psi(f)\bigr)\| \le \|\Psi(f)\|.$$

Since (π, U) was arbitrary, $\|f\| \le \|\Psi(f)\|$ and Ψ is isometric. We can complete the proof by showing that Ψ has dense range.

If $s \in G$, let

$$B := \{\, \Psi(f)(s) : f \in C_c(G, A, \tau) \,\} \subset C_c(M, A, \tau).$$

The first part of Lemma 7.29 on page 217 implies that $\{\, f(ms) : f \in C_c(G, A, \tau) \,\}$ is dense in A for each $m \in M$. Since B is closed under pointwise multiplication by elements in $C_c(M/N_\tau)$,[9] B is dense in $A \rtimes_\alpha^\tau M$ by the second part of Lemma 7.29.

[9] If $\varphi \in C_c(M/N_\tau)$ and if $\bar\varphi$ is an extension of φ to $C_c(G/N_\tau)$, then since N_τ is normal, we can define $\psi(rN_\tau) := \bar\varphi(rs^{-1}N_\tau)$. If $g(r) := \psi(rN_\tau)f(r)$, we have $g(ms) = \varphi(mN_\tau)f(ms)$.

On the other hand, the range of Ψ is closed under multiplication by elements of $C_c(G/M)$. Therefore the range of Ψ is dense by Lemma 7.29. Thus Ψ is an isomorphism. □

7.5 The Type Structure of Regular Crossed Products

C^*-algebras are naturally divided into types based on their structure and the behavior of their representation theory. Such matters are discussed in detail in the standard references on C^*-algebras such as [110, §5.6] or [126, Chap. 6]. In this section, we want to look at CCR and GCR algebras. A C^*-algebra is called *CCR* or *liminary*[10] if $\pi(A) = \mathcal{K}(\mathcal{H}_\pi)$ for every irreducible representation $\pi \in \hat{A}$. In this text, we define a C^*-algebra A to be *GCR* or *postliminary* if $\pi(A) \supset \mathcal{K}(\mathcal{H}_\pi)$ for all $\pi \in \hat{A}$ [110, p. 169]. This definition is best suited for the results below, and it has a pleasing resemblance to the CCR definition. However, we should acknowledge that it is not the classical one — although it is equivalent to the classical definition. The essential feature that a GCR algebra A must possess is that it have a composition series $\{ I_\alpha \}_{0 \le \alpha \le \alpha_0}$ of ideals such that $I_0 = \{ 0 \}$, $I_{\alpha_0} = A$, and such that $I_{\alpha+1}/I_\alpha$ is CCR. (Thus the "G" in GCR is probably meant to suggest that GCR algebras are "generalized" CCR algebras.) Many classical texts [2, 28, 126] say that A is GCR if every quotient A/J (with $J \ne A$) contains a nonzero CCR ideal. This version of the definition has the advantage that it follows without too much difficulty that it is equivalent to A having a composition series as above [2, Theorem 1.5.5]. To see that our definition is equivalent to either of the above conditions requires invoking some of the outstanding results in the subject. In the separable case, these conditions, and many others, are shown to be equivalent in the centerpiece results [126, Theorem 6.8.7] or [28, Theorem 9.1]. In the possibly nonseparable case, it is shown in [157] that A has a composition series as above if and only if A is a type I C^*-algebra. This means that $\pi(A)''$ is a type I von Neumann algebra for all representations of A.[11] In [158], it is shown that A is of Type I if and only if every irreducible representation of A contains the compacts in its image (as in our definition for a GCR algebra). Thus our definition is equivalent to the classical ones.

If A is GCR, then all ideals and quotients of A are GCR [110, Theorem 5.6.2]. Furthermore, if A is GCR and $P \in \operatorname{Prim} A$, then A/P is a primitive GCR algebra.[12] Consequently, it must contain a canonical minimal ideal $\mathcal{K}(P)$ isomorphic to the compact operators on some Hilbert space. Since a C^*-algebra isomorphic to the

[10] The acronym "CCR" is supposed to suggest "completely continuous representations" as compact operators once went by the term completely continuous operator. The term "liminary" is meant to be an English equivalent of the French word "liminaire" meaning "preliminary". The spelling "liminal" is probably more common (cf., [28, 110]), but I prefer to follow Pedersen's spelling in [126]. In these notes, we'll stick to the CCR and GCR designations despite Pedersen's admonition in [126, Remark 6.2.13].

[11] The terminology is classical, but unfortunate. A type I von Neumann algebra is usually not a type I C^*-algebra.

[12] A C^*-algebra is called primitive if it has a faithful irreducible representation.

compacts on some Hilbert space is called *elementary*, we call $\mathcal{K}(P)$ the *elementary ideal determined by* P.

In general, trying to determine conditions on a crossed product that force it to be CCR or GCR is very difficult. However, if we assume that (A, G, α) is separable and regular and that A is GCR, then we can say something.[13] Even with these restrictions, the discussion will get fairly complicated. To make the general discussion a little easier to digest, it will be helpful to first look at the situation for regular transformation groups. (We will look at possibly non-regular transformation groups in Section 8.3.) We call a group G either CCR or GCR when the group C^*-algebra $C^*(G)$ is either CCR or GCR, respectively.

Let $\bigl(C_0(X), G, \mathrm{lt}\bigr)$ be a regular system in which each orbit is locally closed.[14] These hypotheses are automatically met if (G, X) is second countable and the orbit space $G\backslash X$ is a T_0 topological space (Theorem 6.2 on page 173). Let $L = \pi \rtimes u$ be an irreducible representation of $C_0(X) \rtimes_{\mathrm{lt}} G$. Since $\bigl(C_0(X), G, \mathrm{lt}\bigr)$ is quasi-regular, there is a $x \in X$ such that $\ker \pi = J(\overline{G \cdot x})$, were $J(F)$ is the ideal of functions in $C_0(X)$ vanishing on F. By Lemma 6.16 on page 182, L factors though a representation $\bar{L} = \bar{\pi} \rtimes u$ of $C_0(\overline{G \cdot x}) \rtimes_{\mathrm{lt}} G$. By assumption, $G \cdot x$ is an open subset of $\overline{G \cdot x}$. Therefore, we can view $C_0(G \cdot x) \rtimes_{\mathrm{lt}} G$ as an ideal in $C_0(\overline{G \cdot x}) \rtimes_{\mathrm{lt}} G$ (Corollary 3.20 on page 94). Since $\ker \bar{\pi} = \{0\}$, $\bar{L}$ does not vanish on $C_0(G \cdot x) \rtimes_{\mathrm{lt}} G$, and it follows that the restriction $\bar{L}'$ of $\bar{L}$ to $C_0(G \cdot x) \rtimes_{\mathrm{lt}} G$ is irreducible. Furthermore, $\bar{L}$ is the canonical extension of $\bar{L}'$. Since $\bigl(C_0(X), G, \mathrm{lt}\bigr)$ is regular, the map $sG_x \mapsto s \cdot x$ is an equivariant homeomorphism of G/G_x onto $G \cdot x$. Therefore $C_0(G \cdot x) \rtimes_{\mathrm{lt}} G$ is isomorphic to $C_0(G/G_x) \rtimes_{\mathrm{lt}} G$. Thus, by Green's Imprimitivity Theorem, $C_0(G \cdot x) \rtimes_{\mathrm{lt}} G$ is Morita equivalent to $C^*(G_x)$.[15]

Proposition 7.30. *Suppose that* (G, X) *is a second countable transformation group and that* $G\backslash X$ *is a* T_0 *topological space. Then* $C_0(X) \rtimes_{\mathrm{lt}} G$ *is GCR if and only if* G_x *is GCR for all* $x \in X$.

Proof. We adopt the notation from the preceding discussion. Suppose that G_x is GCR for all x and that L is an irreducible representation of $C_0(X) \rtimes_{\mathrm{lt}} G$. Since Morita equivalence preserves the property of being GCR (Proposition I.44 on page 507), it follows that $C_0(G \cdot x) \rtimes_{\mathrm{lt}} G$ is GCR for all x. Hence if L is an irreducible representation of $C_0(X) \rtimes_{\mathrm{lt}} G$, then the representation $\bar{L}'$ described above contains the compacts in its image. Hence so does $\bar{L}$, and therefore L. This shows that $C_0(X) \rtimes_{\mathrm{lt}} G$ is GCR.

Suppose that $C_0(X) \rtimes_{\mathrm{lt}} G$ is GCR, and that $x \in X$. Let $U := X \smallsetminus \overline{G \cdot x}$. Then $C_0(\overline{G \cdot x}) \rtimes_{\mathrm{lt}} G$ is isomorphic to the quotient of $C_0(X) \rtimes_{\mathrm{lt}} G$ by the ideal $C_0(U) \rtimes_{\mathrm{lt}} G$ (Corollary 3.20 on page 94). Therefore $C_0(\overline{G \cdot x}) \rtimes_{\mathrm{lt}} G$ is GCR as is its ideal $C_0(G \cdot x) \rtimes_{\mathrm{lt}} G$ ([110, Theorem 5.6.2]). As we saw in the preceding discussion, $C_0(G \cdot x) \rtimes_{\mathrm{lt}} G$ is Morita equivalent to $C^*(G_x)$. Therefore $C^*(G_x)$ is GCR by Proposition I.44 on page 507. □

[13] The case where (A, G, α) is not regular and A is not GCR is rather mysterious, and very few general criteria are known. Certainly sweeping characterizations seem well out of reach.

[14] Recall that a subset of a topological space is locally closed if it is open in its closure (Lemma 1.25 on page 6).

[15] In fact, by Theorem 4.30 on page 138, $C_0(G \cdot x) \rtimes_{\mathrm{lt}} G$ is isomorphic to $C^*(G_x) \otimes \mathcal{K}(L^2(G/G_x))$.

Notice that if instead of insisting that orbits are merely locally closed, we insist that orbits be closed, then in the above analysis, each irreducible representation L of $C_0(X) \rtimes_{\mathrm{lt}} G$ actually factors through some $C_0(G \cdot x) \rtimes_{\mathrm{lt}} G$ — rather than possibly being lifted from an essential ideal. Since Morita equivalence also preserves the property of being CCR (Proposition I.43 on page 507), we have the following analogue of Proposition 7.30 on the facing page for CCR transformation group C^*-algebras.

Proposition 7.31. *Suppose that (G, X) is a second countable transformation group and that each orbit is closed in X. Then $C_0(X) \rtimes_{\mathrm{lt}} G$ is CCR if and only if every G_x is CCR.*

We now expand our discussion to allow for possibly noncommutative coefficient algebras A. We still insist that A be GCR, and even then we will sketch some of the details. As above, we also require that (A, G, α) be a regular dynamical system and that each G-orbit is locally closed in $\operatorname{Prim} A$. It will be helpful if we review how closed, open and locally closed subsets of $\operatorname{Prim} A$ correspond to quotients, ideals and subquotients of A, respectively, and verify that these identifications are equivariant with respect to a strongly continuous action on A. It is well known, see [139, Proposition A.27] for example, that open subsets of $\operatorname{Prim} A$ are in one-to-one correspondence with ideals in A. If $U \subset \operatorname{Prim} A$ is open, then the corresponding ideal is

$$I_U = \ker(\operatorname{Prim} A \smallsetminus U) := \bigcap \{\, P \in \operatorname{Prim} A : P \notin U \,\}.$$

Furthermore, there is a homeomorphism

$$h_U : U \to \operatorname{Prim} I_U \quad \text{given by} \quad h_U(P) = P \cap I_U. \tag{7.32}$$

If V is also open in $\operatorname{Prim} A$, if $V \subset U$ and if $\pi : I_U \to I_U/I_V$ is the quotient map, then there is a homeomorphism

$$h_{U,V} : U \smallsetminus V \to \operatorname{Prim}\bigl(I_U/I_V\bigr) \quad \text{given by} \quad h_{U,V}(Q) = \pi(Q). \tag{7.33}$$

Often, we'll suppress the map π and write Q/I_V in place of $\pi(Q)$.

Lemma 7.32. *Suppose that (A, G, α) is a dynamical system and that U and V are G-invariant open subsets of* $\operatorname{Prim} A$ *with $V \subset U$. Then the maps h_U and $h_{U,V}$ defined in* (7.32) *and* (7.33) *are G-equivariant for the induced G-actions on* $\operatorname{Prim} A$, $\operatorname{Prim} I_U$ *and* $\operatorname{Prim} I_U/I_V$.

Proof. Note that I_U and I_V are G-invariant ideals so that α induces actions on I_U and I_U/I_V. If $s \in G$ and $P \in U \subset \operatorname{Prim} A$, then

$$\begin{aligned} s \cdot h_U(P) &= \alpha_s\bigl(h_U(P)\bigr) \\ &= \alpha_s\bigl(P \cap I_U\bigr) \\ &= \alpha_s(P) \cap I_U \\ &= h_U(s \cdot P). \end{aligned}$$

On the other hand, if $Q \in U \smallsetminus V$, then

$$\begin{aligned} s \cdot h_{U,V}(Q) &= \alpha_s^{I_V}\big(\pi(Q)\big) \\ &= \pi\big(\alpha_s(Q)\big) \\ &= \pi(s \cdot Q) \\ &= h_{U,V}(s \cdot Q). \end{aligned}$$

□

Recall that a subset $F \subset \operatorname{Prim} A$ is locally closed if F is open in its closure $\overline{F}$. Therefore $\overline{F} \smallsetminus F$ is closed and we can form the ideals corresponding the the complements of $\overline{F}$ and $\overline{F} \smallsetminus F$:

$$J := \ker(\overline{F}) = \bigcap\{\, P \in \operatorname{Prim} A : P \in \overline{F}\,\} \quad \text{and}$$
$$I := \ker(\overline{F} \smallsetminus F) = \bigcap\{\, P \in \operatorname{Prim} A : P \in \overline{F} \smallsetminus F\,\}.$$

Then the subquotient I/J has primitive ideal space naturally identified with F as in (7.33). It will be convenient to denote I/J by $A(F)$. For example, if $P \in \operatorname{Prim} A$ and $G\cdot P$ is locally closed, then we can let $A(G\cdot P)$ be the corresponding subquotient of A. Then in the above discussion, $J := \bigcap_{s\in G} P$, and $Q \mapsto Q/J$ is a G-equivariant homeomorphism of $G \cdot P$ onto $\operatorname{Prim}\big(A(G \cdot P)\big)$. Similarly, $A\big(\overline{G \cdot P}\big)$ is the quotient A/J, and $Q \mapsto Q/J$ is an equivariant homeomorphism of $\overline{G \cdot P} = \{\, Q \in \operatorname{Prim} A : Q \supset \bigcap_{s\in G} s \cdot P\,\}$ onto $\operatorname{Prim}\big(A(\overline{G \cdot P})\big)$.

With these conventions and notations in place, let $L = \pi \rtimes u$ be an irreducible representation of $A \rtimes_\alpha G$. By quasi-regularity, there is a $P \in \operatorname{Prim} A$ such that

$$\ker \pi = \bigcap_{s\in G} s \cdot P.$$

Note that $A\big(\overline{G \cdot P}\big) = A/\ker\pi$ is a GCR algebra and that $Q \mapsto Q/\ker\pi$ is a homeomorphism of $\overline{G \cdot P}$ onto $\operatorname{Prim}\big(A(\overline{G \cdot P})\big)$. By Lemma 6.16 on page 182 and Corollary 3.20 on page 94, L factors through a representation $\bar{L} = \bar{\pi} \rtimes u$ of $A\big(\overline{G \cdot P}\big) \rtimes_{\alpha^{\ker\pi}} G$ with $\ker\bar{\pi} = \{\,0\,\}$. By assumption $G \cdot P$ is open in $\overline{G \cdot P}$ so $A(G \cdot P)$ is an invariant ideal in $A\big(\overline{G \cdot P}\big)$. Thus $A(G \cdot P) = I/\ker\pi$ where

$$I := \bigcap\{\, P \in \operatorname{Prim} A : P \in \overline{G \cdot P} \smallsetminus G \cdot P\,\}.$$

Now we can view $A(G\cdot P) \rtimes_{\alpha^{\ker\pi}} G$ as an ideal in $A\big(\overline{G \cdot P}\big) \rtimes_{\alpha^{\ker\pi}} G$ (Proposition 3.19 on page 93). Since $\bar{\pi}$ is faithful, $\bar{L}$ does not vanish on $A(G \cdot P) \rtimes_{\alpha^{\ker\pi}} G$, and it follows that $\bar{L}$ is the canonical extension of an irreducible representation $\bar{L}'$ of $A(G \cdot P) \rtimes_{\alpha^{\ker\pi}} G$. Since $Q \mapsto Q/\ker\pi$ is an equivariant homeomorphism of $G \cdot P$ onto $\operatorname{Prim}\big(A(G \cdot P)\big)$, and since by assumption $sG_P \to s \cdot P$ is a homeomorphism of G/G_P onto $G \cdot P$, there is a continuous equivariant bijection of $\operatorname{Prim}\big(A(G \cdot P)\big)$ onto G/G_P matching up eG_P with $P/\ker\pi$. Since

$$\frac{A(G \cdot P)}{P/\ker\pi} = \frac{I/\ker\pi}{P/\ker\pi} \cong I/P,$$

Corollary 4.18 on page 130 implies that $A(G \cdot P) \rtimes_{\alpha^{\ker \pi}} G$ is Morita equivalent to $I/P \rtimes_{\alpha^P} G_P$. Since $G \cdot P$ is Hausdorff (it is homeomorphic to G/G_P), I/P is simple and must be the canonical elementary ideal $\mathcal{K}(P)$ in the primitive GCR algebra A/P. Theorem 7.21 on page 205 implies that $I/P \rtimes_{\alpha^P} G_P$ is Morita equivalent to $C^*(G'_P, \tau)$ where G'_P is an extension of G_P by $\mathbf{T}$ determined by the projective representation $\alpha : G_P \to \operatorname{Aut} \mathcal{K}(P)$.[16]

Now we can state analogues of Proposition 7.30 and Proposition 7.31. The proofs follow from the the above discussion and reasoning similar to the transformation group case. Proposition 7.33 is part of [162, Theorem 6.1].

Proposition 7.33. *Suppose that (A, G, α) is a separable dynamical system with A GCR, and $G \backslash \operatorname{Prim} A$ a T_0 topological space. For each $P \in \operatorname{Prim} A$, let $\mathcal{K}(P)$ be the canonical elementary ideal of A/P as described above, and let $C^*(G'_P, \tau)$ be the corresponding twisted group C^*-algebra associated to the projective representation of G_P on $\mathcal{K}(P)$ as above. Then $A \rtimes_\alpha G$ is GCR if and only if each $C^*(G'_P, \tau)$ is GCR.*

Proposition 7.34. *Suppose that (A, G, α) is a separable dynamical system such that A is CCR and such that the G-orbits in $\operatorname{Prim} A$ are closed. Let $C^*(G'_P, \tau)$ be as in Proposition 7.33. Then $A \rtimes_\alpha G$ is CCR if and only if each $C^*(G'_P, \tau)$ is CCR.*

Remark 7.35. As is clear from the discussion preceding Proposition 7.33, it would suffice to assume merely that A is GCR in Proposition 7.34. However, the hypotheses of Proposition 7.34 imply that each orbit is closed and homeomorphic to G/G_P. This implies that points are closed in $\operatorname{Prim} A$. Since A is GCR, this implies that A is actually CCR. Hence we have assumed A is CCR in Proposition 7.34 rather than GCR. Assuming the later gives an extra generality which could none-the-less be misleading.

Notes and Remarks

Theorem 7.1 is due to Takai [161]. The proof given here is adapted from Raeburn's proof in [133] and some comments of Echterhoff's. Reduced crossed products, and Theorem 7.13, were first introduced by Zeller-Meier for discrete groups in [173]. The extension to general crossed products is due to Takai [161]. Theorem 7.21 is from Green's [66] as is most of Section 7.4. Propositions 7.30 and 7.33 are due to Gootman [62] and Takesaki [162], while Propositions 7.31 and 7.34 are due to the author [169]. (The converses of these results will be considered in Section 8.3.)

[16] In Appendix D.3, we remark that G'_P is determined by a class $[\omega_P] \in H^2(G_P, \mathbf{T})$ called the Mackey obstruction for α at P. Furthermore, $C^*(G'_P, \tau) \cong C^*(G_P, \omega_P)$ whose representations correspond to ω_P-representations of G_P. There has been lots of work done on the structure of these twisted group algebras. For example, see [42].

Chapter 8

Ideal Structure

The theme of this chapter, and indeed of the book, is that the fundamental structure of a crossed product $A \rtimes_\alpha G$ is reflected in the structure of the orbit space for the G-action on $\operatorname{Prim} A$ together with the subsystems $(A, G_P, \alpha|_{G_P})$, where G_P is the stability group at $P \in \operatorname{Prim} A$. In Section 8.1, we see that this paradigm is particularly effective for regular (separable) systems. It is a remarkable and deep result that we can push the envelope to include arbitrary separable systems with G amenable. The principle difference is that we have to work with primitive ideals rather than irreducible representations. This result was conjectured by Effros and Hahn — hence systems for which the conjecture holds are known as EH-regular. The proof that the Effros-Hahn conjecture holds is the work of many hands, but the final step is due to Gootman and Rosenberg — building on work of Sauvageot. The proof of the GRS-Theorem (Theorem 8.21 on page 241) is rather lengthy and we devote all of Chapter 9 to it. In Section 8.2 we look at EH-regular systems and investigate some of the powerful implications of EH-regularity. In Section 8.3, we give a detailed discussion of the implications of the GRS-Theorem for transformation group C^*-algebras. In particular, we give a complete description of the primitive ideal space and its topology for the transformation group C^*-algebra $C_0(X) \rtimes_{\mathrm{lt}} G$ when G is abelian. We also give a fairly complete characterization of when $C_0(X) \rtimes_{\mathrm{lt}} G$ is GCR or CCR for abelian G which completes the preliminary discussion in Section 7.5. One of the key tools required in Section 8.3 is the Fell topology on the closed subgroups of a locally compact group G as developed in Appendix H. With this material in hand, it is straightforward to introduce the Fell subgroup crossed product in Section 8.4 which gives a coherent way of working with the subsystems $(A, H, \alpha|_H)$ as H varies. The Fell subgroup crossed product is used in the proof of the GRS-Theorem in Chapter 9.

8.1 Fibering Over the Orbit Space

Many C^*-algebras are naturally fibred over a base space, and as a consequence, a good deal of information about the algebra can be obtained from an understanding of the structure of the fibres and how the fibres are "glued" together. A straightforward example is $A = C\big(X, \mathcal{K}(\mathcal{H})\big)$ for a compact Hausdorff space X and a complex Hilbert space $\mathcal{H}$. Certainly A can be thought of as being fibred over X with constant fibre $\mathcal{K}(\mathcal{H})$. Since $\mathcal{K}(\mathcal{H})$ has a unique class of irreducible representation, it is not surprising to see that the spectrum of A can be identified with X as a topological space (cf. [139, Example A.24]). An essential ingredient in proving this is the existence of a nondegenerate homomorphism of $C(X)$ into the center $ZM(A)$ of the multiplier algebra $M(A)$ of A. It should be kept in mind that a clear picture of the spectrum or primitive ideal space, together with its topology, gives a complete description of the ideal structure of the algebra (cf. [139, Remark A.29]).

In general, a C^*-algebra is called a $C_0(X)$-algebra when there is a nondegenerate homomorphism Φ_A of $C_0(X)$ into the center $ZM(A)$ of the multiplier algebra (Definition C.1 on page 354). A summary of the properties of $C_0(X)$-algebras needed here is given in Appendix C. We'll just recall some of the important terminology and notation. In particular, A is a $C_0(X)$-algebra if and only if there is a continuous map $\sigma_A : \operatorname{Prim} A \to X$ (Proposition C.5 on page 355). If $\Psi_A : C^b(\operatorname{Prim} A) \to ZM(A)$ is the Dauns-Hofmann isomorphism (cf. [139, Theorem A.34]), then Φ_A and σ_A are related by

$$\Phi_A(\varphi) = \Psi_A(\varphi \circ \sigma_A). \tag{8.1}$$

The assertion that $C_0(X)$-algebras are fibred over X is justified by the result stating that A is a $C_0(X)$-algebra if and only if there is an upper semicontinuous C^*-bundle $p : \mathcal{A} \to X$ (Definition C.16 on page 360) such that A is $C_0(X)$-isomorphic to the algebra $\Gamma_0(\mathcal{A})$ of sections vanishing at infinity (Theorem C.26 on page 367). Thus elements in $a \in A$ are (up to isomorphism) functions on X with $a(x)$ taking values in the fibre algebra $\mathcal{A}_x := p^{-1}(x)$ of $\mathcal{A}$ over x. We can identify $\mathcal{A}_x$ with the quotient $A(x) := A/I_x$, where $I_x := \{\, \varphi \cdot a : a \in A \text{ and } \varphi(x) = 0 \,\} = \{\, a \in A : a(x) = 0 \,\}$ is the ideal in A of sections vanishing at x. An upper semicontinuous C^*-bundle is called simply a C^*-bundle (or a continuous C^*-bundle when we want to emphasize the difference) if the map $a \mapsto \|a\|$ is continuous (cf. Definition C.16 on page 360). Although C^*-bundles are more common in the literature than upper semicontinuous C^*-bundles, the properties of upper semicontinuous bundles will be quite sufficient for most of what we want to say here. In particular, the main property we want to invoke is that every irreducible representation must live on a fibre $A(x)$. Specifically, if ρ is an irreducible representation with kernel $P \in \operatorname{Prim} A$ and if $\sigma_A : \operatorname{Prim} A \to X$ is the continuous map satisfying (8.1), then $\sigma_A(P) = x$ if and only if $P \supset I_x$ so that ρ "factors through" $A(x) = A/I_x$ (Proposition C.5 on page 355). When working with $C_0(X)$-algebras, we will normally suppress the map Φ_A and write $\varphi \cdot a$ in place of $\Phi_A(\varphi)(a)$ and view A as a $C_0(X)$-bimodule such that

$$\varphi \cdot a = a \cdot \varphi, \quad (a \cdot \varphi)b = a(\varphi \cdot b) \quad \text{and} \quad (\varphi \cdot a)^* = \bar{\varphi} \cdot a^*.$$

Throughout this section it will be important to keep in mind that if (A, G, α) is a dynamical system, then both the spectrum $\hat{A}$ and the primitive ideal space $\operatorname{Prim} A$ are (not necessarily Hausdorff) locally compact G-spaces (Lemma 2.8 on page 44). If $[\pi]$ is the class of an irreducible representation π in $\hat{A}$, then $s \cdot [\pi] := [s \cdot \pi]$, where $s \cdot \pi := \pi \circ \alpha_s^{-1}$. If $P \in \operatorname{Prim} A$, then $s \cdot P := \alpha_s(P) = \{\, \alpha_s(a) : a \in P \,\}$.

Lemma 8.1. *Suppose that (A, G, α) is a dynamical system. Let A be a $C_0(X)$-algebra with associated map $\sigma : \operatorname{Prim} A \to X$. Then the following statements are equivalent.*

(a) *σ is G-invariant; that is, $\sigma(s \cdot P) = \sigma(P)$ for all $P \in \operatorname{Prim} A$ and $s \in G$.*
(b) *For all $\varphi \in C_0(X)$, $s \in G$ and $a \in A$, $\alpha_s(\varphi \cdot a) = \varphi \cdot \alpha_s(a)$.*
(c) *Each ideal $I_x \subset A$ is α-invariant.*

Proof. Let $\Psi : C^b(\operatorname{Prim} A) \to ZM(A)$ be the Dauns-Hofmann isomorphism, and recall that

$$\varphi \cdot a = \Psi(\varphi \circ \sigma)(a) \quad \text{for all } \varphi \in C_0(X) \text{ and } a \in A.$$

Moreover, for all $f \in C^b(\operatorname{Prim} A)$, [139, Lemma 7.1] implies that

$$\alpha_s\big(\Psi(f)(a)\big) = \Psi\big(\tau_s(f)\big)\alpha_s(a),$$

where $\tau_s(f)(P) := f(s^{-1} \cdot P)$. If (a) holds, then

$$\begin{aligned} \alpha_s(\varphi \cdot a) &= \alpha_s\big(\Psi(\varphi \circ \sigma)(a)\big) \\ &= \Psi\big(\tau_s(\varphi \circ \sigma)\big)\alpha_s(a) \\ &= \Psi(\varphi \circ \sigma)\alpha_s(a) \\ &= \varphi \cdot \alpha_s(a). \end{aligned}$$

Therefore (a) $\Longrightarrow$ (b).

Since I_x is the closed linear span of elements of the form $\varphi \cdot a$ for $a \in A$ and $\varphi \in C_0(X)$ satisfying $\varphi(x) = 0$, we clearly have (b) $\Longrightarrow$ (c).

If (a) fails, then there is a $s_0 \in G$ and a $P_0 \in \operatorname{Prim} A$ such that $\sigma(s_0 \cdot P_0) \neq \sigma(P_0)$. Then there is a $f \in C_0(X)$ such that $f\big(\sigma(P_0)\big) = 1$ while $f\big(\sigma(s_0 \cdot P_0)\big) = 0$. Proposition C.5 on page 355 implies that prop-gen-arv2.2.1

$$\big\|a\big(\sigma(P)\big)\big\| = \sup_{\substack{\pi \in \hat{A} \\ \sigma(\ker \pi) = \sigma(P)}} \|\pi(a)\|.$$

Furthermore, the Dauns-Hofmann Theorem implies that for every $\pi \in \hat{A}$,

$$\pi\big(\Psi(f)(a)\big) = f(\ker \pi)\pi(a) \quad \text{for all } f \in C^b(\operatorname{Prim} A).$$

Thus if $\pi \in \hat{A}$ is such that $\ker \pi = s_0 \cdot P_0$ and a is such that $\pi\big(\alpha_{s_0}(a)\big) \neq 0$, then

$$\begin{aligned} \pi\big(\alpha_{s_0}(f \cdot a)\big) &= \pi\big(\alpha_{s_0}\big(\Psi(f \circ \sigma)(a)\big)\big) \\ &= \pi\big(\Psi\big(\tau_{s_0}(f \circ \sigma)\big)\alpha_{s_0}(a)\big) \\ &= f\big(\sigma(P_0)\big)\pi\big(\alpha_{s_0}(a)\big) \\ &= \pi\big(\alpha_{s_0}(a)\big) \neq 0 \end{aligned}$$

Therefore $\alpha_{s_0}(f \cdot a) \notin I_{\sigma(s_0 \cdot P_0)}$ while $f \cdot a \in I_{\sigma(s_0 \cdot P_0)}$. Thus

$$\alpha_{s_0}\big(I_{\sigma(s_0 \cdot P_0)}\big) \not\subset I_{\sigma(s_0 \cdot P_0)},$$

and (c) fails. Thus (c) $\Longrightarrow$ (a) and the proof is complete. □

Definition 8.2. If A is a $C_0(X)$-algebra, then a dynamical system (A, G, α) is said to be *$C_0(X)$-linear* if $\alpha_s(\varphi \cdot a) = \varphi \cdot \alpha_s(a)$ for all $\varphi \in C_0(X)$, $a \in A$ and $s \in G$.

Recall that we think of $M(A)$ as the collection $\mathcal{L}(A_A)$ of adjointable operators on A viewed as a right Hilbert module over itself. Thus A sits inside $M(A)$ as left multiplication operators: $L_a(b) = ab$.

Lemma 8.3. *Suppose that A is a C^*-algebra and $T \in M(A)$ is such that $T(ab) = aT(b)$ for all $a, b \in A$ (alternatively, $TL_a = L_aT$ for all $a \in A$). Then $T \in ZM(A)$.*

Proof. Let $S \in M(A)$ and $a, b \in A$. Then

$$\begin{aligned} ST(ab) &= S\big(aT(b)\big) \\ &= S(a)T(b) \\ &= T\big(S(a)b\big) \\ &= TS(ab). \end{aligned}$$

This suffices as A^2 is dense in A. □

Theorem 8.4. *Suppose that A is a $C_0(X)$-algebra and that (A, G, α) is a $C_0(X)$-linear dynamical system. Then for each $x \in X$ there are dynamical systems $\big(A(x), G, \alpha^x\big)$ where*

$$\alpha_s^x\big(a(x)\big) = \alpha_s(a)(x).$$

Furthermore, $A \rtimes_\alpha G$ is a $C_0(X)$-algebra such that

$$\varphi \cdot i_A(a) = i_A(\varphi \cdot a) \quad \text{for all } a \in A \text{ and } \varphi \in C_0(X), \tag{8.2}$$

and we have

$$(A \rtimes_\alpha G)(x) \cong A(x) \rtimes_{\alpha^x} G.$$

Proof. Since α is $C_0(X)$-linear, Lemma 8.1 on the previous page implies each I_x is α-invariant. Then we can form the quotient system $\big(A(x), G, \alpha^{I_x}\big)$ as described in Section 3.4. Thus we can let $\alpha^x = \alpha^{I_x}$. Then Proposition 3.19 on page 93 implies that $A(x) \rtimes_{\alpha^x} G$ is isomorphic to the quotient of $A \rtimes_\alpha G$ by $I_x \rtimes_\alpha G$. Let $\Phi_A : C_0(X) \to ZM(A)$ be the structure map $\Phi_A(\varphi)(a) := \varphi \cdot a$. Define $\Phi_{A \rtimes_\alpha G} : C_0(X) \to M(A \rtimes_\alpha G)$ by $\Phi_{A \rtimes_\alpha G}(\varphi) := \bar{\imath}_A\big(\Phi_A(\varphi)\big)$. Since Φ_A and i_A are nondegenerate, so is $\Phi_{A \rtimes_\alpha G}$. Thus to see that $A \rtimes_\alpha G$ is a $C_0(X)$-algebra satisfying (8.2), it suffices to see that $\Phi_{A \rtimes_\alpha G}\big(C_0(X)\big) \subset ZM(A \rtimes_\alpha G)$. Let $f \in C_c(G, A)$ and recall that Corollary 2.36 on page 57 implies

$$f = \int_G i_A\big(f(s)\big) i_G(s) \, d\mu(s). \tag{8.3}$$

It follows that $\Phi_{A \rtimes_\alpha G}(\varphi)(f) \in C_c(G, A)$ with

$$\Phi_{A \rtimes_\alpha G}(\varphi)(f)(s) = \varphi \cdot \big(f(s)\big),$$

and if $f, g \in C_c(G, A)$, then

$$\begin{aligned}\Phi_{A\rtimes_\alpha G}(\varphi)(f*g)(s) &= \Phi_A(\varphi)\int_G f(r)\alpha_r\big(g(r^{-1}s)\big)\,d\mu(r)\\ &= \int_G \varphi\cdot f(r)\alpha_r\big(g(r^{-1}s)\big)\,d\mu(r)\\ &= f*\Phi_{A\rtimes_\alpha G}(\varphi)(g)(s).\end{aligned}$$

Thus $\Phi_{A\rtimes_\alpha G}: C_0(X)\to ZM(A\rtimes_\alpha G)$ by Lemma 8.3 on the facing page.

Now $(A\rtimes_\alpha G)(x)$ is the quotient of $A\rtimes_\alpha G$ by the ideal K_x generated by elements of the form $\varphi\cdot f$ such that $f\in A\rtimes_\alpha G$ and $\varphi\in C_0(X)$ with $\varphi(x)=0$. We can restrict to f belonging to any subset of $A\rtimes_\alpha G$ which spans a dense subspace of $A\rtimes_\alpha G$. Letting $f\in C_c(G,A)$, we clearly have $K_x\subset \operatorname{Ex}(I_x)$. Now let f be of the form $i_A(a)i_G(z)$ with $a\in A$ and $z\in C_c(G)$. Thus, as an element of $C_c(G,A)$, $f=z\otimes a$ where $z\otimes a(s)=z(s)a$. Since $\varphi\cdot i_A(a)=i_A(\varphi\cdot a)$, and since elements of the form $z\otimes\varphi\cdot a$ are dense in $\operatorname{Ex}(I_x)=I_x\rtimes_\alpha G$ by Lemma 1.87 on page 29, it follows that $K_x=I_x\rtimes_\alpha G$. Thus $(A\rtimes_\alpha G)(x)\cong A(x)\rtimes_{\alpha^x} G$ as claimed. □

Remark 8.5. Condition (8.2) is equivalent to

$$(\varphi\cdot f)(s)=\varphi\cdot\big(f(s)\big)\quad\text{for all }\varphi\in C_0(X)\text{ and }f\in C_c(G,A).$$

Corollary 8.6. *Suppose that A is a $C_0(X)$-algebra and that (A,G,α) is a $C_0(X)$-linear dynamical system. Then $A\rtimes_\alpha G$ is the section algebra of an upper semicontinuous C^*-bundle $\mathcal{B}$ over X with fibres $\mathcal{B}_x$ isomorphic to $A(x)\rtimes_{\alpha^x} G$ (where α^x is as defined in Theorem 8.4 on the facing page). If in addition, A is the section algebra of a continuous C^*-bundle $\mathcal{A}$ over X and* if G is amenable, *then $\mathcal{B}$ is also a continuous C^*-bundle over X.*

Proof. The first assertion follows immediately from Theorem 8.4 on the preceding page and Theorem C.26 on page 367. The second assertion will follow once we show that the associated map $\sigma:\operatorname{Prim} A\rtimes_\alpha G\to X$ is open. Recall that $\sigma(P)=x$ if and only if $P\supset I_x\rtimes_\alpha G$. Let U be an open neighborhood in $\operatorname{Prim} A\rtimes_\alpha G$ and suppose that $x_n\to x$ in X with $x_n\notin\sigma(U)$ for all n. It will suffice to see that $x\notin\sigma(U)$. Since $x\in\overline{\{x_n\}}$ and since $\mathcal{A}$ is a C^*-bundle, $I_x\supset\bigcap_{n=1}^\infty I_{x_n}$. It follows from Proposition 5.23 on page 169 (see Remark 5.24) that if G is amenable, then $\operatorname{Ex}(I_y):=I_y\rtimes_\alpha G$ is equal to $\operatorname{Ind}_e^G I_y:=\operatorname{Ind} I_y$. Thus Lemma 5.19 on page 166 implies that

$$\bigcap_{n=1}^\infty (I_{x_n}\rtimes_\alpha G)=\bigcap_{n=1}^\infty \operatorname{Ind} I_{x_n}=\operatorname{Ind}\Big(\bigcap_{n=1}^\infty I_{x_n}\Big)\subset \operatorname{Ind} I_x=I_x\rtimes_\alpha G. \tag{8.4}$$

If we had $x\in\sigma(U)$, then there would be a $P\in U$ such that $P\supset I_x\rtimes_\alpha G$ and a $J\in\mathcal{I}(A\rtimes_\alpha G)$ such that

$$P\in\mathcal{O}_J=\{\,Q\in\operatorname{Prim} A\rtimes_\alpha G: Q\not\supset J\,\}\subset U.$$

Since $P \not\supset J$, it follows from (8.4) that there is a n such that

$$I_{x_n} \rtimes_\alpha G \not\supset J.$$

Then there is a $Q \in \operatorname{Prim} A \rtimes_\alpha G$ such that $Q \supset I_{x_n} \rtimes_\alpha G$ and $Q \not\supset J$. Thus $Q \in \mathcal{O}_J$ and $\sigma(Q) = x_n \in \sigma(U)$ which is a contradiction. □

Naturally, Theorem 8.4 on page 230 and Corollary 8.6 on the preceding page are most useful when we can say something specific about the fibres. We will show that we can often reduce to the case that the action of G on $\operatorname{Prim} A$ is transitive and then invoke Corollary 4.18 on page 130.

Proposition 8.7. *Suppose that (A, G, α) is a dynamical system and that $G\backslash\operatorname{Prim} A$ is Hausdorff. Then $A \rtimes_\alpha G$ is the section algebra of an upper semicontinuous C^*-bundle $\mathcal{B}$ over $G\backslash\operatorname{Prim} A$ with fibres $A(G \cdot P) \rtimes_{\alpha^{G\cdot P}} G$. If G is amenable, then $\mathcal{B}$ is a continuous C^*-bundle. If (A, G, α) is regular (Definition 6.22 on page 186), then each fibre $A(G \cdot P) \rtimes_{\alpha^{G\cdot P}} G$ is Morita equivalent to $A/P \rtimes_{\alpha^P} G_P$, where $G_P = \{\, s \in G : s \cdot P = P \,\}$ is the stability group at P.*

Proof. The natural map $\sigma : \operatorname{Prim} A \to G\backslash\operatorname{Prim} A$ is continuous and open. Since $G\backslash\operatorname{Prim} A$ is Hausdorff, Lemma 8.1 on page 229 and Theorem C.26 on page 367 imply that A is a $C_0(G\backslash\operatorname{Prim} A)$-algebra and that (A, G, α) is a $C_0(G\backslash\operatorname{Prim} A)$-linear dynamical system. Thus $A \rtimes_\alpha G$ is a $C_0(G\backslash\operatorname{Prim} A)$-algebra with fibres $A(G\cdot P) \rtimes_{\alpha^{G\cdot P}} G$ by Theorem 8.4 on page 230. Since $A(G\cdot P) = A/I_{G\cdot P}$, it follows from Lemma 7.32 on page 223 that the map

$$Q \mapsto Q/I_{G\cdot P}$$

is a G-equivariant homeomorphism of $G \cdot P \subset \operatorname{Prim} A$ onto $\operatorname{Prim} A(G \cdot P)$. Since (A, G, α) is regular, there is a G-invariant homeomorphism of $\operatorname{Prim} A(G \cdot P)$ onto G/G_P taking $Q := P/I_{G\cdot P}$ to eG_P. Since

$$A(G \cdot P)/Q = (A/I_{G\cdot P})/(P/I_{G\cdot P})$$

is equivariantly isomorphic to A/P (with respect to the induced G_P-actions), Corollary 4.18 on page 130 implies that $A(G \cdot P) \rtimes_{\alpha^{G\cdot P}} G$ is Morita equivalent to $A/P \rtimes_{\alpha^P} G_P$. The final assertion follows from Corollary 8.6 on the preceding page. □

Notice that if A is commutative, then each primitive quotient A/P is isomorphic to the complex numbers and each fibre is actually Morita equivalent to the group C^*-algebra $C^*(G_P)$ of a stability group. Since the irreducible representations of a $C_0(X)$-algebra come from the fibres, it follows that in set-ups like that of Proposition 8.7, the irreducible representations are determined by the irreducible representations of the stability groups. We want to look at this process in greater detail.

We get representations of $A \rtimes_\alpha G$ from representations associated to the stability groups via induction. Fix $P \in \operatorname{Prim} A$. Rather than start with an arbitrary representation L of $A \rtimes_{\alpha|_{G_P}} G_P$, we want to assume that

$$\operatorname{Res}(\ker L) \supset P. \tag{8.5}$$

Then if $L = \pi \rtimes W$, this implies $\ker \pi \supset P$ and that L factors through $A/P \rtimes_{\alpha^P} G_P$. That is, $L = L' \circ q \rtimes \mathrm{id}$ where $q : A \to A/P$ is the natural map, $q \rtimes \mathrm{id} : A \rtimes_{\alpha|_{G_P}} G_P \to A/P \rtimes_{\alpha^P} G_P$ is the induced map (Proposition 3.19 on page 93) and L' is a representation of $A/P \rtimes_{\alpha^P} G_P$. Notice that if A is abelian, then A/P is the complex numbers, so that the representations satisfying (8.5) correspond exactly to representations of the stability group G_P. In any event, we can form

$$\mathrm{Ind}_{G_P}^G(L)$$

as defined in Definition 5.1 on page 152, which is the representation on $\mathsf{X} \otimes_{A \rtimes G_P} \mathcal{H}_L$ given by

$$\mathrm{Ind}_{G_P}^G(L)(f)(g \otimes h) = N \rtimes v(f)(g) \otimes h = f * g \otimes h.$$

We can also induce L' to a representation of $\mathrm{Ind}_{G_P}^G(A/P, \alpha^P) \rtimes_{\mathrm{lt}} G$ via the imprimitivity bimodule Z of Corollary 4.17 on page 129. If (A, G, α) is regular and $G\backslash \mathrm{Prim}\, A$ is Hausdorff, then we get a representation $R_{G_P}^G(L)$ of $A \rtimes_\alpha G$ by composing this induced representation with the quotient map $\bar{q} \rtimes \mathrm{id}$ from $A \rtimes_\alpha G$ to $A(G\cdot P) \rtimes_{\alpha^{G\cdot P}} G$ with the isomorphism $\Phi \rtimes \mathrm{id} : A(G\cdot P) \rtimes_{\alpha^{G\cdot P}} G \to \mathrm{Ind}_{G_P}^G(A/P, \alpha^P) \rtimes_{\mathrm{lt}} G$ induced by the covariant isomorphism $\Phi : A(G\cdot P) \to \mathrm{Ind}_{G_P}^G(A/P, \alpha^P)$ of Proposition 3.53 on page 105 (where we have identified $A(G\cdot P)/I$ with A/P as above). Thus $R_{G_P}^G(L)$ acts on $\mathsf{Z} \otimes_{A/P \rtimes_{\alpha^P} G_P} \mathcal{H}_{L'}$ by

$$R_{G_P}^G(L)(f)(z \otimes h) = (\Phi \circ \bar{q}) \rtimes \mathrm{id}(f) \cdot z \otimes h, \tag{8.6}$$

where $(\Phi \circ \bar{q}) \rtimes \mathrm{id}(f) \cdot z$ is given by (4.46). A little untangling shows that $\Phi \circ \bar{q}(a)(r) = q\big(\alpha_r^{-1}(a)\big)$, where $q : A \to A/P$ is the natural map. Note that $R_{G_P}^G(L)$ is essentially the representation of $A(G\cdot P) \rtimes_{\alpha^{G\cdot P}} G$ corresponding to L' under the Morita equivalence of Corollary 4.18 on page 130.

Lemma 8.8. *Suppose that (A, G, α) is a regular dynamical system with $G\backslash \mathrm{Prim}\, A$ Hausdorff. Let $P \in \mathrm{Prim}\, A$. If L is a nondegenerate representation of $A \rtimes_{\alpha|_{G_P}} G_P$ with* $\mathrm{Res}(\ker L) \supset P$. *Then $\mathrm{Ind}_{G_P}^G L$ and $R_{G_P}^G L$ are equivalent representations of $A \rtimes_\alpha G$.*

Proof. Let $\mathsf{X}_0 = C_c(G, A)$ and $\mathsf{Z}_0 = C_c(G, A/P)$ be dense subspaces of the imprimitivity bimodules X and Z, respectively. Define

$$u : \mathsf{X}_0 \to \mathsf{Z}_0 \quad \text{by} \quad u(f)(s) = \Delta_G(s)^{\frac{1}{2}} q\big(\alpha_s^{-1}\big(f(s)\big)\big),$$

where $q : A \to A/P$ is the natural map. Define $U : \mathsf{X}_0 \odot \mathcal{H}_L \to \mathsf{Z}_0 \odot \mathcal{H}_L$ by $U(f \otimes h) = u(f) \otimes h$. To keep the notation from becoming distracting, let $B_1 := A/P \rtimes_{\alpha^P} G_P$ and $B_2 := A \rtimes_{\alpha|_{G_P}} G_P$. Let $L = \pi \rtimes W = L' \circ q \rtimes \mathrm{id}$ with $L' = \pi' \rtimes W$

and $\pi = \pi' \circ q$. Therefore

$$\begin{aligned}
\big(U(f\otimes h) \mid U(g\otimes k)\big)_{\mathsf{Z}\otimes_{B_1}\mathcal{H}_L} &= \big(L'(\langle u(g)\,,\,u(f)\rangle_{_{B_1}})h \mid k\big) \\
&= \int_{G_P} \big(\pi'(\langle u(g)\,,\,u(f)\rangle_{_{B_1}}(s))W_s(h) \mid k\big)\,d\mu_{G_P}(s) \\
&= \int_{G_P}\int_G \gamma(s)\Delta_G(t^{-1})\big(\pi'\big(q\big(\alpha_t\big(g(t^{-1})\big)\big)\big)^* \\
&\qquad \alpha^P_s\big(q\big(\alpha_{s^{-1}t}\big(f(t^{-1}s)\big)\big)\big)\big)W_s(h) \mid k\big)\,d\mu_G(t)\,d\mu_{G_P}(s)
\end{aligned}$$

which, after sending $t \mapsto t^{-1}$ and recalling that $\pi = \pi' \circ q$ while $\alpha^P_s \circ q = q \circ \alpha_s$, is

$$\begin{aligned}
&= \int_{G_P}\Big(\pi\Big(\gamma(s)\int_G \alpha_t^{-1}\big(g(t)^*f(ts)\big)\,d\mu_G(t)\Big)W_s(h) \mid k\Big)\,d\mu_{G_P}(s) \\
&= \int_{G_P}\big(\pi(\langle g\,,\,f\rangle_{_{B_2}}(s))W_s(h) \mid k\big)\,d\mu_{G_P}(s) \\
&= \big(f\otimes h \mid g\otimes k\big)_{\mathsf{X}\otimes_{B_2}\mathcal{H}_L}.
\end{aligned}$$

Since $1\otimes q : C_c(G,A) \to C_c(G,A/P)$ is surjective by Lemma 3.18 on page 93, so is u. It follows that U extends to a unitary from $\mathsf{X}\otimes_{B_2}\mathcal{H}_L$ onto $\mathsf{Z}\otimes_{B_1}\mathcal{H}_L$.

Now we compute that

$$\begin{aligned}
(\Phi\circ\bar q)\rtimes \mathrm{id}(f)\cdot z(r) &= \int_G (\Phi\circ\bar q)\rtimes\mathrm{id}(f)(t,r)z(t^{-1}r)\Delta_G(t)^{\frac12}\,d\mu_G(t) \\
&= \int_G q(\alpha_r^{-1}(f(t)))z(t^{-1}r)\Delta_G(t)^{\frac12}\,d\mu_G(t).
\end{aligned}$$

While

$$\begin{aligned}
&u\big(N\rtimes v(f)(g)\big)(r) = u(f*g)(r) \\
&\quad = \Delta_G(r)^{\frac12}q\big(\alpha_r^{-1}(f*g(r))\big) \\
&\quad = \Delta_G(r)^{\frac12}q\circ\alpha_r^{-1}\Big(\int_G f(t)\alpha_t\big(g(t^{-1}r)\big)\,d\mu_G(t)\Big) \\
&\quad = \int_G q\big(\alpha_r^{-1}(f(t))\big)\Delta_G(r)^{\frac12}q\big(\alpha_{r^{-1}t}\big(g(t^{-1}r)\big)\big)\,d\mu_G(t) \\
&\quad = \int_G q\big(\alpha_r^{-1}(f(t))\big)\Delta_G(t^{-1}r)^{\frac12}q\big(\alpha^{-1}_{t^{-1}r}\big(g(t^{-1}r)\big)\big)\Delta_G(t)^{\frac12}\,d\mu_G(t) \\
&\quad = (\Phi\circ\bar q)\rtimes\mathrm{id}(f)\cdot u(g)(r).
\end{aligned}$$

Thus U intertwines $\operatorname{Ind}_{G_P}^G L$ and $R^G_{G_P}L$. □

We can summarize much of the preceding as follows.

Corollary 8.9. *Suppose that (A,G,α) is a regular dynamical system with $G\backslash\operatorname{Prim} A$ Hausdorff. Then every irreducible representation of $A\rtimes_\alpha G$ is equivalent to one of the form $\operatorname{Ind}_{G_P}^G L$ where L is an irreducible representation of $A\rtimes_{\alpha|_{G_P}} G_P$ with $\operatorname{Res}(\ker L)\supset P$.*

Proof. Proposition 8.7 on page 232 implies that $A \rtimes_\alpha G$ is the section algebra of an upper semicontinuous C^*-bundle with fibres $A(G \cdot P) \rtimes_{\alpha^{G \cdot P}} G$ Morita equivalent to $A/P \rtimes_{\alpha^P} G_P$. Therefore every irreducible representation is lifted from a fibre (cf., Proposition C.5 on page 355 and Theorem C.26 on page 367). The Morita equivalence of $A(G \cdot P) \rtimes_{\alpha^{G \cdot P}} G$ with $A/P \rtimes_{\alpha^P} G_P$ implies that any irreducible representation of $A \rtimes_\alpha G$ must be form $R^G_{G_P} L$ as in (8.6). The result follows from Lemma 8.8 on page 233. □

The previous lemma and the preceding discussion motivates the following terminology.

Definition 8.10. Suppose that (A, G, α) is a dynamical system. A representation R of $A \rtimes_\alpha G$ is *induced from a stability group* if there is a primitive ideal $P \in \operatorname{Prim} A$ and a representation L of $A \rtimes_{\alpha|_{G_P}} G_P$ satisfying $\operatorname{Res}(\ker L) \supset P$ such that $\operatorname{Ind}^G_{G_P} L$ is equivalent to R. We say that a primitive ideal $J \in \operatorname{Prim}(A \rtimes_\alpha G)$ is *induced from a stability group* if there is a $P \in \operatorname{Prim} A$ and a $K \in \operatorname{Prim}(A \rtimes_{\alpha|_{G_P}} G_P)$ such that $\operatorname{Res} K = P$ and $\operatorname{Ind}^G_{G_P} K = J$.

We want to generalize Corollary 8.9 on the facing page to deal with regular crossed products where the orbit space $G\backslash \operatorname{Prim} A$ is only *almost* Hausdorff. Although the extension to this case is actually fairly straightforward, the argument is obscured by a number of unfortunate technicalities. Nevertheless, the result, and especially some of its special cases, make it worth expending the effort to sort through the constructions. The first tool we have to recall is the notion of a composition series in a C^*-algebra A.[1] Although the idea is rather simple, the general statement is a bit much to take at one sitting. Since the finite case is what usually occurs in practice, we can start there. Let

$$I_1 \subsetneq I_2 \subsetneq \cdots \subsetneq I_{n-1}$$

be nested proper ideals in a C^*-algebra A with $I_0 = \{0\}$ and $I_n = A$. Then we call $\{I_k\}_{k=0}^n$ a finite composition series for A. The philosophy is that the subquotients

$$I_{k+1}/I_k \quad \text{for } k = 0, 1, \ldots, n-1$$

carry lots of information about A. For example, if $n = 2$, then A is simply an extension of $I = I_1$ by A/I. Notice that if $I \subsetneq J$ are ideals in A and if ρ is an irreducible representation of the subquotient J/I and if $q : J \to J/I$ is the quotient map, then the canonical extension π of $\rho \circ q$ to A is an irreducible representation of A which is said to *live on* J/I. Now if π is any irreducible representation of A and if $\{I_k\}$ is a finite composition series, then there is a $k < n$ such that $\pi(I_k) = \{0\}$ and $\pi(I_{k+1}) \neq \{0\}$. That is, π must live on some subquotient I_{k+1}/I_k and we can recover the spectrum $\hat{A}$ of A from the spectra of the subquotients.

In what follows, it will be necessary to look at possibly infinite composition series. That is, we want to allow for a possibly infinite increasing family of ideals

[1] Composition series usually arise in the classification of GCR or type I C^*-algebras. For example, see Theorems 6.2.6 and 6.8.7 in [126], or [28, Proposition 4.3.4]

inside our algebra A. It will be necessary to index this family a bit more subtly than by nonnegative integers. In fancy terms, our index set Λ will be a segment $0 \le \alpha \le \alpha_0$ of *ordinals.*[2]

Definition 8.11. A *composition series* in a C^*-algebra A is a family $\{ I_\alpha \}_{\alpha\in\Lambda}$ if ideals $I_\alpha \in \mathcal{I}(A)$ indexed by a segment Λ of ordinals $0 \le \alpha \le \alpha_0$ such that

(a) $I_0 = \{0\}$ and $I_{\alpha_0} = A$,
(b) $\alpha < \gamma$ implies $I_\alpha \subsetneq I_\gamma$, and if
(c) γ is a limit ordinal, then

$$I_\gamma = \overline{\bigcup_{\alpha<\gamma} I_\alpha}.$$

We say that the composition series if finite or countable, respectively, if α_0 is finite or countable, respectively.

Remark 8.12. Although any composition series that arises in practice is likely to be finite, the general theory requires that we put up with the possibility of infinite ones. Fortunately, if A is separable, then any composition series $\{ I_\alpha \}_{\alpha\in\Lambda}$ for A can be at most countable. To see this, note that if $\alpha < \alpha_0$, then $J_{\alpha+1}/J_\alpha \ne \{0\}$ and there is a $a_\alpha \in J_{\alpha+1}$ such that $\|a_{\alpha+1} - a\| \ge 1$ for all $a \in J_\alpha$. Since $\beta \ne \alpha$ implies either $\alpha < \beta$ or $\beta < \alpha$, we must have $\|a_\alpha - a_\beta\| \ge 1$. Therefore $\{ a_\alpha \}$ must be countable, and α_0 must be countable. (However, we can still have limit ordinals $\lambda < \alpha_0$.)

Lemma 8.13. *Suppose that* $\{ I_\alpha \}_{\alpha\in\Lambda}$ *is a composition series for a* C^**-algebra* A*. Then every irreducible representation of* A *lives on a subquotient* $I_{\alpha+1}/I_\alpha$ *for some* $\alpha \in \Lambda$*.*

Proof. Let π be an irreducible representation of A. Let S be the set of $\alpha \in \Lambda$ such that $\pi(I_\alpha) \ne \{0\}$. If $\beta = \min S$ is a limit ordinal, then it follows from part (c) of Definition 8.11 that $\pi(I_\beta) = \{0\}$. But this contradicts that fact that $\beta \in S$. Thus β has an immediate predecessor α and π clearly lives on $I_{\alpha+1}/I_\alpha$. □

A subset $F \subset \operatorname{Prim} A$ is locally closed if F is open in its closure $\overline{F}$ (Lemma 1.25 on page 6). Therefore $\overline{F} \setminus F$ is closed and as in Section 7.5 we can form the ideals corresponding to the complements of $\overline{F}$ and $\overline{F} \setminus F$:

$$J := \ker(\overline{F}) = \bigcap \{ P \in \operatorname{Prim} A : P \in \overline{F} \} \quad \text{and}$$
$$I := \ker(\overline{F} \setminus F) = \bigcap \{ P \in \operatorname{Prim} A : P \in \overline{F} \setminus F \}.$$

Then the subquotient I/J has primitive ideal space naturally identified with F as in (7.33) on page 223. As in Section 7.5, it will be convenient to denote I/J by $A(F)$. Note that if $G\backslash\operatorname{Prim} A$ is Hausdorff — as in Corollary 8.9 on page 234, for example — then we can think of $G \cdot P$ as either an element of $G\backslash\operatorname{Prim} A$ or as a

[2]There is a brief description of the properties of ordinals needed here preceding Lemma 6.3 on page 173.

closed subset of $\operatorname{Prim} A$. Thus in the first case, $A(G\cdot P)$ is the fibre of the associated upper semicontinuous C^*-bundle, and in the second, $A(G\cdot P)$ is the quotient of A by the ideal $I_{G\cdot P}$ corresponding to the complement of $G\cdot P$ in $\operatorname{Prim} A$. Thus both interpretations for $A(G\cdot P)$ coincide in this case, so our notation should not be misleading.

More generally, if $G\cdot P$ is locally closed as a subset of $\operatorname{Prim} A$ — or equivalently, as a point in $G\backslash\operatorname{Prim} A$ — then $A(G\cdot P)$ is the quotient by a G-invariant ideal and admits a strongly continuous G-action which we denote by $\alpha^{G\cdot P}$. Let Q_P be the primitive ideal in $A(G\cdot P)$ corresponding to P under the identification of $\operatorname{Prim} A(G\cdot P)$ with $G\cdot P$. Then $A(G\cdot P)/Q_P$ corresponds to an ideal in A/P which can be proper when A/P is not simple.

Example 8.14. Suppose that $A = C_0(X)$ and that $F\subset X$ is locally closed. Then using the notation above, $J = \ker(\overline{F})$ is the ideal of functions vanishing on $\overline{F}$ and $I = \ker(\overline{F}\setminus F)$ is the ideal of functions vanishing on $\overline{F}\setminus F$. The map $g\mapsto g|_F$ is a surjection of I onto $C_0(F)$ with kernel J. Thus $A(F) = I/J$ is naturally identified with $C_0(F)$ in this case.

Remark 8.15. Suppose the C and F are locally closed subsets of $\operatorname{Prim} A$ with $F\subset C$. Then $A(C)(F)\cong A(F)$. To see this, suppose that $A(C) = I/J$. Then the image F' of F in $\operatorname{Prim} A(C)\cong C$ is a locally closed subset and

$$A(C)(F) = I'/J'$$

where I' and J' are ideals in $A(C)$ given by

$$J' = \bigcap\{\, Q\in\operatorname{Prim} A(C) : Q\in\overline{F'}\,\}\quad\text{and}$$
$$I' = \bigcap\{\, Q\in\operatorname{Prim} A(C) : Q\in\overline{F'}\setminus F'\,\}.$$

Thus $I' = I''/J$ and $J' = J''/J$ where $A(F) = I''/J''$. But then

$$A(C)(F) = I'/J' = \frac{I''/J}{J''/J}\cong I''/J'' = A(F).$$

With the terminology and notation developed above, we can now state a basic decomposition result, and also verify that in the regular case, all irreducible representations of $A\rtimes_\alpha G$ are induced from stability groups.

Theorem 8.16. *Suppose that (A,G,α) is a dynamical system with $G\backslash\operatorname{Prim} A$ almost Hausdorff. Then each orbit is locally closed and $A\rtimes_\alpha G$ has a composition series $\{\, I_\lambda\,\}_{\lambda\in\Lambda}$ such that each subquotient $I_{\lambda+1}/I_\lambda$ is the section algebra of an upper semicontinuous C^*-bundle with fibres equal to $A(G\cdot P)\rtimes_{\alpha^{G\cdot P}} G$ for $G\cdot P\in G\backslash\operatorname{Prim} A$. If, in addition, (A,G,α) is regular, then let Q_P be the primitive ideal in $A(G\cdot P)$ corresponding to P under the identification of $\operatorname{Prim} A(G\cdot P)$ with $G\cdot P$. Then $A(G\cdot P)\rtimes_{\alpha^{G\cdot P}} G$ is Morita equivalent to $A(G\cdot P)/Q_P\rtimes_{\alpha^{Q_P}} G_P$, and every irreducible representation of $A\rtimes_\alpha G$ is induced from a stability group.*

Proof. Since $G\backslash\operatorname{Prim} A$ is almost Hausdorff, part (d) of Lemma 6.3 on page 173 implies that there are open sets $\{\, U_\lambda\,\}_{\lambda\in\Lambda}$ satisfying properties (i)–(iv) on page 173.

It follows that points in $G\backslash\operatorname{Prim} A$ are locally closed. Thus each orbit is locally closed and the subquotients $A(G \cdot P)$ are defined for each P. Let $q : \operatorname{Prim} A \to G\backslash\operatorname{Prim} A$ be the orbit map, and let J_λ be the ideal in A corresponding to $q^{-1}(U_\lambda) \subset \operatorname{Prim} A$. Since $q^{-1}(U_\lambda)$ is G-invariant, so is J_λ and we can form

$$I_\lambda := J_\lambda \rtimes_\alpha G.$$

Clearly, $\{ I_\lambda \}$ satisfies axioms (a) & (b) of Definition 8.11 on page 236. Thus we need to see that

$$\bigcup_{\lambda<\beta} J_\lambda \rtimes_\alpha G \quad \text{is dense in} \quad J_\beta \rtimes_\alpha G \tag{8.7}$$

for all limit ordinals $0 < \beta \le \alpha_0$. Since the J_λ are totally ordered by inclusion,

$$\overline{\bigcup_{\lambda<\beta} J_\lambda}$$

is a G-invariant ideal in A. Since a primitive ideal P contains J_λ if and only if $P \notin q^{-1}(U_\lambda)$ and since

$$q^{-1}(U_\beta) = \bigcup_{\lambda<\beta} q^{-1}(U_\lambda),$$

it follows that

$$J_\beta = \overline{\bigcup_{\lambda<\beta} J_\lambda}.$$

Thus $\bigcup J_\lambda$ is dense in J_β, and

$$\operatorname{span}\{ z \otimes a \in C_c(G, J_\beta) : a \in \bigcup J_\lambda \text{ and } z \in C_c(G) \} \tag{8.8}$$

is dense in $J_\beta \rtimes_\alpha G$ by Lemma 1.87 on page 29. But (8.8) is contained in the left-hand side of (8.7). This proves that $\{ I_\lambda \}$ is a composition series as claimed.

Clearly, J_λ is a G-invariant ideal in $J_{\lambda+1}$, and Proposition 3.19 on page 93 implies that

$$I_{\lambda+1}/I_\lambda = \frac{J_{\lambda+1} \rtimes_\alpha G}{J_\lambda \rtimes_\alpha G} \cong (J_{\lambda+1}/J_\lambda) \rtimes_{\alpha^{J_\lambda}} G. \tag{8.9}$$

We can identify $\operatorname{Prim} J_{\lambda+1}/J_\lambda$ with $q^{-1}(U_{\lambda+1} \smallsetminus U_\lambda)$ and $G\backslash\operatorname{Prim}(J_{\lambda+1}/J_\lambda)$ with the Hausdorff space $U_{\lambda+1} \smallsetminus U_\lambda$ (using Lemma 7.32 on page 223). Thus Proposition 8.7 on page 232 implies that (8.9) is the section algebra of an upper semicontinuous C^*-bundle over $U_{\lambda+1} \smallsetminus U_\lambda$ with fibres

$$(I_{\lambda+1}/I_\lambda)(G \cdot P) \cong A(G \cdot P) \rtimes_{\alpha^{G\cdot P}} G \quad \text{for } G \cdot P \in U_{\lambda+1} \smallsetminus U_\lambda$$

(see Remark 8.15 on the previous page).

Now suppose that (A, G, α) is regular, and that ρ is a irreducible representation of $A \rtimes_\alpha G$. Then Lemma 8.13 on page 236 implies that ρ lives on some $I_{\lambda+1}/I_\lambda$. Thus ρ is the natural extension $\overline{R'}$ for an irreducible representation R' on $I_{\lambda+1}$ lifted from an irreducible representation R of $I_{\lambda+1}/I_\lambda$. If (A, G, α) is regular, then so is $(J_{\lambda+1}/J_\lambda) \rtimes_{\alpha^{J_\lambda}} G$ (the orbit space is Hausdorff, so quasi-regularity follows). As in Proposition 8.7 on page 232 and Corollary 8.9 on page 234, $A(G\cdot P) \rtimes_{\alpha^{G\cdot P}} G$ is Morita equivalent to $A(G\cdot P)/Q_P \rtimes_{\alpha^{Q_P}} G_P$, and R must be equivalent to

$$\mathsf{X}'\text{–}\operatorname{Ind}_{G_K}^{G} L$$

where $K \in \operatorname{Prim} J_{\lambda+1}/J_\lambda$, L is an irreducible representation of $(J_{\lambda+1}/J_\lambda) \rtimes_{\alpha^{J_\lambda}} G_K$ with $\operatorname{Res}(\ker L) \supset K$. Here, X' is Green's imprimitivity bimodule between $E^G_{G_K}(J_{\lambda+1}/J_\lambda)$ and $(J_{\lambda+1}/J_\lambda) \rtimes_{\alpha^{J_\lambda}} G_K$.

Of course, K corresponds to a $Q \in \operatorname{Prim} J_{\lambda+1}$ and Lemma 7.32 on page 223 implies G_K coincides with the stability group G_Q for the G-action on $\operatorname{Prim} J_{\lambda+1} \cong q^{-1}(U_{\lambda+1})$. If L' is the lift of L to $J_{\lambda+1} \rtimes_\alpha G_Q$, then $\operatorname{Res}(\ker L') \supset Q$ and Proposition 5.23 on page 169 implies that R' is equivalent to $\mathsf{X}''\text{–}\operatorname{Ind}_{G_Q}^G L'$ where X'' is Green's imprimitivity bimodule corresponding to $E^G_{G_Q}(J_{\lambda+1})$. Furthermore, Q corresponds to a $P \in q^{-1}(U_{\lambda+1} \setminus U_\lambda) \subset \operatorname{Prim} A$ such that $P \cap J_{\lambda+1} = Q$. Again, Lemma 7.32 implies that G_P and G_Q coincide and Proposition 5.21 on page 167 implies that $\overline{R'}$ is equivalent to

$$\mathsf{X}\text{–}\operatorname{Ind}_{G_P}^{G} \overline{L'}$$

where $\overline{L'}$ is the extension of L' to $A \rtimes_\alpha G_P$, and X is Green's imprimitivity bimodule corresponding to $E^G_{G_P}(A)$. Since $\operatorname{Res}(\ker \overline{L'}) \supset P$, we're done. □

Remark 8.17. The statement of Theorem 8.16 on page 237 simplifies significantly with some additional assumptions. The most significant is that we want A be type I or GCR. Secondly, we want A to be separable and G second countable. The type I hypothesis allows us to identify $\operatorname{Prim} A$ with $\hat{A}$ [110, Theorem 5.6.4]. Thus if (A, G, α) is regular and $G\backslash\operatorname{Prim} A$ is almost Hausdorff, then $A(G\cdot P)$ is type I with Hausdorff spectrum and each primitive quotient $A(G\cdot P)/Q_P$ is an elementary C^*-algebra. Thus each fibre $A(G\cdot P) \rtimes_{\alpha^{G\cdot P}} G$ is actually Morita equivalent to a crossed product of the form $\mathcal{K}(\mathcal{H}) \rtimes G_P$ for some complex Hilbert space $\mathcal{H}$. It can be shown that the irreducible representations of $\mathcal{K}(\mathcal{H}) \rtimes G_P$ correspond exactly to certain irreducible projective representations of G_P (cf. Section 7.3). Furthermore, since every quotient of A is also type I, every closed subset of $\operatorname{Prim} A$ has an open dense Hausdorff subset (cf., [126, Theorem 6.2.11] or [28, Theorem 4.4.5]). Thus $\operatorname{Prim} A$ is almost Hausdorff and we can apply the Mackey-Glimm dichotomy (Theorem 6.2 on page 173) to $(G, \operatorname{Prim} A)$. Thus (A, G, α) is regular with $G\backslash\operatorname{Prim} A$ almost Hausdorff if and only if $G\backslash\operatorname{Prim} A$ is a T_0 topological space (or any of the other conditions (a)–(e) of Theorem 6.2 on page 173 are satisfied). See [171] for more details.

8.2 EH-Regularity and the GRS-Theorem

In the previous section, we showed that in the case of separable, regular dynamical systems, every irreducible representation is induced from a stability group (Theorem 8.16 on page 237). Of course this means that every primitive ideal is induced, and we are well on our way to one of our main goals: describing the ideal structure of a crossed product. However, in the non-regular case it is reasonable to expect non type I behavior. Therefore, we do not expect to be able to describe all irreducible representations in a coherent manner. However, it still may be possible to describe the primitive ideals. Consequently, we make the following definition. (The terminology will be justified shortly.)

Definition 8.18. A dynamical system (A, G, α) is EH-*regular* if given $K \in \operatorname{Prim}(A \rtimes_\alpha G)$, then there is a $P \in \operatorname{Prim} A$ and a $J \in \operatorname{Prim}(A \rtimes_\alpha G_P)$ such that $\operatorname{Res} J = P$ and such that $\operatorname{Ind}_{G_P}^G J = K$. That is, (A, G, α) is EH-regular if every primitive ideal in $A \rtimes_\alpha G$ is induced from a stability group.

Of course, Theorem 8.16 on page 237 shows that if the G action on $\operatorname{Prim} A$ is nice — in particular, if (A, G, α) is a separable regular dynamical system — then (A, G, α) is EH-regular. A few examples should suffice to show that EH-regular systems are the ones were "Mackey-Machine"-type methods are likely to uncover significant information about the structure of the corresponding crossed products. The following observation is due to Echterhoff.

Theorem 8.19. *Suppose that (A, G, α) is* EH*-regular, and that the stability group G_P for the action of G on* $\operatorname{Prim} A$ *is amenable for all $P \in \operatorname{Prim} A$. Then the reduced norm $\|\cdot\|_r$ on $C_c(G, A)$ coincides with the universal norm $\|\cdot\|$ and $A \rtimes_{\alpha,r} G = A \rtimes_\alpha G$. In particular,* $\operatorname{Ind}_e^G \pi$ *is a faithful representation of $A \rtimes_\alpha G$ for any faithful representation π of A.*

Proof. Let $\rho \rtimes V$ be an irreducible representation of $A \rtimes_\alpha G$ and let π be a faithful representation of A. By assumption (A, G, α) is EH-regular so we have

$$\ker(\rho \rtimes V) = \operatorname{Ind}_{G_P}^G J$$

for some $P \in \operatorname{Prim} A$ and some $J \in \operatorname{Prim}(A \rtimes_\alpha G_P)$. Since G_P is amenable we have

$$J \supset \{0\} = \ker\bigl(\operatorname{Ind}_e^{G_P} \pi\bigr).$$

Therefore

$$\begin{aligned} \ker(\rho \rtimes V) &\supset \operatorname{Ind}_{G_P}^G \bigl(\ker \operatorname{Ind}_e^{G_P} \pi\bigr) \\ &= \ker\bigl(\operatorname{Ind}_{G_P}^G \operatorname{Ind}_e^{G_P} \pi\bigr) \end{aligned}$$

which, by induction in stages (Theorem 5.9 on page 157), is

$$= \ker\bigl(\operatorname{Ind}_e^G \pi\bigr).$$

Since this holds for any irreducible representation $\rho \rtimes V$, we have $\ker\bigl(\operatorname{Ind}_e^{G_P} \pi\bigr) = \{0\}$, and the result follows. □

As another example, consider the most basic of ideal structure questions: when is $A \rtimes_\alpha G$ simple? This question was the inspiration for a significant amount of work [114–116] — even before the interest in simple C^*-algebras inspired by the work of Kirchberg and Phillips on the classification of separable purely infinite nuclear simple C^*-algebras by their K-theory [89, 90, 128]. However, the simplicity of crossed products is extremely subtle, and we won't be able to do more than consider special cases here.

Note that if I is a non-trivial G-invariant ideal in A, then $I \rtimes_\alpha G$ a non-trivial ideal in $A \rtimes_\alpha G$, and $A \rtimes_\alpha G$ is certainly not simple. Therefore if we want $A \rtimes_\alpha G$ to be simple, a necessary condition is that A be *G-simple*; that is, we assume that A has no non-trivial G-invariant ideals. Assume also that the G-action on $\operatorname{Prim} A$ is free. If A is G-simple, for each $P \in \operatorname{Prim} A$ we must have

$$\bigcap_{s \in G} s \cdot P = \{0\}. \tag{8.10}$$

Note that (8.10) is equivalent to saying that the G-orbit of each $P \in \operatorname{Prim} A$ is dense in $\operatorname{Prim} A$. Thus if P and Q are primitive ideals in $\operatorname{Prim} A$, then there are $s_i \in G$ such that $s_i \cdot P \to Q$ in $\operatorname{Prim} A$. Since Ind_e^G is continuous (Lemma 5.16 on page 164), $\operatorname{Ind}_e^G s_i \cdot P \to \operatorname{Ind}_e^G Q$. On the other hand, $\operatorname{Ind}_e^G s \cdot P$ is equivalent to $\operatorname{Ind}_e^G P$ by Lemma 5.8 on page 157, and it follows that $\operatorname{Ind}_e^G Q \in \overline{\{\operatorname{Ind}_e^G P\}}$; or more simply, $\operatorname{Ind}_e^G P \subset \operatorname{Ind}_e^G Q$. By symmetry, $\operatorname{Ind}_e^G P = \operatorname{Ind}_e^G Q$. Since (A, G, α) is presumed to be EH-regular, every primitive ideal is induced; thus, $\operatorname{Prim}(A \rtimes_\alpha G)$ is a single point and $A \rtimes_\alpha G$ must be simple. We can summarize the above discussion as follows. (Recall that a transformation group is called *minimal* if each orbit is dense.)

Theorem 8.20. *Suppose that (A, G, α) is a EH-regular dynamical system such that the induced action on $\operatorname{Prim} A$ is free. Then $A \rtimes_\alpha G$ is simple if and only if the G-action on $\operatorname{Prim} A$ is minimal.*

If (G, X) is a second countable locally compact transformation group and if the G-action is minimal, then either the action is transitive or the action fails to be regular (Theorem 6.2 on page 173). Therefore in the latter case, the results in the previous section do not apply. Naturally, we want to see what can be said in the non-regular case.

In their 1967 *Memoir* [49], Effros and Hahn conjectured that if (G, X) was a second countable locally compact transformation group with G amenable, $C_0(X) \rtimes_{\mathrm{lt}} G$ should be EH-regular. In Chapter 9, we give the proof that their conjecture, and its natural generalization to dynamical systems (the *generalized Effros-Hahn conjecture*), is true:

Theorem 8.21 (Gootman-Rosenberg-Sauvageot). *Suppose that (A, G, α) is a separable dynamical system with G amenable. Then (A, G, α) is EH-regular.*

Then a fundamental corollary of Theorem 8.20 and Theorem 8.21 is the following.

Corollary 8.22. *Suppose that (G, X) is a second countable locally compact transformation group with G amenable and freely acting. Then $C_0(X) \rtimes_{\mathrm{lt}} G$ is simple if and only if G acts minimally on X.*

Remark 8.23. A direct proof of Corollary 8.22 on the preceding page in the case where $G = \mathbf{Z}$ is given in [21, Theorem VIII.3.9].

As a particularly poignant application of EH-regularity, we can generalize Corollary 8.22 on the previous page for abelian G, and give a tidy description of the primitive ideal space of $C_0(X) \rtimes_{\mathrm{lt}} G$ for any second countable locally compact transformation group (G, X). The proof of this result requires a bit of technology of its own, and is given in Section 8.3.

8.3 The Ideal Structure of $C_0(X) \rtimes_{\mathrm{lt}} G$

It is clear from Sections 7.5, 8.1 and 8.2 that the stability groups G_P for the action of G on $\operatorname{Prim} A$ play a significant role in the theory of crossed products. Naturally, the way the G_P vary with P is often an important consideration. Fortunately, there is a natural compact Hausdorff topology, called the *Fell topology*, on the set Σ of closed subgroups of a locally compact group G. We give a detailed discussion of the Fell topology in Appendix H.1. For now, it suffices to have just an intuitive notion of what it means for a net $\{ H_i \} \subset \Sigma$ to converge to H. If $H_i \to H$ and if $t_i \in H_i$ is such that $t_i \to t$, then we must have $t \in H$. Furthermore, if $t \in H$, then after passing to a subnet and relabeling, we can assume that there are $t_i \in H_i$ such that $t_i \to t$. These two properties actually characterize convergence (Lemma H.2 on page 454). By looking at simple examples, it quickly becomes evident that the map $P \mapsto G_P$ is often not continuous — even in situations where the action is otherwise extremely well behaved. Proposition H.41 on page 477 implies that $P \mapsto G_P$ is always a Borel map, and this will be important in our proof of the Gootman-Rosenberg-Sauvageot result in Chapter 9. In this section we want to say something significant about the ideal structure of transformation group C^*-algebras $C_0(X) \rtimes_{\mathrm{lt}} G$ with G abelian. A key idea is to see that the stability map $x \mapsto G_x$ is "continuous enough" for our purposes.

Throughout this discussion, (G, X) will be a locally compact transformation group, and $\bigl(C_0(X), G, \mathrm{lt}\bigr)$ the associated dynamical system. We will shortly restrict to the situation where G is abelian, but for the moment, we allow G to be arbitrary.

Fix $x \in X$. Let H be a subgroup of the stability group $G_x = \{ s \in G : s \cdot x = x \}$. Using Green's Imprimitivity Theorem (page 132), we can make $C_c(G)$ into a $C_0(G/H) \rtimes_{\mathrm{lt}} G$–$C^*(H)$-pre-imprimitivity bimodule. Let $\mathsf{X} = \mathsf{X}_H^G\bigl(C_0(G/H)\bigr)$ be the completion. Let ω be a representation of H on $\mathcal{H}_\omega$. (As usual, we use the same letter to denote a unitary representation or its integrated form on $C_c(H)$ or $C^*(H)$.) The induced representation $R = \mathsf{X}\text{–}\operatorname{Ind}_H^G \omega$, or simply $\mathsf{X}\text{–}\operatorname{Ind} \omega$, acts on the the completion $\mathcal{H}_R$ of the algebraic tensor product $C_c(G) \odot \mathcal{H}_\omega$ with respect to the inner product

$$
\begin{aligned}
(\varphi \otimes h \mid \psi \otimes k) &:= \bigl(\omega\bigl(\langle \psi , \varphi \rangle_{C^*(H)}\bigr) h \mid k\bigr) \\
&= \int_H \langle \psi , \varphi \rangle_{C^*(H)}(t) \bigl(\omega(t) h \mid k\bigr) \, d\mu_H(t) \\
&= \int_H \int_G \gamma_H(t) \overline{\psi(r)} \varphi(rt) \bigl(\omega(t) h \mid k\bigr) \, d\mu_G(r) \, d\mu_H(t).
\end{aligned}
$$

Then X–Ind ω is the integrated form of (κ, U) where

$$\kappa(\xi)[\varphi \otimes h] = [M(\xi)\varphi \otimes h] \quad \text{with} \quad M(\xi)\varphi(s) = \xi(sH)\varphi(s), \quad \text{and} \qquad (8.11)$$
$$U(r)[\varphi \otimes h] = [\mathrm{lt}_r(\varphi) \otimes h].$$

After identifying $C^*(G)$ with $\mathbf{C} \rtimes G$, the representation of G induced from H, $\mathrm{Ind}_H^G \omega$, is given by the unitary part, U, of X–Ind ω; that is $\mathrm{Ind}_H^G(\omega)(s)[\varphi \otimes h] = [\mathrm{lt}_s(\varphi) \otimes h]$ and $\mathrm{Ind}_H^G(\omega)(\psi)[\varphi \otimes h] = [\psi * \varphi \otimes h]$.

Let ev_x be "evaluation at x" on $C_0(X)$: $\mathrm{ev}_x(\xi) := \xi(x)$. Let $\rho_x = \mathrm{ev}_x \otimes 1$ be the corresponding representation on $\mathcal{H}_\omega$: $\rho_x(\xi)h := \xi(x)h$. Then (ρ_x, ω) is a covariant representation of $\big(C_0(X), H, \mathrm{lt}\big)$, and we can form the induced representation $L := \mathrm{Ind}_H^G(\rho_x \rtimes \omega)$ on $\mathcal{H}_L$ which is the completion of $C_c(G \times X) \odot \mathcal{H}_\omega$ with respect to the inner product[3]

$$\begin{aligned}(f \otimes h \mid g \otimes k) &= \int_H \mathrm{ev}_x\big(\langle g\,,\, f\rangle_{C_0(X)\rtimes_{\mathrm{lt}} H}(t)\big)\big(\omega(t)h \mid k\big)\, d\mu_H(t) \\ &= \int_H \int_G \gamma_H(t)\overline{g(r, r\cdot x)} f(rt, r\cdot x)\big(\omega(t)h \mid k\big)\, d\mu_G(r)\, d\mu_H(t).\end{aligned}$$

Then $L = \pi_L \rtimes U_L$ where, by definition,

$$\pi_L(\xi)[f \otimes h] = [i_{C_0(X)}(\xi) f \otimes h] \quad \text{and} \quad U_L(r)[f \otimes h] = [i_G(r) f \otimes h].$$

(Recall that $i_{C_0(X)}(\xi)f(s,x) = \xi(x)f(s,x)$ and $i_G(r)f(s,x) = f(r^{-1}s, r^{-1}\cdot x)$.)

We can define $U : C_c(G \times X) \odot \mathcal{H}_\omega \to C_c(G) \odot \mathcal{H}_\omega$ by $U(f \otimes h) = \psi(f) \otimes h$, where $\psi(f)(s) := f(s, s\cdot x)$. Then

$$\begin{aligned}&\big(U(f \otimes h) \mid U(g \otimes k)\big)_{\mathcal{H}_R} \\ &\qquad = \int_H \int_G \gamma_H(t)\overline{\psi(g)(r)}\psi(f)(rt)\big(\omega(t)h \mid k\big)\, d\mu_G(r)\, d\mu_H(t) \\ &\qquad = (f \otimes h \mid g \otimes k)_{\mathcal{H}_L}.\end{aligned}$$

Therefore U is isometric, and since U clearly has dense range, it extends to a unitary of $\mathcal{H}_L$ and $\mathcal{H}_R$. Therefore $\mathrm{Ind}_H^G(\rho_x \rtimes \omega)$ is equivalent to a representation $L_H^G(x, \omega)$ on the space of $\mathrm{Ind}_H^G \omega$. We can summarize this discussion as follows.

Proposition 8.24. *Suppose that (G, X) is a locally compact transformation group, that $H \subset G_x$ and that ω is a representation of H. Then $\mathrm{Ind}_H^G(\rho_x \rtimes \omega)$ is equivalent to a representation $L_H^G(x, \omega) = \pi \rtimes \mathrm{Ind}_H^G \omega$ where*

$$\pi(\xi)[\varphi \otimes h] = [\tilde{M}(\xi)\varphi \otimes h],$$

with $\tilde{M}(\xi)\varphi(s) := \xi(s\cdot x)\varphi(s)$.

Remark 8.25. If $j_{(x,H)} : C_0(X) \to C^b(G/H)$ is given by $j_{(x,H)}(\xi)(sH) := \xi(s\cdot x)$, then $\tilde{M} = \bar{M} \circ j_{(x,H)}$ where M is given in (8.11).

[3] Formally, $\mathcal{H}_L$ is the completion of $C_c(G, C_0(X)) \odot \mathcal{H}_\omega$, but it is not hard to see that $C_c(G \times X) \odot \mathcal{H}_\omega$ is dense in $C_c(G, C_0(X)) \odot \mathcal{H}_\omega$.

Proof of Proposition 8.24. Since $\psi\big(i_{C_0(X)}(\xi)(f)\big)(s) = i_{C_0(X)}(\xi)f(s, s\cdot x) = \xi(s\cdot x)\psi(f)(s)$, U clearly intertwines π_L and π. Similarly, we compute that $\psi\big(i_G(r)f\big)(s) = \mathrm{lt}_r\big(\psi(f)\big)(s)$, and U intertwines U_L with $\operatorname{Ind}_H^G \omega$. □

Let $(j_{C_0(G/H)}, j_G)$ be the natural covariant homomorphism of $\big(C_0(G/H), G, \mathrm{lt}\big)$ into $M\big(C_0(G/H) \rtimes_{\mathrm{lt}} G\big)$. Let $k = \bar{j}_{C_0(G/H)} \circ j_{(x,H)} : C_0(X) \to M\big(C_0(G/H) \rtimes_{\mathrm{lt}} G\big)$; that is, $k(\xi)f(s,tH) = \xi(t\cdot x)f(s,tH)$. Then it is easy to check that (k, j_G) is a covariant homomorphism of $\big(C_0(X), G, \mathrm{lt}\big)$ into $M\big(C_0(G/H) \rtimes_{\mathrm{lt}} G\big)$.

Lemma 8.26. *Suppose that (G, X) is a locally compact transformation group, and that ω is a representation of $H \subset G_x$ for some $x \in X$. Then the diagram*

$$\begin{array}{ccc} C_0(X) \rtimes_{\mathrm{lt}} G & \xrightarrow{\;k \rtimes j_G\;} & M\big(C_0(G/H) \rtimes_{\mathrm{lt}} G\big) \\ & {\scriptstyle L_H^G(x,\omega)} \searrow \qquad \swarrow {\scriptstyle \textsf{X}\text{–}\operatorname{Ind}\omega} & \\ & B(\mathcal{H}_R) & \end{array}$$

commutes. In particular, if $\ker\omega \subset \ker\tau$ *(as representations of $C^*(H)$), then*

$$\ker\big(\operatorname{Ind}_H^G(\rho_x \rtimes \omega)\big) \subset \ker\big(\operatorname{Ind}_H^G(\rho_x \rtimes \tau)\big).$$

Proof. On the one hand, $L_H^G(x,\omega) = \pi \rtimes \operatorname{Ind}_H^G \omega$. On the other hand, $L' := (\textsf{X}\text{–}\operatorname{Ind}\omega) \circ (k \rtimes j_G) = \pi_{L'} \rtimes U_{L'}$. Then

$$\begin{aligned} \pi_{L'}(\xi) &= \bar{L}'\big(i_{C_0(X)}(\xi)\big) \\ &= \textsf{X}\text{–}\operatorname{Ind}^{-}\big(\bar{j}_{C_0(G/H)}\big(j_{(x,H)}(\xi)\big)\big) \\ &= \pi(\xi), \end{aligned}$$

where checking the last equality is facilitated by Remark 8.25 on the preceding page. Similarly, $U_{L'}(s) = L'\big(i_G(s)\big) = \textsf{X}\text{–}\operatorname{Ind}\big(j_G(s)\big) = \operatorname{Ind}_H^G \omega(s)$. Therefore $L' = L_H^G(x,\omega)$ as claimed.

Since $\textsf{X}$ is an imprimitivity bimodule, $\textsf{X}$–Ind preserves containment ([139, Theorem 3.22]). Thus $\ker\omega \subset \ker\tau$ implies the kernel of $L_H^G(x,\omega) = (\textsf{X}\text{–}\operatorname{Ind}\omega) \circ (k \rtimes j_G)$ is contained in the kernel of $L_H^G(x,\tau) = (\textsf{X}\text{–}\operatorname{Ind}\tau) \circ (k \rtimes j_G)$. Therefore $\ker L_H^G(x,\omega) \subset \ker L_H^G(x,\tau)$, and the last assertion now follows from Proposition 8.24 on the previous page. □

Proposition 8.27. *Let (G, X) be a locally compact transformation group. Suppose that ω is an irreducible representation of G_x. Then $\operatorname{Ind}_{G_x}^G(\rho_x \rtimes \omega)$ is an irreducible representation of $C_0(X) \rtimes_{\mathrm{lt}} G$.*

Proof. It will suffice to see that $L_{G_x}^G(x,\omega) = \pi \rtimes \operatorname{Ind}_{G_x}^G \omega$ is irreducible. For this, it suffices to see that any operator that commutes with $\pi(\xi)$ for each $\xi \in C_0(X)$ and that also commutes with $\operatorname{Ind}_{G_x}^G(\omega)(s)$ for each $s \in G$ must be a scalar operator. However, since $\textsf{X}$ is an imprimitivity bimodule, $\textsf{X}\text{–}\operatorname{Ind}\omega = \kappa \rtimes \operatorname{Ind}_{G_x}^G \omega$ is irreducible ([139, Corollary 3.32]). Therefore any operator commuting with each $\kappa(\zeta)$ and each

$\operatorname{Ind}_{G_x}^G(\omega)(s)$ is a scalar operator. Hence it will suffice to show that each $\kappa(\zeta)$, with $\zeta \in C_0(G/G_x)$, can be approximated in the weak operator topology by operators of the form $\pi(\xi)$, with $\xi \in C_0(X)$.

Let $C \subset G/G_x$ be a compact set. Since $sG_x \mapsto s \cdot x$ is a continuous injection of G/G_x onto $G \cdot x$, it restricts to a *homeomorphism* of C onto a compact set (since X is Hausdorff). The Tietze Extension Theorem implies that there is a $\xi_C \in C_c(X)$ whose image (under $j_{(x,G_x)}$) in $C^b(G/G_x)$ agrees with ζ on C. Then $\{\pi(\xi_C)\}$ is a net in $B(\mathcal{H}_R)$, and it suffices to see that $\pi(\xi_C) \to \kappa(\zeta)$ in the weak operator topology. Since $\{\pi(\xi_C)\}$ is bounded, it suffices to show convergence on a dense subset of $\mathcal{H}_R$. In particular, we can look at elements of the form $\varphi \otimes h$ in $C_c(G) \odot \mathcal{H}_\omega$. If $\varphi \otimes h$ and $\varphi' \otimes h'$ are in $C_c(G) \odot \mathcal{H}_\omega$ with $\operatorname{supp} \varphi' \subset K$, then provided $C \supset K$,

$$\begin{aligned}
&\big(\pi(\xi_C)(\varphi \otimes h) \mid (\varphi' \otimes h')\big) - \big(\kappa(\zeta)(\varphi \otimes h) \mid (\varphi' \otimes h')\big) \\
&= \int_{G_x} \int_G \gamma_{G_x}(t) \overline{\varphi'(r)} \big(\tilde{M}(\xi_C)\varphi(rt) - M(\zeta)\varphi(rt)\big) \big(\omega(t)h \mid h'\big)\, d\mu_G(r)\, d\mu_{G_x}(t) \\
&= \int_{G_x} \int_G \gamma_{G_x}(t) \overline{\varphi'(r)} \big(\xi_C(r \cdot x) - \zeta(rG_x)\big)\varphi(rt) \big(\omega(t)h \mid h'\big)\, d\mu_G(r)\, d\mu_{G_x}(t),
\end{aligned}$$

which equals zero since $\xi_C(r \cdot x) = \zeta(rG_x)$ on the support of φ'. This completes the proof. □

Using Proposition 8.27 on the facing page, we aim to prove the following.

Proposition 8.28. *Let (X, G) be a locally compact transformation group* with G abelian. *Define $\Phi : X \times \widehat{G} \to \operatorname{Prim}\big(C_0(X) \rtimes_{\mathrm{lt}} G\big)$ by*

$$\Phi(x, \tau) := \ker\big(\operatorname{Ind}_{G_x}^G(\rho_x \rtimes \tau|_{G_x})\big).$$

Then Φ is continuous.

We'll need a bit of machinery to prove Proposition 8.28. In particular, if we let Σ be the compact Hausdorff space of closed subgroups of G endowed with the Fell topology (Corollary H.4 on page 455), then we have to cope with the fact that the stability map $x \mapsto G_x$ is rarely continuous. The following simple observation is sometimes a useful substitute for this lack of continuity.[4]

Lemma 8.29. *If (G, X) is a locally compact transformation group, then*

$$P = \{\, (x, H) \in X \times \Sigma : H \subset G_x \,\}$$

is closed in $X \times \Sigma$.

[4]The proof of this result, and a number of others in the sequel, require that convergent nets in X have unique limits. Of course, this is true since X is Hausdorff. But in the case of dynamical systems (A, G, α) where A is not commutative, and A fails to have Hausdorff spectrum or primitive ideal space, then these techniques fail. This is just one of many reasons that the ideal structure of general crossed products is considerably more mysterious than that of transformation group C^*-algebras.

Proof. Suppose that (x_i, H_i) is in P for all i, and that $(x_i, H_i) \to (x, H)$. If $t \in H$, then we can pass to subnet, relabel, and assume that there are $t_i \in H_i$ such that $t_i \to t$ (Lemma H.2 on page 454). Then $x_i = t_i \cdot x_i \to t \cdot x$. Since X is Hausdorff, $t \cdot x = x$, and $t \in G_x$. Then $H \subset G_x$, and P is closed. □

We begin by defining $\Psi : P \times \widehat{G} \to \mathcal{I}\bigl(C_0(X) \rtimes_{\mathrm{lt}} G\bigr)$ by

$$\Psi(x, H, \tau) := \ker\bigl(\mathrm{Ind}_H^G(\rho_x \rtimes \tau|_H)\bigr).$$

Recall that a *subbasis* for the topology on $\mathcal{I}\bigl(C_0(X) \rtimes_{\mathrm{lt}} G\bigr)$ is given by sets $\mathcal{O}_J$ where J is an ideal in $C_0(X) \rtimes_{\mathrm{lt}} G$, and

$$\mathcal{O}_J := \{\, I \in \mathcal{I}\bigl(C_0(X) \rtimes_{\mathrm{lt}} G\bigr) : I \not\supset J \,\}.$$

Lemma 8.30. *Ψ is continuous.*

Remark 8.31. It seems natural that Ind_H^G should depend continuously on H (see Remark 8.58 on page 260). To work out the details the tool needed is a *continuous choice of Haar measures* on Σ as defined in Lemma H.8 on page 458. If $f_0 \in C_c^+(G)$ is such that $f_0(e) > 0$, then we can fix a Haar measure μ_H on H such that

$$\int_H f_0(s)\, d\mu_H(s) = 1.$$

Then Lemma H.8 on page 458 implies that

$$H \mapsto \int_H f(s)\, d\mu_H(s)$$

is continuous for all $f \in C_c(G)$.

Proof. Let $\{\, (x_i, H_i, \tau_i) \,\}$ be a net in $P \times \widehat{G}$ converging to (x, H, τ). It will suffice to see that if $\Psi(x_i, H_i, \tau_i) \supset J$ for all i, then $\Psi(x, H, \tau) \supset J$. Fix a continuous choice of Haar measures μ_H on Σ as in Lemma H.8 on page 458. By Proposition 8.24 on page 243, $\mathrm{Ind}_{H_i}^G(\rho_{x_i} \rtimes \tau_i|_{H_i})$ is equivalent to $L_{H_i}^G(x_i, \tau_i|_{H_i})$ which acts on the completion of $C_c(G)$ with respect to the inner product[5]

$$(\varphi \mid \psi)_i = \int_{H_i} \psi^* * \varphi(t)\tau_i(t)\, d\mu_{H_i}(t). \tag{8.12}$$

(Let $(\cdot \mid \cdot)_0$ be the inner product for $L_H^G(x, \tau|_H) \cong \mathrm{Ind}_H^G(\rho_x \rtimes \tau|_H)$.) Furthermore, $L_{H_i}^G(x_i, \tau_i|_{H_i}) = \pi_i \rtimes V_i$, where $V_i = \mathrm{Ind}_{H_i}^G(\tau_i|_{H_i})$ and π_i is given by $\pi_i(\xi)\varphi(s) = \xi(s \cdot x_i)\varphi(s)$. By definition of the topology on $\widehat{G}$, τ_i converges to τ uniformly on compact subsets of G. Therefore part (b) of Lemma H.9 on page 460 implies that $(\varphi \mid \psi)_i \to (\varphi \mid \psi)_0$ for all φ and ψ in $C_c(G)$.

[5]Since we're assuming G is abelian, there are no modular functions to worry about. In particular, γ_{H_i} is identically one and has been omitted from our formulas.

Claim. Let $A \subset C_0(X) \rtimes_{\mathrm{lt}} G$ be dense. Then it will suffice to show that

$$\big(L^G_{H_i}(x_i, \tau_i|_{H_i})(F)\varphi \mid \psi\big)_i \to \big(L^G_H(x, \tau|_H)(F)\varphi \mid \psi\big)_0 \tag{8.13}$$

for all $F \in A$ and all $\varphi, \psi \in C_c(G)$.

Proof of Claim. Let $F' \in J$. Since $\operatorname{Ind}_H^G(\rho_x \rtimes \tau|_H)$ is equivalent to $L^G_H(x, \tau|_H)$ and since $C_c(G)$ is dense in the space of $L^G_H(x, \tau|_H)$, it suffices to see that $\big(L^G_H(x, \tau|_H)(F')\varphi \mid \psi\big)_0 = 0$ for all φ and ψ in $C_c(G)$. Fix $\epsilon > 0$. Since $J \subset \ker\big(\operatorname{Ind}_{H_i}^G(\rho_{x_i} \rtimes \tau_i|_{H_i})\big)$ and since $\operatorname{Ind}_{H_i}^G(\rho_{x_i} \rtimes \tau_i|_{H_i})$ is equivalent to $L^G_{H_i}(x_i, \tau_i|_{H_i})$, we have $\big(L^G_{H_i}(x_i, \tau_i|_{H_i})(F')\varphi \mid \psi\big)_i = 0$ for all i. Since $(\varphi \mid \psi)_i \to (\varphi \mid \psi)_0$ implies that $\|\varphi\|_i = (\varphi \mid \varphi)_i^{\frac{1}{2}}$ converges to $\|\varphi\|_0$ for all $\varphi \in C_c(G)$, we can find $F \in A$ such that

$$\big|\big(L^G_{H_i}(x_i, \tau_i|_{H_i})(F)\varphi \mid \psi\big)_i\big| < \frac{\epsilon}{2} \quad \text{for large } i\text{, and such that}$$
$$\big|\big(L^G_H(x, \tau|_H)(F)\varphi \mid \psi\big)_0 - \big(L^G_H(x, \tau|_H)(F')\varphi \mid \psi\big)_0\big| < \frac{\epsilon}{2}.$$

It follows from (8.13) that $\big|\big(L^G_H(x, \tau|_H)(F')\varphi \mid \psi\big)_0\big| < \epsilon$. Since ϵ was arbitrary, the claim follows. □

Let $A = \operatorname{span}\{\, \varphi \otimes \xi : \varphi \in C_c(G) \text{ and } \xi \in C_0(X)\,\}$, where $\varphi \otimes \xi(s, x) := \varphi(s)\xi(x)$. Then A is dense in $C_0(X) \rtimes_{\mathrm{lt}} G$ (Lemma 1.87 on page 29). Note that

$$\big(L^G_{H_i}(x_i, \tau_i|_{H_i})(\varphi \otimes \xi)\varphi' \mid \psi\big)_i = \big(\pi_i(\xi)V_i(\varphi)\varphi' \mid \psi\big)_i = \big(\pi_i(\xi)\varphi * \varphi' \mid \psi\big)_i.$$

Therefore it will be enough to show that

$$\big(\pi_i(\xi)\varphi \mid \psi\big)_i \to \big(\pi(\xi)\varphi \mid \psi\big)_0,$$

where $L^G_H(x, \tau|_H) = \pi \rtimes V$.

Recall that

$$\big(\pi_i(\xi)\varphi \mid \psi\big)_i = \int_{H_i} \tau_i(t)\psi^* * \pi_i(\xi)\varphi(t)\, d\mu_{H_i}(t).$$

Since, by definition of the topology on $\widehat{G}$, $\tau_i \to \tau$ uniformly on compact sets, it will suffice, by Lemma H.9 on page 460, to see that $\psi * \pi_i(\xi)\varphi$ converges to $\psi * \pi(\xi)\varphi$ in the inductive limit topology on $C_c(G)$. Clearly, we just need to see that $\pi_i(\xi)\varphi \to \pi(\xi)\varphi$ in the inductive limit topology. Since the supports are all contained in $\operatorname{supp}\varphi$, it will suffice to see that the convergence is uniform. If not, then after passing to subnet and relabeling, there are $s_i \in \operatorname{supp}\varphi$ and an $\epsilon > 0$ such that

$$\big|\xi(s_i \cdot x_i)\varphi(s_i) - \xi(s_i \cdot x)\varphi(s_i)\big| \geq \epsilon > 0 \quad \text{for all } i.$$

Since we can pass to another subnet, relabel, and assume that $s_i \to s$, we arrive at a contradiction. Thus, Ψ must be continuous as claimed. □

Proof of Proposition 8.28 on page 245. Suppose that $(x_i, \tau_i) \to (x, \tau)$ in $X \times \widehat{G}$, and that J is an ideal in $C_0(X) \rtimes_{\mathrm{lt}} G$ such that $\Phi(x_i, \tau_i) \supset J$ for all i. Then as in Lemma 8.30 on page 246, it will suffice to see that $\Phi(x, \tau) \supset J$. Since Σ is compact, we can pass to a subnet, relabel, and assume that $G_{x_i} \to H \in \Sigma$. If $t \in H$, then (possibly after passing to another subnet and relabeling) there are $t_i \in G_{x_i}$ such that $t_i \to t$. Then $x_i = t_i \cdot x_i \to t \cdot x$. Since X is Hausdorff, $x = t \cdot x$ and $t \in G_x$. Therefore $H \subset G_x$ and $(x, H) \in P$. Since $\Phi(x_i, \tau_i) = \Psi(x_i, G_{x_i}, \tau_i)$ and since Ψ is continuous, we have

$$\Psi(x, H, \tau) \supset J.$$

However, using "Induction by Stages" (Theorem 5.9 on page 157), we have

$$\operatorname{Ind}_H^G(\rho_x \rtimes \tau|_H) \cong \operatorname{Ind}_{G_x}^G\big(\operatorname{Ind}_H^{G_x}(\rho_x \rtimes \tau|_H)\big).$$

Using Proposition 8.24 on page 243 (with $G = G_x$), we have

$$\operatorname{Ind}_H^{G_x}(\rho_x \rtimes \tau|_H) \cong \rho_x \rtimes \big(\operatorname{Ind}_H^{G_x} \tau|_H\big).$$

Therefore

$$\operatorname{Ind}_H^G(\rho_x \rtimes \tau|_H) \cong \operatorname{Ind}_{G_x}^G\big(\rho_x \rtimes \big(\operatorname{Ind}_H^{G_x} \tau|_H\big)\big).$$

On the other hand, Corollary 5.15 on page 162 implies that $\ker\big(\operatorname{Ind}_H^{G_x} \tau|_H\big) \subset \ker \tau|_{G_x}$. Therefore the last part of Lemma 8.26 on page 244 implies that

$$\begin{aligned}\Phi(x, \tau) = \ker\big(\operatorname{Ind}_{G_x}^G(\rho_x \rtimes \tau|_{G_x})\big) &\supset \ker\big(\operatorname{Ind}_{G_x}^G\big(\rho_x \rtimes \operatorname{Ind}_H^{G_x} \tau|_H\big)\big) \\ &= \ker\big(\operatorname{Ind}_H^G(\rho_x \rtimes \tau|_H)\big) \\ &= \Psi(x, H, \tau) \supset J.\end{aligned}$$

Therefore, $\Phi(x, \tau) \supset J$, and we're done. $\square$

Lemma 8.32. *Suppose that (G, X) is a locally compact transformation group with G abelian, that $H \subset G_x$ and that $\tau \in \widehat{G}$. Then $\operatorname{Ind}_H^G(\rho_x \rtimes \tau|_H)$ is equivalent to $\operatorname{Ind}_H^G(\rho_{s \cdot x} \rtimes \tau|_H)$.*

Remark 8.33. Lemma 8.32 is a special case of Lemma 5.8 on page 157, but it seems reasonable to give a separate proof in this special case.

Proof. In view of Proposition 8.24 on page 243, it suffices to show that $L_H^G(x, \tau|_H)$ is equivalent to $L_H^G(s \cdot x, \tau|_H)$. Both these representations act on the completion of $C_c(G)$ with respect to the inner product

$$(\varphi \mid \psi) = \int_H \psi^* * \varphi(t)\tau(t)\, d\mu_H(t) = \int_H \int_G \overline{\psi(r)}\varphi(rt)\tau(t)\, d\mu_G(r)\, d\mu_H(t).$$

Furthermore, $L_H^G(x, \tau|_H)(f)\varphi(r) = \int_G f(v, r \cdot x)\varphi(v^{-1}r)\, d\mu_G(v)$. Define

$U : C_c(G) \to C_c(G)$ by $U(\varphi)(r) = \varphi(sr)$. Then, recalling that G is abelian,

$$\begin{aligned} U\big(L^G_H(x,\tau|_H)(f)\varphi\big)(r) &= L^G_H(x,\tau|_H)(f)\varphi(sr) \\ &= \int_G f(v, sr\cdot x)\varphi(v^{-1}sr)\,d\mu_G(v) \\ &= \int_G f(v, rs\cdot x)U(\varphi)(v^{-1}r)\,d\mu_G(v) \\ &= L^G_H(s\cdot x,\tau|_H)U(\varphi)(r). \end{aligned}$$ □

It follows immediately from Lemma 8.32 that $\Phi(x,\tau) = \Phi(s\cdot x,\tau)$ for all $s \in G$. The continuity of Φ implies a bit more.

Lemma 8.34. *Suppose that (G, X) is a locally compact transformation group with G abelian, and that $\overline{G\cdot x} = \overline{G\cdot y}$. Then $G_x = G_y$, and if τ and σ are in $\widehat{G}$ and are such that $\tau\overline{\sigma} \in G_x^\perp$, then $\Phi(x,\tau) = \Phi(y,\sigma)$.*

Proof. Since $y \in \overline{G\cdot x}$, there are $s_i \in G$ such that $s_i\cdot x \to y$. But if $t \in G_x$, then since G is abelian, $s_i\cdot x = ts_i\cdot x \to t\cdot y$. Since X is Hausdorff, $t\cdot y = y$ and $t \in G_y$. This shows that $G_x \subset G_y$. By symmetry, $G_y \subset G_x$ and the first assertion follows.

Let $s_i\cdot x \to y$, as above. By Lemma 8.34, $\Phi(s_i\cdot x,\tau) = \Phi(x,\tau)$. Since Φ is continuous, we have $\Phi(y,\tau) \in \overline{\Phi(x,\tau)}$. Thus, by definition of the hull-kernel topology, $\Phi(x,\tau) \subset \Phi(y,\tau)$. By symmetry, we must have $\Phi(x,\tau) = \Phi(y,\tau)$, and a moments reflection shows that $\Phi(y,\tau) = \Phi(y,\sigma)$. □

The continuity assertions in Corollary 6.15 on page 182 and Lemma 5.16 on page 164 are also a consequence of the following general observation from [66, Proposition 9].

Lemma 8.35. *Let B and D be C^*-algebras and let $\varphi : B \to M(D)$ be a homomorphism. Define $\varphi^* : \mathcal{I}(D) \to \mathcal{I}(B)$ by*

$$\varphi^*(J) = \{\, b \in B : \varphi(b)d \in J \text{ for all } d \in D \,\}.$$

Then φ^* is continuous.

Proof. Since D and J are an ideals in $M(D)$, we clearly have $\varphi^*(J) \in \mathcal{I}(B)$. Suppose $K \in \mathcal{I}(B)$ and let $\mathcal{O}_K = \{\, I \in \mathcal{I}(B) : I \not\supset K \,\}$. We need to see that $(\varphi^*)^{-1}(\mathcal{O}_K)$ is open.

If $\varphi^*(J) \not\supset K$, then there is a $b \in K$ and a $d \in D$ such that $\varphi(b)d \notin J$. Therefore $\varphi(K)D \not\subset J$. Let K' be the ideal in D generated by $\varphi(K)D$. Then $K' \not\subset J$. On the other hand, if $K' \not\subset J$, then as J is an ideal, it follows that $\varphi(K)D \not\subset J$, and hence $\varphi^*(J) \not\supset K$. Summarizing,

$$\begin{aligned} (\varphi^*)^{-1}(\mathcal{O}_K) &= \{\, J \in \mathcal{I}(D) : \varphi^*(J) \not\supset K \,\} \\ &= \{\, J \in \mathcal{I}(D) : \varphi(K)D \not\subset J \,\} \\ &= \mathcal{O}_{K'}. \end{aligned}$$

Therefore φ^* is continuous. □

Lemma 8.36. *Suppose that (A, G, α) is a dynamical system. There is a continuous map $i_G^* : \mathcal{I}(A \rtimes_\alpha G) \to \mathcal{I}\big(C^*(G)\big)$ such that given a representation $L = \pi \rtimes u$ of $A \rtimes_\alpha G$, $i_G^*(\ker L) = \ker u$ (where u is viewed as a representation of $C^*(G)$).*

Proof. Let $i_G : C^*(G) \to M(A \rtimes_\alpha G)$ be the integrated form of the natural map. Then i_G^* is continuous by Lemma 8.35. Moreover,

$$i_G^*(\ker L) = \{\, f \in C^*(G) : i_G(f)d \in \ker L \text{ for all } d \in A \rtimes_\alpha G \,\}.$$

Since $L\big(i_G(f)d\big) = u(f)L(d)$ and since L is nondegenerate, $u(f) = 0$ if and only if $f \in i_G^*(\ker L)$. □

Lemma 8.37. *Suppose that (G, X) is a locally compact transformation group, and that $J(F)$ is the ideal in $C_0(X)$ corresponding to the closed set $F \subset X$. Let $(G\backslash X)^\sim$ be the T_0-ization of $G\backslash X$ (see Definition 6.9 on page 180). Define $\pi : (G\backslash X)^\sim \to \mathcal{I}\big(C_0(X)\big)$ by $\pi(\overline{G \cdot x}) := J(\overline{G \cdot x})$. Then π is a homeomorphism onto its range.*

Proof. Define $\bar\pi : X \to \mathcal{I}\big(C_0(X)\big)$ by $\bar\pi(x) = \pi\big(\overline{G \cdot x}\big)$ Since Lemma 6.19 on page 183 implies that $\bar\pi(x) = \operatorname{Res}\big(\Phi(x, 1)\big)$, $\bar\pi$ is continuous by Proposition 8.28 on page 245. Since $\bar\pi$ is clearly injective, it will suffice to see that $\bar\pi$ is open as a map onto its range.

Let U be open in X, and let $K := J(X \smallsetminus G \cdot U)$. Then if $x \in U$,

$$J(\overline{G \cdot x}) = \bigcap_{s \in G} J(s \cdot x) \not\supset K.$$

Thus, $\bar\pi(x) \in \mathcal{O}_K = \{\, I \in \mathcal{I}\big(C_0(X)\big) : I \not\supset K \,\}$.

On the other hand, suppose that $\bar\pi(x) \in \mathcal{O}_K$. If $s \cdot x \notin U$ for all $s \in G$, then $x \notin G \cdot U$, and since $G \cdot U$ is open, $\overline{G \cdot x} \cap G \cdot U = \emptyset$. This would imply that

$$\bigcap_{s \in G} J(s \cdot x) = J(\overline{G \cdot x}) \supset K,$$

which contradicts our assumption that $\bar\pi(x) \in \mathcal{O}_K$. Therefore there must be a s such that $s \cdot x \in U$. Since $\bar\pi(x) = \bar\pi(s \cdot x)$, we have $\bar\pi(x) \in \bar\pi(U)$. In other words, $\bar\pi(U) = \mathcal{O}_K \cap \bar\pi(X)$. Since $\mathcal{O}_K$ is a subbasic open set in $\mathcal{I}\big(C_0(X)\big)$, this proves that $\bar\pi$ is open onto its range. The result is proved. □

There is a one-to-one correspondence between closed subsets $F \subset \operatorname{Prim} A$ and ideals $I \in \mathcal{I}(A)$:

$$I(F) := \bigcap_{P \in F} P$$

for a closed set $F \subset \operatorname{Prim} A$ [139, Proposition A.27]. The following lemma is helpful when using the topology on $\mathcal{I}(A)$.[6]

[6]It is possible to describe the topology on $\mathcal{I}(A)$ as a topology on the closed subsets of $\operatorname{Prim} A$. However, this is definitely not Fell's compact Hausdorff topology discussed in Appendix H, and it seemed best not to introduce a second topology on the closed subsets of $\operatorname{Prim} A$. In any event, it is interesting to contrast Lemma 8.38 with Lemma H.2 on page 454.

Lemma 8.38. *Suppose that A is a C^*-algebra. For each closed set $F \subset \operatorname{Prim} A$, let $I(F)$ be the corresponding ideal in $\mathcal{I}(A)$. Then a net $\{ I(F_j) \}$ converges to $I(F)$ in $\mathcal{I}(A)$ if and only if given $P \in F$, there is a subnet $\{ I(F_{j_k}) \}$ and $P_k \in F_{j_k}$ such that $P_k \to P$ in $\operatorname{Prim} A$.*

Proof. Suppose that $I(F_j) \to I(F)$ in $\mathcal{I}(A)$ and that $P \in F$. Let U be a neighborhood of P in $\operatorname{Prim} A$, and let $J = I(U^c)$ be the ideal corresponding to the complement of U. Then $I(F) \not\supset J$. Therefore

$$\mathcal{O}_J = \{ I \in \mathcal{I}(A) : I \not\supset J \}$$

is a neighborhood of $I(F)$. Thus there is a j_0 such that $j \geq j_0$ implies that $I(F_j) \in \mathcal{O}_J$. In particular, if $j \geq j_0$, then $U \cap F_j \neq \emptyset$. Then if we let

$$M := \{ (U, j) : U \text{ is a neighborhood of } P \text{ and } U \cap F_j \neq \emptyset \},$$

then M is directed by decreasing U and increasing j. Let $F_{(U,j)} = F_j$. For each $m = (U, j) \in M$, we can pick $P_m \in F_j \cap U$. Then $\{ P_m \}$ converges to P as required.

For the converse, suppose that $\{ I(F_j) \}$ has the property given in the lemma, and that $I(F_j) \not\to I(F)$. After passing to a subnet, and relabeling, we can assume that there is an open set $U \subset \operatorname{Prim} A$ such that $U \cap F \neq \emptyset$ and such that $F_j \cap U = \emptyset$ for all j. But if $P \in F \cap U$, then we can pass to a subnet, relabel, and find $P_j \in F_j$ such that $P_j \to P$. Then P_j must eventually be in U, which is a contradiction. This completes the proof. □

We can form the quotient space

$$X \times \widehat{G}/\!\sim,$$

where $(x, \tau) \sim (y, \sigma)$ if $\overline{G \cdot x} = \overline{G \cdot y}$ and $\tau\overline{\sigma} \in G_x^{\perp}$. We give $X \times \widehat{G}/\!\sim$ the quotient topology.

Theorem 8.39. *Let (G, X) be a locally compact transformation group with G abelian. Then $\Phi : X \times \widehat{G} \to \operatorname{Prim}\bigl(C_0(X) \rtimes_{\mathrm{lt}} G\bigr)$ is open as a map onto its range, and factors through $X \times \widehat{G}/\!\sim$. Furthermore, Φ defines a homeomorphism of $X \times \widehat{G}/\!\sim$ onto its range. If $\bigl(C_0(X), G, \mathrm{lt}\bigr)$ is* EH*-regular, which is automatic if (G, X) is second countable, then Φ defines a homeomorphism of $X \times \widehat{G}/\!\sim$ onto* $\operatorname{Prim}\bigl(C_0(X) \rtimes_{\mathrm{lt}} G\bigr)$.

Remark 8.40. The openness of Φ is very important. It implies that the quotient map of $X \times \widehat{G}$ onto $X \times \widehat{G}/\!\sim$ is open, and as a result, it is very easy to describe the quotient topology. For example, the forward image of a basis for the topology in $X \times \widehat{G}$ is a basis for the quotient topology in $X \times \widehat{G}/\!\sim$.

Remark 8.41. We are working with the primitive ideal space because we are not assuming that $C_0(X) \rtimes_{\mathrm{lt}} G$ is GCR, so that there is little hope of describing the spectrum in any meaningful way. However, since the topology on the spectrum is pulled back from the topology on the primitive ideal space, the map

$$\widehat{\Phi}(x, \tau) := \operatorname{Ind}_{G_x}^G (\rho_x \rtimes \tau|_{G_x})$$

is always a continuous map into the spectrum $(C_0(X) \rtimes_{\mathrm{lt}} G)^\wedge$. If the orbit space is T_0, then Proposition 7.30 on page 222 implies that $C_0(X) \rtimes_{\mathrm{lt}} G$ is GCR,[7] and the primitive ideal space is homeomorphic to the spectrum. In this case, we actually get a complete description of the spectrum, and every irreducible representation is equivalent to one of the form $\widehat{\Phi}(x, \tau)$.

Proof of Theorem 8.39. If $\bigl(C_0(X), G, \mathrm{lt}\bigr)$ is EH-regular, every primitive ideal in $\mathrm{Prim}\bigl(C_0(X) \rtimes_{\mathrm{lt}} G\bigr)$ is of the form $\ker\bigl(\mathrm{Ind}_{G_x}^G(\rho \rtimes \omega\bigr)$ for an irreducible representation $\rho \rtimes \omega$ of $C_0(X) \rtimes_{\mathrm{lt}} G_x$, where $\ker \rho$ is the ideal in $C_0(X)$ of functions vanishing at x. In particular, $\rho = \rho_x$ and ω must be an irreducible representation of G_x. But every irreducible representation of G_x is a character, and every character of G_x is the restriction of a character on G [56, Corollary 4.41]. Therefore Φ is surjective when $\bigl(C_0(X), G, \mathrm{lt}\bigr)$ is EH-regular.

If (G, X) is second countable, then $\bigl(C_0(X), G, \mathrm{lt}\bigr)$ is a separable dynamical system with G amenable. Therefore, $\bigl(C_0(X), G, \mathrm{lt}\bigr)$ is EH-regular by Theorem 8.21 on page 241. Also, Φ factors through $X \times \widehat{G}/{\sim}$ in view of Lemma 8.34 on page 249. Thus it suffices to see that Φ is an open map onto its range, and that it is injective on $X \times \widehat{G}/{\sim}$.

If F is a closed subset of X, let $J(F)$ be the corresponding ideal in $C_0(X)$. Lemma 6.19 on page 183 implies that $\mathrm{Res}\bigl(\Phi(x,\tau)\bigr) = \bigcap_{s \in G} J(s \cdot x) = J\bigl(\overline{G \cdot x}\bigr)$. Similarly, if $I(C)$ is the ideal in $C_0(\widehat{G})$ corresponding to a closed subset $C \subset \widehat{G}$, then Lemma 8.36 on page 250 and Proposition 5.14 on page 162 imply that $i_G^*\bigl(\Phi(x,\tau)\bigr) = I(\tau G_x^\perp)$. It follows that if $\Phi(x,\tau) = \Phi(y,\sigma)$, then $\overline{G \cdot x} = \overline{G \cdot y}$ and $\tau G_x^\perp = \sigma G_y^\perp$. Since Lemma 8.34 on page 249 implies $G_x = G_y$, Φ is injective on $X \times \widehat{G}/{\sim}$.

To show that Φ is open, we will use Proposition 1.15 on page 4. Thus if $\{\, \Phi(x_i, \tau_i) \,\}$ is a net converging to $\Phi(x, \tau)$, then it will suffice to see that, after possibly passing to a subnet and relabeling, that there is a net $\{\, (y_i, \sigma_i) \,\}$ converging to (x, τ) (in $X \times \widehat{G}$) such that $\overline{G \cdot x_i} = \overline{G \cdot y_i}$ and such that $\tau_i \overline{\sigma_i} \in G_{x_i}^\perp$. Since Res is continuous (Corollary 6.15 on page 182), $J(\overline{G \cdot x_i}) \to J(\overline{G \cdot x})$ in $\mathcal{I}\bigl(C_0(X)\bigr)$. Lemma 8.37 on page 250 implies that $\overline{G \cdot x_i} \to \overline{G \cdot x}$ in $(G\backslash X)^\sim$. Since the natural map of X onto $(G\backslash X)^\sim$ is open (Lemma 6.12 on page 180), we can pass to a subnet, relabel, and assume that there is a net $y_i \to x$ such that $\overline{G \cdot y_i} = \overline{G \cdot x_i}$ (Proposition 1.15 on page 4).

At the same time, the continuity of i_G^* (Lemma 8.36 on page 250) implies that $I(\tau_i G_{x_i}^\perp) \to I(\tau G_x^\perp)$. Then, after possibly passing to another subnet and relabeling, Lemma 8.38 on the preceding page implies there are $\sigma_i \in G_{x_i}^\perp$ such that $\tau_i \sigma_i \to \tau$ in $\widehat{G}$. Then $(y_i, \sigma_i \tau_i) \to (x, \tau)$, and this shows that Φ is open onto its range which completes the proof. □

Remark 8.42 (Alternate description of $\mathrm{Prim}\bigl(C_0(X) \rtimes_{\mathrm{lt}} G\bigr)$). Let $(G\backslash X)^\sim$ be the "T_0-ization" of the orbit space $G\backslash X$ (Definition 6.9 on page 180). We will abuse notation slightly and use $\overline{G \cdot x}$ for both the closure of $G \cdot x$ in X, and for the class of $G \cdot x$ in $(G\backslash X)^\sim$. (Since $G \cdot x$ and $G \cdot y$ define the same class in $(G\backslash X)^\sim$ if

[7] Also see Theorem 8.43 on the next page.

and only if they have the same closure, this should cause little harm.) In view of Lemma 8.34 on page 249, we can also define the quotient space

$$(G\backslash X)^{\sim} \times \widehat{G}/{\sim},$$

where $(\overline{G\cdot x},\tau) \sim (\overline{G\cdot y},\sigma)$ if $\overline{G\cdot x} = \overline{G\cdot y}$ and $\tau\overline{\sigma} \in G_x^{\perp}$. Of course there is a bijection θ on $X \times \widehat{G}/{\sim}$ onto $(G\backslash X)^{\sim} \times \widehat{G}/{\sim}$. We want to observe that $\theta : X \times \widehat{G}/{\sim} \to (G\backslash X)^{\sim} \times \widehat{G}/{\sim}$ is a homeomorphism for the respective quotient topologies.[8] Consider the commutative diagram

$$\begin{array}{ccc} X \times \widehat{G} & \xrightarrow{\;q\times \mathrm{id}\;} & (G\backslash X)^{\sim} \times \widehat{G} \\ {\scriptstyle q_1}\big\downarrow & \searrow & \big\downarrow{\scriptstyle q_2} \\ X \times \widehat{G}/{\sim} & \xrightarrow{\;\theta\;} & (G\backslash X)^{\sim} \times \widehat{G}/{\sim}, \end{array}$$

where q, q_1 and q_2 are the natural quotient maps. Since q and q_2 are continuous, the dotted diagonal arrow is continuous which shows that θ is continuous. Suppose that $V \subset X \times \widehat{G}/{\sim}$ is open. Since θ is a bijection, $q \times \mathrm{id}\big(q_1^{-1}(V)\big) = q_2^{-1}\big(\theta(V)\big)$. Therefore it will suffice to see that $q\times\mathrm{id}\big(q_1^{-1}(V)\big)$ is open. However, $q_1^{-1}(V)$ is open, and therefore is the union of basic open rectangles. Since q is open (Lemma 6.12 on page 180), it follows that $q \times \mathrm{id}\big(q_1^{-1}(V)\big)$ is open, and the assertion follows. Therefore, when convenient, and when $\big(C_0(X), G, \mathrm{lt}\big)$ is EH-regular, we can identify $\mathrm{Prim}\big(C_0(X) \rtimes_{\mathrm{lt}} G\big)$ with $(G\backslash X)^{\sim} \times \widehat{G}/{\sim}$.

With Theorem 8.39 on page 251 in hand, we can provide the promised converses to Proposition 7.30 on page 222 and Proposition 7.31 on page 223.

Theorem 8.43. *Suppose that (G,X) is a locally compact transformation group with G abelian. If $C_0(X) \rtimes_{\mathrm{lt}} G$ is GCR, then the orbit space $G\backslash X$ is T_0. In particular, if (G,X) is second countable, then $C_0(X) \rtimes_{\mathrm{lt}} G$ is GCR if and only if $G\backslash X$ is T_0.*

Proof. Suppose that $C_0(X) \rtimes_{\mathrm{lt}} G$ is GCR, and that $\overline{G\cdot x} = \overline{G\cdot y}$. We want to see that $G\cdot x = G\cdot y$. Let $H = G_x = G_y$, and let ι denote the trivial character. The continuity of Φ (Lemma 8.34 on page 249) implies that

$$\ker\big(\mathrm{Ind}_H^G(\rho_x \rtimes \iota)\big) = \ker\big(\mathrm{Ind}_H^G(\rho_y \rtimes \iota)\big).$$

Since $C_0(X) \rtimes_{\mathrm{lt}} G$ is GCR, this implies that $\mathrm{Ind}_H^G(\rho_x \rtimes \iota)$ and $\mathrm{Ind}_H^G(\rho_y \rtimes \iota)$ are equivalent. Using Proposition 5.4 on page 153, we can realize $\mathrm{Ind}_H^G(\rho_x \rtimes \iota) = M_x \rtimes U$ on $L^2_1(G, \mu_{G/H}; \mathbf{C}) \cong L^2(G/H)$ where

$$M_x(\xi)h(\dot{s}) = \xi(s\cdot x)h(\dot{s}).$$

Of course, we must have M_x equivalent to M_y. If $G\cdot x \neq G\cdot y$, then $G\cdot x \cap G\cdot y = \emptyset$. Let $Y = Z = G/H$, and define $i : Y \to X$ by $i(sH) = s\cdot x$ and $j : Z \to X$ by

[8] Since Theorem 8.39 on page 251 implies that latter quotient embeds into a primitive ideal space, this description helps to explain the reason that the quasi-orbit space $(G\backslash X)^{\sim}$ plays a key role in the proof of Theorem 8.39 on page 251

$j(sH) = s \cdot y$. Since i and j have disjoint images, Lemma I.42 on page 505 implies that M_x is not equivalent to M_y. This is a contradiction and completes the proof of the first assertion. The second assertion follows from the first and Proposition 7.30 on page 222 (together with the observation that, since G is abelian, G_x is CCR for all x). □

Theorem 8.44. *Suppose that (G, X) is a locally compact transformation group with G abelian. If $C_0(X) \rtimes_{\mathrm{lt}} G$ is CCR, then the G-orbits in X are closed. In particular, if (G, X) is second countable, then $C_0(X) \rtimes_{\mathrm{lt}} G$ is CCR if and only if the G-orbits in X are closed.*

Proof. Suppose that $C_0(X) \rtimes_{\mathrm{lt}} G$ is CCR. Then Theorem 8.43 on the previous page implies that $G\backslash X$ is T_0. Thus $(G\backslash X)^\sim = G\backslash X$. Then Theorem 8.39 on page 251 (together with Remark 8.42 on page 252) implies that $x \mapsto \Phi(x, \iota) = \ker\bigl(\mathrm{Ind}_{G_x}^G(\rho_x \rtimes \iota)\bigr)$ defines a continuous injection of $G\backslash X$ into $\mathrm{Prim}(C_0(X) \rtimes_{\mathrm{lt}} G)$. Since $C_0(X) \rtimes_{\mathrm{lt}} G$ is CCR, points in $\mathrm{Prim}(C_0(X) \rtimes_{\mathrm{lt}} G)$ are closed. Therefore points in $G\backslash X$ are closed; that is, each G-orbit in X must be closed as required.

The last statement now follows from Proposition 7.31 on page 223. □

Example 8.45 (Irrational Rotation Algebras). Let A_θ be the rotation algebra $C(\mathbf{T}) \rtimes_\alpha \mathbf{Z}$ where α is determined by rotation through the angle $2\pi\theta$ with θ *irrational.* Then the action of $\mathbf{Z}$ on $\mathbf{T}$ is free. Thus Theorem 8.39 on page 251 implies that $\mathrm{Prim}\, A_\theta$ is homeomorphic to the T_0-ization $(\mathbf{Z}\backslash\mathbf{T})^\sim$ of the orbit space. However, by Lemma 3.29 on page 96, each $\mathbf{Z}$-orbit is dense. Therefore $\mathbf{Z}\backslash\mathbf{T}$ is not a T_0-space, and $(\mathbf{Z}\backslash\mathbf{T})^\sim$ is reduced to a point. It follows that A_θ is simple.[9] By Theorem 8.43 on the preceding page, A_θ is not GCR. Therefore if θ is irrational, then the irrational rotation C^*-algebra A_θ is a simple NGCR algebra.[10]

Example 8.46 (Rational Rotation Algebras). Let A_θ be the rotation algebra $C(\mathbf{T}) \rtimes_\alpha \mathbf{Z}$ where α is determined by rotation through the angle $2\pi\theta$ with θ *rational.* Say $\theta = p/q$ with p and q relatively prime. Then the stability group at each point of $\mathbf{T}$ is $q\mathbf{Z}$. The orbits are finite, and hence closed. Furthermore,

$$\mathbf{Z}\backslash\mathbf{T} \cong (\frac{p}{q}\mathbf{Z})\backslash(\mathbf{R}/\mathbf{Z}) \cong \mathbf{R}/\frac{1}{q}\mathbf{Z} \cong \mathbf{T}.$$

Therefore Theorem 8.39 on page 251 implies that

$$\mathrm{Prim}\, A_\theta \cong \mathbf{T} \times (\widehat{\mathbf{Z}}/(q\mathbf{Z})^\perp) \cong \mathbf{T} \times \widehat{q\mathbf{Z}} \cong \mathbf{T}^2.$$

Since the orbits are closed, A_θ is CCR by Proposition 7.31 on page 223 (or Theorem 8.44). Furthermore, every irreducible representation of A_θ is of the form $\mathrm{Ind}_{G_x}^G(\rho_z \rtimes \omega)$ for some $z \in \mathbf{T}$ and $\omega \in (q\mathbf{Z})^\wedge$ (Remark 8.41 on page 251). Since we can realize $\mathrm{Ind}_{G_x}^G(\rho_z \rtimes \omega)$ on the q-dimensional Hilbert space $L^2(\mathbf{Z}/q\mathbf{Z})$, it follows

[9] It is interesting to compare this "short" proof of the simplicity of A_θ with our earlier proof in Proposition 2.56 on page 68. Of course, this proof is only "short" because we are able to invoke the machinery of this section — which includes the GRS-Theorem.

[10] A C^*-algebra is called NGCR if it has no nonzero CCR ideals.

that when θ is rational, A_θ is a q-homogeneous C^*-algebra with spectrum $\mathbf{T}^2$. (As we pointed out in Remark 2.60 on page 71, the rational rotation algebras are even Morita equivalent to $C(\mathbf{T}^2)$.[11]

Example 8.47. Let $\mathbf{R}^+_\times$ be the group of positive real numbers under multiplication. Let $\mathbf{R}^+_\times$ act on $\mathbf{R}^2$ by

$$s \cdot (a,b) := (a/s, b/s).$$

The origin is a fixed points and all other orbits are rays originating at the origin. The action is free except at the origin. Since the orbits are all locally closed, but not all closed, $C_0(\mathbf{R}^2) \rtimes_\alpha \mathbf{R}^+_\times$ is GCR, but not CCR (Theorems 8.43 and 8.44 on the facing page). Thus the primitive ideal space can be identified with the spectrum, and the induced representations are (up to equivalence) given by inducing from stability groups. The spectrum can be identified with the the set $\mathbf{T} \cup \mathbf{R}^+$. Since the natural map of $\mathbf{R}^+_\times \times \mathbf{R}^2$ onto $\mathbf{T} \cup \mathbf{R}^+$ is open, the open sets in $\mathbf{T} \cup \mathbf{R}^+$ are given by

$$\{\, U \subset \mathbf{T} : U \text{ is open in } \mathbf{T} \,\} \cup \{\, \mathbf{T} \cup V : V \text{ is open in } \mathbf{R}^+ \,\}.$$

In particular, $\mathbf{T}$ is open and corresponds to the ideal $I := C_0(\mathbf{R}^2 \smallsetminus \{\, (0,0) \,\}) \rtimes_\alpha \mathbf{R}^+_\times$, and $\mathbf{R}^+$ is closed and corresponds to the quotient $C_0(\mathbf{R}^2) \rtimes_\alpha \mathbf{R}^+_\times / I \cong C^*(\mathbf{R}^+_\times)$.

Example 8.48. The situation in Example 8.47 changes dramatically if we alter the action so that

$$s \cdot (a,b) := (a/s, sb).$$

We described the orbit space F in Example 3.32 on page 96. Since the $\mathbf{R}^+_\times$-action is free everywhere except at the origin, we may identify the spectrum of $C_0(\mathbf{R}^2) \rtimes_\beta \mathbf{R}^+_\times$ with a set identical to F except that the origin has been replaced by $\mathbf{R}^+$. The open sets in the spectrum are given by

$$\begin{aligned} &\{\, U : U \subset F \smallsetminus \{\, (0,0) \,\} \text{ and } U \text{ open in } F \,\} \cup \\ &\qquad \{\, U \cup V : U \text{ is a open neighborhood of } (0,0) \text{ in } F \text{ with } (0,0) \text{ deleted}, \\ &\qquad\qquad\qquad\qquad \text{and } V \text{ is open in } \mathbf{R}^+ \,\}. \end{aligned}$$

Example 8.49 (Semidirect Products of Abelian Groups). Suppose that $G = N \rtimes_\varphi H$ is the semidirect product of two locally compact abelian groups N and K. As in Example 3.16 on page 91, $C^*(G)$ is isomorphic to the transformation group C^*-algebra $C_0(\widehat{N}) \rtimes H$, where $h \cdot \omega(\cdot) := \omega(h^{-1} \cdot h)$. By Theorem 8.39 on page 251, the primitive ideal space of $C^*(G)$ is homeomorphic to

$$(H \backslash \widehat{N})^\sim \times \widehat{H}/\!\sim$$

as described above.

[11] Since q-homogeneous C^*-algebras are always locally trivial [50], A_θ must have continuous trace. Since its spectrum is $\mathbf{T}^2$ and since $H^3(\mathbf{T}^2) = \{\, 0 \,\}$, it follows that A_θ has trivial Dixmier-Douady class. Therefore A_θ is Morita equivalent to $C(\mathbf{T}^2)$ by [139, Proposition 5.33].

Example 8.50 (The $ax + b$ group). To continue with the previous example, we consider the specific case of the $ax+b$ group. Then, as in Example 3.15 on page 91, $C^*(G)$ is isomorphic to the transformation group C^*-algebra $C_0(\mathbf{R}) \rtimes_\tau \mathbf{R}^+_\times$ where the action of $\mathbf{R}^+_\times$ on $\mathbf{R}$ is given by

$$s \cdot a = a/s.$$

There are just three orbits: the origin, and the two half lines $p_{-1} := \mathbf{R}^+_\times \cdot (-1)$ and $p_2 := \mathbf{R}^+_\times \cdot 1$. The orbits are all locally closed, but not all closed. Hence $C^*(G)$ is GCR (but not CCR). The action is free except at the origin — which is a fixed point. Therefore we can identify the spectrum of $C^*(G)$ with the set $\{ p_{-1} \} \cup \mathbf{R}^+ \cup \{ p_1 \}$. The sets $\{ p_i \}$ are open (for $i = -1, 1$) as are sets of the form $U \cup \{ p_{-1}, p_1 \}$, when U is open in $\mathbf{R}^+$. Since the irreducible representations are all equivalent to an induced representation, it is a straightforward matter to write down representatives for the infinite-dimensional representations corresponding to each p_i, as well as the complex homomorphisms corresponding to each point in $\mathbf{R}^+$.

8.4 The Fell Subgroup Crossed Product

If (A, G, α) is a dynamical system and if Σ is the compact Hausdorff space of closed subgroups of G, then for each $H \in \Sigma$, we have a crossed product $A \rtimes_{\alpha|_H} H$. Our study of EH-regular systems, and regular systems in particular, suggests that a good deal of the ideal structure of the crossed product $A \rtimes_\alpha G$ is contained in the crossed products by subgroups. One way to encode this information is in an analogue of Fell's Subgroup Algebra [52], which we call *Fell's subgroup crossed product.* Fell used the subgroup algebra to study the continuity of the induction process, and the subgroup crossed product can be used similarly (cf., for example, [170]). However, we introduce its construction here primarily so that we can use it in the proof of the Gootman-Rosenberg-Sauvageot Theorem on page 241 in Chapter 9.

Let $\{ \mu_H \}$ be a continuous choice of Haar measures on Σ (Lemma H.8 on page 458), and let

$$G * \Sigma = \{ (s, H) \in G \times \Sigma : s \in H \}.$$

Using Lemma H.2 on page 454, it is easy to see that $G * \Sigma$ is closed in $G \times \Sigma$, and therefore it is locally compact.

Proposition 8.51. *Suppose that (A, G, α) is a dynamical system. Then $C_c(G * \Sigma, A)$ is a $*$-algebra with respect to the operations defined by*

$$f * g(s, H) := \int_H f(r, H)\alpha_r\big(g(r^{-1}s, H)\big)\, d\mu_H(r) \quad \textit{and}$$
$$f^*(s, H) := \Delta_H(s^{-1})\alpha_s\big(f(s^{-1}, H)^*\big).$$

Most of the proof of Proposition 8.51 are routine; after all, for $H \in \Sigma$, the "fibre" over H is $C_c(H, A)$ with the $*$-algebra structure coming from $A \rtimes_{\alpha|_H} H$. We do have to work a bit to see that the above formulas define elements in $C_c(G*\Sigma, A)$. For this, the following lemma is useful.

Lemma 8.52. *Let $G * G * \Sigma = \{ (s, r, H) \in G \times G \times \Sigma : s, r \in H \}$. Then $G * G * \Sigma$ is locally compact and if $f \in C_c(G * G * \Sigma, A)$, then*

$$F_f(s, H) := \int_H f(s, r, H) \, d\mu_H(r)$$

defines an element $F_f \in C_c(G * \Sigma, A)$.

Proof. Clearly, $G * G * \Sigma$ is closed in $G \times G \times \Sigma$. Hence it is locally compact, and given f as above, there is a compact set $K \subset G$ such that $f(s, r, H) = 0$ if $s \notin K$ or $r \notin K$. Let $n(K) \in \mathbf{R}^+$ be such that $\mu_H(K \cap H) \leq n(K)$ for all $H \in \Sigma$ (Lemma H.9 on page 460). Then $F_f(s, H)$ is a well-defined element of A and

$$\|F_f(s, H)\| \leq \|f\|_\infty n(K).$$

It follows that if $f_i \to f$ in the inductive limit topology on $C_c(G * G * \Sigma, A)$, then $F_{f_i} \to F_f$ uniformly. If $\varphi \in C_c(G * G * \Sigma)$ and if $a \in A$, then we can define $\varphi \otimes a \in C_c(G * G * \Sigma, A)$ by $\varphi \otimes a(s, r, H) = \varphi(s, r, H)a$. Since $G * G * \Sigma$ is closed in $G \times G \times \Sigma$, there is a $\Phi \in C_c(G \times G \times \Sigma)$ extending φ. Since

$$F_{\varphi \otimes a}(s, H) = \int_H \Phi(s, r, H) \, d\mu_H(r)a,$$

$F_{\varphi \otimes a} \in C_c(G * \Sigma, A)$ by Lemma H.9 on page 460. Since the span of elements of the form $\varphi \otimes a$ is dense in $C_c(G * \Sigma, A)$ in the inductive limit topology (Lemma 1.87 on page 29), this suffices. □

Proof of Proposition 8.51 on the preceding page. The continuity of f^* follows immediately from the continuity of $(s, H) \mapsto \Delta_H(s)$ which is proved in Lemma H.9 on page 460. On the other hand,

$$(s, r, H) \mapsto f(r, H)\alpha_r\big(g(r^{-1}s, H)\big)$$

is in $C_c(G * G * \Sigma, A)$. Therefore the continuity of $f * g$ follows from Lemma 8.52. The rest is routine. □

We can define a norm $\| \cdot \|_\Sigma$ on $C_c(G * \Sigma, A)$ by

$$\|f\|_\Sigma := \sup_{H \in \Sigma} \int_H \|f(s, H)\| \, d\mu_H(s).$$

A nondegenerate $*$-homomorphism $L : C_c(G * \Sigma, A) \to B(\mathcal{H}_L)$ is called a $\| \cdot \|_\Sigma$-decreasing representation if $\|L(f)\| \leq \|f\|_\Sigma$ for all $f \in C_c(G * \Sigma, A)$. The universal norm on $C_c(G * \Sigma, A)$ is given by

$$\|f\| := \sup\{ \|L(f)\| : L \text{ is a } \| \cdot \|_\Sigma\text{-decreasing representation} \}.$$

Remark 8.53. As usual in this sort of construction, the class of $\| \cdot \|_\Sigma$-representations is not necessarily a set. But the values $\|L(f)\|$ form a subclass of the set $\mathbf{R}$ of real numbers. The separation axioms of set theory imply that a subclass of a set is a set. Hence the supremum is well-defined [84, §1.1]. Alternatively, we could restrict to the set of cyclic representations on a Hilbert space of suitably large dimension, but this hardly seems to clarify anything.

The completion of $C_c(G*\Sigma, A)$ in the universal norm is denoted by $A \rtimes_\alpha \Sigma$ and is called *Fell's subgroup crossed product.*

Lemma 8.54. *Suppose that A is a C^*-algebra and that F is a closed subset of a locally compact space X. Then the restriction map $r : C_0(X, A) \to C_0(F, A)$ is surjective. In particular, if $g \in C_c(F, A)$, then there is a $f \in C_c(X, A)$ such that $r(f) = g$.*

Proof. To see that r is surjective, it suffices to see that it has dense range (since r is a homomorphism between C^*-algebras). However, $C_c(F, A)$ is dense in $C_0(F, A)$, and the span of functions of the form $\varphi \otimes a(s) := \varphi(s)a$, with $\varphi \in C_c(F)$ and $a \in A$, are dense in $C_c(F, A)$ in the inductive limit topology (and hence in the $\|\cdot\|_\infty$-norm) by Lemma 1.87 on page 29. If $\varphi \in C_c(F)$, then the Tietze extension theorem implies that there is a $\psi \in C_c(X)$ which extends φ. Since $r(\psi \otimes a) = \varphi \otimes a$, it follows that r has dense range.

If $g \in C_c(F, A)$, then by the above, there is a $f \in C_0(X, A)$ such that $r(f) = g$. If $\psi \in C_c(X)$ is such that $\psi(x) = 1$ for all $x \in \operatorname{supp} g$, then $r(\psi \cdot f) = g$ and $\psi \cdot f \subset C_c(X, A)$ as required. $\square$

Lemma 8.54 implies that the restriction map

$$\kappa_H : C_c(G * \Sigma, A) \to C_c(H, A)$$

is surjective for each $H \in \Sigma$. If R is a representation of $A \rtimes_{\alpha|_H} H$, then $R \circ \kappa_H$ is a $\|\cdot\|_\Sigma$-decreasing representation of $C_c(G*\Sigma, A)$. Thus κ_H extends to a surjective homomorphism, also called κ_H, of $A \rtimes_\alpha \Sigma$ onto $A \rtimes_{\alpha|_H} H$.

Proposition 8.55. *Suppose that (A, G, α) is a dynamical system and that Σ is the compact Hausdorff space of closed subgroups of G. Then $A \rtimes_\alpha \Sigma$ is a $C(\Sigma)$-algebra. The $C(\Sigma)$-action on $f \in C_c(G * \Sigma, A)$ is given by $\varphi \cdot f(s, H) = \varphi(H)f(s, H)$. The fibre $A \rtimes_\alpha \Sigma(H)$ over H is (isomorphic to) $A \rtimes_{\alpha|_H} H$, and every irreducible representation of $A \rtimes_\alpha \Sigma$ is (equivalent to one) of the form $R \circ \kappa_H$, where $H \in \Sigma$ and R is an irreducible representation of $A \rtimes_{\alpha|_H} H$.*

Proof. In view of Lemma 8.3 on page 230, to see that $A \rtimes_\alpha \Sigma$ is a $C(\Sigma)$-algebra it suffices to see that $\|\varphi \cdot f\| \le \|\varphi\|_\infty \|f\|$ for $\varphi \in C(\Sigma)$ and $f \in C_c(G * \Sigma, A)$. Let $\psi := \big(\|\varphi\|_\infty^2 1_\Sigma - |\varphi|^2\big)^{\frac{1}{2}}$. Then $\psi \in C(\Sigma)$, and

$$\begin{aligned}(\varphi \cdot f)^* * (\varphi \cdot f) &= |\varphi|^2 \cdot f^* * f \\ &= \|\varphi\|_\infty^2 f^* * f - \big(\|\varphi\|_\infty^2 1_\Sigma - |\varphi|^2\big) \cdot f^* * f \\ &= \|\varphi\|_\infty^2 f^* * f - (\psi \cdot f)^* * (\psi \cdot f).\end{aligned}$$

Therefore $(\varphi \cdot f)^* * (\varphi \cdot f) \le \|\varphi\|_\infty^2 f^* * f$ in $A \rtimes_\alpha \Sigma$, and

$$\|\varphi \cdot f\|^2 = \|(\varphi \cdot f)^* * (\varphi \cdot f)\| \le \|\varphi\|_\infty^2 \|f * f\| = \|\varphi\|_\infty^2 \|f\|^2.$$

It follows that $A \rtimes_\alpha \Sigma$ is a $C(\Sigma)$-algebra as claimed.

Let I_H be the ideal of $A \rtimes_\alpha \Sigma$ spanned by $C_{0,H}(\Sigma) \cdot C_c(G * \Sigma, A)$. Then $A \rtimes_\alpha \Sigma(H) = A \rtimes_\alpha \Sigma / I_H$. Clearly $I_H \subset \ker \kappa_H$. To see that $A \rtimes_\alpha \Sigma(H)$ is isomorphic to $A \rtimes_{\alpha|_H} H$, it will suffice to prove that $I_H = \ker \kappa_H$. Let L be a representation of $A \rtimes_\alpha \Sigma$ with $I_H \subset \ker L$. It follows that if $\varphi \in C(\Sigma)$ is such that $\varphi(H) = 1$, then $L(f) = L(\varphi \cdot f)$ for all $f \in C_c(G * \Sigma, A)$. On the other hand,

$$H \mapsto \int_H \|f(s,H)\| \, d\mu_H(s) = \|\kappa_H(f)\|_1$$

is continuous on Σ (since $\{\, \mu_H \,\}$ is a continuous choice of Haar measures). Therefore, given $f \in C_c(G * \Sigma, A)$ and $\epsilon > 0$, there is a $\varphi \in C(\Sigma)$ such that $\varphi(H) = 1$ and such that

$$\|\varphi \cdot f\|_\Sigma \le \|\kappa_H(f)\|_1 + \epsilon.$$

Since ϵ is arbitrary, it follows that

$$\|L(f)\| \le \|\kappa_H(f)\|_1.$$

Since $\kappa_H\bigl(C_c(G * \Sigma, A)\bigr) = C_c(H, A)$, L defines a $\|\cdot\|_1$-decreasing representation L' on $C_c(H,A)$ by

$$L'\bigl(\kappa_H(f)\bigr) = L(f).$$

Therefore, $\|L(f)\| \le \|\kappa_H(f)\|_1$. Since L is any representation with I_H in its kernel, we must have $I_H = \ker \kappa_H$ as required.

The last assertion follows as every irreducible representation of a $C(\Sigma)$-algebra must factor through a fibre (Proposition C.5 on page 355). □

Remark 8.56. If $r = (\pi, u)$ is a representation of $A \rtimes_{\alpha|_H} H$, then there is a natural representation $r' := r \circ \kappa_H$ of $A \rtimes_\alpha \Sigma$. Often, we will not distinguish between r and r' and trust that the meaning will be clear from context.

The following observations will be needed in Chapter 9. Suppose that (π, u) is a covariant representation of $(A, H, \alpha|_H)$. If $s \in G$, let $s \cdot \pi$ and $s \cdot u$ be defined by

$$s \cdot \pi(s) := \pi\bigl(\alpha_s^{-1}(a)\bigr) \quad \text{for } a \in A \quad \text{and} \quad s \cdot u(t) := u(s^{-1}ts) \quad \text{for } t \in s \cdot H.$$

Then a short computation shows that $(s \cdot \pi, s \cdot u)$ is a covariant representation of $(A, s \cdot H, \alpha|_{s \cdot H})$. Let $L := (\pi, u)$ and $s \cdot L := (s \cdot \pi, s \cdot u)$. If $\Phi \in C_c(G * \Sigma)$ and if we define

$$s \cdot \Phi(t, H) := \omega(s, s^{-1} \cdot H)\alpha_s\bigl(\Phi(s^{-1}ts, s^{-1} \cdot H)\bigr), \tag{8.14}$$

where ω is as in Lemma H.10 on page 461, then

$$\begin{aligned} L(s^{-1} \cdot \Phi) &= \int_H \pi\bigl(s^{-1} \cdot \Phi(t,H)\bigr)u(t)\, d\mu_H(t) \\ &= \omega(s^{-1}, s \cdot H) \int_H \pi\bigl(\alpha_s^{-1}\bigl(\Phi(s^{-1}ts, s^{-1} \cdot H)\bigr)\bigr)u(t)\, d\mu_H(t) \\ &= \int_{s \cdot H} s \cdot \pi\bigl(\Phi(t, s \cdot H)\bigr) s \cdot u(t)\, d\mu_{s \cdot H}(t) \\ &= s \cdot L(\Phi). \end{aligned}$$

Lemma 8.57. *Suppose that* (A, G, α) *is a dynamical system and that* $A \rtimes_\alpha \Sigma$ *is Fell's subgroup algebra. Then there is a dynamical system*

$$\beta : G \to \operatorname{Aut} A \rtimes_\alpha \Sigma \tag{8.15}$$

such that $\beta_s(\Phi) = s \cdot \Phi$ *for all* $\Phi \in C_c(G * \Sigma, A)$.

Proof. The preceding discussion shows that β_s is isometric on $C_c(G * \Sigma, A)$ so that (8.15) is a homomorphism. However, if $s_i \to e$, then it is not hard to check that $s_i \cdot \Phi \to \Phi$ in the inductive limit topology. It now follows easily that β is strongly continuous and therefore a dynamical system. □

Remark 8.58. As mentioned at the beginning of this section, we have introduced $A \rtimes_\alpha \Sigma$ primarily because it is needed in the proof of the Gootman-Rosenberg-Sauvageot Theorem in Chapter 9. Unfortunately, there isn't space to develop some of its other applications to crossed products. Nevertheless, we want to at least describe Fell's original motivation for introducing the subgroup algebra $C^*_\Sigma(G) := \mathbf{C} \rtimes_{\mathrm{id}} \Sigma$ in [52]. Let

$$\mathscr{S}(G) = \{\, (H, \pi) : H \in \Sigma \text{ and } \pi \text{ is a representation of } C^*(H) \,\}.$$

We can, and do, view $\mathscr{S}(G)$ as a collection of representation of $C^*_\Sigma(G)$, and topologize it by pulling back the topology from $\mathcal{I}\bigl(C^*_\Sigma(G)\bigr)$. Therefore $(H_i, \pi_i) \to (H, \pi)$ in $\mathscr{S}(G)$ if $\ker(\pi_i \circ \kappa_{H_i}) \to \ker(\pi \circ \kappa_H)$ in $\mathcal{I}\bigl(C^*_\Sigma(G)\bigr)$. We already know that $\pi \mapsto \operatorname{Ind}_H^G \pi$ is continuous in this sense; that is, if $\ker \pi_i \to \ker \pi$ in $\mathcal{I}\bigl(C^*(H)\bigr)$, then Lemma 5.16 on page 164 implies that $\ker\bigl(\operatorname{Ind}_H^G \pi\bigr) \to \ker\bigl(\operatorname{Ind}_H^G \pi\bigr)$ in $\mathcal{I}\bigl(C^*(G)\bigr)$. More generally, let

$$X := \{\, (K, H, \pi) \in \Sigma \times \mathscr{S}(G) : K \supset H \,\},$$

and give X the relative topology as a subset of $\Sigma \times \mathscr{S}(G)$. Then we get a map of X into $\mathscr{S}(G)$ defined by

$$(K, H, \pi) \mapsto \operatorname{Ind}_H^K \pi.$$

Fell shows that this map is continuous [52, Theorem 4.2]. More colloquially, $\operatorname{Ind}_H^K \pi$ is continuous in all three variables H, K and π. Naturally, these sorts of considerations extend to Fell's subgroup crossed product, but we leave the details for another day.

Notes and Remarks

Theorem 8.16 is a "$C_0(X)$-algebra" version of the main result in [171] (as is Corollary 8.6 and much of the material in Section 8.1). Theorem 8.19 was shown to me by Echterhoff. Of course, Theorem 8.21 is due to Gootman & Rosenberg [64], but the contribution of Sauvageot should not be underestimated [159, 160]. Section 8.3 is taken primarily from [169] (although Theorem 8.43 is a special case of Gootman's [62] with a different proof). Section 8.4 is a straightforward variation on Fell's [52].

Chapter 9

The Proof of the Gootman-Rosenberg-Sauvageot Theorem

The proof of the Gootman-Rosenberg-Sauvageot Theorem (on page 241) was the focus of intense work in the 1970s. The method of attack was developed by Sauvageot in [159] and [160]. The complete solution was given by Gootman and Rosenberg in [64]. The proof given here follows these ground breaking papers and also borrows heavily from Renault's proof of the analogue of the GRS-Theorem for groupoids [144]. The proof is involved and necessitates the introduction of sophisticated technology such as direct integrals and the theory of standard Borel spaces. This will force us to up the difficulty level a fair bit. However, as EH-regularity and the proof of the validity of the generalized Effros-Hahn conjecture is so important, it seems worth the effort to present the argument here at a level of detail comparable with that in the previous chapters. Much of the additional technology alluded to above has been relegated to appendices. While this allows us to get right to the matter at hand, those unfamiliar with topics such as direct integrals, homogeneous representations of C^*-algebras and Borel structures will have to make frequent detours to the appropriate appendices.

Summary of the Proof

We have to show that given $K \in \operatorname{Prim}(A \rtimes_\alpha G)$ there is a $P \in \operatorname{Prim} A$ and a $J \in \operatorname{Prim} A \rtimes_\alpha G_P$ such that $\operatorname{Res} J = P$ and such that $\operatorname{Ind}_{G_P}^G J = K$.

Step I: We would like to show that if $J \in \operatorname{Prim}(A \rtimes_\alpha G_P)$ with $\operatorname{Res} J = P$, then $\operatorname{Ind}_{G_P}^G J$ is a primitive ideal in $A \rtimes_\alpha G$. Unfortunately, this seems to be difficult to prove.[1] Instead, we follow Sauvageot and work with homogeneous representations. (Homogeneous representations are treated in Appendix G.1.) Suppose that $J = \ker L$, and that we also assume that $L = (\pi, u)$ is a homogeneous representation of $A \rtimes_\alpha G_P$ such that π is homogeneous (with kernel P). The "extra" hypothesis on the "π part" of L allows us to show that $\operatorname{Ind}_{G_P}^G L$ is homogeneous. Since we're assuming (A, G, α) is separable throughout, it follows that $\operatorname{Ind}_{G_P}^G J = \ker\bigl(\operatorname{Ind}_{G_P}^G L\bigr)$ is primitive. This is Proposition 9.5 on page 268.

Step II: Fix $K \in \operatorname{Prim} A \rtimes_\alpha G$. We need to produce an induced primitive ideal which is related to K. Let $R = (\pi, V)$ be a factorial representation of $A \rtimes_\alpha G$ with kernel K. Sauvageot's idea was to employ Effros's ideal center decomposition (treated in Appendix G). Although it might seem peculiar not be simply take R irreducible, we want to assume that R has infinite multiplicity (cf., Remark 9.6 on page 271), then as in Remark G.23 on page 444, we can form an ideal center decomposition of π, say

$$\pi = \int_{\operatorname{Prim} A}^{\oplus} \pi_P \, d\mu(P)$$

where the associated Borel Hilbert bundle $\operatorname{Prim} A * \mathfrak{H}$ is trivial. This allows us to assume that π acts on $L^2(\operatorname{Prim} A, \mu, \mathcal{H})$, and thus to remove one layer of direct integrals. This will reduce, but unfortunately not eliminate, some of the technical detail of direct integrals in the sequel. (This simplification will be especially useful in Step III below.) Then, by decomposing V, we obtain for μ-almost every P a covariant representation (π_P, σ_P) of (A, G_P, α) which determines a representation r_P of the subgroup crossed product algebra $A \rtimes_\alpha \Sigma$. (Fell's subgroup crossed product, $A \rtimes_\alpha \Sigma$, is defined in Section 8.4.) Then it is possible to construct a representation of $A \rtimes_\alpha \Sigma$,

$$r := \int_{\operatorname{Prim} A}^{\oplus} r_P \, d\mu(P),$$

which is called the *restriction of R to the stability groups.* Once we have r properly defined, we need to examine an ideal center decomposition

$$\tilde{r} := \int_{\operatorname{Prim}(A \rtimes_\alpha \Sigma)}^{\oplus} \tilde{r}_Q \, d\nu(Q)$$

for r. There are a number of technicalities to worry about. For example, we want to see that ν is quasi-invariant and ergodic[2] for the natural G-action on $\operatorname{Prim} A \rtimes_\alpha \Sigma$ (Proposition 9.8 on page 274). Also, we show that the ideal center $\mathcal{IC}(\pi)$ of π is contained in the ideal center $\mathcal{IC}(r)$ of r (Lemma 9.7 on page 272). This gives an inclusion of $L^\infty(\operatorname{Prim} A, \mu)$ in $L^\infty\bigl(\operatorname{Prim}(A \rtimes_\alpha \Sigma), \nu\bigr)$, and it is well-known that such inclusions are given by a map $\tau : \operatorname{Prim}(A \rtimes_\alpha \Sigma) \to \operatorname{Prim} A$ of the underlying

[1] If A is type I, then the above statement can be proved as written. In fact, more is true. If $L = (\pi, u)$ is an *irreducible* representation of $A \rtimes_\alpha G_P$ such that $\ker \pi = P$ and if A is type I, then $\operatorname{Ind}_{G_P}^G L$ is irreducible [44].

[2] We say that ν is *ergodic* if every G-invariant Borel set is either null or conull.

spaces (Lemma I.11). Using this, we find that $\tilde{r}$ decomposes as a direct integral over $\operatorname{Prim} A$ in a useful way. In particular, it follows that, up to a null set of course, each $\tilde{r}_Q$ is determined by a covariant representation $(\tilde{\pi}_Q, \tilde{u}_Q)$ of $(A, G_{\tau(Q)}, \alpha)$ with $\tilde{\pi}_Q$ homogeneous with kernel $\tau(Q)$.

Step III: In this step, we form a representation $\operatorname{ind} r$ of $A \rtimes_\alpha G$ which we view as being induced from r. We take advantage of having π act on $L^2(\operatorname{Prim} A, \mu, \mathcal{H})$ to explicitly realize $\operatorname{ind} r$ on a space of equivalence classes of $\mathcal{H}$-valued functions on $\operatorname{Prim} A$. After making the above precise, there is still hard work to do in order to show that $K' := \ker(\operatorname{ind} r)$ is an induced primitive ideal as in Step I. This is proved in Proposition 9.14 on page 283.

Step IV: If G is amenable, then we show in Proposition 9.22 on page 292 that

$$\ker(\operatorname{ind} r) = K' \subset K := \ker R.$$

Step V: Steps I, II, III and IV are contained in Sauvageot's [160], and the proof here is taken from his work. The final, and most technical, piece of the proof was accomplished by Gootman and Rosenberg in [64], where it is shown that we always have

$$\ker R \subset \ker(\operatorname{ind} r).$$

Since $\ker(\operatorname{ind} r)$ is induced, this completes the proof. Our proof is strongly influenced by Renault's generalization of the GRS-Theorem to locally compact groupoids in [144]. Our result is Proposition 9.24 on page 298.

Preliminaries

In this chapter, (A, G, α) will be a *separable* dynamical system. We write Σ for the compact Hausdorff space of closed subgroups of G (equipped with the Fell topology as in Corollary H.4 on page 455), and we fix a continuous choice of Haar measures μ_H on Σ (as in Lemma H.8 on page 458). Since G is second countable, Proposition H.17 on page 466 implies that there is a continuous function $\mathfrak{b} : G \times \Sigma \to [0, \infty)$, called a *generalized Bruhat approximate cross section*, such that

(a) $\displaystyle\int_H \mathfrak{b}(st, H)\, d\mu_H(t) = 1$ for all $H \in \Sigma$ and $s \in G$, and

(b) if $K \subset G$ is compact, then

$$\operatorname{supp} \mathfrak{b} \cap \{\, (st, H) \in G \times \Sigma : s \in K \text{ and } t \in H \,\}$$

is compact.

We let $\omega : G \times \Sigma \to (0, \infty)$ be as defined in Lemma H.10 on page 461, and we define $\rho : G \times \Sigma \to (0, \infty)$ as in Equation (H.4) on page 461. Then we let β^{G_P} be the quasi-invariant measure on G/G_P corresponding to $s \mapsto \rho(s, G_P)$ as in Equation (H.5) on page 462. We almost always write β^P in place of β^{G_P}. Since A is separable, $\operatorname{Prim} A$ is a standard Borel space (Theorem H.40 on page 477). In fact, we can equip Prim A with its regularized topology, which is a Polish topology finer than the hull-kernel topology, and which generates the same (standard) Borel structure. In its Polish topology, $(G, \operatorname{Prim} A)$ is still a topological transformation group (Theorem H.39 on page 476).

If μ is a finite quasi-invariant measure on Prim A, then we let $d : G \times \operatorname{Prim} A \to (0, \infty)$ be a Borel choice of Radon-Nikodym derivatives as in Corollary D.34 on page 389. In particular, if f is a bounded Borel function on Prim A, then

$$\int_{\operatorname{Prim} A} f(P) d(s, P)\, d\mu(P) = \int_{\operatorname{Prim} A} f(s \cdot P)\, d\mu(P).$$

Recall that if $\tau : X \to Y$ is a Borel map and if ν is a measure on X, then the *push-forward* of ν by τ is the measure $\tau_* \nu$ on Y given by $\tau_*(E) := \nu\big(\tau^{-1}(E)\big)$ (see Lemma H.13 on page 463).

Lemma 9.1. *Let μ be a finite quasi-invariant measure on* Prim A. *For all nonnegative Borel functions f on* $\operatorname{Prim} A \times \operatorname{Prim} A$, *the function*

$$P \mapsto \int_{G/G_P} f(P, s \cdot P)\, d\beta^P(\dot{s})$$

is Borel, and there is a σ-finite Borel measure γ on $\operatorname{Prim} A \times \operatorname{Prim} A$ *such that*

$$\gamma(f) = \int_{\operatorname{Prim} A} \int_{G/G_P} f(P, s \cdot P)\, d\beta^P(\dot{s})\, d\mu(P). \tag{9.1}$$

(a) *If $\sigma : \operatorname{Prim} A \times \operatorname{Prim} A \to \operatorname{Prim} A \times \operatorname{Prim} A$ is the flip, $\sigma(P, Q) = (Q, P)$, then $\sigma_* \gamma$ is equivalent to γ. In particular, there is a Borel function $D : \operatorname{Prim} A \times \operatorname{Prim} A \to (0, \infty)$ such that $D(P, Q)^{-1} = D(Q, P)$ for all (P, Q) and such that for all nonnegative Borel functions f on* $\operatorname{Prim} A \times \operatorname{Prim} A$ *we have*

$$\int_{\operatorname{Prim} A} \int_{G/G_P} f(P, s \cdot P) D(P, s \cdot P)\, d\beta^P(\dot{s})\, d\mu(P) = \int_{\operatorname{Prim} A} \int_{G/G_P} f(s \cdot P, P)\, d\beta^P(\dot{s})\, d\mu(P). \tag{9.2}$$

(b) *Let $d : G \times \operatorname{Prim} A$ be a Borel choice of Radon-Nikodym derivatives as above. Then*

$$D(P, s \cdot P) = \rho(s, G_P)^{-1} d(s^{-1}, P) \tag{9.3}$$

for $\mu \times \mu_G$-almost all $(P, s) \in (\operatorname{Prim} A) \times G$.

Proof. Since the stabilizer map $P \mapsto G_P$ is Borel by Proposition H.41 on page 477, $(P, s) \mapsto \mathfrak{b}(s, G_P)\rho(s, G_P)$ is a Borel function F on $(\operatorname{Prim} A) \times G$, and $\nu := F \cdot d(\mu \times \mu_G)$ is a Borel measure on $(\operatorname{Prim} A) \times G$ such that

$$\nu(g) = \iint_{(\operatorname{Prim} A) \times G} g(P, s) \mathfrak{b}(s, G_P) \rho(s, G_P)\, d(\mu \times \mu_G)(P, s).$$

Let $\tau : \operatorname{Prim} A \times G \to \operatorname{Prim} A \times \operatorname{Prim} A$ be given by $\tau(P, s) := (P, s \cdot P)$, and define $\gamma := \tau_* \nu$. By Lemma H.13 on page 463, if f is a nonnegative Borel function on $\operatorname{Prim} A \times \operatorname{Prim} A$, then

$$\gamma(f) = \int_{\operatorname{Prim} A} \int_G f(P, s \cdot P) \mathfrak{b}(s, G_P) \rho(s, G_P) \, d\mu_G(s) \, d\mu(P)$$

which, by Proposition H.11 on page 462, is

$$= \int_{\operatorname{Prim} A} \int_{G/G_P} f(P, s \cdot P) \int_{G_P} \mathfrak{b}(st, G_P) \, d\mu_{G_P}(t) \, d\beta^P(\dot{s}) \, d\mu(P)$$
$$= \int_{\operatorname{Prim} A} \int_{G/G_P} f(P, s \cdot P) \, d\beta^P(\dot{s}) \, d\mu(P).$$

Note that

$$P \mapsto \int_G f(P, s \cdot P) \mathfrak{b}(s, G_P) \rho(s, G_P) \, d\mu_G(s) = \int_{G/G_P} f(P, s \cdot P) \, d\beta^P(\dot{s}).$$

is Borel by Fubini's Theorem.

To see that γ is σ-finite we note that it suffices to show that γ is equivalent to a finite measure on $\operatorname{Prim} A \times \operatorname{Prim} A$. But ν is σ-finite and therefore equivalent to a finite measure $\tilde{\nu}$. Then $\tau_* \tilde{\nu}$ is finite and equivalent to $\tau_* \nu = \gamma$.

Lemma H.10 on page 461 implies that $\rho(sr, H) = \rho(r, H) \rho(s, r \cdot H)$. In particular, $\rho(s, G_P)^{-1} = \rho(s^{-1}, s \cdot G_P) = \rho(s^{-1}, G_{s \cdot P})$. Let ν' be the measure on $\operatorname{Prim} A \times G$ given by

$$\nu'(g) := \iint_{(\operatorname{Prim} A) \times G} g(P, s) d(s^{-1}, P) \mathfrak{b}(s, G_P) \, d(\mu \times \mu_G)(P, s).$$

Since d and ρ are everywhere nonzero, ν' is equivalent to ν. Furthermore,

$$\tau_* \nu'(f) = \int_G \int_{\operatorname{Prim} A} f(P, s \cdot P) d(s^{-1}, P) \mathfrak{b}(s, G_P) \, d\mu(P) \, d\mu_G(s)$$

which, since $P \mapsto d(s^{-1}, P)$ is a Radon-Nikodym derivative, and after using Fubini's Theorem again, is

$$= \int_{\operatorname{Prim} A} \int_G f(s^{-1} \cdot P, P) \mathfrak{b}(s, G_{s^{-1} \cdot P}) \, d\mu_G(s) \, d\mu(P)$$
$$= \int_{\operatorname{Prim} A} \int_G f(s \cdot P, P) \mathfrak{b}(s^{-1}, G_{s \cdot P}) \Delta_G(s^{-1}) \, d\mu_G(s) \, d\mu(P)$$

which, since $\rho(s, G_P)^{-1} \Delta_G(s^{-1}) = \rho(s^{-1}, G_{s \cdot P}) \Delta_G(s^{-1}) = \omega(s^{-1}, G_{s \cdot P})$ (where ω is defined in Lemma H.10), is

$$= \int_{\operatorname{Prim} A} \int_{G/G_P} f(s \cdot P, P) \int_{G_P} \omega(t^{-1} s^{-1}, G_{s \cdot P}) \mathfrak{b}(t^{-1} s^{-1}, G_{s \cdot P})$$
$$d\mu_{G_P}(t) \, d\beta^P(\dot{s}) \, d\mu(P)$$

which, since $\omega(t^{-1}s^{-1}, G_{s\cdot P}) = \omega(st, G_P)^{-1} = \Delta_{G_P}(t^{-1})\omega(s, G_P)^{-1}$, is

$$\begin{aligned}
&= \int_{\operatorname{Prim} A} \int_{G/G_P} f(s\cdot P, P)\omega(s, G_P)^{-1} \int_{G_P} \mathfrak{b}(ts^{-1}, G_{s\cdot P}) \\
&\qquad\qquad\qquad\qquad d\mu_{G_P}(t)\, d\beta^P(\dot s)\, d\mu(P) \\
&= \int_{\operatorname{Prim} A} \int_{G/G_P} f(s\cdot P, P) \int_{G_{s\cdot P}} \mathfrak{b}(s^{-1}t, G_{s\cdot P})\, d\mu_{G_{s\cdot P}}(t)\, d\beta^P(\dot s)\, d\mu(P) \\
&= \int_{\operatorname{Prim} A} \int_{G/G_P} f(s\cdot P, P)\, d\beta^P(\dot s)\, d\mu(P) \\
&= \sigma_*\gamma(f).
\end{aligned}$$

Thus γ and $\sigma_*\gamma$ are equivalent as claimed. Let $\widetilde{D} : \operatorname{Prim} A \times \operatorname{Prim} A \to (0,\infty)$ be a Radon-Nikodym derivative for σ so that (9.2) holds for $\widetilde{D}$ in place of D. Then $\widetilde{D}(P,Q)^{-1} = \widetilde{D}(Q,P)$ for γ-almost all (P,Q). Therefore

$$D(P,Q) := \widetilde{D}(P,Q)^{\frac{1}{2}}\widetilde{D}(Q,P)^{-\frac{1}{2}}$$

is also a Radon-Nikodym derivative and satisfies all the requirements of part (a).

To establish part (b), let $f \in C_c(G)$, and define

$$H(P, s\cdot P) := \int_{G_P} f(st)\rho(st, G_P)^{-1}\, d\mu_{G_P}(t).$$

(As usual, we can define $H(P,Q)$ to be zero if $Q \neq s\cdot P$ for some s.) Then we can calculate as above that

$$\begin{aligned}
&\iint_{(\operatorname{Prim} A)\times G} f(s)D(s\cdot P, P)\rho(s, G_P)^{-1}d(s^{-1}, P)\, d(\mu\times\mu_G)(P,s) \\
&\quad= \int_{\operatorname{Prim} A}\int_G f(s^{-1})\Delta_G(s^{-1})D(P, s\cdot P)\rho(s, G_P) \\
&\qquad\qquad\qquad\qquad d\mu_G(s)\, d\mu(P) \\
&\quad= \int_{\operatorname{Prim} A}\int_{G/G_P} D(P, s\cdot P)\int_{G_P} f(t^{-1}s^{-1})\Delta_G(t^{-1}s^{-1}) \\
&\qquad\qquad\qquad\qquad d\mu_{G_P}(t)\, d\beta^P(\dot s)\, d\mu(P) \\
&\quad= \int_{\operatorname{Prim} A}\int_{G/G_P} D(P, s\cdot P) \\
&\qquad\qquad \omega(s, G_P)\int_{s\cdot G_P} f(s^{-1}t^{-1})\Delta_G(s^{-1}t^{-1}) \\
&\qquad\qquad\qquad\qquad d\mu_{s\cdot G_P}(t)\, d\beta^P(\dot s)\, d\mu(P) \\
&\quad= \int_{\operatorname{Prim} A}\int_{G/G_P} D(P, s\cdot P)\int_{G_{s\cdot P}} f(s^{-1}t)\rho(s^{-1}t, G_{s\cdot P})^{-1} \\
&\qquad\qquad\qquad\qquad d\mu_{G_{s\cdot P}}(t)\, d\beta^P(\dot s)\, d\mu(P)
\end{aligned}$$

which, since $H(s\cdot P,P) = H(s\cdot P, s^{-1}s\cdot P)$, is

$$= \int_{\operatorname{Prim} A}\int_{G/G_P} D(P,s\cdot P)H(s\cdot P,P)\,d\beta^P(\dot s)\,d\mu(P)$$
$$= \int_{\operatorname{Prim} A}\int_{G/G_P} H(P,s\cdot P)\,d\beta^P(\dot s)\,d\mu(P)$$
$$= \iint_{(\operatorname{Prim} A)\times G} f(s)\,d(\mu\times\mu_G)(P,s).$$

Since this holds for all $f\in C_c(G)$, the assertion in part (b) follows. □

Recall if $H\in\Sigma$, then the homomorphism $\gamma_H : H\to \mathbf{R}^+_\times$ is given by

$$\gamma_H(t) = \Bigl(\frac{\Delta_G(t)}{\Delta_H(t)}\Bigr)^{\frac12}.$$

Thus, for example, $\rho(st,G_P) = \gamma_{G_P}(t)^{-2}\rho(s,G_P)$ for all $s\in G$ and $t\in G_P$.

Lemma 9.2. *Suppose that μ is a finite quasi-invariant measure on* Prim A *and that $d : G\times \operatorname{Prim} A\to(0,\infty)$ is a Borel choice of Radon-Nikodym derivatives as in Corollary D.34 on page 389. Then for μ-almost all $P\in\operatorname{Prim} A$,*

$$d(t,P) = \gamma_{G_P}(t)^2 \quad \textit{for all } t\in G_P.$$

The proof of Lemma 9.2 seems as if it should be a simple consequence of (9.3) (together with $D(P,P)=1$ for all P). But (9.3) holds only almost everywhere. To make matters worse, usually G_P will have measure zero in G. Fortunately, there are selection results, due to Ramsay, that allow us to give a proper proof. Since this material is only used in the proof of Lemma 9.11 on page 279, we have exiled the proof of Lemma 9.2 to Appendix G.2.1 on page 452 where the necessary technology is discussed.

We will also need the following observation concerning the β^P's.

Lemma 9.3. *Suppose that φ is a nonnegative Borel function on* Prim A. *Then for all $r\in G$,*

$$\int_{G/G_P}\varphi(s\cdot P)\,d\beta^P(\dot s) = \int_{G/G_{r\cdot P}}\varphi(sr\cdot P)\,d\beta^{r\cdot P}(\dot s).$$

Remark 9.4. This result has a more elegant formulation in terms of the measure β_P which we define to be the measure on $G\cdot P$ equal to the push-forward of β^P via the natural continuous bijection $sG_P\mapsto s\cdot P$ of G/G_P onto $G\cdot P$. Thus β_P is given by

$$\int_{G\cdot P}\varphi(Q)\,d\beta_P(Q) = \int_{G/G_P}\varphi(s\cdot P)\,d\beta^P(\dot s)$$

for all nonnegative Borel functions on $G\cdot P$ (see Lemma H.13 on page 463). Using this formalism, the Lemma states that $\beta_P = \beta_{r\cdot P}$ for all r. Since G is σ-compact and since $K\cdot P$ is closed in Prim A for $K\subset G$ compact, $G\cdot P$ is Borel in Prim A. Thus there is no harm is considering β_P to be a Borel measure on all of Prim A.

Proof. Suppose that φ is nonnegative and that $\mathfrak{b}$ is a Bruhat approximate cross-section for G over G/G_P. Then, using the properties of ρ, ω and the β^P as developed on pages 461–462,

$$\begin{aligned}\int_{G/G_P} \varphi(s\cdot P)\,d\beta^P(\dot s) &= \int_G \varphi(s\cdot P)\mathfrak{b}(s)\rho(s,G_P)\,d\mu_G(s)\\ &= \int_G \varphi(sr\cdot P)\mathfrak{b}(sr)\rho(sr,G_P)\Delta_G(r)\,d\mu_G(s)\\ &= \int_G \varphi(sr\cdot P)\mathfrak{b}(sr)\omega(r,G_P)\rho(s,G_{r\cdot P})\,d\mu_G(s)\\ &= \int_{G/G_{r\cdot P}} \varphi(sr\cdot P)\omega(r,G_P)\int_{G_{r\cdot P}} \mathfrak{b}(str)\,d\mu_{G_{r\cdot P}}(t)\,d\beta^{r\cdot P}(\dot s)\\ &= \int_{G/G_{r\cdot P}} \varphi(sr\cdot P)\int_{G_P} \mathfrak{b}(srt)\,d\mu_{G_P}(t)\,d\beta^{r\cdot P}(\dot s)\\ &= \int_{G/G_{r\cdot P}} \varphi(sr\cdot P)\,d\beta^{r\cdot P}(\dot s).\end{aligned}$$

□

9.1 Step I: Induced Primitive Ideals

As outlined above, our goal in this section is to prove the following.

Proposition 9.5 ([160, Proposition 2.1]). *Let (A,G,α) be a separable dynamical system. Suppose that $L=\pi\rtimes u$ is a homogeneous representation of $A\rtimes_{\alpha|_{G_P}} G_P$ on $\mathcal{H}$, and that π is homogeneous with kernel P. Then $\operatorname{Ind}_{G_P}^G L$ is homogeneous. In particular, the kernel of $\operatorname{Ind}_{G_P}^G L$ is primitive.*

To begin with, suppose that $L=(\pi,u)$ is a covariant representation of (A,H,α). As in Proposition 5.4 on page 153, we'll realize $\operatorname{Ind}_H^G L$ on the space $L^2_u(G,\beta^H,\mathcal{H})$. We want to view $\operatorname{Ind}_H^G L$ as a direct integral. This will allow us to use the technology from Appendices F and G; in particular, we want to invoke Proposition G.20 on page 441 which will imply that any operator in the commutant of the induced representation is decomposable. To accomplish this, it will be convenient to realize $L^2_u(G,\beta^H,\mathcal{H})$ as equivalence classes of functions on G. Recall that $L^2_u(G,\beta^H,\mathcal{H})$ was defined as the completion of the set $\mathcal{V}_c$ of $h\in C_b(G,\mathcal{H})$ such that

$$h(rt)=u_t^{-1}\big(h(r)\big)\quad\text{for all } t\in H \text{ and } \in G \tag{9.4}$$

and such that $r\mapsto \|h(r)\|$ is in $C_c(G/H)$. Let $\mathcal{L}^2_u(G,\beta^H,\mathcal{H})$ be the set of Borel functions $h:G\to\mathcal{H}$ satisfying (9.4) and such that

$$\int_{G/H}\|h(r)\|\,d\beta^H(\dot r)<\infty.$$

If $h,k\in\mathcal{L}^2_u(G,\beta^H,\mathcal{H})$, then $r\mapsto \big(h(r)\mid k(r)\big)$ is constant on H-cosets and belongs to $\mathcal{L}^1(G/H,\beta^H)$. Therefore we can define a sesquilinear form on $\mathcal{L}^2_u(G,\beta^H,\mathcal{H})$ by

$$(h\mid k):=\int_{G/H}\big(h(r)\mid k(r)\big)\,d\beta^H(\dot r). \tag{9.5}$$

Since saturated sets in G are μ_G-null if and only if their image in G/H is β^H-null (Lemma H.14 on page 463), (9.5) induces a bona fide inner product on the μ_G-equivalence classes $\mathcal{L}^2_u(G,\beta^H,\mathcal{H})/\sim$ in $\mathcal{L}^2_u(G,\beta^H,\mathcal{H})$. It is not hard to see that $\mathcal{L}^2_u(G,\beta^H,\mathcal{H})/\sim$ is complete and therefore a Hilbert space. Inclusion gives an isometric injection of $\mathcal{V}_c$ into $\mathcal{L}^2_u(G,\beta^H,\mathcal{H})/\sim$. It is not hard to see that the image is dense. Therefore we can, and do, identify $L^2_u(G,\beta^H,\mathcal{H})$ with $\mathcal{L}^2_u(G,\beta^H,\mathcal{H})/\sim$.[3]

Since G is second countable, there is a Borel cross section $c : G/H \to G$ for the quotient map $q : G \to G/H$. Let $b : G \to H$ be the Borel map determined by $s = c(\dot{s})b(s)$ for all $s \in G$. Note that $b(st) = b(s)t$ for all $s \in G$ and $t \in H$. If $f \in \mathcal{L}^2(G/H,\beta^H,\mathcal{H})$, then

$$H_f(s) := u^{-1}_{b(s)}\big(f(\dot{s})\big)$$

belongs to $\mathcal{L}^2_u(G,\beta^H,\mathcal{H})$ and $\|H_f\|_2 = \|f\|_{L^2(G/H)}$.

Conversely, if $h \in \mathcal{L}^2_u(G,\beta^H,\mathcal{H})$, then

$$F_h(\dot{s}) := h\big(c(\dot{s})\big)$$

is in $\mathcal{L}^2(G/H,\beta^H,\mathcal{H})$. The maps $f \mapsto H_f$ and $h \mapsto F_h$ are inverses, and we obtain unitaries between $L^2_u(G,\beta^H,\mathcal{H})$ and $L^2(G/H,\beta^H,\mathcal{H})$. Thus we can view $L^2_u(G,\beta^H,\mathcal{H})$ as the L^2-sections of the trivial Borel Hilbert bundle $G/H\times\mathcal{H}$. The diagonal operators $\Delta(G/H\times\mathcal{H},\beta^H)$ are naturally identified with $L^\infty(G/H)$, and the decomposable operators can be identified with equivalence classes of bounded weak-operator Borel functions $F : G/H \to B(\mathcal{H})$ via

$$T_F(h)(s) := u^{-1}_{b(s)}F(\dot{s})u_{b(s)}\big(h(s)\big) = \widetilde{F}(s)\big(h(s)\big),$$

where $\widetilde{F}(s) := u^{-1}_{b(s)}F(\dot{s})u_{b(s)}$. Note that $\widetilde{F}$ satisfies

$$\widetilde{F}(st) = u_t^{-1}\widetilde{F}(s)u_t \quad \text{for all } s \in G \text{ and } t \in H. \tag{9.6}$$

If $\widetilde{F} : G \to B(\mathcal{H})$ is any weak-operator Borel function satisfying (9.6), then $F(\dot{s}) := \widetilde{F}(c(\dot{s}))$ is a weak-operator Borel function on G/H such that

$$T_F h(s) = \widetilde{F}(s)\big(h(s)\big).$$

Thus we can identify the decomposable operators on $L^2_u(G,\beta^H,\mathcal{H})$ with the weak-operator Borel functions $F : G \to B(\mathcal{H})$ such that transform as in (9.6).

Proof of Proposition 9.5. We realize $\operatorname{Ind}_{G_P}^G L = \tilde{\pi} \rtimes \tilde{u}$ on $L^2_u(G,\beta^P,\mathcal{H})$, where

$$\tilde{\pi}(a)h(s) = \pi\big(\alpha_s^{-1}(a)\big)\big(h(s)\big) \text{ and}$$
$$\tilde{u}(r)h(s) = \Big(\frac{\rho(r^{-1}s,G_P)}{\rho(s,G_P)}\Big)^{\frac12} h(r^{-1}s) = \rho(r^{-1},G_{s\cdot P})^{\frac12}h(r^{-1}s).$$

[3]See Remark 9.21 on page 292.

Notice that
$$\pi\big(\alpha_{st}^{-1}(a)\big) = u_t^{-1}\pi\big(\alpha_s^{-1}(a)\big)u_t.$$
Therefore we can express $\tilde{\pi}$ as a direct integral
$$\tilde{\pi} = \int_{G/G_P}^{\oplus} \pi_{\dot{s}}\, d\beta^P(\dot{s}),$$
where
$$\pi_{\dot{s}}(a)h(s) = \pi\big(\alpha_s^{-1}(a)\big)\big(h(s)\big). \tag{9.7}$$
Clearly, each $\pi_{\dot{s}}$ is homogeneous with kernel $s \cdot P$. Thus if $\dot{s} \neq \dot{r}$, then $\ker \pi_{\dot{s}} \neq \ker \pi_{\dot{r}}$. Thus (9.7) is essentially an ideal center decomposition, and $\mathcal{IC}(\pi)$ coincides with the diagonal operators $\Delta(G/G_P, \beta^P)$ by Proposition G.20 on page 441. Since $\Delta(G/G_P, \beta^P) = \mathcal{IC}(\tilde{\pi}) \subset \tilde{\pi}(A)'' \cap \tilde{\pi}(A)'$, any operator in $\tilde{\pi}(A)'$ commutes with $\Delta(G/G_P, \beta^P)$ and is decomposable (Theorem F.21 on page 418).

Let E be a nonzero projection in the commutant of $\operatorname{Ind}_{G_P}^G L$. Then $E \in \tilde{\pi}(A)'$ and there is a weak-operator Borel function $E : G \to B(\mathcal{H})$ such that $Eh(s) = E(s)\big(h(s)\big)$ and such that $E(st) = u_t^{-1}E(s)u_t$ for all $s \in G$ and $t \in G_P$. Since A is separable and since the set of s for which $E(s)$ is a projection is saturated, we can change E on a null set and assume that $E(s)$ is a projection in the commutant of π for each s. Since $E \in \tilde{u}(G)'$, for each $r \in G$,
$$E(s)h(r^{-1}s) = E(r^{-1}s)h(r^{-1}s) \quad \text{for } \mu_G\text{-almost all } s.$$
Equivalently,
$$E(rs)h(s) = E(s)h(s) \quad \text{for } \mu_G\text{-almost all } s.$$
Thus
$$N(h) = \{\, (s, r) : E(rs)h(s) \neq E(s)h(s) \,\}$$
is a $\mu_G \times \mu_G$-null set. Let $\{\, h_i \,\}$ be a fundamental sequence for $\mathcal{L}_u^2(G, \beta^P, \mathcal{H})$. Then $\{\, h_i(s) \,\}$ is total in $\mathcal{H}$ for all $s \in G$. Let $N = \bigcup N(h_i)$. Then N is a $\mu_G \times \mu_G$-null set such that $(s, r) \notin N$ implies that $E(rs) = E(s)$. Thus there is a $s_0 \in G$ such that $e := E(s_0) \neq 0$ and such that $e = E(s_0) = E(rs_0)$ for μ_G-almost all $r \in G$. Notice that $e \in \pi(A)' \cap u(G)'$. Thus $e \in L(A \rtimes_{\alpha|_{G_P}} G_P)'$. Since L is homogeneous, $\ker L = \ker L^e$. It is not hard to check that
$$\operatorname{Ind}_{G_P}^G L^e = \big(\operatorname{Ind}_{G_P}^G L\big)^E.$$
Thus
$$\begin{aligned}
\ker\big(\operatorname{Ind}_{G_P}^G L\big)^E &= \ker\big(\operatorname{Ind}_{G_P}^G L^e\big) \\
&= \operatorname{Ind}_{G_P}^G \big(\ker L^e\big) \\
&= \operatorname{Ind}_{G_P}^G (\ker L) \\
&= \ker\big(\operatorname{Ind}_{G_P}^G L\big).
\end{aligned}$$

Since E was arbitrary, $\operatorname{Ind}_{G_P}^G L$ is homogeneous. The last statement follows from Corollary G.9 on page 434. □

9.2 Step II: Restricting to the Stability Groups

Remark 9.6. In our proof of the Effros-Hahn conjecture, we want to work with a factor representation $R = (\pi, V)$ of $A \rtimes_\alpha G$ with a particular primitive kernel P_0. Note that $n \cdot R$ is a factor representation for any $1 \le n \le \aleph_0$. This follows, for example, from [28, Proposition 5.3.4 and Definition 5.3.1(iv)]. Alternatively, we can recall that $n \cdot \pi$ is equivalent to $\pi \otimes 1_{\mathcal{H}_n}$ where $\mathcal{H}_n$ is a fixed Hilbert space of dimension n (cf., [139, §B.4]). But $\pi \otimes 1_{\mathcal{H}_n}(A)' = \pi(A)' \otimes B(\mathcal{H}_n)$ and $\pi \otimes 1_{\mathcal{H}_n}(A)'' = \pi(A)'' \otimes 1_{\mathcal{H}_n}$ by [29, I.2.4 Proposition 4]. Thus $n \cdot R$ is factorial if R is. Furthermore, $n \cdot R$ still has kernel P_0. Thus for our purposes we can assume that R has infinite multiplicity.

Let $R = (\pi, V)$ be a separable covariant representation with infinite multiplicity of a separable dynamical system (A, G, α). Since π has infinite multiplicity, Remark G.23 on page 444 implies that π has an ideal center decomposition (Theorem G.22 on page 444) on $L^2(\operatorname{Prim} A, \mu, \mathcal{H})$ for a finite Borel measure μ on the standard Borel space $\operatorname{Prim} A$. Therefore, we can replace R by an equivalent representation acting on $L^2(\operatorname{Prim} A, \mu, \mathcal{H})$ such that

$$\pi = \int_{\operatorname{Prim} A}^{\oplus} \pi_P \, d\mu(P) \tag{9.8}$$

with each π_P homogeneous with kernel P. We apply Proposition G.24 on page 445 to $R = (\pi, V)$, taking advantage of the simplification that $\operatorname{Prim} A * \mathfrak{H} = \operatorname{Prim} A \times \mathcal{H}$ is trivial. Therefore μ is quasi-invariant, and if $d : G \times \operatorname{Prim} A \to (0, \infty)$ is a Borel choice of Radon-Nikodym derivatives as in Corollary D.34 on page 389, then $W(s)f(P) := d(s, P)^{\frac{1}{2}} f(s^{-1} \cdot P)$ is a unitary on $L^2(\operatorname{Prim} A, \mu, \mathcal{H})$ such that $U(s) := V(s)W(s^{-1})$ is decomposable for all s. Let $X \subset \operatorname{Prim} A$ be a Borel μ-conull set and

$$E := \{ (s, P) \in G \times \operatorname{Prim} A : P \in X \text{ and } s^{-1} \cdot P \in X \}$$

be as in Proposition G.24. Thus if $U(\mathcal{H})$ is the unitary group of $\mathcal{H}$ (with the weak operator topology), then there is a Borel map $(s, P) \mapsto U(s, P)$ from E into $U(\mathcal{H})$ such that the following statements hold.

(A) For all $s \in G$,

$$U(s) = \int_{\operatorname{Prim} A}^{\oplus} U(s, P) \, d\mu(P).$$

(B) For all $s, r \in G$ and all $P \in X$ such that $s^{-1} \cdot P \in X$ and $r^{-1}s^{-1} \cdot P \in X$, we have

$$U(sr, P) = U(s, P)U(r, s^{-1} \cdot P).$$

(C) If $(s, P) \in E$, then

$$\pi_P(a) = U(s, P)\pi_{s^{-1} \cdot P}\big(\alpha_s^{-1}(a)\big)U(s, P)^* \quad \text{for all } a \in A.$$

(D) If $P \in X$, then $\sigma_P(t) := U(t, P)$ defines a unitary representation of G_P and $r_P := (\pi_P, \sigma_P)$ is a covariant representation of (A, G_P, α).

We will treat r_P as a representation of $A \rtimes_\alpha \Sigma$ on $\mathcal{H}$. Since $P \mapsto G_P$ is Borel by Proposition H.41 on page 477, for each $\Phi \in C_c(G*\Sigma, A)$ and $h, k \in L^2(\operatorname{Prim} A, \mu, \mathcal{H})$

$$(s, P) \mapsto \big(\pi_P\big(\Phi(s, G_P)\big)\sigma_P(s)h(P) \mid k(P)\big)$$

is Borel.[4] Thus Lemma H.30 on page 473 implies that

$$P \mapsto \big(r_P(\Phi)h(P) \mid k(P)\big) = \int_{G_P} \big(\pi_P\big(\Phi(s, G_P)\big)\sigma_P(s)h(P) \mid k(P)\big)\, d\mu_{G_P}(s)$$

is Borel, and hence $\{ r_P \}$ is a Borel field of representations on $L^2(\operatorname{Prim} A, \mu, \mathcal{H})$. Thus we can form the direct integral

$$r := \int_{\operatorname{Prim} A}^{\oplus} r_P \, d\mu(P) \tag{9.9}$$

on $L^2(\operatorname{Prim} A, \mu, \mathcal{H})$. We call r the *restriction of R to the stability groups.* (In groupoid language, r is the restriction of R to the isotropy group bundle.)

Since (9.8) is an ideal center decomposition, Lemma G.19 on page 441 implies that the ideal center $\mathcal{IC}(\pi)$ of π is the algebra of diagonal operators $\Delta(\operatorname{Prim} A \times \mathcal{H}, \mu)$, which is isomorphic to $L^\infty(\operatorname{Prim} A, \mu)$. If φ is a bounded Borel function on $\operatorname{Prim} A$, then the corresponding diagonal operator is denoted by T_φ. Since r acts on $L^2(\operatorname{Prim} A, \mu, \mathcal{H})$, its ideal center $\mathcal{IC}(r)$ is also a von Neumann subalgebra of $B\big(L^2(\operatorname{Prim} A, \mu, \mathcal{H})\big)$.

Lemma 9.7. *Let $R = (\pi, V)$ be a factorial representation with infinite multiplicity of $A \rtimes_\alpha G$ on $L^2(\operatorname{Prim} A, \mu, \mathcal{H})$ such that π has an ideal center decomposition given in* (9.8)*. Let r be the restriction of R to the stability groups. Then $\mathcal{IC}(\pi) \subset \mathcal{IC}(r)$.*

Proof. Let $e(I)$ be the projection onto the essential subspace of $\pi|_I$. Then, almost by definition,

$$\mathcal{IC}(\pi) = \{ e(I) : I \in \mathcal{I}(A) \}''.$$

Now if $I \in \mathcal{I}(A)$, then we let $\tilde{I}$ be the ideal of $A \rtimes_\alpha \Sigma$ which is the closure of

$$\tilde{I}_0 := \{ \Phi \in C_c(G * \Sigma, A) : \Phi(s, H) \in I \cap \alpha_s(I) \text{ for all } (s, H) \in G * \Sigma \}.$$

Notice that if $\Phi \in C_c(G * \Sigma, A)$, then

$$\begin{aligned} r(\Phi)h(P) &= \int_{G_P} \pi_P\big(\Phi(s, G_P)\big)\sigma_P(s)h(P)\, d\mu_{G_P}(s) \\ &= \int_{G_P} \sigma_P(s)\pi_P\big(\alpha_s^{-1}\big(\Phi(s, G_P)\big)\big)h(P)\, d\mu_{G_P}(s). \end{aligned} \tag{9.10}$$

[4] We will prove a considerably stronger statement in Lemma 9.9 on page 277. Using an orthonormal basis for $\mathcal{H}$ and Parseval's identity, it suffices to see that $F_\Phi(s, H, P) := \big(\pi_P\big(\Phi(s, H)\big)h(P) \mid k(P)\big)$ is Borel on $G * \Sigma \times \operatorname{Prim} A$. However, this is clear if Φ is an elementary tensor $(s, H) \mapsto \varphi(s, H)a$. On the other hand, the span of elementary tensors is dense in the inductive limit topology, and if $\Phi_i \to \Phi$ in the inductive limit topology, then $F_{\Phi_i} \to F_\Phi$ pointwise. This justifies the assertion.

Lemma G.10 on page 434 implies that

$$e(I) = \int_{\operatorname{Prim} A}^{\oplus} e_{\pi_P}(I)\, d\mu(P)$$

where $e_{\pi_P}(I)$ is the projection onto the essential subspace of $\pi_P|_I$. Since $\mathcal{IC}(\pi)$ is the collection of diagonal operators on $L^2(\operatorname{Prim} A, \mu, \mathcal{H})$, there is a Borel set $Y \subset \operatorname{Prim} A$ such that $e(I) = T_{\mathbb{1}_Y}$. Thus $e_{\pi_P}(I)$ is zero μ-almost everywhere off Y, and the identity μ-almost everywhere on Y. Thus if $a \in I$, then $\pi_P(a) = 0$ for μ-almost every $P \notin Y$. If $h \in \ker e(I)$, then $h(P) = 0$ μ-almost everywhere on Y. Thus it follows from (9.10) that $r(\Phi)h = 0$ for all $\Phi \in \tilde{I}_0$. Thus $h \in \ker e_r(\tilde{I})$ and we have

$$e(I) \geq e_r(\tilde{I}). \tag{9.11}$$

On the other hand, let $\varphi \in C_c(G * \Sigma)$ and let $a, b \in I$. Define

$$\Phi(s, H) = \varphi(s, H) a \alpha_s(b).$$

Then $\Phi \in \tilde{I}_0$ and

$$\begin{aligned} r_P(\Phi) &= \int_{G_P} \varphi(s, G_P) \pi_P\big(a\alpha_s(b)\big) \sigma_P(s)\, d\mu_{G_P}(s) \\ &= \pi_P(a) \int_{G_P} \varphi(s, G_P) \sigma_P(s)\, d\mu_{G_P}(s) \pi_P(b) \\ &= \pi_P(a) \sigma_P\big(\varphi(\cdot, G_P)\big) \pi_P(b). \end{aligned}$$

If $h \in \ker e_r(\tilde{I})$, then there is a μ-null set $N(a, b, \varphi)$ such that for all $P \notin N(a, b, \varphi)$ we have

$$\pi_P(a) \sigma_P\big(\varphi(\cdot, G_P)\big) \pi_P(b) h(P) = 0. \tag{9.12}$$

Since I and $C_c(G * \Sigma)$ have countable dense subsets, there is a μ-null set N such that (9.12) holds for all $a, b \in I$ and $\varphi \in C_c(G * \Sigma)$ provided $P \notin N$. However, if $a \in I$ and if $\pi_P(a)h(P) \neq 0$ then, using an approximate identity for $C_c(G * \Sigma)$, we can find $\varphi \in C_c(G * \Sigma)$ such that $\pi_P(a)\sigma_P\big(\varphi(\cdot, G_P)\big)h(P) \neq 0$. Similarly, using an approximate identity for I and the fact that we must have $e_{\pi_P}(I) \neq 0$, we can find $b \in I$ such that

$$\pi_P(a) \sigma_P\big(\varphi(\cdot, G_P)\big) \pi_P(b) h(P) \neq 0. \tag{9.13}$$

Thus $P \in N$. It follows that $\pi(a)h = 0$ for all $a \in I$. Thus $e(I)h = 0$. It follows from this and (9.11) that $e_r(\tilde{I}) = e(I)$. Since I was arbitrary, the lemma is proved. □

Let

$$\tilde{r} := \int_{\operatorname{Prim}(A \rtimes_\alpha \Sigma)}^{\oplus} \tilde{r}_Q\, d\nu(Q) \tag{9.14}$$

be an ideal center decomposition for r on $L^2\big(\operatorname{Prim}(A \rtimes_\alpha \Sigma) * \mathfrak{H}, \nu\big)$. Therefore each $\tilde{r}_Q$ is homogeneous with kernel Q, and there is a unitary M^r from $L^2(\operatorname{Prim} A, \mu, \mathcal{H})$ onto $L^2\big(\operatorname{Prim}(A \rtimes_\alpha \Sigma) * \mathfrak{H}, \nu\big)$ intertwining r and $\tilde{r}$.

Proposition 9.8. *Suppose that $R = \pi \rtimes V$ is a factor representation of $A \rtimes_\alpha G$ with infinite multiplicity. Let r be the restriction of $R = (\pi, V)$ to $A \rtimes_\alpha \Sigma$ as above. Then the measure ν in the ideal center decomposition* (9.14) *of r is quasi-invariant and ergodic for the natural action of G on $\operatorname{Prim}(A \rtimes_\alpha \Sigma)$ induced by the action $\beta : G \to \operatorname{Aut}(A \rtimes_\alpha \Sigma)$ defined in Lemma 8.57 on page 260. In particular, $\tilde{r}_{s\cdot Q}$ is equivalent to $s \cdot \tilde{r}_Q$ for all $s \in G$ and $Q \in \operatorname{Prim} A \rtimes_\alpha \Sigma$.*

Since the proof of the proposition will require some technical gyrations, we will break it up into two bits; first, we'll consider the quasi-invariance. Then after developing some additional structure, we'll turn to the proof of ergodicity on page 281. We retain the notation and set-up described in and before items (A)–(D) on page 271.

Proof of quasi-invariance. Suppose that $\Phi \in C_c(G * \Sigma, A)$, that $s \in G$ and that $h \in L^2(\operatorname{Prim} A, \mu, \mathcal{H})$. Let $P \in X \cap s \cdot X$. Then $(s, P) \in E$, and item (A) implies that we can also find a null sets N_1 and N_2 such that $P \notin N_1$ implies that $V(s)W(s^{-1})\bigl(W(s)r(\Phi)h\bigr)(P) = U(s,P)\bigl(W(s)r(\Phi)h(P)\bigr)$, and that $P \notin N_2$ implies that $V(s)W(s^{-1})\bigl(W(s)h\bigr)(P) = U(s,P)\bigl(W(s)h(P)\bigr)$. Then, if $P \in (X \cap s \cdot X) \smallsetminus (N_1 \cup N_2)$, we compute that

$$\begin{aligned} V(s)r(\Phi)h(P) &= V(s)W(s^{-1})W(s)r(\Phi)h(P) \\ &= d(s,P)^{\frac{1}{2}}U(s,P)r(\Phi)h(s^{-1}\cdot P) \\ &= d(s,P)^{\frac{1}{2}}U(s,P)\int_{G_{s^{-1}\cdot P}} \pi_{s^{-1}\cdot P}\bigl(\Phi(t, G_{s^{-1}\cdot P})\bigr)\sigma_{s^{-1}\cdot P}(t)h(s^{-1}\cdot P) \\ &\qquad\qquad d\mu_{G_{s^{-1}\cdot P}}(t) \end{aligned}$$

which, since $G_{s^{-1}\cdot P} = s^{-1}\cdot G_P$, and since $(s,P) \in E$ implies that $U(s,P)\pi_{s^{-1}\cdot P}(a) = \pi_P\bigl(\alpha_s(a)\bigr)U(s,P)$, is

$$\begin{aligned} &= \int_{s^{-1}\cdot G_P} \pi_P\bigl(\alpha_s\bigl(\Phi(t, s^{-1}\cdot G_P)\bigr)\bigr)U(s,P)U(t, s^{-1}\cdot P)W(s)h(P) \\ &\qquad\qquad d\mu_{s^{-1}\cdot G_P}(t) \end{aligned}$$

which, since P and $s^{-1}\cdot P$ are in X, is

$$\begin{aligned} &= \int_{s^{-1}\cdot G_P} \pi_P\bigl(\alpha_s\bigl(\Phi(t, s^{-1}\cdot G_P)\bigr)\bigr)U(st,P)W(s)h(P)\,d\mu_{s^{-1}\cdot G_P}(t) \\ &= \int_{G_P} \omega(s, s^{-1}\cdot G_P)\pi_P\bigl(\alpha_s\bigl(\Phi(s^{-1}ts, s^{-1}\cdot G_P)\bigr)\bigr)U(ts,P)W(s)h(P) \\ &\qquad\qquad d\mu_{G_P}(t) \\ &= \int_{G_P} \pi_P\bigl(\beta_s(\Phi)(t, G_P)\bigr)U(t,P)U(s,P)W(s)h(P)\,d\mu_{G_P}(t) \end{aligned}$$

which, since $P \notin N_2$ is

$$= r\bigl(\beta_s(\Phi)\bigr)V(s)h(P).$$

Since the set of P such that $P \in X$, $s^{-1} \cdot P \in X$ and $P \notin N_1 \cup N_2$ is conull, the above computation shows that

$$s \cdot r(\Phi) := r\big(\beta_s^{-1}(\Phi)\big) = V(s)^* r(\Phi) V(s) \tag{9.15}$$

for all $\Phi \in A \rtimes_\alpha \Sigma$ and $s \in G$. In particular, r and $s \cdot r$ are equivalent.

To see that ν and $s \cdot \nu$ are equivalent measures, we proceed as in the proof of Proposition G.24 on page 445 using (9.15). In particular, M^r implements an equivalence between $s \cdot r$ on $L^2(\operatorname{Prim} A, \mu, \mathcal{H})$ and

$$\int_{\operatorname{Prim}(A \rtimes_\alpha \Sigma)}^{\oplus} s \cdot \tilde{r}_Q \, d\nu(Q)$$

on $L^2\big(\operatorname{Prim}(A \rtimes_\alpha \Sigma) * \mathfrak{H}, \nu\big)$. Then $L(s)h(P) := h(s^{-1} \cdot P)$ defines a unitary from $L^2\big(\operatorname{Prim}(A \rtimes_\alpha \Sigma) * \mathfrak{H}, \nu\big)$ onto $L^2\big(\operatorname{Prim}(A \rtimes_\alpha \Sigma) * \mathfrak{H}, s \cdot \nu\big)$ intertwining (9.15) with

$$\int_{\operatorname{Prim}(A \rtimes_\alpha \Sigma)}^{\oplus} \tilde{r}'_Q \, d(s \cdot \nu)(Q), \tag{9.16}$$

where $\tilde{r}'_Q := \tilde{r}_{s^{-1} \cdot Q} \circ \beta_s^{-1}$. Since $\tilde{r}'_Q$ is homogeneous with kernel Q, the direct integral decomposition in (9.16) is an ideal center decomposition, and then our uniqueness result (Proposition G.21 on page 443) implies that ν and $s \cdot \nu$ are equivalent.[5] Since s is arbitrary, we have shown that ν is quasi-invariant. Proposition G.21 also implies that $\tilde{r}'_Q$ and $\tilde{r}_Q$ are equivalent. Therefore $\tilde{r}_{s \cdot Q}$ and $s \cdot \tilde{r}_{s \cdot Q}$ are equivalent for all s. This completes the proof of the proposition with the exception of the ergodicity of ν. □

To see that ν is also ergodic requires that we look a bit more closely at the operator $M^r V(s) M^{r*} L(s)^* : L^2\big(\operatorname{Prim}(A \rtimes_\alpha \Sigma) * \mathfrak{H}, s \cdot \nu\big) \to L^2\big(\operatorname{Prim}(A \rtimes_\alpha \Sigma) * \mathfrak{H}, \nu\big)$ which implements an equivalence between the ideal center decompositions given by (9.14) and by (9.16). The uniqueness of ideal center decompositions (Proposition G.21 on page 443) implies that $M^r V(s) M^{r*} L(s)^*$ commutes with the diagonal operators. More precisely, if φ is a bounded Borel function on $\operatorname{Prim}(A \rtimes_\alpha \Sigma)$ and if T_φ and T_φ^s are the corresponding diagonal operators on $L^2\big(\operatorname{Prim}(A \rtimes_\alpha \Sigma) * \mathfrak{H}, \nu\big)$ and $L^2\big(\operatorname{Prim}(A \rtimes_\alpha \Sigma) * \mathfrak{H}, s \cdot \nu\big)$, respectively, then

$$M^r V(s) M^{r*} L(s)^* T_\varphi^s = T_\varphi M^r V(s) M^{r*} L(s)^*.$$

Therefore,

$$M^r V(s) M^{r*} T_\varphi L(s)^* = M^r V(s) M^{r*} L(s)^* T_{s \cdot \varphi}^s = T_{s \cdot \varphi} M^r V(s) M^{r*} L(s)^*,$$

where $s \cdot \varphi(P) = \operatorname{lt}_s(\varphi)(P) := \varphi(s^{-1} \cdot P)$. Thus,

$$M^r V(s) M^{r*} T_\varphi = T_{s \cdot \varphi} M^r V(s) M^{r*}. \tag{9.17}$$

[5] It is worth noting that the special uniqueness of the ideal center decomposition is used here. Normally, one has to worry about the algebras of diagonal operators when making statements about direct integrals. See for example, Proposition F.33 on page 423. This will also be important in the proof of ergodicity to follow.

Therefore if $\varphi = \mathbb{1}_B$ for a G-invariant Borel set $B \subset \operatorname{Prim}(A \rtimes_\alpha \Sigma)$, then $E := T_\varphi$ commutes with $\widetilde{V}(s) := M^r V(s) M^{r*}$. Of course, E also commutes with $\tilde{r}(A \rtimes_\alpha \Sigma)$. If $j_A : A \to M(A \rtimes_\alpha \Sigma)$ is the natural map, then

$$r\big(j_A(a)\Phi)\big) = \pi(a)r(\Phi).$$

Thus E also commutes with $\tilde{\pi}(A)$ where $\tilde{\pi}(a) = M^r \pi(a) M^{r*}$. Therefore $E \in \tilde{R}(A \rtimes_\alpha G)'$ where $\tilde{R} = (\tilde{\pi}, \tilde{V})$. If R were irreducible, this would force E to be 0 or I and ν would be ergodic. Unfortunately, we are assuming only that R is factorial (so that we can make some of the initial constructions a little less technical). Here we have to pay for that assumption. Since we have $E \in \mathcal{IC}(\tilde{r}) \subset \tilde{r}(A \rtimes_\alpha \Sigma)''$, if we can show that

$$\tilde{r}(A \rtimes_\alpha \Sigma)'' \subset \tilde{R}(A \rtimes_\alpha G)'', \quad \text{or equivalently} \quad r(A \rtimes_\alpha \Sigma)'' \subset R(A \rtimes_\alpha G)'',$$

then E belongs to the center of $\tilde{R}(A \rtimes_\alpha G)''$. Since $\tilde{R}(A \rtimes_\alpha G)''$ is a von Neumann factor by assumption, its center is trivial and this will prove that ν is ergodic and finish the proof of Proposition 9.8 on page 274. Showing this requires some involved constructions. Fortunately, these constructions will be of use later on, so we will work out the details here.

Observe that if $f \in C_c(G, A)$ and $h, k \in L^2(\operatorname{Prim} A, \mu, \mathcal{H})$, then

$$\begin{aligned}\big(R(f)h \mid k\big) &= \int_G \big(\pi\big(f(s)\big)V(s)h \mid k\big)\, d\mu_G(s)\\ &= \int_G \int_{\operatorname{Prim} A} \big(\pi\big(f(s)\big)V(s)W(s^{-1})W(s)(h)(P) \mid k(P)\big)\, d\mu(P)\, d\mu_G(s)\end{aligned}$$

which, since $V(s)W(s^{-1}) = \int_{\operatorname{Prim} A}^{\oplus} U(s, P)\, d\mu(P)$ and $\pi = \int_{\operatorname{Prim} A}^{\oplus} \pi_P\, d\mu(P)$, is

$$= \int_G \int_{\operatorname{Prim} A} \big(\pi_P\big(f(s)\big)U(s, P)h(s^{-1} \cdot P) \mid k(P)\big) d(s, P)^{\frac{1}{2}}\, d\mu(P)\, d\mu_G(s). \tag{9.18}$$

To take advantage of (9.18), we want to consider the collection $\mathscr{B}^b_{cc}(G \times \operatorname{Prim} A, A)$ of bounded Borel functions $F : G \times \operatorname{Prim} A \to A$ such that there is compact set $K \subset G$, depending on F, such that $F(s, P) = 0$ if $s \notin K$. If $F \in \mathscr{B}^b_{cc}(G \times \operatorname{Prim} A, A)$, then we define

$$\begin{aligned}\|F\|_{I,r} &:= \sup_P \int_G \|F(s, P)\|\, d\mu_G(s),\\ \|F\|_{I,s} &= \sup_P \int_G \|F(s, s \cdot P)\|\, d\mu_G(s) \text{ and}\\ \|F\|_I &= \max\big\{\|F\|_{I,r}, \|F\|_{I,s}\big\}.\end{aligned}$$

Notice that if K is such that $F(s, P) = 0$ if $s \notin K$, then all the above quantities are bounded by $\|F\|_\infty \mu_G(K)$.[6] It is also worth remarking that $\|F\|_{I,s} = \|F^*\|_{I,r}$,

[6]These norms, and the notation, come from treating $G \times \operatorname{Prim} A$ as a groupoid. Good references for the theory of groupoids and their C^*-algebras are [143] or [108].

where $F^*(s,P) := \Delta_G(s^{-1})F(s^{-1}, s^{-1}\cdot P)^*$. Moreover, we want to observe that $\mathscr{B}^b_{cc}(G\times \operatorname{Prim} A, A)$ is a $*$-algebra. Suppose that $F, F' \in \mathscr{B}^b_{cc}(G\times \operatorname{Prim} A, A)$. Clearly, $r \mapsto F(r,P)\alpha_r\big(F'(r^{-1}s, r^{-1}\cdot P)\big)$ is Borel. Making use of vector-valued integrals such as defined in Appendix B, we can use Proposition B.34 on page 343 to define $F * F'(s,P) \in A$ by

$$F * F'(s,P) := \int_G F(r,P)\alpha_r\big(F'(r^{-1}s, r^{-1}\cdot P)\big)\,d\mu_G(s).$$

Let $\varphi \in A^*$. Then

$$\varphi\big(F * F'(s,P)\big) = \int_G \varphi\big(F(r,P)\alpha_r\big(F'(r^{-1}s, r^{-1}\cdot P)\big)\big)\,d\mu_G(s). \tag{9.19}$$

Since $(s,r,P) \mapsto F(r,P)\alpha_r\big(F'(r^{-1}s, r^{-1}\cdot P)\big)$ is Borel, Fubini's Theorem implies that the left-hand side of (9.19) is a Borel function of (s,P). Thus, $(s,P) \mapsto F * F'(s,P)$ is weakly Borel, and hence Borel by Lemma H.32 on page 474. Thus $F * F' \in \mathscr{B}^b_{cc}(G\times \operatorname{Prim} A, A)$. The other required properties, such as associativity, follow from similar considerations.

Lemma 9.9. *If $F \in \mathscr{B}^b_{cc}(G\times \operatorname{Prim} A, A)$, then*

$$(s,P) \mapsto \big(\pi_P\big(F(s,P)\big)U(s,P)h(s^{-1}\cdot P) \mid k(P)\big)$$

is a Borel function on $G\times \operatorname{Prim} A$ for all $h, k \in L^2(\operatorname{Prim} A, \mu, \mathcal{H})$. In particular, there is a bounded operator $R'(F)$ on $L^2(\operatorname{Prim} A, \mu, \mathcal{H})$ such that $\|R'(F)\| \le \|F\|_I$ given by

$$\begin{aligned}\big(R'(F)h \mid k\big) :=&\\ &\int_G\int_{\operatorname{Prim} A} \big(\pi_P\big(F(s,P)\big)U(s,P)h(s^{-1}\cdot P) \mid k(P)\big)\\ &\qquad\qquad d(s,P)^{\frac12}\,d\mu(P)\,d\mu_G(s).\end{aligned} \tag{9.20}$$

In fact, R' defines a $*$-homomorphism of $\mathscr{B}^b_{cc}(G\times \operatorname{Prim} A, A)$ into the bounded operators on $L^2(\operatorname{Prim} A, \mu, \mathcal{H})$.

Proof. Let $\{\, e_k \,\}$ be a countable orthonormal basis for $\mathcal{H}$. Then

$$\begin{aligned}&\big(\pi_P\big(F(s,P)\big)U(s,P)h(s^{-1}\cdot P) \mid k(P)\big)\\ &\quad= \sum_k \big(U(s,P)h(s^{-1}\cdot P) \mid e_k\big)\big(\pi_P\big(F(s,P)\big)e_k \mid k(P)\big).\\ &\quad= \sum_{km} \big(U(s,P)h(s^{-1}\cdot P) \mid e_k\big)\big(\pi_P\big(F(s,P)\big)e_k \mid e_m\big)\big(e_m \mid k(P)\big)\end{aligned}$$

Thus it suffices to see that

$$(s,P) \mapsto \big(\pi_P\big(F(s,P)\big)e_k \mid e_m\big)$$

is Borel for all k and m. Since A is separable, given $\epsilon > 0$, there is a countable partition $\{ B_i \}$ of A by nonempty Borel sets of diameter at most ϵ. Let $E_i' := F^{-1}(B_i)$ and fix $a_i \in B_i$. If $K \subset G$ is a compact set such that $F(s,P) = 0$ if $s \notin K$, then let $E_i := E_i' \cap (K \times \operatorname{Prim} A)$. Define

$$F_\epsilon(s,P) := \sum_i a_i \mathbb{1}_{E_i}(s,P).$$

Then $F_\epsilon \in \mathscr{B}^b_{cc}(G \times \operatorname{Prim} A, A)$ and $\|F - F_\epsilon\|_\infty \le \epsilon$. Moreover

$$(s,P) \mapsto \big(\pi_P\big(F_\epsilon(s,P)\big)e_k \mid e_m\big) = \sum_i \mathbb{1}_{E_i}(s,P)\big(\pi_P(a_i)e_k \mid e_m\big)$$

is Borel. Since $F_{\frac{1}{n}} \to F$ pointwise, the first assertion follows.

Now we can use the Cauchy-Schwarz inequality on $L^2(\mu \times \mu_G)$ to calculate that

$$\begin{aligned}
&\big|\big(R'(F)h \mid k\big)\big| \\
&\quad\le \iint_{(\operatorname{Prim} A)\times G} \|F(s,P)\|\|h(s^{-1}\cdot P)\|\|k(P)\| d(s,P)^{\frac{1}{2}}\, d(\mu_G \times \mu)(s,P) \\
&\quad\le \Big(\iint_{(\operatorname{Prim} A)\times G} \|F(s,P)\|\|k(P)\|^2\, d(\mu_G \times \mu)(s,P)\Big)^{\frac{1}{2}} \\
&\qquad\qquad \Big(\iint_{(\operatorname{Prim} A)\times G} \|F(s,s\cdot P)\|\|h(P)\|^2\, d(\mu_G \times \mu)(s,P)\Big)^{\frac{1}{2}} \\
&\quad\le \|F\|_{I,r}^{\frac{1}{2}}\|k\|_2\|F\|_{I,s}^{\frac{1}{2}}\|h\|_2 \\
&\quad\le \|F\|_I\|h\|_2\|k\|_2
\end{aligned}$$

Therefore $R'(F)$ is a bounded operator with $\|R'(F)\| \le \|F\|_I$. □

Notice that each $f \in C_c(G,A)$ defines an element F_f of $\mathscr{B}^b_{cc}(G \times \operatorname{Prim} A, A)$, given by $F_f(s,P) = f(s)$; of course, $R'(F_f) = R(f)$. Thus we should view R' as an extension of R to a larger class of functions.

Lemma 9.10. *Let R and R' be as above.*

(a) *If $T \in R(A \rtimes_\alpha G)'$, then T is decomposable. If*

$$T = \int_{\operatorname{Prim} A}^{\oplus} T(P)\, d\mu(P), \tag{9.21}$$

then for all $s \in G$,

$$T(P)U(s,P) = U(s,P)T(s^{-1}\cdot P) \quad \textit{for } \mu\textit{-almost all } P.$$

(b) *If $F \in \mathscr{B}^b_{cc}(G \times \operatorname{Prim} A, A)$, then $R'(F) \in R(A \rtimes_\alpha G)''$.*

Proof. If $T \in R(A \rtimes_\alpha G)'$, then T commutes with both $\pi(A)$ and $V(G)$. Since $\mathcal{IC}(\pi)$ is in the center of $\pi(A)'$, $T \in \mathcal{IC}(\pi)' = \Delta(\operatorname{Prim} A \times \mathcal{H}, \mu)'$, and hence, T is decomposable. If T is given by (9.21), then for almost all P,

$$\begin{aligned}TV(s)f(P) &= d(s,P)^{\frac{1}{2}}T(P)V(s)W(s^{-1})f(s^{-1}\cdot P),\\ &= d(s,p)^{\frac{1}{2}}T(P)U(s,P)f(s^{-1}\cdot P).\end{aligned}$$

On the other hand, since T commutes with $V(s)$, we have, for almost all P,

$$\begin{aligned}TV(s)f(P) &= V(s)Tf(P)\\ &= V(s)W(s^{-1})W(s)Tf(P)\\ &= d(s,P)^{\frac{1}{2}}U(s,P)T(s^{-1}\cdot P)f(s^{-1}\cdot P).\end{aligned}$$

If $\{ e_m \}$ is an orthonormal basis for $\mathcal{H}$ and $f_m(P) := e_m$ for all P, then applying the above to f_m shows that there is a μ-null set N_m such that

$$U(s,P)T(s^{-1}\cdot P)e_m = T(P)U(s,P)e_m \quad \text{if } P \notin N_m.$$

If we let $N = \bigcup N_m$, then $T(P)U(s,P) = U(s,P)T(s^{-1}\cdot P)$ for all $P \notin N$. This proves part (a).

Using part (a) and (9.20), it is not hard to see that if $T \in R(A \rtimes_\alpha G)'$ and $F \in \mathscr{B}^b_{cc}(G \times \operatorname{Prim} A, A)$, then T and $R'(F)$ commute. Therefore $R'(F) \in R(A \rtimes_\alpha G)''$ as required. □

Lemma 9.11. *If $F \in \mathscr{B}^b_{cc}(G \times \operatorname{Prim} A, A)$ and $\Phi \in C_c(G * \Sigma, A)$, then*

$$\Phi \cdot F(s,P) := \int_{G_P} \Phi(t, G_P)\alpha_t\bigl(F(t^{-1}s,P)\bigr)\gamma_{G_P}(t)^{-1}\,d\mu_{G_P}(t) \tag{9.22}$$

defines an element of $\mathscr{B}^b_{cc}(G \times \operatorname{Prim} A, A)$ such that $R'(\Phi \cdot F) = r(\Phi)R'(F)$. In particular, if $f \in C_c(G, A)$, then $R'(\Phi \cdot F_f) = r(\Phi)R(f)$.

Proof. Suppose $F(s,P)$ vanishes if $s \notin K$ and that $\Phi(s,H)$ vanishes if $s \notin K'$. Then $\Phi \cdot F(s,P)$ vanishes for $s \notin K'K$. Furthermore,

$$\|\Phi \cdot F(s,P)\| \le \|\Phi\|_\infty \|F\|_\infty \|\varphi\|_\infty,$$

where

$$\varphi(H) := \int_H \psi(s)|\gamma_{G_P}(s)^{-1}|\,d\mu_H(s),$$

and $\psi \in C_c^+(G)$ is such that $\psi(s) = 1$ for all $s \in K'K$. Thus to see that $\Phi \cdot F \in \mathscr{B}^b_{cc}(G \times \operatorname{Prim} A, A)$, we just have to see that it is a Borel function. But

$$\tilde{F}(P,s,t,H) := \Phi(t,H)\alpha_t\bigl(F(t^{-1}s,H)\bigr)\gamma_H(t)^{-1}$$

is Borel on $\operatorname{Prim} A \times G \times G * \Sigma$, and it vanishes if $s \notin K'K$. Thus Lemma H.30 on page 473 (with $X = \operatorname{Prim} A \times G$) implies that

$$\tilde{\varphi}(P,s,H) = \int_H \tilde{F}(P,s,t,H)\,d\mu_H(t)$$

is weakly Borel, and therefore Borel by Lemma H.32 on page 474. Now

$$(s,P) \mapsto \Phi \cdot F(s,P) = \tilde{\varphi}(P,s,G_P)$$

is Borel because $P \mapsto G_P$ is (Proposition H.41 on page 477).

Using (9.20), we have[7]

$$\big(R'(\Phi \cdot F)h \mid k\big) = \int_G \int_{\operatorname{Prim} A} \big(\pi_P(\Phi \cdot F(s,P))U(s,P)h(s^{-1}\cdot P) \mid k(P)\big) \\ d(s,P)^{\frac{1}{2}}\, d\mu(P)\, d\mu_G(s)$$

which, since each π_P is a representation of A, is

$$= \int_G \int_{\operatorname{Prim} A} \int_{G_P} \big(\pi_P\big(\Phi(t,G_P)\alpha_t\big(F(t^{-1}s,P)\big)\big)U(s,P)h(s^{-1}\cdot P) \mid k(P)\big) \\ \gamma_{G_P}(t)^{-1} d(s,P)^{\frac{1}{2}}\, d\mu_{G_P}(t)\, d\mu(P)\, d\mu_G(s)$$

which, since $d(ts,P) = d(t,P)d(s,P) = \gamma_{G_P}(t)^2 d(s,P)$, is

$$= \int_{\operatorname{Prim} A} \int_{G_P} \int_G \big(\pi_P\big(\Phi(t,G_P)\alpha_t\big(F(s,P)\big)\big)U(ts,P)h(s^{-1}\cdot P) \mid k(P)\big) \\ d(s,P)^{\frac{1}{2}}\, d\mu_G(s)\, d\mu_{G_P}(t)\, d\mu(P)$$

which, since for all $s \in G$ and $t \in G_P$, $U(ts,P) = U(t,P)U(s,P) = \sigma_P(t)U(s,P)$ for μ-almost all P, is

$$= \int_G \int_{\operatorname{Prim} A} \int_{G_P} \big(\pi_P\big(\Phi(t,G_P)\alpha_t\big(F(s,P)\big)\big)\sigma_P(t)U(s,P)h(s^{-1}\cdot P) \mid k(P)\big) \\ d(s,P)^{\frac{1}{2}}\, d\mu_{G_P}(t)\, d\mu(P)\, d\mu_G(s)$$

which, since (π_P, σ_P) is covariant, is

$$= \int_G \int_{\operatorname{Prim} A} \int_{G_P} \big(\sigma_P(t)\pi_P\big(\alpha_t^{-1}\big(\Phi(t,G_P)\big)F(s,P)\big)U(s,P)h(s^{-1}\cdot P) \mid k(P)\big) \\ d(s,P)^{\frac{1}{2}}\, d\mu_{G_P}(t)\, d\mu(P)\, d\mu_G(s)$$

$$= \int_G \int_{\operatorname{Prim} A} \int_{G_P} \big(\pi_P\big(F(s,P)\big)U(s,P)h(s^{-1}\cdot P) \mid$$

[7]This computation can be simplified by realizing that we can define $R'(F)$ pointwise via a vector-valued integral:

$$R'(F)h(P) = \int_G \pi_P\big(F(s,P)\big)U(s,P)h(s^{-1}\cdot P)d(s,P)^{\frac{1}{2}}\, d\mu_G(s).$$

We've given the proof as above simply to reduce the use of vector-valued integrals of Borel functions.

$$
\begin{aligned}
&\pi_P\bigl(\alpha_{t^{-1}}\bigl(\Phi(t,G_P)^*\bigr)\bigr)\sigma_P(t^{-1})k(P)\bigr)d(s,P)^{\frac12}\,d\mu_{G_P}(t)\,d\mu(P)\,d\mu_G(s)\\
&= \int_G\int_{\operatorname{Prim}A}\bigl(\pi_P\bigl(F(s,P)\bigr)U(s,P)h(s^{-1}\cdot P)\mid r(\Phi^*)k(P)\bigr)d(s,P)^{\frac12}\,d\mu(P)\,d\mu_G(s)\\
&= \bigl(R'(F)h\mid r(\Phi^*)k\bigr)\\
&= \bigl(r(\Phi)R'(F)h\mid k\bigr).
\end{aligned}
$$

Thus $R'(\Phi\cdot F)=r(\Phi)R'(F)$ and the final statement is an easy consequence of this. $\square$

Proof of Proposition 9.8 on page 274 continued. We still need to see that ν is ergodic. But it suffices to see that $r(A\rtimes_\alpha\Sigma)\subset R(A\rtimes_\alpha G)''$. Let $\{f_i\}$ be an approximate identity for $A\rtimes_\alpha G$. Then $R(f_i)\to I$ in the strong operator topology. Therefore if $\Phi\in A\rtimes_\alpha\Sigma$, then $r(\Phi)$ is the strong operator limit of $r(\Phi)R(f_i)=R'(\Phi\cdot F_{f_i})$. Since each of the latter is in $R(A\rtimes_\alpha G)''$ by Lemma 9.10 on page 278, so is $r(\Phi)$. $\square$

Since (9.14) is an ideal center decomposition for r, $\mathcal{IC}(r)\cong\Delta(\operatorname{Prim}(A\rtimes_\alpha\Sigma)*\mathfrak{H},\nu)$, and $L^\infty(\operatorname{Prim}(A\rtimes_\alpha\Sigma),\nu)$ is isomorphic to the latter via $\psi\mapsto T^\Sigma_\psi$. Since $\mathcal{IC}(\pi)\subset\mathcal{IC}(r)$ (Lemma 9.7 on page 272), $M^r: L^2(\operatorname{Prim}A,\mu,\mathcal{H})\to L^2\bigl(\operatorname{Prim}(A\rtimes_\alpha\Sigma)*\mathfrak{H},\nu\bigr)$ maps $\mathcal{IC}(\pi)=\Delta(\operatorname{Prim}A\times\mathcal{H},\mu)$ onto a von Neumann subalgebra $\mathscr{L}$ of $\Delta(\operatorname{Prim}(A\rtimes_\alpha\Sigma)*\mathfrak{H},\nu)$. Let $\varphi\mapsto T_\varphi$ be the natural isomorphism of $L^\infty(\operatorname{Prim}(A),\mu)$ onto the diagonal operators $\Delta(\operatorname{Prim}(A)\times\mathcal{H},\mu)$. Lemma I.11 on page 489 implies that there is a Borel map $\tau:\operatorname{Prim}(A\rtimes_\alpha\Sigma)\to\operatorname{Prim}A$ which implements the induced isomorphism of $L^\infty(\operatorname{Prim}A,\mu)$ onto its range in $L^\infty(\operatorname{Prim}(A\rtimes_\alpha\Sigma),\nu)$. Thus $M^rT_\varphi=T^\Sigma_{\varphi\circ\tau}M^r$.

Remark 9.12. Notice that if $\varphi_s(P):=\varphi(s\cdot P)$, then since

$$U(s):=V(s)W(s^{-1})=\int^\oplus_{\operatorname{Prim}A}U(s,P)\,d\mu(P)$$

commutes with the diagonal operators,

$$
\begin{aligned}
V(s)T_\varphi &= V(s)W(s^{-1})W(s)T_\varphi\\
&= V(s)W(s^{-1})T_{s\cdot\varphi}W(s)\\
&= T_{s\cdot\varphi}V(s)W(s^{-1})W(s)\\
&= T_{s\cdot\varphi}V(s).
\end{aligned}
$$

As we showed in deriving (9.17) on page 275, if $\widetilde V(s):=M^rV(s)M^{r*}$, then we have

$$\widetilde V(s)T^\Sigma_\psi=T^\Sigma_{s\cdot\psi}\widetilde V(s),$$

where $\psi_s(Q)=\psi(s\cdot Q)$. Now, on the one hand,

$$
\begin{aligned}
M^rV(s)T_\varphi &= M^rT_{s\cdot\varphi}(s)\\
&= T^\Sigma_{(s\cdot\varphi)\circ\tau}M^rV(s).
\end{aligned}
$$

On the other hand,

$$\begin{aligned} M^r V(s) T_\varphi &= \widetilde{V}(s) M^r T_\varphi \\ &= \widetilde{V}(s) T^\Sigma_{\varphi\circ\tau} M^r \\ &= T^\Sigma_{s\cdot(\varphi\circ\tau)} \widetilde{V}(s) M^r \\ &= T^\Sigma_{s\cdot(\varphi\circ\tau)} M^r V(s). \end{aligned}$$

Thus for all $s \in G$ and all $\varphi \in L^\infty(\operatorname{Prim} A, \mu)$, $\varphi\bigl(\tau(s\cdot Q)\bigr)$ and $\varphi\bigl(s\cdot\tau(Q)\bigr)$ define the same element of $L^\infty(\operatorname{Prim}(A \rtimes_\alpha \Sigma), \nu)$. Therefore for each $s \in G$, $\tau(s\cdot Q) = s\cdot\tau(Q)$ for ν-almost all Q. Thus τ is essentially equivariant. Since we can replace τ by any τ' which agrees with τ almost everywhere, Theorem D.46 on page 397 implies that we may assume that there is a G-invariant Borel ν-conull set $Y_0 \subset \operatorname{Prim}(A \rtimes_\alpha \Sigma)$ such that $\tau(s \cdot Q) = s \cdot \tau(Q)$ for all $Q \in Y_0$ and all $s \in G$.

Since τ implements an injection of $L^\infty(\operatorname{Prim} A, \mu)$ into $L^\infty(\operatorname{Prim}(A \rtimes_\alpha \Sigma), \nu)$, it follows that $\tau_*\nu \ll \mu$. Therefore, we can apply Corollary I.9 on page 487 to disintegrate ν with respect to μ. This means that there are finite measures ν_P on $\operatorname{Prim}(A \rtimes_\alpha \Sigma)$ such that if φ is a bounded Borel function on $\operatorname{Prim}(A \rtimes_\alpha \Sigma)$, then

$$\int_{\operatorname{Prim}(A\rtimes_\alpha\Sigma)} \varphi(Q)\, d\nu(Q) = \int_{\operatorname{Prim} A} \int_{\operatorname{Prim}(A\rtimes_\alpha\Sigma)} \varphi(Q)\, d\nu_P(Q)\, d\mu(P). \tag{9.23}$$

Since $Q \mapsto \tilde{r}_Q$ is a Borel field of representations, we can form the direct integral representation

$$\hat{r}_P := \int^\oplus_{\operatorname{Prim}(A\rtimes_\alpha\Sigma)} \tilde{r}_Q \, d\nu_P(Q) \tag{9.24}$$

on $\mathcal{V}_P := L^2(\operatorname{Prim}(A \rtimes_\alpha \Sigma) * \mathfrak{H}, \nu_P)$. Let $\operatorname{Prim} A * \mathscr{V}$ be the Borel Hilbert bundle induced from $L^2\bigl(\operatorname{Prim}(A \rtimes_\alpha \Sigma) * \mathfrak{H}, \nu\bigr)$ via the disintegration of ν with respect to μ (see Example F.19 on page 416). Since we can identify $L^2\bigl(\operatorname{Prim}(A \rtimes_\alpha \Sigma) * \mathfrak{H}, \nu\bigr)$ with $L^2(\operatorname{Prim} A * \mathscr{V}, \mu)$ we have $\tilde{r}$ equivalent to

$$\hat{r} = \int^\oplus_{\operatorname{Prim} A} \hat{r}_P \, d\mu(P).$$

If φ is a bounded Borel function on $\operatorname{Prim} A$ and if T'_φ is the corresponding diagonal operator on $L^2(\operatorname{Prim} A * \mathscr{V}, \mu)$, then

$$M^r T_\varphi = T'_\varphi M^r.$$

Thus Corollary F.34 on page 426 implies r_P and $\hat{r}_P$ are equivalent for μ-almost all P.

Since $\ker \hat{r}_P$ is separable, a simple argument applied to (9.24) shows that there is a ν_P-null set $N(P)$ such that

$$\ker \hat{r}_P \subset \ker \tilde{r}_Q \quad \text{if } Q \notin N(P). \tag{9.25}$$

Since $\operatorname{supp} \nu_P \subset \tau^{-1}(P)$, we can rewrite (9.25) as

$$\ker \hat{r}_{\tau(Q)} \subset \ker \tilde{r}_Q. \tag{9.26}$$

Since (9.26) holds for ν_P-almost all Q and for all P, it follows that (9.26) holds for ν-almost all Q. Thus off a ν-null set N, $\tilde{r}_Q$ factors through $A \rtimes_\alpha G_{\tau(Q)}$. Therefore $\tilde{r}_Q = (\tilde{\pi}_Q, \tilde{u}_Q)$ is a covariant representation of $(A, G_{\tau(Q)}, \alpha)$ for all $Q \notin N$. If $j : A \to M(A \rtimes_\alpha \Sigma)$ is the natural map, then $\tilde{\pi}_Q(a) = \tilde{r}_Q^{\,-}\big(j(a)\big)$. Thus if $\{ \Phi_i \}$ is a countable approximate identity for $A \rtimes_\alpha \Sigma$, then off N,

$$\big(\tilde{\pi}_Q(a)h(P) \mid k(P)\big) = \lim_i \big(\tilde{r}_Q\big(j(a)\Phi_i\big)h(P) \mid k(P)\big).$$

Therefore $Q \mapsto \tilde{\pi}_Q$ is a Borel field of representations.[8] Since $\hat{r}_P$ is equivalent to r_P for μ-almost all P, we have $\hat{r}_P = (\hat{\pi}_P, \hat{\sigma}_P)$ with $\hat{\pi}_P$ homogeneous with kernel P (for almost all P). Since $\hat{r}_P$ is the direct integral of the the $\tilde{r}_Q$'s (see (9.24)), it follows that

$$\hat{\pi}_P = \int_{\operatorname{Prim}(A \rtimes_\alpha \Sigma)}^{\oplus} \tilde{\pi}_Q \, d\nu_P(Q).$$

Since $\hat{\pi}_P$ is homogeneous with kernel P, Corollary G.17 on page 440 implies that $\tilde{\pi}_Q$ is homogeneous with kernel P for ν_P-almost all Q. Thus $\tilde{\pi}_Q$ is homogeneous with kernel $\tau(Q)$ for ν-almost all Q. To summarize, we have the following lemma.

Lemma 9.13. *Let r be the restriction of R to the stability groups, and let (9.14) be the ideal center decomposition of r on $L^2\big(\operatorname{Prim}(A \rtimes_\alpha \Sigma) * \mathfrak{H}, \nu\big)$. Then there is a G-invariant Borel set $Y_0 \subset \operatorname{Prim}(A \rtimes_\alpha \Sigma)$ and a Borel map $\tau : \operatorname{Prim}(A \rtimes_\alpha \Sigma) \to \operatorname{Prim} A$ such that $\tau(s \cdot Q) = s \cdot \tau(Q)$ for all $s \in G$ and all $Q \in Y_0$. Furthermore, for ν-almost all Q, $\tilde{r}_Q$ is given by a homogeneous covariant representation $(\tilde{\pi}_Q, \tilde{u}_Q)$ of $(A, G_{\tau(Q)}, \alpha)$, where $\tilde{\pi}_Q$ is homogeneous with kernel $\tau(Q)$.*

9.3 Step III: Sauvageot's Induced Representation

In this step, we want to "induce" the restriction r to a representation $\operatorname{ind} r$ of $A \rtimes_\alpha G$, and then to see that the kernel of $\operatorname{ind} r$ is an induced primitive ideal. We will define $\operatorname{ind} r$ in an *ad hoc* manner (explaining the quotations marks above) in order to avoid the overhead of developing a general theory for induction from $A \rtimes_\alpha \Sigma$ to $A \rtimes_\alpha G$.

Proposition 9.14. *The kernel of the representation* $\operatorname{ind} r$ *induced from the restriction to the stability groups of a factorial representation R of $A \rtimes_\alpha G$ with infinite multiplicity is an induced primitive ideal of $A \rtimes_\alpha G$.*

We'll need a very tidy description of the space of the induced representation. (This will also be helpful when we get around to defining $\operatorname{ind} r$ in Lemma 9.19 on page 286.) Let $\mathscr{V}_1$ be the set of Borel functions $\xi : \operatorname{Prim} A \times G \to \mathcal{H}$ such that

$$\xi(P, st) = \sigma_P(t)^{-1}\big(\xi(P, s)\big) \quad \text{for all } s \in G \text{ and } t \in G_P.$$

[8] We can let π_Q be the zero representation off N.

Let $\mathcal{L}^2_\sigma(\operatorname{Prim} A \times G, \mu, \mathcal{H})$ be the set of $\xi \in \mathscr{V}_1$ such that

$$\int_{\operatorname{Prim} A} \int_{G/G_P} \|\xi(P,s)\|^2 \, d\beta^P(\dot{s}) \, d\mu(P) < \infty.$$

We let $L^2_\sigma(\operatorname{Prim} A \times G, \mu, \mathcal{H})$ be the quotient of $\mathcal{L}^2_\sigma(\operatorname{Prim} A \times G, \mu, \mathcal{H})$ where we identify two functions which agree $\mu \times \mu_G$-almost everywhere.

Example 9.15. Suppose that $\mathcal{H} = \mathbf{C}$ so that $\sigma_P(t) = 1$ for all $t \in G_P$. Then $\mathscr{L}^2_\sigma(\operatorname{Prim} A \times G, \mu, \mathbf{C})$ is the set of Borel functions $h : \operatorname{Prim} A \times G \to \mathbf{C}$ such that $\xi(P, st) = \xi(P, s)$ for all $s \in G$ and $t \in G_P$, and such that

$$\int_{\operatorname{Prim} A} \int_{G/G_P} |\xi(P,s)| \, d\beta^P(\dot{s}) \, d\mu(P) < \infty.$$

Suppose that γ is the measure on $\operatorname{Prim} A \times \operatorname{Prim} A$ given in Lemma 9.1 on page 264. Then in this case, the map φ sending $f \in \mathcal{L}^2(\operatorname{Prim} A \times \operatorname{Prim} A, \gamma)$ to $\varphi(f)(P,s) := f(P, s \cdot P)$ defines an isomorphism of $L^2(\operatorname{Prim} A \times \operatorname{Prim} A, \gamma)$ onto $L^2_\sigma(\operatorname{Prim} A \times G, \mu, \mathbf{C})$.

Lemma 9.16. *If* $\xi, \eta \in \mathcal{L}^2_\sigma(\operatorname{Prim} A \times G, \mu, \mathcal{H})$, *then*

$$P \mapsto \int_{G/G_P} \big(\xi(P,s) \mid \eta(P,s)\big) \, d\beta^P(\dot{s})$$

is in $\mathcal{L}^1(\operatorname{Prim} A, \mu)$, *and*

$$(\xi \mid \eta) := \int_{\operatorname{Prim} A} \int_{G/G_P} \big(\xi(P,s) \mid \eta(P,s)\big) \, d\beta^P(\dot{s}) \, d\mu(P) \tag{9.27}$$

defines an inner product on $L^2_\sigma(\operatorname{Prim} A \times G, \mu, \mathcal{H})$.

Proof. Let γ be the measure on $\operatorname{Prim} A \times \operatorname{Prim} A$ described in Lemma 9.1 on page 264. Let

$$\hat{\xi}(P, s \cdot P) := \|\xi(P,s)\| \quad \text{and} \quad \hat{\eta}(P, s \cdot P) = \|\eta(P,s)\|,$$

and extend $\hat{\xi}$ and $\hat{\eta}$ to all of $\operatorname{Prim} A \times \operatorname{Prim} A$ by setting them to zero elsewhere.[9] Then $\hat{\xi}$ and $\hat{\eta}$ are in $\mathcal{L}^2(\gamma)$, and their pointwise product is in $\mathcal{L}^1(\gamma)$. Thus

$$\int_{\operatorname{Prim} A} \int_{G/G_P} \|\xi(P,s)\| \|\eta(P,s)\| \, d\beta^P(\dot{s}) \, d\mu(P) < \infty.$$

It follows that the inner integral in (9.27), as a function of P, is in $\mathcal{L}^1(\mu)$ and the first assertion follows.

Suppose $\xi \in \mathcal{L}^2_\sigma(\operatorname{Prim} A \times G, \mu, \mathcal{H})$. Let $D = \{\, (P,s) : \xi(P,s) \neq 0 \,\}$ and $D_P = \{\, s : (s, P) \in D \,\}$. Then $(\xi \mid \xi) = 0$ if and only if for μ-almost all P, the image of D_P in G/G_P is a β^P-null set. Since D_P is saturated, the image of D_P is β^P null if and only if it is μ_G-null (Lemma H.14 on page 463). Thus $(\xi \mid \xi) = 0$ if and only if D is $\mu \times \mu_G$-null. The rest is straightforward. □

[9]Naturally, we want $\hat{\xi}$ and $\hat{\eta}$ to be Borel functions. But $S = \{\, (P, s \cdot P) : s \in G \,\}$ is σ-compact in the product Jacobson topology on $\operatorname{Prim} A \times \operatorname{Prim} A$. Thus S is Borel and $\hat{\xi}$ and $\hat{\eta}$ are Borel.

In order to show that $L^2_\sigma(\operatorname{Prim} A \times G, \mu, \mathcal{H})$ is nontrivial, we will extend the key construct in the proof of Proposition 5.4 on page 153. We aim to define a map $\mathcal{W}$ on the algebraic tensor product $C_c(G, A) \odot \mathcal{L}^2(\operatorname{Prim} A, \mu, \mathcal{H})$ which takes values in $\mathcal{L}^2_\sigma(\operatorname{Prim} A \times G, \mu, \mathcal{H})$. Therefore we define

$$\begin{aligned}\mathcal{W}(f \otimes h)(P, s)&\\ &:= \int_{G_P} \rho(st, G_P)^{-\frac{1}{2}} \pi_P\big(\alpha_s^{-1}\big(f(st)\big)\big)\sigma_P(t)\big(h(P)\big)\, d\mu_{G_P}(t). \qquad (9.28)\end{aligned}$$

The map $(P, s) \mapsto \mathcal{W}(f \otimes h)(P, s)$ is Borel by Lemma H.30 on page 473, and a calculation shows that it transforms correctly. To see that $\mathcal{W}(f \otimes h)$ is appropriately square integrable, and therefore belongs to $\mathcal{L}^2_\sigma(\operatorname{Prim} A \times G, \mu, \mathcal{H})$, requires a bit of work. However, it follows from the next observation which we will need later.

Lemma 9.17. *Define $\mathscr{P} : C_c(G, A) \to C_c(G * \Sigma, A)$ by*

$$\mathscr{P}(f)(s, H) = \gamma_H(s) f(s).$$

Then for all $f, g \in C_c(G, A)$ and $h, k \in \mathcal{L}^2(\operatorname{Prim} A, \mu, \mathcal{H})$ we have

$$\big(\mathcal{W}(f \otimes h) \mid \mathcal{W}(g \otimes k)\big) = \big(r\big(\mathscr{P}(g^* * f)\big)h \mid k\big).$$

In particular, $\mathcal{W}(f \otimes h) \in \mathcal{L}^2_\sigma(\operatorname{Prim} A \times G, \mu, \mathcal{H})$ and $f \mapsto \mathcal{W}(f \otimes h)$ is continuous from $C_c(G, A)$ with the inductive limit topology into $L^2_\sigma(\operatorname{Prim} A \times G, \mu, \mathcal{H})$.

Proof. To prove the first assertion, we compute as follows:

$$\begin{aligned}&\big(\mathcal{W}(f \otimes h) \mid \mathcal{W}(g \otimes k)\big)\\
&= \int_{\operatorname{Prim} A} \int_{G/G_P} \big(W(f \otimes h)(P, s) \mid W(g \otimes k)(P, s)\big)\, d\dot{s}\, dP\\
&= \int_{\operatorname{Prim} A} \int_{G/G_P} \int_{G_P} \int_{G_P} \big(\pi_P\big(\alpha_s^{-1}\big(f(st)\big)\big)\sigma_P(t)h(P) \mid\\
&\qquad \pi_P\big(\alpha_s^{-1}\big(g(sv)\big)\big)\sigma_P(v)k(P)\big)\rho(st, G_P)^{-\frac{1}{2}}\rho(sv, G_P)^{-\frac{1}{2}}\, dt\, dv\, d\dot{s}\, dP\end{aligned}$$

which, since (π_P, σ_P) is covariant μ-almost everywhere, is

$$\begin{aligned}&= \int_{\operatorname{Prim} A} \int_{G/G_P} \int_{G_P} \int_{G_P} \big(\pi_P\big(\alpha_{sv}^{-1}\big(g(sv)^* f(st)\big)\big)\sigma_P(v^{-1}t)h(P) | k(P)\big)\\
&\qquad\qquad \rho(st, G_P)^{-\frac{1}{2}}\rho(sv, G_P)^{-\frac{1}{2}}\, dt\, dv\, d\dot{s}\, dP\\
&= \int_{\operatorname{Prim} A} \int_{G/G_P} \int_{G_P} \int_{G_P} \big(\pi_P\big(\alpha_{sv}^{-1}\big(g(sv)^* f(svt)\big)\big)\sigma_P(t)h(P) | k(P)\big)\\
&\qquad\qquad \gamma_{G_P}(t)\rho(sv, G_P)^{-1}\, dt\, dv\, d\dot{s}\, dP\\
&= \int_{\operatorname{Prim} A} \int_G \int_{G_P} \big(\pi_P\big(\alpha_s^{-1}\big(g(s)^* f(st)\big)\big)\sigma_P(t)h(P) | k(P)\big)\\
&\qquad\qquad \gamma_{G_P}(t)\, dt\, ds\, dP\end{aligned}$$

$$= \int_{\operatorname{Prim} A} \int_{G_P} \big(\pi_P(g^* * f(t))\gamma_{G_P}(t)\sigma_P(t)h(P) \mid k(P)\big)\,dt\,dP$$
$$= \big(r(\mathscr{P}(g^* * f))h \mid k\big).$$

The second assertion follows since $\|\mathcal{W}(f \otimes h)\|^2 = \big(\mathcal{W}(f \otimes h) \mid \mathcal{W}(f \otimes h)\big)$. For the final assertion, note that if $g_i \to 0$ in the inductive limit topology, then $\mathscr{P}(g_i^* * g_i) \to 0$ in the inductive limit topology on $C_c(G * \Sigma, A)$, and therefore in the C^*-norm in $A \rtimes_\alpha \Sigma$. The result follows from this, the first assertion and the observation that $\mathcal{W}(f_i \otimes h) - \mathcal{W}(f \otimes h) = \mathcal{W}(f_i - f \otimes h)$. □

In Proposition 5.4, we showed that we could realize $\operatorname{Ind}_{G_P}^G r_P$ on $\mathcal{K}(P) := L^2_{\sigma_P}(G, \beta^P, \mathcal{H})$ as $\Pi_P \rtimes W_P$ where

$$\Pi_P(a)h(s) = \pi_P\big(\alpha_s^{-1}(a)\big)\big(h(s)\big), \text{ and}$$
$$W_P(r)h(s) = \Big(\frac{\rho(r^{-1}s, G_P)}{\rho(s, G_P)}\Big)^{\frac{1}{2}} h(r^{-1}s) = \rho(r^{-1}, G_{s\cdot P})^{\frac{1}{2}} h(r^{-1}s).$$

Moreover if $\bar{v}$ denotes the constant function $P \mapsto v$ for each $v \in \mathcal{H}$, then the proof of Proposition 5.4 shows that

$$\{\,\mathcal{W}(f \otimes \bar{v})(P, \cdot) : f \in C_c(G, A) \text{ and } v \in \mathcal{H}\,\}$$

spans a dense subspace of $\mathcal{K}(P)$. Let $\mathfrak{K} = \{\,\mathcal{K}(P)\,\}_{P \in \operatorname{Prim} A}$. Notice that each $\xi \in \mathcal{L}^2_\sigma(\operatorname{Prim} A \times G, \mu, \mathcal{H})$ defines a section of $\operatorname{Prim} A * \mathfrak{K}$ in the obvious way:[10] $\check{\xi}(P)(s) := \xi(P, s)$.

Proposition 9.18. *There is a unique Borel structure on* $\operatorname{Prim} A * \mathfrak{K}$ *such that it is an analytic Borel Hilbert bundle over* $\operatorname{Prim} A$*, and such that* $\check{\xi} \in B(\operatorname{Prim} A * \mathfrak{K})$ *for all* $\xi \in \mathcal{L}^2_\sigma(\operatorname{Prim} A \times G, \mu, \mathcal{H})$*. Furthermore,* $L^2_\sigma(\operatorname{Prim} A \times G, \mu, \mathcal{H})$ *is a separable Hilbert space, and* $\xi \mapsto \check{\xi}$ *is a unitary isomorphism of* $L^2_\sigma(\operatorname{Prim} A \times G, \mu, \mathcal{H})$ *onto* $L^2(\operatorname{Prim} A * \mathfrak{K}, \mu)$.

In order to concentrate on the task at hand, we will postpone the proof of Proposition 9.18 to the end of this section (on page 290).

When convenient, we will identify $L^2_\sigma(\operatorname{Prim} A \times G, \mu, \mathcal{H})$ and $L^2(\operatorname{Prim} A * \mathfrak{K}, \mu)$. In particular, we will usually not distinguish between ξ and $\check{\xi}$.

Lemma 9.19 (The Definition of $\operatorname{ind} r$). *The formulas*

$$\Pi(a)\xi(P, s) := \pi_P\big(\alpha_s^{-1}(a)\big)\big(\xi(P, s)\big), \text{ and} \tag{9.29}$$
$$W(r)\xi(P, s) := \rho(r^{-1}, G_{s\cdot P})^{\frac{1}{2}} \xi(P, r^{-1}s) \tag{9.30}$$

define a covariant representation $\operatorname{ind} r := (\Pi, W)$ *of* (A, G, α) *on* $L^2_\sigma(\operatorname{Prim} A \times G, \mu, \mathcal{H})$*. After identifying* $L^2_\sigma(\operatorname{Prim} A \times G, \mu, \mathcal{H})$ *with* $L^2(\operatorname{Prim} A * \mathfrak{K}, \mu)$*, we have*

$$\operatorname{ind} r = \int_{\operatorname{Prim} A}^{\oplus} \operatorname{Ind}_{G_P}^G r_P \, d\mu(P). \tag{9.31}$$

---

[10] Since $s \mapsto \|\xi(P, s)\|$ can fail to be square integrable off a μ-null set, we can let $\check{\xi}(P) = 0$ in those cases.

Proof. Algebraically, the final statement simply amounts to untangling definitions. The primary issues are to see that the right-hand side of (9.29) defines a Borel function, and that $r \mapsto W(r)\xi$ is continuous for all $\xi \in L^2_\sigma(\operatorname{Prim} A \times G, \mu, \mathcal{H})$. For the first of these, it suffices to see that

$$(P, s) \mapsto \big(\pi_P\big(\alpha_s^{-1}(a)\big)\big(\xi(P, s)\big) \mid v\big)$$

is Borel (Lemma D.43 on page 395). But if $\{ e_i \}$ is a countable orthonormal basis for $\mathcal{H}$, then

$$\big(\pi_P\big(\alpha_s^{-1}(a)\big)\xi(P, s)\big) \mid v\big) = \sum_i \big(\xi(P, s) \mid e_k\big)\big(\pi_P\big(\alpha_s^{-1}(a)\big)e_k \mid v\big).$$

Thus, it suffices to see that

$$(P, s) \mapsto \big(\pi_P\big(\alpha_s^{-1}(a)\big)v \mid w\big)$$

is Borel. Since G is second countable, there is a countably-valued Borel function $b_\epsilon : G \to A$ such that $\|b_\epsilon(s) - \alpha_s^{-1}(a)\| < \epsilon$ for all $s \in G$. Since b_ϵ is a countable sum of characteristic functions,

$$(P, s) \mapsto \big(\pi_P\big(b_\epsilon(s)\big)v \mid w\big)$$

is Borel. Since

$$\big(\pi_P\big(b_{\frac{1}{n}}(s)\big)v \mid w\big) \to \big(\pi_P\big(\alpha_s^{-1}(a)\big)v \mid w\big),$$

the right-hand side of (9.29) is Borel and Π is a representation of A.

To show that W is strongly continuous, it suffices to see that $s \mapsto W(s)\xi$ is continuous for a dense set of ξ. In the proof of Proposition 9.18, we will show that vectors of the form $\mathcal{W}(f \otimes h)$ for $f \in C_c(G, A)$ and $k \in L^2(\operatorname{Prim} A, \mu, \mathcal{H})$ span a dense subspace of $L^2_\sigma(\operatorname{Prim} A \times G, \mu, \mathcal{H})$. Hence, it suffices to consider ξ of the form $\mathcal{W}(f \otimes h)$. Note that $W(s)\mathcal{W}(f \otimes h) = \mathcal{W}(i_G(s)(f) \otimes h)$. Since $s \mapsto i_G(s)f$ is continuous from G into $C_c(G, A)$ with the inductive limit topology, the strong continuity of $\mathcal{W}$ follows from Lemma 9.17 on page 285. □

We showed in Section 9.2 that r has an ideal center decomposition

$$\tilde{r} = \int^\oplus_{\operatorname{Prim}(A \rtimes_\alpha \Sigma)} \tilde{r}_Q \, d\nu,$$

where for ν-almost all Q, $\tilde{r}_Q = \tilde{\pi}_Q \rtimes \tilde{u}_Q$ is a covariant representation of $(A, G_{\tau(Q)}, \alpha)$. Equation (9.31) suggests that $\operatorname{ind} r$ ought to be equivalent to a representation $\operatorname{ind} \tilde{r}$ which is a direct integral of representations of the form $\operatorname{Ind}_{G_{\tau(Q)}}^G \tilde{r}_Q$. Since we have only an *ad hoc* means of inducing from $A \rtimes_\alpha \Sigma$ to $A \rtimes_\alpha G$, making sense of this requires a bit of fussing.

Let $\widetilde{\mathcal{K}}(Q)$ be the "usual" space for $\operatorname{Ind}_{G_{\tau(Q)}}^G \tilde{r}_Q$ as defined in Section 5.1. Thus $\widetilde{\mathcal{K}}(Q)$ is the completion of the algebraic tensor product $C_c(G, A) \odot \mathcal{H}(Q)$ with respect to the pre-inner product

$$(f \otimes v \mid g \otimes w) = \big(\tilde{r}_Q\big(\langle g \, , f \rangle_{A \rtimes G_{\tau(Q)}}\big)v \mid w\big).$$

Each $f\otimes\tilde h$ in $C_c(G,A)\odot\mathcal{L}^2(\operatorname{Prim}(A\rtimes_\alpha\Sigma)*\mathfrak{H},\nu)$ defines a section of $\operatorname{Prim}(A\rtimes_\alpha\Sigma)*\widetilde{\mathfrak{K}}$ is an obvious way: $(f\otimes\tilde h)(Q):=f\otimes\tilde h(Q)$. Clearly

$$Q\mapsto\big((f\otimes\tilde h)(Q)\mid(g\otimes\tilde k)(Q)\big)$$

is Borel, and it is not hard to see that $\operatorname{Prim}(A\rtimes_\alpha\Sigma)*\widetilde{\mathfrak{K}}$ has a unique Borel structure making it into an analytic Borel Hilbert bundle such that each $f\otimes\tilde h$ defines a Borel section.[11] If we define $\mathscr{P}:C_c(G,A)\to C_c(G*\Sigma,A)$ as in Lemma 9.17 on page 285, then

$$\langle f\,,g\rangle_{A\rtimes G_{\tau(Q)}}(s)=\gamma_{G_{\tau(Q)}}(s)g^*\ast f(s).$$

It follows that the inner product in $L^2(\operatorname{Prim}(A\rtimes_\alpha\Sigma)*\widetilde{\mathfrak{K}},\nu)$ is given by

$$(f\otimes\tilde h\mid g\otimes\tilde k)=\big(\tilde r\big(\mathscr{P}(g^*\ast f)\big)\tilde h\mid\tilde k\big).$$

Since

$$\big(\operatorname{Ind}_{G_{\tau(Q)}}^G\tilde r_Q\big)(g)(f\otimes\tilde h(Q))=g\ast f\otimes\tilde h(Q),$$

it is clear that $\{\operatorname{Ind}_{G_{\tau(Q)}}^G\tilde r_Q\}$ is a Borel field of representations of $A\rtimes_\alpha G$ and we can define the desired direct integral:

$$\operatorname{ind}\tilde r:=\int_{\operatorname{Prim}(A\rtimes_\alpha\Sigma)}^{\oplus}\operatorname{Ind}_{G_{\tau(Q)}}^G\tilde r_Q\,d\nu(Q).\tag{9.32}$$

Proposition 9.20. *The representations* $\operatorname{ind}r$ *defined in Lemma 9.19 on page 286, and* $\operatorname{ind}\tilde r$ *defined in* (9.32) *above are equivalent representations of* $A\rtimes_\alpha G$.

Proof. Just as above, we can view $C_c(G,A)\odot\mathcal{L}^2(\operatorname{Prim}A,\mu,\mathcal{H})$ as a dense subspace of square integrable Borel sections of an analytic Borel Hilbert bundle $\operatorname{Prim}A*\widetilde{\mathfrak{H}}$ where $\widetilde{\mathcal{H}}(P)$ is the completion of $C_c(G,A)\odot\mathcal{H}$ with respect to the inner product

$$(f\otimes h\mid g\otimes k)=\big(r_P(\langle g\,,f\rangle_{A\rtimes G_P})h\mid k\big).$$

The inner product on $L^2(\operatorname{Prim}A*\widetilde{\mathfrak{H}},\mu)$ is given by

$$(f\otimes h\mid g\otimes k)=\big(r\big(\mathscr{P}(g^*\ast f)\big)h\mid k\big).$$

Recall that r and $\tilde r$ are equivalent via a unitary $M^r:L^2(\operatorname{Prim}A,\mu,\mathcal{H})\to L^2(\operatorname{Prim}(A\rtimes_\alpha\Sigma)*\mathfrak{H},\nu)$. Then we can compute that

$$\begin{aligned}\big(f\otimes M^r(h)\mid g\otimes M^r(k)\big)&=\big(\tilde r\big(\mathscr{P}(g^*\ast f)\big)M^r(h)\mid M^r(k)\big)\\&=\big(M^r\big(r\big(\mathscr{P}(g^*\ast f)\big)h\big)\mid M^r(k)\big)\end{aligned}$$

[11]For example, if $\{f_i\}$ is dense in the inductive limit topology on $C_c(G,A)$ and if $\{\tilde v_i\}$ is a special orthogonal fundamental sequence for $\operatorname{Prim}(A\rtimes_\alpha\Sigma)*\mathfrak{H}$, then $\{f_i\otimes\tilde v_j\}$ is a set of sections which satisfy properties (ii) and (iii) of part (b) of Proposition F.8 on page 412.

which, since M^r is unitary, is

$$\begin{aligned} &= \big(r\big(\mathscr{P}(g^* * f)\big)h \mid k\big) \\ &= (f \otimes h \mid g \otimes k). \end{aligned}$$

Thus, we obtain a well-defined unitary $\mathcal{M} : L^2(\operatorname{Prim} A * \widetilde{\mathfrak{H}}, \mu) \to L^2(\operatorname{Prim}(A \rtimes_\alpha \Sigma) * \widetilde{\mathfrak{K}}, \nu)$ characterized by $\mathcal{M}(f \otimes h) = f \otimes M^r(h)$.

In view of Lemma 9.17 on page 285, $\mathcal{W} : C_c(G, A) \odot \mathcal{L}^2(\operatorname{Prim} A, \mu, \mathcal{H}) \to \mathcal{L}^2_\sigma(\operatorname{Prim} A \times G, \mu, \mathcal{H})$ defined by (9.28) defines an isometry of $L^2(\operatorname{Prim} A * \widetilde{\mathfrak{H}}, \mu)$ into $L^2_\sigma(\operatorname{Prim} A \times G, \mu, \mathcal{H})$. Moreover, if we use Proposition 9.18 on page 286 to identify $L^2_\sigma(\operatorname{Prim} A \times G, \mu, \mathcal{H})$ with $L^2(\operatorname{Prim} A * \mathfrak{K}, \mu)$, then $\mathcal{W}$ is implemented by a id-isomorphism $\{\,\mathcal{W}(P)\,\}$ where $\mathcal{W}(P)$ is the isomorphism of $\widetilde{\mathcal{H}}(P)$ onto $\mathcal{K}(P) = L^2_{\sigma_P}(G, \beta^P, \mathcal{H})$ given by

$$\mathcal{W}(P)(f \otimes h(P))(s) = \mathcal{W}(f \otimes h)(P, s).$$

(We proved that $\mathcal{W}(P)$ is a unitary in Proposition 5.4 on page 153.) In particular, $\mathcal{W}$ is surjective (see Lemma F.32 on page 422), and defines a unitary from $L^2(\operatorname{Prim} A * \widetilde{\mathfrak{H}}, \mu)$ onto $L^2_\sigma(\operatorname{Prim} A \times G, \mu, \mathcal{H})$.

If $\operatorname{ind} r = \Pi \rtimes W$, as in Lemma 9.19 on page 286, then it is routine to check that

$$\begin{aligned} \Pi(a)\mathcal{W}(f \otimes h) &= \mathcal{W}(i_A(a)(f) \otimes h) \quad \text{and} \\ W(r)\mathcal{W}(f \otimes h) &= \mathcal{W}(i_G(r)(f) \otimes h). \end{aligned}$$

Therefore $(\operatorname{ind} r)(g)\mathcal{W}(f \otimes h) = \mathcal{W}(g * f \otimes h)$. Then $\mathcal{W}\mathcal{M}^{-1}$ is a unitary and

$$\begin{aligned} \mathcal{W}\mathcal{M}^{-1}(\operatorname{ind} \tilde{r})(g)(f \otimes \tilde{h}) &= \mathcal{W}\mathcal{M}^{-1}(g * f \otimes h) \\ &= \mathcal{W}\big(g * f \otimes (M^r)^{-1}(\tilde{h})\big) \\ &= (\operatorname{ind} r)(g)\mathcal{W}\mathcal{M}^{-1}(f \otimes \tilde{h}). \end{aligned}$$

Thus $\operatorname{ind} r$ and $\operatorname{ind} \tilde{r}$ are equivalent as claimed. □

In view of the previous result, we can compute the kernel of $\operatorname{ind} r$ by examining the kernel of (9.32). Lemma 9.13 on page 283 implies that each $\tilde{r}_Q = \tilde{\pi}_Q \rtimes \tilde{u}_Q$ is homogeneous with $\tilde{\pi}_Q$ also homogeneous and having kernel $\tau(Q)$. Proposition 9.5 on page 268 implies that

$$I_Q := \ker \operatorname{Ind}_{G_{\tau(Q)}}^G (\tilde{\pi}_Q \rtimes \tilde{u}_Q)$$

is an induced primitive ideal of the form $\operatorname{Ind}_{G_{\tau(Q)}}^G J_Q$ with $\operatorname{Res} J_Q = \tau(Q)$. Lemma F.28 on page 420 implies that $Q \mapsto I_Q$ is a Borel map of $\operatorname{Prim}(A \rtimes_\alpha \Sigma)$ to $\operatorname{Prim} A$, and Lemma 9.13 on page 283 implies that there is a ν-conull set $Y_0 \subset \operatorname{Prim}(A \rtimes_\alpha \Sigma)$ such that $\tau(s \cdot Q) = s \cdot \tau(Q)$ for all $Q \in Y_0$ and all $s \in G$. Recall that

$$\ker\big(\operatorname{Ind}_{G_{\tau(Q)}}^G \tilde{r}_Q\big) = \operatorname{Ind}_{G_{\tau(Q)}}^G \big(\ker \tilde{r}_Q\big)$$

by Lemma 5.16 on page 164, and that $\tilde{r}_{s\cdot Q}$ is equivalent to $\ker s\cdot\tilde{r}_Q$ by Proposition 9.8 on page 274. Thus for all $Q\in Y_0$ we have

$$\begin{aligned} I_{s\cdot Q} &:= \ker\bigl(\operatorname{Ind}_{G_{\tau(s\cdot Q)}}^G \tilde{r}_{s\cdot Q}\bigr) \\ &= \ker\bigl(\operatorname{Ind}_{G_{s\cdot\tau(Q)}}^G s\cdot\tilde{r}_Q\bigr) \\ &= \ker\bigl(\operatorname{Ind}_{s\cdot G_{\tau(Q)}}^G s\cdot\tilde{r}_Q\bigr) \end{aligned}$$

which, since $\operatorname{Ind}_{s\cdot G_{\tau(Q)}}^G s\cdot\tilde{r}_Q$ is equivalent to $\operatorname{Ind}_{G_{\tau(Q)}}^G \tilde{r}_Q$ by Lemma 5.8 on page 157, is

$$\begin{aligned} &= \ker\bigl(\operatorname{Ind}_{G_{\tau(Q)}}^G \tilde{r}_Q\bigr) \\ &= I_Q. \end{aligned}$$

Thus $Q\mapsto I_Q$ is G-invariant from $Y_0\subset\operatorname{Prim}(A\rtimes_\alpha\Sigma)\to\operatorname{Prim}A$. Thus $Q\mapsto I_Q$ is essentially constant by Lemma D.47 on page 398, and there is a $P_0:=\tau(Q_0)$ such that $\ker(\operatorname{ind}r)=I_{Q_0}$ is an induced ideal of the form $\operatorname{Ind}_{G_{P_0}}^G J$ with $\operatorname{Res}J=P_0$.

This completes the proof of Proposition 9.14 except for proving Proposition 9.18 on page 286.

Proof of Proposition 9.18. Suppose that ξ and η are in $\mathcal{L}^2_\sigma(\operatorname{Prim}A\times G,\mu,\mathcal{H})$. If $\mathfrak{b}$ is a generalized Bruhat approximate cross section, then

$$\int_{G/G_P}\bigl(\xi(P,s)\mid\eta(P,s)\bigr)\,d\beta^P(\dot{s})=\int_G\bigl(\xi(P,s)\mid\eta(P,s)\bigr)\mathfrak{b}(s,G_P)\rho(s,G_P)\,d\mu_G(s).$$

It follows that $P\mapsto\bigl(\check{\xi}(P)\mid\check{\eta}(P)\bigr)$ is Borel for all ξ and η. In particular, let $\{f_i\}$ be dense in $C_c(G,A)$ in the inductive limit topology, and let $\{e_i\}$ be an orthonormal basis for $\mathcal{H}$. If $\bar{e}_i$ is the corresponding constant function on $\operatorname{Prim}A$, then

$$\{\,\mathcal{W}(f_i\otimes\bar{e}_j)(P,\cdot)\,\}$$

spans a dense subspace of $\mathcal{K}(P)$ for each P (by the proof of Proposition 5.4 on page 153). It follows from Proposition F.8 on page 412, that there is a unique Borel structure on $\operatorname{Prim}A*\mathfrak{K}$ making the latter into an analytic Borel Hilbert bundle such that $\{\check{\mathcal{W}}(f_i\otimes\bar{e}_j)\}$ is a fundamental sequence. Furthermore, each $\check{\xi}$ is in $B(\operatorname{Prim}A*\mathfrak{K})$.

Since that map $\xi\mapsto\check{\xi}$ is isometric and since (9.27) on page 284 defines a definite inner-product on $L^2_\sigma(\operatorname{Prim}A\times G,\mu,\mathcal{H})$, once we show that $L^2_\sigma(\operatorname{Prim}A\times G,\mu,\mathcal{H})$ is complete, it will follow that $L^2_\sigma(\operatorname{Prim}A\times G,\mu,\mathcal{H})$ and $L^2(\operatorname{Prim}A*\mathfrak{K},\mu)$ are isomorphic. Since the latter is separable by Lemma F.17 on page 415, this will suffice to prove the proposition.

Let $\{\xi_n\}$ be a Cauchy sequence in $\mathcal{L}^2_\sigma(\operatorname{Prim}A\times G,\mu,\mathcal{H})$. We can pass to a subsequence, relabel, and assume that

$$\|\xi_{n+1}-\xi_n\|_2\le\frac{1}{2^n}\quad\text{for all } n.$$

Define extended real-valued functions z_n and z on $\operatorname{Prim} A \times \operatorname{Prim} A$ by

$$z_n(P, s \cdot P) = \sum_{k=1}^{n} \|\xi_{k+1}(P, s) - \xi_k(P, s)\|, \text{ and}$$
$$z(P, s \cdot P) = \sum_{k=1}^{\infty} \|\xi_{k+1}(P, s) - \xi_k(P, s)\|.$$

(We let z and z_n be zero at (P, Q) if $Q \neq s \cdot P$ for some $s \in G$ — see footnote 9 on page 284.) Using the triangle inequality for $\|\cdot\|_\gamma = \|\cdot\|_{L^2(\gamma)}$, and the observation that $\|\hat{\xi}\|_\gamma = \|\xi\|_2$ (where $\hat{\xi}(P, s \cdot P) = \|\xi(P, s)\|$), we have

$$\|z_n\|_\gamma \leq \sum_{k=1}^{n} \|\xi_{k+1} - \xi_k\|_2 \leq 1.$$

Since

$$\|z_n\|_\gamma^2 = \iint_{\operatorname{Prim} A \times \operatorname{Prim} A} z_n(P, Q)^2 \, d\gamma(P, Q),$$

the Monotone Convergence Theorem implies that

$$\|z\|_\gamma^2 = \iint_{\operatorname{Prim} A \times \operatorname{Prim} A} z(P, Q)^2 \, d\gamma(P, Q) \leq 1.$$

Therefore there is a γ-null set N such that $(P, s \cdot P) \notin N$ implies

$$\sum_{k=1}^{\infty} \xi_{k+1}(P, s) - \xi(P, s)$$

is absolutely convergent in $\mathcal{H}$. Thus the series converges to some $\xi'(P, s) \in \mathcal{H}$. Furthermore,

$$\begin{aligned} \xi'(P, s) &= \lim_{n \to \infty} \sum_{k=1}^{n} \xi_{k+1}(P, s) - \xi_k(P, s) \\ &= \lim_{n \to \infty} \xi_{n+1}(P, s) - \xi_1(P, s). \end{aligned}$$

Thus $\xi(P, s) := \xi'(P, s) + \xi_1(P, s)$ satisfies

$$\xi(P, s) = \lim_{n \to \infty} \xi_n(P, s) \quad \text{for all } (P, s \cdot P) \notin N.$$

If we define $\xi(P, s) = 0$ if $(P, s \cdot P) \notin N$, then, since $(P, s \cdot P) \in N$ if and only if $(P, st \cdot P) \in N$ for all $t \in G_P$,

$$\xi(P, st) = \sigma_P(t)^{-1}\big(\xi(P, s)\big) \quad \text{for } \textit{all } s \in G \text{ and } t \in G_P.$$

Furthermore,

$$\{ (P, s) : (P, s \cdot P) \in N \}$$

is a $\mu \times \mu_G$-null set. Thus $\xi_n \to \xi$ almost everywhere.

Given $\epsilon > 0$, there is a N such that $n, m \geq N$ implies that

$$\|\xi_n - \xi_m\|_2 < \epsilon.$$

If $(P, s \cdot P) \notin N$,

$$\|\xi(P, s) - \xi_k(P, s)\| = \lim_n \|\xi_n(P, s) - \xi_k(P, s)\|.$$

Thus, if $k \geq N$, Fatou's Lemma implies that

$$\|\xi - \xi_k\|_2^2 \leq \liminf_n \|\xi_n - \xi_k\|_2^2 \leq \epsilon^2.$$

Also,

$$\begin{aligned}\|\xi(P, s)\|^2 &\leq \big(\|\xi(P, s) - \xi_k(P, s)\| + \|\xi_k(P, s)\|\big)^2 \\ &\leq 4\|\xi(P, s) - \xi_k(P, s)\|^2 + 4\|\xi_k(P, s)\|^2.\end{aligned}$$

Thus $\xi \in \mathcal{L}^2_\sigma(\operatorname{Prim} A \times G, \mu, \mathcal{H})$ and $\xi_k \to \xi$ in $L^2_\sigma(\operatorname{Prim} A \times G, \mu, \mathcal{H})$. This completes the proof of completeness. □

Remark 9.21. If in the preceding discussion we replace A by $\mathbf{C}$ (and therefore $\operatorname{Prim} A$ by a point) and σ by a unitary representation of a subgroup H on $\mathcal{H}$. Then we get a Hilbert space $\mathcal{V}$ whose elements are μ_G-almost everywhere equivalence classes of Borel functions $h : G \to \mathcal{H}$ which satisfy $h(st) = u(t)^{-1}\big(h(s)\big)$ for all $s \in G$ and $t \in H$ for which

$$\int_{G/H} \|h(s)\|^2 \, d\beta^H(\dot{s}) < \infty.$$

Since the functions

$$h(s) := \mathcal{W}(f \otimes \bar{v})(s) := \int_H \rho(st, H)^{-\frac{1}{2}} \varphi(st) u(t)(v) \, d\mu_H(t) \tag{9.33}$$

span a dense subset and also lie in the space $\mathcal{V}_c$ defined prior to Proposition 5.4 on page 153, we see that the space $L^2_u(G, \beta^H, \mathcal{H})$ in that Proposition can be identified with $\mathcal{V}$.

9.4 Step IV: G Amenable

Proposition 9.22 (Sauvageot). *Suppose that G is amenable and that R is a factor representation of $A \rtimes_\alpha G$ with infinite multiplicity. If* $\operatorname{ind} r$ *is the representation of $A \rtimes_\alpha G$ induced from Sauvageot's restriction of R to the stability groups, then*

$$\ker(\operatorname{ind} r) \subset \ker R.$$

Let $\theta : \operatorname{Prim} A \times \operatorname{Prim} A \to [0,\infty)$ be a Borel function such that

$$\int_{G/G_P} \theta(P, s\cdot P)^2 \, d\beta^P(\dot s) = 1 \quad \text{for all } P \in \operatorname{Prim} A. \tag{9.34}$$

(We will produce such a function in Lemma 9.23 on page 295.)

We are still using the notation and set up described in the beginning of Section 9.2. In particular, let $X \subset \operatorname{Prim} A$ be as specified there. Thus if $P \in X$, $s \cdot P \in X$ and $t \in G_P$, then

$$U(t^{-1}s^{-1}, P) = U(t^{-1}, P)U(s^{-1}, P) = \sigma_P(t)^{-1}U(s^{-1}, P).$$

Let $N := \operatorname{Prim} A \smallsetminus X$ and set

$$N' := \{\, (P, s) : s \cdot P \in N \,\}.$$

Since μ is quasi-invariant, for each $s \in G$ we have

$$\mu\bigl(\{\, P : s \cdot P \in N \,\}\bigr) = \mu(s^{-1} \cdot N) = 0.$$

It follows that N' is $\mu \times \mu_G$-null. Let $h \in \mathcal{L}^2(\operatorname{Prim} A, \mu, \mathcal{H})$ with $\|h\|_2 = 1$, and define $\xi^{\gamma,h}(P, s)$ to be

$$D(P, s\cdot P)^{\frac12}\theta(s\cdot P, P)U(s^{-1}, P)\bigl(h(s\cdot P)\bigr) \quad \text{if } (P, s) \notin N', \tag{9.35}$$

and zero otherwise. Note that if $(P, s) \notin N'$ and $t \in G_P$ then $(P, st) \notin N'$, and we have

$$\begin{aligned}
\xi^{\gamma,h}(P, st) &= D(P, s\cdot P)^{\frac12}\theta(s\cdot P, P)U(t^{-1}s^{-1}, P)\bigl(h(s\cdot P)\bigr) \\
&= D(P, s\cdot P)^{\frac12}\theta(s\cdot P, P)U(t^{-1}, P)U(s^{-1}, P)\bigl(h(s\cdot P)\bigr) \\
&= \sigma_P(t)^{-1}\bigl(\xi^{\gamma,h}(P, s)\bigr).
\end{aligned}$$

Thus $\xi^{\gamma,h}$ transforms properly for all (P, s). Furthermore,

$$\begin{aligned}
\|\xi^{\gamma,h}\|_2^2 &= \iint_{\operatorname{Prim} A\times \operatorname{Prim} A} D(P, Q)\theta(Q, P)^2 \|h(Q)\|^2 \, d\gamma(P, Q) \\
&= \iint_{\operatorname{Prim} A\times \operatorname{Prim} A} \theta(P, Q)^2 \|h(P)\|^2 \, d\gamma(P, Q) \\
&= \int_{\operatorname{Prim} A} \|h(P)\|^2 \int_{G/G_P} \theta(P, s\cdot P)^2 \, d\beta^P(\dot s)\, d\mu(P) \\
&= 1.
\end{aligned}$$

Thus $\xi^{\gamma,h} \in \mathcal{L}^2_\sigma(\operatorname{Prim} A \times G, \mu, \mathcal{H})$ with norm one.

If $\varphi \in C_c(G, A)$, then we compute as follows:

$$\big(\operatorname{ind} r(\varphi)\xi^{\gamma,h} \mid \xi^{\gamma,h}\big) = \int_{\operatorname{Prim} A} \int_{G/G_P} \big(\operatorname{ind} r(\varphi)\xi^{\gamma,h}(P,s) \mid \xi^{\gamma,h}(P,s)\big) \, d\beta^P(\dot s)\, d\mu(P)$$

$$= \int_{\operatorname{Prim} A} \int_{G/G_P} \int_G \rho(r^{-1}, G_{s\cdot P})^{\frac12} \big(\pi_P\big(\alpha_s^{-1}(\varphi(r))\big)\big(\xi^{\gamma,h}(P, r^{-1}s)\big) \mid \xi^{\gamma,h}(P,s)\big) \, d\mu_G(r)\, d\beta^P(\dot s)\, d\mu(P)$$

which, since (9.35) holds $\mu \times \mu_G$-almost everywhere, is

$$= \iint_{\operatorname{Prim} A \times \operatorname{Prim} A} \int_G n(P,Q,r)\Theta(P,Q,r)I(P,Q,r)\, d\mu_G(r)\, d\gamma(P,Q), \tag{9.36}$$

where n, Θ and I are defined by

$$\begin{aligned} n(P, s\cdot P, r) &:= \rho(r^{-1}, G_{s\cdot P})^{\frac12} D(P, r^{-1}s\cdot P)^{\frac12} D(P, s\cdot P)^{\frac12} \\ \Theta(P, s\cdot P, r) &:= \theta(r^{-1}s\cdot P, P)\theta(s\cdot P, P) \text{ and} \\ I(P, s\cdot P, r) &:= \big(\pi_P\big(\alpha_s^{-1}(\varphi(r))\big)U(s^{-1}r, P)\big(h(r^{-1}s\cdot P)\big) \mid \\ &\qquad\qquad U(s^{-1}, P)\big(h(s\cdot P)\big)\big). \end{aligned}$$

Using (9.3) of Lemma 9.1 on page 264, we have, $\mu \times \mu_G \times \mu_G$-almost everywhere,

$$\begin{aligned} D(P, r^{-1}s\cdot P) &= \rho(r^{-1}s, G_P)^{-1} d(s^{-1}r, P) \\ &= \rho(s, G_P)^{-1}\rho(r^{-1}, G_{s\cdot P})^{-1} d(s^{-1}, P) d(r, s\cdot P) \\ &= D(P, s\cdot P) D(s\cdot P, r^{-1}s\cdot P). \end{aligned}$$

Therefore we have, $\mu \times \mu_G \times \mu_G$-almost everywhere,

$$\begin{aligned} n(P, s\cdot P, r) &= \rho(r^{-1}, G_{s\cdot P})^{\frac12} D(s\cdot P, r^{-1}s\cdot P)^{\frac12} D(P, s\cdot P) \\ &= \rho(r^{-1}, G_{s\cdot P})^{\frac12} \big(\rho(r^{-1}, G_{s\cdot P})^{-1} d(r, s\cdot P)\big)^{\frac12} D(P, s\cdot P) \\ &= d(r, s\cdot P)^{\frac12} D(P, s\cdot P). \end{aligned} \tag{9.37}$$

Thus (9.36) equals

$$\iint_{\operatorname{Prim} A \times \operatorname{Prim} A} \int_G d(r,Q)^{\frac12} \Theta(P,Q,r) I(P,Q,r)\, d\mu_G(r) D(P,Q)\, d\gamma(P,Q)$$

Furthermore, for each $s, r \in G$, the following hold μ-almost everywhere:

$$\begin{aligned} \pi_P\big(\alpha_s^{-1}(\varphi(r))\big) &= U(s^{-1}, P)\pi_{s\cdot P}\big(\varphi(r)\big)U(s^{-1}, P)^{-1} \\ U(s^{-1}, P)^{-1} &= U(s, s\cdot P) \\ U(r, s\cdot P) &= U(s, s\cdot P)U(s^{-1}r, P). \end{aligned}$$

Therefore, $\gamma \times \mu_G$-almost everywhere,

$$I(P,Q,r) = \big(\pi_Q\big(\varphi(r)\big)U(r,Q)\big(h(r^{-1}\cdot Q)\big) \mid h(Q)\big). \tag{9.38}$$

Now the definition of D and equations (9.36) and (9.38) show that (9.36) is equal to

$$\iint_{\operatorname{Prim} A\times\operatorname{Prim} A}\int_G d(r,P)^{\frac12}\theta(r^{-1}\cdot P,Q)\theta(P,Q) \\ \big(\pi_P\big(\varphi(r)\big)U(r,P)\big(h(r^{-1}\cdot P)\big) \mid h(P)\big)\,d\mu_G(r)\,d\gamma(P,Q). \tag{9.39}$$

On the other hand, Equation (9.18) implies that

$$\big(R(\varphi)h \mid h\big) = \\ \int_{\operatorname{Prim} A}\int_G d(r,P)^{\frac12}\big(\pi_P\big(\varphi(r)\big)U(r,P)h(r^{-1}\cdot P) \mid h(P)\big)\,d\mu_G(r)\,d\mu(P). \tag{9.40}$$

Lemma 9.23. *If G is amenable, then, given $\epsilon > 0$, we can choose a Borel function $\theta : \operatorname{Prim} A \times \operatorname{Prim} A \to [0,\infty)$ satisfying* (9.34), *and such that*

$$\Big|\int_{G/G_P}\theta(r^{-1}\cdot P, s\cdot P)\theta(P,s\cdot P)\,d\beta^P(\dot s) - 1\Big| < \epsilon \tag{9.41}$$

for all $r \in \operatorname{supp}\varphi$ and all $P \in \operatorname{Prim} A$.

Proof. For the moment, assume that $g \in C_c^+(G)$ with $\|g\|_2 = 1$. Then, if we define

$$\theta(P,s\cdot P) = \Big(\int_{G_P}\rho(st,G_P)^{-1}g(st)^2\,d\mu_{G_P}(t)\Big)^{\frac12}, \tag{9.42}$$

we have

$$\int_{G/G_P}\theta(P,s\cdot P)^2\,d\beta^P(\dot s) = \int_{G/G_P}\int_{G_P}\rho(st,G_P)^{-1}g(st)^2\,d\mu_{G_P}(t)\,d\beta^P(\dot s) \\ = \|g\|_2^2 = 1,$$

and θ satisfies (9.34). Furthermore

$$\begin{aligned}\theta(r^{-1}\cdot P,s\cdot P)^2 &= \theta(r^{-1}\cdot P, sr(r^{-1}\cdot P))^2 \\ &= \int_{G_{r^{-1}\cdot P}}\rho(srt,G_{r^{-1}\cdot P})^{-1}g(srt)^2\,d\mu_{G_{r^{-1}\cdot P}}(t) \\ &= \omega(r,G_{r^{-1}\cdot P})\int_{G_P}\rho(str,G_{r^{-1}\cdot P})^{-1}g(str)^2\,d\mu_{G_P}(t) \\ &= \omega(r,G_{r^{-1}\cdot P})\rho(r,G_{r^{-1}\cdot P})^{-1}\int_{G_P}\rho(st,G_P)^{-1}g(str)^2\,d\mu_{G_P}(t)\end{aligned}$$

which, since $\rho(r, G_{r^{-1}\cdot P}) = \Delta_G(r^{-1})\omega(r, G_{r^{-1}\cdot P})$, is

$$= \int_{G_P} \Delta_G(r)\rho(st, G_P)^{-1} g(str)^2 \, d\mu_{G_P}(t).$$

Notice that $\theta(P, s\cdot P) = \|a\|_{L^2(G_P)}$ where $a(t) := \rho(st, G_P)^{-\frac{1}{2}} g(st)$ and $\theta(r^{-1}\cdot P, s\cdot P) = \|b\|_{L^2(G_P)}$ where $b(t) := \Delta_G(r)^{\frac{1}{2}}\rho(st, G_P)^{-\frac{1}{2}} g(str)$. Since $\big(\|a\|_2 - \|b\|_2\big)^2 \le \|a-b\|_2^2$, we conclude that

$$\begin{aligned} &\big|\theta(r^{-1}\cdot P, s\cdot P) - \theta(P, s\cdot P)\big|^2 \\ &\qquad\qquad \le \int_{G_P} \rho(st, G_P)^{-1}\big|\Delta_G(r)^{\frac{1}{2}} g(str) - g(st)\big|^2 \, d\mu_{G_P}(t). \end{aligned}$$

Therefore

$$\begin{aligned} &\int_{G/G_P} \big|\theta(r^{-1}\cdot P, s\cdot P) - \theta(P, s\cdot P)\big|^2 \, d\beta^P(\dot{s}) \\ &\qquad \le \int_{G/G_P}\int_{G_P} \rho(st, G_P)^{-1}\big|\Delta_G(r)^{\frac{1}{2}} g(str) - g(st)\big|^2 \, d\mu_{G_P}(t)\, d\beta^P(\dot{s}). \end{aligned} \tag{9.43}$$

It follows that the right-hand side of (9.43) is

$$\int_G \big|\Delta_G(r)^{\frac{1}{2}} g(sr) - g(s)\big|^2 \, d\mu_G(s) = \|g^r - g\|_{L^2(G)}^2,$$

where $g^r(s) := \Delta(r)^{\frac{1}{2}} g(sr)$.[12] Since $\|g^r\|_{L^2(G)} = 1$, we have

$$\begin{aligned} \|g^r - g\|_{L^2(G)}^2 &= 2 - 2\operatorname{Re}\big(g^r \mid g\big)_{L^2(G)} \\ &= 2 - 2\operatorname{Re}\int_G \Delta_G(r)^{\frac{1}{2}} g(sr)\overline{g(s)}\, d\mu_G(s) \end{aligned}$$

which, since $g \in C_c^+(G)$, is

$$= 2\Big(1 - \int_G \Delta_G(r)^{\frac{1}{2}} g(sr) g(s)\, d\mu_G(s)\Big).$$

Furthermore, we have

$$\begin{aligned} &\Big|\int_{G/G_P} \theta(r^{-1}\cdot P, s\cdot P)\theta(P, s\cdot P)\, d\beta^P(\dot{s}) - 1\Big|^2 \\ &\qquad = \Big|\int_{G/G_P} \theta(P, s\cdot P)\big(\theta(r^{-1}\cdot P, s\cdot P) - \theta(P, s\cdot P)\big)\, d\beta^P(\dot{s})\Big|^2 \end{aligned}$$

[12]Usually $\rho(g) = g^r$ is called the *right-regular representation* of G on $L^2(G)$. But 'ρ' is already employed here, so we have not used this "standard notation".

which, using the Cauchy-Schwarz inequality, is

$$
\begin{aligned}
&\le \Bigl(\int_{G/G_P} \theta(P, s\cdot P)^2 \, d\beta^P(\dot s)\Bigr)\cdot \\
&\qquad \Bigl(\int_{G/G_P} \bigl|\theta(r^{-1}\cdot P, s\cdot P) - \theta(P, s\cdot P)\bigr|^2 \, d\beta^P(\dot s)\Bigr) \\
&\le 2\Bigl(1 - \int_G \Delta(r)^{\frac12} g(sr)g(s)\, d\mu_G(s)\Bigr). \qquad (9.44)
\end{aligned}
$$

If G is amenable, then Proposition A.17 on page 325 and its proof show that given a compact set $K := \operatorname{supp}\varphi$, and $\epsilon > 0$, there is a $f \in C_c^+(G)$ such that $\|f\|_{L^2(G)} = 1$ and such that

$$\bigl|1 - f * \tilde f(r)\bigr| < \epsilon/2 \quad \text{for all } r \in K,$$

where $\tilde f(r) = \overline{f(r^{-1})} = f(r^{-1})$. But

$$
\begin{aligned}
f * \tilde f(r) &= \int_G f(s)\tilde f(s^{-1}r)\, d\mu_G(s) \\
&= \int_G f(s) f(r^{-1}s)\, d\mu_G(s) \\
&= \int_G f(rs) f(s)\, d\mu_G(s).
\end{aligned}
$$

Now let $g(s) := \Delta_G(s)^{-\frac12} f(s^{-1})$. Then $g \in C_c^+(G)$ and $\|g\|_{L^2(G)} = 1$. Furthermore,

$$
\begin{aligned}
\int_G \Delta(r)^{\frac12} g(sr)g(s)\, d\mu_G(s) &= \int_G \Delta_G(s)^{-1} f(r^{-1}s^{-1}) f(s^{-1})\, d\mu_G(s) \\
&= \int_G f(r^{-1}s) f(s)\, d\mu_G(s) \\
&= \int_G f(s) f(rs)\, d\mu_G(s).
\end{aligned}
$$

Using this g to define θ gives (9.41). □

Proof of Proposition 9.22 on page 292. Choose θ as in Lemma 9.23 on page 295. Then the above gymnastics imply that

$$
\begin{aligned}
&\Bigl|\bigl(\operatorname{ind} r(\varphi)\xi^{\gamma,h} \mid \xi^{\gamma,h}\bigr) - \bigl(R(\varphi)h \mid h\bigr)\Bigr| \\
&\quad = \Bigl|\int_G \int_{\operatorname{Prim} A} d(r,P)^{\frac12} \bigl(\pi_P\bigl(\varphi(r)\bigr)U(r,P)\bigl(h(r^{-1}\cdot P)\bigr) \mid h(P)\bigr) \\
&\qquad\qquad \Bigl(\int_{G/G_P} \theta(r^{-1}\cdot P, s\cdot P)\theta(P, s\cdot P)\, d\beta^P(\dot s) - 1\Bigr)\, d\mu(P)\, d\mu_G(r)\Bigr| \\
&\quad \le \epsilon \int_G \int_{\operatorname{Prim} A} d(r,P)^{\frac12} \|\varphi(r)\| \|h(r^{-1}\cdot P)\| \|h(P)\|\, d\mu(P)\, d\mu_G(r) \\
&\quad \le \epsilon \int_G \|\varphi(r)\| \int_{\operatorname{Prim} A} d(r,P)^{\frac12} \|h(r^{-1}\cdot P)\| \|h(P)\|\, d\mu(P)\, d\mu_G(r)
\end{aligned}
$$

which, by Hölder's inequality, is

$$\leq \epsilon \|d(r,\cdot)^{\frac{1}{2}} h(r^{-1}\cdot)\|_2 \|h\|_2 \int_G \|\varphi(r)\| \, d\mu_G(r)$$
$$= \epsilon \|\varphi\|_1.$$

Thus we can find states ω_n of the form $\omega_{\xi^{\gamma,h},\xi^{\gamma,h}}$ associated to $\operatorname{ind} r$ such that for all $\varphi \in C_c(G, A)$, $\omega_n(\varphi) \to \omega(\varphi)$, where $\omega = \omega_{h,h}$ is the state associated to R. Therefore $\omega_n \to \omega$ weakly. But if $a \in \ker(\operatorname{ind} r)$, then $\omega_n(a) = 0$ for all n. Thus $\omega(a) = 0$. Since h was arbitrary, with conclude that $a \in \ker R$. This completes the proof. □

9.5 Step V: Gootman-Rosenberg *a la* Renault

Proposition 9.24 (Gootman-Rosenberg). *If R is a factor representation of $A \rtimes_\alpha G$ with infinite multiplicity and if* $\operatorname{ind} r$ *is the representation of $A \rtimes_\alpha G$ induced from Sauvageot's restriction of R to the stability groups, then*

$$\ker R \subset \ker(\operatorname{ind} r).$$

The two key difficulties to surmount in proving the reverse containment to Proposition 9.22 on page 292 is the lack of continuity of the map $P \to G_P$ from $\operatorname{Prim} A$ to Σ, and the possibility that the orbit structure of $\operatorname{Prim} A/G$ may be pathological. To overcome the first problem, we use a version of Lusin's Theorem that applies to Polish spaces. (Recall that in our set-up, where (A, G, α) is separable, both $\operatorname{Prim} A$ and Σ are Polish spaces in the regularized and Fell topologies, respectively. See Theorem H.39 on page 476 and Theorem H.5 on page 455.).

Theorem 9.25 (Lusin's Theorem). *Suppose that X and Y are Polish spaces and that μ is a probability measure on X. If $f : X \to Y$ is Borel, then given $\epsilon > 0$, there is a compact subset $L \subset X$ such that $\mu(X \setminus L) < \epsilon$ and the restriction of f to L is continuous.*[13]

This result is proved, for example, in [124, Corollary 24.22], and is a consequence of the fact that finite measures on Polish spaces are tight (Lemma D.38 on page 392).

To deal with badly behaved orbits, we need to see that the orbits are at least locally well-behaved. To do this, Gootman and Rosenberg used a "local cross-section" result ([64, Theorem 1.4]) based on some clever selection theorems. Instead, we follow Renault who used a result about equivalence relations [144, Proposition 1.11] which is based on techniques due to P. Forrest. In both cases, the local compactness of G is crucial. As always in this chapter, *we give* $\operatorname{Prim} A$ *its regularized topology so that* $\operatorname{Prim} A$ *is a Polish G-space* (Theorem H.39 on page 476).

[13]Notice that the result definitely does not say that f is continuous at each point of L. The characteristic function of the rationals is nowhere continuous, but its restriction to the irrationals is constant and therefore continuous.

Proposition 9.26 (Renault). *Let K be a compact symmetric neighborhood of e in G, let M be a neighborhood of the diagonal $\Delta \subset \operatorname{Prim} A \times \operatorname{Prim} A$ and let $L \subset \operatorname{Prim} A$ be such that the restriction of $P \mapsto G_P$ to L is continuous. If $P, Q \in \operatorname{Prim} A$, then define $P \sim_K Q$ if there is a $s \in K$ such that $Q = s \cdot P$. Then each $P_0 \in L$ has a neighborhood V in L with the property that $\sim_K$ is an equivalence relation on V such that $P \sim_K Q$ implies $(P, Q) \in M$. Furthermore, if $U \subset V$ is open, then its $\sim_K$-saturation $[U] := \{ R \in V : R \sim_K P \text{ with } P \in U \}$ is open in V. (That is, $\sim_K$ is an open equivalence relation on V.)*

Proof. Let

$$G(P_0) := \{ (t, P_0) : t \in G_{P_0} \}.$$

For clarity, we'll break the proof up into a number of steps.

Claim 1. Let

$$K' := \{ (ts, P) \in G \times L : s \in K \text{ and } t \in G_P \}.$$

Then K' is a neighborhood of $G(P_0)$ in $G \times L$.

Proof of Claim 1. Let $(t, P_0) \in G(P_0)$ and suppose that (r_i, P_i) is a sequence in $G \times L$ converging to (t, P_0). It will suffice to prove that $\{ (r_i, P_i) \}$ has a subsequence which is eventually in K'. Since $\{ P_i \} \subset L$, by assumption, $G_{P_i} \to G_{P_0}$ in Σ. Thus we can pass to a subsequence, relabel, and find $t_i \in G_{P_i}$ such that $t_i \to t$. Then $s_i := t_i^{-1} r_i \to e$, and s_i is eventually in K. Thus $(r_i, P_i) = (t_i s_i, P_i)$ is eventually in K'. □

Claim 2. There is a neighborhood M_0 of Δ such that (P, Q) and (Q, J) in M_0 imply that (P, J) belongs to M.

Proof of Claim 2. Let d be a metric on $\operatorname{Prim} A$ which is compatible with its Polish topology. Let

$$D_\epsilon(P) := \{ (Q, Q') \in \operatorname{Prim} A \times \operatorname{Prim} A : d(P, Q) < \epsilon \text{ and } d(P, Q') < \epsilon \}.$$

For each $P \in \operatorname{Prim} A$, choose ϵ_P such that $D_{3\epsilon_P}(P) \subset M$, and define

$$M_0 := \bigcup_{P \in \operatorname{Prim} A} D_{\epsilon_P}(P).$$

If $(P, Q), (Q, J) \in M_0$, then there are $P_i \in \operatorname{Prim} A$ such that $(P, Q) \in D_{\epsilon_{P_1}}(P_1)$ and $(Q, J) \in D_{\epsilon_{P_2}}(P_2)$. We can assume that $\epsilon_{P_2} \le \epsilon_{P_1}$. Then

$$\begin{aligned} d(P_1, J) &\le d(P_1, Q) + d(Q, J) \\ &\le d(P_1, Q) + d(Q, P_2) + d(P_2, J) \\ &< \epsilon_{P_1} + \epsilon_{P_2} + \epsilon_{P_2} \\ &\le 3\epsilon_{P_1} \end{aligned}$$

Thus $(P, J) \in D_{3\epsilon_{P_1}}(P_1) \subset M$. This proves claim 2. □

Let $\pi : G \times L \to \operatorname{Prim} A \times \operatorname{Prim} A$ be the continuous map given by $\pi(s, P) := (P, s^{-1} \cdot P)$. Thus $V := \pi^{-1}(M)$ is a neighborhood of $G(P_0)$.

Claim 3. There is a neighborhood N of $G(P_0)$ in $G \times L$ such that (s, P) and $(r, s^{-1} \cdot P)$ in N implies that (sr, P) belongs to $K' \cap V$.

Proof of Claim 3. Let M_0 be as in Claim 2. Let $V_0 := \pi^{-1}(M_0)$ and pick a compact neighborhood K_0 of e in G such that $K_0^2 \subset K$. If K_0' is as in claim 1, then $N := K_0' \cap V_0$ is a neighborhood of $G(P_0)$. If (s, P) and $(r, s^{-1} \cdot P)$ both belong to N, then $s = ts'$ and $r = v'r'$ with $s', r' \in K_0$, $t \in G_P$ and $v' \in G_{s^{-1} \cdot P}$. Since $s^{-1} \cdot P = (s')^{-1} \cdot P$, $v' = (s')^{-1} v s'$ with $v \in G_P$. Thus $sr = tvs'r'$ and $(sr, P) \in K'$. On the other hand, we have $(P, s^{-1} \cdot P) \in M_0$ and $(s^{-1} \cdot P, r^{-1}s^{-1} \cdot P) \in M_0$. By Claim 2, $(sr, P) \in V$ as well. This proves the claim. □

Claim 4. There is a neighborhood V of P_0 in L such that $s \in K$, $P \in V$ and $s^{-1} \cdot P \in V$ implies that $(s, P) \in N$.

Proof of claim 4. Let $\{ V_i \}$ be a neighborhood basis at P_0 with $V_{i+1} \subset V_i$ for all i. If the claim where false, then for each i there is a $s_i \in K$ and $P_i \in V_i$ such that $s_i^{-1} \cdot P_i \in V$ and $(s_i, P_i) \notin N$. Since K is compact, we can pass to a subsequence, relabel, and assume that $s_i \to s$. Since $P_i \to P_0$, $s_i^{-1} P_i \to s^{-1} \cdot P_0 = P_0$. Thus $s \in G_{P_0}$ and $(s, P_0) \in G(P_0)$. Thus (s_i, P_i) is eventually in N. This contradicts our assumptions and proves the claim. □

Fix P_0 and V as in Claim 4. Consider $\sim_K$ restricted to V. Since K is symmetric and $e \in K$, $\sim_K$ is always reflexive and and symmetric. To see that $\sim_K$ is an equivalence relation on V, we need to see that it is transitive. Suppose that $P, Q, J \in V$, that $P \sim_K Q$ and that $Q \sim_K J$. Then there are $s, r \in K$ such that $Q = s^{-1} \cdot P$ and $J = r^{-1} \cdot Q = r^{-1}s^{-1} \cdot P$. Claim 4 implies that (s, P) and $(r, s^{-1} \cdot P)$ belong to N. Thus claim 3 implies $(sr, P) \in K' \cap V$. Thus $sr = tk$ with $k \in K$ and $t \in G_P$. Thus $J = k^{-1} \cdot P$ and $P \sim_K J$. This proves that $\sim_K$ is an equivalence relation on V.

If $P, Q \in V$ and $P \sim_K Q$, then $Q = s^{-1} \cdot P$ with $s \in K$. Furthermore, $(s, P) \in N$ (by claim 4). Thus $(P, Q) = (P, s^{-1} \cdot P) \in M$.

If $U \subset V$ is open, then

$$[U] = \Big(\bigcup_{s \in K} s^{-1} \cdot U \Big) \cap V$$

is open as each $s^{-1} \cdot U$ is open. This completes the proof of the proposition. □

Next we require a technical lemma which will be used to guarantee that certain integrals are nonzero. Here and in the sequel, we'll adopt notation so that if V is a subset of G, then V/G_P will denotes the image of V in G/G_P via the natural map. Also recall that if μ is a Radon measure on a locally compact space X, then the *support of* μ, $\operatorname{supp} \mu$, is the closed set which is the complement of

$$N := \bigcup \{ U \subset X : U \text{ is open and } \mu(U) = 0 \}.$$

Since μ is regular, we have $\mu(N) = 0$. Moreover, $x_0 \in \operatorname{supp}\mu$ if and only if

$$\int_X f(x)\,d\mu(x) > 0$$

for every $f \in C_c^+(X)$ with $f(x_0) > 0$.

Lemma 9.27. *Suppose that $L \subset \operatorname{Prim} A$ is compact. Then there is a Borel set $L' \subset L$ such that $\mu(L \setminus L') = 0$ and such that if $P \in L'$ and V is a neighborhood of e in G, then*

$$\int_{V/G_P} \psi(s \cdot P)\,d\beta^P(\dot{s}) > 0$$

provided $\psi \in C^+(L)$ and $\psi(P) > 0$. (We view ψ as a Borel function on Prim A *which vanishes off L.)*

Proof. The map $sG_P \mapsto s \cdot P$ is a continuous bijection of G/G_P onto $G \cdot P$. Thus we can give $G \cdot P$ the locally compact topology coming from the homogeneous space G/G_P. We'll call this the quotient topology to distinguish it from the relative topology coming from Prim A (which is often coarser than than the quotient topology). Let β_P be the measure on $G \cdot P$ given by the push-forward of β^P. Then we define $\mathfrak{C}(P)$ to be the support, with respect to the quotient topology on $G \cdot P$, of the measure β_P restricted to $L \cap G \cdot P$. Define

$$L' := \{\, P \in L : P \in \mathfrak{C}(P) \,\}.$$

To see that L' is Borel, let $\{ K_i \}$ be a neighborhood basis at e of compact sets in G. Let $M_i := \{\, (P, s \cdot P) : s \in K_i \text{ and } s \cdot P \in L \,\}$. Then M_i is closed in $L \times L$, and $\mathbb{1}_{M_i}$ is Borel. Let $\mathfrak{b}$ be a generalized Bruhat approximate cross section. Since

$$\int_{V/G_P} \mathbb{1}_{M_i}(P, s \cdot P)\,d\beta^P(\dot{s}) = \int_G \mathbb{1}_{VG_P}(s)\mathbb{1}_{M_i}(P, s \cdot P)\mathfrak{b}(s, G_P)\rho(s, G_P)\,d\mu_G(s)$$

is a Borel function of P by Fubini's Theorem,

$$B_i := \Big\{\, P \in L : \int_{V/G_P} \mathbb{1}_{M_i}(P, s \cdot P)\,d\beta^P(\dot{s}) > 0 \,\Big\}$$

is Borel. Since $K_i \cdot P$ is a neighborhood of P in the quotient topology on $G \cdot P$, we clearly have $L' \subset B_i$.

On the other hand, if $P \in L \setminus L'$, then there is a neighborhood U of e in G such that $\beta_P(U \cdot P \cap L) = 0$. If i is such that $K_i \subset U$, then $P \notin B_i$. Thus

$$L' = \bigcap B_i$$

is Borel.

Let

$$A = \{\, (P, s \cdot P) \in L \times L : s \cdot P \in \mathfrak{C}(P) \,\}.$$

Since $\beta_P = \beta_{r\cdot P}$ for all $r \in G$ (Remark 9.4 on page 267), we must have $\mathfrak{C}(P) = \mathfrak{C}(r \cdot P)$ for all r. Thus

$$A = \{ (P, s \cdot P) \in L \times L : s \cdot P \in L' \}. \tag{9.45}$$

Let γ be the measure on $\operatorname{Prim} A \times \operatorname{Prim} A$ defined in Lemma 9.1 on page 264. Then

$$\gamma\big(\operatorname{Prim} A \times (L \smallsetminus L')\big) = \int_{\operatorname{Prim} A} \int_{G/G_P} \mathbb{1}_{L\smallsetminus L'}(Q)\, d\beta_P(Q)\, d\mu(P)$$

which, in view of (9.45), is

$$= \int_{\operatorname{Prim} A} \beta_P(L \smallsetminus \mathfrak{C}(P))\, d\mu(P),$$

which vanishes since $\beta_P(L \smallsetminus \mathfrak{C}(P)) = 0$ by the definition of $\mathfrak{C}(P)$.

Lemma 9.1 on page 264 implies that the flip preserves γ-null sets, so

$$\begin{aligned} 0 &= \gamma\big((L \smallsetminus L') \times \operatorname{Prim} A\big) \\ &= \int_{L\smallsetminus L'} \beta^P(G/G_P)\, d\mu(P). \end{aligned}$$

Thus $\mu(L \smallsetminus L') = 0$.

Now suppose that $\psi \in C^+(L)$ and that $P \in L'$. Then $U := \{ Q \in L : \psi(Q) > 0 \}$ is a neighborhood of P in L. Since $P \in \mathfrak{C}(P)$, $\beta_P(U) > 0$. The final assertion follows. □

Fix a compact symmetric neighborhood K of e in G, a neighborhood M of the diagonal Δ in $\operatorname{Prim} A \times \operatorname{Prim} A$ and a compact set $L \subset \operatorname{Prim} A$ such that the restriction of $P \to G_P$ to L is continuous on L. For each such triple, we will define a linear map

$$\mathscr{Q} = \mathscr{Q}(K, M, L) : C_c(G, A) \to \mathscr{B}^b_{cc}(G \times \operatorname{Prim} A, A).$$

We proceed as follows. Using Proposition 9.26 on page 299, let $\{ V_1, \dots, V_n \}$ be a cover of L by open sets such that $\sim_K$ is an open equivalence relation on each V_i and such that $P \sim_K Q$ in V_i implies that $(P, Q) \in M$. Let $\{ \psi_1, \dots, \psi_n \}$ be a partition of unity in $C(L)$ subordinate to the V_i. We'll view each ψ_i as a Borel function on $\operatorname{Prim} A$ which vanishes off L. Using Lemma 9.27 on the preceding page, there is a Borel set $L' \subset L$ such that $\mu(L \smallsetminus L') = 0$ and such that

$$\varphi_i(P) := \int_{K/G_P} \psi_i(s \cdot P)\, d\beta^P(\dot{s}) > 0 \quad \text{if } P \in L' \text{ and } \psi_i(P) > 0.$$

The special properties of the V_i and L' now allow us to perform the following construction.

Lemma 9.28. *Given K, M, L, L', $\{V_i\}$ and $\{\psi_i\}$ as above, there are Borel functions $e_1, \dots, e_n$ on* Prim A *such that each e_i vanishes off V_i and such that for all $P \in L'$*

$$1 = \sum_{i=1}^{n} e_i(P) \int_{K/G_P} e_i(s \cdot P)\, d\beta^P(\dot{s}). \tag{9.46}$$

Proof. We claim that if $r \in K$, $P \in V_i$ and $r \cdot P \in V_i$, then

$$K \cdot P \cap V_i = Kr \cdot P \cap V_i. \tag{9.47}$$

If $s \in K$ and if $s \cdot P \in V_i$, then $r \cdot P \sim_K P$ and $s \cdot P \sim_K P$, thus $r \cdot P \sim_K s \cdot P$ in V_i and $s \cdot P \in Kr \cdot P$. Thus the left-hand side of (9.47) is contained in the right-hand side. For the other containment, suppose that $sr \cdot P \in V_i$ with $s \in K$. Then $r \cdot P \sim_K sr \cdot P$, and since $r \cdot P \sim_K P$, we have $sr \cdot P \in K \cdot P$. This proves the claim. Thus if $P \in V_i$, $r \cdot P \in V_i$ and $r \in K$, then, since $\operatorname{supp} \psi_i \subset V_i$,

$$\begin{aligned} \varphi_i(P) &= \int_{K/G_P} \psi_i(s \cdot P)\, d\beta^P(\dot{s}) \\ &= \int_{K \cdot P \cap V_i} \psi_i(Q)\, d\beta_P(Q) \end{aligned}$$

which, by Remark 9.4 on page 267 and (9.47), is

$$\begin{aligned} &= \int_{Kr \cdot P \cap V_i} \psi_i(Q)\, d\beta_{r \cdot P}(Q) \\ &= \varphi_i(r \cdot P). \end{aligned}$$

Using a generalized bruhat approximate cross section, as in the proof of Lemma 9.27 on page 301, it is not hard to see that φ_i is Borel. Therefore $\{ P \in \operatorname{Prim} A : \varphi_i(P) > 0 \}$ is Borel, and we can define Borel functions by

$$e_i(P) := \begin{cases} \psi_i(P)\varphi_i(P)^{-\frac{1}{2}} & \text{if } \varphi_i(P) > 0, \text{ and} \\ 0 & \text{otherwise.} \end{cases}$$

If $s \in K$, $P \in L'$ and $\psi_i(P) > 0$, then we must have $P \in V_i$ and $\varphi_i(P) > 0$. Then we claim that

$$e_i(s \cdot P) = \varphi_i(P)^{-\frac{1}{2}} \psi_i(s \cdot P). \tag{9.48}$$

The claim is clearly true if $\psi_i(s \cdot P) = 0$. Otherwise, $s \cdot P \in V_i$ and (9.48) follows as we just observed that $\varphi_i(P) = \varphi_i(s \cdot P)$ in this case. Thus if $\psi_i(P) > 0$ and $P \in L'$, then

$$\begin{aligned} e_i(P) \int_{K/G_P} e_i(s \cdot P)\, d\beta^P(\dot{s}) &= \psi_i(P)\varphi_i(P)^{-1} \int_{K/G_P} \psi_i(s \cdot P)\, d\beta^P(\dot{s}) \\ &= \psi_i(P). \end{aligned}$$

This equality is clear if $\psi_i(P) = 0$. The result follows as $\sum_i \psi_i(P) = 1$ for all $P \in L$ by assumption. □

We define $\mathscr{Q} = \mathscr{Q}(K, M, L)$ by

$$\mathscr{Q}(f)(r, P) := \sum_{i=1}^{n} e_i(P) f(r) e_i(r^{-1} \cdot P). \tag{9.49}$$

Of course, each $\mathscr{Q}$ also depends on our choices of the V_i and ψ_i, but this will not be an issue in the sequel: we only require that the e_i are as constructed in Lemma 9.28 on the previous page.

If $\varphi \in C_c(G)$, then we can define a function Φ on $G \times \Sigma$ by

$$\Phi(s, H) := \int_{s \cdot H} \varphi(s^{-1}t)\, d\mu_{s \cdot H}(t).$$

If $h \in H$, then $(sh)^{-1} = s^{-1}(sh^{-1}s^{-1})$ and $sh^{-1}s^{-1} \in s \cdot H$. Thus $\Phi(sh, H) = \Phi(s, H)$ for all $s \in G$ and $h \in H$. Let $C = \operatorname{supp} \varphi$. If Φ were not bounded, then there would be a sequence $\{ (s_i, H_i) \} \subset C \times \Sigma$ such that $\lim_i \Phi(s_i, H_i) = \infty$. Since $C \times \Sigma$ is compact, we may as well assume that $(s_i, H_i) \to (s, H)$. But Φ is continuous (Lemma H.9 on page 460), so this leads to a contradiction. Thus if $f \in C_c(G, A)$ and M is a neighborhood of the diagonal in $\operatorname{Prim} A \times \operatorname{Prim} A$, then we can define a finite constant

$$C(f, M) := \sup\Bigl\{ \int_{G_{s \cdot P}} \|f(s^{-1}t)\|\, d\mu_{G_{s \cdot P}}(t) : (P, s \cdot P) \in M \Bigr\}.$$

When M is all of $\operatorname{Prim} A \times \operatorname{Prim} A$, we'll write simply $C(f)$.

Lemma 9.29. *Let $\mathscr{Q} = \mathscr{Q}(K, M, L)$ be as above. Then if $f \in C_c(G, A)$ and if $\operatorname{supp} f \subset K$, we have*

$$\|\mathscr{Q}(f)\|_I \le \max\bigl\{ C(f, M), C(f^*, M) \bigr\},$$

where $\|\cdot\|_I$ is the norm on $\mathscr{B}^b_{cc}(G \times \operatorname{Prim} A, A)$ defined on page 276.

Proof. We compute

$$\begin{aligned}
\|\mathscr{Q}(f)\|_{I,r} &= \sup_{P \in \operatorname{Prim} A} \int_G \|\mathscr{Q}(f)(s, P)\|\, d\mu_G(s) \\
&\le \sup_P \sum_{i=1}^{n} e_i(P) \int_G e_i(s^{-1} \cdot P) \|f(s)\|\, d\mu_G(t) \\
&= \sup_P \sum_{i=1}^{n} e_i(P) \int_G e_i(s \cdot P) \|f(s^{-1})\| \Delta_G(s^{-1}) \rho(s, G_P)^{-1} \\
&\qquad\qquad \rho(s, G_P)\, d\mu_G(s)
\end{aligned}$$

which, since K is symmetric and $\operatorname{supp} f \subset K$, is

$$\begin{aligned}
&= \sup_P \sum_{i=1}^{n} e_i(P) \int_{K/G_P} e_i(s \cdot P) \int_{G_P} \|f(t^{-1}s^{-1})\| \\
&\qquad\qquad \Delta_G(t^{-1}s^{-1}) \rho(t^{-1}s^{-1}, G_{s \cdot P})\, d\mu_{G_P}(t)\, d\beta^P(\dot{s})
\end{aligned}$$

which, since we have $\rho(t^{-1}s^{-1}, G_{s\cdot P})$ equal to $\Delta_G(ts)\omega(t^{-1}s^{-1}, G_{s\cdot P})$ and since $\omega(t^{-1}s^{-1}, G_{s\cdot P})$ is the same as $\Delta_{G_P}(t)^{-1}\omega(s, G_P)^{-1}$, is

$$= \sup_P \sum_{i=1}^n e_i(P) \int_{K/G_P} e_i(s\cdot P) \omega(s,G_P)^{-1} \int_{G_P} \|f(ts^{-1})\| \, d\mu_{G_P}(t)\, d\beta^P(\dot s)$$

$$= \sup_P \sum_{i=1}^n e_i(P) \int_{K/G_P} e_i(s\cdot P) \int_{G_{s\cdot P}} \|f(s^{-1}t)\| \, d\mu_{G_{s\cdot P}}(t)\, d\beta^P(\dot s)$$

which, since $s \in K$, $e_i(P) > 0$ and $e_i(s\cdot P) > 0$ imply that $(P, s\cdot P) \in M$, is

$$\begin{aligned} &\le C(f,M) \sup_P \sum_{i=1}^n e_i(P) \int_{K/G_P} e_i(s\cdot P)\, d\beta^P(\dot s) \\ &\le C(f,M). \end{aligned}$$

Since a straightforward computation shows that $\|\mathscr{Q}(f^*)\|_{I,s} = \|\mathscr{Q}(f)\|_{I,r}$, the above shows that $\|\mathscr{Q}(f)\|_{I,s} \le C(f^*, M)$. This completes the proof. □

The next result uses the intricate properties of Proposition 9.26 on page 299, and it is the key to Renault's approach. Recall that R' is defined in Lemma 9.9 on page 277.

Lemma 9.30. *Suppose that $f \in C_c(G,A)$ and that we define $\mathscr{P}(f) \in C_c(G*\Sigma, A)$ by*

$$\mathscr{P}(f)(s,H) := \gamma_H(s)f(s).$$

Let $\mathfrak{Q}$ be the set of maps $\mathscr{Q} = \mathscr{Q}(K,M,L)$ constructed as in (9.49), *where K is a symmetric compact neighborhood of e in G containing* $\operatorname{supp} f$, *M is a neighborhood of the diagonal Δ in* $\operatorname{Prim} A \times \operatorname{Prim} A$ *and L is compact subset of* $\operatorname{Prim} A$ *such that the restriction of the stability map $P \mapsto G_P$ is continuous as a function from L to Σ. Then $r(\mathscr{P}(f))$ is in the weak operator closure of $\{ R'(\mathscr{Q}(f)) : \mathscr{Q} \in \mathfrak{Q} \}$.*

Proof. Let $h_1', h_2', \dots, h_m'$ and $k_1, k_2, \dots, k_m$ be vectors in $L^2(\operatorname{Prim} A, \mu, \mathcal{H})$ and fix $\epsilon' > 0$. It will suffice to produce $\mathscr{Q} \in \mathfrak{Q}$ such that

$$\bigl|\bigl(R'(\mathscr{Q}(f))h_i' \mid k_i\bigr) - \bigl(r(\mathscr{P}(f))h_i' \mid k_i\bigr)\bigr| < \epsilon' \quad \text{for } i = 1,2,\dots,m. \tag{9.50}$$

Since R is nondegenerate and since $\{ R'(\mathscr{Q}(f)) : \mathscr{Q} \in \mathfrak{Q} \}$ is bounded (Lemma 9.29 on the preceding page), we can assume that $h_i' = R(g_i)h_i$ for $g_i \in C_c(G,A)$ and $h_i \in L^2(\operatorname{Prim} A, \mu, \mathcal{H})$. Then Lemmas 9.9 on page 277 and 9.11 on page 279 imply that if we view g_i as an element of $\mathscr{B}^b_{cc}(G \times \operatorname{Prim} A, A)$, then

$$R'(\mathscr{Q}(f))h_i' = R'(\mathscr{Q}(f) * g_i)h_i \quad \text{and} \quad r(\mathscr{P}(f))h_i' = R'(\mathscr{P}(f)\cdot g_i)h_i.$$

Fix an $\epsilon > 0$. Use Lusin's Theorem 9.25 on page 298 to find a compact subset $L \subset \operatorname{Prim} A$ such that $P \mapsto G_P$ is continuous when restricted to L, and such that

$$\int_{\operatorname{Prim} A \setminus L} \|k_i(P)\|^2 \, d\mu(P) < \epsilon^2 \quad \text{for all } i.$$

Define

$$\begin{aligned}\tilde{\xi}_i(P, r, s) &:= \int_{G_{r \cdot P}} f(r^{-1}t)\alpha_{r^{-1}t}\big(g_i(t^{-1}rs)\big)\, d\mu_{G_{r\cdot P}}(t) \\ &= \omega(r, G_P)^{-1} \int_{G_P} f(tr^{-1})\alpha_{tr^{-1}}\big(g_i(rt^{-1}s)\big)\, d\mu_{G_P}(t).\end{aligned}$$

Our choice of L, Lemma H.9 on page 460 and Lemma H.10 on page 461 imply that $\tilde{\xi}$ is continuous on $L \times G \times G$. Furthermore, $\tilde{\xi}_i(P, rt, s) = \tilde{\xi}_i(P, r, s)$ if $t \in G_P$. Thus we can define $\xi(P, r \cdot P, s) := \tilde{\xi}(P, r, s)$. If we let K be a compact symmetric neighborhood of e containing $\operatorname{supp} f$ and $(\operatorname{supp} f)(\operatorname{supp} g_i)$ for all i, then $\xi(P, rP, s)$ vanishes if $s \notin K$ or if $r \notin KG_P$. We claim that there is a neighborhood M of the diagonal such that $P \in L$, $(P, r \cdot P) \in M$ and $r \in K$ imply that

$$\|\xi_i(P, r \cdot P, s) - \xi_i(P, P, s)\| < \epsilon \quad \text{for all } i. \tag{9.51}$$

If the claim fails, then for some i we can find a neighborhood basis $\{ M_j \}$ of the diagonal and $(P_j, r_j, s_j) \in L \times K \times K$ such that

$$\|\xi_i(P_j, r_j \cdot P_j, s_j) - \xi_i(P_j, P_j, s_j)\| \geq \epsilon. \tag{9.52}$$

But, passing to a subsequence and relabeling, we may as well assume that $P_i \to P$, $r_i \to r \in G_P$ and $s_i \to s$. This, and the continuity of $\tilde{\xi}_i$ eventually contradicts (9.52).

Now we let $\mathscr{Q} = \mathscr{Q}(K, M, L)$. Then we observe that

$$\begin{aligned}\big(R'(\mathscr{Q}(f))h_i' \mid k_i\big) - \big(r(\mathscr{P}(f))h_i' \mid k_i\big) &= \big(R'(\mathscr{Q}(f) * g_i - \mathscr{P}(f) \cdot g_i)h_i \mid k_i\big) \\ &= \int_{\operatorname{Prim} A} \big(R'(\mathscr{Q}(f) * g_i - \mathscr{P}(f) \cdot g_i)h_i(P) \mid k_i(P)\big)\, d\mu(P)\end{aligned}$$

This last expression is the sum of the value A of the integral over $\operatorname{Prim} A \setminus L$, and the value B of the integral over L. It is not hard to check that

$$\|\mathscr{Q}(f) * g_i\|_I \leq \|\mathscr{Q}(f)\|_I \|g_i\|_I \leq C(f, M)\|g_i\|_1, \tag{9.53}$$

and similarly that

$$\|\mathscr{P}(f) \cdot g_i\|_I \leq \sup_P \|f|_{G_P}\|_{L^1(A \rtimes G_P)} \|g_i\|_1. \tag{9.54}$$

Therefore

$$|A| \leq \|R'(\mathscr{Q}(f) * g_i - \mathscr{P}(f) \cdot g_i)\| \|h_i\| \Big(\int_{\operatorname{Prim} A \setminus L} \|k_i(P)\|^2 \, d\mu(P) \Big)^{\frac{1}{2}}$$

which, by our choice of L, is

$$\leq \|\mathscr{Q}(f) * g_i - \mathscr{P}(f) \cdot g_i\|_I \|h_i\|\epsilon$$

and, in view of (9.53) and (9.54), there is a constant C, depending only on f and the g_i, such that the above is

$$\leq C\|h_i\|\epsilon.$$

On the other hand, since $\mu(L \smallsetminus L') = 0$,

$$|B| \leq \|\mathbb{1}_{L' \times G}\big(\mathscr{Q}(f) * g_i - \mathscr{P}(f) \cdot g_i\big)\|_I \|h_i\|\|k_i\|.$$

Since $\mathbb{1}_{L' \times G}\big(\mathscr{Q}(f) * g_i - \mathscr{P}(f) \cdot g_i\big)$ has support in $L' \times K$,

$$\begin{aligned}&\|\mathbb{1}_{L' \times G}\big(\mathscr{Q}(f) * g_i - \mathscr{P}(f) \cdot g_i\big)\|_I \\ &\qquad \leq \mu_G(K) \sup\{\, \|\mathscr{Q}(f) * g_i(s,P) - \mathscr{P}(f) \cdot g_i(s,P)\| : (s,P) \in K \times L' \,\}\end{aligned}$$

Thus it will suffice to show that the supremum is small. But calculations show that

$$\mathscr{Q}(f) * g_i(s,P) = \sum_{j=1}^n e_j(P) \int_{K/G_P} e_j(r \cdot P) \xi_i(P, r \cdot P, s) \, d\beta^P(\dot{r}),$$

while

$$\mathscr{P}(f) \cdot g_i(s,P) = \xi_i(P,P,s).$$

Combining these with the fundamental property enjoyed by the e_j's (Lemma 9.28 on page 303), we get

$$\begin{aligned}&\|\mathscr{Q}(f) * g_i(s,P) - \mathscr{P}(f) \cdot g_i(s,P)\| \\ &\qquad \leq \sum_{j=1}^n e_j(P) \int_{K/G_P} e_j(r \cdot P) \|\xi_i(P, r \cdot P, s) - \xi_i(P,P,s)\| \, d\beta^P(\dot{r}). \qquad (9.55)\end{aligned}$$

But if $e_j(P) > 0$ and $e_j(r \cdot P) > 0$, then our choice of V_j guarantees that $(P, r \cdot P) \in M$. Thus (9.55) is less than ϵ. Thus we can choose ϵ sufficiently small so that (9.50) holds. This completes the proof. □

To complete Renault's proof, we want to realize $\operatorname{ind} r$ as an induced representation acting on the completion of $C_c(G,A) \odot L^2(\operatorname{Prim} A, \mu, \mathcal{H})$ with respect to the pre-inner product given on elementary tensors by

$$(f \otimes h \mid g \otimes k) := \big(r\big(\mathscr{P}(g^* * f)\big)h \mid k\big), \qquad (9.56)$$

as we did for $\operatorname{ind} \tilde{r}$ in the proof of Proposition 9.20 on page 288. Thought of in this way, $(\operatorname{ind} r)(g)$ sends the class of $f \otimes \xi$ to the class of $g * f \otimes \xi$.

Recall that if $f \in C_c(G,A)$ and if $h \in L^2(\operatorname{Prim} A, \mu, \mathcal{H})$, then we defined $\mathcal{W}(f \otimes h) \in L^2_\sigma(\operatorname{Prim} A \times G, \mu, \mathcal{H})$ by (9.28). We also saw that

$$(\operatorname{ind} r)(g)\mathcal{W}(f \otimes h) = \mathcal{W}(g * f \otimes h), \text{ and} \tag{9.57}$$

$$\big(\mathcal{W}(f \otimes h) \mid \mathcal{W}(g \otimes k)\big) = \big(r\big(\mathscr{P}(g^* * f)\big)h \mid k\big). \tag{9.58}$$

Combining (9.57) and (9.58), we see that

$$\big((\operatorname{ind} r)(f)\mathcal{W}(g_1 \otimes h) \mid \mathcal{W}(g_2 \otimes k)\big) = \big(r\big(\mathscr{P}(g_2^* * f * g_1)\big)h \mid k\big). \tag{9.59}$$

If $a \in A \rtimes_\alpha G$ and if $\epsilon > 0$, then we can find $f \in C_c(G,A)$ such that

$$\Big|\big((\operatorname{ind} r)(a)\mathcal{W}(g_1 \otimes h) \mid \mathcal{W}(g_2 \otimes k)\big) - \big((\operatorname{ind} r)(f)\mathcal{W}(g_1 \otimes h) \mid \mathcal{W}(g_2 \otimes k)\big)\Big| < \frac{\epsilon}{3}$$

and such that

$$C(g_1^* * g_1)^{\frac{1}{2}} C(g_2^* * g_2)^{\frac{1}{2}} \|f - a\| \|h\| \|k\| < \frac{\epsilon}{3},$$

where $C(\cdot)$ is the constant defined prior to Lemma 9.29 on page 304. Lemma 9.30 on page 305 allows us to approximate (9.59) by terms of the form

$$\big(R'\big(\mathscr{Q}(g_2^* * f * g_1)\big)h \mid k\big). \tag{9.60}$$

Recall that if e is a bounded Borel function on $\operatorname{Prim} A$, then the corresponding diagonal operator T_e is given by $T_e h(P) = e(P)h(P)$ for $h \in L^2(\operatorname{Prim} A, \mu, \mathcal{H})$. Thus (9.60) is given by

$$\begin{aligned}
&\int_G \int_{\operatorname{Prim} A} \big(\pi_P\big(\mathscr{Q}(g_2^* * f * g_1)\big)(s,P) U(s,P) h(s^{-1}\cdot P) \mid k(P)\big) d(s,P)^{\frac{1}{2}}\, dP\, ds \\
&\quad = \sum_{i=1}^n \int_G \int_{\operatorname{Prim} A} e_i(P) e_i(s^{-1}\cdot P) \\
&\qquad\qquad \big(\pi_P\big(g_2^* * f * g_1(s)\big) U(s,P) h(s^{-1}\cdot P) \mid k(P)\big) d(s,P)^{\frac{1}{2}}\, dP\, ds \\
&\quad = \sum_{i=1}^n \int_G \int_{\operatorname{Prim} A} \big(\pi_P\big(g_2^* * f * g_1(s)\big) U(s,P) T_{e_i} h(s^{-1}\cdot P) \mid \\
&\qquad\qquad\qquad\qquad T_{e_i} k(P)\big) d(s,P)^{\frac{1}{2}}\, dP\, ds \\
&\quad = \sum_{i=1}^n \big(R(g_2^* * f * g_1) T_{e_i} h \mid T_{e_i} k\big) \\
&\quad = \sum_{i=1}^n \big(R(f) h_i \mid k_i\big),
\end{aligned}$$

where

$$h_i := R(g_1) T_{e_i} h \quad \text{and} \quad k_i := R(g_2) T_{e_i} k.$$

Note that

$$\sum_{i=1}^{n} \|h_i\|^2 = \big(R'\big(\mathscr{Q}(g_1^* * g_1)\big)h \mid h\big)$$
$$\leq C(g_1^* * g_1)\|h\|^2,$$

where the inequality follows from Lemma 9.29 on page 304. Similarly, $\sum_i \|k_i\|^2 \leq C(g_2^* * g_2)\|k\|^2$. Thus if $a \in A \rtimes_\alpha G$, then

$$\Big|\sum_{i=1}^{n}\big(R(a)h_i \mid k_i\big)\Big| \leq \|a\| \sum_{i=1}^{n} \|h_i\|\|k_i\|$$
$$\leq \|a\|\Big(\sum \|h_i\|^2\Big)^{\frac{1}{2}}\Big(\sum \|k_i\|^2\Big)^{\frac{1}{2}}$$
$$\leq \|a\| C(g_1^* * g_1)^{\frac{1}{2}} C(g_2^* * g_2)^{\frac{1}{2}} \|h\|\|k\|.$$

Thus, with $f \in C_c(G, A)$ as above,

$$\Big|\sum_{i=1}^{n}\big(R(a)h_i \mid k_i\big) - \sum_{i=1}^{n}\big(R(f)h_i \mid k_i\big)\Big| < \frac{\epsilon}{3}.$$

Thus for any $a \in A \rtimes_\alpha G$ and any $\epsilon > 0$, there are vectors h_i and k_i such that

$$\Big|\big((\operatorname{ind} r)(a)\mathcal{W}(g_1 \otimes h) \mid \mathcal{W}(g_2 \otimes k)\big) - \sum_{i=1}^{n}\big(R(a)h_i \mid k_i\big)\Big| < \epsilon.$$

Thus if $a \in \ker R$, $(\operatorname{ind} r)(a)$ is zero on the span $\mathcal{S}$ of the $\mathcal{W}(f \otimes h)$. We showed in the proof of Proposition 9.20 on page 288 that $\mathcal{S}$ is dense. Therefore $(\operatorname{ind} r)(a)$ is zero. This completes the proof of Proposition 9.24 on page 298.

This result together with Proposition 9.22 on page 292 gives the Sauvageot-Gootman-Rosenberg-Renault proof of the Effros-Hahn conjecture (Theorem 8.21 on page 241).

Notes and Remarks

Lemma 9.7 and Propositions 9.5, 9.8, 9.14 and 9.22 are all due to Sauvageot [160]. Proposition 9.24 is due to Gootman & Rosenberg [64]; however, the proof given here is modeled on Renault's proof in [144]. In particular, Proposition 9.26 is due to Renault [144].

Appendix A

Amenable Groups

In this chapter we want to collect a few basic results concerning amenable groups. In particular, we want to arrive at the classical result stating that a group G is amenable if and only if the left-regular representation is a faithful representation of $C^*(G)$ [82]. This result could serve as a definition of amenability were it not for the fact that we want to apply some of the details of the argument to the theory of crossed products. Furthermore, we prefer to have a definition that easily leads to results that allows us to recognize an amenable group when we see one. For example, we'd like to know that closed subgroups and quotients of amenable groups are amenable, etc. (see Remark A.15 on page 322). I have tried to pare down the results reproduced here to a bare minimum. Much more complete treatments are available elsewhere. Most of the ideas here were stolen from [126, §§7.1–7.2]. Other good sources are [68], [125] and [72].

A.1 States and Positive Definite Functions

If φ is a state on $C^*(G)$, then the usual GNS-construction ([139, Proposition A.6]) gives us a nondegenerate representation π^φ and a cyclic vector h_φ such that

$$\varphi(f) = \bigl(\pi^\varphi(f)h_\varphi \mid h_\varphi\bigr) \quad \text{for all } f \in C_c(G). \tag{A.1}$$

Of course, π^φ is the integrated form of a unitary representation, also denoted π^φ, and if we define $\Phi = \Phi_\varphi$ to be the bounded continuous function on G given by

$$\Phi(s) := (\pi^\varphi_s h_\varphi \mid h_\varphi), \tag{A.2}$$

then we have

$$\varphi(f) = \bigl(\pi^\varphi(f)h_\varphi \mid h_\varphi\bigr) = \int_G f(s)(\pi^\varphi_s h_\varphi \mid h_\varphi)\,d\mu(s) = \int_G f(s)\Phi(s)\,d\mu(s).$$

One goal of this section is to characterize those functions on G having the form given in (A.2). Then it is possible to study the states of $C^*(G)$ in terms of these functions. Specifically, we want to prove the following.

Proposition A.1. *Let Φ be a bounded* continuous *function on a locally compact group G. Then the following statements are equivalent.*

(a) *There is a unitary representation $\pi : G \to U(\mathcal{H}_\pi)$ such that*

$$\Phi(s) = (\pi_s h \mid h)$$

for a vector $h \in \mathcal{H}_\pi$.

(b) *For any finite set $\{ s_1, \dots, s_n \} \subset G$, the matrix*

$$\big(\Phi(s_i^{-1}s_j)\big)$$

is positive in M_n.

(c) *For all $f \in C_c(G)$,*

$$\int_G f^* * f(r)\Phi(r)\,d\mu(r) \geq 0. \tag{A.3}$$

(d) *There is a positive functional φ of norm $\|\Phi\|_\infty$ on $C^*(G)$ such that*

$$\varphi(f) = \int_G f(s)\Phi(s)\,d\mu(s) \quad \textit{for all } f \in C_c(G). \tag{A.4}$$

Definition A.2. A function Φ, continuous or not, on a group G is called *positive definite* if it satisfies the matrix condition in part (b) of Proposition A.1. The set of *continuous* positive definite functions on G is denoted by $\mathcal{P}(G)$.

Example A.3. Of course, functions such as those defined in (A.2) are positive definite. More generally, if V is any complex inner product space and $s \mapsto u_s$ is a homomorphism of G into the group of unitary operators on V, then

$$\Phi(s) := (u_s v \mid v)$$

is positive definite for any $v \in V$. To see this, let $s_1, \dots, s_n$ be elements of G. We need to show that for any $(c_1, \dots, c_n) \in \mathbf{C}^n$ we have

$$\sum_{i,j} \bar{c}_i c_j \Phi(s_i^{-1}s_j) \geq 0. \tag{A.5}$$

However, the left-hand side is

$$\begin{aligned}\sum_{i,j} \bar{c}_i c_j (u_{s_i^{-1}s_j} v \mid v) &= \sum_{i,j} (c_j u_{s_j} v \mid c_i u_{s_i} v) \\ &= \Big(\sum_i c_i u_i v \Bigm| \sum_i c_i u_i v\Big),\end{aligned}$$

which is certainly nonnegative.

In view of Example A.3, it is clear that a positive definite function need not be continuous or even measurable. For example, any character on G, continuous or not, is positive definite. But if a positive definite function is continuous at e, then it's actually uniformly continuous [72, Theorem VIII §32.4]. (We won't need this fact as the functions we'll want to consider will obviously be continuous.)

Positive definite functions are rather special, and enjoy lots of properties. A few of these are summarized in the next lemma. It will be convenient to introduce the following notation. If f is any function on G, then we define $\tilde{f}$ and f° by

$$\tilde{f}(s) := \overline{f(s^{-1})} \quad \text{and} \quad f^\circ(s) := f(s^{-1}). \tag{A.6}$$

Lemma A.4. *Suppose that Φ is a positive definite function on G. Then Φ is bounded with $\|\Phi\|_\infty = \Phi(e)$. We also have $\Phi(r^{-1}) = \overline{\Phi(r)}$ so that $\tilde{\Phi} = \Phi$. Furthermore, Φ° is positive definite, and if Ψ is also positive definite, then so is the pointwise product $\Phi\Psi$.*

Proof. The matrix

$$\begin{pmatrix} \Phi(e) & \Phi(r) \\ \Phi(r^{-1}) & \Phi(e) \end{pmatrix}$$

is positive by assumption. Thus $\Phi(e) \geq 0$, $\overline{\Phi(r)} = \Phi(r^{-1})$ and $0 \leq \Phi(e)^2 - \Phi(r)\Phi(r^{-1}) = \Phi(e)^2 - |\Phi(r)|^2$.

It is straightforward to check that Φ° is positive definite. For the last statement, it suffices to check that the Schur product $(a_{ij}b_{ij})$ of two positive matrices (a_{ij}) and (b_{ij}) is again positive. But since (b_{ij}) is positive, we can find c_{ij} such that $b_{ij} = \sum_k \bar{c}_{ki}c_{kj}$. Then for any scalars d_i,

$$\sum_{ij} \bar{d}_i d_j a_{ij} b_{ij} = \sum_{ijk} \bar{d}_i d_j a_{ij} \bar{c}_{ki} c_{kj} = \sum_k \sum_{ij} \overline{(c_{ki}d_i)}(c_{kj}d_j)a_{ij},$$

which is positive since (a_{ij}) is positive. □

Remark A.5. It follows from Lemma 1.72 on page 23 that $C_c(G)$ has an approximate identity for the inductive limit topology and therefore for the L^1-norm or universal norm. In fact, any net $\{ u_V \}$ indexed by decreasing neighborhoods of the identity in G of functions in $C_c^+(G)$ satisfying $\operatorname{supp} u_V \subset V$, $u_V^* = u_V$ and having integral one will do. In the sequel, we'll refer to this sort of approximate identity as an *approximate identity for $C^*(G)$ in $C_c(G)$*.

Proof of Proposition A.1 on the facing page. (a) $\implies$ (b): This implication is proved in Example A.3 on the preceding page.

(b) $\implies$ (c): Suppose that $f \in C_c(G)$ with $K = \operatorname{supp} f$. Then $F(s,r) := \overline{f(s)}f(r)\Phi(s^{-1}r)$ is in $C_c(G \times G)$ with support in $K \times K$. In particular, F is uniformly continuous (Lemma 1.62 on page 19) and there is a neighborhood V of e in G such that

$$|F(s,r) - F(s',r')| < \epsilon$$

provided both (s,r) and (s',r') are in the same right translate $Vs_0 \times Vr_0$ of $V \times V$. We can cover K by finitely many right translates of V and form a partition of K by nonempty Borel sets $E_1, \dots, E_n$ such that each E_i is contained in a right translate of V. Let $s_i \in E_i$. Then

$$\begin{aligned}\int_G f^* * f(r)\Phi(r)\,d\mu(r) &= \int_G\int_G f^*(s)f(s^{-1}r)\Phi(r)\,d\mu(s)\,d\mu(r) &\text{(A.7)}\\ &= \int_G\int_G \overline{f(s)}f(r)\Phi(s^{-1}r)\,d\mu(r)\,d\mu(s)\\ &= \sum_{i,j}\int_{E_i}\int_{E_j} \overline{f(s)}f(r)\Phi(s^{-1}r)\,d\mu(r)\,d\mu(s)\\ &= \sum_{i,j}\overline{f(s_i)\mu(E_i)}f(s_j)\mu(E_j)\Phi(s_i^{-1}s_j) + R, &\text{(A.8)}\end{aligned}$$

where

$$R = \sum_{i,j}\int_{E_i}\int_{E_j}\big(F(s,r) - F(s_i,s_j)\big)\,d\mu(r)\,d\mu(s).$$

Since the uniform continuity of F implies $|R| \le \epsilon\mu(K)^2$ and since the first term of (A.8) is nonnegative by assumption, it follows that the left-hand side of (A.7) is bounded below by $-\epsilon\mu(K)^2$. Since ϵ is arbitrary, the left-hand side of (A.7) is nonnegative as claimed.

(c) $\Longrightarrow$ (d): If Φ satisfies (A.3), then (A.4) defines a positive linear functional on $C_c(G)$ with norm $\|\Phi\|_\infty$ with respect to the L^1-norm on $C_c(G)$. We can mimic the usual GNS-construction (see Proposition A.19 on page 328 in Appendix A.3 at the end of this Chapter) and conclude that there is a L^1-norm decreasing, nondegenerate $*$-homomorphism $\pi_\varphi : C_c(G) \to B(\mathcal{H}_\varphi)$ and a cyclic vector h_φ such that

$$\varphi(f) = \big(\pi_\varphi(f)h_\varphi \mid h_\varphi\big).$$

Since π_φ is bounded with respect to the L^1-norm, it is bounded with respect to the universal norm (Lemma 2.45 and Example 2.44 on page 61). Thus φ extends to a positive functional on $C^*(G)$.

(d) $\Longrightarrow$ (a): If φ is a positive linear functional, then the argument at the beginning of this section shows that φ is given by integration against $\Psi(s) = \big(\pi_s^\varphi h_\varphi \mid h_\varphi\big)$, where $(H_\varphi, \pi_\varphi, h_\varphi)$ is the GNS representation associated to φ. But then Φ and Ψ have to agree almost everywhere on every compact set K in G. Since both Φ and Ψ are continuous, they must agree everywhere. □

Remark A.6. Since $\|\Phi\|_\infty = \Phi(e)$ if Φ is positive definite, it follows from Proposition A.1 on page 312 that (A.4) establishes a one-to-one correspondence between states $\varphi \in \mathcal{S}\big(C^*(G)\big)$ on $C^*(G)$ and continuous positive definite functions Φ satisfying $\Phi(e) = 1$.

Example A.7. Although we have been at pains to avoid measure theory, there is one crucial example of positive definite functions that requires we consider arbitrary functions in $\mathcal{L}^2(G)$. If $f, g \in \mathcal{L}^2(G)$ and λ is the left-regular representation, then

$$\Phi_{f,g}(r) := (f \mid \lambda_r g) \tag{A.9}$$

is a bounded continuous function on G. If $f = g$, then (A.9) is positive definite in view of Proposition A.1 on page 312 and Lemma A.4 on page 313. Thus $\Phi_{f,f}$ is a continuous positive definite function on G with $\|\Phi_{f,f}\|_\infty = \Phi_{f,f}(e) = \|f\|_2^2$ for all $f \in \mathcal{L}^2(G)$. Since

$$(f \mid \lambda_r g) = \int_G f(s)\overline{g(r^{-1}s)}\,d\mu(s) = \int_G f(s)\tilde{g}(s^{-1}r)\,d\mu(s),$$

the more suggestive notation $f * \tilde{g}$ is used in place of $\Phi_{f,g}$. If f is in the unit ball of $L^2(G)$, then we'll write $\omega_{f,f}$ for the state

$$\omega_{f,f}(g) := \big(\lambda(g)f \mid f\big).$$

If $g \in C_c(G)$, then

$$\begin{aligned}\omega_{f,f}(g) &= \int_G g(r)(\lambda_r f \mid f)\,d\mu(r)\\ &= \int_G g(r) f * \tilde{f}(r^{-1})\,d\mu(r).\end{aligned}$$

Thus the state $\omega_{f,f}$ corresponds to the positive definite function $(f * \tilde{f})^o$.

Lemma A.8. *Suppose that $\{ u_i \}$ is an approximate identity for $C^*(G)$ in $C_c(G)$ (as described in Remark A.5 on page 313). Then for any continuous function φ on G, $\lim_{i,j} u_i * \varphi * \tilde{u}_j(s) = \varphi(s)$.*

Proof. Since each u_i is non-negative with integral one,

$$\begin{aligned}u_i * \varphi * \tilde{u}_j(s) - \varphi(s) &= \int_G u_i * \varphi(r)\tilde{u}_j(r^{-1}s)\,d\mu(r) - \varphi(s)\\ &= \int_G\int_G u_i(t)\varphi(t^{-1}r)u_j(s^{-1}r)\,d\mu(r)\,d\mu(t) - \varphi(s)\\ &= \int_G\int_G u_i(t)u_j(r)\big(\varphi(t^{-1}sr) - \varphi(s)\big)\,d\mu(r)\,d\mu(t).\end{aligned}$$

Since φ is continuous at s, we can make the last term as small as we please by choosing i and j large enough to make the supports of u_i and u_j sufficiently small. □

Example A.7 shows that if $f \in \mathcal{L}^2(G)$, then $f * \tilde{f} \in \mathcal{P}(G)$ with $\|f\|_2^2 = f * \tilde{f}(e)$. The next result gives a partial converse.

Lemma A.9. *If Φ is a continuous positive definite function with compact support on a locally compact group G, then there is a $f \in \mathcal{L}^2(G)$ such that $\Phi = f * \tilde{f}$.*

Proof. Define a map $\alpha(\Phi) : C_c(G) \to C_c(G)$ by $\alpha(\Phi)(f) := f * \Phi$. Notice that

$$\begin{aligned}\alpha(\Phi)(f)(s) &= \int_G f(r)\Phi(r^{-1}s)\,d\mu(r)\\ &= \int_G \Phi(r^{-1})f(sr)\,d\mu(r)\end{aligned}$$

which, if we let ρ denote the right-regular representation and define $\Psi(r) := \Phi(r^{-1})\Delta(r)^{-\frac{1}{2}}$, is

$$= \int_G \Psi(r)\rho(r)f(s)\,d\mu(r)$$
$$= \rho(\Psi)(f)(s).$$

It follows that $\alpha(\Phi)$ is a bounded operator on $C_c(G) \subset L^2(G)$, and extends to an element in $B\big(L^2(G)\big)$. Moreover

$$\begin{aligned}
\big(\alpha(\Phi)f \mid f\big) &= \int_G f * \Phi(s)\overline{f(s)}\,d\mu(s) \\
&= \int_G \int_G f(r)\Phi(r^{-1}s)\overline{f(s)}\,d\mu(r)\,d\mu(s) \\
&= \int_G \int_G \Phi(r^{-1})f(sr)\overline{f(s)}\,d\mu(r)\,d\mu(s) \\
&= \int_G \Phi(r^{-1}) \int_G f^*(s)f(s^{-1}r)\,d\mu(s)\,d\mu(r) \\
&= \int_G \Phi^\circ(r)f^* * f(r)\,d\mu(r), \qquad \text{(A.10)}
\end{aligned}$$

and (A.10) is nonnegative by Proposition A.1 on page 312 (since Φ° is positive definite by Lemma A.4 on page 313). This shows that $\alpha(\Phi)$ is a positive operator on $L^2(G)$ and so has a positive square root $\alpha(\Phi)^{\frac{1}{2}}$ which commutes with any operator commuting with $\alpha(\Phi)$. Since $\lambda_s(f * \Phi) = (\lambda_s f) * \Phi$, λ_s commutes with $\alpha(\Phi)$, and therefore with $\alpha(\Phi)^{\frac{1}{2}}$.

Let $\{u_i\}$ be an approximate identity for $C^*(G)$ in $C_c(G)$ (Remark A.5 on page 313). Define

$$f_i := \alpha(\Phi)^{\frac{1}{2}}u_i.$$

Since

$$\begin{aligned}
\|f_i - f_j\|_2^2 &= \big(\alpha(\Phi)(u_i - u_j) \mid u_i - u_j\big) \\
&= \big((u_i - u_j) * \Phi \mid u_i - u_j\big) \\
&= u_i * \Phi * \tilde{u}_i(e) - u_i * \Phi * \tilde{u}_j(e) - u_j * \Phi * \tilde{u}_i(e) + u_j * \Phi * \tilde{u}_j(e),
\end{aligned}$$

it follows from Lemma A.8 on the preceding page that $\{f_i\}$ is Cauchy in $L^2(G)$. Thus there must be a $f \in \mathcal{L}^2(G)$ such that $f_i \to f$ in $L^2(G)$. We have

$$\begin{aligned}
f * \tilde{f}(s) &= (f \mid \lambda_s f) \\
&= \lim_i \big(\alpha(\Phi)^{\frac{1}{2}}u_i \mid \lambda_s\alpha(\Phi)^{\frac{1}{2}}u_i\big) \\
&= \lim_i (u_i * \Phi \mid \lambda_s u_i) \\
&= \lim_i u_i * \Phi * \tilde{u}_i(s) \\
&= \Phi(s).
\end{aligned}$$

□

Theorem A.10 ([126, Proposition 7.1.11]). *Let $\mathcal{P}_1(G)$ be the continuous positive definite functions on G with $\Phi(e) = 1$ equipped with the topology of uniform convergence on compacta. Then $\mathcal{P}_1(G)$ is closed in $C(G)$ in the topology of uniform convergence on compacta. If we give the state space $\mathcal{S}\bigl(C^*(G)\bigr)$ of $C^*(G)$ the the weak-$*$ topology (coming from $C^*(G)^*$), then the bijection $\varphi \mapsto \Phi$ of Remark A.6 on page 314 is a homeomorphism of $\mathcal{S}\bigl(C^*(G)\bigr)$ onto $\mathcal{P}_1(G)$.*

Remark A.11. If φ is a state on $C^*(G)$, then we can use the GNS-construction to realize φ as a vector state as in (A.1). Since π_φ is nondegenerate, it extends to all of $M\bigl(C^*(G)\bigr)$, and we can extend φ to a state $\tilde{\varphi}$ on $M\bigl(C^*(G)\bigr)$ by the formula

$$\tilde{\varphi}(m) = \bigl(\bar{\pi}_\varphi(m)h_\varphi \mid h_\varphi\bigr).$$

Although we don't need to use it here, $\tilde{\varphi}$ is the unique state extending φ by [126, Proposition 3.1.6].

Proof of Theorem A.10. Let Φ_i be a net in $\mathcal{P}_1(G)$ converging to Φ in the compact open topology. Then $\Phi \in \mathcal{P}_1(G)$ by Proposition A.1 on page 312. Let φ_i and φ be the states corresponding to Φ_i and Φ, respectively. If $f \in C_c(G)$, then we certainly have

$$\varphi_i(f) = \int_G f(s)\Phi_i(s)\,d\mu(s) \to \int_G f(s)\Phi(s)\,d\mu(s) = \varphi(f).$$

Since $C_c(G)$ is dense in $C^*(G)$ and each φ_i has norm one, it follows that $\varphi_i \to \varphi$ in the weak-$*$ topology. Thus we have proved the first statement and half of the second.

Now suppose $\varphi_i \to \varphi$ in the state space of $C^*(G)$. Let Φ_i and Φ be the corresponding elements of $\mathcal{P}_1(G)$. Fix $\epsilon > 0$ and a compact set $C \subset G$. Let $\{u_i\}$ be a approximate identity for $C^*(G)$ in $C_c(G)$ as in Remark A.5 on page 313. Now $\|\varphi\| = 1 = \lim_i \varphi(u_i) = \lim_i \varphi(u_i^*u_i)$ by [139, Lemma A.7] and part (d) of Lemma A.21 on page 328. Since

$$(1-u_i)^*(1-u_i) = 1 - u_i - u_i + u_i^*u_i,$$

there is a $u \in C_c^+(G)$ such that

$$\tilde{\varphi}\bigl((1-u)^*(1-u)\bigr) < \epsilon^2. \tag{A.11}$$

Since $\tilde{\varphi}_i(1) = 1$ for all i and since $\varphi_i \to \varphi$ in the weak-$*$ topology, there is a i_0 such that $i \geq i_0$ implies that

$$\tilde{\varphi}_i\bigl((1-u)^*(1-u)\bigr) < \epsilon^2. \tag{A.12}$$

If $i_G : G \to M(C^*(G))$ is the canonical homomorphism, then it follows from Remark A.11 that $\tilde{\varphi}\bigl(i_G(r)\bigr) = \Phi(r)$. Thus we have

$$\begin{aligned}
|\Phi(r) - \Phi * \tilde{u}(r)|^2 &= \Bigl|\Phi(r) - \int_G \Phi(s)\tilde{u}(s^{-1}r)\,d\mu(s)\Bigr|^2 \\
&= \Bigl|\Phi(r) - \int_G \Phi(s)u(r^{-1}s)\,d\mu(s)\Bigr|^2 \\
&= \bigl|\tilde{\varphi}\bigl(i_G(r)\bigr) - \varphi\bigl(i_G(r)u\bigr)\bigr|^2 \\
&= \bigl|\tilde{\varphi}\bigl(i_G(r)(1-u)\bigr)\bigr|^2
\end{aligned}$$

which, using the Cauchy-Schwarz inequality for states, is

$$\leq \tilde{\varphi}(1)\tilde{\varphi}\big((1-u)^*(1-u)\big) < \epsilon^2,$$

where the last inequality comes from (A.11). A similar calculation using (A.12) shows that for all $i \geq i_0$

$$\big|\Phi_i(r) - \Phi_i * \tilde{u}(r)\big| < \epsilon.$$

From the above calculation, we also see that

$$\Phi * \tilde{u}(r) - \Phi_i * \tilde{u}(r) = \varphi\big(i_G(r)u\big) - \varphi_i\big(i_G(r)u\big).$$

Since $r \mapsto i_G(r)u$ is continuous from G into $C^*(G)$, since each φ_i has norm 1, and since C is compact, there is an $i_1 \geq i_0$ such that $i \geq i_1$ implies

$$\big|\Phi * \tilde{u}(r) - \Phi_i * \tilde{u}(r)\big| < \epsilon \quad \text{for all } r \in C.$$

But then $i \geq i_1$ implies that

$$\big|\Phi(r) - \Phi_i(r)\big| < 3\epsilon \quad \text{for all } r \in C,$$

and this completes the proof. □

A.2 Amenability

If G is a locally compact group, then $\ell^\infty(G)$ is the set of *all* bounded functions on G equipped with the supremum norm $\|\cdot\|_\infty$. Note that $\ell^\infty(G) = L^\infty(G_d)$, where G_d denotes the group G equipped with the discrete topology.

Definition A.12. If X is a closed subspace of $\ell^\infty(G)$ containing the constant functions and closed under complex conjugation, then a linear functional $m : X \to \mathbf{C}$ is called a *mean* on X if

(a) $m(\bar{f}) = \overline{m(f)}$,
(b) $m(1) = 1$, and
(c) $f \geq 0$ implies $m(f) \geq 0$.

A mean is called *left-invariant* if

(d) $m\big(\lambda_s(f)\big) = m(f)$ for all $s \in G$ and $f \in X$.

The term mean is justified by the observation that conditions (b) and (c) are equivalent to

(e) $\inf_{s\in G} f(s) \leq m(f) \leq \sup_{s\in G} f(s)$ for all real-valued $f \in X$.

It follows from condition (a) that if $f \in X$, then

$$m(f) = m\big(\operatorname{Re} f + i \operatorname{Im} f\big) = \operatorname{Re} m(f) + i \operatorname{Im} m(f).$$

Then $\|f\|_\infty = 1$ implies $1 - (\operatorname{Re} f)^2 - (\operatorname{Im} f)^2 \geq 0$. Thus $1 - \big(\operatorname{Re} m(f)\big)^2 - \big(\operatorname{Im} m(f)\big)^2 \geq 0$, which implies $|m(f)| \leq 1$. Therefore if X is a C^*-subalgebra of $\ell^\infty(G)$, then a mean is simply a state on X.

A *discrete* group G is called *amenable* if there is a left-invariant mean on $X = \ell^\infty(G)$. In [68, Theorem 1.2.1], Greenleaf shows that G_d is amenable if G is an abelian group. He also shows that the free group on two generators $\mathbb{F}_2$ is not amenable [68, Example 1.2.3]. Since subgroups of discrete amenable groups are amenable [68, Theorem 1.2.5], any discrete group — such as $\mathbb{F}_n$ — containing $\mathbb{F}_2$ as a subgroup cannot be amenable. It was an open question for many years whether any nonamenable discrete group had to contain $\mathbb{F}_2$ as a subgroup. But there are now known to be nonamenable groups which do not contain $\mathbb{F}_2$ as a subgroup [118].

When G is not discrete, then $\ell^\infty(G)$ is too big and must be replaced by a suitable subspace X which is more closely tied to the underlying topology of G.[1] One candidate is the bounded continuous functions $C^b(G)$ on G. A function $f \in C^b(G)$ is called *left-uniformly continuous* if for all $\epsilon > 0$ there is a symmetric neighborhood V of e in G such that

$$|f(r) - f(s)| < \epsilon \tag{A.13}$$

provided $rs^{-1} \in V$. We say that f is *right-uniformly continuous* if given $\epsilon > 0$ there is a symmetric neighborhood V such that (A.13) holds whenever $r^{-1}s \in V$. Those $f \in C^b(G)$ which are both right- and left-uniformly continuous are called *uniformly continuous*, and form a sub-C^*-algebra $C^b_u(G)$ of $\ell^\infty(G)$. Then we have

$$C^b_u(G) \subset C^b(G) \subset L^\infty(G), \tag{A.14}$$

where $L^\infty(G)$ is the usual Banach space of equivalence classes of essentially bounded functions on G when G is σ-compact. In general, $L^\infty(G)$ consists of equivalences classes of bounded measurable functions $\mathcal{L}^\infty(G)$ on G which agree locally almost everywhere with norm given by

$$\|\varphi\|_\infty := \inf\{\, \alpha \in \mathbf{R} : \{\, s \in G : |f(s)| > \alpha \,\} \text{ is locally null} \,\}$$

(see Appendix I.5). Then $L^\infty(G)$ is a Banach space which is the dual of $L^1(G)$ in the usual way (Proposition I.27 on page 497). Naturally, we say a state m on $L^\infty(G)$ is a left-invariant mean if $m(\lambda_r\varphi) = m(\varphi)$ for all r and $\varphi \in L^\infty(G)$.

As we shall show, the existence of a left-invariant mean on any of the subalgebras in (A.14) implies the existence of left-invariant means on the others (as well as a stronger property we'll need down the road). To show all this, we'll have to take a short detour into measure theory.

Now suppose that $f \in C_c(G)$ and $\varphi \in \mathcal{L}^\infty(G)$. If $s \in G$, then $r \mapsto \varphi(r^{-1}s)$ is a bounded measurable function with $\|\cdot\|_\infty$-norm equal to $\|\varphi\|_\infty$. Thus Hölder's inequality implies we can define a function on G by

$$f * \varphi(s) := \int_G f(r)\varphi(r^{-1}s)\, d\mu(r), \tag{A.15}$$

and

$$\|f * \varphi\|_\infty \le \|f\|_1 \|\varphi\|_\infty.$$

[1] As we shall see, compact groups must be amenable. But the real orthogonal group $O(3)$ contains $\mathbb{F}_2$ as a non-closed subgroup. Thus there can be no left-invariant mean on $\ell^\infty(O(3))$.

Note that

$$|f*\varphi(s)-f*\varphi(r)| = \Bigl|\int_G f(t)\varphi(t^{-1}s)\,d\mu(t) - \int_G f(t)\varphi(t^{-1}r)\,d\mu(t)\Bigr|$$
$$= \Bigl|\int_G \bigl(f(st)-f(rt)\bigr)\varphi(t^{-1})\,d\mu(t)\Bigr|$$

which, since $t\mapsto\varphi(t^{-1})$ is bounded with $\|\cdot\|_\infty$-norm equal to $\|\varphi\|_\infty$, is

$$\le \|\lambda_{s^{-1}}f-\lambda_{r^{-1}}f\|_1\|\varphi\|_\infty = \|\lambda_{rs^{-1}}f - f\|_1\|\varphi\|_\infty.$$

Since translation is continuous in L^1, it follows that $f*\varphi$ is left-uniformly continuous.

On the other hand, $r\mapsto \tilde f(r^{-1}s)=\overline{f(s^{-1}r)}$ has L^1-norm $\|f\|_1$, and we can define a function on G by

$$\varphi*\tilde f(s) := \int_G \varphi(r)\tilde f(r^{-1}s)\,d\mu(r)$$
$$= \int_G \varphi(r)\overline{f(s^{-1}r)}\,d\mu(r),$$

such that $\|\varphi*\tilde f\|_\infty \le \|\varphi\|_\infty\|f\|_1$. This time we can show that $\varphi*\tilde f$ is right-uniformly continuous.[2]

If $f,g\in C_c(G)$, $\varphi\in\mathcal{L}^\infty(G)$ and $s\in G$, then both

$$(r,t)\mapsto f(rt)\varphi(t^{-1})\overline{g(s^{-1}r)} \quad\text{and}\quad (r,t)\mapsto f(sr)\varphi(t)\overline{g(rt)}$$

are measurable and vanish off a compact set in $G\times G$. This will allow us to apply Fubini's Theorem in the following calculation:

$$(f*\varphi)*\tilde g(s) = \int_G f*\varphi(r)\overline{g(s^{-1}r)}\,d\mu(r)$$
$$= \int_G\int_G f(rt)\varphi(t^{-1})\overline{g(s^{-1}r)}\,d\mu(t)\,d\mu(r)$$
$$= \int_G\int_G f(rt)\varphi(t^{-1})\overline{g(s^{-1}r)}\,d\mu(r)\,d\mu(t)$$
$$= \int_G\int_G f(r)\varphi(t^{-1})\overline{g(s^{-1}rt^{-1})}\Delta(t^{-1})\,d\mu(r)\,d\mu(t)$$
$$= \int_G\int_G f(r)\varphi(t)\overline{g(s^{-1}rt)}\,d\mu(r)\,d\mu(t)$$
$$= \int_G\int_G f(sr)\varphi(t)\overline{g(rt)}\,d\mu(r)\,d\mu(t)$$
$$= \int_G\int_G f(sr)\varphi(t)\tilde g(t^{-1}r^{-1})\,d\mu(t)\,d\mu(r)$$

[2]It is more natural to write $\tilde f$ here as it is possible to make sense of these formulas when $f\in L^1(G)$. Then $\tilde f$ may not be in $L^1(G)$ even though $\|\varphi*\tilde f\|_\infty\le\|\varphi\|_\infty\|f\|_1$.

$$= \int_G f(sr)\varphi * \tilde{g}(r^{-1})\,d\mu(r)$$
$$= f * (\varphi * \tilde{g})(s).$$

Thus

$$f * (\varphi * \tilde{g}) = (f * \varphi) * \tilde{g}, \tag{A.16}$$

and we can drop the parentheses from now on. Since the left-hand side of (A.16) is left-uniformly continuous and the right-hand side is right-uniformly continuous, we have

$$f * \varphi * \tilde{g} \in C_u^b(G) \quad \text{for all } f, g \in C_c(G) \text{ and } \varphi \in \mathcal{L}^\infty(G).$$

Similar considerations imply that

$$(h * f) * \varphi * \tilde{g} = h * (f * \varphi * \tilde{g})$$

for $h, f, g \in C_c(G)$ and $\varphi \in \mathcal{L}^\infty(G)$.

Now we're ready to prove a result which will allow us to give a general definition of amenability. We've taken this from [68, Theorem 2.2.1] and [126, Proposition 7.3.4]. It will prove convenient for this result and the next to introduce the notation $\mathcal{S}_c(G)$ for the functions in $C_c^+(G)$ with integral one. (We're thinking of $\mathcal{S}_c(G)$ as a subset of the state space of $L^\infty(G)$: f is associated to the state $\varphi \mapsto (\varphi \mid f)$.)

Theorem A.13. *Let G be a locally compact group. The restriction of a left-invariant mean on $\ell^\infty(G)$ is a left-invariant mean on $C^b(G)$. Moreover, the following statements are equivalent.*

(a) *There is a left invariant mean on $C_u^b(G)$.*
(b) *There is a left invariant mean on $C^b(G)$.*
(c) *There is a left invariant mean on $L^\infty(G)$.*
(d) *There is a state m on $L^\infty(G)$ such that*

$$m(f * \varphi) = \int_G f(s)\,d\mu(s) \cdot m(\varphi) \tag{A.17}$$

for all $f \in C_c(G)$ and $\varphi \in \mathcal{L}^\infty(G)$.

Proof. The first statement is obvious, as is (c) $\Longrightarrow$ (b) and (b) $\Longrightarrow$ (a). If $f \in \mathcal{S}_c(G)$, then if (d) holds, $m(\varphi) = m(f * \varphi)$ for all $\varphi \in L^\infty(G)$. Since $f * \lambda_s\varphi = f_1 * \varphi$ where $f_1(r) := f(rs^{-1})\Delta(s^{-1})$ and since $\int_G f_1(r)\,d\mu(r) = \int_G f(r)\,d\mu(r) = 1$, we have $m(\lambda_s\varphi) = m(f * \lambda_s\varphi) = m(f_1 * \varphi) = m(\varphi)$ for all $\varphi \in \mathcal{L}^\infty(G)$. Thus we also have (d) $\Longrightarrow$ (c).

So we need to prove that (a) $\Longrightarrow$ (d). Assume that m is a mean on $C_u^b(G)$. Fix a nonnegative function φ in $C_u^b(G)$. If $f \in C_c(G)$, then $f * \varphi$ is left-uniformly continuous. But $f * \varphi(sr) = f * \varphi^r(s)$, were $\varphi^r(s) := \varphi(sr)$. Since $|f * \varphi(sr) - f * \varphi(s)| \le \|f\|_1 \|\varphi^r - \varphi\|_\infty$, it follows from the uniform continuity of φ that $f * \varphi$ is also right-uniformly continuous. That is $f * \varphi$ is in $C_u^b(G)$ and we can define $J : C_c(G) \to \mathbf{C}$ by

$$J(f) := m(f * \varphi).$$

Since $\lambda_s f * \varphi = \lambda_s(f * \varphi)$, J is a left-invariant positive functional on $C_c(G)$. Thus there is a scalar α such that

$$m(f * \varphi) = \alpha \int_G f(s)\,d\mu(s).$$

But if $\{ u_i \}$ is an approximate identity for $C^*(G)$ in $C_c(G)$, then, using the uniform continuity of φ, it is easy to compute that

$$\|u_i * \varphi - \varphi\|_\infty \to 0.$$

Thus

$$\alpha = J(u_i) = m(u_i * \varphi) \to m(\varphi).$$

Since each φ in $C_u^b(G)$ is a linear combination of four nonnegative functions in $C_u^b(G)$,[3] it follows that (A.17) holds for all $f \in C_c(G)$ and $\varphi \in C_u^b(G)$.

Now fix $f \in \mathcal{S}_c(G)$. Then for each $\varphi \in \mathcal{L}^\infty(G)$, $f * \varphi * \tilde{f} \in C_u^b(G)$ and we can define

$$\bar{m}(\varphi) := m(f * \varphi * \tilde{f}).$$

Clearly $\bar{m}$ is a positive functional on $\mathcal{L}^\infty(G)$. Since $\|f*\varphi*\tilde{f}\|_\infty \le \|f\|_1\|\varphi\|_\infty\|f\|_1 = \|\varphi\|_\infty$, $\bar{m}$ defines a functional on $L^\infty(G)$ of norm at most one, and since an easy computation shows that $\bar{m}(1) = 1$, $\bar{m}$ is a state. We still have to show that $\bar{m}$ satisfies (A.17). Let $\{ u_i \}$ be an approximate unit in $C_c(G)$ and $h' \in C_c(G)$. Note that $\|h' * \varphi * \tilde{f}\|_\infty \le \|h'\|_1\|\varphi * \tilde{f}\|_\infty$. Since $f * h' * u_i \to f * h'$ in $L^1(G)$,

$$\begin{aligned}
\bar{m}(h * \varphi) &:= m(f * h * \varphi * \tilde{f}) \\
&= \lim_i m(f * h * u_i * \varphi * \tilde{f}) \\
&= \int_G f * h(s)\,d\mu(s) \lim_i m(u_i * \varphi * \tilde{f}) \\
&= \int_G h(s)\,d\mu(s) \lim_i \int_G f(s)\,d\mu(s) m(u_i * \varphi * \tilde{f}) \\
&= \int_G h(s)\,d\mu(s) \lim_i m(f * u_i * \varphi * \tilde{f}) \\
&= \int_G h(s)\,d\mu(s) m(f * \varphi * \tilde{f}) \\
&= \int_G h(s)\,d\mu(s) \bar{m}(\varphi).
\end{aligned}$$

□

Definition A.14. A locally compact group is *amenable* if any of the equivalent conditions (a)–(d) in Theorem A.13 on the previous page are satisfied.

Remark A.15. As we noted, every abelian group has a left-invariant mean on $\ell^\infty(G)$, and is amenable by the first part of Theorem A.13 on the preceding page. Since

[3] If $\varphi \in C_u^b(G)$, then $2\operatorname{Re}\varphi = \varphi + \overline{\varphi}$ shows that $\operatorname{Re}\varphi$ is still uniformly continuous. Similarly, $2\varphi^+ = |f| + f$ and $2\varphi^- = |f| - f$ show that $\operatorname{Re} f = (\operatorname{Re} f)^+ - (\operatorname{Re} f)^-$ with both $(\operatorname{Re} f)^+$ and $(\operatorname{Re} f)^-$ nonnegative and uniformly continuous.

Haar measure gives a left invariant mean on $C^b(G)$ for any compact group, compact groups are amenable. Furthermore, closed subgroups and quotients (by closed subgroups) of amenable groups are amenable by [68, Theorems 2.3.1 and 2.3.2]. If H and G/H are amenable, then G is too [68, Theorem 2.3.3]. In particular, any solvable or nilpotent group is amenable.

The next theorem is from [68, §2.4] and [126, Proposition 7.3.7]. For the statement note that functions in $\mathcal{S}_c(G)$ can be viewed as a convex set of states on $L^\infty(G)$ which is closed under convolution. It follows from Lemma I.28 on page 497 that $\mathcal{S}_c(G)$ is weak-$*$ dense in the state space of $L^\infty(G)$.

Proposition A.16. *Suppose that G is a locally compact group. Then the following statements are equivalent.*

(a) *G is amenable.*

(b) *There is net $\{ g_i \} \subset \mathcal{S}_c(G)$ such that $h * g_i - g_i$ converges to 0 in the weak-$*$ topology on $L^\infty(G)^*$ for all $h \in \mathcal{S}_c(G)$.*

(c) *There is net $\{ g_i \} \subset \mathcal{S}_c(G)$ such that $\|h * g_i - g_i\|_1 \to 0$ for all $h \in \mathcal{S}_c(G)$.*

(d) *For each compact set $C \subset G$ and each $\epsilon > 0$, there is a $g \in \mathcal{S}_c(G)$ such that $\|\lambda_s g - g\|_1 < \epsilon$ for every $s \in C$.*

(e) *For each compact set $C \subset G$ and each $\epsilon > 0$, there is a g in the unit ball of $L^1(G)$ with $g \geq 0$ almost everywhere such that $\|\lambda_s g - g\|_1 < \epsilon$ for every $s \in C$.*

Proof. (a) $\Longrightarrow$ (b): Since G is amenable, there is a state m on $L^\infty(G)$ satisfying (A.17) in part (d) of Theorem A.13 on page 321. Since $\mathcal{S}_c(G)$ is weak-$*$ dense, there is net $\{ g_i \} \subset \mathcal{S}_c(G)$ such that $(\varphi \mid g_i) \to m(\varphi)$ for all $\varphi \in \mathcal{L}^\infty(G)$. Since h and g_i each have compact support, there is no difficulty in applying Fubini's Theorem in the following computation:

$$\begin{aligned}
(\varphi \mid h * g_i) &= \int_G \varphi(s)\overline{h * g_i(s)}\, d\mu(s) \\
&= \int_G \int_G \varphi(s)\overline{h(sr)g_i(r^{-1})}\, d\mu(r)\, d\mu(s) \\
&= \int_G \int_G h^*(rs)\varphi(s^{-1})\overline{g_i(r)}\, d\mu(s)\, d\mu(r) \\
&= \int_G h^* * \varphi(r)\overline{g_i(r)}\, d\mu(r) \\
&= (h^* * \varphi \mid g_i).
\end{aligned}$$

Since m satisfies (A.17) and h^* has integral one, we have

$$(\varphi \mid h * g_i) = (h^* * \varphi \mid g_i) \to m(h^* * \varphi) = m(\varphi).$$

It follows that $h * g_i - g_i \to 0$ in the weak-$*$ topology.

(b) $\Longrightarrow$ (c): Equipped with the L^1-norm, $C_c(G)$ is a normed vector space with dual $C_c(G)^* = L^1(G)^* \cong L^\infty(G)$.[4] Fix $h_1, \dots, h_n \in \mathcal{S}_c(G)$ and let

$$C := \{ (h_1 * g - g, \dots, h_n * g - g) \in \bigoplus_{k=1}^{n} C_c(G) : g \in \mathcal{S}_c(G) \}.$$

Then C is a *convex* subset of $\bigoplus_{k=1}^n C_c(G)$. Since the dual of $\bigoplus_{k=1}^n C_c(G)$ can be identified with $\bigoplus_{k=1}^n L^\infty(G)$, part (b) implies that there is a net $\{ g_i \}$ in $\mathcal{S}_c(G)$ such that $(h_k * g_i - g_i)$ converges weakly to $(0, \dots, 0)$. Since C is convex, the Hahn-Banach Separation Theorem [126, Theorem 2.4.7] implies the weak and norm closures of C coincide. Thus, there is a $g \in \mathcal{S}_c(G)$ such that

$$\|h_k * g - g\|_1 < \epsilon \quad \text{for } k = 1, 2, \dots, n.$$

Let J be the set of pairs (F, ϵ) where F is a finite subset of $\mathcal{S}_c(G)$ and $\epsilon > 0$. Then J is a directed set with $(F, \epsilon) \geq (F', \epsilon')$ when $F \supset F'$ and $\epsilon < \epsilon'$. For each $j = (F, \epsilon) \in J$, we just showed that there is a $g_j \in \mathcal{S}_c(G)$ such that

$$\|h * g_j - g_j\| < \epsilon \quad \text{for all } h \in F.$$

Then $\{ g_j \}_{j \in J}$ is the required net.

(c) $\Longrightarrow$ (d): Let $C \subset G$ be compact and $\epsilon > 0$. Fix $h \in \mathcal{S}_c(G)$ and a net $\{ g_i \} \subset \mathcal{S}_c(G)$ satisfying (c). Then for all $s \in G$,

$$\begin{aligned} \|\lambda_s(h * g_i) - h * g_i\|_1 &\leq \|\lambda_s(h * g_i) - g_i\|_1 + \|g_i - h * g_i\|_1 \\ &= \|\lambda_s h * g_i - g_i\|_1 + \|h * g_i - g_i\|_1. \end{aligned} \tag{A.18}$$

Since h and $\lambda_s h$ are in $\mathcal{S}_c(G)$, it follows that (A.18) goes to 0 for large i. Since $s \mapsto \lambda_s h$ is continuous from G to $C_c(G) \subset L^1(G)$, a compactness argument shows that for large i

$$\|\lambda_s(h * g_i) - h * g_i\|_1 < \epsilon \quad \text{for all } s \in C.$$

Thus we can let $g = h * g_i$ for a suitably large i.

(d) $\Longrightarrow$ (e) is trivial.

(e) $\Longrightarrow$ (a): Let $I = \{ (C, \epsilon) : C \subset G \text{ is compact and } \epsilon > 0 \}$ directed by $(C, \epsilon) \geq (C', \epsilon')$ if $C \supset C'$ and $\epsilon < \epsilon'$. For each $i = (C, \epsilon) \in I$, we can choose g_i in the unit ball of $L^1(G)$ with $g_i \geq 0$ such that

$$\|\lambda_s g_i - g_i\|_1 < \epsilon \quad \text{for all } s \in C.$$

For each $i \in I$, we can define a state m_i on $C^b(G)$ by

$$m_i(\varphi) = \int_G \varphi(s) g_i(s) \, d\mu(s).$$

[4]If G is σ-compact, then it is straightforward to see that $L^1(G)^*$ can be identified with $L^\infty(G)$. For the general case see Appendix I.5.

If $i = (C, \epsilon)$ and $s \in C$, then

$$\begin{aligned}\big|m_i(\lambda_{s^{-1}}\varphi - \varphi)\big| &= \Big|\int_G \big(\varphi(sr) - \varphi(r)\big)g_i(r)\,d\mu(r)\Big| \\ &= \Big|\int_G \varphi(r)\big(g_i(s^{-1}r) - g_i(r)\big)\,d\mu(r)\Big| \\ &\le \|\varphi\|_\infty \|\lambda_s g_i - g_i\|_1 \\ &\le \|\varphi\|_\infty \epsilon.\end{aligned}$$

Since the state space of $C^b(G)$ is compact in the weak-$*$ topology, the net $\{ m_i \}$ must have a weak-$*$ convergent subnet $\{ m_{i_j} \}$ converging to a state m. Clearly, m is a left-invariant mean on $C^b(G)$, and G is amenable by definition. □

The next result is the key which connects our basic characterizations of amenability (Proposition A.16 on page 323) with positive definite functions on G (recall Example A.7 on page 314). The proof is stolen from Pedersen's book [126, Proposition 7.3.8].

Proposition A.17. *A locally compact group G is amenable if and only if there is a net $\{ \Phi_i \}$ of* compactly supported *positive definite functions in $\mathcal{P}_1(G)$ converging to the constant function* 1 *in $\mathcal{P}_1(G)$.*

Proof. Suppose that G is amenable. If $f \in C_c(G)$, then Example A.7 on page 314 implies that $f * \tilde{f}$ is a compactly supported positive definite function with $f * \tilde{f}(e) = \|f\|_2^2$. Thus it will suffice to produce, for each compact set C and $\epsilon > 0$, a function $f \in C_c^+(G)$ with $\|f\|_2 = 1$ such that

$$|1 - f * \tilde{f}(s)| < \epsilon \quad \text{for all } s \in C.$$

However, by part (d) of Proposition A.16 on page 323, there is a $g \in \mathcal{S}_c(G)$ such that

$$\|\lambda_s g - g\|_1 < \epsilon^2 \quad \text{for all } s \in C. \tag{A.19}$$

Define f by $f(s) := \sqrt{g(s)}$. Then $\|f\|_2 = 1$, and

$$\begin{aligned}\big|1 - f * \tilde{f}(s)\big|^2 &= \big|f * \tilde{f}(e) - f * \tilde{f}(s)\big|^2 \\ &= \Big|\int_G f(r)\big(\overline{f(r)} - \overline{f(s^{-1}r)}\big)\,d\mu(r)\Big|^2 \\ &= \big|(f \mid f - \lambda_s f)\big|^2 \\ &\le 1\|f - \lambda_s f\|_2^2 \\ &= \int_G \big|\sqrt{g(r)} - \sqrt{g(s^{-1}r)}\big|^2\,d\mu(r) \\ &\le \int_G \big|\sqrt{g(r)} - \sqrt{g(s^{-1}r)}\big|\big|\sqrt{g(r)} + \sqrt{g(s^{-1}r)}\big|\,d\mu(r) \\ &= \int_G \big|g(r) - g(s^{-1}r)\big|\,d\mu(r) \\ &= \|g - \lambda_s g\|_1 < \epsilon^2.\end{aligned}$$

This proves "only if" implication.

For the second implication, first consider any $f \in \mathcal{L}^2(G)$ with $\|f\|_2 = 1$. Let $g = |f|^2$ and note that g is in the unit ball of $L^1(G)$ and that $g \geq 0$. For each $s \in G$, we have

$$\begin{aligned}
\|g - \lambda_s g\|_1 &= \int_G \big| |f(r)|^2 - |f(s^{-1}r)|^2 \big| \, d\mu(r) \\
&= \int_G \big| |f(r)^2| - |f(s^{-1}r)^2| \big| \, d\mu(r) \\
&\leq \int_G |f(r)^2 - f(s^{-1}r)^2| \, d\mu(r) \\
&= \int_G |f(r) + f(s^{-1}r)| |\overline{f(r)} - \overline{f(s^{-1}r)}| \, d\mu(r) \\
&= \big(|f + \lambda_s f| \mid |f - \lambda_s f| \big) \\
&\leq \|f + \lambda_s f\|_2 \|f - \lambda_s f\|_2 \\
&\leq 2\|f - \lambda_s f\|_2 \\
&= 2\big(2 - (f \mid \lambda_s f) - (\lambda_s f \mid f)\big)^{\frac{1}{2}} \\
&= 2\big(2 - 2\operatorname{Re}(f \mid \lambda_s f)\big)^{\frac{1}{2}} \\
&\leq 2\sqrt{2}\big|1 - (f \mid \lambda_s f)\big|^{\frac{1}{2}} \\
&= 2\sqrt{2}\big|1 - f * \tilde{f}(s)\big|^{\frac{1}{2}}.
\end{aligned}$$

Therefore if $\{ \Phi_i \}$ is net converging to 1, then given $i = (C, \epsilon)$ with C compact and $\epsilon > 0$, we can choose Φ_i such that

$$2\sqrt{2}\big|1 - \Phi_i(s)\big|^{\frac{1}{2}} < \epsilon \quad \text{for all } s \in C.$$

By Lemma A.9 on page 315, there is a f_i in the unit ball of $L^2(G)$ such that $\Phi_i = f_i * \tilde{f}_i$. Therefore if $g_i := |f_i|^2$, then

$$\|g_i - \lambda_s g_i\|_1 < \epsilon \quad \text{for all } s \in C.$$

Since C and ϵ are arbitrary, G must be amenable by part (e) of Proposition A.16 on page 323. □

And now for the main result. Let $\iota : G \to \mathbf{C}$ be the *trivial representation* $\iota_s = 1$ for all $s \in G$.

Theorem A.18. *Let G be a locally compact group. Then the following statements are equivalent.*

(a) *G is amenable.*

(b) *The left-regular representation is faithful on $C^*(G)$.*

(c) *The kernel of the left-regular representation is contained in the kernel of the trivial representation ι (as representations of $C^*(G)$).*

Proof. (a) $\Longrightarrow$ (b): Since $\lambda(b) = 0$ implies $\lambda(b^*b) = 0$, it will suffice to show that every positive element a in $\ker \lambda$ is zero. Let ω be a state on $C^*(G)$ with

$$\omega(a) = \|a\|$$

(cf., e.g., [139, Lemma A.10]). Let Ψ be the positive definite function corresponding to ω via Theorem A.10 on page 317. Let $\Phi = \Psi^\circ$ which is positive definite by Lemma A.4 on page 313. By Proposition A.17 on page 325, there are compactly supported Φ_i in $\mathcal{P}_1(G)$ such that $\Phi_i \to 1$ in $\mathcal{P}_1(G)$. Since each Φ_i has compact support, $\Phi_i\Phi$ is a positive definite function with compact support by Lemma A.4 on page 313, and $\Phi_i\Phi$ converges to Φ in $\mathcal{P}_1(G)$. By Lemma A.9 on page 315, there are $\{ f_i \}$ in the unit ball of $L^2(G)$ such that $\Phi_i\Phi = f_i * \tilde{f}_i$. It follows that $(f_i * \tilde{f}_i)^\circ \to \Psi$. Example A.7 on page 314 shows that $(f_i * \tilde{f}_i)^\circ$ corresponds to the vector state $g \mapsto \big(\lambda(g)f_i \mid f_i\big)$, and Theorem A.10 on page 317 implies

$$\big(\lambda(a)f_i \mid f_i\big) \to \omega(a) = \|a\|.$$

Since $a \in \ker \lambda$ by assumption, $a = 0$ and λ must be faithful.

The implication (b) $\Longrightarrow$ (c) is trivial.

(c) $\Longrightarrow$ (a): Since $\ker \lambda \subset \ker \iota$, ι factors through λ and there is a state ι' on the concrete C^*-algebra $\lambda\big(C^*(G)\big)$ such that $\iota'\big(\lambda(g)\big) = \iota(g)$. It follows from [139, Proposition B.5], for example, that ι' is the weak-$*$ limit of convex combinations ω_i of vector states of the form

$$\omega'_{f,f}\big(\lambda(g)\big) = \big(\lambda(g)f \mid f\big) \quad (f \in C_c(G)). \tag{A.20}$$

But the positive definite function corresponding to $\omega_{f,f} := \omega'_{f,f} \circ \lambda$ is the compactly supported function $(f * \tilde{f})^o$ (Example A.7 on page 314). Thus the positive definite function Φ_i corresponding to $\omega_i := \omega'_i \circ \lambda$ is a finite convex combination of compactly supported functions and is therefore compactly supported itself. Since $\omega_i \to \iota$ in the weak-$*$ topology and since the positive definite function corresponding to ι is the constant function 1, $\Phi_i \to 1$ in $\mathcal{P}_1(G)$ (Theorem A.10 on page 317). Thus G is amenable by Proposition A.17 on page 325. □

A.3 Another GNS Construction

In this section, a *normed $*$-algebra* B will mean a normed algebra with an involution $a \mapsto a^*$ satisfying $\|ab\| \le \|a\|\|b\|$ and $\|a^*\| = \|a\|$ for all $a, b \in B$. We say that B has an approximate identity $\{ u_i \}$ if $\|u_i\| \le 1$ for all i and

$$\lim_i \|b - bu_i\| = 0 = \lim_i \|b - u_ib\| \quad \text{for all } b \in B.$$

Since the involution is isometric, a similar statement holds for $\{ u_i^* \}$. In addition to C^*-algebras, the sort of algebras we have in mind are $C_c(G)$ and $C_c(G, A)$ when equipped with the L^1-norm.

A linear functional $\varphi : B \to \mathbf{C}$ is called *positive* if $\varphi(b^*b) \ge 0$ for all $b \in B$. Of course, φ is called bounded if its norm $\|\varphi\| := \sup_{\|b\|=1} |\varphi(b)|$ is finite.

Proposition A.19. *If φ is a positive linear functional of norm 1 on a normed $*$-algebra B with an approximate identity $\{ u_i \}$, then there is a Hilbert space $\mathcal{H}_\varphi$, a norm decreasing $*$-homomorphism $\pi_\varphi : B \to B(\mathcal{H}_\varphi)$, and a unit vector $h_\varphi \in \mathcal{H}_\varphi$ such that*

$$\varphi(b) = \big(\pi_\varphi(b)h_\varphi \mid h_\varphi\big),$$

and $\{ \pi_\varphi(b)h_\varphi : b \in B \}$ is dense in $\mathcal{H}_\varphi$.

Remark A.20. The proof is based on the usual GNS-construction for states on C^*-algebras as outlined in §A.1 of [139]. Some finesse is required to show that left multiplication is bounded since we can't assume B is even a pre-C^*-algebra. Of course, we could pass to the Banach $*$-algebra completion of B, and there is a good deal of literature on positive functionals on Banach algebras with approximate identities (e.g., [28, §2.1] and [72, Chap. VIII §32]). However, we have chosen to work directly with the potentially incomplete algebra B.

Lemma A.21. *Let B be a normed $*$-algebra with approximate identity $\{ u_i \}$ as above. If φ is a bounded positive linear functional on B then the following hold for all $a, b \in B$.*

(a) $\varphi(b^*) = \overline{\varphi(b)}$.
(b) $\varphi(b^*a) = \overline{\varphi(a^*b)}$.
(c) *(Cauchy-Schwarz inequality)* $|\varphi(b^*a)|^2 \le \varphi(b^*b)\varphi(a^*a)$.
(d) $\|\varphi\| = \lim_i \varphi(u_i^*u_i)$.
(e) $|\varphi(b^*ab)| \le \|a\|\varphi(b^*b)$.
(f) $|\varphi(b)|^2 \le \|\varphi\|\varphi(b^*b)$.

Proof. Parts (b) and (c) follow from [139, Lemma A.4] together with the observation that the proof in [139] only uses that A is a $*$-algebra. Then (a) follows from

$$\varphi(b^*) = \lim_i \varphi(b^*u_i) = \lim_i \overline{\varphi(u_i^*b)} = \overline{\varphi(b)}.$$

Let $L := \limsup_i \varphi(u_i^*u_i)$. The Cauchy-Schwarz inequality implies that

$$\begin{aligned} |\varphi(b)|^2 &= \lim_i |\varphi(u_i^*b)|^2 \\ &\le \limsup_i \varphi(u_i^*u_i)\varphi(b^*b) \qquad \text{(A.21)} \\ &\le \|\varphi\|\|b\|^2 L. \end{aligned}$$

This forces $L \ge \|\varphi\|$. But the opposite inequality is obvious, and this proves $\limsup_i \varphi(u_i^*u_i) = \|\varphi\|$ and (f) follows from (A.21). Since any subnet of $\{ u_i \}$ is again an approximate identity, it follows that $\lim_i \varphi(u_i^*u_i) = \|\varphi\|$, and (d) is proved.

To prove part (e), notice that $\varphi_b : B \to \mathbf{C}$ defined by $\varphi_b(a) := \varphi(b^*ab)$ is again a bounded positive linear functional on B. Thus

$$\begin{aligned} \|\varphi_b\| &= \lim_i \varphi_b(u_i^*u_i) \\ &= \lim_i \varphi\big((u_ib)^*u_ib\big) \\ &= \varphi(b^*b). \end{aligned}$$

□

Proof of Proposition A.19. It follows from (b) and (c) of Lemma A.21 on the preceding page that

$$(a \mid b) := \varphi(b^*a)$$

is a sesquilinear form on B, and we can form the Hilbert space completion $\mathcal{H}_\varphi$. Let $q : B \to \mathcal{H}_\varphi$ be the obvious map. Now part (e) of Lemma A.21 implies that

$$\begin{aligned}(ba \mid ba) &= \varphi(a^*b^*ba)\\ &\le \|b^*b\|\varphi(a^*a)\\ &\le \|b\|^2(a \mid a).\end{aligned}$$

Therefore we can define a bounded operator $\pi_\varphi(b)$ on $\mathcal{H}_\varphi$ which satisfies $\pi_\varphi(b)\big(q(a)\big) = q(ba)$. It is not hard to see that π_φ is a norm-decreasing $*$-homomorphism.

If B has an identity, then $h_\varphi := q(1_B)$ is a cyclic vector for π_φ representing φ. If B does not have an identity, then we proceed along the lines of [139, Proposition A.6]. We make $B^1 = B \oplus \mathbf{C}$ into a normed $*$-algebra by giving it the obvious $*$-algebra structure and norm $\|b + \lambda 1\| := \|b\| + |\lambda|$. We can extend φ to a linear functional τ on B^1 by the formula

$$\tau(b + \lambda 1) := \varphi(b) + \lambda.$$

Since

$$\begin{aligned}|\tau(b + \lambda 1)| &\le |\varphi(b)| + |\lambda|\\ &\le \|b\| + |\lambda|\\ &= \|b + \lambda 1\|,\end{aligned}$$

τ has norm 1. Using Lemma A.21 on the facing page we have

$$\begin{aligned}\tau\big((b + \lambda 1)^*(b + \lambda 1)\big) &= \tau(b^*b + \lambda b^* + \bar{\lambda} b + |\lambda|^2 1)\\ &= \varphi(b^*b) + \lambda\overline{\varphi(b)} + \bar{\lambda}\varphi(b) + |\lambda|^2\\ &= \varphi(b^*b) + 2\operatorname{Re}\big(\bar{\lambda}\varphi(b)\big) + |\lambda|^2\\ &\ge \varphi(b)|^2 - 2|\lambda||\varphi(b)| + |\lambda|^2\\ &= \big(|\varphi(b)| - |\lambda|\big)^2\\ &\ge 0,\end{aligned}$$

and it follows that τ is positive. Let $(\pi_\tau, \mathcal{H}_\tau)$ be the representation corresponding to τ as above. If $q^1 : B^1 \to \mathcal{H}_\tau$ is the natural map, then $q(b) \mapsto q^1(b)$ induces an isometry V of $\mathcal{H}_\varphi$ into $\mathcal{H}_\tau$ such that

$$V\pi_\varphi(b) = \pi_\tau(b)V \quad \text{for all } b \in B.$$

Thus we can identify $\mathcal{H}_\varphi$ with the subspace $V\mathcal{H}_\varphi$ of $\mathcal{H}_\tau$. Note that $\mathcal{H}_\varphi$ is the essential subspace $\overline{\operatorname{span}}\{\, \pi_\tau(b)h : b \in B \text{ and } h \in \mathcal{H}_\tau \,\}$ of $\pi_\tau|_B$, and $\pi_\tau|_B = \pi_\varphi \oplus 0$ on $\mathcal{H}_\varphi \oplus \mathcal{H}_\varphi^\perp$. If h_φ is the projection of $q^1(1)$ onto $\mathcal{H}_\varphi$, then

$$\pi_\varphi(b)h_\varphi = \pi_\tau(b)q^1(1) = q^1(b).$$

Thus h_φ is cyclic for π_φ and

$$\big(\pi_\varphi(b)h_\varphi \mid h_\varphi\big) = \big(\pi_\tau(b)q^1(1) \mid q^1(1)\big) = \tau(b) = \varphi(b).$$

Since

$$\begin{aligned} 1 = \|\varphi\| &= \lim_i \varphi(u_i^* u_i) \\ &= \lim_i \|\pi_\varphi(u_i)h_\varphi\|^2 \\ &= \|h_\varphi\|^2, \end{aligned}$$

h_φ is a unit vector as required. □

Appendix B

The Banach $*$-Algebra $L^1(G, A)$

In classical treatments of the crossed product — such as [30, 66, 162] — $A \rtimes_\alpha G$ is defined to be the enveloping C^*-algebra of the Banach $*$-algebra $L^1(G, A)$. The latter is usually quickly disposed of with the phrase "$L^1(G, A)$ is the collection of Bochner integrable functions from G to A equipped with the convolution product ...." I have avoided this approach, and there are a number of good reasons for this. Certainly not the least of which is that I find the theory of vector-valued integration a bit formidable. Nevertheless, in this section, we make an attempt at sketching some of the necessary background for those interested in persevering.

For vector-valued integration, there are a number of references available. Among these are Dunford and Schwarz [32, Chap. III], Bourbaki [11], Fell and Doran [54, Chap. II] and Hille and Phillips [73, Part I, §III.1]. It is hardly clear at first glance that these four references are talking about theories which have a nontrivial intersection. In fact, Bourbaki has chosen to focus only on Radon measures on locally compact spaces (as defined below), and has presented the theory in what has fortunately become a completely nonstandard order. Nevertheless, the presentation here is closest in spirit to Bourbaki's as it treats the density of $C_c(G, A)$ in $L^1(G, A)$ as the central feature. Thus, one could avoid measure theory altogether by defining $L^1(G, A)$ as the completion of $C_c(G, A)$ in the L^1-norm where it not that one of the primary reasons to pass to the L^1-algebra is to be able to use functions which are not necessarily continuous or compactly supported.

B.1 Vector-Valued Integration

In the discussion here, X will be a locally compact Hausdorff space and μ a *Radon measure* on a measurable space $(X, \mathscr{M})$. A Radon measure is a measure associated to a positive linear functional $I : C_c(X) \to \mathbf{C}$ via the Riesz Representation Theorem. More precisely, μ is the unique measure defined on a σ-algebra $\mathscr{M}$ containing the Borel sets of X such that

(a) $I(f) = \int_X f(x)\, d\mu(x)$ for all $f \in C_c(X)$,

(b) for each open set $V \subset X$, $\mu(V) = \sup\{\, \mu(K) : K \subset V$ is compact$\,\}$, and

(c) For all $E \in \mathscr{M}$, $\mu(E) = \inf\{\, \mu(V) : E \subset V$ is open$\,\}$.

Measures satisfying (b) and (c) are called *regular*. As shown in [156, Theorem 2.14] or [71, Theorem 11.32], any regular measure must also satisfy

$$\mu(E) = \sup\{\, \mu(C) : C \subset E \text{ is compact}\,\} \quad \text{if } \mu(E) < \infty. \tag{B.1}$$

Remark B.1. A measure which satisfies (B.1) without also requiring that $\mu(E) < \infty$ is called *inner regular*. A measure satisfying (c) is called *outer regular*. As Arveson points out in [3], one can require measures arising from linear functionals to be either outer regular, as in [156], or inner regular, as in [3]. As we shall see, it is not possible to have both conditions hold in general.

Although any σ-algebra $\mathscr{M}$ containing the Borel sets in X would do, our choice for a σ-algebra will coincide with that in [156, Theorem 2.14]. One important consequence of this choice is that our measures are *complete* in that every subset of a null[1] set is measurable. Another more subtle consequence is that μ is *saturated* in that *locally measurable* sets are measurable. That is,

$$A \in \mathscr{M} \iff A \cap K \in \mathscr{M} \quad \text{for all } K \subset X \text{ compact.} \tag{B.2}$$

Remark B.2. Saturation and local measurability are annoyances that arise only because we have not insisted that our space X be σ-compact or second countable. Normally, nonseparable examples such as these don't play a particularly important role, but it seems silly to make blanket assumptions that can be avoided with just a bit more effort. This is especially true when X is a locally compact group G, and Folland describes the situation there rather succinctly in [56, §2.3]. There he also gives a classic example that shows exactly the sort of pathology we have to beware of if we're going to allow large spaces such as $X = \mathbf{R} \times \mathbf{R}_d$; that is, X is the locally compact group which is the product of $\mathbf{R}$ with its usual second countable topology with $\mathbf{R}$ equipped with the discrete topology. Then $\mathbf{R} \times \mathbf{R}_d$ is far from σ-compact. Now let μ be a Haar measure on $\mathbf{R} \times \mathbf{R}_d$. Thus μ is (a multiple of) the product of Lebesgue measure and counting measure. Consider $S = \{\,0\,\} \rtimes \mathbf{R}_d$. Then S is closed and hence measurable. Furthermore, any compact subset of S is finite and has measure zero. On the other hand, any open set containing S is the uncountable disjoint union of nonempty open sets. Since any nonempty open set has strictly positive Haar measure, regularity forces $\mu(S) = \infty$. Thus S is an example of a set which is *locally null*, but not null. In particular, Haar measure on $\mathbf{R} \times \mathbf{R}_d$ fails to be inner regular.

The first task here is to define what it means for a function $f : X \to B$ to be measurable if B is a Banach space (which need not be separable). The definition has to be finessed so that the image of a measurable function is not too large. For

[1] Unless we're dealing with more than one measure, we'll just write "null set" in place of "μ-null set".

example, if f is to be the limit of a sequence of simple functions, then the range of f has to be contained in the closure of the images of these simple functions which is a separable subspace of B.

Definition B.3. Let B be a Banach space. A function $f : X \to B$ is *essentially separately-valued* on a set S if there is a countable set $D \subset B$ and a null set $N \subset S$ such that $f(x) \in \overline{D}$ for all $x \in S \setminus N$.

Since will often be convenient to deal with separable subspaces rather than separable subsets, the next lemma will often be invoked without comment.

Lemma B.4. *Let D be a countable set in a Banach space B. Let B_D be the closed subspace of B generated by D. Then B_D is separable. In particular, we can replace the countable* set *D in Definition B.3 with a separable subspace.*

Proof. Note that B_D is the closure of the linear span $S(D)$ of D in B. But the countable set $S_{\mathbf{Q}}(D)$ of rational linear combinations of elements of D is certainly countable and dense in $S(D)$. Thus $S_{\mathbf{Q}}(D)$ is also dense in B_D. □

Another apparent complication is that there are at least two reasonable notions of measurability.

Definition B.5. Let B be a Banach space and $f : X \to B$ a function. Then f is *strongly measurable* if

(a) $f^{-1}(V)$ is measurable for all open sets $V \in B$, and
(b) f is essentially separately valued on every compact subset of X.

Definition B.6. Let B be a Banach space and $f : X \to B$ a function. Then f is *weakly measurable* if

(a) $\varphi \circ f$ is a measurable function from X to $\mathbf{C}$ for all $\varphi \in B^*$, and
(b) f is essentially separately-valued on each compact subset of X.

This complication is not a complication at all as the two notions coincide.

Lemma B.7. *Let B be a Banach space. A function $f : X \to B$ is weakly measurable if and only if it is strongly measurable.*

For the proof, it will be convenient to record the following standard result.

Lemma B.8. *Suppose that B_0 is a separable subspace of a Banach space B. Then there is a countable set F of linear functionals in the unit ball of B^* such that*

$$\|b\| = \sup_{\varphi \in F} |\varphi(b)| \quad \textit{for all } b \in B_0.$$

Proof. Let D be a countable dense subset of B_0. Using the Hahn-Banach Theorem, for each $d \in D$, there is a norm one linear functional $\varphi_d \in B^*$ such that $\varphi_d(d) = \|d\|$. Now for any $b \in B_0$ and $d \in D$,

$$\begin{aligned} \|b\| \geq |\varphi_d(b)| &= |\varphi_d(d) - \varphi_d(d-b)| \geq |\varphi_d(d)| - |\varphi_d(d-b)| \\ &\geq \|d\| - \|d-b\|. \end{aligned}$$

Given $\epsilon > 0$, we can choose $d \in D$ such that $\|d - b\| < \frac{\epsilon}{2}$. Then $\|d\| \geq \|b\| - \frac{\epsilon}{2}$ and

$$\|b\| \geq \sup_{d \in D} |\varphi_d(b)| \geq \|b\| - \epsilon.$$

Since ϵ is arbitrary, the result follows. □

Proof of Lemma B.7. Since each $\varphi \in B^*$ is continuous, it is straightforward to check that strong measurability implies weak measurability.

So, suppose that f is weakly measurable. Clearly, it suffices only to check that $f^{-1}(V)$ is measurable for all open sets $V \subset B$. Since μ is saturated, it will then suffice to check that $f^{-1}(V) \cap K$ is measurable for all $K \subset X$ compact. Since f is essentially separately-valued on K, there is a null set $N \subset K$ and a separable subspace B_0 such that $f(x) \in B_0$ for all $x \in K \setminus N$ (Lemma B.4 on the preceding page). Define $f_0 : X \to B$ by

$$f_0(x) := \begin{cases} f(x) & \text{if } x \in X \setminus N \text{, and} \\ 0 & \text{if } x \in N. \end{cases}$$

Since μ is complete and

$$f^{-1}(V) \cap K = \big(f^{-1}(V) \cap N\big) \cup \big(f_0^{-1}(V) \cap (K \setminus N)\big),$$

it will suffice to see that $f_0^{-1}(V) \cap K \setminus N$ is measurable. But

$$f_0^{-1}(V) \cap K \setminus N = f_0^{-1}(V \cap B_0) \cap K \setminus N,$$

and since B_0 is separable, there are countably many closed balls $B_n := \{\, b \in B_0 : \|b - b_n\| \leq \epsilon_n \,\}$ with $b_n \in B_0$ such that

$$V \cap B_0 = \bigcup_{n=1}^{\infty} B_n.$$

Thus

$$f_0^{-1}(V) \cap K \setminus N = \bigcup_{n=1}^{\infty} f_0^{-1}(B_n) \cap K \setminus N,$$

and it will suffice to see that $f_0^{-1}(B_n)$ is measurable. But Lemma B.8 on the previous page implies there is a countable set $\{\, \varphi_m \,\}_{m=1}^{\infty}$ in B^* such that for all $b \in B_0$

$$\|b - b_n\| \leq \epsilon_n \iff |\varphi_m(b - b_n)| \leq \epsilon_n \quad \text{for all } m \geq 1.$$

Therefore

$$f_0^{-1}(B_n) = \bigcap_{m=1}^{\infty} \{\, x \in X : |\varphi_m\big(f(x) - b_n\big)| \leq \epsilon_n \,\},$$

which is measurable since each of the sets in the countable intersection on the right-hand side is measurable by assumption. □

Remark B.9. In view of the above, we'll normally just say a Banach space valued function is measurable if it satisfies either of Definitions B.5 or B.6 on page 333.

Remark B.10. Since our measures are saturated, it follows, as in the proof of Lemma B.7 on page 333, that $f : X \to B$ is measurable if and only if $f|_K$ is measurable for all $K \subset X$ compact.

Lemma B.11. *A linear combination of measurable B-valued functions is measurable, and every element of $C_c(X, B)$ is measurable. If f is a B-valued function on X with the property that for each compact set $K \subset X$, there is a sequence $\{ f_n \}$ of measurable B-valued functions such that $f_n|_K$ converges to $f|_K$ almost everywhere, then f is measurable.*

Proof. Each of the assertions is well-known for scalar-valued functions. Since it suffices to check weak measurability, it only remains to verify that the functions in question are essentially separately-valued on compact sets. However this is clear for a linear combination, and any element of $C_c(X, B)$ has compact image which is necessarily separable. So suppose that $f_n(x) \to f(x)$ for all $x \in K \setminus N_0$, where N_0 is a null set. If K is compact, then by assumption there are null sets N_n and separable subspaces B_n such that $f_n(x) \in B_n$ provided $x \in K \setminus N_n$. Let $N = \bigcup_{n=0}^{\infty} N_n$, and let $\bigvee B_n$ be the subspace generated by the B_n. Thus for all $x \in K \setminus N$,

$$f(x) \in \overline{\bigvee_{n=1}^{\infty} B_n},$$

which is a separable subspace. □

Lemma B.12 (Egoroff's Theorem). *Suppose that $\mu(E) < \infty$ and that $\{ f_n \}$ are measurable functions on X such that $f_n(x) \to f(x)$ for almost all $x \in E$. Then for each $\epsilon > 0$, there is a subset $F \subset E$ such that $\mu(E \setminus F) < \epsilon$ and $f_n \to f$* uniformly *on F.*

Proof. Let

$$S(n, k) := \bigcap_{m \geq n} \{ x \in E : \|f_m(x) - f(s)\| < \frac{1}{k} \}.$$

Thus $S(n + 1, k) \supset S(n, k)$ and $\mu\bigl(\bigcup_{n=1}^{\infty} S(n, k)\bigr) = \mu(E)$. Therefore for each $k \geq 1$, we have $\lim_n \mu\bigl(S(n, k)\bigr) = \mu(E)$. Since $\mu(E) < \infty$, this implies $\lim_n \mu\bigl(E \setminus S(n, k)\bigr) = 0$. Thus we can inductively choose an increasing sequence n_k such that

$$\mu\bigl(E \setminus S(n_k, k)\bigr) < \frac{\epsilon}{2^k}.$$

Let $F := \bigcap_{k=1}^{\infty} S(n_k, k)$. Then $\mu(E \setminus F) < \epsilon$ and if $x \in F$, then $x \in S(n_k, k)$ so that $n \geq n_k$ implies that $\|f_n(x) - f(x)\| < \frac{1}{k}$. □

Now we want to give another criterion for measurability which is very convenient and also helps to connect with the treatment in Bourbaki, where a number of the arguments below can be found. The term "C-measurable" will only be used until we can prove it is equivalent to (weak and strong) measurability.

Definition B.13. A B-valued function is called *C-measurable* if given any compact set $K \subset X$ and $\epsilon > 0$, there is a compact set $K' \subset K$ such that $\mu(K \setminus K') < \epsilon$ and such that the restriction $f|_{K'}$ of f to K' is continuous.

Remark B.14. It is important to keep in mind that the assertion in the above definition is that $f|_{K'}$ is continuous on K', not that f is continuous on K'. For example, the characteristic function of the irrationals restricts to a constant function on the irrationals, but it is not continuous at even single point.

Lemma B.15. *If f is C-measurable and $K \subset X$ is compact, then there are disjoint compact sets $K_1, K_2, \dots \subset K$ such that $\mu\big(K \setminus \bigcup_{n=1}^{\infty} K_n\big) = 0$ and $f|_{K_n}$ is continuous for all n. (Some of the K_n can be empty.)*

Proof. By definition, we can choose $K_1 \subset K$ such that $\mu(K \setminus K_1) < 1$ and $f|_{K_1}$ is continuous. Since $K \setminus K_1$ has finite measure, there is, by (B.1), a compact set $K_2' \subset K \setminus K_1$ such that $\mu(K \setminus K_1 \setminus K_2') < \frac{1}{2}$. By assumption, there must be a $K_2 \subset K_2'$ such that $f|_{K_2}$ is continuous and $\mu(K \setminus K_1 \setminus K_2) < \frac{1}{2}$. Continuing, we get pairwise disjoint compact sets $K_1, K_2, \dots$ such that $\mu\big(K \setminus \bigcup_{i=1}^{n} K_i\big) < \frac{1}{n}$. Thus $\mu\big(K \setminus \bigcup_{i=1}^{\infty} K_i\big) = 0$. □

Lemma B.16. *If f is C-measurable, then f is measurable.*

Proof. We'll use Definition B.5 on page 333. To verify part (a), it will suffice to see that $f^{-1}(A)$ is measurable for all closed sets $A \subset B$. Since μ is saturated, it will suffice to show that $f^{-1}(A) \cap K$ is measurable for any compact set K. By Lemma B.15, we can partition $K = N \cup \bigcup_{n=1}^{\infty} K_n$ with each K_n compact, N null, and $f|_{K_n}$ continuous. Then

$$f^{-1}(A) \cap K = f^{-1}(A) \cap N \cup \bigcup_{n=1}^{\infty} f^{-1}(A) \cap K_n.$$

Since $f|_{K_n}$ is continuous, $f^{-1}(A) \cap K_n$ is closed, and $f^{-1}(A) \cap N$ is null. This suffices.

To establish part (b) of Definition B.5 on page 333, let $K = N \cup \bigcup K_n$ as above. Then each $f(K_n)$ is compact and therefore separable. Thus the subspace generated by $\bigcup f(K_n)$ is separable. This suffices as N is null. □

Lemma B.17. *Suppose that each f_n is a B-valued C-measurable function and that $f_n(x) \to f(x)$ for almost all x. Then f is C-measurable.*

Proof. Let $K \subset X$ be compact and $\epsilon > 0$. Choose $K_n \subset K$ such that $\mu(K \setminus K_n) < \frac{\epsilon}{2^{n+1}}$ and such that $f_n|_{K_n}$ is continuous. Egoroff's Theorem and (B.1) imply there is a compact set $K_0 \subset K$ such that $\mu(K \setminus K_0) < \frac{\epsilon}{2}$ and $f_n \to f$ uniformly on K_0. Notice that the restriction of each f_n to $K' := \bigcap_{n=0}^{\infty} K_n$ is continuous and that $\mu(K \setminus K') < \epsilon$. Since $f_n \to f$ uniformly on K', $f|_{K'}$ is continuous. □

Even though we want to work with continuous compactly supported functions whenever possible, the fundamental tool of measure theory is the simple function.

We will have to do some work to fully integrate continuous functions and measurable functions. It should be noted that our definition of simple function is a bit restrictive — we insist that they be measurable and take nonzero values only on sets of finite measure.

Definition B.18. A measurable function $s : X \to B$ is called *simple* if it takes only finitely many values $b_1, \dots, b_n$ and $\mu(\{ x \in X : f(x) = b_i \}) < \infty$ if $b_i \neq 0$.

Lemma B.19. *Every simple function is C-measurable.*

Proof. It suffice to see that $\mathbb{1}_E \otimes b$ is C-measurable for every measurable set E with finite measure and every $b \in B$. Suppose that K is compact. Using regularity and (B.1), there is a compact set K_1 and an open set U such that

$$K_1 \subset K \cap E \subset U \quad \text{and} \quad \mu(U \setminus K_1) < \frac{\epsilon}{3}.$$

Also there is a compact set K_2 such that $K_2 \subset K \setminus U$ with $\mu\big((K \setminus U) \setminus K_2\big) < \epsilon/3$. Let $K' := K_1 \cup K_2$. Then

$$\begin{aligned}\mu(K \setminus K') &\le \mu\big(K \cap E) \setminus K_1\big) + \mu\big((U \setminus E) \cap K\big) + \mu\big((K \setminus U) \setminus K_2\big)\\ &< \epsilon.\end{aligned}$$

Furthermore, $\mathbb{1}_E \otimes b$ is constant on each K_i. Since K_1 and K_2 are disjoint, $\mathbb{1}_E \otimes b|_{K'}$ is continuous as required. □

Proposition B.20. *A B-valued function is C-measurable if and only if it is (strongly) measurable.*

Proof. In view of Lemma B.16 on the preceding page, it suffices to show that strong measurability implies C-measurability. So suppose that f is strongly measurable and that $K \subset X$. There is null set $N \subset K$ and a separable subspace B_0 such that $f(x) \in B_0$ if $x \in K \setminus N$. Let $\{ d_n \}$ be a countable dense subset of B_0, and define

$$B_n(d_m) := \{ b \in B : \|b - d_m\| < \frac{1}{2n} \}.$$

By assumption each

$$X_{n,m} := f^{-1}\big(B_n(d_m)\big)$$

is measurable. For each n,

$$K \setminus N \subset \bigcup_{m=1}^{\infty} X_{n,m}.$$

Let $\{ Y_i \}$ be a measurable partition of $K \setminus N$ (some of which may be empty) such that $Y_i \subset X_{n,i}$. If Y_i is empty, let x_i be any point in K, otherwise choose $x_i \in Y_i$. Then for each k,

$$s_k := \sum_{i=1}^{k} f(x_i) \mathbb{1}_{Y_i}$$

is a simple function. Thus

$$p_n := \sum_{i=1}^{\infty} f(x_i)\mathbb{1}_{Y_i}$$

is the pointwise limit of the s_k and is C-measurable by Lemmas B.19 on the preceding page and B.17 on page 336. Furthermore,

$$\|p_n(x) - f(x)\| \le \frac{1}{n} \quad \text{for all } x \in K \setminus N.$$

Thus $p_n(x) \to f(x)$ for all $x \in K \setminus N$. Thus f is C-measurable by Lemma B.17. □

Corollary B.21. *Suppose that B is a Banach algebra and that f and g are B-valued measurable functions on X. Then the pointwise product $h(x) := f(x)g(x)$ is measurable from X to B.*

Remark B.22. This result is not as "obvious" as it might seem at first blush. The usual proof in the scalar case takes advantage of the fact that open sets in $\mathbf{C} \times \mathbf{C}$ are the *countable* union on open rectangles. This need not be the case in $B \times B$ and the usual proof would have to be modified to make use of the fact that measurable functions are essentially separately valued on compact sets. Another straightforward proof of this corollary can be made using the characterization of measurability to be given in Proposition B.24 on the facing page.

Proof. This follows immediately from Proposition B.20 on the previous page since the product of continuous functions is continuous. □

Lemma B.23. *Suppose that the restriction of $f : X \to B$ to a compact set K is continuous. Then for all $\epsilon > 0$ there is a simple function s vanishing outside K such that for all $x \in K$,*

$$\|s(x)\| \le \|f(x)\| \quad \textit{and} \quad \|s(x) - f(x)\| < \epsilon.$$

Proof. Since K is compact, $f|_K$ is uniformly continuous. Thus there are open sets $U_1, \dots, U_n$ such that $K \subset \bigcup U_i$ and $x, y \in U_i \cap K$ implies $\|f(x) - f(y)\| < \frac{\epsilon}{2}$. Thus there is a measurable partition $E_1, \dots, E_m$ of K by nonempty sets such that each E_i is contained in some U_j. Let $x_i \in E_i$ and define

$$b_i := \begin{cases} 0 & \text{if } \|f(x_i)\| \le \frac{\epsilon}{2}, \text{ and} \\ f(x_i)\Big(1 - \dfrac{\epsilon}{2\|f(x_i)\|}\Big) & \text{if } \|f(x_i)\| > \frac{\epsilon}{2}. \end{cases}$$

Let $s := \sum_{i=1}^{m} b_i \mathbb{1}_{E_i}$. Then if $x \in K$, it belongs to exactly one E_i. Assume $\|f(x_i)\| > \frac{\epsilon}{2}$. Then

$$\begin{aligned} \|f(x) - s(x)\| = \|f(x) - b_i\| &= \Big\| f(x) - f(x_i)\Big(1 - \frac{\epsilon}{2\|f(x_i)\|}\Big)\Big\| \\ &\le \|f(x) - f(x_i)\| + \frac{\epsilon}{2} < \epsilon. \end{aligned}$$

On the other hand,

$$\begin{aligned}\|f(x)\| - \|s(x)\| &= \|f(x)\| - \|f(x_i)\|\Big(1 - \frac{\epsilon}{2\|f(x_i)\|}\Big)\\ &= \|f(x)\| - \|f(x_i)\| + \frac{\epsilon}{2} \geq 0,\end{aligned}$$

since $\big|\|f(x)\| - \|f(x_i)\|\big| \leq \|f(x) - f(x_i)\| < \frac{\epsilon}{2}$.

The case where $\|f(x_i)\| \leq \frac{\epsilon}{2}$ is even easier. □

Now we can give another characterization of a measurable function.

Proposition B.24. *A B-valued function f is measurable if and only if for each compact set $K \subset X$, there is a sequence of simple functions $\{ s_n \}$ such that for almost all $x \in K$*

$$\|s_n(x)\| \leq \|f(x)\| \quad \textit{and} \quad s_n(x) \to f(x).$$

Proof. The "if" direction follows from Lemma B.11 on page 335. For the "only if" direction, we can invoke Lemma B.15 on page 336 (and Proposition B.20 on page 337) to partition a compact set K as $N \cup \bigcup K_n$ where N is null and $f|_{K_n}$ is continuous. Using Lemma B.23 on the preceding page, for each n we can find a sequence of simple functions $\{ t^n_m \}$ supported in K_n such that $\|t^n_m(s)\| \leq \|f(x)\|$ and $\lim_m t^n_m(x) = f(x)$ for all $x \in K_n$. Define $s_m = t^1_m + t^2_m + \cdots + t^m_m$. Then s_m is a simple function, and since $s_m(x) = t^n_m(x)$ if $x \in K_n$ and $m \geq n$, the $\{ s_m \}$ have the required properties. □

Lemma B.25. *If E is a σ-finite subset of X, then there is a σ-compact subset $F \subset E$ such that $\mu(E \setminus F) = 0$.*

Proof. Since the countable union of null sets is null, it clearly suffices to prove the assertion when $\mu(E) < \infty$. But then the assertion follows immediately from (B.1). □

Lemma B.26. *If f is a B-valued measurable function which vanishes off a σ-finite subset, then there are simple functions s_n and a null set N such that $\|s_n(x)\| \leq \|f(x)\|$ and $s_n(x) \to f(x)$ for all $x \in X \setminus N$.*

Proof. Using Lemma B.26 and repeated application of Lemma B.15 on page 336, we can partition X as $Z \cup N_0 \cup \bigcup K_n$ such that $f(x) = 0$ if $x \in Z$, N_0 is null, K_n is compact and $f|_{K_n}$ is continuous. Lemma B.23 on the facing page implies that for each n, there is a sequence $\{ t^n_m \}$ of simple functions and a null set $N_n \subset K_n$ such that $\|t^n_m(x)\| \leq \|f(x)\|$ and $\lim_m t^n_m(x) = f(x)$ for $x \in K_n \setminus N_n$. Now define $s_m = t^1_m + \cdots + t^m_m$ and $N = \bigcup_{n=0}^{\infty} N_n$. Then $\{ s_m \}$ is the required sequence. □

The next result is [54, Theorem II.14.8], and is a vector-valued version of the usual Tietze Extension Theorem. Unlike the scalar case, there is no assertion that the extension preserves the sup-norm of the original function.

Proposition B.27 (Tietze Extension Theorem). *Suppose that $K \subset X$ is compact and that $g \in C(K, B)$. Then there is a $f \in C_c(X, B)$ such that $f|_K = g$.*

Proof. Fix $\epsilon > 0$ and cover K by precompact open sets $V_1, \dots, V_n$ such that $x, y \in V_i \cap K$ implies $\|g(x) - g(y)\| < \epsilon$. We can assume that $V_i \cap K \neq \emptyset$ and choose $x_i \in V \cap K$. By Lemma 1.43 on page 12, there are $\{\varphi_i\}_{i=1}^n \subset C_c^+(X)$ such that $\operatorname{supp} \varphi_i \subset V_i$ and such that

$$\sum_{i=1}^n \varphi_i(x) \begin{cases} = 1 & \text{if } x \in K, \text{ and} \\ \leq 1 & \text{if } x \notin K. \end{cases}$$

Then $f := \sum_{i=1}^n g(x_i)\varphi_i$ satisfies

$$\|f(x) - g(x)\| = \Big\| \sum_{i=1}^n \varphi_i(x)\big(g(x_i) - g(x)\big) \Big\| \leq \epsilon \quad \text{for all } x \in K.$$

Therefore there are $f_i \in C_c(X, B)$ such that $f_i|_K \to g$ uniformly on K. Passing to a subsequence and relabeling, we can assume that for all $n \geq 2$,

$$\|f_n(x) - f_{n-1}(x)\| < \frac{1}{2^n} \quad \text{for all } x \in K.$$

For $n \geq 2$ let $h_n' := f_n - f_{n-1}$ and define

$$h_n(x) := \begin{cases} h_n'(x) & \text{if } \|h_n'(x)\| \leq \frac{1}{2^n}, \text{ and} \\ \dfrac{h_n'(x)}{2^n \|h_n'(x)\|} & \text{otherwise.} \end{cases}$$

Notice that each h_n is continuous on X, vanishes off $U := \bigcup V_i$, and satisfies $\|h_n(x)\| \leq \frac{1}{2^n}$. Thus

$$f_1(x) + \sum_{n=2}^\infty h_n(x)$$

converges absolutely and uniformly to a continuous function f with support in $\overline{U}$. Thus $f \in C_c(X, B)$, and since $h_n(x) = f_n(x) - f_{n-1}(x)$ if $x \in K$, $f(x) = g(x)$ for all $x \in K$. □

Corollary B.28 (Lusin's Theorem). *Let f be a B-valued measurable function and K a compact subset of X. If $\epsilon > 0$ then there is a $g \in C_c(X, B)$ such that*

$$\mu\big(\{x \in K : g(x) \neq f(x)\}\big) < \epsilon.$$

Lemma B.29. *If $f : X \to B$ is measurable, then $x \mapsto \|f(x)\|$ is measurable.*

Proof. This follows easily from the definition of strong measurability and the continuity of $b \mapsto \|b\|$. □

Definition B.30. A measurable function $f : X \to B$ is *integrable* if

$$\|f\|_1 := \int_X \|f(x)\|\,d\mu(x) < \infty.$$

We call $\|\cdot\|_1$ the L^1-norm. The collection of all B-valued integrable functions on X is denoted $\mathcal{L}^1(X,B)$, and the set of equivalence classes of functions in $\mathcal{L}^1(X,B)$ where two functions which agree almost everywhere are identified is denoted by $L^1(X,B)$.

With the scalar-valued case as a guide, it is routine to check that both $\mathcal{L}^1(X,B)$ and $L^1(X,B)$ are vector spaces, and that $\|\cdot\|_1$ is a seminorm on $\mathcal{L}^1(X,B)$ which defines a norm on $L^1(X,B)$. Thus to obtain the expected result that $L^1(X,B)$ is a Banach space, we only need to prove completeness.

Proposition B.31. *Suppose that μ is a Radon measure on a locally compact space X, and that B is a Banach space. Then $L^1(X,B)$ is a Banach space. In particular, if $\{f_n\}$ is a Cauchy sequence in $\mathcal{L}^1(X,B)$, then there is a $f \in \mathcal{L}^1(X,B)$ and a subsequence $\{f_{n_k}\}$ such that $f_{n_k} \to f$ in $L^1(X,B)$ and such that $f_{n_k}(x) \to f(x)$ for almost all $x \in X$.*

Proof. Since a Cauchy sequence with convergent subsequence is convergent, it suffice to prove the last statement. Therefore we can pass to a subsequence, relabel, and assume that

$$\|f_{n+1} - f_n\|_1 \le \frac{1}{2^n} \quad \text{for } n \ge 1.$$

Now define

$$z_n(x) := \sum_{k=1}^{n} \|f_{k+1}(x) - f_k(x)\| \quad \text{and} \quad z(x) := \sum_{k=1}^{\infty} \|f_{k+1}(x) - f_k(x)\|,$$

with z taking values in $[0,\infty]$. Then

$$\|z_n\|_1 = \sum_{k=1}^{n} \|f_{k+1} - f_k\|_1 \le 1.$$

In other words,

$$\|z_n\|_1 = \int_X z_n(x)\,d\mu(x) \le 1 \quad \text{for all } n,$$

and the Monotone Convergence Theorem implies

$$\int_X z(x)\,d\mu(x) \le 1 < \infty.$$

Therefore z is finite almost everywhere, and there is null set $N \subset X$ such that $\sum_{k=1}^{\infty} f_{k+1}(x) - f_k(x)$ is absolutely convergent in B for all $x \in X \setminus N$. Since B is complete, the series converges and there is a $f'(x)$ such that

$f'(x) = \lim_{n\to\infty} \sum_{k=1}^{n} f_{k+1}(x) - f_k(x) = \big(\lim_{n\to\infty} f_{n+1}(x)\big) - f_1(x)$. Thus we can define f on $X \setminus N$ by

$$f(x) := \lim_n f_n(x) = f'(x) + f_1(x).$$

We can define f to be identically 0 on N, and then f is measurable by Lemma B.11 on page 335.

We still have to see that $f \in \mathcal{L}^1(X, B)$ and that $f_n \to f$ in $L^1(X, B)$. Let $\epsilon > 0$. By assumption, we can choose N so that $n, m \geq N$ implies

$$\|f_n - f_m\|_1 < \epsilon.$$

For all $s \in X \setminus N$, we have $\|f(x) - f_m(x)\| = \lim_n \|f_n(x) - f_m(x)\|$. Thus if $m \geq N$, Fatou's Lemma implies

$$\|f - f_m\|_1 \leq \liminf_n \|f_n - f_m\|_1 \leq \epsilon. \tag{B.3}$$

Since

$$\|f(x)\| \leq \big(\|f(x) - f_m(x)\| + \|f_m(x)\|\big),$$

it follows from (B.3) that $f \in \mathcal{L}^1(X, B)$ and that $f_n \to f$ in $L^1(X, B)$. □

Proposition B.32 (Dominated Convergence Theorem). *Let (X, μ) be a Radon Measure and B a Banach space. Suppose that $\{ f_n \}$ is a sequence of measurable B-valued functions on X such that there is nonnegative integrable function $g \in \mathcal{L}^1(X)$ such that $\|f_n(x)\| \leq g(x)$ for almost all x and such that $f_n(x) \to f(x)$ for almost all x. Then $f_n \to f$ in $L^1(X, B)$.*

Proof. For almost all x, $\|f_n(x) - f(x)\| \leq 2g(x)$. Since $\|f_n(x) - f(x)\|$ converges to 0 for almost all x and $2g$ is integrable, the scalar-valued Dominated Convergence Theorem implies

$$\|f_n - f\|_1 = \int_X \|f_n(x) - f(x)\| \, d\mu(x) \to 0 \quad \text{as } n \to \infty.$$ □

Proposition B.33. *Both the collection of simple functions and the set $C_c(X, B)$ are dense in $L^1(X, B)$.*

Proof. If $f \in \mathcal{L}^1(X, B)$, then f vanishes off a σ-finite set. Thus the density of simple functions follows from Lemma B.26 on page 339 and the Dominated Convergence Theorem. To see that $C_c(X, B)$ is dense, it suffices to see that we can approximate $\mathbb{1}_E \otimes b$ by an element of $C_c(X, B)$ provided E has finite measure and $b \in B$. Let $\epsilon > 0$. The regularity of μ implies there is an open set V and a compact set K such that $K \subset E \subset V$ and $\mu(V \setminus K) < \epsilon$. Then Urysohn's Lemma (Lemma 1.41 on page 11) implies there is a $z \in C_c(X)$ such that $0 \leq z(x) \leq 1$ for all $x \in X$, $z(x) = 1$ if $x \in K$, and z vanishes off V. Then $z \otimes b \in C_c(X, B)$ and

$$\|z \otimes b - \mathbb{1}_E \otimes b\|_1 < \epsilon \|b\|.$$

Since ϵ is arbitrary, we're done. □

Recall from Lemma 1.91 on page 32 that there is a linear map $I : C_c(X,B) \to B$ given by

$$I(f) = \int_X f(x)\,d\mu(x), \tag{B.4}$$

and $\|I(f)\| \le \|f\|_1$. Since $C_c(X,B)$ is dense in $L^1(X,B)$ and since $L^1(X,B)$ is complete, we can extend I to all of $L^1(X,B)$ and continue to write I as an integral as in (B.4) even when f is not in $C_c(X,B)$. If $f \in \mathcal{L}^1(X,B)$, then we can find $\{f_i\} \subset C_c(X,B)$ such that $f_i \to f$ in $L^1(X,B)$. Then the definition of I implies that

$$\int_X f_i(x)\,d\mu(x) \to \int_X f(x)\,d\mu(x)$$

in B. Thus if $\varphi \in B^*$,

$$\varphi\Big(\int_X f_i(x)\,d\mu(x)\Big) \to \varphi\Big(\int_X f(x)\,d\mu(x)\Big).$$

Since it is easy to check that $\varphi \circ f_i \to \varphi \circ f$ in $\mathcal{L}^1(X)$, we conclude that

$$\varphi\Big(\int_X f(x)\,d\mu(x)\Big) = \int_X \varphi\big(f(x)\big)\,d\mu(x) \quad \text{for all } f \in \mathcal{L}^1(X,B) \text{ and } \varphi \in B^*.$$

Thus we've essentially proved the following (compare with Lemma 1.91 on page 32).

Proposition B.34. *If μ is a Radon measure on X and if B is a Banach space, then*

$$f \mapsto \int_X f(x)\,d\mu(x)$$

is a linear map satisfying

$$\Big\|\int_X f(x)\,d\mu(x)\Big\| \le \|f\|_1 \quad \textit{for all } f \in \mathcal{L}^1(X,B).$$

The integral is characterized by

$$\varphi\Big(\int_X f(x)\,d\mu(x)\Big) = \int_X \varphi\big(f(x)\big)\,d\mu(x) \quad \textit{for all } \varphi \in B^*. \tag{B.5}$$

More generally, if L is any bounded linear map from B to B_1, then

$$L\Big(\int_X f(x)\,d\mu(x)\Big) = \int_X L\big(f(x)\big)\,d\mu(x).$$

Remark B.35. It follows from (B.5) that if z is a scalar-valued integrable function on X, then

$$\int_X (z \otimes b)(x)\,d\mu(x) = \int_X z(x)\,d\mu(x) b. \tag{B.6}$$

In particular, if $s = \sum_{i=1}^n b_i \mathbb{1}_{E_i}$ is a simple function with each E_i of finite measure, then

$$\int_X s(x)\,d\mu(x) = \sum_{i=1}^n \mu(E_i) b_i. \tag{B.7}$$

Remark B.36. If A is a C^*-algebra, it is now straightforward to extend Lemma 1.92 on page 32 to $L^1(G,A)$.

B.1.1 The Literature

We should pause to see how the treatments and assumptions here intersect with the standard treatments mentioned above. Using (B.1), it follows that $A \in \mathscr{M}$ if and only if for all compact sets K, $A \cap K$ is the union of a σ-compact set and a null set. Since every null set is a subset of a G_δ-null set, any two complete and saturated Radon measures corresponding the the same functional have the same σ-algebra of definition. Since the measures defined in [156, Theorem 2.14] and [71, §III.11] are both complete and saturated, the notion of measurability in the two treatments coincide — even though this might not be evident on the first read through. Since the treatments in [32, 54, 73] are written for general measure spaces, we can only expect agreement when we use Radon measures with the above σ-algebra. To make contact with Bourbaki, we need to notice that there the σ-algebra of measurable sets is specified in an indirect way. To see that it coincides with Rudin's, and hence with ours, there is a bit of work to do. Even to see that the collection of measurable sets $\mathscr{M}_B$ in Bourbaki is a σ-algebra requires looking up [11, VI §5 no. 1 Corollaire to Proposition 3, IV §5 no. 3 Corollaire 4, and IV §5 no. 5 Corollaire 2]. To see that $\mathscr{M}_B = \mathscr{M}$, it suffices to consider $A \subset K$ for K compact by [156, p. 42] and [11, IV §5 no. 1 Proposition 3]. But $A \in \mathscr{M}_B \iff A \in \mathscr{M}$ follows from [11, IV §4 no. 6 Théorème 4] together with the observation that Rudin's and Bourbaki's extensions of μ coincide on integrable sets (which follows from [11, IV §1 no. 1 Definition 1, IV §1 no. 4 Proposition 19, IV §4 no. 5 Definition 2, IV §4 no. 2 Proposition 1]).

Our definition of (strong) measurability is clearly consistent with the usage in Dunford and Schwartz — see [32, Theorem III.6.10] and recall that locally measurable implies measurable for Radon measures — and with Bourbaki after consulting [11, IV §5 no. 5 Proposition 8 and IV §5 no. 5 Théorème 4]. To see that this is the same definition as used in Fell and Doran and Hill and Phillips, we need Lemma B.7 on page 333. In view of [54, Proposition II.5.12] and [73, Theorem I.3.5.3], we see that all four authorities are in agreement.

B.2 Product Measures

Since iterated integrals play an important role in the theory, we should spend a little time on product measures. In this section, μ and ν will be Radon measures on locally compact spaces X and Y, respectively. We get a positive linear functional

$$J : C_c(X \times Y) \to \mathbf{C}$$

given by[2]

$$J(f) := \int_X \int_Y f(x, y)\, d\nu(y)\, d\mu(x) = \int_Y \int_X f(x, y)\, d\mu(x)\, d\nu(y).$$

[2] It is not necessary to appeal to Fubini's Theorem to see the two iterated integrals below are equal. This can be accomplished by direct estimates such as [71, Theorem III.13.2]. Alternatively, any version of Fubini gives us a product measure allowing us to define J.

The Radon measure associated to J is what we'll mean by the product of μ and ν and it will be denoted by $\mu \times \nu$. We will assume the basic results about product measures and iterated integrals usually referred to as Fubini's and Tonelli's Theorems as stated below. Good references for Fubini's Theorem are [154, Theorem 12.4.19] or [71, Theorem III.13.8]. Our version of Tonelli's Theorem is proved in [71, III.13.9], and also follows from [154, 12.4.20] together with the observation that a function vanishing off a σ-finite subset of $X \times Y$ is equal almost everywhere to a function vanishing of the product of two σ-compact sets.

Theorem B.37 (Fubini's Theorem). *Suppose that μ and ν are Radon measures on locally compact spaces X and Y, respectively. If $f \in \mathcal{L}^1(X \times Y)$, then the following statements hold.*

(a) *For almost all x, $y \mapsto f(x,y)$ belongs to $\mathcal{L}^1(Y)$.*

(b) *For almost all y, $x \mapsto f(x,y)$ belongs to $\mathcal{L}^1(X)$.*

(c) *The function*

$$x \mapsto \int_Y f(x,y)\,d\nu(y)$$

is defined almost everywhere and defines a class in $L^1(X)$.

(d) *The function*

$$y \mapsto \int_X f(x,y)\,d\mu(x)$$

is defined almost everywhere and defines a class in $L^1(Y)$.

(e) *The iterated integrals*

$$\int_X \int_Y f(x,y)\,d\nu(y)\,d\mu(x) = \int_Y \int_X f(x,y)\,d\mu(x)\,d\nu(y)$$

are equal, and the common value is

$$\int_{X\times Y} f(x,y)\,d\mu \times \nu(x,y).$$

Remark B.38. When dealing with functions

$$x \mapsto \int_Y f(x,y)\,d\nu(y), \tag{B.8}$$

as in part (c) of Fubini's Theorem above, it can be awkward not to have an everywhere defined function. Hence we'll adopt the usual convention that (B.8) takes the value 0 at x when $y \mapsto f(x,y)$ fails to be integrable. Of course so long as $y \mapsto f(x,y)$ is integrable for almost all x, the class in $L^1(X)$ is unchanged.

In practice there are usually two obstacles to applying Fubini's Theorem to a function f on a product space. First one must verify that f is measurable with respect to the product measure. Second, one has to check that f is integrable. The first of these is the more formidable. Once measurability is established, integrability can usually be handled with Tonelli's Theorem and some care about the support of the function.

Theorem B.39 (Tonelli's Theorem). *Let μ and ν be Radon measures on locally compact spaces X and Y, respectively. Suppose that $f : X \times Y \to [0, \infty]$ is measurable with respect the product measure $\mu \times \nu$, and that f vanishes off a σ-finite set. Then the following statements hold.*

(a) *For almost all x, $y \mapsto f(x, y)$ is measurable.*

(b) *For almost all y, $x \mapsto f(x, y)$ is measurable.*

(c) *The function*

$$x \mapsto \int_Y f(x, y)\, d\nu(y)$$

is defined almost everywhere and defines a measurable $[0, \infty]$-valued function on X.

(d) *The function*

$$y \mapsto \int_X f(x, y)\, d\mu(x)$$

is defined almost everywhere and defines a measurable $[0, \infty]$-valued function on Y.

(e) *The iterated integrals*

$$\int_X \int_Y f(x, y)\, d\nu(y)\, d\mu(x) = \int_Y \int_X f(x, y)\, d\mu(x)\, d\nu(y)$$

are equal, and the common value is

$$\int_{X \times Y} f(x, y)\, d\mu \times \nu(x, y).$$

Having accepted these two results, we'd like to establish their vector-valued counterparts. However the counterpart to Tonelli's Theorem is an immediate consequence of the scalar-valued version.

Lemma B.40. *Suppose that $f : X \times Y \to B$ is a measurable function which vanishes off a σ-finite set. Then the iterated integrals*

$$\int_X \int_Y \|f(x, y)\|\, d\nu(y)\, d\mu(x) = \int_Y \int_X \|f(x, y)\|\, d\mu(x)\, d\nu(y)$$

coincide. The common value is finite if and only if f is integrable.

Proof. By definition, f is integrable if and only if the scalar valued function $(x, y) \mapsto \|f(x, y)\|$ is integrable. Now we can apply Tonelli to the scalar-valued integral. □

The proof of a vector-valued Fubini Theorem based on the scalar-valued case is a bit more subtle. We've used the same approach as in [54, §II.16] and take advantage of the density of $C_c(X, B)$ in $L^1(X, B)$ (Proposition B.33 on page 342).

Theorem B.41 (Fubini's Theorem). *Suppose that μ and ν are Radon measures on locally compact spaces X and Y, respectively. If $f \in \mathcal{L}^1(X \times Y, B)$ for a Banach space B, then the following statements hold.*

(a) *For almost all* x*,* $y \mapsto f(x,y)$ *belongs to* $\mathcal{L}^1(Y,B)$*.*
(b) *For almost all* y*,* $x \mapsto f(x,y)$ *belongs to* $\mathcal{L}^1(X,B)$*.*
(c) *The function*

$$x \mapsto \int_Y f(x,y)\,d\nu(y)$$

is defined almost everywhere and defines a class in $L^1(X,B)$*.*
(d) *The function*

$$y \mapsto \int_X f(x,y)\,d\mu(x)$$

is defined almost everywhere and defines a class in $L^1(Y,B)$*.*
(e) *The iterated integrals*

$$\int_X \int_Y f(x,y)\,d\nu(y)\,d\mu(x) = \int_Y \int_X f(x,y)\,d\mu(x)\,d\nu(y)$$

are equal, and the common value is

$$\int_{X\times Y} f(x,y)\,d\mu\times\nu(x,y).$$

Proof. Since $C_c(X\times Y,B)$ is dense in $L^1(X\times Y,B)$, Lemma B.31 on page 341 implies there is a sequence $\{\, f_i \,\} \subset C_c(X\times Y,B)$ such that $f_i \to f$ in $L^1(X\times Y,B)$ and such that there is a $\mu\times\nu$-null set N such that $f_i(x,y) \to f(x,y)$ for all $(x,y) \notin N$. The scalar-valued Fubini Theorem applied to $\mathbb{1}_N$ implies that there is a μ-null set N_X such that

$$N^x := \{\, y : (x,y) \in N \,\} \tag{B.9}$$

is a ν-null set for all $x \in X \smallsetminus N_X$. Therefore if $x \in X \smallsetminus N_X$ is fixed, then $f_i(x,y) \to f(x,y)$ for all $y \in Y \smallsetminus N^x$. Thus if $x \in X \smallsetminus N_X$ then $y \mapsto f(x,y)$ is the almost everywhere limit of continuous functions, and is therefore measurable. On the other hand, $(x,y) \mapsto \|f(x,y)\|$ is in $\mathcal{L}^1(X\times Y)$, the the scalar-valued Fubini Theorem implies $y \mapsto \|f(x,y)\|$ is integrable for almost all x. Since, by definition, $y \mapsto f(x,y)$ is integrable whenever it measurable and $y \mapsto \|f(x,y)\|$ is integrable, part (a) follows. Part (b) is proved similarly.

To prove (c), we'll apply the scalar-valued Fubini Theorem to the functions

$$\varphi_i(x,y) := \|f_i(x,y) - f(x,y)\| \quad\text{and}\quad \psi_i(x) := \int_Y \varphi_i(x,y)\,d\nu(y).$$

By our choice of $\{\, f_i \,\}$, $\varphi_i \to 0$ in $L^1(X\times Y)$. Since

$$\begin{aligned}\int_X |\psi_i(x)|\,d\mu(x) &= \int_X \psi_i(x)\,d\mu(x)\\ &= \int_X\int_Y \varphi_i(x,y)\,d\nu(y)\,d\mu(x)\\ &= \int_{X\times Y} \varphi_i(x,y)\,d\mu\times\nu(x,y)\\ &= \int_{X\times Y} |\varphi_i(x,y)|\,d\mu\times\nu(x,y),\end{aligned}$$

we also have $\psi_i \to 0$ in $L^1(X)$. Passing to a subsequence and relabeling, we can assume that $\psi_i(x) \to 0$ for almost all x in X. But if we let

$$g_i(x) := \int_Y f_i(x,y)\,d\nu(y) \quad \text{and} \quad g(x) := \int_Y f(x,y)\,d\nu(y),$$

then each $g_i \in C_c(X, B)$ by Lemma 1.102 on page 36. (We're using the convention established in Remark B.38 on page 345 to view g as a function on all of X.) Thus each g_i is certainly measurable, and

$$\begin{aligned}\psi_i(x) &= \int_Y \|f_i(x,y) - f(x,y)\|\,d\nu(y)\\ &\geq \Big\|\int_Y f_i(x,y) - f(x,y)\,d\nu(y)\Big\|\\ &= \|g_i(x) - g(x)\|,\end{aligned}$$

and it follows that $g_i(x) \to g(x)$ for almost all x in X. Thus g is measurable. Since

$$\int_X \|g(x)\|\,d\mu(x) \leq \int_X \int_Y \|f(x,y)\|\,d\nu(y)\,d\mu(x) = \|f\|_1 < \infty,$$

$g \in \mathcal{L}^1(X)$ as claimed. This proves part (c) and part (d) is proved similarly.

To prove (e), it suffices, by symmetry, to see that one of the iterated integrals coincides with the integral with respect to the product measure. However, this follows by applying a state $\varphi \in B^*$ to both sides and applying the scalar-valued version of the theorem. □

B.2.1 The Convolution Product

In this section, we'll replace the Banach space B with a C^*-algebra A, the space X with a locally compact group G and μ will be a left-Haar measure on G. Let (A, G, α) be a dynamical system. To turn $L^1(G, A)$ into a Banach $*$-algebra, we have to extend the convolution product to $L^1(G, A)$. At one level, this is easy. If $f, g \in C_c(G, A)$, then $f*g$ as defined in (2.16) is in $C_c(G, A)$ and $\|f*g\|_1 \leq \|f\|_1\|g\|_1$. Similarly, $\|f^*\|_1 = \|f\|_1$. Since $C_c(G, A)$ is dense in $L^1(G, A)$ by Lemma B.33 on page 342, the product and involution on $C_c(G, A)$ extend to all of $L^1(G, A)$ by continuity. This solution is not very satisfactory as it leaves us with no decent formula for a representative of either $f * g$ or f^* in $\mathcal{L}^1(G, A)$. Of course, we expect these representatives to be given by the same formulas as for $C_c(G, A)$: namely (2.16) and (2.17) on page 48. We'll give the proof for $f * g$ here. Showing that f^* represented by the $\mathcal{L}^1(G, A)$ function $s \mapsto \alpha_s\big(f(s^{-1})\big)^*\Delta(s^{-1})$ is considerably easier.

Proposition B.42. *Suppose (A, G, α) is a dynamical system with $f, g \in \mathcal{L}^1(G, A)$. Then there is a null set $M \subset G$ such that $s \notin M$ implies $r \mapsto f(r)\alpha_r\big(g(r^{-1}s)\big)$ is in $\mathcal{L}^1(G, A)$. Furthermore, if we define*

$$c(s) := \int_G f(r)\alpha_r\big(g(r^{-1}s)\big)\,d\mu(r) \quad \text{if } s \notin M, \tag{B.10}$$

and let $c(s) := 0$ *if* $s \in M$*, then* c *is an element of* $\mathcal{L}^1(G, A)$ *representing the product* $f * g$ *defined above.*

The proof of Proposition B.42 on the facing page boils down to an application of the vector-valued Fubini Theorem B.41 on page 346. However, as is often the case when trying to apply a Fubini type theorem, the hard part is establishing that the function of two variables in question is measurable and supported on a reasonable set. Here measurability is a bit thorny since we've insisted that our Radon measures, and therefore Haar measure, be defined on σ-algebras strictly larger than the Borel sets and there is no *a priori* reason to suspect that $(r, s) \mapsto g(r^{-1}s)$ is measurable if g is (however, see Lemma B.45 on page 351). One solution is to prove that all elements of $\mathcal{L}^1(G, A)$ are equivalent to a Borel function so that $(r, s) \mapsto g(r^{-1}s)$ is measurable, and then define the convolution only for Borel f and g [71, Corollary 11.11.41]. Here we'll prove the proposition as stated.

First, to deal with support issues, we'll take advantage of the following observation which seems special to locally compact groups (compare with Lemma B.25 on page 339).

Lemma B.43. *If* $U \subset G$ *is open and* $\mu(U) < \infty$*, then* U *is contained in a* σ*-compact set. In particular, if* C *is a* σ*-finite subset of* G*, then* C *is contained in a* σ*-compact set.*

Proof. Since Haar measure is regular, every set of finite measure is contained in an open set of finite measure. Thus, it suffices to prove the first assertion. However, G is the disjoint union

$$\bigcup_{i \in I} G_i$$

of clopen σ-compact sets (Lemma 1.38 on page 10). Thus U is the disjoint union of open sets

$$\bigcup_{i \in I} G_i \cap U,$$

and since an open set is a Haar null set if and only if it is empty, $G_i \cap U = \emptyset$ if and only if $\mu(G_i \cap U) > 0$. However, for each n, there can be at most finitely many i such that $\mu(G_i \cap U) > \frac{1}{n}$. Thus, at most countably many $G_i \cap U$ are nonempty. The assertion follows. □

Proof of Proposition B.42. Let $h : G \times G \to A$ be defined by

$$h(r, s) := f(r)\alpha_r\big(g(r^{-1}s)\big).$$

Since f and g are in $\mathcal{L}^1(G, A)$, they vanish off a σ-finite set. In view of Lemma B.43, there is a σ-compact set C such that f and g vanish off C. Since C^2 is still σ-compact, there is a σ-compact set S such that h vanishes off $S \times S$.

We claim that to prove c as defined in (B.10) is in $\mathcal{L}^1(G, A)$, it suffices to see that h is measurable. For if h is measurable, then the scalar-valued Tonelli's Theorem B.39 on page 346 implies $\|h\| \in \mathcal{L}^1(G \times G)$. Thus $h \in \mathcal{L}^1(G \times G, A)$, and that (B.10) defines an element of $\mathcal{L}^1(G, A)$ now follows our vector-valued Fubini Theorem B.41 on page 346.

To prove measurability, we follow [55, Proposition VIII.5.4]. Choose $\{ f_n \}$ and $\{ g_n \}$ in $C_c(G, A)$ such that $f_n \to f$ and $g_n \to g$ in $L^1(G, A)$. In view of Lemma B.31 on page 341, we can pass to a subsequence, relabel, and assume that off a *Borel* null set N, we have

$$f_n(s) \to f(s) \quad \text{and} \quad g_n(s) \to g(s).$$

Define

$$h_n(r, s) := f_n(r)\alpha_r\big(g_n(r^{-1}s)\big).$$

Then $h_n \in C_c(G \times G, A)$, and by enlarging S if necessary, we can assume that each h_n has support in $S \times S$. Then

$$h_n(r, s) \to h(r, s)$$

for all $(r, s) \in G \times G \setminus D$ where

$$D \subset N \times S \cup \{ (r, s) \in S \times S : r^{-1}s \in N \}.$$

Since the almost everywhere limit of measurable functions is measurable, to see that h is measurable, it will suffice to see that D is a $\mu \times \mu$-null set. However, $N \times S$ is certainly null, so it remains to see that

$$D' := \{ (r, s) \in S \times S : r^{-1}s \in N \}$$

is a null set. However, D' is certainly measurable (Borel in fact) since we were careful to pick N Borel. Thus $\mathbb{1}_{D'}$ is a measurable function vanishing off a σ-compact set. So Tonelli's Theorem B.39 on page 346 implies that

$$\mu \times \mu(D') = \int_G \mu(D'_r) \, d\mu(r),$$

where

$$D'_r := \{ s \in G : (r, s) \in D' \}.$$

Since $D'_r \subset rN$, $\mu(D'_r) = 0$ for all r, and we've shown D' is a $\mu \times \mu$-null set. Hence h is measurable, and therefore belongs to $\mathcal{L}^1(G \times G)$.

We still want to see that the element of $L^1(G, A)$ defined by the function c defined in the proposition is equal to the limit of $f_n * g_n$. For this, it suffices to see that

$$\begin{aligned} \|f_n * g_n - c\|_1 &= \int_G \|f_n * g_n(s) - c(s)\| \, d\mu(s) \\ &\le \int_G \int_G \|h_n(r, s) - h(r, s)\| \, d\mu(r) \, d\mu(s) \\ &= \|h_n - h\|_1 \end{aligned}$$

tends to zero as $n \to \infty$. But this follows from

$$\begin{aligned}\|h_n - h\|_1 &= \int_G \int_G \|f_n(r)\alpha_r\big(g_n(r^{-1}s)\big) - f(r)\alpha_r\big(g(r^{-1}s)\big)\| \, d\mu(r) \, d\mu(s) \\ &\le \int_G \int_G \|f_n(r) - f(r)\| \|g_n(r^{-1}s)\| \, d\mu(r) \, d\mu(s) \\ &\qquad + \int_G \int_G \|f(r)\| \|g_n(r^{-1}s) - g(r^{-1}s)\| \, d\mu(r) \, d\mu(s) \\ &= \|f_n - f\|_1 \|g_n\|_1 + \|f\|_1 \|g_n - g\|_1,\end{aligned}$$

where we have invoked Tonelli's Theorem B.39 on page 346 in the first line and the last. □

Remark B.44. One advantage of the approach to defining the convolution product via $C_c(G, A)$ is that associativity in $L^1(G, A)$ follows from associativity of convolution in $C_c(G, A)$. However, it is still likely that it will be necessary to deal with measurability issues such as the following. Suppose that that if f, g and h are in $\mathcal{L}^1(G, A)$, then we certainly expect that

$$(r, t, s) \mapsto f(r)\alpha_r\big(g(r^{-1}t)\big)\alpha_t\big(h(t^{-1}s)\big) \tag{B.11}$$

is a measurable function on $G \times G \times G$. To prove this, the following lemma suffices.

Lemma B.45. *Let (A, G, α) be a dynamical system. Suppose that $f \in \mathcal{L}^1(G, A)$ and $h(r, s) := \alpha_r\big(f(r^{-1}s)\big)$. Then $h : G \times G \to A$ is measurable.*

Proof. Since Haar measure is saturated, it suffices to see that $h|_C$ is measurable for each compact set $C \subset G \times G$. Therefore it will suffice to show that $h|_{K\times K}$ is measurable for each compact set $K \subset G$. To do this, we'll produce measurable functions h_n such that $h_n \to h$ almost everywhere on $K \times K$. This will suffice in view of Lemma B.11 on page 335.

Since $f \in \mathcal{L}^1(G, A)$, there are $f_n \in C_c(G, A)$ such that $f_n \to f$ in $L^1(G, A)$. Passing to a subsequence and relabeling, we can assume that there is a *Borel* null set N such that $f_n(s) \to f(s)$ for all $s \notin N$. Since f_n is continuous, $h_n(r, s) := \alpha_r\big(f_n(r^{-1}s)\big)$ certainly defines a measurable function (continuous in fact), and

$$h_n(r, s) \to h(r, s)$$

for all $(r, s) \in K \times K \setminus D$, where

$$D = \{\, (r, s) \in K \times K : r^{-1}s \in N \,\}.$$

Since N is Borel, D is a measurable subset of $K \times K$, and Tonelli's Theorem B.39 on page 346 implies that

$$\mu \times \mu(D) = \int_G \mu(D_r) \, d\mu(r),$$

where $D_r := \{\, s : (r, s) \in D \,\}$. Since $D_r \subset rN$, it follows that D is a null set. This completes the proof. □

Remark B.46. Notice that once we know that (B.11) is measurable, it follows from Tonelli's Theorem that (B.11) is in $\mathcal{L}^1(G \times G \times G, A)$. Then Fubini's Theorem implies for almost all $s \in G$,

$$(r, t) \mapsto f(r)\alpha_r\big(g(r^{-1}t)\big)\alpha_t\big(h(t^{-1}s)\big) \tag{B.12}$$

is in $\mathcal{L}^1(G \times G, A)$. It is interesting to note that the integrability of (B.12) for any given s seems hard to prove directly when considered as a function of r and t alone.

Appendix C

Bundles of C^*-algebras

It is often repeated that a C^*-algebra should be thought of as a noncommutative analogue of the space of continuous functions vanishing at infinity on a locally compact Hausdorff space X. This analogy can be very powerful and underlies much recent progress subsumed by the terms noncommutative algebraic topology and noncommutative geometry. A very modest step in this direction is to try to view a given C^*-algebra A as a set of sections of some sort of bundle. For example, $C_0(X)$ is the family of sections of the trivial bundle over X, and $C_0(X, M_n)$ is the family of sections of $X \times M_n$ viewed as a bundle over X. The natural candidate for the base space X is the primitive ideal space $\operatorname{Prim} A$ of A. However, from the point of view of bundle theory, $\operatorname{Prim} A$ is rarely sufficiently well behaved as a topological space to be of much use. Instead, the usual thing to do is to find a locally compact *Hausdorff* space X — which will turn out to be a continuous image of $\operatorname{Prim} A$ — over which A fibres in a nice way. Such algebras now go by the name of $C_0(X)$-algebras, and were introduced by Kasparov in [87]. We prove here that they all can be realized as section algebras of an *upper semicontinuous C^*-bundle over X*. The adjective "upper semicontinuous" is used to distinguish these bundles from the more classical C^*-bundles which have been studied extensively. C^*-bundles were originally studied by Fell under the name "continuous fields of C^*-algebras", and there is an extensive literature starting with [34, 50, 53]. A up to date survey can be found in [54, §II.13]. Upper semicontinuous C^*-bundles were introduced by Hofmann and further references and a discussion of both sorts of bundles can be found in [35]. Lee's Theorem [99, Theorem 4], which characterizes which C^*-algebras can be realized as section algebras of C^*-bundles is part of the main result in this chapter where it is also proved that a C^*-algebra is a $C_0(X)$-algebra if and only if it is isomorphic to the section algebra of a upper semicontinuous C^*-bundle over X (Theorem C.26 on page 367). Much of the material developed here on $C_0(X)$-algebras, and a good deal more, can be found in [9, 43, 113]. The correspondence between $C_0(X)$-algebras and bundles was explored in [113] but without explicitly exhibiting a total space.

It is interesting to note that even though there are $C_0(X)$-algebras which are not the section algebras of C^*-bundles over X, the technology used here is nearly identical to that developed by Fell for C^*-bundles.

C.1 $C_0(X)$-algebras

Definition C.1. Suppose that A is a C^*-algebra and that X is a locally compact Hausdorff space. Then A is a *$C_0(X)$-algebra* if there is a homomorphism Φ_A from $C_0(X)$ into the center $ZM(A)$ of the multiplier algebra $M(A)$ which is nondegenerate in that the ideal

$$\Phi_A\big(C_0(X)\big) \cdot A = \operatorname{span}\{\, \Phi_A(f)a : f \in C_0(X) \text{ and } a \in A \,\} \tag{C.1}$$

is dense in A.

One reason to study $C_0(X)$-algebras is that they are fibred over X in a natural way. Note that if J is an ideal in $C_0(X)$ (or even a subalgebra), then the closure of $\Phi_A(J) \cdot A$ is an ideal in A. If J_x is the ideal of functions vanishing at $x \in X$, then we denote the ideal $\overline{\Phi_A(J_x) \cdot A}$ by I_x and think of the quotient $A(x) := A/I_x$ as the *fibre of A over x*. If $a \in A$, then we write $a(x)$ for the image of a in $A(x)$, and we think of a as a function from X to $\coprod_{x \in X} A(x)$.

Example C.2. If D is any C^*-algebra, then $A = C_0(X, D)$ is a $C_0(X)$-algebra in a natural way:

$$\Phi_A(f)(a)(x) := f(x)a(x) \quad \text{for } f \in C_0(X) \text{ and } a \in A.$$

In this case, each fibre $A(x)$ is easily identified with D, and the identification of elements of A with functions on X is the obvious one.

Example C.3. A degenerate example is the following. Let A be any C^*-algebra and x_0 any point in X. Then A becomes a $C_0(X)$-algebra via

$$\Phi_A(f)(a) = f(x_0)a.$$

Here $A(x)$ is the zero C^*-algebra unless $x = x_0$ in which case $A(x_0) = A$.

To avoid pathological examples such as Example C.3, one could insist that Φ_A be injective. However, the general theory goes through smoothly enough without this assumption. Of course, in most examples of interest, it will turn out that Φ_A will be injective.

Example C.4. Suppose that X and Y are locally compact spaces and that $\sigma : Y \to X$ is continuous. Then $C_0(Y)$ becomes a $C_0(X)$-algebra with respect to the map $\Phi_{C_0(Y)}$ defined by

$$\Phi_{C_0(Y)}(f)(g)(y) := f\big(\sigma(y)\big)g(y).$$

In this example, the fibres $C_0(Y)(x)$ are isomorphic to $C_0\big(\sigma^{-1}(x)\big)$. If $f \in C_0(Y)$, then $f(x)$ is just the restriction of f to $\sigma^{-1}(x)$.

Proof. It is clear that $\Phi_{C_0(Y)}(f) \in ZM\big(C_0(Y)\big) = M\big(C_0(Y)\big) \cong C^b(Y)$. The only issue is nondegeneracy. But (C.1) is an ideal I in $C_0(Y)$ without any common zeros. It then follows from the Stone-Weierstrass Theorem ([127, Corollary 4.3.5]) that I is dense in $C_0(Y)$. □

In fact, Example C.4 on the preceding page is very instructive and is not nearly as specialized as it might seem at first. To explain this comment, recall that the Dauns-Hofmann Theorem ([139, Theorem A.34]) allows us to identify $ZM(A)$ with $C^b(\operatorname{Prim} A)$. More precisely, if $a(P)$ denotes the image of $a \in A$ in the primitive quotient A/P, then there is an isomorphism Ψ of $C^b(\operatorname{Prim} A)$ onto $ZM(A)$ such that

$$\big(\Psi(f)(a)\big)(P) = f(P)a(P) \quad \text{for all } f \in C^b(\operatorname{Prim} A),\ a \in A \text{ and } P \in \operatorname{Prim} A.$$

Alternatively, we can also characterize Ψ in terms of irreducible representations. If $\pi \in \hat{A}$, then

$$\pi\big(\Psi(f)a\big) = f(\ker \pi)\pi(a) \quad \text{for all } f \in C^b(\operatorname{Prim} A) \text{ and } a \in A.$$

It is common practice to suppress the Ψ and write $f \cdot a$ in place of $\Psi(f)(a)$. Thus if $\sigma : \operatorname{Prim} A \to X$ is any continuous map, we get a homomorphism $\Phi_A : C_0(X) \to C^b(\operatorname{Prim} A) \cong ZM(A)$ by defining

$$\Phi_A(f) = f \circ \sigma.$$

To see that Φ_A is nondegenerate, recall that (C.1) is an ideal I. If $\pi \in \hat{A}$, then

$$\pi\big(\Phi_A(f)(a)\big) = f\big(\sigma(\ker \pi)\big)\pi(a),$$

and it follows that $\pi(I) \neq \{0\}$. Since π is arbitrary, I must be dense in A. Thus, we have proved the first statement in the following proposition.

Proposition C.5. *Suppose that A is a C^*-algebra and that X is a locally compact space. If there is a continuous map $\sigma_A : \operatorname{Prim} A \to X$, then A is a $C_0(X)$-algebra with*

$$\Phi_A(f) = f \circ \sigma_A \quad \textit{for all } f \in C_0(X). \tag{C.2}$$

Conversely, if $\Phi_A : C_0(X) \to C^b(\operatorname{Prim} A) \cong ZM(A)$ is a $C_0(X)$-algebra, then there is a continuous map $\sigma_A : \operatorname{Prim} A \to X$ such that (C.2) holds.

In particular, every irreducible representation of A is lifted from a fibre $A(x)$ for some $x \in X$. More precisely, if $\pi \in \hat{A}$, then the ideal $I_{\sigma_A(\ker \pi)}$ is contained in $\ker \pi$, and π is lifted from an irreducible representation of $A\big(\sigma(\ker \pi)\big)$. Thus we can identify $\hat{A}$ with the disjoint union $\coprod_{x \in X} A(x)^\wedge$.

As we shall see in the proof of this result, it is often easier to work with the spectrum $\hat{A}$ of A rather than the primitive ideal space $\operatorname{Prim} A$. Because the topology on $\hat{A}$ is simply pulled back from that on $\operatorname{Prim} A$, any continuous map $f : \operatorname{Prim} A \to Y$ has a continuous lift $\bar{f}$ to $\hat{A}$. If points are closed in Y, then every continuous map from $\hat{A}$ to Y is a lift.

Lemma C.6. *Suppose that Y is a topological space in which points are closed and suppose that $\bar{f} : \hat{A} \to Y$ is continuous. Then $\bar{f}$ factors through $\operatorname{Prim} A$; that is, there is continuous function $f : \operatorname{Prim} A \to Y$ such that $f(\ker \pi) = \bar{f}(\pi)$ for all $\pi \in \hat{A}$.*

Proof. It is easy to see that if such an f exists, then it is continuous. Thus it suffices to see that if $\ker \pi = \ker \rho$, then we must have $\bar{f}(\pi) = \bar{f}(\rho)$. Since points in Y are closed, $\bar{f}^{-1}(\bar{f}(\pi))$ is closed and therefore contains $\overline{\{\pi\}}$. If $\ker \rho = \ker \pi$, then $\rho \in \overline{\{\pi\}}$. □

Lemma C.7. *Suppose that* $\pi_i \to \pi$ *in* $\hat{A}$ *and that* $\pi_i(a) = 0$ *for all* i. *Then* $\pi(a) = 0$.

Proof. Let $J = \bigcap \ker \pi_i$. Then $a \in J$, and $\{\rho \in \hat{A} : \ker \rho \subset J\}$ is a closed subset of $\hat{A}$ containing each π_i. Thus it must also contain π and $\ker \pi \supset J$. This implies $\pi(a) = 0$ as required. □

As with the Dauns-Hofmann Theorem, it is common practice to suppress mention of the homomorphism Φ_A and simply write $f \cdot a$ in place of $\Phi_A(f)(a)$. Then A becomes a nondegenerate $C_0(X)$-bimodule satisfying

$$f \cdot a = a \cdot f \quad \text{and} \quad (f \cdot a)^* = a^* \cdot \bar{f}. \tag{C.3}$$

In particular, it is a consequence of the Cohen Factorization Theorem that every element in A is of the form $f \cdot a$ [139, Proposition 2.33].

Proof of Proposition C.5. The first assertion was proved above. Assume that $\Phi_A : C_0(X) \to C^b(\operatorname{Prim} A) \cong ZM(A)$ is a nondegenerate homomorphism. Let $\pi \in \hat{A}$. Since $\Phi_A(f)$ is in the center of $M(A)$, $\bar{\pi}(\Phi_A(f))$ commutes with every operator in $\pi(A)$. Since π is irreducible, $\bar{\pi}(\Phi_A(f))$ is a scalar multiple $\omega(f)$ of the identity, and $\omega : C_0(X) \to \mathbf{C}$ is a complex homomorphism. Since Φ_A is nondegenerate, ω is nonzero and is given by $\omega(f) = f(\bar{\sigma}(\pi))$ for some point $\bar{\sigma}(\pi) \in X$ (cf. [127, Example 4.2.5]). In view of Lemma C.6 on the preceding page, to prove the second assertion it suffices to prove that $\bar{\sigma}$ is continuous.

To that end, suppose $\pi_i \to \pi$, and suppose to the contrary that $\bar{\sigma}(\pi_i) \not\to \bar{\sigma}(\pi)$. Then we can pass to a subnet, relabel, and assume there is an open neighborhood V of $\bar{\sigma}(\pi)$ which is disjoint from each $\bar{\sigma}(\pi_i)$. But then there is a $f \in C_c(X)$ such that $f(\bar{\sigma}(\pi)) = 1$ while $f(\bar{\sigma}(\pi_i)) = 0$ for all i. Then there must be an $a \in A$ such that $\pi(f \cdot a) \neq 0$ while $\pi_i(f \cdot a) = 0$ for all i. Together with Lemma C.7, this is a contradiction.

If $\pi \in \hat{A}$, then the definition of $\bar{\sigma}$ implies that

$$\pi(f \cdot a) = f(\bar{\sigma}(\pi))\pi(a) \quad \text{for all } f \in C_0(X) \text{ and } a \in A.$$

In particular, $I_{\bar{\sigma}(\pi)} := \operatorname{span}\{f \cdot a : f(\bar{\sigma}(\pi)) = 0\}$ is contained in $\ker \pi$ and π is lifted from an irreducible representation of the quotient $A(\bar{\sigma}(\pi))$. The remaining assertions follow easily. □

It will be helpful to keep in mind a straightforward special case of Example C.4 on page 354.

Example C.8. Let Y be the subset of $\mathbf{R}^2$ consisting of the line segments from $(0,0)$ to $(2,1)$ to $(1,2)$ to $(3,3)$. Let $X = [0,3] \subset \mathbf{R}$ and let $p : Y \to X$ be the projection onto the first factor. Then, as in Example C.4 on page 354, $A = C(Y)$ is a $C(X)$-algebra. Note that

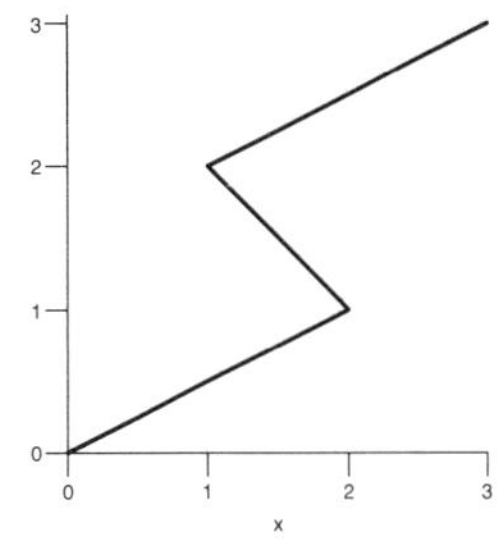

$$A(x) \cong \begin{cases} \mathbf{C} & \text{if } x \in [0,1) \text{ or } x \in (2,3], \\ \mathbf{C}^2 & \text{if } x = 1 \text{ or } x = 2, \\ \mathbf{C}^3 & \text{if } x \in (1,2). \end{cases}$$

Let $f \in A$ be defined by

$$f(x,y) = \begin{cases} 1 & \text{if } y \le \frac{3}{2}, \\ 4 - 2y & \text{if } \frac{3}{2} \le y \le 2, \text{ and} \\ 0 & \text{if } y \ge 2. \end{cases}$$

If we write $f(x,\cdot)$ for the image of f in $C(Y)(x)$, then $x \mapsto \|f(x,\cdot)\|$ is the indicator function of the closed interval $[0,2]$ in $[0,3]$. This is the prototypical example of an *upper semicontinuous function.*

Definition C.9. If X is a topological space, then $f : X \to \mathbf{R}$ is *upper semicontinuous* if

$$\{\, x \in X : f(x) < r \,\}$$

is open for each $r \in \mathbf{R}$.

Proposition C.10. *Suppose that A is a $C_0(X)$-algebra.*

(a) *For each $a \in A$, the function $x \mapsto \|a(x)\|$ is upper semicontinuous, and vanishes at infinity on X in the sense that $\{\, x \in X : \|a(x)\| \ge k \,\}$ is compact for each $k > 0$.*

(b) *The functions $x \mapsto \|a(x)\|$ are continuous for all $a \in A$ if and only if the map $\sigma_A : \operatorname{Prim} A \to X$ is open.*

(c) *For each $a \in A$,*
$$\|a\| = \sup_{x \in X} \|a(x)\|.$$

(d) *If $f \in C_0(X)$ and $a \in A$, then $(f \cdot a)(x) = f(x)a(x)$.*

Proof. Let $\bar{\sigma}_A$ be the lift of σ_A to $\hat{A}$. Since the irreducible representations of $A(x)$ are all of the form $a(x) \mapsto \pi(a)$ for some $\pi \in \hat{A}$ satisfying $\bar{\sigma}_A(\pi) = x$, and since given $a(x) \in A(x)$, there is an irreducible representation of $A(x)$ preserving the norm of $a(x)$ ([139, Theorem A.14]), it follows that for all $k > 0$

$$\{\, x \in X : \|a(x)\| \ge k \,\} = \bar{\sigma}_A\big(\{\, \pi \in \hat{A} : \|\pi(a)\| \ge k \,\}\big). \tag{C.4}$$

However, for any C^*-algebra A and any $k > 0$, $\{\,\pi \in \hat{A} : \|\pi(a)\| \ge k\,\}$ is compact ([139, Lemma A.30(b)]). Since σ_A, and hence $\bar{\sigma}_A$, is continuous, the left-hand side of (C.4) is compact and therefore closed in the Hausdorff space X. It now follows easily that $\{\,x \in X : \|a(x)\| < k\,\}$ is open for all k. This proves (a).

Notice that

$$\{\,x \in X : \|a(x)\| > k\,\} = \bar{\sigma}_A\bigl(\{\,\pi \in \hat{A} : \|\pi(a)\| > k\,\}\bigr). \tag{C.5}$$

Suppose that σ_A is an open map. Then $\bar{\sigma}_A$ is open, and since $\{\,\pi \in \hat{A} : \|\pi(a)\| > k\,\}$ is open in any C^*-algebra, it follows that the left-hand side of (C.5) is open. Together with (a), this shows that $x \mapsto \|a(x)\|$ is continuous when σ_A is open.

For the converse, assume that $x \mapsto \|a(x)\|$ is continuous for all $a \in A$. To see that σ_A is open, it suffices to check that $\bar{\sigma}_A$ is open. For this it is enough to show that $\bar{\sigma}_A(O_J)$ is open, where $O_J = \{\,\rho \in \hat{A} : \rho(J) \ne \{\,0\,\}\,\}$ for some ideal J in A. But

$$O_J = \bigcup_{a\in J}\bigcup_{n\in\mathbf{N}} \{\,\rho \in \hat{A} : \|\rho(a)\| > \frac{1}{n}\,\}.$$

Using (C.5),

$$\bar{\sigma}_J(O_J) = \bigcup_{a\in J}\bigcup_{n\in\mathbf{N}} \{\,x \in X : \|a(x)\| > \frac{1}{n}\,\}.$$

Since the latter is open, this completes the proof of (b).

Part (c) is straightforward. To prove part (d), note that if $a = g \cdot b$, then $f(x)a - f \cdot a = \bigl(f(x)g - fg\bigr) \cdot a \in I_x$. Thus $f(x)a(x) = (f \cdot a)(x)$ in this case. Since every element of A is of this form by [139, Proposition 2.33],[1] this completes the proof. □

Since any ideal I of A is also an ideal in $M(A)$, a multiplier $m \in M(A)$ defines a multiplier m_I of A/I: $m_I(a+I) := m(a) + I$. Thus if A is a $C_0(X)$-algebra, then each multiplier $m \in M(A)$ determines a map $x \mapsto m(x)$ with each $m(x) \in M\bigl(A(x)\bigr)$ and $m(a)(x) = m(x)\bigl(a(x)\bigr)$. It is natural to ask when such a map determines a multiplier.

Lemma C.11 (cf. [99, Lemma 2]). *Suppose that A is a $C_0(X)$-algebra and that for each $x \in X$, m_x is an element in $M\bigl(A(x)\bigr)$. If for each $a \in A$ there are elements $b, c \in A$ such that for all $x \in X$,*

$$b(x) = m_x a(x) \quad \textit{and} \quad c(x) = m_x^* a(x), \tag{C.6}$$

then

(a) *there is a $m \in M(A)$ such that $m(x) = m_x$ for all $x \in X$, and*
(b) $\sup_{x\in X} \|m(x)\| = \|m\| < \infty$.

Remark C.12. Condition (C.6) is equivalent to requiring that for all $a \in A$ there are elements $b, b' \in A$ such that $b(x) = m_x a(x)$ and $b'(x) = a(x)m_x$ for all $x \in X$.

[1] Instead of invoking [139, Proposition 2.33], we could simply observe that elements of the form $g \cdot b$ span a dense subspace.

Proof. We define a map $m : A \to A$ by $m(a)(x) := m_x a(x)$. Then m is clearly adjointable with $m^*(a) = m_x^* a(x)$, and so defines an element $m \in \mathcal{L}(A_A) := M(A)$. This establishes (a).

Let $L := \sup_x \|m(x)\|$. Then Proposition C.10 implies that

$$\begin{aligned}\|m(a)\| &= \sup_x \|m(a)(x)\| \\ &\le \sup_x \|m(x)\|\|a(x)\| \\ &\le \sup_x \|m(x)\|\|a\|.\end{aligned}$$

Thus $\|m\| \le L$. Fix $\epsilon > 0$ and $x \in X$. We can find $b \in A(x)$ of norm one such that $\|m(x)b(x)\| \ge \|m(x)\| - \epsilon$. Since the norm on $A(x)$ is the quotient norm, there is an $a \in A$ with $a(x) = b$ and $\|a\| \le 1+\epsilon$. But then $\|m(a)\| \ge \|m(a)(x)\| \ge \|m(x)\| - \epsilon$. It follows that

$$\|m\| \ge \frac{\|m(x)\| - \epsilon}{1+\epsilon}.$$

Since ϵ is arbitrary, $\|m\| \ge \|m(x)\|$ for all $x \in X$, and $L \le \|m\|$. □

Example C.13. It would be easy to speculate that, if in the statement of Lemma C.11 on the facing page, we also require X to be compact (or more generally, if we require $x \mapsto \|m(x)\|$ to vanish at infinity) and $m_x \in A$ for all x, then the multiplier m should actually belong to A. However, this is not the case. Consider $X = \{0\} \cup \{\frac{1}{n} : n \ge 1\}$ with the relative topology from $\mathbf{R}$. Let $A = C\big(X, \mathcal{K}(\mathcal{H})\big)$ where $\mathcal{H}$ is an infinite-dimensional Hilbert space with orthonormal basis $\{e_n\}$. Then $M(A)$ is $C(X, B(\mathcal{H})_s)$ where $B(\mathcal{H})_s$ denotes $B(\mathcal{H})$ equipped with the $*$-strong topology ([139, Proposition 2.57]). In particular, if p_n is the rank-one projection onto the space spanned by e_n, then $p_n \to 0$ strongly (and, since $p_n = p_n^*$, $*$-strongly), and

$$m(x) = \begin{cases} 0 & \text{if } x = 0, \text{ and} \\ p_n & \text{if } x = \frac{1}{n} \end{cases}$$

defines a multiplier of A which is not in A.

Remark C.14. Example C.13 also shows that $x \mapsto \|m(x)\|$ need not be upper semicontinuous; the multiplier m defined there has the property that $\{x : \|m(x)\| < 1\} = \{0\}$ which is not open. At first glance, this might appear to be a little surprising since $M(A)$ misses being a $C_0(X)$-algebra only because Φ_A need not be nondegenerate as a map into $M(A)$.

Corollary C.15. *Suppose that A is a $C_0(X)$-algebra, that $d_x \in A(x)$ for all $x \in X$ and that for all $a \in A$ there are $b, c \in A$ such that for all $x \in X$*

$$b(x) = d_x a(x) \quad \textit{and} \quad c(x) = d_x^* a(x).$$

If for each $k > 0$, $\{x \in X : \|d_x\| \ge k\}$ has compact closure, then the following are equivalent.

(a) *There is a* $d \in A$ *such that* $d(x) = d_x$ *for all* $x \in X$.
(b) *The functions* $x \mapsto \|d_x - a(x)\|$ *are upper semicontinuous for all* $a \in A$.
(c) *For each* $x \in X$ *and* $\epsilon > 0$ *there is a neighborhood* U *of* x *and an element* $a \in A$ *such that* $\|d_y - a(y)\| < \epsilon$ *for all* $y \in U$.

Proof. That (a) $\Longrightarrow$ (b) follows immediately from Proposition C.10 on page 357.

Given $x \in X$ there is an $a \in A$ such that $a(x) = d_x$. If (b) holds, then $y \mapsto \|d_y - a(y)\|$ is upper semicontinuous, and has the value zero at x. Thus (b) $\Longrightarrow$ (c).

We still need to show that (c) $\Longrightarrow$ (a). However, a partition of unity argument implies that for each n there is an $a_n \in A$ such that

$$\sup_{x \in X} \|d_x - a_n(x)\| < \frac{1}{n}.$$

However, Lemma C.11 on page 358 implies that there is a $d \in M(A)$ such that $d(x) = d_x$, and that $\|d - a_n\| < \frac{1}{n}$. Thus $a_n \to d$ in norm, and $d \in A$. □

C.2 Upper Semicontinuous C^*-bundles

In this section we introduce the notion of an upper semicontinuous C^*-bundle over a locally compact space X. Our definition is a minor modification of what many authors have called a C^*-bundle or a continuous C^*-bundle. It is a bit surprising to see that the theory for upper semicontinuous C^*-bundles is so similar to that for ordinary C^*-bundles. Many of the arguments and proofs here are virtually identical to those for C^*-bundles as given in [54, §II.13]. These observations on upper semicontinuous-bundles go back to work of Hofmann [75–77], and many of the details can be found in [35].

Recall that in general, a *bundle* is simply a surjective map p from one set $\mathcal{A}$ onto another, say X. Then X is the *base space* and $\mathcal{A}$ is called the *total space*. The set $p^{-1}(x)$ is called the fibre over x. Various additional requirements are added depending on what category one is working in. For topological bundles, it is standard to require $\mathcal{A}$ and X to be topological spaces and to require p to be both continuous and open. Often there are requirements for "local triviality", but here the bundles we are looking at will rarely be locally trivial or even have isomorphic fibres. Although a bundle formally consists of the triple $(\mathcal{A}, p, X)$, it is standard to refer to the bundle by either the map p or, if p is understood, by simply the total space $\mathcal{A}$.

Definition C.16. An *upper semicontinuous C^*-bundle over X* is a continuous, open, surjection $p : \mathcal{A} \to X$ together with operations and norms making each fibre $\mathcal{A}_x = p^{-1}(x)$ into a C^*-algebra such that the following axioms hold.

A1: The map $a \mapsto \|a\|$ is upper semicontinuous from $\mathcal{A}$ to $\mathbf{R}$.
A2: The involution $a \mapsto a^*$ is continuous from $\mathcal{A}$ to $\mathcal{A}$.
A3: The maps $(a, b) \mapsto ab$ and $(a, b) \mapsto a+b$ are continuous from $\mathcal{A}^{(2)} := \{ (a, b) \in \mathcal{A} \times \mathcal{A} : p(a) = p(b) \}$ to $\mathcal{A}$.

A4: For each $\lambda \in \mathbf{C}$, the map $a \mapsto \lambda a$ is continuous from $\mathcal{A}$ to $\mathcal{A}$.

A5: If $\{ a_i \}$ is a net in $\mathcal{A}$ such that $\|a_i\| \to 0$ and $p(a_i) \to x$ in X, then $a_i \to 0_x$, where 0_x is the zero element in $\mathcal{A}_x$.

If "upper semicontinuous" is replaced by "continuous" in axiom A1, then $p : \mathcal{A} \to X$ is called a *C^*-bundle over X*.

The axioms in Definition C.16 on the facing page are more generous that one might think at first glance. For example, we have the following strengthening of axiom A4.

Proposition C.17 (cf. [54, Proposition II.13.10]). *Suppose that $p : \mathcal{A} \to X$ is an upper semicontinuous C^*-bundle over X. Then scalar multiplication $(\lambda, a) \mapsto \lambda a$ is continuous from $\mathbf{C} \times \mathcal{A}$ to $\mathcal{A}$.*

For the proof we need another observation.

Lemma C.18. *If $a_i \to 0_x$ in $\mathcal{A}$, then $\|a_i\| \to 0$.*

Proof. By axiom A1, the set $\{ a \in \mathcal{A} : \|a\| < \epsilon \}$ is open for all $\epsilon > 0$. Thus we eventually have $\|a_i\| < \epsilon$ for all ϵ and the result is proved. $\square$

Proof of Proposition C.17. Suppose that $\lambda_i \to \lambda$ and $a_i \to a$. Notice that $p(\lambda_i a_i - \lambda a_i) = p(a_i) \to p(a)$. Since $\|a_i\| \leq \|a_i - a\| + \|a\|$ and $\|a_i - a\| \to 0$ by Lemma C.18, $\{ \|a_i\| \}$ is eventually bounded and $\|\lambda_i a_i - \lambda a_i\| = |\lambda_i - \lambda| \|a_i\| \to 0$. Thus $\lambda_i a_i - \lambda a_i \to 0_{p(a)}$ by axiom A1. But $\lambda a_i \to \lambda a$ by axiom A4, and

$$\lambda_i a_i = \lambda_i a_i - \lambda a_i + \lambda a_i \to 0_{p(a)} + \lambda a = \lambda a. \qquad \square$$

Remark C.19. As we shall see (Example C.27 on page 368), the total space of an upper semicontinuous C^*-bundle $p : \mathcal{A} \to X$ need not be Hausdorff. However, if p is a (continuous) C^*-bundle, then $\mathcal{A}$ is Hausdorff. To see this, suppose $\{ a_i \}$ converges to both b and c. Since X is Hausdorff, $p(b) = p(c)$, and axiom A3 implies that $0_{p(a_i)} = a_i - a_i \to b - c$. If $a \mapsto \|a\|$ is continuous, then $\|b - c\| = 0$ and $b = c$. However, for any upper semicontinuous C^*-bundle $p : \mathcal{A} \to X$, we at least know that the relative topology on each fibre $\mathcal{A}_x$ is just the norm topology. This is proved exactly as in [54, Proposition II.13.11]: suppose that $a_i \to a$ in $\mathcal{A}$ with $p(a_i) = p(a)$ for all i. Then $a_i - a \to 0_{p(a)}$ by axiom A3, and $\|a_i - a\| \to 0$ by Lemma C.18. Conversely, if $\|a_i - a\| \to 0$, then $a_i - a \to 0_{p(a)}$ by axiom A5. Thus $a_i \to a$ in $\mathcal{A}$ by axiom A3.

Although the statement of the next result seems awkward, it is nevertheless very valuable in practice as it allows us to pin down the topology on upper semicontinuous C^*-bundles.

Proposition C.20 (cf. [54, Proposition II.13.12]). *Suppose that $p : \mathcal{A} \to X$ is an upper semicontinuous C^*-bundle over X. Let $\{ a_i \}$ be a net in $\mathcal{A}$ such that $p(a_i) \to p(a)$ for some $a \in \mathcal{A}$. Suppose that for all $\epsilon > 0$ there is net $\{ u_i \}$ (indexed by the same set) in $\mathcal{A}$ and $u \in \mathcal{A}$ such that*

(a) *$u_i \to u$ in $\mathcal{A}$,*

(b) $p(u_i) = p(a_i)$,

(c) $\|a - u\| < \epsilon$, *and*

(d) $\|a_i - u_i\| < \epsilon$ *for large* i.

Then $a_i \to a$.

Proof. Since X is Hausdorff, we must have $p(u) = p(a)$ (and so (c) at least makes sense). Since it suffices to show that every subnet of $\{ a_i \}$ has a subnet converging to a, we can pass to subnets at will. In particular, since p is open, we can pass to a subnet, relabel, and find $c_i \in \mathcal{A}$ such that $c_i \to a$ and $p(c_i) = p(a_i)$. Now fix $\epsilon > 0$ and choose u_i as above. Since addition is continuous, $c_i - u_i \to a - u$ in $\mathcal{A}$. Since $\|a - u\| < \epsilon$ by assumption and since $\{ b \in \mathcal{A} : \|b\| < \epsilon \}$ is open, we eventually have $\|c_i - u_i\| < \epsilon$. The triangle inequality then implies that we eventually have $\|a_i - c_i\| < 2\epsilon$. As ϵ is arbitrary, we've shown that $\|a_i - c_i\| \to 0$. Then axiom A5 implies $a_i - c_i \to 0_{p(a)}$. Therefore

$$a_i = (a_i - c_i) + c_i \to 0_{p(a)} + a = a. \qquad \square$$

Definition C.21. If $p : \mathcal{A} \to X$ is any surjection, then $f : X \to \mathcal{A}$ is called a *section* if $p\big(f(x)\big) = x$ for all $x \in X$. The set of continuous sections of an upper semicontinuous C^*-bundle is denoted $\Gamma(\mathcal{A})$. The collection of continuous sections which vanish at infinity is denoted $\Gamma_0(\mathcal{A})$.[2]

It is not obvious that an upper semicontinuous C^*-bundle has any continuous sections other than the zero section. In general a bundle $p : \mathcal{A} \to X$ is said to have *enough sections* if given any $x \in X$ and any $a \in \mathcal{A}_x$, then there is a continuous section f such that $f(x) = a$. A result of Douady and dal Soglio-Hérault implies that C^*-bundles over locally compact spaces X have enough sections [54, Appendix C]. Hofmann has noted that the same is true for upper semicontinuous C^*-bundles over locally compact spaces in [76, Proposition 3.4], but the details have not been published [75]. In the examples we're interested in — namely bundles arising from $C_0(X)$-algebras — it will be easy to see that there will always be enough sections. Thus, in the sequel, we will assume that all our bundles do have enough sections. The next lemma is an immediate consequence of the axioms and Proposition C.17 on the previous page.

Lemma C.22. *Suppose that* $p : \mathcal{A} \to X$ *is an upper semicontinuous* C^**-bundle over* X. *Then* $\Gamma(\mathcal{A})$ *is a* $*$*-algebra with respect to the natural pointwise operations. Moreover, if* $\varphi : X \to \mathbf{C}$ *is continuous and* $f \in \Gamma(\mathcal{A})$, *then* $\varphi \cdot f \in \Gamma(\mathcal{A})$ *where* $\varphi \cdot f(x) := \varphi(x) f(x)$.

Proposition C.23. *Suppose that* $p : \mathcal{A} \to X$ *is an upper semicontinuous* C^**-bundle over* X. *Then* $\Gamma_0(\mathcal{A})$ *is a* $C_0(X)$*-algebra with respect to the sup-norm and the natural action of* $C_0(X)$ *on sections. For each* $x \in X$, *we have* $\Gamma_0(\mathcal{A})(x) = \mathcal{A}_x$.

[2] If $f \in \Gamma(\mathcal{A})$, then $\{ x : \|f\| \geq \epsilon \}$ is always closed in X. If it is compact for all $\epsilon > 0$, then we say that f vanishes at infinity.

Proof. Since $\|f\| := \sup_x \|f(x)\|$ clearly defines a norm on $\Gamma_0(\mathcal{A})$, and in view of Lemma C.22 on the facing page, in order to show that $\Gamma_0(\mathcal{A})$ is a C^*-algebra, we just need to see that it is complete. Let $\{ f_n \}$ be a Cauchy sequence in $\Gamma_0(\mathcal{A})$. Since each $\mathcal{A}_x$ is a C^*-algebra, there is a (not necessarily continuous) section $f : X \to \mathcal{A}$ such that $f_n(x) \to f(x)$ for all $x \in X$. Standard arguments imply that $\|f_n - f\| \to 0$. It will suffice to see that f is continuous. Towards this end, let $x_i \to x$ in X, and we'll show that $f(x_i) \to f(x)$ using Proposition C.20. Fix $\epsilon > 0$, and choose N so that $\|f_N - f\| < \epsilon$. Since $f_N(x_i) \to f_N(x)$, we can let $a_i = f(x_i)$ and $u_i = f_N(x_i)$, and then Proposition C.20 on page 361 implies that $f(x_i) \to f(x)$.

We still have to see that $\Phi_A(\varphi)$ is in $ZM\big(\Gamma_0(\mathcal{A})\big)$ where $\Phi_A(\varphi)(f) = \varphi \cdot f$. But $\Phi_A(\varphi)$ is clearly an adjointable operator with adjoint $\Phi_A(\bar{\varphi})$. Since $f(\varphi{\cdot}g) = \varphi{\cdot}(fg)$, it follows that $\Phi_A(\varphi) \in ZM(A)$, and nondegeneracy is an easy exercise. Thus we have established that $\Gamma_0(\mathcal{A})$ is a $C_0(X)$-algebra.

It is not hard to check that $I_x = \{ f \in \Gamma_0(\mathcal{A}) : f(x) = 0 \}$. Then $f + I_x \mapsto f(x)$ is clearly an isomorphism of $\Gamma_0(\mathcal{A})(x)$ onto $\mathcal{A}_x$. (Note that after identifying $\Gamma_0(\mathcal{A})(x)$ with $\mathcal{A}_x$, we can view $f(x)$ as either the value of f at x or the image of f in $\Gamma_0(\mathcal{A})(x)$.) □

Proposition C.24. *Suppose that $p : \mathcal{A} \to X$ is an upper semicontinuous C^*-bundle over X and that Γ is a subspace of $\Gamma_0(\mathcal{A})$ such that*

(a) *$f \in \Gamma$ and $\varphi \in C_0(X)$ implies $\varphi \cdot f \in \Gamma$, and*

(b) *for each $x \in X$, $\{ f(x) : f \in \Gamma \}$ is dense in $\mathcal{A}_x$.*

Then Γ is dense in $\Gamma_0(\mathcal{A})$.

Proof. Fix $f \in \Gamma_0(\mathcal{A})$ and $\epsilon > 0$. We need to find $g \in \Gamma$ such that $\sup_{x \in X} \|f(x) - g(x)\| \le \epsilon$. Let K be the compact set $\{ x \in X : \|f(x)\| \ge \frac{\epsilon}{3} \}$. Given $x \in K$, there is a $g \in \Gamma$ such that $\|f(x) - g(x)\| < \frac{\epsilon}{3}$. Using upper semicontinuity, there is a neighborhood U of x such that

$$\|f(y) - g(y)\| < \frac{\epsilon}{3} \quad \text{if } y \in U.$$

Since K is compact, there is a cover $\{ U_i \}_{i=1}^n$ of K by open sets with compact closure, and $g_i \in \Gamma$ such that

$$\|f(y) - g_i(y)\| < \frac{\epsilon}{3} \quad \text{if } y \in U_i.$$

Using a partition of unity as in Lemma 1.43 on page 12, there are functions $\{ \varphi_i \} \subset C_c^+(X)$ such that $\operatorname{supp} \varphi_i \subset U_i$ and

$$\sum_{i=1}^n \varphi_i(x) \begin{cases} = 1 & \text{if } x \in K, \text{ and} \\ \le 1 & \text{if } x \notin K. \end{cases}$$

By assumption, $\sum_i \varphi_i \cdot g_i \in \Gamma$. On the other hand, if $x \in K$, then

$$\Big\| f(x) - \sum_i \varphi_i(x) g_i(x) \Big\| = \Big\| \sum_i \varphi_i(x)\big(f(x) - g_i(x)\big) \Big\| \\ \leq \sum_i \varphi_i(x) \| f(x) - g_i(x) \| \\ \leq \frac{\epsilon}{3} < \epsilon.$$

But if $x \in U_i \setminus K$, then $\|g_i(x)\| < \frac{2\epsilon}{3}$. Since $\operatorname{supp}\varphi_i \subset U_i$, for any x in the complement of K, $\varphi_i(x)\|g_i(x)\| \leq \frac{2\epsilon}{3}\varphi_i(x)$. Thus if $x \notin K$, we still have

$$\Big\| f(x) - \sum_i \varphi_i(x) g_i(x) \Big\| \leq \|f(x)\| + \sum_i \varphi_i(x) \|g_i(x)\| \\ < \frac{\epsilon}{3} + \frac{2\epsilon}{3} = \epsilon.$$

Therefore $\sup_x \big\| f(x) - \big(\sum_i \varphi_i \cdot g_i\big)(x) \big\| \leq \epsilon$ as required. □

Since it is not at all obvious how to find an upper semicontinuous C^*-bundle, it is significant that the converse of Proposition C.23 on page 362 holds: every $C_0(X)$-algebra is the section algebra of an upper semicontinuous C^*-bundle. It is worth repeating that ideas and proof are virtually unchanged from Fell's proof for C^*-bundles (cf. [54, Theorem II.13.18]).

Theorem C.25 (Fell). *Suppose that $\mathcal{A}$ is a set and $p : \mathcal{A} \to X$ is a surjection onto a locally compact space X such that each $\mathcal{A}_x := p^{-1}(x)$ is a C^*-algebra. Let Γ be a $*$-algebra of sections of $\mathcal{A}$ such that*

(a) *for each $f \in \Gamma$, $x \mapsto \|f(x)\|$ is upper semicontinuous, and*

(b) *for each $x \in X$, $\{\, f(x) : f \in \Gamma \,\}$ is dense in $\mathcal{A}_x$.*

Then there is a unique topology on $\mathcal{A}$ such that $p : \mathcal{A} \to X$ is an upper semicontinuous C^*-bundle over X with $\Gamma \subset \Gamma(\mathcal{A})$.

If we replace "upper semicontinuous" by "continuous" in (a), *then $p : \mathcal{A} \to X$ is a C^*-bundle over X.*

Proof. If τ is any topology on $\mathcal{A}$ such that $p : \mathcal{A} \to X$ is an upper semicontinuous C^*-bundle with Γ consisting of continuous sections satisfying (a) and (b), then Proposition C.20 on page 361 implies that $a_i \to a$ in $(\mathcal{A}, \tau)$ if and only if $p(a_i) \to p(a)$ and for each $\epsilon > 0$ there is an $f \in \Gamma$ such that $\big\| f\big(p(a)\big) - a \big\| < \epsilon$ and we eventually have $\big\| f\big(p(a_i)\big) - a_i \big\| < \epsilon$. Thus, τ, if it exists, is uniquely determined by Γ.

To show that an appropriate topology exists, we'll proceed in three steps. First, we'll nominate a family of sets and prove it constitutes a topology on $\mathcal{A}$. Second, we'll show that $p : \mathcal{A} \to X$ is an upper semicontinuous C^*-bundle with respect to this topology. Thirdly, we'll need to see that each $f \in \Gamma$ is continuous.

Step I: Let τ be the collection of unions of sets of the form

$$W(f, U, \epsilon) = \big\{\, a \in \mathcal{A} : p(a) \in U \text{ and } \big\| a - f\big(p(a)\big) \big\| < \epsilon \,\big\}$$

where $f \in \Gamma$, U is an open set in X and $\epsilon > 0$. To see that τ is a topology, it will suffice to see that it is closed under intersections. For this it will suffice to see that if $a \in W(f, U, \epsilon) \cap W(g, V, \delta)$, then there is a triple (h, Z, σ) such that

$$a \in W(h, Z, \sigma) \subset W(f, U, \epsilon) \cap W(g, V, \delta).$$

Note that $\|a - f(p(a))\| < \epsilon$ and $\|a - g(p(a))\| < \delta$. Thus there is a $\sigma > 0$ so that

$$\|a - f(p(a))\| < \epsilon - 2\sigma \quad \text{and} \quad \|a - g(p(a))\| < \delta - 2\sigma.$$

Choose $h \in \Gamma$ such that $\|a - h(p(a))\| < \sigma$. Since $h - f \in \Gamma$, $x \mapsto \|h(x) - f(x)\|$ is upper semicontinuous, as is $x \mapsto \|h(x) - g(x)\|$. Since

$$\|h(p(a)) - f(p(a))\| \le \|h(p(a)) - a\| + \|a - f(p(a))\| < \epsilon - \sigma \quad \text{and}$$
$$\|h(p(a)) - g(p(a))\| \le \|h(p(a)) - a\| + \|a - g(p(a))\| < \delta - \sigma,$$

it follows that there is a neighborhood $Z \subset U \cap V$ of $p(a)$ such that

$$\|h(y) - f(y)\| < \epsilon - \sigma \quad \text{and} \quad \|h(y) - g(y)\| < \delta - \sigma$$

for all $y \in Z$.

Now by our choice of h, $a \in W(h, Z, \sigma)$ and if $b \in W(h, Z, \sigma)$, then $p(b) \in Z \subset U$ and

$$\begin{aligned}\|b - f(p(b))\| &\le \|b - h(p(b))\| + \|h(p(b)) - f(p(b))\| \\ &< \sigma + (\epsilon - \sigma) = \epsilon.\end{aligned}$$

That is, $W(h, Z, \sigma) \subset W(f, U, \epsilon)$. Similarly, $W(h, Z, \sigma) \subset W(g, V, \delta)$ and we've completed Step I.

Step II: We need to see that $p : (\mathcal{A}, \tau) \to X$ is an upper semicontinuous C^*-bundle as in Definition C.16 on page 360. In particular, we have to check that p is continuous and open. Suppose that $U \subset X$ is open and that $a \in p^{-1}(U)$. Choose $f \in \Gamma$ such that $\|a - f(p(a))\| < 1$. Then $W(f, U, 1)$ is a neighborhood of a contained in $p^{-1}(U)$, and p is continuous. Given $W(g, V, \delta)$ and $x \in V$, we have $g(x) \in W(g, V, \delta)$. Thus $p(W(g, V, \delta)) = V$ and it follows that p is open. So it only remains to verify axioms A1–A5.

Axiom A1: Suppose that $\|a\| < \alpha$. Let $\delta > 0$ be such that $\|a\| < \alpha - 2\delta$, and choose $f \in \Gamma$ such that $\|f(p(a)) - a\| < \delta$. Then $\|f(p(a))\| < \alpha - \delta$ and the upper semicontinuity of $x \mapsto \|f(x)\|$ implies that there is a neighborhood U of $p(a)$ such that $\|f(y)\| < \alpha - \delta$ for all $y \in U$. Therefore if $b \in W(f, U, \delta)$, then

$$\|b\| \le \|b - f(p(b))\| + \|f(p(b))\| < \alpha,$$

and it follows that $\{ a \in \mathcal{A} : \|a\| < \alpha \}$ is open. In other words, axiom A1 is satisfied.

Axiom A2: Since $a \in W(f, U, \epsilon)$ if and only if $a^* \in W(f^*, U, \epsilon)$, it follows that $a \mapsto a^*$ is continuous.

Axiom A3: Let $(a,b) \in \mathcal{A}^{(2)}$ with $ab \in W(f,U,\epsilon)$. For convenience, let x be the common value of $p(a)$ and $p(b)$. Let $\delta > 0$ be such that $\|ab - f(x)\| < \epsilon - 2\delta$. Choose g and h in Γ such that

$$\|g(x) - a\| < \sigma_a := \min\Big\{\frac{\delta}{2(\|b\|+1)}, \frac{1}{2}\Big\} \quad \text{and}$$
$$\|h(x) - b\| < \sigma_b := \min\Big\{\frac{\delta}{2(\|a\|+1)}, \frac{1}{2}\Big\}.$$

Note that $\|g(x)\| < \|a\| + \frac{1}{2}$ and $\|h(x)\| < \|b\| + \frac{1}{2}$ and consequently our choices imply that

$$\|ab - g(x)h(x)\| < \delta. \tag{C.7}$$

Therefore,

$$\|g(x)h(x) - f(x)\| < \epsilon - \delta.$$

Since g, h and $gh - f$ are in Γ, there is a neighborhood $V \subset U$ of x such that for all $y \in V$,

$$\|g(y)\| < \|a\| + \frac{1}{2}, \quad \|h(y)\| < \|b\| + \frac{1}{2} \quad \text{and} \quad \|g(y)h(y) - f(y)\| < \epsilon - \delta.$$

Thus if $c \in W(g,V,\sigma_a)$ and $d \in W(h,V,\sigma_b)$ are such that $p(c) = y = p(d)$, then $\|c\| \le \|c - g(y)\| + \|g(y)\| \le \|a\| + 1$ (similarly, $\|d\| \le \|b\| + 1$), and our choices imply that

$$\begin{aligned}\|cd - f(y)\| &\le \|cd - g(y)h(y)\| + \|g(y)h(y) - f(y)\| \\ &< \delta + (\epsilon - \delta) = \epsilon.\end{aligned}$$

Therefore $W(g,V,\sigma_a)W(h,V,\sigma_b) \subset W(f,U,\epsilon)$, and we proved that multiplication is continuous.[3] The proof that addition is continuous is similar, but easier.

Axiom A4: We just have to notice that if $\lambda \neq 0$, then $a \in W(\frac{1}{\lambda}f, U, \frac{\epsilon}{|\lambda|})$ implies that $\lambda a \in W(f,U,\epsilon)$.

Axiom A5: Suppose that $\|a_i\| \to 0$ and that $p(a_i) \to x$. We need to check that $a_i \to 0_x$. Let $0_x \in W(f,U,\epsilon)$ Then $x \in U$ and there is a $\delta > 0$ such that $\|f(x)\| < \epsilon - \delta$. Eventually, we have $p(a_i) \in U$, $\|f(p(a_i))\| < \epsilon - \delta$ and $\|a_i\| < \delta$. But then

$$\|a_i - f(p(a_i))\| \le \|a_i\| + \|f(p(a_i))\| < \epsilon.$$

Thus we eventually have $a_i \in W(f,U,\epsilon)$, and since the $W(f,U,\epsilon)$'s form a basis for τ, $a_i \to 0_x$ as required. We've shown $p : \mathcal{A} \to X$ is an upper semicontinuous C^*-bundle over X.

Step III: We need to see that each $f \in \Gamma$ is continuous from X to $\mathcal{A}$. So suppose that $x_i \to x$ in X, and suppose, to the contrary, that $f(x_i) \not\to f(x)$. Passing to a subnet and relabeling, we may as well assume that there is a neighborhood

[3] If C and D are subsets of a bundle $\mathcal{A}$, then CD is defined to be $\{\, cd : c \in C,\ d \in D \text{ and } p(c) = p(d)\,\}$.

$W(g,U,\epsilon)$ of $f(x)$ which is disjoint from $\{\, f(x_i) \,\}$. Since $p\big(f(x_i)\big) = x_i$, we must eventually have

$$\|f(x_i) - g(x_i)\| \geq \epsilon.$$

Since $f - g \in \Gamma$, $y \mapsto \|f(y) - g(y)\|$ is upper semicontinuous, and we must have

$$\|f(x) - g(x)\| \geq \epsilon.$$

Therefore $f(x) \notin W(g,U,\epsilon)$ which contradicts our choice of $W(g,U,\epsilon)$, and completes proof in the upper semicontinuous case.

To prove the final assertion, we just have to show that if $x \mapsto \|f(x)\|$ is continuous for all $f \in \Gamma$, then $a \mapsto \|a\|$ is continuous from $\mathcal{A}$ to $\mathbf{R}$. However, since we already have shown $a \mapsto \|a\|$ is upper semicontinuous, it will suffice to show that $\{\, a \in A : \|a\| > \alpha \,\}$ is open. So let a be such that $\|a\| > \alpha$. Let δ be such that $\|a\| > \alpha + 2\delta$. Choose $f \in \Gamma$ such that $\|f(p(a)) - a\| < \delta$. Then $\|f\big(p(a)\big)\| > \alpha + \delta$. Continuity implies there is a neighborhood U of $p(a)$ such that $y \in U$ implies that $\|f(y)\| > \alpha + \delta$. Now if $b \in W(f,U,\delta)$, we have

$$\|b\| \geq \big\|f\big(p(b)\big)\big\| - \big\|f\big(p(b)\big) - b\big\| > (\alpha + \delta) - \delta = \alpha,$$

and it follows that $\{\, a \in \mathcal{A} : \|a\| > \alpha \,\}$ is open. □

We can summarize much of this chapter in the following theorem. The final assertion is often called Lee's Theorem (cf. [99, Theorem 4]).

Theorem C.26. *Suppose that A is a C^*-algebra. Then the following statements are equivalent.*

(a) *A is a $C_0(X)$-algebra.*
(b) *There is a continuous map $\sigma_A : \operatorname{Prim} A \to X$.*
(c) *There is an upper semicontinuous C^*-bundle $p : \mathcal{A} \to X$ over X and a $C_0(X)$-linear isomorphism of A onto $\Gamma_0(\mathcal{A})$.*

Moreover, if A is a $C_0(X)$-algebra, then $\sigma_A : \operatorname{Prim} A \to X$ is open if and only if $p : \mathcal{A} \to X$ is actually a C^*-bundle over X.

Proof. Proposition C.5 on page 355 implies (a) and (b) are equivalent. Proposition C.23 on page 362 shows that (c) implies (a). Thus it will suffice to prove that (a) $\Longrightarrow$ (c). Let $\mathcal{A} = \coprod_{x \in X} A(x)$, let $p : \mathcal{A} \to X$ be the canonical map and let $\Gamma = \{\, x \mapsto a(x) : a \in A \,\}$. Now Fell's Theorem C.25 on page 364 implies that $\mathcal{A}$ has a topology making $p : \mathcal{A} \to X$ an upper semicontinuous C^*-bundle with $\Gamma \subset \Gamma(\mathcal{A})$. In view of Proposition C.10 on page 357, $\Gamma \subset \Gamma_0(\mathcal{A})$, and Γ is dense in $\Gamma_0(\mathcal{A})$ by Proposition C.24 on page 363. Of course, the map from A to Γ is a $C_0(X)$-linear isomorphism in view of parts (c) and (d) of Proposition C.10. Therefore, Γ is closed and equal to $\Gamma_0(\mathcal{A})$.

Now suppose A is a $C_0(X)$-algebra. If $\sigma : \operatorname{Prim} A \to X$ is continuous and open, then $x \mapsto \|a(x)\|$ is continuous for all $a \in A$ all by Proposition C.10 on page 357, and Theorem C.25 on page 364 implies the existence of $\mathcal{A}$ as required.

The converse follows from Proposition C.23 on page 362 and Proposition C.10 on page 357. □

It is a worthwhile, albeit tedious, exercise to work out the topology on the total space corresponding to Example C.8 on page 357. In particular, we note the following.

Example C.27. Let $\mathcal{A}$ be the total space for the upper semicontinuous C^*-bundle corresponding over X so that $C(Y) \cong \Gamma_0(\mathcal{A})$ in Example C.8 on page 357. Consider the section $f(x, \cdot)$ defined there. We have $f(x, \cdot) = 0_x$ if $x > 2$. But as $x \searrow 2$, we must have $0_x = f(x, \cdot) \to f(2, \cdot) \neq 0$. Since 0_2 must also be a limit point, $\mathcal{A}$ is not Hausdorff.

Appendix D

Groups

Locally compact groups and Polish groups have rich Borel structures that often have important implications. One important example of this are results which imply that Borel and/or measurable homomorphisms are necessarily continuous. We prove a couple of these "automatic continuity" results in Appendix D.1. In Appendix D.2 we prove an old result of Mackey's that if a Borel group has an invariant measure and a suitably rich Borel structure, then it also has a locally compact topology compatible with the given Borel structure. In Appendix D.3, we employ some of the machinery developed here to discuss projective and multiplier representations of locally compact groups. Appendices D.4 and D.5 record a number of technical results needed elsewhere in this book.

D.1 Group homomorphisms

When working in the category of topological groups, it goes without saying that the natural notion of morphisms are continuous group homomorphisms. Nevertheless there are situations, particularly when working with projective and multiplier representations (cf. Appendix D.3), when the continuity of naturally arising homomorphisms is in doubt. The purpose of this section is to show that under reasonable hypotheses on the groups involved, all but the most pathological homomorphisms are continuous.

Example D.1 (Discontinuous homomorphism). In one sense, it's easy to give examples of discontinuous homomorphisms; for example, the identity map from **R** with its usual topology to itself with the discrete topology is not continuous. However if the groups in question are second countable locally compact groups, then exhibiting examples is harder. Nevertheless, such examples exist — even from **R** to **R**. To produce such an example, note that **R** is a vector space over the field **Q** of rational numbers. Therefore **R** has an uncountable vector space basis $\{ r_\alpha \}_{\alpha\in A}$ over **Q**.[1] Then we can prescribe a **Q**-linear map $\varphi : \mathbf{R} \to \mathbf{R}$ by setting the $\varphi(r_\alpha)$ to whatever

[1]Such a basis is sometimes called a *Hamel basis*, and its existence requires the Axiom of Choice in the form of Zorn's Lemma. If you're bothered by that, you're reading the wrong book.

values we want. For example, let $B = \{\alpha_i\}_{i=1}^{\infty}$ be a countable subset of A, and choose $q_i \in \mathbf{Q}$ so that

$$\left|\frac{r_{\alpha_i}}{q_i}\right| \leq \frac{1}{i}.$$

Then we can define φ by

$$\varphi(r_\alpha) = \begin{cases} q_i & \text{if } \alpha = \alpha_i \in B, \text{ and} \\ 0 & \text{if } \alpha \notin B. \end{cases}$$

Now we have $r_{\alpha_i}/q_i \to 0$ in $\mathbf{R}$, but $\varphi(r_{\alpha_i}/q_i) = 1$ for all i. Thus φ is not continuous.

As we shall see, any discontinuous homomorphism from $\mathbf{R}$ to itself, such as the one constructed in Example D.1, will fail to be measurable (cf. Theorem D.3 or Theorem D.11 on page 372). Some hint of this dichotomy is given by the following remark.

Remark D.2. If a homomorphism $\varphi : G \to H$ between topological groups fails to be continuous at a single point, then it fails to be continuous everywhere. If φ is not continuous at s, then there is a net $\{s_\alpha\}$ in G such that $s_\alpha \to s$ and $\varphi(s_\alpha) \not\to \varphi(s)$. But if $r \in G$, then $rs^{-1}s_\alpha \to r$. If $\varphi(rs^{-1}s_\alpha) \to \varphi(r) = \varphi(rs^{-1})\varphi(s)$, then after multiplying by $\varphi(rs^{-1})^{-1}$, we would have a contradiction.

However, we still need some topological hypotheses to prove anything of substance. A case of particular interest occurs when all the groups in sight are locally compact. Then measurability is enough to ensure continuity! (Of course, measurability means with respect to the Haar measures on the groups involved.) The full result [94] is bit too demanding to reproduce. However we can prove a nice result which includes most cases of interest. Since any second countable locally compact group is σ-compact, our result will hold for all second countable locally compact groups.

Theorem D.3. *Suppose that G and H are locally compact groups, and that H is σ-compact. Then every measurable homomorphism $\varphi : G \to H$ is continuous.*

We need the following lemma which is of interest in its own right.

Lemma D.4. *Suppose that G is a locally compact group with Haar measure μ, and that A is a measurable subset with $0 < \mu(A) < \infty$. Then $A^{-1}A$ contains an open neighborhood of the identity e of G.*

Remark D.5. Of course the Lemma applies equally well to any measurable set of nonzero measure provided it contains a non-null measurable subset of finite measure. If G is σ-compact, then every set of infinite measure contains such a subset. But in general, this can fail to be the case. For example, let $\mathbf{R}_d$ be the reals with the discrete topology. Then $S = \mathbf{R}_d \times \{0\}$ is a subgroup of $\mathbf{R}_d \times \mathbf{R}$ with infinite measure, and yet it has no non-null subsets of finite measure (see Remark B.2 on page 332). Clearly, $S = S^{-1}S$ contains no open subset. Thus S is an example of a *locally null set* which is not null. (More importantly, S is also an example of why many people restrict to second countable or at least σ-compact groups.)

Proof of Lemma D.4 on the facing page. Since Haar measure is regular on sets of finite measure, we can find a compact set $K \subseteq A$ with $\mu(K) > 0$. Then the characteristic function f of K is in $L^1(G) \cap L^\infty(G)$. Let $f^*(s) = f(s^{-1})\Delta(s^{-1})$. Then it follows that $f^* * f$ is continuous; for example, see the computations on page 319. Certainly it is positive and compactly supported with $\operatorname{supp} f^* * f \subset K^{-1}K$. Since $f^* * f(e) = \mu(K)$, the result follows. □

Proof of Theorem D.3 on the preceding page. Since it suffices to see that φ is continuous at e (Remark D.2 on the facing page), it suffices to show that $\varphi^{-1}(V)$ is open whenever V is an open neighborhood of e in H.

Next I claim that it suffices to show that $\varphi^{-1}(W)$ contains interior for every open neighborhood W of e in H. To see this, choose $g \in \varphi^{-1}(V)$. Since $\varphi(g) \in V$, there is a neighborhood W of e in H such that $\varphi(g)W^2 \subseteq V$ and such that $W = W^{-1}$. By assumption, $\varphi^{-1}(W^2) \supset \varphi^{-1}(W)\varphi^{-1}(W)^{-1}$ contains a neighborhood of e in G. Since $g\varphi^{-1}(W^2)$ is a neighborhood of g contained in $\varphi^{-1}(V)$, the claim follows.

Now let W be a neighborhood of e in H. Let U be a neighborhood of e in H such that $U^2 \subseteq W$ and $U = U^{-1}$. Since H is σ-compact, there is a *sequence* $\{ h_n \}_{n=1}^\infty \subseteq \varphi(G) \subseteq H$ such that $\bigcup_{n=1}^\infty h_n U$ covers $\varphi(G)$. Choose $g_n \in G$ so that $\varphi(g_n) = h_n$. Notice that $\varphi^{-1}(h_n U) = g_n \varphi^{-1}(U)$. In particular, $\bigcup_{n=1}^\infty g_n \varphi^{-1}(U)$ covers G. Since Haar measure is left invariant, and since some $g_n\varphi^{-1}(U)$ must contain a non-null subset of finite measure, we conclude that $\varphi^{-1}(U)$ has the same property. But then Lemma D.4 on the preceding page implies that $\varphi^{-1}(U^2) = \varphi^{-1}(U)\varphi^{-1}(U)^{-1}$ contains interior. Since this last set is contained in $\varphi^{-1}(W)$, we're done. □

Recall that a character on a group G is a homomorphism from G to the circle $\mathbf{T}$.

Corollary D.6. *Every measurable character on a locally compact group is continuous.*

Although Theorem D.3 on the facing page is striking, it is not quite what we need to handle the sorts of questions arising from projective representations. To study these, we have to consider homomorphisms into the unitary group $U(\mathcal{H})$ of a complex Hilbert space equipped with the strong operator topology. Note that unless $\mathcal{H}$ is finite dimensional, $U(\mathcal{H})$ is never locally compact. However, when $\mathcal{H}$ is separable, $U(\mathcal{H})$ admits a complete metric which is compatible with the strong operator topology (Lemma D.42 on page 395).

Recall that a metric m on the underlying set of a topological space X is compatible with the topology on X if the topology induced by m coincides with the given topology on X. If such a metric exists, X is called *metrizable.* If X admits a compatible metric which is complete (that is, one in which every Cauchy sequence is convergent in X), then X is called *completely metrizable*. It is important to keep in mind that complete metrizabilty is a property of the underlying topology rather than any other structure — even a metric — naturally presented with the space.

Example D.7. Let X be $(0,1)$ together with its natural topology as a subset of $\mathbf{R}$. Of course, $(0,1)$ is metrizable as the usual metric on $\mathbf{R}$ restricts to a compatible metric. However, $(0,1)$ is homeomorphic with $\mathbf{R}$ and hence admits a complete metric compatible with its topology. Thus $(0,1)$ is completely metrizable even thought the "natural" metric is not complete on $(0,1)$. More generally, a subset A of a completely metrizable space X is completely metrizable if and only if A is a G_δ subset of X [168, Theorem 24.12].

Recall that a topological space is called *separable* if it has a countable dense subset. Note that any second countable space is separable, and that a separable *metrizable* space is necessarily second countable. There are separable spaces which are not second countable.

Definition D.8. A topological group is called (completely) *metrizable* if and only if the underlying set admits a (complete) metric compatible with the given topology. Such a group is called a *metric group*. A separable completely metrizable group is called a *Polish group*.

Lemma D.9. *Any second countable locally compact group is Polish.*

Proof. This is a special case of Lemma 6.5 on page 175. □

Remark D.10. Any metrizable space is *first countable* in that every point must have a countable neighborhood basis. There are certainly first countable spaces that are not metrizable, but every topological group with a first countable topology is metrizable. In fact, the metric can be taken to be left-invariant [71, Theorem 8.3].

Theorem D.11. *Suppose that G and H are separable metric groups with G polish. Then any Borel homomorphism $\varphi : G \to H$ is continuous.*

In view of Remark D.2 on page 370, we just have to show that φ has a single point of continuity. We'll proceed by showing that the set of discontinuities in G can't be large enough to fill up all of G. To quantify "small" and "large" we'll use the classical notions of first and second category, respectively.

Definition D.12. A subset A in a topological space X is said to be *meager* or of *first category* in X if it can be written as a countable union $\bigcup F_n$ with each $\overline{F_n}$ having empty interior. (One says that F_n is nowhere dense.) A set which is not of first category is said to be of *second category*.

Remark D.13. Any Baire space is of second category in itself; that is, a Baire space can't be written as a countable union of nowhere dense subsets. Since a complete metric space is a Baire space, every Polish group is of second category. The hypothesis that G is Polish in Theorem D.11 is only used to guarantee that G is of second category.

Definition D.14. A subset B in a topological space X has the *Baire property* if there is an open set U in X such that the symmetric difference $B \triangle U := B \setminus U \cup U \setminus B$ is of first category in X.

Lemma D.15. *If X is a topological space, then every Borel set in X has the Baire property.*

Proof. Since every open set trivially has the Baire property, it will suffice to see that the collection $\mathcal{S}$ of subsets of X with the Baire property is a σ-algebra. Since $X \in \mathcal{S}$ we have to see that $\mathcal{S}$ is closed under taking complements and countable unions. But if $S \in \mathcal{S}$ then there is an open set U such that $F := S \triangle U$ is of first category. Let U^c denote the complement $X \smallsetminus U$, and let $\operatorname{int}(U^c)$ be its interior. Since $\operatorname{int}(U^c)^c \subset \overline{U}$, a bit of fussing shows that

$$S^c \triangle \operatorname{int}(U^c) \subset F \cup (\overline{U} \smallsetminus U).$$

Since $\overline{U} \smallsetminus U$ is closed with empty interior, $F \cup \overline{U} \smallsetminus U$ is of first category. It follows that $S^c \in \mathcal{S}$.

Now suppose that $\{ A_i \}_{i=1}^\infty \subset \mathcal{S}$. Then there are open sets U_i such that $A_i \smallsetminus U_i$ and $U_i \smallsetminus A_i$ are both of first category. But then

$$\Bigl(\bigcup A_i\Bigr) \triangle \Bigl(\bigcup U_i\Bigr) \subset \bigcup (A_i \smallsetminus U_i) \cup \bigcup (U_i \smallsetminus A_i).$$

Thus $\bigcup A_i \in \mathcal{S}$. □

The next result explains all the attention to first and second category sets above. Keep in mind that if $f : X \to Y$ is a function between topological spaces, then saying that the restriction $f|_A$ of f to a subset $A \subset X$ is continuous does not imply that f is continuous on A. For example, the characteristic function of the rationals in $\mathbf{R}$ is nowhere continuous, but the restriction of f to either the rationals or irrationals is continuous.

Lemma D.16. *If M and N are separable metrizable spaces and if $f : M \to N$ is Borel, then there is a set $P \subset M$ of first category such that $f|_{M \smallsetminus P}$ is continuous.*

Proof. Let $\{ O_i \}$ be a countable basis for the topology on N. Since f is Borel, $f^{-1}(O_i)$ is Borel and has the Baire property by Lemma D.15 on the preceding page. Thus there is an open set U_i such that $P_i := f^{-1}(O_i) \triangle U_i$ is first category in M. Let $P = \bigcup P_i$. Then P is first category. Since

$$\begin{aligned}(f|_{M \smallsetminus P})^{-1}(O_i) &= f^{-1}(O_i) \cap (M \smallsetminus P) \\ &\subset f^{-1}(O_i) \cap P_i^c \\ &\subset f^{-1}(O_i) \cap U_i \\ &\subset U_i,\end{aligned}$$

we have $(f|_{M \smallsetminus P})^{-1}(O_i) \subset U_i \cap (M \smallsetminus P)$. On the other hand, a similar argument shows that $U_i \cap (M \smallsetminus P) \subset f^{-1}(O_i) \cap (M \smallsetminus P)$. Thus

$$(f|_{M \smallsetminus P})^{-1}(O_i) = U_i \cap (M \smallsetminus P)$$

which is open in $M \smallsetminus P$. Since every open set in N is a union of the O_i, we've shown that $f|_{M \smallsetminus P}$ is continuous. □

Proof of Theorem D.11 on page 372. In view of Remark D.2 on page 370, it suffices to see that φ is continuous at e. Since G is a metric group, it suffices to show that given a *sequence* $s_n \to e$ in G, then we have $\varphi(s_n) \to e$ in H. By Lemma D.16 on the preceding page, there is a first category set $P \subset G$ such that $\varphi|_{G \setminus P}$ is continuous. But

$$F := P \cup \bigcup_{n=1}^{\infty} s_n^{-1} P$$

is still of first category. Since G is completely metrizable, it is not of first category in itself ([168, Corollary 25.4]), and there must be a $s \in G \setminus F$. But then for all $n \geq 1$, $s_n s \notin P$. Thus $s_n s \to s$ in $G \setminus P$. Since φ is continuous when restricted to $G \setminus P$, $\varphi(s_n s) \to \varphi(s)$. Thus $\varphi(s_n) \to e$ and we're done. □

D.2 Borel Groups and Invariant Measures

In certain circumstances, we may be presented with a group without a natural topology. If we have an invariant measure on the group, it is natural to ask if there is a locally compact topology on the group so that the measure is a Haar measure for the group. This turns out to be possible in a very general setting. Weil proved [167, Appendice I] that any such group G could be viewed as a dense subgroup of a locally compact group G'. (A nice treatment of Weil's theorem can be found in [70, p. 266].) Later Mackey showed that under some weak measure theoretic hypotheses, which we'll state below, the two groups G and G' coincide [105, Theorem 7.1] so that Weil's topology exactly what we're looking for. (Another proof appears in [142, Theorem 1.3].) In order to state and prove Mackey's result (Theorem D.23 on page 376), we need a brief primer on *Borel structures*. Although much of this material Borel structure also had it origins in Mackey's original work [105], our basic reference is Chapter 3 of Arveson's book [2]. Arveson's treatment is an excellent resource, and well worth a thorough reading.

If X is a set and $\mathscr{C}$ is a collection of subsets of X, then the smallest σ-algebra containing $\mathscr{C}$ is called the *σ-algebra generated by $\mathscr{C}$*. For example, if X is a topological space, then the *field of Borel sets in X* is the σ-algebra $\mathscr{B}(X)$ generated by the open subsets of X, and a subset of X is called *Borel* if it belongs to $\mathscr{B}(X)$. Following the terminology in [2, 105], we call a pair $(X, \mathscr{B})$ consisting of a set X and a σ-algebra $\mathscr{B}$ of subsets of X a *Borel space*. Many authors prefer the term *measurable space* to emphasize that we definitely don't want to assume that X has a topology, or even if it does, that $\mathscr{B}$ has anything to do with Borel subsets of X. When we do want to work with the Borel subsets of a topological space X, we'll retain the notation $\mathscr{B}(X)$ to emphasize the fact.

A Borel space $(X, \mathscr{B})$ is said to be *countably separated* if there is a countable subset $\mathscr{D} \subset \mathscr{B}$ which separates points of X. This means that given distinct points $x, y \in X$, there is a $D \in \mathscr{D}$ such that either $x \in D$ and $y \notin D$ or $y \in D$ and $x \notin D$. We say $(X, \mathscr{B})$ is *countably generated* if there is a countable set $\mathscr{D}$ as above which separates points and which generates $\mathscr{B}$.

Example D.17. If X is a separable metrizable space, then the topology on X is second countable. Thus a countable basis for the topology will separate points and generate $\mathscr{B}(X)$. Thus $\big(X, \mathscr{B}(X)\big)$ is a countably generated Borel space.

If $(X, \mathscr{B})$ and $(Y, \mathscr{M})$ are Borel spaces, then a map $f : X \to Y$ is called *Borel* if $f^{-1}(M) \in \mathscr{B}$ for all $M \in \mathscr{M}$. We say that f is a *Borel isomorphism* if f is bijective and both f and f^{-1} are Borel maps. We say that $(X, \mathscr{B})$ and $(Y, \mathscr{M})$ are *Borel isomorphic* if there is a Borel isomorphism from X to Y.

A separable completely metrizable space is called a *Polish space*. Polish spaces, or rather their Borel structures, play a central role in the theory.

Definition D.18. A Borel space $(X, \mathscr{B})$ is called *standard* if it is Borel isomorphic to a Borel subset of some Polish space P with the relative Borel structure.

A remarkable result, due to Kuratowski [96, p. 451 remark (i)], implies that any uncountable standard Borel space is isomorphic to $[0, 1]$ with the usual Borel structure (Theorem I.40 on page 503). Thus standard Borel spaces are like Hilbert spaces in the sense that once you've seen one, of a given dimension, you've seen them all. Another important, and far more varied, class of Borel spaces arise from analytic subsets of Polish spaces. Recall that a subset A in a Polish space P is called *analytic* if there is a continuous map f of a Polish space Q into P such that $f(Q) = A$. While every Borel subset is analytic [2, Theorem 3.2.1], the converse if false [96, p. 479]. Nevertheless, given an analytic subset A, we can equip it with the *relative Borel structure* coming from $\mathscr{B}(P)$: $\mathscr{M} = \{\, A \cap B : B \in \mathscr{B}(P) \,\}$.

Definition D.19. A Borel space $(X, \mathscr{B})$ is called *analytic* if it is Borel isomorphic to $(A, \mathscr{M})$ where A is an analytic subset of a Polish space and $\mathscr{M}$ is the relative Borel structure on A.

We will need two fundamental results about analytic Borel spaces from [2, Chap. 3] that we won't reproduce here. The first is a *generalized Souslin Theorem* which says that a Borel bijection $f : X \to Y$ between analytic Borel spaces is necessarily a Borel isomorphism [2, Corollary 2 of Theorem 3.3.4]. The second is the *Unique Structure Theorem* which says that if $(X, \mathscr{B})$ is an analytic Borel space and if $\mathscr{B}_0$ is a countably generated sub-σ-algebra of $\mathscr{B}$ which separates points of X, then $\mathscr{B} = \mathscr{B}_0$ [2, Theorem 3.3.5]. As an example of the Unique Structure Theorem, we give the following.

Lemma D.20. *Suppose that $(X, \mathscr{B})$ is an analytic Borel space and that $\{\, f_n \,\}$ is a sequence of (complex-valued) Borel functions on X which separate points. Then $\mathscr{B}$ is the smallest σ-algebra in X such that each f_n is Borel. In particular, $g : Y \to X$ is Borel if and only if $f_n \circ g$ is Borel for all n*

Proof. Let $\mathscr{B}_0$ be the smallest σ-algebra such that each f_n is Borel. If U_n is a countable basis for $\mathbf{C}$, then $\{\, f_m^{-1}(U_n) \,\}$ is a countable family that generates $\mathscr{B}_0$ and which separates points. Thus $\mathscr{B} = \mathscr{B}_0$ by the Unique Structure Theorem.

If $f_n \circ g$ is Borel for all n, then $g^{-1}\big(f_m^{-1}(U_n)\big)$ is Borel for all n and m. Since $\{\, f_m^{-1}(U_n) \,\}$ generate $\mathscr{B}$, g is Borel. □

If $(X, \mathscr{B})$ and $(Y, \mathscr{M})$ are Borel spaces, then the *product Borel structure* on $X \times Y$ is the σ-algebra $\mathscr{B} \times \mathscr{M}$ generated by the measurable rectangles $A \times B$ with $A \in \mathscr{B}$ and $B \in \mathscr{M}$. If X and Y are topological spaces, then it is possible for $\mathscr{B}(X) \times \mathscr{B}(Y)$ to be a proper subset of $\mathscr{B}(X \times Y)$ even if X and Y are compact groups [7, §2]. (However, if X and Y are second countable locally compact spaces then equality holds — see Lemma 4.44 on page 141.)

Definition D.21. A *Borel group* is a group G together with a σ-algebra $\mathscr{B}$ of subsets of G such that

$$(r, s) \mapsto r^{-1}s$$

is a Borel map from $(G \times G, \mathscr{B} \times \mathscr{B})$ to $(G, \mathscr{B})$. We say that G is an analytic (resp., standard) if the underlying Borel structure is analytic (resp., standard).

Example D.22. Suppose that G is a second countable locally compact topological group. Then every open set in $G \times G$ is a countable union of open rectangles which are clearly in $\mathscr{B}(G) \times \mathscr{B}(G)$. Therefore $\big(G, \mathscr{B}(G)\big)$ is a standard Borel group (Lemma D.9 on page 372).

If X is a topological space, then the natural Borel structure on X is given by the Borel sets. However, if we start with a Borel space $(X, \mathscr{B})$, there may or may not be a topology on X such that $\mathscr{B}(X) = \mathscr{B}$. Such a topology is said to be *compatible with the Borel structure.*

Theorem D.23 (Mackey). *Suppose that G is an analytic Borel group with a nonzero left-invariant σ-finite measure μ. Then there is a unique second countable locally compact topology on G which is compatible with the Borel structure on G. With respect to this topology, μ is a Haar measure on G.*

Proof. Suppose that τ is a second countable locally compact topology compatible with the given Borel structure. Since G must have a σ-finite Haar measure ν, it follows from Lemma D.40 on page 393, that μ is a multiple of ν. Therefore, μ is a Haar measure on G.[2] Suppose that τ' is another second countable locally compact topology on G which is compatible with the given Borel structure. The identity map from (G, τ) to (G, τ') is a Borel isomorphism, and it must be a homeomorphism by Theorem D.3 on page 370. Thus $\tau = \tau'$, and a compatible second countable locally compact topology, once we show that it exists, is unique.

To show that such a topology exists, let $\mathcal{H} = L^2(G, \mu)$. Since G is analytic, G is countably generated as a Borel space. Thus $\mathcal{H}$ is separable by Lemma D.41 on page 394, and $U(\mathcal{H})$ is a Polish group by Lemma D.42 on page 395. For each $s \in G$ and $h \in \mathcal{H}$, define

$$L_s h(r) := h(s^{-1}r) \quad \text{for all } r \in G.$$

Since μ is left-invariant, $L_s h \in \mathcal{H}$, and L_s is is a unitary operator with $L_s^* = L_s^{-1} = L_{s^{-1}}$. In fact, it is easy to check that $s \mapsto L_s$ is a homomorphism of G onto a subgroup $L(G)$ of the Polish group $U(\mathcal{H})$.

[2] In order to conclude that μ is a Haar measure, we need to know that it is a Radon measure. As this is not immediately clear, we appeal to Lemma D.40.

We'll proceed as follows. First we'll show that L is an injective Borel map of G onto $L(G)$. The next, and hardest step, is to show that $L(G)$ is closed in $U(\mathcal{H})$. Then we can pull-back the (necessarily second countable) Polish topology on $L(G)$ to G, and see that it generates the given Borel structure. Then we use a result which implies that measures on Polish spaces are nearly supported on compact sets to show that this topology is actually locally compact.

To see that L is a Borel map, notice that

$$(L_s h \mid k) = \int_G h(s^{-1}r)\overline{k(r)}\,d\mu(r) \quad \text{for all } s \in G \text{ and } h, k \in \mathcal{H}.$$

Since G is a Borel group, $(s, r) \mapsto h(s^{-1}r)\overline{k(r)}$ is Borel (with respect to the product Borel structure $\mathscr{B} \times \mathscr{B}$) and Fubini's Theorem ([156, Theorem 8.8] or [57, Theorem 2.37]) implies that $s \mapsto (L_s h \mid k)$ is Borel for all $h, k \in \mathcal{H}$. Thus $s \mapsto L_s$ is Borel in view of Lemma D.43 on page 395.

To see that L is one-to-one, it suffice to see that $L_s = 1_{\mathcal{H}}$ implies that $s = e$. Since G is countably separated, Lemma D.44 on page 395 implies there is an injective Borel map $\psi : G \to [0, 1]$ which is square integrable. Thus if $L_s = 1_{\mathcal{H}}$, then $L_s\psi = \psi$ and

$$\psi(r) = \psi(s^{-1}r) \quad \text{for almost all } r \in G.$$

Therefore there is some $r_0 \in G$, such that $\psi(r_0) = \psi(s^{-1}r_0)$. Since ψ is one-to-one, this implies $s = e$ as desired.

Next we want to see that the range $L(G)$ of L is closed in $U(\mathcal{H})$. Suppose that $L_{s_n} \to U$ in $U(\mathcal{H})$. If E and F are Borel sets in G of finite measure, then $\mathbb{1}_E$, $\mathbb{1}_F$ and $\mathbb{1}_E \cdot \mathbb{1}_F$ are all in $\mathcal{L}^2(G)$. Since multiplication is continuous in $U(\mathcal{H})$,

$$U(\mathbb{1}_E\mathbb{1}_F) = \lim_n L_{s_n}(\mathbb{1}_E)L_{s_n}(\mathbb{1}_F) = U(\mathbb{1}_E)U(\mathbb{1}_F).$$

In particular, $U(\mathbb{1}_E)^2 = U(\mathbb{1}_E)$. Thus there is a Borel set F of finite measure such that $U(\mathbb{1}_E) = \mathbb{1}_F$ in $L^2(G)$, and F is, up to a null set, uniquely determined by U. Thus we obtain a map Φ from the Borel sets of G with finite measure into the boolean σ-algebra $\mathscr{B}/\mathscr{N}$ of Borel sets in G modulo μ-null sets. (See Appendix I.6 for a discussion of boolean σ-algebras and operations such as $\bigvee$ below.) If $F \in \mathscr{B}$ is of finite measure and if F is the disjoint union $\bigcup_{i=1}^{\infty} F_i$, then

$$\mathbb{1}_F = \sum_{i=1}^{\infty} \mathbb{1}_{F_i} \quad \text{in } L^2(G), \text{ and} \quad U(\mathbb{1}_F) = \sum_{i=1}^{\infty} U(\mathbb{1}_{F_i}).$$

It follows that

$$\Phi(F) = \bigvee_{i=1}^{\infty} \Phi(F_i).$$

Since μ is σ-finite, any $B \in \mathscr{B}$ can be written as the countable disjoint union of sets of finite measure. If

$$\bigcup_{i=1}^{\infty} F_i = B = \bigcup_{i=1}^{\infty} E_i$$

are two such decompositions, then for each i, $F_i = \bigcup_{j=1}^{\infty} F_i \cap E_j$. Thus if $K \in \mathscr{B}/\mathscr{N}$ is such that $K \geq \Phi(F_i \cap E_j)$ for all i and j, then

$$K \geq \bigvee_{i=1}^{\infty} \bigvee_{j=1}^{\infty} \Phi(F_i \cap E_j) = \bigvee_{i=1}^{\infty} \Phi(F_i).$$

Thus

$$\bigvee_{j=1}^{\infty} \Phi(E_j) \geq \bigvee_{i=1}^{\infty} \Phi(F_i). \tag{D.1}$$

By symmetry, we have equality in (D.1), and it follows that we can extend Φ to a σ-homomorphism of $\mathscr{B}$ into $\mathscr{B}/\mathscr{N}$ with the property that $E \triangle F \in \mathscr{N}$ implies $\Phi(E) = \Phi(F)$. Lemma I.36 on page 501 implies that there is a Borel map $\varphi : G \to G$ such that

$$\Phi(B) = [\varphi^{-1}(B)],$$

where $[\varphi^{-1}(B)]$ denotes the class of $\varphi^{-1}(B)$ in $\mathscr{B}/\mathscr{N}$. In particular, if E has finite measure, then

$$U(\mathbb{1}_E)(s) = \mathbb{1}_{\varphi^{-1}(E)}(s) = \mathbb{1}_E\bigl(\varphi(s)\bigr) \quad \text{for almost all } s \in G.$$

It follows that for all $h \in \mathcal{H} = L^2(G)$,

$$U(h)(s) = h\bigl(\varphi(s)\bigr) \quad \text{for almost all } s \in G.$$

For each $r \in G$, let $\rho(r) \in U(\mathcal{H})$ be given by

$$\rho(r)(h)(s) = \Delta^{\frac{1}{2}}(r)h(sr).$$

Then $\rho(r)$ commutes with L_s for all $s \in G$. Therefore, $\rho(r)$ must also commute with U. Thus for all $h \in \mathcal{H}$ and $r \in G$, we have

$$h\bigl(\varphi(sr)\bigr) = h\bigl(\varphi(s)r\bigr) \quad \text{for almost all } s \in G.$$

As above, let $\psi : G \to [0,1]$ be injective and square integrable. Since $\psi\bigl(\varphi(sr)\bigr) = \psi\bigl(\varphi(s)r\bigr)$ for almost all s, we have

$$\varphi(sr) = \varphi(s)r \quad \text{for almost all } s.$$

Using Fubini's Theorem, there must be a $s_0 \in G$ and a null set N such that

$$\varphi(s_0 r) = \varphi(s_0)r \quad \text{for } r \notin N.$$

Thus if $s \notin s_0 N$, we have

$$\varphi(s) = \varphi(s_0 s_0^{-1} s) = \varphi(s_0)s_0^{-1}s = \bigl(s_0\varphi(s_0)^{-1}\bigr)^{-1}s.$$

Thus if $r := s_0\varphi(s_0)^{-1}$, then for almost all s,

$$U(h)(s) = h\bigl(\varphi(s)\bigr) = L_r(h)(s),$$

and $U = L_r$. It follows that $L(G)$ is a closed subgroup of the Polish group $U(\mathcal{H})$.

Since G is analytic, the generalized Souslin Theorem ([2, Corollary 2 of Theorem 3.3.4]) implies that L is a Borel isomorphism of G onto $L(G)$ in its relative Borel structure. Since the relative Borel structure on $L(G)$ is that generated by the relative topology on $L(G)$, we can pull back the Polish topology on $L(G)$ to G and obtain a *second countable* Polish topology on G which is compatible with the given Borel structure on G.

It only remains to show that the topology on $L(G)$ is locally compact. To do this, we'll take advantage of the fact that the unit ball $B(\mathcal{H})_1$ of $B(H)$ is a compact Polish space in the weak operator topology (Lemma D.37 on page 391). Therefore the closure, C, of $L(G)$ in $B(H)_1$ is a second countable compact space. Thus it will suffice to show that

$$L(G) = C \setminus \{0\}.$$

Suppose that $L_{s_n} \to T \neq 0$ in the weak operator topology. Since $T \neq 0$, there must be Borel sets E' and F' of finite measure such that

$$(L_{s_n} \mathbb{1}_{E'} \mid \mathbb{1}_{F'}) \to a := (T\mathbb{1}_{E'} \mid \mathbb{1}_{F'}) > 0.$$

It follows that for large n,

$$\mu(F' \cap s_n E') \geq a/2.$$

It follows from Corollary D.39 on page 393 that there are compact sets E and F such that

$$\mu(F' \setminus F) < a/12 \quad \text{and} \quad \mu(E' \setminus E) < a/12.$$

But then

$$\begin{aligned}\mu(F' \cap s_n E') &\leq \mu\big((F \cup F' \setminus F) \cap s_n(E \cup E' \setminus E)\big) \\ &\leq \mu(F \cap s_n E) + \frac{a}{4}.\end{aligned}$$

Thus for large n, we have $\mu(F \cap s_n E) \geq a/4 > 0$. Thus we eventually have $F \cap s_n E \neq \emptyset$ and $s_n \in FE^{-1}$. Since FE^{-1} is compact, we can pass to a subsequence, relabel, and assume that $s_n \to s$ in G. But then $L_{s_n} \to L_s$ in the weak operator topology and we must have $T = L_s$. This completes the proof. □

D.3 Projective Representations

Let $\mathcal{H}$ be a complex Hilbert space. Then the center of the unitary group $U(\mathcal{H})$ is easily seen to be the scalar operators and we have a homeomorphism i from the circle $\mathbf{T}$ onto the center given by $i(z) = z1_{\mathcal{H}}$. The quotient $P(\mathcal{H}) = U(\mathcal{H})/i(\mathbf{T})$ is called the *projective unitary group*. A *projective representation* of a group G is a homomorphism $\alpha : G \to P(\mathcal{H})$ which is continuous when $P(\mathcal{H})$ is given the quotient topology coming from the strong operator topology on $U(\mathcal{H})$. Of course each $U \in U(\mathcal{H})$ determines an automorphism in $\operatorname{Aut}\mathcal{K}(\mathcal{H})$ given by $\operatorname{Ad}U(T) := UTU^*$. As discussed in [139, Chap. 1], Ad factors through $P(\mathcal{H})$ and induces a homeomorphism of $P(\mathcal{H})$ with $\operatorname{Aut}\mathcal{K}(\mathcal{H})$ with the point norm topology [139, Proposition 1.6].

Although much of what is being reproduced here holds in great generality (cf. [5,93]), we'll restrict to the case in which $\mathcal{H}$ is separable and G is second countable. In particular, if $\mathcal{H}$ is separable, we have the following useful observation.

Lemma D.24. *Suppose that $\mathcal{H}$ is separable complex Hilbert space and that $\alpha_0 \in \operatorname{Aut}\mathcal{K}(\mathcal{H})$ with $\operatorname{Ad} U_0 = \alpha_0$. Then there is a Borel map $c : \operatorname{Aut}\mathcal{K}(\mathcal{H}) \to U(\mathcal{H})$ and a neighborhood N of α_0 such that $\operatorname{Ad} c(\beta) = \beta$ for all $\beta \in \operatorname{Aut}\mathcal{K}(\mathcal{H})$, c is continuous on N and $c(\alpha_0) = U_0$.*

Proof. Since $\mathcal{H}$ is separable, $\operatorname{Aut}\mathcal{K}(\mathcal{H})$ is Polish (cf. [139, Lemma 7.18]) and therefore second countable. Lemma 1.6 of [139] implies that for each $\alpha \in \operatorname{Aut}\mathcal{K}(\mathcal{H})$, there is a neighborhood N_α of α and a continuous map $c_\alpha : N_\alpha \to U(\mathcal{H})$ such that $\operatorname{Ad} c_\alpha(\beta) = \beta$ for all $\beta \in N_\alpha$. Since $\operatorname{Aut}\mathcal{K}(\mathcal{H})$ is second countable, we can find a countable cover $\{ N_i \}_{i=1}^\infty$ of $\operatorname{Aut}\mathcal{K}(\mathcal{H})$ by open sets and continuous maps $c_i : N_i \to U(\mathcal{H})$ such that $\operatorname{Ad} c_i(\beta) = \beta$ for all $\beta \in N_i$. We can assume that $\alpha_0 \in N_1$, and define Borel sets B_k such that

$$B_1 := N_1 \quad \text{and} \quad B_k := N_k \setminus \bigcup_{i=1}^{k-1} B_i \quad \text{for } k \geq 2.$$

Then we can define c' by $c'(\beta) = c_i(\beta)$ if $\beta \in B_i$. Then c' satisfies all the requirements of the lemma with the possible exception of $c'(\alpha_0) = U_0$. But $\operatorname{Ad} c'(\alpha_0) = \operatorname{Ad} U_0$, so there is a $z \in T$ such that $c'(\alpha_0) = zU_0$. But then we can define $c := \bar{z}c'$. □

If $\alpha : G \to \operatorname{Aut}\mathcal{K}(\mathcal{H})$ is a projective representation of G and if c is as in Lemma D.24, then we get a Borel map

$$\pi : G \to U(\mathcal{H})$$

given by $\pi(s) := c(\alpha_s)$. Since α is a homomorphism, there is a function $\omega : G \times G \to \mathbf{T}$ such that

$$\pi(s)\pi(r) = \omega(s,r)\pi(sr) \quad \text{for all } s, r \in G. \tag{D.2}$$

If $h \in \mathcal{H}$ with $\|h\| = 1$, then

$$\omega(s,r) = \big(\pi(s)\pi(r)\pi(sr)^{-1}h \mid h\big),$$

and the right-hand side is the composition of the maps

(a) $(s,r) \mapsto (s,r,sr)$ from $G \times G$ to $G \times G \times G$,
(b) $(s,r,t) \mapsto (\pi(s),\pi(r),\pi(t))$ from $G \times G \times G$ to $U(\mathcal{H}) \times U(\mathcal{H}) \times U(\mathcal{H})$,
(c) $(U,V,W) \mapsto UVW^{-1}$ from $U(\mathcal{H}) \times U(\mathcal{H}) \times U(\mathcal{H})$ to $U(\mathcal{H})$ and
(d) $U \mapsto (Uh \mid h)$ from $U(\mathcal{H}) \to \mathbf{C}$.

Since (a), (c) and (d) are continuous, it will follow that ω is Borel provided we see that (b) is a Borel map. But this follows because $s \mapsto \pi(s)$ is Borel and since every open set in $U(\mathcal{H}) \times U(\mathcal{H}) \times U(\mathcal{H})$ is the *countable* union of open rectangles $V_1 \times V_2 \times V_3$. Thus ω is Borel. Further (D.2) and associativity imply that

$$\omega(s,t)\omega(st,r) = \omega(s,tr)\omega(t,r) \quad \text{for all } s,t,r \in G. \tag{D.3}$$

Definition D.25. A Borel function $\omega : G \times G \to \mathbf{T}$ is called a 2-*cocycle* with values in $\mathbf{T}$ if it satisfies (D.3). We say that ω is *normalized* if $\omega(e,s) = \omega(s,e) = 1$ for all $s \in G$. A normalized 2-cocycle is called a *multiplier*. We say that ω is trivial or a 2-*boundary* if there is a Borel function $b : G \to \mathbf{T}$ such that

$$\omega(s,r) = b(s)b(r)b(sr)^{-1} \quad \text{for all } s, r \in G.$$

The collection of all 2-cocycles is denoted by $Z^2(G,\mathbf{T})$, and the 2-boundaries are denoted by $B^2(G,\mathbf{T})$.

It is immediate that $Z^2(G,\mathbf{T})$ is an abelian group under pointwise multiplication and that $B^2(G,\mathbf{T})$ is a subgroup. We let $H^2(G,\mathbf{T})$ be the quotient group. We'll denote the image of $\omega \in Z^2(G,\mathbf{T})$ in $H^2(G,\mathbf{T})$ by $[\omega]$. Two elements ω and ω' that differ by a boundary are called similar. (As the notation suggests, $H^2(G,\mathbf{T})$ is part of a group cohomology for G with coefficients in $\mathbf{T}$. See §7.4 of [139] for more details and references.)

Note that if we choose $c : \operatorname{Aut}\mathcal{K}(\mathcal{H}) \to U(\mathcal{H})$ as in Lemma D.24 on the preceding page with the property that $c(1_{\mathcal{K}(\mathcal{H})}) = 1_{\mathcal{H}}$, then the resulting cocycle (D.2) is normalized. If ω is any cocycle, then ω is similar to a normalized cocycle. Let $b(s) = \overline{\omega(s,e)}$, and $\delta b(s,r) := b(s)b(r)b(sr)^{-1}$. Then $\omega' := \delta b \cdot \omega$ does the trick.

If $\pi : G \to U(\mathcal{H})$ is a Borel map such that $\operatorname{Ad}\pi$ — defined by $(\operatorname{Ad}\pi)_s = \operatorname{Ad}\pi(s)$ — is a homomorphism, then there is a function ω such that (D.2) holds. Just as before, ω is Borel and is easily seen to be a 2-cocycle.

Definition D.26. If ω is a multiplier on G, then a ω-*representation* of G on $\mathcal{H}$ is a Borel map $\pi : G \to U(\mathcal{H})$ such that

$$\pi(s)\pi(r) = \omega(s,r)\pi(sr) \quad \text{for all } s, r \in G.$$

Notice that if π is an ω-representation of G on $\mathcal{H}$, then $\operatorname{Ad}\pi$ is a Borel homomorphism of G into $\operatorname{Aut}\mathcal{K}(\mathcal{H})$. If G is second countable and $\mathcal{H}$ is separable, then G and $\operatorname{Aut}\mathcal{K}(\mathcal{H})$ are Polish groups (Lemmas D.9 and [139, Lemma 7.18]), and $\operatorname{Ad}\pi$ is continuous by Theorem D.11 on page 372. Therefore $\operatorname{Ad}\pi$ is a projective representation of G. If π and π' are ω- and ω'-representations, respectively, of G on $\mathcal{H}$ such that $\operatorname{Ad}\pi = \operatorname{Ad}\pi'$ then for each $s \in G$ there is a scalar $b(s) \in \mathbf{T}$ such that $\pi(s) = b(s)\pi'(s)$. It is easy to see that b is Borel and that $\omega(s,r) = b(s)b(r)b(sr)^{-1}\omega'(s,r)$. That is, ω and ω' are similar.

Of course if $\alpha : G \to \operatorname{Aut}\mathcal{K}(\mathcal{H})$ is equal to $\operatorname{Ad}\pi$ for a ordinary representation of G — that is, α is unitarily implemented — then the corresponding cocycle is identically one and has trivial image in $H^2(G,\mathbf{T})$. On the other hand, if α is implemented by a ω-representation with ω trivial, then α is unitarily implemented. To see this, suppose $\omega(s,r) = b(s)b(r)b(sr)^{-1}$. Then we can define $\pi'(s) = \overline{b(s)}\pi(s)$. Then π' is a Borel homomorphism, and therefore continuous. But $\alpha = \operatorname{Ad}\pi'$.

We can summarize the above discussion as follows.

Proposition D.27. *Suppose that G is a second countable locally compact group and that $\mathcal{H}$ is a separable Hilbert space. If $\alpha : G \to \operatorname{Aut}\mathcal{K}(\mathcal{H})$ is a projective representation on $\mathcal{H}$, then there is a multiplier ω in $Z^2(G,\mathbf{T})$ and a ω-representation*

π such that $\alpha = \operatorname{Ad}\pi$. The class $[\omega] \in H^2(G, \mathbf{T})$ depends only on α and is called the Mackey obstruction for α. Furthermore, α is unitarily implemented if and only if its Mackey obstruction is trivial.

Conversely, if π is an ω-representation for some multiplier ω, then $\operatorname{Ad}\pi$ is a projective representation with Mackey obstruction $[\omega]$.

Now we want to see that multiplier representations of a second countable locally compact group G are in one-to-one correspondence with certain ordinary representations of a locally compact group which is an extension of $\mathbf{T}$ by G. If $\omega \in Z^2(G, \mathbf{T})$ is a multiplier, then we let G^ω be the set $\mathbf{T} \times G$ equipped with the operations

$$(z, s)(z', r) := \big(zz'\omega(s, r), sr\big) \quad \text{and}$$
$$(z, s)^{-1} := \big(\overline{z\omega(s, s^{-1})}, s^{-1}\big).$$

It is not hard to check that G^ω is a group, and that we have an (algebraic) exact sequence

$$1 \longrightarrow \mathbf{T} \xrightarrow{\ i\ } G^\omega \xrightarrow{\ j\ } G \longrightarrow e \tag{D.4}$$

where $i(z) = (z, e)$, $j(z, s) = s$ and $i(\mathbf{T})$ is central in G^ω. If ω is continuous, then we can give G^ω the product topology, and then (D.4) becomes a short exact sequence of locally compact groups (Proposition 1.53 on page 15). Note that the product measure $\mu_\mathbf{T} \times \mu_G$ is a Haar measure on G^ω in this case. On the other hand, if ω is merely Borel, then we have to work a little harder. Since G and $\mathbf{T}$ are second countable, G and $\mathbf{T}$ are Borel groups (Example D.22 on page 376), and it is not hard to see that G^ω is a Borel group when we give it the Borel structure $\mathscr{B}(\mathbf{T} \times G) = \mathscr{B}(\mathbf{T}) \times \mathscr{B}(G)$ coming from the Polish topology on $\mathbf{T} \times G$ (Lemma D.9 on page 372). Furthermore, the product measure $\mu_\mathbf{T} \times \mu_G$ is invariant under left multiplication. So all that's missing is a locally compact topology on G^ω. It is a nontrivial result due to Mackey (Theorem D.23 on page 376) that G^ω has a unique second countable locally compact topology such that $\mathscr{B}(G^\omega) = \mathscr{B}(\mathbf{T} \times G)$ and such that $\mu_\mathbf{T} \times \mu_G$ is a Haar measure. Since i and j are clearly Borel homomorphisms, it follows from Theorem D.3 on page 370 that they are continuous. Proposition 1.53 on page 15 then implies that (D.4) becomes a short exact sequence of locally compact groups. Thus there is a unique locally compact topology on G^ω such that (D.4) is again a short exact sequence of locally compact groups.

Notice that if π' is an ω-representation of G on $\mathcal{H}$, then

$$\pi(z, s) := z\pi'(s) \tag{D.5}$$

satisfies

$$\begin{aligned} \pi(z, s)\pi(z', r) &= zz'\pi'(s)\pi'(r) \\ &= zz'\omega(s, r)\pi'(sr) \\ &= \pi\big((z, s)(z', r)\big). \end{aligned}$$

Thus π is a Borel homomorphism of G^ω into $U(\mathcal{H})$. Since Theorem D.11 on page 372 implies that π is strongly continuous, π is a representation satisfying

$$\pi\big(i(z)\big) = z1_\mathcal{H}. \tag{D.6}$$

Conversely, if π is a representation of G^ω on $\mathcal{H}$ satisfying (D.6), then

$$\pi'(s) := \pi(1, s) \tag{D.7}$$

is a Borel map of G into $U(\mathcal{H})$ such that

$$\begin{aligned}\pi'(s)\pi'(r) &= \pi(1, s)\pi(1, r) \\ &= \pi\big((\omega(s, r), sr)\big) \\ &= \omega(s, r)\pi'(sr).\end{aligned}$$

We've proved the following.

Proposition D.28. *Suppose that ω is a multiplier on a second countable locally compact group G and that G^ω is the corresponding extension* (D.4). *Then there is a one-to-one correspondence between ω-representations of G on separable Hilbert spaces and representations of G^ω on separable Hilbert spaces satisfying* (D.6)

Thus to understand ω-representations, we'll want to get a handle on representations of G^ω satisfying (D.6). We'll do this by showing that there is a quotient $C^*(G, \omega)$ of $C^*(G^\omega)$ whose representations are exactly those which satisfy (D.6). We'll do this in slightly more generality than necessary here to stress the connection with twisted crossed products as discussed in Section 7.4.

For the moment, G can be any locally compact group (second countable or not), and let

$$1 \longrightarrow \mathbf{T} \stackrel{i}{\longrightarrow} E \stackrel{j}{\longrightarrow} G \longrightarrow e \tag{D.8}$$

be any central extension of $\mathbf{T}$ by G. (That is, we assume that i and j are continuous homomorphisms, that i is a homeomorphism onto $\ker j$, that j is an open surjection and that $i(\mathbf{T})$ lies in the center of E.) Let $\tau : \mathbf{T} \to \mathbf{T}$ be any character of $\mathbf{T}$. (The example of interest for multiplier representations is $E = G^\omega$ and $\tau(z) = z$ for all $z \in \mathbf{T}$.) Since $i(\mathbf{T})$ is central, if $f \in C_c(E)$ and dz is normalized Lebesgue measure on $\mathbf{T}$, then

$$\int_{\mathbf{T}} f\big(si(z)\big)\, dz$$

depends only on $j(s) := \dot{s}$ in G. Thus we can define a Haar integral on E by

$$\int_E f(s)\, d\mu_E(s) := \int_G \int_{\mathbf{T}} f\big(si(z)\big)\, dz\, d\mu_G(\dot{s})$$

where μ_G is any fixed Haar measure on G. A unitary representation π of E is said to *preserve* τ if

$$\pi\big(i(z)\big) = \tau(z) 1_{\mathcal{H}_\pi} \quad \text{for all } z \in \mathbf{T}.$$

If I is the ideal of $M\big(C^*(E)\big)$ generated by $\{\, i_E\big(i(z)\big) - \tau(z)1_E : z \in \mathbf{T} \,\}$, then let

$$I_\tau := I \cap C^*(E).$$

If π preserves τ, then $\bar{\pi}(I) = \{0\}$ and $I_\tau \subset \ker \pi$. On the other hand, if $I_\tau \subset \ker \pi$, then $\bar{\pi}(I) = \{0\}$, and

$$C^*(E, \tau) := C^*(E)/I_\tau$$

can be characterized as the quotient of $C^*(E)$ such that a representation of E preserves τ if and only if it factors through $C^*(E,\tau)$.

Example D.29. In the example at hand — G second countable, $E = G^\omega$ and $\tau(z) = z$ — a representation of G^ω preserves τ if and only if it satisfies (D.6). Thus representations satisfying (D.6) are exactly those factoring through $C^*(G^\omega,\tau)$, and representations of $C^*(G^\omega,\tau)$ on separable Hilbert spaces are in one-to-one correspondence with ω-representations of G. We usually write $C^*(G,\omega)$ in place of $C^*(G^\omega,\tau)$ in this case.

As with ordinary group C^*-algebras, it will be important to have an underlying (dense) function space inside of $C^*(E,\tau)$. This will be fairly straightforward since we can take advantage of the facts that $i(\mathbf{T})$ is central and dz is normalized. Let

$$C_c(E,\tau) := \{\, f \in C_c(E) : f\big(i(z)s\big) = \overline{\tau(z)}f(s)\,\},$$

and for each $f \in C_c(E)$, let

$$Q_\tau(f)(s) := \int_{\mathbf{T}} f\big(si(z)\big)\tau(z)\,dz.$$

We have $Q_\tau(f) \in C_c(E)$ by Corollary 1.103 on page 36. Since $i(\mathbf{T})$ is in the center of E, it follows that $Q_\tau(f) \in C_c(E,\tau)$. Since we're working with normalized Lebesgue measure on $\mathbf{T}$, the $\mathbf{T}$-integrals can often disappear from formulas entirely. For example, if $f,g \in C_c(E,\tau)$, then $(r,s) \mapsto f(r)g(r^{-1}s)$ depends only on $(\dot r, s)$. In particular,

$$\begin{aligned} f * g(s) &= \int_G \int_{\mathbf{T}} f\big(ri(z)\big)g\big(i(\bar z)r^{-1}s\big)\,dz\,d\mu_G(\dot r) \\ &= \int_G \int_{\mathbf{T}} f(r)g(r^{-1}s)\,dz\,d\mu_G(\dot r) \\ &= \int_G f(r)g(r^{-1}s)\,d\mu_G(\dot r). \end{aligned}$$

In a similar vein, if π preserves τ and if $f \in C_c(E,\tau)$, then $s \mapsto f(s)\pi(s)$ depends only on $\dot s$, and

$$\pi(f) = \int_G f(s)\pi(s)\,d\mu_G(\dot s).$$

If $f,g \in C_c(E)$, then applying Fubini's Theorem as needed,

$$\begin{aligned} Q_\tau(f*g)(s) &= \int_{\mathbf{T}} f*g\big(si(z)\big)\tau(z)\,dz \\ &= \int_{\mathbf{T}}\int_G\int_{\mathbf{T}} f\big(ri(w)\big)g\big(i(\bar w)r^{-1}si(z)\big)\tau(z)\,dw\,d\mu_G(\dot r)\,dz \\ &= \int_G \Big(\int_{\mathbf{T}} f(ri(w))\tau(w)\,dw\Big)\Big(\int_{\mathbf{T}} g\big(r^{-1}si(z)\big)\tau(z)\,dz\Big)\,d\mu_G(\dot r) \\ &= Q_\tau(f) * Q_\tau(g)(s). \end{aligned}$$

Since we easily check that $Q_\tau(f^*) = Q_\tau(f)^*$ and that $Q_\tau(f) = f$ if $f \in C_c(E,\tau)$, it follows that $C_c(E,\tau)$ is a $*$-subalgebra of $C_c(E)$ and that Q_τ is a homomorphism of $C_c(E)$ onto $C_c(E,\tau)$. Furthermore if $f \in C_c(E)$ and if π preserves τ, then

$$\begin{aligned}\pi(f) &= \int_G \int_{\mathbf{T}} f\big(si(z)\big)\pi\big(si(z)\big)\,dz\,d\mu_G(\dot s)\\ &= \int_G \int_{\mathbf{T}} f\big(si(z)\big)\tau(z)\,dz\pi(s)\,d\mu_G(\dot s)\\ &= \int_G Q_\tau(f)(s)\pi(s)\,d\mu_G(\dot s)\\ &= \pi\big(Q_\tau(f)\big).\end{aligned}$$

We have essentially proved the following.

Proposition D.30. *Suppose that E is a central extension of $\mathbf{T}$ by a locally compact group G as in* (D.8)*, and that $\tau : \mathbf{T} \to \mathbf{T}$ is a character. Then Q_τ defines an isomorphism of $C^*(E,\tau)$ onto the completion of $C_c(E,\tau)$ with respect to the norm*

$$\|f\|_\tau := \sup\{\, \|\pi(f)\| : \pi \text{ is a representation of } E \text{ which preserves } \tau \,\}.$$

Proposition D.30 gives us a powerful method for studying ω-representations of a second countable locally compact group G. If ω is a multiplier of G, then we can form the locally compact central extension G^ω as in (D.4). Then $C^*(G,\omega) := C^*(G^\omega,\tau)$ (with $\tau(z) = z$ for all z) is the completion of $C_c(G^\omega,\tau)$ as above. Nevertheless, some people prefer to work with an underlying structure for $C^*(G,\omega)$ which is closer to the original data — namely G and ω — rather than the extension G^ω and its Mackey topology. In general, this can't be done without working with measurable functions and $L^1(G)$ rather than $C_c(G)$. If ω is a multiplier of G and if $f, g \in \mathcal{L}^1(G)$, then using some care (as in Proposition B.42 on page 348) we can show that

$$(r,s) \mapsto f(r)g(r^{-1}s)\omega(r,r^{-1}s)$$

is measurable and belongs to $\mathcal{L}^1(G \times G)$. Then Fubini's Theorem implies that

$$f * g(s) = \int_G f(r)g(r^{-1}s)\omega(r,r^{-1}s)\,d\mu(r) \tag{D.9}$$

is defined for almost all s, and in fact, defines an element in $L^1(G)$ which depends only on the classes of f and g in $L^1(G)$. As in Remark B.44 on page 351, we can use Fubini's Theorem to justify interchanging the order of integration in the following computation:

$$\begin{aligned}(f*g)*h(s) &= \int_G f*g(r)h(r^{-1}s)\omega(r,r^{-1}s)\,d\mu(r)\\ &= \int_G\int_G f(t)g(t^{-1}r)h(r^{-1}s)\omega(t,t^{-1}r)\omega(r,r^{-1}s)\,d\mu(t)\,d\mu(r)\\ &= \int_G\int_G f(t)g(r)h(r^{-1}t^{-1}s)\omega(t,r)\omega(tr,r^{-1}t^{-1}s)\,d\mu(r)\,d\mu(t)\end{aligned}$$

which, since $\omega(t,r)\omega(tr,r^{-1}t^{-1}s)=\omega(t,t^{-1}s)\omega(r,r^{-1}t^{-1}s)$, equals

$$\begin{aligned}&=\int_G f(t)\int_G g(r)h(r^{-1}t^{-1}s)\omega(r,r^{-1}t^{-1}s)\,d\mu(r)\omega(t,t^{-1}s)\,d\mu(t)\\&=\int_G f(t)g*h(t^{-1}s)\omega(t,t^{-1}s)\,d\mu(t)\\&=f*(g*h)(s).\end{aligned}$$

Thus (D.9) defines an associative product on $L^1(G)$. Similar computations show that

$$f^*(s):=\overline{\omega(s,s^{-1})}\Delta(s^{-1})\overline{f(s^{-1})} \tag{D.10}$$

is an involution on $L^1(G)$ making $L^1(G)$ into a Banach $*$-algebra that we denote by $L^1(G,\omega)$ to distinguish it from $L^1(G)$ with the usual group C^*-algebra structure.

If π' is an ω-representation of G on $\mathcal{H}$ and $f\in\mathcal{L}^1(G)$, then $s\mapsto f(s)\pi'(s)h$ is in $\mathcal{L}^1(G,\mathcal{H})$ for each $h\in\mathcal{H}$.[3] Thus we can define a linear operator $\pi'(f)$ on $\mathcal{H}$ by

$$\pi'(f)h=\int_G f(s)\pi'(s)h\,d\mu(s)\quad\text{for all } h\in\mathcal{H}.$$

Clearly $\|\pi'(f)\|\le\|f\|_1$, and more gyrations with Fubini's Theorem justify

$$\begin{aligned}\pi'(f*g)h&=\int_G\int_G f(r)g(r^{-1}s)\omega(r,r^{-1}s)\pi'(s)h\,d\mu(r)\,d\mu(s)\\&=\int_G\int_G f(r)g(s)\omega(r,s)\pi'(rs)h\,d\mu(s)\,d\mu(r)\\&=\int_G\int_G f(r)g(s)\pi'(r)\pi'(s)h\,d\mu(s)\,d\mu(r)\\&=\pi'(f)\pi'(g)h.\end{aligned}$$

Similarly, $\pi'(f^*)h=\pi'(f)^*h$, and π' extends to a $*$-homomorphism of $L^1(G,\omega)$ into $B(\mathcal{H})$. We can define a C^*-norm on $L^1(G,\omega)$ by

$$\|f\|_\omega:=\sup\{\,\|\pi'(f)\|:\pi'\text{ is an }\omega\text{-representation of }G\,\}\le\|f\|_1. \tag{D.11}$$

We want to see that the completion of $L^1(G,\omega)$ in the $\|\cdot\|_\omega$-norm is (isomorphic to) $C^*(G,\omega):=C^*(G^\omega,\tau)$ with $\tau(z)=z$ for all $z\in\mathbf{T}$. However, we can define

$$\Phi:C_c(G^\omega,\tau)\to L^1(G,\omega) \tag{D.12}$$

[3] Vector valued integration is discussed in Appendix B. We could avoid vector-valued integrals here by defining $\pi'(f)$ weakly:

$$(\pi'(f)h\mid k):=\int_G f(s)(\pi'(s)h\mid k)\,d\mu(s).$$

by $\Phi(f)(s) := f(1,s)$. Note that

$$\begin{aligned}\Phi(f*g)(s) &= f*g(1,s) \\ &= \int_G f(1,r)g\big((1,r)^{-1}(1,s)\big)\,d\mu(r) \\ &= \int_G f(1,r)g\big(\overline{\omega(r,r^{-1})}\omega(r^{-1},s),r^{-1}s\big)\,d\mu(r) \\ &= \int_G f(1,r)g(1,r^{-1}s)\omega(r,r^{-1}s)\,d\mu(r) \\ &= \Phi(f)*\Phi(g)(s).\end{aligned}$$

Since we easily check that $\Phi(f^*) = \Phi(f)^*$, Φ is a $*$-homomorphism. If $g \in C_c(G)$, then $f(z,s) := \bar{z}g(s)$ defines an element of $C_c(G^\omega,\tau)$ such that $\Phi(f) = g$. Therefore Φ is has dense range. If π' is an ω-representation of G and if π is the corresponding representation of G^ω, then

$$\pi(f) = \pi'\big(\Phi(f)\big) \quad \text{for all } f \in C_c(G^\omega,\tau).$$

Therefore $\|f\|_\tau = \|\Phi(f)\|_\omega$ and Φ is isometric. We've summarize the above discussion in the following proposition.

Proposition D.31. *If ω is a multiplier on a second countable locally compact group G, then* (D.12) *defines an isomorphism of $C^*(G,\omega)$ onto the completion of $L^1(G,\omega)$ with respect to the norm $\|\cdot\|_\omega$ defined in* (D.11).

D.4 Quasi-invariant Measures

A Borel space $(X,\mathscr{B})$ is called a *Borel G-space* if G acts on X in such a way that $(s,x) \mapsto s\cdot x$ is a Borel map from $\big(G\times X,\mathscr{B}(G)\times\mathscr{B}\big)$ to $(X,\mathscr{B})$. If μ is a measure on X, then $s\cdot\mu$ is, by definition, the measure given by $s\cdot\mu(E) := \mu(s^{-1}\cdot E)$. We say that μ is *quasi-invariant* if μ and $s\cdot\mu$ are equivalent for all $s\in G$. Recall that if $\tau : X\to Y$ is a Borel map and if μ is a measure on X, then the *push-forward* of μ by τ is the measure $\tau_*\mu$ on Y given by $\tau_*(E) := \mu\big(\tau^{-1}(E)\big)$ (see Lemma H.13 on page 463). Notice that if τ is a Borel isomorphism, then $\tau_*\mu$ is σ-finite whenever μ is.

Lemma D.32. *Suppose that G is a second countable locally compact group. Let μ be a finite quasi-invariant measure on a Borel G-space X and define $\sigma : G\times X\to G\times X$ by $\sigma(s,x) = (s,s\cdot x)$. Then σ is a Borel isomorphism, and if ν is any σ-finite Borel measure on G, then $\sigma_*(\nu\times\mu)$ is equivalent to $\nu\times\mu$.*

Proof. Clearly, σ is Borel with Borel inverse $(s,x)\mapsto(s,s^{-1}\cdot x)$, so σ is a Borel isomorphism as claimed. Thus we only have to show that $\sigma_*(\nu\times\mu)$ is equivalent to $\nu\times\mu$. We can replace ν by an equivalent finite measure and assume that ν is finite. Let D be a Borel subset of $G\times X$ and let $f = \mathbb{1}_D$. Since Fubini's Theorem implies that

$$\sigma_*(\nu\times\mu)(D) = \int_G\int_X f(s,s\cdot x)\,d\mu(x)\,d\nu(s), \tag{D.13}$$

it suffices to show that (D.13) is zero if and only if $\nu \times \mu(D) = 0$. As usual, let $D_s := \{ x \in X : (s, x) \in D \}$. Then (D.13) vanishes if and only if for ν-almost all s,

$$\mu(\{ x \in X : s \cdot x \in D_s \}) = \mu(s^{-1}D_s) = 0.$$

Since μ is quasi-invariant, $\mu(s^{-1}D_s) = 0$ if and only if $\mu(D_s) = 0$. Thus $\sigma_*(\nu \times \mu)(D) = 0$ if and only if for ν-almost all s, $\mu(D_s) = 0$. But by Fubini's Theorem, $\mu(D_s) = 0$ for ν-almost all s if and only if $\nu \times \mu(D) = 0$. □

Since $\sigma_*(\nu \times \mu)$ is equivalent to $\nu \times \mu$ we can assume that its Radon-Nikodym derivative strictly positive everywhere.

Remark D.33 (Measured Groupoids). Let $\kappa : G \times X \to G \times X$ be given by $k(s, x) := (s^{-1}, s^{-1} \cdot x)$. Then κ is a Borel isomorphism and for appropriate f, and we can use Lemma D.32 on the preceding page to compute that

$$\begin{aligned}\iint_{G\times X} f(s,x)\kappa_*(\mu_G \times \mu)(s,x) &= \int_G \int_X f(s^{-1}, s^{-1} \cdot x)\, d\mu(x)\, d\mu_G(s) \\ &= \int_G \int_X f(s, s \cdot x)\Delta(s^{-1})\, d\mu(x)\, d\mu_G(s) \\ &= \int_G \int_X f(s,x)\Delta(s^{-1}) \frac{d\sigma_*(\mu_G \times \mu)}{d(\mu_G \times \mu)}(s,x)\, d\mu(x)\, d\mu_G(s).\end{aligned}$$

Since can assume that $\Delta(s^{-1})\frac{d\sigma_*(\mu_G\times\mu)}{d(\mu_G\times\mu)}(s, x)$ is strictly positive almost everywhere, it follows that $\kappa_*(\mu_G \times \mu)$ is equivalent to $\mu_G \times \mu$. In order to apply some selection results, we will need to realize that these observations amount to showing that $\mathfrak{G} := G \times X$ together with the measure $\mu_G \times \mu$ is what is called a *measured groupoid* in [140] or [142]. The groupoid operations on $\mathfrak{G}$ are given by $(s, x)^{-1} := (s^{-1}, s^{-1} \cdot x)$ and $(s, x)(r, s^{-1} \cdot x) := (sr, x)$. For more on groupoids and the selection results below we need see Remark G.26 on page 450.

Let $s \cdot \mu$ be the push-forward of μ by left-translation by s; thus $s \cdot \mu(E) = \mu(s^{-1} \cdot E)$. Since μ is quasi-invariant, for each s there is a Radon-Nikodym derivative $\frac{d(s\cdot\mu)}{d\mu}$ which is determined only almost everywhere. We would like to select representatives for $\frac{d(s\cdot\mu)}{d\mu}$ so that

$$(s, x) \mapsto \frac{d(s \cdot \mu)}{d\mu}(x) \tag{D.14}$$

is Borel on $G \times X$. Although at first blush it seems "obvious" that we should be able to do so, our proof requires a very subtle selection theorem in measured groupoid theory due to Ramsay. (But see Remark D.35 on page 391.) To use Ramsay's result, we'll need to assume that ν is equivalent to Haar measure on G. Then $\big((G \times X), \nu \times \mu\big)$ is a measured groupoid (Remark D.33). In the sequel, we'll abuse notation a bit and write

$$f(x) = \frac{d(s \cdot \mu)}{d\mu}(x) \quad \text{or} \quad f = \frac{d(s \cdot \mu)}{d\mu}$$

to mean that f is a specific choice of Radon-Nikodym derivative.

Corollary D.34. *Suppose that G is a second countable locally compact group. Let μ be a finite quasi-invariant finite Borel measure on an analytic Borel G-space X, let σ be as in Lemma D.32 on page 387 and let μ_G be Haar measure. Then there is a Borel function $d : G \times X \to (0, \infty)$ such that*

(a) *d is a Radon-Nikodym derivative for $\sigma_*(\mu_G \times \mu)$ with respect to $\mu_G \times \mu$,*

(b) *such that for all $s, r \in G$,*

$$d(sr, x) = d(s, x)d(r, s^{-1} \cdot x) \quad \text{for } \mu\text{-almost all } x, \text{ and} \tag{D.15}$$

(c) *such that for all $s \in G$, $d(s, \cdot)$ is a Radon-Nikodym derivative for $s \cdot \mu$ with respect to μ. That is,*

$$d(s, \cdot) = \frac{ds \cdot \mu}{d\mu} \tag{D.16}$$

for all $s \in G$.

We call d a Borel choice of Radon-Nikodym derivatives.

Proof. For convenience, we can assume that μ is a probability measure and let ν be a finite Borel measure equivalent to μ_G. By Lemma D.32 on page 387, there is a Borel function $d_0 : G \times X \to (0, \infty)$ which is a Radon-Nikodym derivative for $\sigma_*(\nu \times \mu)$ with respect to $\nu \times \mu$.

Note that for any bounded Borel function f,

$$\iint_{G\times X} f(s, x)\, d\sigma_*(\nu \times \mu)(s, x) = \int_G \int_X f(s, s \cdot x)\, d\mu(x)\, d\nu(s) \tag{D.17}$$

$$= \int_G \int_X f(s, x)\, d(s \cdot \mu)(x)\, d\nu(s). \tag{D.18}$$

On the other hand, the definition of d_0 implies that (D.17) is equal to

$$\int_G \int_X f(s, x) d_0(s, x)\, d\mu(x)\, d\nu(s) = \int_G \int_X f(s, x)\, d\mu_s(x)\, d\nu(s), \tag{D.19}$$

where $\mu_s := d_0(s, \cdot)\mu$. In both cases, Fubini's theorem implies that

$$s \mapsto \int_X f(s, x)\, d(s \cdot \mu)(x) = \int_X f(s, s \cdot x)\, d\mu(x)$$

and

$$s \mapsto \int_X f(s, x)\, d\mu_s(x) = \int_X f(s, x) d_0(s, x)\, d\mu(x)$$

are Borel for all bounded Borel functions f on $G \times X$. If q is the projection of $G \times X$ onto G, then it follows from (D.17) and the fact that μ is a probability measure that $q_*\big(\sigma_*(\nu \times \mu)\big) = \nu$. Thus (D.18) and (D.19) are both disintegrations of $\sigma_*(\nu \times \mu)$ with respect to ν as in Theorem I.5 on page 482. Thus the uniqueness assertion in that result implies that there is a μ_G-conull set Y such that $s \cdot \mu = \mu_s$ for all $s \in Y$. The uniqueness of the Radon-Nikodym derivative gives

$$d_0(s, \cdot) = \frac{d(s \cdot \mu)}{d\mu} \quad \text{for all } s \in Y.$$

Using Fubini's Theorem, it is easy to check that

$$A := \{\, (s,r) \in G \times G : s \notin Y,\, r \notin Y \text{ and } sr \notin Y \,\}$$

is a $\mu_G \times \mu_G$-null set. If $(s,r) \notin A$ and φ is a bounded Borel function on G, then on the one hand

$$\int_X \varphi(x) d_0(sr,x)\, d\mu(x) = \int_X \varphi(sr \cdot x)\, d\mu(x).$$

On the other hand,

$$\begin{aligned}\int_X \varphi(x) d_0(s,x) d_0(r,s^{-1}\cdot x)\, d\mu(x) &= \int_X \varphi(s\cdot x) d_0(r,x)\, d\mu(x) \\ &= \int_X \varphi(sr\cdot x)\, d\mu(x).\end{aligned} \tag{D.20}$$

It follows that for each $(s,r) \notin A$,

$$d_0(sr,x) = d_0(s,x) d_0(r,s^{-1}\cdot x) \quad \text{for } \mu\text{-almost all } x. \tag{D.21}$$

Since the set of $(s,r,x) \in G \times G \times X$ where (D.21) holds is Borel, it follows that (D.21) holds $\mu_G \times \mu_G \times \mu$-almost everywhere.

Since X is analytic and μ is quasi-invariant, $(G \times X, \mu_G \times \mu)$ is an example of what Ramsay calls a *virtual group* in [140, page 274] and a *measured groupoid* in [142] (Remark D.33 on page 388). Using that language, the fact that (D.21) holds $\mu_G \times \mu_G \times \mu$-almost everywhere implies that d_0 defines an *almost everywhere homomorphism* of $G \times X$ into the multiplicative group $\mathbf{R}^+_\times$ of positive reals. Ramsay's selection theorem (Remark G.26 on page 450) implies that there is a Borel function $d : G \times X \to \mathbf{R}^+_\times$ and a μ-conull set X_0 such that $d = d_0$ $\mu_G \times \mu$-almost everywhere and such that if x, $s^{-1}\cdot x$ and $r^{-1}s^{-1}\cdot x$ all lie in X_0, then

$$d(sr,x) = d(s,x) d(r,s^{-1}\cdot x). \tag{D.22}$$

Since $d = d_0$ almost everywhere, we can shrink Y a bit if necessary so that $d(s,\cdot) = \frac{d(s\cdot\mu)}{d\mu}$ for $s \in Y$. In particular, the set

$$\Sigma := \{\, s \in G : d(s,\cdot) \text{ is a Radon-Nikodym derivative for } s\cdot\mu \text{ with respect to } \mu \,\}$$

is a μ_G-measurable[4] conull subset of G. For fixed $s, r \in \Sigma$, the quasi-invariance of μ implies that

$$\{\, x \in X_0 : s^{-1}\cdot x \in X_0 \text{ and } r^{-1}s^{-1}\cdot x \in X_0 \,\}$$

is μ-conull. Thus for all $s, r \in \Sigma$, we have $d(sr,x) = d(s,x)d(r,s^{-1}\cdot x)$ for μ-almost all x. Repeating the calculation in (D.20) with d in place of d_0 shows that $d(sr,\cdot)$ is a Radon-Nikodym derivative and hence that Σ is a conull subsemigroup of G. Lemma D.36 on the facing page implies that $\Sigma = G$. This completes the proof. □

[4]Recall that a subset is measurable if it a union of a Borel set and a subset of a Borel null set.

Remark D.35. It should be pointed out that one can omit the use of Ramsay's Theorem in the proof at the expense of having (D.15) hold only for almost all (s, r) and having (D.16) hold for μ_G almost all s. (Then we can even replace Haar measure by any σ-finite Borel measure on G.) Since we'll have to invoke Ramsay's result in the proof of the Effros-Hahn conjecture we used it here as well. A proof without invoking Ramsay's result can be found in [176, Theorem B9].

D.5 Technical Results

In this section, we collect a number of technical results needed elsewhere but which either did not seem of sufficient import to include at the time, or which would have been distracting to the matter at hand.

Lemma D.36. *Suppose that G is a locally compact group with left-Haar measure μ_G. If Σ is a measurable conull subsemigroup of G (that is, $\mu_G(G \setminus \Sigma) = 0$), then $\Sigma = G$.*

Proof. Let $H := \Sigma \cap \Sigma^{-1}$. Then H is a conull subgroup of G. Let K be a compact neighborhood of e. Since $K \cap (G \setminus H)$ is a null set, $0 < \mu_G(K \cap H) < \infty$. Let $A := H \cap K$. By Lemma D.4 on page 370, $A^{-1}A \subset H^2 = H$ is a neighborhood of e. Thus H is open and G is the disjoint union of open sets each of which is a translate of H. Since H is conull, $H = G$ and the result follows as $H \subset \Sigma$. □

Lemma D.37. *If H is a separable complex Hilbert space, then*

$$B(\mathcal{H})_1 := \{\, T \in B(\mathcal{H}) : \|T\| \le 1 \,\}$$

is a compact Polish space in the weak-operator topology, and a Polish space in the $*$-strong operator topology.

Proof. Let $\{\, h_k \,\}$ be a countable dense subset of $\mathcal{H}$, and let

$$C(h_k) := \{\, z \in \mathbf{C} : |z| \le \|h_k\|^2 \,\}.$$

Then $\prod_k C(h_k)$ is a compact Polish space in the product topology. The map $\varphi : B(\mathcal{H}_1) \to \prod C(h_k)$ given by $\varphi(T)_k := (Th_k \mid h_k)$ is a continuous injection. Suppose that $\varphi(T_n)$ converges to f in $\prod C(h_k)$. Since $\{\, h_k \,\}$ are dense in $\mathcal{H}$, it follows that $\{\, (T_n h \mid h) \,\}$ converges for all $h \in \mathcal{H}$. Since

$$(T_n h \mid k) = \sum_{j=1}^{3} i^j \big(T_n(h + i^j k) \mid h + i^j k\big),$$

$\{\, (T_n h \mid k) \,\}$ converges for all $h, k \in \mathcal{H}$. The Principle of Uniform boundedness implies that there is an operator $T \in B(\mathcal{H})_1$ such that $T_n \to T$ in the weak operator topology, and that $\varphi(T_n) \to \varphi(T)$. Thus the range of φ is closed and φ is a homeomorphism of $B(\mathcal{H})_1$ onto its range. This proves the first assertion.

The second assertion is proved similarly. Let $\mathcal{H}_k := \{ h \in \mathcal{H} : \|h\| \leq \|h_k\| \}$ with $\{ h_k \}$ as above. Then

$$P := \prod_{k=1}^{\infty} \mathcal{H}_k \times \mathcal{H}_k$$

is a Polish space in the product topology. Define $\varphi : B(\mathcal{H}_1) \to P$ by $\varphi(T)_k = (Th_k, T^*h_k)$. Then φ is clearly injective and continuous in the $*$-strong operator topology. If $\varphi(T_n) \to f$ in P, then since $\{ h_k \}$ is dense in $\mathcal{H}$, $T_n h$ and $T_n^* h$ converge for all $h \in \mathcal{H}$. Just as above, there are operators $T, S \in B(\mathcal{H})_1$ such that $T_n \to T$ and $T_n^* \to S$ in the strong operator topology. It follows that $S = T^*$ so that $T_n \to T$ in the $*$-strong operator topology. Therefore φ has closed range and is a homeomorphism onto its range. Thus $B(\mathcal{H})_1$ is Polish in the $*$-strong operator topology. □

If $(X, \mathscr{B})$ is a Borel space, then by a measure on X we mean a measure on the σ-algebra $\mathscr{B}$. So in the following, our measures are not necessarily Radon measures, nor are they even necessarily complete. In fact, our model Borel space is a topological space equipped with its Borel field $\mathscr{B}(X)$.

In the literature, our next result says that finite measures on Polish spaces are *tight*.

Lemma D.38 ([70, §9(10)]). *Suppose that μ is a finite Borel measure on a Polish space P. If $\epsilon > 0$, then there is a compact set $C \subset P$ such that $\mu(C) \geq \mu(P) - \epsilon$.*

Proof. We may as well assume that $\mu(P) = 1$. Let $0 < \epsilon < 1$, and let $D = \{ x_n \}$ be a countable dense subset of P. Suppose that d is a compatible complete metric on P. Define

$$C(n,k) = \{ x \in P : d(x, x_n) \leq \frac{1}{k} \} \quad \text{and} \quad F(m,k) = \bigcup_{n=1}^{m} C(n,k).$$

Then $C(n,k)$ and $F(m,k)$ are closed sets and for each $k \geq 1$,

$$\bigcup_{m=1}^{\infty} F(m,k) = P$$

Thus $\lim_{m \to \infty} \mu\big(F(m,k)\big) = 1$. Thus there is a m_1 such that $\mu\big(F(m_1,1)\big) > 1 - \epsilon$. But

$$F(m_1, 1) = \bigcup_{m=1}^{\infty} F(m,2) \cap F(m_1,1).$$

Thus there is a m_2 such that

$$\mu\Big(\bigcap_{i=1}^{2} F(m_i, i)\Big) > 1 - \epsilon.$$

Thus we can find a sequence $\{ m_i \}$ such that

$$\mu\Big(\bigcap_{i=1}^{k} F(m_i, i)\Big) > 1 - \epsilon \quad \text{for all } k \geq 1.$$

Let

$$C := \bigcap_{i=1}^{\infty} F(m_i, i).$$

Since we clearly have $\mu(C) \geq 1 - \epsilon$, we just have to check that C is compact. But C is clearly closed and therefore complete. But construction, C is covered by finitely many closed balls of radius $\frac{1}{i}$ for each $i \geq 1$. Thus C is totally bounded and therefore compact by [57, Theorem 0.25]. □

Corollary D.39. *Suppose that μ is a Borel measure on a Polish space P. If E is any subset of P with finite μ-measure and if $\epsilon > 0$, then there is a compact set C such that $\mu(E \setminus C) < \epsilon$.*

Proof. Let μ_E be the finite Borel measure on P given by $\mu_E(B) := \mu(B \cap E)$. Then apply the previous lemma to μ_E. □

Lemma D.40. *Suppose that G is a Borel group with a left-invariant σ-finite measure μ. Then μ is unique up to a positive scalar and has the same null sets as any right-invariant measure on G. In particular, if N is a μ-null set, then so is Ns for all $s \in G$. Moreover, there is a Borel homomorphism $\Delta : G \to (0, \infty)$ such that*

$$\Delta(r) \int_G f(sr)\, d\mu(s) = \int_G f(s)\, d\mu(s) \quad \textit{for all } f \in L^1(G).$$

Proof. Suppose that N is a μ-null set. By assumption, sN is μ-null for all $s \in G$. Let

$$E = \{ (r, s) \in G \times G : r^{-1}s \in N^{-1} \}.$$

Then $E^s := \{ r \in G : r^{-1}s \in N^{-1} \} = sN$ is null for all s, and Fubini's Theorem implies that E is a $\mu \times \mu$-null set. Thus there is a $r_0 \in G$ such that

$$E_{r_0} := \{ s \in G : r^{-1}s \in N^{-1} \} = r_0 N^{-1}$$

is a null set. Since μ is left invariant, N^{-1} is a null set. It follows that $Ns = (s^{-1}N^{-1})^{-1}$ is null for all $s \in G$ whenever N is a null set.

Now suppose that ν is also a left-invariant σ-finite measure on G. We want to see that ν is a multiple of μ. Note that $\lambda = \mu + \nu$ is also left-invariant, and by symmetry it suffices to show that μ and λ are multiples.

Since $\mu \ll \lambda$, we can let $f = d\mu/d\lambda$ be the Radon-Nikodym derivative [57, Proposition 3.8]. Thus for all $\varphi \in L^1(\mu)$,

$$\int_G \varphi(s)\, d\mu(s) = \int_G \varphi(s) f(s)\, d\lambda(s).$$

Since both μ and λ are left-invariant, for each $r \in G$ we have

$$\int_G \varphi(s)f(s)\,d\lambda(s) = \int_G \varphi(s)f(rs)\,d\lambda(s).$$

Since the Radon-Nikodym derivative is uniquely defined up to a λ-null set, for each $r \in G$, $f(s) = f(rs)$ off a λ-null set. But then Fubini's Theorem implies that

$$E := \{\,(s,r) \in G \times G : f(s) \neq f(rs)\,\}$$

is a $\lambda \times \lambda$-null set. Thus there is a $s_0 \in G$ and a λ-null set N such that

$$f(rs_0) = f(s_0) \quad \text{for all } r \notin N.$$

In other words, $f(s) = f(s_0)$ for all $s \notin Ns_0$. Since Ns_0 must be a λ-null set by the above remarks, f is constant almost everywhere. Thus μ and λ are multiples as claimed.

It follows that for each $r \in G$, $\nu(E) := \mu(Er)$ is a left-invariant measure on G and there is a positive scalar $\Delta(r)$ such that $\mu(Er) = \Delta(r)\mu(E)$ for every measurable set E. By considering characteristic functions, we have

$$\int_G f(sr^{-1})\,d\mu(s) = \Delta(r)\int_G f(s)\,d\mu(s) \quad \text{for all } f \in L^1(\mu).$$

Fubini's Theorem implies that

$$r \mapsto \int_G f(sr^{-1})\,d\mu(s)$$

is Borel, and it follows that $r \mapsto \Delta(r)$ is Borel. Since Δ is easily seen to be multiplicative, Δ is a homomorphism.

Since any right-invariant measure is transformed into a left-invariant measure via inversion, the remaining statements follow. □

Lemma D.41. *Suppose that μ is a σ-finite measure on a countably generated Borel space $(X, \mathscr{B})$. Then $L^2(X,\mu)$ is separable.*

Proof. Since μ is σ-finite, it is equivalent to finite measure ν and $L^2(X,\nu)$ is isomorphic to $L^2(X,\mu)$. Thus we may as well assume that μ is finite. Let $\mathscr{D} = \{\,E_i\,\}_{i=1}^{\infty}$ be a countable generating set for $\mathscr{B}$ with $E_1 = X$. Let $\mathscr{A}_n$ be the algebra of sets in X generated by $\{\,E_i\,\}_{i=1}^{n}$. Then the algebra $\mathscr{A}$ generated by $\mathscr{D}$ is the union of the $\mathscr{A}_n$. Since each $\mathscr{A}_n$ is finite,[5] $\mathscr{A}$ is countable. We will show that the set of characteristic functions of sets in $\mathscr{A}$ (each of which is in L^2 since μ is finite) is dense. Thus it will suffice to see that any f orthogonal to each such function is zero. Let $\mathscr{M}$ be the set of Borel sets E such that

$$\int_E f(x)\,d\mu(x) = 0.$$

[5] To see that $\mathscr{A}_n$ is finite, note that it consists of finite unions of its "atoms": $E_x := \bigcap\{\,E_i : 1 \leq i \leq n \text{ and } x \in E_i\,\}$.

The dominated convergence theorem implies that $\mathscr{M}$ is closed under countable increasing unions and countable decreasing intersections. Since $\mathscr{A} \subset \mathscr{M}$, $\mathscr{M}$ must contain the σ-algebra generated by $\mathscr{A}$ by the Monotone Class Lemma ([57, Lemma 2.35]). Since $\mathscr{D} \subset \mathscr{A}$, we must have $\mathscr{M} = \mathscr{B}$. It follows that $f = 0$ in L^2. □

Lemma D.42. *Let $U(\mathcal{H})$ be the group of unitary operators on a complex Hilbert space $\mathcal{H}$. Then the weak and strong operator topologies coincide on $U(\mathcal{H})$, and $U(\mathcal{H})$ is a topological group with respect to this topology. If $\mathcal{H}$ is separable, then $U(\mathcal{H})$ is a Polish group.*

Proof. Since the the strong operator topology is finer than the weak operator topology, in order to see that the two topologies coincide on $U(\mathcal{H})$, it is enough to see that whenever $U_\alpha \to U$ weakly, then $\|U_\alpha h - Uh\| \to 0$ for all $h \in \mathcal{H}$. But

$$\begin{aligned}\|U_\alpha h - Uh\|^2 &= (U_\alpha h - Uh \mid U_\alpha h - Uh) \\ &= 2\|h\|^2 - (U_\alpha h \mid Uh) - (Uh \mid U_\alpha h)\end{aligned}$$

which tends to 0 as α increases. Since multiplication on bounded subsets is continuous in the strong operator topology and since taking adjoints is continuous in the weak operator topology, it follows that $(U, V) \to UV^{-1} = UV^*$ is continuous in $U(\mathcal{H})$. Thus $U(\mathcal{H})$ is a topological group. Since $U(\mathcal{H})$ is clearly closed in $B(\mathcal{H})_1$ in the $*$-strong operator topology, it follows from Lemma D.37 on page 391 that $U(\mathcal{H})$ is Polish when $\mathcal{H}$ is separable. □

Lemma D.43. *Suppose that G is a locally compact group and that $\mathcal{H}$ is a separable Hilbert space. Let $\varphi : G \to U(\mathcal{H})$ be a function. Then the following are equivalent.*

(a) *φ is Borel.*

(b) *$s \mapsto \varphi(s)h$ is Borel from G to $\mathcal{H}$ for all $h \in \mathcal{H}$.*

(c) *$s \mapsto \big(\varphi(s)h \mid k\big)$ is Borel from G to $\mathbf{C}$ for all $h, k \in \mathcal{H}$.*

Proof. Since the composition of a continuous map with a Borel map is Borel, we clearly have (a) $\Longrightarrow$ (b) $\Longrightarrow$ (c). To show that (c) $\Longrightarrow$ (a), we note that the strong and weak operator topologies coincide on $U(\mathcal{H})$ (Lemma D.42). Thus it will suffice to show that $\varphi^{-1}(V)$ is Borel for a weak operator open set V in $U(\mathcal{H})$. Since $\mathcal{H}$ is separable, every such set is a countable union of finite intersections of sets of the form

$$V(U_0, h, k, \epsilon) := \{\, U \in U(\mathcal{H}) : |(Uh \mid k) - (U_0 h \mid k)| < \epsilon \,\}$$

for $U_0 \in U(\mathcal{H})$; $h, k \in \mathcal{H}$; and $\epsilon > 0$. Since (c) implies that $\varphi^{-1}\big(V(U_0, h, k, \epsilon)\big)$ is Borel, this suffices. □

Lemma D.44. *Suppose that $(X, \mathscr{B})$ is a countably separated Borel space. Then there is an injective Borel map $\varphi : X \to [0, 1]$. If μ is a σ-finite measure on $(X, \mathscr{B})$, then we may choose $\varphi \in L^2(X, \mu)$.*

Proof. Suppose that $\{E_i\}_{i=1}^{\infty} \subset \mathscr{B}$ separates points in X. Let $\sigma(x)_i := \mathbb{1}_{E_i}(x)$. Then

$$\sigma : X \to \prod_{i=1}^{\infty} \mathbf{Z}_2$$

is injective and Borel. We can define a continuous map

$$\beta : \prod_{i=1}^{\infty} \mathbf{Z}_2 \to [0,1] \quad \text{by} \quad \beta(x) := \sum_{i=1}^{\infty} \frac{x(i)}{3^i}.$$

To see that β is injective, suppose $x \neq y$ in $\prod \mathbf{Z}_2$. Then there is a $k \geq 1$ such that $x(i) = y(i)$ for $i \leq k-1$, and $x(k) \neq y(k)$. We may as well assume that $x(k) = 1$ and $y(k) = 0$. If

$$A := \sum_{i=1}^{k-1} \frac{x(i)}{3^i},$$

then

$$\beta(x) \geq A + \frac{1}{3^k},$$

while

$$\beta(y) \leq A + \sum_{i=k+1}^{\infty} \frac{1}{3^i} = A + \frac{1}{2} \cdot \frac{1}{3^k} < \beta(x).$$

Thus β is injective and we can set $\varphi := \beta \circ \sigma$.

If μ is a σ-finite measure, then there is a sequence of disjoint sets X_i such that $X = \bigcup X_i$ and $\mu(X_i) < \infty$ for all i. We can choose integers n_i such that

$$\frac{\mu(X_i)}{n_i} \leq \frac{1}{i} \quad \text{and} \quad n_{i+1} > n_i \geq i. \tag{D.23}$$

If φ is defined as above, then define

$$\psi(x) = \frac{1}{\frac{1}{4} + \frac{1}{2}\varphi(x) + n_i} \quad \text{for } x \in X_i.$$

Then ψ is clearly Borel. Since $x \in X_i$ implies that

$$\frac{1}{n_{i+1}} \leq \frac{1}{n_i + 1} < \psi(x) < \frac{1}{n_i}, \tag{D.24}$$

$\psi(x) = \psi(y)$ implies that $x, y \in X_i$ for some i. Since φ is one-to-one, this implies $x = y$ and ψ is injective. In view of (D.23) and (D.24), we also have

$$\begin{aligned} \int_X |\psi(x)|^2 \, d\mu(x) &= \sum_{i=1}^{\infty} \int_{X_i} |\psi(x)|^2 \, d\mu(x) \\ &\leq \sum_{i=1}^{\infty} \frac{1}{n_i^2} \mu(X_i) \\ &\leq \sum_{i=1}^{\infty} \frac{1}{i^2} < \infty. \end{aligned}$$

Thus ψ is in $L^2(X, \mu)$ as required. □

Corollary D.45. *If X is a standard Borel space, then there is a Borel isomorphism of X onto a Borel subset of $[0,1]$.*

Proof. Since X is countably separated, Lemma D.44 on page 395 implies there is an injective Borel map $f : X \to [0,1]$. Then [2, Theorem 3.3.2] implies that $f(X)$ is Borel and that f is an isomorphism onto its range. □

If $\tau : X \to Y$ is a Borel map and if μ is a measure on X, then we say that τ is *essentially constant* if there is a $y_0 \in Y$ such that $\tau(x) = y_0$ for μ-almost all x. (Of course, y_0 is unique provided μ is nonzero.) Our next result is subtle, and is as close as we can come to proving that an *essentially equivariant map* is equal almost everywhere to an equivariant map. We first have to throw away a null set. Note that we are not claiming that the restriction of an essentially equivariant map to a conull set is equivariant.

Theorem D.46 ([176, Proposition B.5]). *Suppose that G is a second countable locally compact group and that X and Y are standard Borel G-spaces. Let μ be a σ-finite measure on X. If $\tau : X \to Y$ is a Borel map such that for all $s \in G$, $\tau(s \cdot x) = s \cdot \tau(x)$ for μ-almost all x, then there is a G-invariant μ-conull Borel set $X_0 \subset X$ and a Borel map $\tilde{\tau} : X \to Y$ such that $\tau = \tilde{\tau}$ almost everywhere, and for all $x \in X_0$ and all $s \in G$, $\tilde{\tau}(s \cdot x) = s \cdot \tilde{\tau}(x)$.*

Proof. Let

$$X_0 = \{\, x \in X : s \mapsto s^{-1} \cdot \tau(s \cdot x) \text{ is essentially constant.} \,\}$$

Let ν be a probability measure on G which is equivalent to Haar measure on G. In view of Corollary D.45, we can assume that Y is a Borel subset of $[0,1]$. Fubini's Theorem implies that

$$\hat{\tau}(x) := \int_G s^{-1} \cdot \tau(s \cdot x)\, d\nu(s)$$

is a Borel function from X to $[0,1]$. Similarly,

$$J(x) := \int_G |s^{-1} \cdot \tau(s \cdot x) - \hat{\tau}(x)|\, d\nu(s)$$

is Borel from X to $[0,1]$ and $X_0 = J^{-1}(\{\,0\,\})$. Thus X_0 is Borel and $\hat{\tau}(X_0) \subset Y \subset [0,1]$. If $y_0 \in Y$, we can define a Borel function $\tilde{\tau} : X \to Y$ by

$$\tilde{\tau}(x) := \begin{cases} \hat{\tau}(x) & \text{if } x \in X_0, \text{ and} \\ y_0 & \text{otherwise.} \end{cases}$$

Let $D = \{\, (s,x) : s^{-1} \cdot \tau(s \cdot x) \neq \tau(x) \,\}$. Then D is Borel and

$$D_s := \{\, x : s^{-1} \cdot \tau(s \cdot x) \neq \tau(x) \,\}$$

is μ-null for all s by assumption. Thus D is $\mu_G \times \mu$-null and Fubini's Theorem implies that there is a μ-conull set X_{00} such that $x \in X_{00}$ implies that

$$D^x := \{\, s : s^{-1} \cdot \tau(s \cdot x) \neq \tau(x) \,\}$$

is μ_G-null. Since $X_{00} \subset X_0$, X_0 is conull.

Suppose that $x \in X_0$ and that $s \in G$. Then there is a μ_G-null set N such that $t \notin N$ implies

$$t^{-1} \cdot \tau(t \cdot x) = \tilde{\tau}(x).$$

Since Ns^{-1} is μ_G-null (Lemma D.40 on page 393),

$$s^{-1}r^{-1} \cdot \tau(rs \cdot x) = \tilde{\tau}(x) \quad \text{for almost all } r.$$

Thus,

$$r^{-1} \cdot \tau(rs \cdot x) = s \cdot \tilde{\tau}(x) \quad \text{for almost all } r.$$

This shows that $s \cdot x \in X_0$ and that $\tilde{\tau}(s \cdot x) = s \cdot \tilde{\tau}(x)$. This completes the proof. □

Lemma D.47. *Suppose that Y is a Borel G-space and that μ is a finite ergodic measure on Y. If $f : Y \to X$ is a G-invariant Borel map into a countably separated Borel space X, then f is essentially constant.*

Proof. Let $\{ A_i \}$ be a separating family of Borel sets in X. Since $f^{-1}(A_i)$ and $f^{-1}(X \setminus A_i)$ are G-invariant, one is μ-null and the other conull. Define

$$B_i := \begin{cases} A_i & \text{if } f^{-1}(A_i) \text{ is conull, and} \\ X \setminus A_i & \text{if } f^{-1}(X \setminus A_i) \text{ is conull.} \end{cases}$$

Note that $\{ B_i \}$ still separates points in X. Furthermore, writing $B^c := Y \setminus B$, we have

$$\mu\Big(f^{-1}\Big(\bigcap B_i\Big)^c\Big) = \mu\Big(\bigcup f^{-1}(B_i)^c\Big) = 0.$$

Thus $\bigcap B_i \neq \emptyset$. Since the B_i separate points, $\bigcap B_i = \{ x_0 \}$ and $f(y) = x_0$ for μ-almost all y. □

Appendix E

Representations of C^*-algebras

In this chapter, which is primarily expository, we want to review some of the well-known (albeit a bit unfashionable these days) basics of representation theory for C^*-algebras. Much of the material in this Appendix is a brutal shortening of Chapters 2 and 4 of Arveson's wonderful little book [2]. Anyone wanting the real story should skip this appendix, and read Arveson's book. In particular, we often refer to Arveson's book for the proofs.

One of the basic goals of any representation theory is to break down general representations into their basic building blocks. Our first instinct is to break representations down into "irreducible" bits. Although this works very well in finite dimensions, it is not very useful for infinite dimensional representations — even in the commutative case. Another approach, using the notion of multiplicity as defined below, is very useful for the class of GCR C^*-algebras — which, of course, includes the commutative ones — and provides the basis for more general approaches as well. However, these more general approaches require that both the C^*-algebra and the space of the representations in question be separable. Consequently, we will shortly be making such assumptions, and we will not complicate matters by holding onto the general situation overly long.

E.1 Multiplicity

Recall that a *representation* of a C^*-algebra A is a *nondegenerate* homomorphism $\pi : A \to B(\mathcal{H}_\pi)$ for some complex Hilbert space $\mathcal{H}_\pi$. Two representations π and σ are *equivalent* if there is a unitary operator $U : \mathcal{H}_\pi \to \mathcal{H}_\sigma$ such that $\sigma(a) = U\pi(a)U^*$ for all $a \in A$. We say that π is separable if $\mathcal{H}_\pi$ is separable, *n-dimensional* if $\mathcal{H}_\pi$ is n-dimensional, *infinite-dimensional* if $\mathcal{H}_\pi$ is infinite dimensional, etc. A subspace $V \subset \mathcal{H}_\pi$ is called *invariant* if $\pi(a)h \in V$ for all $a \in A$ and $h \in V$. Since π is $*$-preserving, it is not hard to see that V invariant implies that $V^\perp$ is also invariant. If V is closed — so that V is a Hilbert space — then the orthogonal projection P

onto V commutes with $\pi(a)$ for all $a \in A$ (we say simply that P commutes with π). Conversely, if P is an orthogonal projection commuting with π, then the space of P is a closed invariant subspace V for π. If P is the orthogonal projection onto the subspace V, and if P commutes with π, then the restriction π^P of π to V is a representation of A called a *subrepresentation* of π.[1] Provided V is closed and P is the corresponding projection, we have

$$\pi = \pi^P \oplus \pi^{I-P}.$$

A representation is called *irreducible* if it has no nontrivial closed invariant subspaces — and hence no nontrivial subrepresentations.[2] Irreducible representations are a critical component in the theory of C^*-algebras, but in the case of infinite-dimensional representations, it is not possible to express every representation as a direct sum of irreducibles (as it is in the finite-dimensional case). For example, the natural representation of $C_0(\mathbf{R})$ on $L^2(\mathbf{R})$ via multiplication has no irreducible subrepresentations at all. (Recall that all irreducible representations of $C_0(\mathbf{R})$ are point-evaluations and therefore 1-dimensional.) Nevertheless, we will have more to say about decomposition of this representation and others into irreducibles latter. First we introduce a concept which does allow a decomposition into direct sums for a large class of representations.

Definition E.1. A representation π of a C^*-algebra A is called *multiplicity-free* if π does not have two equivalent orthogonal nonzero subrepresentations.

Remark E.2. Clearly, π is multiplicity-free provided that there is no nonzero representation σ of A such that $\sigma \oplus \sigma$ is equivalent to a subrepresentation of π.

If σ is a subrepresentation of π, then since the orthogonal projection P onto the space of σ must commute with π, P belongs to the *commutant*

$$\pi(A)' := \{\, T \in B(\mathcal{H}_\pi) : T\pi(a) = \pi(a)T \text{ for all } a \in A \,\}$$

of π. Note that the *double commutant* $\pi(A)''$ (the commutant of $\pi(A)'$) is the closure of $\pi(A)$ in the weak operator topology (by von Neumann's Double Commutant Theorem [110, Theorem 4.2.5]). It coincides with the strong closure of $\pi(A)$ by [110, Theorem 4.2.7]. Thus $\mathcal{Z} := \pi(A)' \cap \pi(A)''$ can be viewed as the center of either the commutant of $\pi(A)$ or of the von Neumann algebra generated by $\pi(A)$. We call σ a *central subrepresentation* of π if the corresponding orthogonal projection belongs to the center of the von Neumann algebra generated by π. If $\pi(A)'$ is commutative, then $\mathcal{Z} = \pi(A)'$ and every subrepresentation is central.

Lemma E.3 ([2, p. 42]). *A representation π of a C^*-algebra A is multiplicity-free if and only if the commutant $\pi(A)'$ is commutative.*

[1] Some care is needed here as we require representations to be nondegenerate. But π is nondegenerate if and only if $\pi(e_i) \to I_{\mathcal{H}_\pi}$ in the strong operator topology for any approximate identity $\{\, e_i \,\}$ of A. Thus $\pi^P(e_i) \to I_V$ and π^P is nondegenerate.

[2] It is known that irreducible representations of C^*-algebras are actually *algebraically* irreducible in that they have no nontrivial invariant subspaces at all ([28, Corollary 2.8.4] or [126, Theorem 3.13.2]).

Proof. Suppose that σ and η equivalent nonzero orthogonal subrepresentations of π. Then there is a partial isometry $U \in \pi(A)'$ such that U^*U is the orthogonal projection onto the space of σ and UU^* the orthogonal projection onto the space of η. Thus U and U^* do not commute and $\pi(A)'$ is not commutative.

Conversely, suppose that $\pi(A)'$ is not commutative. Then there is a self-adjoint operator $T \in \pi(A)'$ that does not commute with all of $\pi(A)'$. Using the spectral theorem, there is a spectral projection $P \in \pi(A)'$ for T with the same property. Thus there is a $S \in \pi(A)'$ such that

$$R := PS(I - P) \neq 0.$$

Let $R = U|R|$ be the polar decomposition of R (see [2, p. 7] or [126, Proposition 2.2.9]). Then

$$U^*U \leq (I - P) \quad \text{and} \quad UU^* \leq P.$$

Thus U is a nonzero partial isometry in $\pi(A)'$ with orthogonal initial and final spaces. Thus π^{U^*U} and π^{UU^*} are equivalent orthogonal subrepresentations of π, and π is not multiplicity-free. □

Definition E.4. Two representations π and σ of a C^*-algebra A are called *disjoint* if no nonzero subrepresentation of π is equivalent to any subrepresentation of σ.

Let X be a *second countable* locally compact Hausdorff space so that $C_0(X)$ is a *separable* commutative C^*-algebra.[3] Let μ be a *finite* Borel measure on X. If $\mathcal{B}^b(X)$ is the set of complex-valued bounded Borel functions on X, then each $f \in \mathcal{B}^b(X)$ determines a bounded operator L_f on $L^2(X, \mu)$ via

$$L_f h(x) := f(x)h(x).$$

(Of course, we could equally well have used $L^\infty(X, \mu)$ in place of $\mathcal{B}^b(X)$ as $L_f = L_g$ if f and g agree μ-almost everywhere. However, we don't gain anything and it will be convenient later, in the vector-valued case, not to be burdened with the niceties of defining L^∞.) As in [2, §2.2], we make the following conventions.

Definition E.5. Let μ be a finite measure on the locally compact space X. We let

$$\mathscr{L} := \{\, L_f \in B\big(L^2(X)\big) : f \in \mathcal{B}^b(X)\,\},$$

and we write π_μ for the representation

$$\pi_\mu : C_0(X) \to B\big(L^2(X)\big)$$

given by $\pi_\mu(f) = L_f$.

The representations π_μ are fundamental to the what follows. Their basic properties are developed in [2, §2.2] and we summarize a few of them here.

[3] Of course, every separable commutative C^*-algebra is isomorphic to $C_0(\hat{A})$ with $\hat{A}$ second countable.

Theorem E.6 ([2, Theorems 2.2.1, 2.2.2 and 2.2.4] — see also Proposition I.41). *Suppose that X is a second countable locally compact Hausdorff space and that μ is a finite measure on X.*

(a) *$\mathscr{L}$ is the closure of $\pi_\mu\big(C_0(X)\big)$ in the strong operator topology and*

(b) *$\mathscr{L}' = \mathscr{L}$.*

If ν is also a finite measure on X, then

(c) *π_μ is equivalent to π_ν if and only if μ and ν are equivalent measures, and*

(d) *π_μ and π_ν are disjoint if and only if μ and ν are disjoint measures.*

Moreover, each π_μ is multiplicity-free. In fact,

(e) *A representation π of $C_0(X)$ is multiplicity-free if and only if there is a finite Borel measure μ on X such that π is equivalent to π_μ.*

Remark E.7. Part (a) implies that $\mathscr{L}$ is a strongly closed subalgebra of $B\big(L^2(X,\mu)\big)$. Thus $\mathscr{L}$ is not only a C^*-algebra, but a von Neumann algebra as well. Part (b) implies that $\mathscr{L}$ is a *maximal abelian subalgebra* of $B\big(L^2(X)\big)$ — that is, there is no strongly closed commutative subalgebra of $B\big(L^2(X)\big)$ which properly contains $\mathscr{L}$. Lemma E.3 on page 400 and part (b) imply that each π_μ is multiplicity-free. The converse (part (e)) follows from the fact that multiplicity-free representations of separable C^*-algebras must have a cyclic vector [2, Lemma 2.2.3]. Consequently, every multiplicity-free representation of a separable C^*-algebra is separable.

We want to suggest that multiplicity-free representations make good "building blocks" for more general representations. For example, if π is any representation of a C^*-algebra A, then we can define its multiple $n \cdot \pi$ for any integer $n \geq 1$: $n \cdot \pi$ denotes the representation $\pi \oplus \cdots \oplus \pi$ on $\mathcal{H}_\pi \oplus \cdots \oplus \mathcal{H}_\pi$ where there are n-summands in each case. We also allow the case $n = \infty$ where there are countably infinite many summands.[4] Note that $n \cdot \pi$ is equivalent to $\pi \otimes 1_{\mathcal{H}}$ on $\mathcal{H}_\pi \otimes \mathcal{H}$ where $\mathcal{H}$ is any n-dimensional Hilbert space [139, p. 254]. Naturally, if $n \geq 2$, then $n \cdot \pi$ is not multiplicity-free and we'd even like to say that $n \cdot \pi$ has "multiplicity n". In order for such a definition to be well defined, we need must first observe that if $n \cdot \pi$ is equivalent to $m \cdot \sigma$, then $n = m$. (Note that we do not claim we also have π and σ equivalent.) Arveson does this in [2, Proposition 2.1.3], and we take advantage of this to make the following fundamental definition.

Definition E.8. A representation π of A has *multiplicity* n ($1 \leq n \leq \infty$) if there is a multiplicity-free representation σ of A such that π is equivalent to $n \cdot \sigma$. Equivalently, π has multiplicity n if π has n orthogonal mutually equivalent subrepresentations $\{ \pi_i \}$ such that $\pi = \bigoplus_i \pi_i$.

Example E.9. Suppose that μ is a finite measure on a locally compact space X and that $\mathcal{H}_n$ is a Hilbert space of dimension $1 \leq n \leq \infty$. Then $n \cdot \pi_\mu$ is equivalent to

[4] There is no reason to restrict to countably many summands for this definition, but as we are ultimately only interested in separable representations, countably infinite suffices for our purposes.

the representation $\pi_\mu \otimes 1_{\mathcal{H}_n}$ on $L^2(X) \otimes \mathcal{H}_n$ which we identify with $L^2(X, \mathcal{H}_n)$.[5] We shall see that every representation of $C_0(X)$ with multiplicity n is equivalent to $n \cdot \pi_\mu$ for a finite measure μ on X.

Since it can be impossible to break infinite-dimensional representations into direct sums of irreducibles, another approach is to try to break a representation up into direct sums of multiplicity-free representations. This is also not going to be possible in general, but there are large classes of C^*-algebras for which it is possible.

Definition E.10. A representation π of a C^*-algebra A is called *type I* if every central subrepresentation of π has a multiplicity-free subrepresentation. A C^*-algebra is said to be of *type I* if every representation of A is of type I.[6]

Remark E.11. Kaplansky [86] proved that all GCR C^*-algebras are of type I (see also [2, Theorem 2.4.1]). That the converse holds for separable C^*-algebras is one of the monumental achievements in the subject and is due to Glimm [60]. Glimm's result was extended to the general case by Sakai [157]. However, for the moment, we will only need to keep in mind that commutative C^*-algebras have only type I representations.

The significance of type I representations to our discussion is illustrated by the following decomposition theorem [2, Theorem 2.1.8].

Theorem E.12 (Decomposition Theorem). *Suppose that π is a type I representation of a C^*-algebra A on a* separable *Hilbert space. Then there is a unique orthogonal family $\{\pi_n\}$ of central subrepresentations of π such that*

(a) *each π_n either has multiplicity n or is the zero representation, and*

(b) $\pi = \bigoplus_n \pi_n$.

Remark E.13. A von Neumann algebra $\mathcal{M} \subset B(\mathcal{H})$ is called a *factor* if it has trivial center: $\mathcal{M} \cap \mathcal{M}' = \mathbf{C}I_{\mathcal{H}}$. A representation π of a C^*-algebra A is called a *factor representation* if $\pi(A)$ generates factor (i.e., $\pi(A)''$ is a factor). Note that every irreducible representation, and in fact every multiple of an irreducible representation is a factor representation [28, Proposition 5.4.11]. Also, any multiplicity-free factor representation is necessarily irreducible (since $\pi(A)'$ is commutative). Using Theorem E.12, it is not too hard to see that a type I factor representation has to be a multiple of an irreducible representation, and a non-type I factor representation can't be a multiple of an irreducible. Another part of Glimm's fundamental paper proving that separable type I C^*-algebras are necessarily GCR [60] asserts that a separable non-type I C^*-algebra does have non-type I factor representations. (In fact, it must have factor representations generating factors both of von Neumann type II and type III.)

[5] To avoid working with vector-valued functions, we can simply *define* $L^2(X, \mu, \mathcal{H})$ to be $L^2(X, \mu) \otimes \mathcal{H}$. Otherwise, we can refer to Appendix I.4, and note that we can replace G there with the second countable locally compact space X.

[6] An alternative formulation that A is of type I if and only if the von Neumann algebra $\pi(A)''$ is a type I von Neumann algebra for all representations π of A. This is standard terminology even though a type I von Neumann algebra is rarely a type I C^*-algebra.

Since commutative C^*-algebras are always of type I, the discussion in this section allows us to give a complete description of the representations of commutative C^*-algebras.

Theorem E.14 ([2, pp. 54–5]). *Suppose that $A = C_0(X)$ is a* separable *commutative C^*-algebra and that π is a* separable *representation of A. Then π is equivalent to a representation of the form*

$$(\pi_{\mu_\infty} \otimes 1_{\mathcal{H}_\infty}) \oplus \pi_{\mu_1} \oplus (\pi_{\mu_2} \otimes 1_{\mathcal{H}_2}) \oplus \cdots$$

where each μ_n is a finite Borel measure on X with μ_n disjoint from μ_m if $n \neq m$. (Some, but not all, of the μ_n can be the zero measure). If

$$\sigma = (\pi_{\nu_\infty} \otimes 1_{\mathcal{H}_\infty}) \oplus \pi_{\nu_1} \oplus (\pi_{\nu_2} \otimes 1_{\mathcal{H}_2}) \oplus \cdots$$

is another such representation, then σ is equivalent to π and if and only if μ_n and ν_n are equivalent measures for all n.

Remark E.15 (Decomposition into irreducibles). Theorem E.14 can also be thought of as a continuous decomposition of π into irreducibles. First recall that the irreducible representations of $A = C_0(X)$ are the point evaluations ev_x given by $\mathrm{ev}_x(f) := f(x)$, and that we can identify the spectrum $\hat{A}$ of A with X [139, Example A.24]. If μ is a finite Borel measure on X, then we can write

$$\pi_\mu(f)h(x) = \mathrm{ev}_x(f)h(x), \tag{E.1}$$

and this suggests we think of π_μ as a continuous sum of the ev_x's. Formally this is called a *direct integral*

$$\int_{\hat{A}}^{\oplus} \mathrm{ev}_x \, d\mu(x), \tag{E.2}$$

where (E.2) is simply a shorthand for (E.1). In general, given a representation π of A and measures μ_n as in Theorem E.14, we get a Borel partition $\{ X_n \}$ of X such that $\mu_n(E) = 0$ if $E \cap X_n = \emptyset$. After possibly multiplying each μ_n by a constant, we get a finite measure μ on X by setting $\mu = \sum_n \mu_n$. We can define a *multiplicity function* $m : X \to \{ 1, 2, \dots, \infty \}$ by $m(x) = n$ if $x \in X_n$ and then use the suggestive notation

$$\int_{\hat{A}}^{\oplus} m(x) \cdot \mathrm{ev}_x \, d\mu(x) \tag{E.3}$$

as a shorthand for the decomposition of π given by Theorem E.14. The point of this shorthand is to suggest how π is built up from the point evaluations. Moreover, with a bit of work, we can see that the equivalence class of π uniquely determines the measure class of μ and the multiplicity function m almost everywhere. We'll have much more to say about direct integrals in Appendix F (cf., Example F.25 on page 419). A complete treatment of the theory of direct integrals and applications to representations of C^*-algebras can be found in Dixmier's books [28, 29].

E.2 Decomposable Operators

Naturally, we also want to decompose representations of noncommutative C^*-algebras in analogy with what we did for abelian algebras in Theorem E.14 on the preceding page. This will require that we look a bit further into the pit of direct integrals. However, in this section, we'll keep following Arveson's treatment, and try to get only a few toes out over the abyss. We take a much more extensive look in Appendix F.

We let $\mathcal{B}^b(X,\mathcal{H})$ be the set of all bounded functions $F : X \to B(\mathcal{H})$ such that

$$x \mapsto \big(F(x)h \mid k\big)$$

is Borel for all $h, k \in \mathcal{H}$.[7] (Such functions are called *weak-operator Borel.*) The usual pointwise operations make $\mathcal{B}^b(X,\mathcal{H})$ into a $*$-algebra with norm

$$\|F\| := \sup_{x\in X} \|F(x)\|.$$

This norm satisfies the C^*-norm identity $\|F^*F\| = \|F\|^2$ and, although we don't require it, $\mathcal{B}^b(X,\mathcal{H})$ is a C^*-algebra. If $F \in \mathcal{B}^b(X,\mathcal{H})$, and if $h \in L^2(X,\mu,\mathcal{H}) := L^2(X)\otimes\mathcal{H}$, then it is not hard to check that $x \mapsto F(x)\big(h(x)\big)$ is in $L^2(X,\mu,\mathcal{H})$ with norm bounded by $\|F\|\|h\|_2$. Thus each $F \in \mathcal{B}^b(X,\mathcal{H})$ defines a bounded operator L_F on $L^2(X,\mathcal{H})$ via the formula

$$L_F(h)(x) := F(x)\big(h(x)\big).$$

The subalgebra

$$\mathscr{L}\otimes 1_{\mathcal{H}} = \{\, L_f \otimes 1_{\mathcal{H}} : f \in \mathcal{B}^b(X)\,\}$$

is called the set of *diagonal operators* in $B\big(L^2(X,\mathcal{H})\big)$ and we see easily that $L_F \in (\mathscr{L}\otimes 1_H)'$ for all $F \in \mathcal{B}^b(X,\mathcal{H})$.

Definition E.16. A bounded operator T on $L^2(X,\mu,\mathcal{H})$ is called *decomposable* if there is a $F \in \mathcal{B}^b(X,\mathcal{H})$ such that $T = L_F$.

In addition to being more easily understood, decomposable operators have a tidy characterization as exactly the operators in the commutant of the algebra $\mathscr{L}\otimes 1_{\mathcal{H}}$ of diagonal operators [2, Theorem 4.2.1]. (Also see Theorem F.21 on page 418.) We summarize some of the facts we need in the next result.

Theorem E.17. *Suppose that $\mathcal{H}$ is a separable Hilbert space, that X is a second countable locally compact Hausdorff space and that μ is a finite Borel measure on X. Then $T \in B\big(L^2(X,\mu,\mathcal{H})\big)$ is decomposable if and only if $T \in (\mathscr{L}\otimes 1_{\mathcal{H}})'$. Furthermore $\mathscr{L}\otimes 1_{\mathcal{H}}$ is an abelian von Neumann algebra, and $\pi_\mu \otimes 1_{\mathcal{H}}\big(C_0(X)\big)$ is dense in $\mathscr{L}\otimes 1_{\mathcal{H}}$ is the strong operator topology.*

[7] As in the scalar case, $\mathcal{B}^b(X,\mathcal{H})$ could be replace by some sort of L^∞ space. However, we are not assuming that the $F \in \mathcal{B}^b(X,\mathcal{H})$ be essentially separately valued (note that $B(\mathcal{H})$ is not separable if $\mathcal{H}$ is infinite dimensional), so the material in Appendix B does not apply directly.

Proof. As mentioned above, the first assertion is [2, Theorem 4.2.1]. To see that $(\pi_\mu \otimes 1_{\mathcal{H}})\big(C_0(X)\big)$ is strongly dense in $\mathscr{L} \otimes 1_{\mathcal{H}}$, fix $g \in \mathcal{B}^b(X)$. Then there is a bounded sequence $\{ f_n \} \subset C_0(X)$ such that $f_n(x) \to g(x)$ for μ-almost every x. Let $h \in L^2(X, \mu, \mathcal{H})$. Then

$$\|f_n(x)h(x) - g(x)h(x)\|^2 \to 0$$

for all x off a null set. The dominated convergence theorem then implies that $\pi_\mu(f_n) \otimes 1_{\mathcal{H}}$ converges strongly to $L_g \otimes 1_{\mathcal{H}}$. This suffices.

Since $\mathscr{L}$ is closed in the strong operator topology by Theorem E.6 on page 402, it is not hard to see directly that $\mathscr{L} \otimes 1_{\mathcal{H}}$ is as well. Alternatively, we can appeal to [29, I.2.4 Proposition 4] to conclude that $\mathscr{L} \otimes 1_{\mathcal{H}} = \big(\mathscr{L} \otimes B(\mathcal{H})\big)'$. □

Now we suppose that we have a C^*-subalgebra $\mathfrak{A}$ of decomposable operators on $L^2(X, \mathcal{H})$. Then, by definition, for each $T \in \mathfrak{A}$, there is a $F_T \in \mathcal{B}^b(X, \mathcal{H})$ such that $T = L_{F_T}$. If $\mathfrak{A}$ and $\mathcal{H}$ are separable, then we can make our choices so that $T \mapsto F_T$ is a $*$-homomorphism. The details are worked out in [2, §4.2].

Corollary E.18. *Suppose that $\mathcal{H}$ is separable Hilbert space and that $\mathfrak{A}$ is a separable C^*-subalgebra of $(\mathscr{L} \otimes 1_{\mathcal{H}})'$. Then there is an isometric $*$-isomorphism $\pi : \mathfrak{A} \to \mathcal{B}^b(X, \mathcal{H})$ such that*

$$L_{\pi(T)} = T$$

for all $T \in \mathfrak{A}$.

Notice that if we are given $\pi : \mathfrak{A} \to \mathcal{B}^b(X, \mathcal{H})$ as in Corollary E.18, then we obtain a family of possibly degenerate representations $\{ \pi_x \}_{x \in X}$ of $\mathfrak{A}$ as follows:

$$\pi_x(T) := \pi(T)(x).$$

These representations will play a crucial role in what follows. The idea will be to show that representations which we wish to study are equivalent to representations ρ having image $\rho(A)$ contained in $(\mathscr{L} \otimes 1_{\mathcal{H}})'$ for appropriate choices of X, μ and $\mathcal{H}$. Then letting $\rho(A)$ play the role of $\mathfrak{A}$, we obtain a "decomposition" of ρ into representations $\rho_x := \pi_x \circ \rho$. To determine the properties of the π_x, and therefore the ρ_x, we need the following technical result from [2, §4.2].

Proposition E.19. *If $\{ F_n \}$ is sequence in $\mathcal{B}^b(X, \mathcal{H})$ such that $L_{F_n} \to L_F$ in the strong operator topology, then there is a subsequence $\{ F_{n_k} \}$ and a μ-null set N such that $F_{n_k}(x) \to F(x)$ in the strong operator topology for all $x \in X \setminus N$.*

Notice that if $\mathfrak{A}$ is a nondegenerate subalgebra — that is, if the identity map is a nondegenerate representation of $\mathfrak{A}$ — and if $\{ E_i \}$ is an approximate identity in $\mathfrak{A}$, then $E_i \to I$ in the strong operator topology. Thus off a null set (and after reindexing) $\pi_x(E_i) \to I_{\mathcal{H}}$ and almost all the π_x are nondegenerate.

Proposition E.20. *Let $\mathcal{H}$ be a separable Hilbert space, X a second countable locally compact Hausdorff space and μ a finite measure on X. Suppose that $\mathfrak{A}$ is a separable C^*-subalgebra of $(\mathscr{L} \otimes 1_H)'$ with $\pi : \mathfrak{A} \to \mathcal{B}^b(X, \mathcal{H})$ as in Corollary E.18.*

(a) *If* $L_F \in \mathfrak{A}''$, *then* $F(x) \in \pi_x(\mathfrak{A})''$ *for* μ*-almost all* x.
(b) *If* $L_F \in \mathfrak{A}'$, *then* $F(x) \in \pi_x(\mathfrak{A})'$ *for* μ*-almost all* x.
(c) *If* $\mathscr{L} \otimes 1_{\mathcal{H}} = \mathfrak{A}' \cap \mathfrak{A}''$, *then* π_x *is a factor representation for* μ*-almost all* x.
(d) *If* $\mathscr{L} \otimes 1_{\mathcal{H}} \subset \mathfrak{A}''$, *then there is a* μ*-null set* N *such that* $x \neq y$ *and* $x, y \in X \setminus N$ *implies that* π_x *and* π_y *are disjoint.*

Proof. Part (a) is Corollary 1 of [2, Proposition 4.2.2].

For part (b), let $\{ A_n \}$ be a countable dense subset of $\mathfrak{A}$. Then L_F commutes with each $L_{\pi(A_n)}$. Thus there is a μ-null set N such that

$$F(x)\pi_x(A_n) = \pi_x(A_n)F(x) \quad \text{for all } x \in X \setminus N.$$

Thus $F(x) \in \pi_x(\mathfrak{A})'$ for almost all x.

Part (d) is Corollary 2 to [2, Proposition 4.2.2].

For part (c), note that if the diagonal operators coincide with the center of $\mathfrak{A}''$, then $\mathfrak{A}'$ and $\mathfrak{A}''$ generate the decomposable operators in the strong operator topology.[8] If $T \in B(\mathcal{H})$ generates $B(\mathcal{H})$ and if $F(x) := T$ for all $x \in X$, then $L_F \in (\mathscr{L} \otimes 1_{\mathcal{H}})'$. Since $\mathcal{H}$ is separable, there is a sequence $\{ T_n \}$ of elements in the complex algebra spanned by $\mathfrak{A}'$ and $\mathfrak{A}''$ such that $T_n \to L_F$ (see [2, Corollary following Theorem 1.2.2]). Using Proposition E.19 on the facing page and replacing $\{ T_n \}$ by a subsequence if necessary, we can assume that there is a μ-null set N such that $x \in X \setminus N$ implies

$$\pi(T_n)(x) \to F(x) = T$$

is the strong operator topology. However, enlarging N if necessary, parts (a) and (b) allow us to assume that $x \notin N$ implies each $\pi(T_n)(x) = \pi_x(T_n)$ is in the algebra generated by $\pi_x(\mathfrak{A})'$ and $\pi_x(\mathfrak{A})''$. Thus T is almost everywhere in the commutant of the center $\pi_x(\mathfrak{A})' \cap \pi_x(\mathfrak{A})''$. Since T generates $B(\mathcal{H})$, the center is trivial for almost every x which completes the proof. □

[8] In general, the commutant of $\mathcal{M} \cap \mathcal{N}$ is the von Neumann algebra generated by $\mathcal{M}'$ and $\mathcal{N}'$.

Appendix F

Direct Integrals

We knocked on the door to the theory of direct integrals of representations of C^*-algebras in Appendix E. However, the proof of the Gootman-Rosenberg-Sauvageot Theorem 8.21 on page 241 requires that we deal with the theory in a more general setting. We need this generality not only for the proof of Theorem 8.21 in Chapter 9, but also to develop the key tool used there: Effros's ideal center decomposition of a representation of a separable C^*-algebra (Theorem G.22 on page 444). Effros's theorem and the consequences we require are developed in Appendix G.

As always when dealing with direct integrals, separability is essential. All the C^*-algebras and Hilbert spaces appearing in this section are assumed to be separable.

F.1 Borel Hilbert Bundles

The classical notion of a direct integral of Hilbert spaces and decomposable operators is developed in detail in [29, Part II, Chap. 1&2]. A brief survey can be found in [56, §7.4]. The point of view taken here is that a direct integral of Hilbert spaces is simply the Banach space of square integrable sections of an appropriate bundle of Hilbert spaces. A decomposable operator is an operator which respects the fibres. I got this approach from [108, §3.1], which is based on [141, §1].

We start with a collection

$$\mathfrak{H} := \{\, \mathcal{H}(x) \,\}_{x\in X}$$

of separable (nonzero) complex Hilbert spaces indexed by an *analytic* Borel space X. Then the total space is the disjoint union

$$X * \mathfrak{H} := \{\, (x,h) : h \in \mathcal{H}(x) \,\},$$

and we let $\pi : X * \mathfrak{H} \to X$ be the obvious map.

Definition F.1. Let $\mathfrak{H} = \{\, \mathcal{H}(x) \,\}_{x\in X}$ be a family of separable Hilbert spaces indexed by an analytic Borel space X. Then $(X * \mathfrak{H}, \pi)$ is a *Borel Hilbert bundle* if $X * \mathfrak{H}$ has a Borel structure such that

(a) π is a Borel map and

(b) there is sequence $\{ f_n \}$ of sections such that

(i) the maps $\tilde{f}_n : X * \mathfrak{H} \to \mathbf{C}$ defined by
$$\tilde{f}_n(x,h) := \bigl(f_n(x) \mid h\bigr),$$
are Borel for each n,

(ii) for each n and m,
$$x \mapsto \bigl(f_n(x) \mid f_m(x)\bigr)$$
is Borel, and

(iii) the functions $\{ \tilde{f}_n \}$ and π separate points of $X * \mathfrak{H}$.

The sequence $\{ f_n \}$ is called a *fundamental sequence* for $(X * \mathfrak{H}, \pi)$. We let $B(X * \mathfrak{H})$ be the set of sections of $X * \mathfrak{H}$ such that
$$x \mapsto \bigl(f(x) \mid f_n(x)\bigr)$$
is Borel for all n.

Remark F.2. We are being a bit sloppy with our notation for sections. A section f of $X * \mathfrak{H}$ is really of the form $f(x) = (x, \hat{f}(x))$, where $\hat{f}(x) \in \mathcal{H}(x)$. Therefore we should really write something like $\bigl(\hat{f}_n(x) \mid \hat{f}_m(x)\bigr)_{\mathcal{H}(x)}$ in place of $\bigl(f_n(x) \mid f_m(x)\bigr)$. Alternatively, we could identify $\mathcal{H}(x)$ with $\{ x \} \times \mathcal{H}(x)$. In any event, we are going to suppress the $\hat{f}$ to reduce clutter.

Remark F.3. If f is a section of $X * \mathfrak{H}$, then, $\bigl(f_n(x) \mid f(x)\bigr) = \tilde{f}_n\bigl(f(x)\bigr)$ (see Remark F.2). Thus if f is Borel (as a map of X into $X * \mathfrak{H}$), then $f \in B(X * \mathfrak{H})$. If in addition, $X * \mathfrak{H}$ is an analytic Borel space, then Lemma D.20 on page 375 (and property (iii) of axiom (b)) implies that the Borel field on $X * \mathfrak{H}$ is that generated by π and the $\tilde{f}_n$. Thus, if $X * \mathfrak{H}$ is analytic, a section $f \in B(X * \mathfrak{H})$ if and only if f is Borel as a map of X into $X * \mathfrak{H}$.

Remark F.4. Suppose that $X * \mathfrak{H}$ is an analytic Borel Hilbert bundle; that is, we assume $X * \mathfrak{H}$ is Borel Hilbert bundle and that $X * \mathfrak{H}$ is an analytic Borel space. We can give X the *largest* Borel structure such that π is Borel. Then with this new, *a priori* larger (but still countably separated), Borel structure, X is still analytic [2, Corollary 1 of Theorem 3.3.5]. But then the Unique Structure Theorem [2, Theorem 3.3.5] implies that the new Borel structure coincides with the original. In particular, for analytic Borel Hilbert bundles, we can replace condition (a) in Definition F.1 on the preceding page with

(a)′ $\pi^{-1}(E)$ is Borel in $X * \mathfrak{H}$ if and only if E is Borel in X.

Therefore our definition of an *analytic* Borel Hilbert bundle coincides with Ramsay's in [141, §1] and Muhly's in [108, Chap 3., §1]. (We are interested *only* in analytic Borel Hilbert bundles. We have not made this a blanket assumption in this section only because in the proof of Proposition F.8 on page 412 we want to deal with bundles which are not *a priori* analytic.)

Example F.5. Let $\mathcal{H}$ be any separable Hilbert space with orthonormal basis $\{ e_n \}$, and let X be an analytic Borel space. If we let $\mathcal{H}(x) := \mathcal{H}$ for all x, then with the product Borel structure on $X * \mathfrak{H} = X \times \mathcal{H}$, $(X \times \mathcal{H}, \pi)$ becomes a Borel Hilbert bundle with respect to the fundamental sequence given by $f_n(x) := e_n$ for all x. If X is an analytic Borel space (respectively, a standard Borel space) then $X * \mathfrak{H}$ is analytic (respectively, standard). More generally, let $X = X_\infty \cup X_1 \cup X_2 \cup \dots$ be a Borel partition of X, and let $\mathcal{H}_d$ be a Hilbert space of dimension d with orthonormal basis $\{ e^d_n \}_{n=1}^{n=d}$ for $d = 1, \dots, \aleph_0$. For each $x \in X_d$, let $\mathcal{H}(x) = \mathcal{H}_d$. Then we can give $X * \mathfrak{H} = \coprod_{d=1}^{d=\infty} X_d \times \mathcal{H}_d$ the obvious Borel structure coming from the product Borel structure on each factor.[1] Then $X * \mathfrak{H}$ is a Borel Hilbert bundle with respect to the fundamental sequence

$$f_n(x) = \begin{cases} e^d_n & \text{if } x \in X_d \text{ and } 1 \le n \le d, \text{ and} \\ 0 & \text{otherwise.} \end{cases}$$

A above, $X * \mathfrak{H}$ is either analytic or standard depending on whether X has the corresponding property.

If we agree that two Borel Hilbert bundles $X * \mathfrak{H}$ and $X * \mathfrak{K}$ over X are isomorphic if there is a bundle map $\varphi : X * \mathfrak{H} \to X * \mathfrak{K}$ which is a Borel isomorphism such that $(x, h) \mapsto \varphi(x, h)$ is a unitary isomorphism of $\mathcal{H}(x)$ onto $\mathcal{K}(x)$,[2] then it is something of a surprise to discover that every bundle is isomorphic to one of the seemingly trivial examples described in Example F.5 (cf., Corollary F.9 on page 413). On the other hand, the isomorphism is often not natural or convenient to introduce in applications. But the result is none-the-less indispensable in the theory.

Proposition F.6. *Let $X * \mathfrak{H}$ be a Borel Hilbert bundle with fundamental sequence $\{ f_n \}$, and let $d(x) := \dim \mathcal{H}(x)$. Then, for each $n = 1, 2, \dots, \aleph_0$,*

$$X_n := \{ x \in X : \dim \mathcal{H}(x) = n \}$$

is Borel and there is a sequence $\{ e_k \}$ in $B(X * \mathfrak{H})$ such that

(a) *for each $x \in X$, $\{ e_k(x) \}_{k=1}^{d(x)}$ is an orthonormal basis for $\mathcal{H}(x)$, and if $d(x) < \infty$, $e_k(x) = 0$ for all $k > d(x)$, and*
(b) *for each k, there is a Borel partition $X = \bigcup_{n=1}^\infty B^k_n$ and for each (n, k) finitely many Borel functions $\varphi^{n,k}_j : B^k_n \to \mathbf{C}$ $\big(1 \le j \le m(n, k)\big)$ such that*

$$e_k(x) = \sum_{j=1}^{m(n,k)} \varphi^{n,k}_j(x) f_j(x) \quad \textit{for all } x \in B^k_n.$$

If $\{ u_k \}$ is any sequence which satisfies (a) *&* (b)*, then each $u_k \in B(X * \mathfrak{H})$ and*

[1] If X and Y are Borel spaces, then the disjoint union $X \coprod Y$ is a Borel space where $B \subset X \coprod Y$ is Borel if and only if $B \cap X$ and $B \cap Y$ are Borel. This generalizes in a straightforward way to *countable* disjoint unions.

[2] We will return to the notion of an isomorphism of Borel Hilbert bundles in Definition F.22 on page 418.

(c) $f \in B(X * \mathfrak{H})$ *if and only if* $x \mapsto \big(f(x) \mid u_k(x)\big)$ *is Borel for all* k*, and*
(d) *if* $f, g \in B(X * \mathfrak{H})$*, then* $x \mapsto \big(f(x) \mid g(x)\big)$ *is Borel.*

Remark F.7. Clearly, $\{ e_k \}$ is also a fundamental sequence for $X * \mathfrak{H}$. We call such a sequence a *special orthogonal fundamental sequence.*

Proof of Proposition F.6. The measurability of the X_n and the existence of $\{ e_k \}$ satisfying (a) and (b) are shown in [56, Proposition 7.27] or [29, Lemma II.1.1] (using a careful application of the usual Gram-Schmidt process). We won't repeat the proof here.

Note that part (b) implies that

$$x \mapsto \big(f(x) \mid u_k(x)\big) \tag{F.1}$$

is Borel for all k provided $f \in B(X * \mathfrak{H})$. Thus $u_k \in B(X * \mathfrak{H})$. If (F.1) is Borel for all k, then Parseval's Identity implies that

$$x \mapsto \big(f(x) \mid f_j(x)\big) = \sum_k \big(f(x) \mid u_k(x)\big)\big(u_k(x) \mid f_j(x)\big)$$

is also Borel. This establishes part (c). Part (d) also follows from Parseval's Identity:

$$\big(f(x) \mid g(x)\big) = \sum_k \big(f(x) \mid u_k(x)\big)\big(u_k(x) \mid g(x)\big). \qquad \square$$

Other than straightforward situations such as Example F.5 on the preceding page, Borel Hilbert bundles do not arise fully formed with a natural Borel structure on the total space. Instead, we rely on the following.

Proposition F.8. *Suppose that* X *is an analytic Borel space and that* $\mathfrak{H} = \{ \mathcal{H}(x) \}_{x \in X}$ *is a family of separable Hilbert spaces. Suppose that* $\{ f_n \}$ *is a countably family of sections of* $X * \mathfrak{H}$ *such that conditions* (ii) *and* (iii) *of axiom* (b) *in Definition F.1 on page 409 are satisfied. Then there is a unique analytic Borel structure on* $X * \mathfrak{H}$ *such that* $(X * \mathfrak{H}, \pi)$ *becomes an analytic Borel Hilbert bundle and* $\{ f_n \}$ *is a fundamental sequence.*

Proof. If $(X * \mathfrak{H}, \pi)$ has an analytic Borel structure such that $\{ f_n \}$ is a fundamental sequence, then as in Remark F.3 on page 410, the Borel structure is that generated by π and the $\tilde{f}_n$. Thus, if $X * \mathfrak{H}$ has an analytic Borel structure, it is unique.

On the other hand, we can let $X * \mathfrak{H}$ have the smallest Borel structure such that the maps $\tilde{f}_n$ and π are Borel. Then $(X * \mathfrak{H}, \pi)$ is a Borel Hilbert bundle and $\{ f_n \}$ is a fundamental sequence. Let $X_d = \{ x \in X : \dim(\mathcal{H}(x)) = d \}$. Then X_d is Borel (Proposition F.6 on the previous page), and it will suffice to see that $X_d * \mathfrak{H}$ is analytic.[3] By Proposition F.6 on the preceding page, there is a special orthogonal fundamental sequence $\{ u_n \}$ for $X_d * \mathfrak{H}$ such that $\{ u_n(x) \}$ is an orthonormal basis for $\mathcal{H}(x)$ for each $x \in X_d$. Let $\mathcal{H}_d$ be a Hilbert space of dimension d and let $\{ e_n \}$

[3]If Y is a Borel subset of X and if $X * \mathfrak{H}$ is a Borel Hilbert bundle over X, then we let $Y * \mathfrak{H}$ denote the obvious Borel Hilbert bundle over Y obtained by restriction.

be an orthonormal basis for $\mathcal{H}_d$. Then $(X_d \times \mathcal{H}_d, \pi_d)$ is an analytic Borel Hilbert bundle and $g_n(x) = e_n$ for all $x \in X_d$ defines a fundamental sequence for $X_d \times \mathcal{H}_d$. Note that π_d and the $\tilde{g}_n$ determine the Borel structure on the analytic Borel Hilbert Bundle $X_d \times \mathcal{H}_d$. Since the $\{ u_n(x) \}$ determine a unitary isomorphism of $\mathcal{H}(x)$ onto $\mathcal{H}_d$, we get a bundle map $\varphi : X_d * \mathfrak{H} \to X_d \times \mathcal{H}_d$ which is a unitary on each fibre. Since

$$\tilde{g}_n\bigl(\varphi(x,h)\bigr) = \tilde{u}_n(x,h),$$

it follows that φ is a Borel map. Reversing the roles of g_n and u_n shows that φ^{-1} is Borel. Since $X_d \times \mathcal{H}_d$ is analytic, this completes the proof. □

Examining the proof of Proposition F.8 and in view of Remark F.3 on page 410, we have the following corollary.

Corollary F.9. *Suppose that $X * \mathfrak{H}$ is an analytic Borel Hilbert bundle over an analytic space X. Let $X_d = \{ x \in X : \dim\bigl(\mathcal{H}(x)\bigr) = d \}$, and let $\mathcal{H}_d$ be a Hilbert space of dimension d. Then $X * \mathfrak{H}$ is isomorphic to $\coprod_{d=1}^{d=\infty} X_d \times \mathcal{H}_d$ (as in Example F.5 on page 411).*

F.2 The Direct Integral of Hilbert Spaces

For the rest of this appendix, and in fact elsewhere in this book with the exception of the previous section, we will be concerned solely with analytic Borel Hilbert bundles over analytic spaces. In particular, let $(X * \mathfrak{H}, \pi)$ be an *analytic* Borel Hilbert bundle over an analytic space X. Then $B(X * \mathfrak{H})$ denotes the set of Borel sections. If $f \in B(X * \mathfrak{H})$, then $x \mapsto \|f(x)\|$ is Borel by Proposition F.6 on page 411. If μ is a Borel measure on X, then let $L^2(X * \mathfrak{H}, \mu)$ be the normed vector space formed by the quotient of

$$\mathcal{L}^2(X * \mathfrak{H}, \mu) = \{ f \in B(X * \mathfrak{H}) : x \mapsto \|f(x)\|^2 \text{ is integrable} \},$$

where functions agreeing μ-almost everywhere are identified. Notice that if $f, g \in \mathcal{L}^2(X * \mathfrak{H}, \mu)$, then the usual Cauchy-Schwarz inequality implies that $x \mapsto \bigl(f(x) \mid g(x)\bigr)$ is integrable, and $L^2(X * \mathfrak{H}, \mu)$ is an inner product space in the obvious way. If $X * \mathfrak{H}$ is the trivial bundle $X \times \mathcal{H}$, then it is immediate that $L^2(X \times \mathcal{H}, \mu)$ is the Hilbert space $L^2(X, \mu, \mathcal{H})$ (see Appendix I.4 and Remark I.12 on page 490). In general, it is not hard to see that $L^2(X * \mathfrak{H}, \mu)$ is a Hilbert space with the above inner product. To prove this, we could either mimic the proof that $L^2(X, \mu, \mathcal{H})$ is complete, or we could invoke Corollary F.9. We'll leave this as an exercise.

We should notice that $L^2(X * \mathfrak{H}, \mu)$ is nothing more than the *direct integral*

$$\int_X^{\oplus} \mathcal{H}(x)\, d\mu(x) \tag{F.2}$$

as defined in [29].

Remark F.10. Although the integral notation, (F.2), is classical, here we'll usually use $L^2(X * \mathfrak{H}, \mu)$ as it makes clear the dependence on the Borel structure on $X * \mathfrak{H}$ induced by the fundamental sequence.

With the above definition is hand, we can record some straightforward corollaries of Proposition F.6 on page 411 and Corollary F.9 on the previous page.

Corollary F.11. *Suppose that $X*\mathfrak{H}$ is an analytic Borel Hilbert bundle and suppose that* $\dim \mathcal{H}(x) = d$ *for all x and $1 \le d \le \aleph_0$. Then, if $\mathcal{H}_d$ is any Hilbert space of dimension d, $X*\mathfrak{H}$ is isomorphic to the trivial bundle $X \times \mathcal{H}_d$. In particular, if μ is any finite Borel measure on X, then $L^2(X*\mathfrak{H},\mu)$ is isomorphic to $L^2(X,\mu)\otimes\mathcal{H}_d \cong L^2(X,\mu,\mathcal{H}_d)$.*

Corollary F.12. *Let $X*\mathfrak{H}$ be an analytic Borel Hilbert bundle and let μ be a finite measure on X. Then there is a Borel partition $X = X_\infty \cup X_1 \cup X_2 \cup \cdots$ such that, if μ_d is the restriction of μ to X_d and if $\mathcal{H}_d$ is a fixed Hilbert space of dimension $1 \le d \le \aleph_0$, then*

$$\begin{aligned} L^2(X*\mathfrak{H},\mu) &\cong L^2(X_\infty,\mu_\infty)\otimes\mathcal{H}_\infty \oplus L^2(X_1,\mu_1) \oplus L^2(X_2,\mu_2)\otimes\mathcal{H}_2 \oplus \cdots \\ &\cong \bigoplus_{d=1}^{d=\infty} L^2(X_d,\mu_d,\mathcal{H}_d). \end{aligned}$$

Definition F.13. Suppose that μ is a finite measure on a Borel space X and that $X*\mathfrak{H}$ is a Borel field of Hilbert spaces. An operator T on $L^2(X*\mathfrak{H},\mu)$ is called *diagonal* if there is a bounded (scalar-valued) Borel function $\varphi \in \mathcal{B}^b(X)$ such that

$$Th(x) = \varphi(x)h(x)$$

for μ-almost every x. The collection of diagonal operators on $L^2(X*\mathfrak{H},\mu)$ is denoted by $\Delta(X*\mathfrak{H},\mu)$. If $\varphi \in \mathcal{B}^b(X)$, then the associated diagonal operator is denoted by T_φ.

Example F.14. If $X*\mathfrak{H}$ is a trivial bundle $X\times\mathcal{H}$, then $\Delta(X\times\mathcal{H},\mu)$ is the collection of operators $\mathscr{L}\otimes 1_{\mathcal{H}}$ as defined on page 405 of Appendix E.2.

Lemma F.15. *Suppose that $X*\mathfrak{H}$ is an analytic Borel Hilbert bundle and that μ is a finite Borel measure on X. Then $\Delta(X*\mathfrak{H},\mu)$ is an abelian von Neumann subalgebra of $B\big(L^2(X*\mathfrak{H},\mu)\big)$, and the map $\varphi\mapsto T_\varphi$ induces an isomorphism of $L^\infty(X,\mu)$ onto $\Delta(X*\mathfrak{H},\mu)$.*

Proof. The map $\varphi\mapsto T_\varphi$ clearly defines an isometric $*$-isomorphism of $L^\infty(X,\mu)$ onto $\Delta(X*\mathfrak{H},\mu)$.[4] Since

$$(T_\varphi f \mid g) = \int_X \varphi(x)\big(f(x)\mid g(x)\big)\,d\mu(x),$$

it follows that $\varphi\mapsto T_\varphi$ is continuous from $L^\infty(X,\mu)$ with the weak-$*$ topology (as the dual of $L^1(X,\mu)$) onto $\Delta(X*\mathfrak{H},\mu)$ with the weak operator topology. Since the unit ball of $L^\infty(X,\mu)$ is weak-$*$ compact, the unit ball of $\Delta(X*\mathfrak{H},\mu)$ is compact, and hence closed, in the weak operator topology. Since the Kaplansky Density Theorem [110, Theorem 4.3.3] implies that the unit ball of $\Delta(X*\mathfrak{H},\mu)$ is weakly dense in the unit ball of its weak closure, this implies that $\Delta(X*\mathfrak{H},\mu)$ is closed in the weak operator topology. This completes the proof. □

[4]We are using the standing assumption that each $\mathcal{H}(x)\ne 0$. Otherwise, we could replace μ by its restriction to the set where $\mathcal{H}(x)\ne 0$, and proceed as above (see [29, II.2.4]).

Lemma F.16. *Suppose that X is an analytic Borel space. Then there is a countable family $\{ \psi_i \}$ of bounded Borel functions on X with the property that given a finite Borel measure μ on X and a function $f \in \mathcal{L}^1(X, \mu)$ such that*

$$\int_X \psi_i(x) f(x) \, d\mu(x) = 0 \quad \textit{for all } i,$$

then $f(x) = 0$ μ-almost everywhere.

Proof. Let $\{ A_i \}$ be a countable family of Borel sets generating the Borel sets in X as a σ-algebra. As in Lemma D.41 on page 394, the algebra $\mathscr{A}$ generated by $\{ A_i \}$ is countable.

Let μ be a finite Borel measure on X, let $f \in \mathcal{L}^1(X, \mu)$ and let $\mathscr{C}$ be the family of Borel sets B such that

$$\int_B f(x) \, d\mu(x) = 0.$$

The Dominated Convergence Theorem implies that $\mathscr{C}$ is closed under countable increasing unions and countable decreasing intersections. If $\mathscr{A} \subset \mathscr{C}$, then the Monotone Class Lemma ([57, Lemma 2.35]) implies that $\mathscr{C}$ contains the σ-algebra generated by $\mathscr{A}$ — and therefore every Borel set. In that case, we must have $f = 0$ almost everywhere. Therefore it will suffice to let the $\{ \psi_i \}$ be the set of characteristic functions of elements of $\mathscr{A}$. □

Lemma F.17. *Suppose that $X * \mathfrak{H}$ is an analytic Borel Hilbert bundle over an analytic space X. Then there is a countable family $\{ \psi_i \} \subset \mathcal{B}^b(X)$ such that given a finite Borel measure μ on X and a square integrable fundamental sequence $\{ f_j \}$ for $X * \mathfrak{H}$, the set*

$$\{ T_{\psi_i} f_j : i \geq 1 \textit{ and } j \geq 1 \}$$

is dense in $L^2(X * \mathfrak{H}, \mu)$. In particular, $L^2(X * \mathfrak{H}, \mu)$ is separable for any finite Borel measure μ on X.

Proof. Let $\{ \psi_i \}$ be as in Lemma F.16. Suppose that f is orthogonal to every $T_{\psi_i} f_j$. Then for each j,

$$\int_X \psi_i(x) \big(f_j(x) \mid f(x) \big) \, d\mu(x) = 0 \tag{F.3}$$

for all i. Thus there is a μ-null set N_j such that $\big(f_j(x) \mid f(x) \big) = 0$ for all $x \notin N_j$. Then $N := \bigcup N_j$ is a null set and $f(x) = 0$ for all $x \notin N$. That is, $f = 0$ in $L^2(X * \mathfrak{H}, \mu)$. □

Example F.18 (Pull-backs). Suppose that $X * \mathfrak{H}$ is an analytic Borel Hilbert bundle with fundamental sequence $\{ f_n \}$ and that $\sigma : Y \to X$ is a Borel map. Then we can form the pull-back Borel Hilbert bundle

$$\sigma^*(X * \mathfrak{H}) := Y * \mathfrak{H}_\sigma := \{ (y, h) : h \in \mathcal{H}\big(\sigma(y)\big) \}$$

with Borel structure compatible with the fundamental sequence $\{ f_n \circ \sigma \}$. Therefore $f \in B(X * \mathfrak{H})$ implies that $f \circ \sigma \in B(Y * \mathfrak{H}_\sigma)$. If ν is finite Borel measure on Y

and if $\sigma_*\nu$ is the push-forward measure on X given by $\sigma_*\nu(E) := \nu\big(\sigma^{-1}(E)\big)$ (see Lemma H.13 on page 463), then

$$\int_Y \big(f\big(\sigma(y)\big) \mid g\big(\sigma(y)\big)\big)\, d\nu(y) = \int_Y \big(f(x) \mid g(x)\big)\, d(\sigma_*\nu)(x).$$

Thus $W(f)(y) := f\big(\sigma(y)\big)$ defines an isometry

$$W : L^2(X * \mathfrak{H}, \sigma_*\nu) \to L^2(Y * \mathfrak{H}_\sigma, \nu)$$

which is an isomorphism if σ is a Borel isomorphism.

Example F.19. Let $Y * \mathfrak{K}$ be a Borel Hilbert bundle over a standard Borel space Y, and let $\tau : Y \to X$ be a Borel surjection onto a standard Borel space X. Let ν be a finite Borel measure on Y and let $\mu = \tau_*\nu$ be the push forward. Using the Disintegration Theorem on page 482, we can disintegrate ν with respect to μ so that we have, for each $x \in X$, finite measures ν_x with $\operatorname{supp}\nu_x \subset \tau^{-1}(x)$ and such that for all bounded Borel functions φ on Y

$$\int_Y \varphi(y)\, d\nu(y) = \int_X \int_Y \varphi(y)\, d\nu_x(y)\, d\mu(x).$$

Let $\{ u_i \}$ be a special orthogonal fundamental sequence for $Y * \mathfrak{K}$, and let

$$\mathcal{H}(x) := L^2(Y * \mathfrak{K}, \nu_x).$$

Let $\{ \psi_j \} \subset \mathcal{B}^b(Y)$ be as in Lemma F.17 on the preceding page. Since each ν_x is a finite measure, $u_i \in \mathcal{L}^2(Y * \mathfrak{K}, \nu_x)$, and hence $\{ T_{\psi_j} u_i \}$ is dense in $L^2(Y * \mathfrak{K}, \nu_x)$ for each x. Define a section g_{ij} of $X * \mathfrak{H}$ by letting $g_{ij}(x)(y) = \psi_j(y)u_i(y)$. If $f \in \mathcal{L}^2(Y * \mathfrak{K}, \nu_x)$ satisfies $(g_{ij}(x) \mid f)_{L^2(\nu_x)} = 0$ for all i and j, then $f = 0$ ν_x-almost everywhere. Thus $\{ g_{ij} \}$ define a Borel structure on $X * \mathfrak{H}$ (Proposition F.8 on page 412). The map sending $f \in \mathcal{L}^2(Y * \mathfrak{K}, \nu)$ to the section $g(x)(y) = f(y)$ is a natural isomorphism of $L^2(Y * \mathfrak{K}, \nu)$ onto $L^2(X * \mathfrak{H}, \mu)$.

F.3 Decomposable Operators

Let $X * \mathfrak{H}$ and $X * \mathfrak{K}$ be analytic Borel Hilbert bundles over the same space X with fundamental sequences $\{ f_n \}$ and $\{ g_k \}$, respectively. A bundle map $\widehat{T} : X * \mathfrak{H} \to X * \mathfrak{K}$ is determined by a family of maps $T(x) : \mathcal{H}(x) \to \mathcal{K}(x)$ such that $\widehat{T}(x, h) = \big(x, T(x)h\big)$. Using Corollary F.9 on page 413 and Lemma D.20 on page 375, it is possible to prove that $\widehat{T}$ is Borel if and only if

$$x \mapsto \tilde{g}_k\big(\widehat{T}\big(f_n(x)\big)\big) = \big(T(x)f_n(x) \mid g_k(x)\big) \quad \text{is Borel for all } n \text{ and } k. \tag{F.4}$$

Rather than sketch a proof here, we will *define* a family of linear maps $T(x) : \mathcal{H}(x) \to \mathcal{K}(x)$ to be a *Borel field of operators* if (F.4) holds. In particular, if $f \in B(X * \mathfrak{H})$, then $x \mapsto T(x)f(x)$ is in $B(X * \mathfrak{K})$.

If each $T(x)$ is a bounded linear operator, then we have

$$\|T(x)\| = \sup_{\substack{h\in S\\ h(x)\neq 0}} \|T(x)h(x)\|\|h(x)\|^{-1},$$

where S is any countable family in $B(X * \mathfrak{H})$ such that $\{ h(x) : h \in S \}$ is dense in $\mathcal{H}(x)$ for all x. (Rational linear combinations of the f_n will do.) It follows that $x \mapsto \|T(x)\|$ is Borel. If μ is a finite Borel measure on X, then we can set

$$\lambda := \operatorname{ess\,sup}_{x\in X} \|T(x)\|. \tag{F.5}$$

If $\lambda < \infty$, then we get a linear operator T on $L^2(X * \mathfrak{H}, \mu)$ with norm at most λ via $Tf(x) := T(x)f(x)$ for all $f \in L^2(X * \mathfrak{H}, \mu)$. In fact, $\|T\| = \lambda$. Thus an essentially bounded Borel field of operators $\{ T(x) \}$ defines a bounded linear operator $T : L^2(X * \mathfrak{H}, \mu) \to L^2(X * \mathfrak{K}, \mu)$ with norm λ given by (F.5).

Operators T determined by a Borel field of operators $\{ T(x) \}$ as above are called *decomposable* and the classical notation is

$$T := \int_X^{\oplus} T(x)\, d\mu(x). \tag{F.6}$$

Lemma F.20. *Suppose that $X * \mathfrak{H}$ and $X * \mathfrak{K}$ are Borel Hilbert bundles over X and that μ is a finite Borel measure on X. Let $\{ T(x) \}_{x\in X}$ be an essentially bounded Borel field of operators $T(x) : \mathcal{H}(x) \to \mathcal{K}(x)$, and let the direct integral*

$$T := \int_X^{\oplus} T(x)\, d\mu(x)$$

be the associated decomposable operator from $L^2(X * \mathfrak{H}, \mu)$ to $L^2(X * \mathfrak{K}, \mu)$.

(a) *If $T = 0$, then $T(x) = 0$ for μ-almost all x.*

(b) *If $\{ T'(x) \}_{x\in X}$ is another essentially bounded Borel family such that*

$$T = \int_X^{\oplus} T'(x)\, d\mu(x),$$

then $T'(x) = T(x)$ for μ-almost all x.

Proof. Clearly, it suffices to prove part (a). Let $\{ f_n \}$ and $\{ g_m \}$ be fundamental sequences in $L^2(X * \mathfrak{H}, \mu)$ and $L^2(X * \mathfrak{K}, \mu)$, respectively. Then for each n and m,

$$\big(T(x)f_n(x) \mid g_m(x)\big) = 0 \quad \text{for } \mu\text{-almost all } x.$$

Therefore $T(x) = 0$ almost everywhere. □

Note that a decomposable operator such as T in (F.6) commutes with diagonal operators: if φ is a bounded Borel function on X, then we can let $T_\varphi^{\mathfrak{H}}$ and $T_\varphi^{\mathfrak{K}}$ be the corresponding diagonal operators on $L^2(X * \mathfrak{H}, \mu)$ and $L^2(X * \mathfrak{K}, \mu)$, respectively, and we have

$$TT_\varphi^{\mathfrak{H}} = T_\varphi^{\mathfrak{K}}T. \tag{F.7}$$

Although we won't prove it here, a bounded operator T from $L^2(X*\mathfrak{H},\mu) \to L^2(X*\mathfrak{K},\mu)$ is decomposable if and only if (F.7) holds for all bounded Borel functions φ [29, II.2.5 Theorem 1].[5] The case where $\mathfrak{H} = \mathfrak{K}$ is a bit more straightforward.

Theorem F.21. *Suppose that $X*\mathfrak{H}$ is a Borel Hilbert bundle and that μ is a finite measure on X. Let $T \in B\big(L^2(X*\mathfrak{H},\mu)\big)$. Then T is decomposable if and only if T is in the commutant of the diagonal operators $\Delta(X*\mathfrak{H},\mu)$.*

Sketch of the Proof. In view of the above discussion, any decomposable operator in $B\big(L^2(X*\mathfrak{H},\mu)\big)$ belongs to $\Delta(X*\mathfrak{H},\mu)'$. Hence it suffices to see that any $T \in \Delta(X*\mathfrak{H},\mu)'$ is decomposable. Since the commutant is a von Neumann algebra, and hence a C^*-algebra, it is spanned by its unitary elements. If T is unitary, the result is a special case of (the proof of) part (c) of Proposition F.33 on page 423.[6] Alternatively, we could use Corollary F.12 on page 414 to reduce to the case where $X*\mathfrak{H}$ is trivial an appeal to [2, Theorem 4.2.1]. A complete proof of the theorem can be found in [29, II.2.5 Theorem 1]. □

We can now formalize the notion of a Hilbert bundle isomorphism employed in Appendices F.1 and F.2.

Definition F.22. Let $X*\mathfrak{H}$ and $X*\mathfrak{K}$ be Borel Hilbert bundles. Then $X*\mathfrak{H}$ and $X*\mathfrak{K}$ are *isomorphic* (as Borel Hilbert bundles) if there is a Borel bundle map $\widehat{V} : X*\mathfrak{H} \to X*\mathfrak{K}$ such that $V(x) : \mathcal{H}(x) \to \mathcal{K}(x)$ is a unitary for each x.

Remark F.23. If $\widehat{V}$ is an isomorphism of $X*\mathfrak{H}$ onto $X*\mathfrak{K}$ then,

$$V := \int_X^{\oplus} V(x)\,d\mu(x)$$

is a unitary from $L^2(X*\mathfrak{H},\mu)$ onto $L^2(X*\mathfrak{K},\mu)$ for any finite Borel measure μ on X.

Remark F.24. In the above remark about isomorphisms, it is critical that the measure μ is used for both $X*\mathfrak{H}$ and $X*\mathfrak{K}$. Even if $\mathfrak{H} = \mathfrak{K}$ and $T(x) = \mathrm{id}_{\mathcal{H}(x)}$ for all x, the corresponding operator need not even map $\mathcal{L}^2(X*\mathfrak{H},\mu)$ into $\mathcal{L}^2(X*\mathfrak{K},\nu)$ for arbitrary finite Borel measures μ and ν — even if μ is equivalent to ν. In particular, the direct integral notation used in (F.6) should be used only in the case where the measures match up. We will discuss how to proceed when we have bundles over different base spaces in Appendix F.4.

Let A be a separable C^*-algebra. A collection of representations $\pi_x : A \to B\big(\mathcal{H}(x)\big)$ is called a *Borel field of representations of* A if $\{\,\pi_x(a)\,\}$ is a Borel field of operators for all $a \in A$. In that case, we get a representation π of A on $L^2(X*\mathfrak{H},\mu)$ given by $\pi(a)f(x) := \pi_x(a)f(x)$ for all $f \in L^2(X*\mathfrak{H},\mu)$. The representation π is called the *direct integral* of the $\{\,\pi_x\,\}$ and is denoted by

$$\int_X^{\oplus} \pi_x\,d\mu(x).$$

[5] In essence, we prove this as part of the proof of Proposition F.33 on page 423; see page 423.
[6] See footnote 8 on page 423.

Example F.25 (Remark E.15 on page 404 — take II). If X is a second countable locally compact space and if π is a separable representation of $C_0(X)$, then Theorem E.14 on page 404 and the discussion in Appendix E.1 implies that there is an analytic Borel Hilbert bundle $X * \mathfrak{H} = \coprod_{d=1}^{d=\infty} X_d \times \mathcal{H}_d$ and a finite Borel measure μ on X such that π is equivalent to the representation ρ_μ on $L^2(X * \mathfrak{H}, \mu)$ given by

$$\rho_\mu(f)h(x) := f(x)h(x).$$

If $m(x) := d$ for $x \in X_d$, then it is a matter of untangling definitions to see that

$$\rho_\mu = \int_X^\oplus m(x) \cdot \mathrm{ev}_x \, d\mu(x).$$

This justifies the assertions made in Remark E.15 on page 404. Furthermore, since ρ_μ is essentially the direct sum of representations of the form $\pi_\nu \otimes 1$, it follows from the second half of Theorem E.17 on page 405 that $\rho_\mu\big(C_0(X)\big)'' = \Delta(X * \mathfrak{H}, \mu)$.

The above example is the key which allows us to find direct integral decompositions with various properties (cf., Proposition E.20 on page 406).

Proposition F.26 ([28, Theorem 8.3.2]). *Suppose that ρ is a separable representation of a separable C^*-algebra A on $\mathcal{H}_\rho$ and that $\mathcal{C}$ is an abelian von Neumann subalgebra of $\rho(A)'$. Then there is a second countable locally compact Hausdorff space X, a Borel Hilbert bundle $X * \mathfrak{H}$, a finite measure μ on X, a Borel family $\{\pi_x\}$ of representations of A on $\mathcal{H}(x)$ and a unitary isomorphism of $\mathcal{H}_\rho$ onto $L^2(X * \mathfrak{H}, \mu)$ intertwining ρ and the direct integral*

$$\pi = \int_X^\oplus \pi_x \, d\mu(x) \tag{F.8}$$

as well as $\mathcal{C}$ and the diagonal operators $\Delta(X * \mathfrak{H}, \mu)$.

Remark F.27. We call (F.8) a *direct integral decomposition of ρ with respect to $\mathcal{C}$.* If

$$\rho' := \int_Y^\oplus \rho_y \, d\nu(y)$$

is another direct integral decomposition of ρ, then (F.8) and ρ' are equivalent via a unitary intertwining the respective diagonal operators. Then Proposition F.33 on page 423 applies.

Sketch of the Proof of Proposition F.26. Since $\mathcal{H}_\rho$ is separable, there is a separable C^*-subalgebra C_0 of $\mathcal{C}$ which is dense in the weak operator topology [2, Proposition 1.2.3]. Then C_0 is isomorphic to $C_0(X)$ for a second countable locally compact space X [2, Theorem 1.1.1]. The identity map on $B(\mathcal{H}_\rho)$ is a representation of $C_0(X)$ and it follows from Example F.25 that $\mathcal{H}_\rho$ is unitarily isomorphic to $L^2(X * \mathfrak{H}, \mu)$ for a suitable Borel Hilbert bundle $X * \mathfrak{H}$ and finite measure μ, and that the unitary intertwines the identity representation of C_0 with the representation ρ_μ from that example. Therefore $\mathcal{C} = C_0''$ is intertwined with the diagonal operators $\Delta(X * \mathfrak{H}, \mu) = \rho_\mu\big(C_0(X)\big)''$. Thus ρ is equivalent to a representation π on $L^2(X * \mathfrak{H}, \mu)$ such that $\pi(A) \subset \Delta(X * \mathfrak{H}, \mu)'$. Using Corollary E.18 on page 406 (together with Corollary F.12 on page 414), it follows that there is a Borel field $\{ \pi_x \}$ of representations on A on $\mathcal{H}(x)$ such that

$$\pi = \int_X^\oplus \pi_x \, d\mu(x). \tag{F.9}$$

This completes the proof. □

Example F.25 on the preceding page can be thought of as a decomposition of π into irreducibles. Furthermore, the equivalence class of π is determined by the class of the measure μ and the μ-almost everywhere equivalence class of the "multiplicity function" m [2, §2.2]. Although this result is very satisfactory, it does not extend to the general case. (Well, it works pretty well in case that A is of GCR — see [2, Chap. 4] for example — but we want to allow for the possibility of non-GCR algebras.) The difficulty is that, in the general case, the uniqueness of a direct integral decomposition of a representation (Proposition F.33) is rather fussy and depends strongly on the abelian subalgebra of $\pi(A)'$ used in the decomposition.[7]

Lemma F.28. *Suppose that*

$$\pi = \int_X^\oplus \pi_x \, d\mu(x)$$

is a direct integral of representations of a separable C^*-algebra A on a Hilbert space $L^2(X * \mathfrak{H}, \mu)$. Then the map $x \mapsto \ker \pi_x$ is a Borel map of X into $\mathcal{I}(A)$. If $\ker \pi_x \in \operatorname{Prim} A$ *for all x, then $x \mapsto \ker \pi_x$ is a Borel map into* $\operatorname{Prim} A$.

Proof. Since the topology on $\operatorname{Prim} A$ is the relative topology coming from $\mathcal{I}(A)$, it suffices to prove the first statement.

Let J be an ideal in A and let

$$\mathcal{O}_J := \{ I \in \mathcal{I}(A) : J \not\subset I \}.$$

Since the $\mathcal{O}_J$'s form a subbasis for the topology on $\mathcal{I}(A)$, it suffice to show that the inverse image of each $\mathcal{O}_J$ is Borel; that is, we need to see that

$$B = \{ x \in X : \pi_x(J) \neq \{ 0 \} \}$$

is Borel. Let $\{ f_i \}$ be a fundamental sequence for $X * \mathfrak{H}$ in $\mathcal{L}^2(X * \mathfrak{H}, \mu)$ and let $\{ a_i \}$ be a countable dense subset of J. Then

$$B_{ijk} := \{ x \in X : \big(\pi_x(a_i) f_j(x) \mid f_k(x)\big) \neq 0 \}$$

[7] For example, to get a direct integral decomposition in which almost all the component representations π_x are irreducible, it is necessary that the abelian subalgebra of $\pi(A)'$ be maximal abelian [28, Lemma 8.5.1]. Unfortunately, in the non-GCR case, $\pi(A)'$ can have lots of maximal abelian subalgebras, and the decompositions can have very little to do with one another [28, §8.5.3 & §18.9.8].

is Borel and

$$B = \bigcup_{ijk} B_{ijk}.$$

Thus B is Borel. $\square$

For later use, we need the following variation on Proposition E.19 on page 406. It can be proved using Proposition E.19 and Corollary F.9 on page 413. It can also be proved directly by mimicking the proof of Proposition E.19 (as found in [2, §4.2]). A complete proof can be found in [29, II.2.3 Proposition 4(i)]. We'll omit the details here.

Proposition F.29. *Suppose that $X * \mathfrak{H}$ is an analytic Borel Hilbert bundle over an analytic Borel space X. Suppose that*

$$T = \int_X^{\oplus} T(x)\,d\mu(x) \quad \text{and} \quad T_i = \int_X^{\oplus} T_i(x)\,d\mu(x) \quad i = 1, 2, 3, \dots$$

are decomposable operators such that the sequence $T_i \to T$ in the strong operator topology. Then there is a subsequence $\{T_{i_j}\}$ and a μ-null set N such that $T_{i_j}(x) \to T(x)$ in the strong operator topology for all $x \in X \setminus N$.

F.4 τ-isomorphisms

We want to expand our discussion at the beginning of Appendix F.3 so that we can consider Hilbert space isomorphisms U of direct integrals $L^2(X * \mathfrak{H}, \mu)$ and $L^2(Y * \mathfrak{K}, \nu)$ over *different* spaces X and Y. We will want U to be "decomposable" in an appropriate sense. For example, if we have a Borel isomorphism $\tau : X \to Y$, then we could look for maps $U(x) : \mathcal{H}(x) \to \mathcal{K}\big(\tau(x)\big)$ such that $y \mapsto U\big(\tau^{-1}(y)\big)f\big(\tau^{-1}(y)\big)$ is Borel if and only if $x \mapsto f(x)$ is Borel. If this map is to be isometric, it will be convenient to assume that $\nu = \tau_*\mu$ so that τ is measure preserving.

Definition F.30. Suppose that $X * \mathfrak{H}$ and $Y * \mathfrak{K}$ are Borel Hilbert bundles and that $\tau : X \to Y$ is a Borel isomorphism. Then a family of unitaries $V(x) : \mathcal{H}(x) \to \mathcal{K}\big(\tau(x)\big)$ is called a *τ-isomorphism* if

$$x \mapsto \big(V(x)f(x) \mid g\big(\tau(x)\big)\big) \tag{F.10}$$

is Borel for all $f \in B(X * \mathfrak{H})$ and $g \in B(Y * \mathfrak{K})$.

Remark F.31. If $\{e_k\}$ and $\{u_j\}$ are special orthogonal fundamental sequences (Remark F.7 on page 412) for $X * \mathfrak{H}$ and $Y * \mathfrak{K}$, respectively, then

$$\begin{aligned}\big(V(x)f(x) \mid g\big(\tau(x)\big)\big) \\ &= \sum_{k,j} \big(f(x)|e_j(x)\big)\big(V(x)e_j(x) \mid u_k\big(\tau(x)\big)\big)\big(u_k\big(\tau(x)\big) \mid g\big(\tau(x)\big)\big),\end{aligned}$$

and it follows that it suffices to check (F.10) for f and g in a fundamental sequence for $X * \mathfrak{H}$ and $Y * \mathfrak{K}$, respectively.

The terminology "τ-isomorphism" is justified by the following easy observation. Note that the choice of measure in the range space is determined by τ.

Lemma F.32. *Suppose that $\tau : X \to Y$ is a Borel isomorphism and that $\{ V(x) \}_{x\in X}$ is a τ-isomorphism of $X * \mathfrak{H}$ onto $Y * \mathfrak{K}$. Then if μ is a finite Borel measure on X, there is a unitary*

$$V : L^2(X * \mathfrak{H}, \mu) \to L^2(Y * \mathfrak{K}, \tau_*\mu)$$

given by $Vf\bigl(\tau(x)\bigr) := V(x)\bigl(f(x)\bigr)$ for all $x \in X$.

Proof. If $f \in \mathcal{L}^2(X * \mathfrak{H}, \mu)$, then $h\bigl(\tau(x)\bigr) = V(x)f(x)$ is in $B(Y * \mathfrak{K})$ since $\{ V(x) \}$ is a τ-isomorphism. Then, using the definition of $\tau_*\mu$, we easily check that $\|h\| = \|f\|$. Thus V is certainly an isometry. If we let $W\bigl(\tau(x)\bigr) := V(x)^*$, then $\{ W(y) \}$ is a τ^{-1}-isomorphism implementing an inverse for V. Thus V is a unitary as claimed. □

Since null sets are invisible to integral formulas, it is not always appropriate to expect to get an everywhere defined isomorphism τ as above. Instead, there will be a ν-null set N and a ν-null set M and an Borel isomorphism $\tau : X \smallsetminus N \to Y \smallsetminus M$. Then we can define U as before with the understanding that $Uf(y)$ or $Uf\bigl(\tau(x)\bigr)$ is zero if $y \in M$ or $x \in N$. Of course in this situation, $\nu := \tau_*\mu$ is formally only a measure on $Y \smallsetminus M$, but we interpret it as a measure on Y in the obvious way. We will also make repeated use of the following observation: if $N' \supset N$ is a μ-null set, then $\tau(N' \smallsetminus N)$ is a ν-null set and τ restricts to a Borel isomorphism of $X \smallsetminus N'$ onto $Y \smallsetminus M'$ where $M' = M \cup \tau(N' \smallsetminus N)$.

Using Lemma F.32, it is clear that a bona fide τ-isomorphism $\{ V(x) \}$ from $X * \mathfrak{H}|_{X\smallsetminus N}$ to $Y * \mathfrak{K}|_{Y\smallsetminus M}$ induces in a simple way a unitary

$$U := \int_X^\oplus U(x)\,d\mu(x) \tag{F.11}$$

from $L^2(X * \mathfrak{H}, \mu)$ to $L^2(Y * \mathfrak{K}, \tau_*\mu)$. We will simply say that this unitary is induced by the τ-isomorphism $\{ V(x) \}$ via Lemma F.32. A straightforward variation on Lemma F.20 on page 417 implies that the U determines the $U(x)$ for μ-almost all x. Thus if a unitary U is determined by a τ-isomorphism, then the unitaries $U(x)$ that appear in the decomposition (F.11) are *essentially unique.*

F.5 Uniqueness of Direct Integrals

If direct integral decompositions of a representation are going to be a useful replacement for the decomposition of finite dimensional representations into irreducible summands, then we are going to want some sort of uniqueness result under unitary equivalence. The basic result is taken from [29, II.6.3 Theorem 4]. The key hypothesis is that the unitary implementing the equivalence *also intertwine the respective diagonal operators.*

Proposition F.33. *Let A be a separable C^*-algebra. Suppose that*

$$\pi := \int_X^\oplus \pi_x \, d\mu(x) \quad \text{and} \quad \rho := \int_Y^\oplus \rho_y \, d\nu(y)$$

are direct integral representations on $L^2(X * \mathfrak{H}, \mu)$ and $L^2(Y * \mathfrak{K}, \nu)$, respectively. Suppose that $U : L^2(X * \mathfrak{H}, \mu) \to L^2(Y * \mathfrak{K}, \nu)$ is an isomorphism implementing an equivalence between π and ρ which also intertwines the respective algebras of diagonal operators $\Delta(X * \mathfrak{H}, \mu)$ and $\Delta(Y * \mathfrak{K}, \nu)$. Then there is

(a) *a μ-null set N and ν-null set M,*
(b) *a Borel isomorphism $\tau : X \setminus N \to Y \setminus M$ such that $\tau_*\mu$ is equivalent to ν,*
(c) *an essentially unique τ-isomorphism from $X * \mathfrak{H}|_{X \setminus N}$ to $Y * \mathfrak{K}|_{Y \setminus M}$ consisting of unitaries $V(x) : \mathcal{H}(x) \to \mathcal{K}(\tau(x))$ for each $x \in X \setminus N$ such that*

$$Vf(\tau(x)) = V(x)f(x) \quad (x \in X \setminus N)$$

induces a unitary $V : L^2(X * \mathfrak{H}, \mu) \to L^2(Y * \mathfrak{K}, \tau_*\mu)$ with the properties that

(d) *$U = WV$, where $W : L^2(Y * \mathfrak{K}, \tau_*\mu) \to L^2(Y * \mathfrak{K}, \nu)$ is the natural isomorphism*

$$Wf(y) = \frac{d(\tau_*\mu)}{d\nu}(y)^{\frac{1}{2}} f(y), \quad \text{and}$$

(e) *such that $V(x)$ implements an equivalence between π_x and $\rho_{\tau(x)}$ for all $x \in X \setminus N$.*

If φ and ψ are bounded Borel functions on X and Y, respectively, and if T_φ and $\widetilde{T}_\psi$ are the corresponding diagonal operators on $L^2(X * \mathfrak{H}, \mu)$ and $L^2(Y * \mathfrak{K}, \tau_*\mu)$, respectively, then $VT_\varphi = \widetilde{T}_{\varphi \circ \tau^{-1}} V$.

Proof. Since U induces an isomorphism of $\Delta(\mu)$ and $\Delta(\nu)$, we obtain an isomorphism Ψ of $L^\infty(X, \mu)$ onto $L^\infty(Y, \nu)$. Von Neumann's Theorem (Corollary I.38 on page 502) implies that there are N, M and τ as in parts (a) and (b) such that $\Psi\varphi(y) = \varphi(\tau^{-1}(y))$ for $y \notin M$.

If $\varphi \in \mathcal{B}^b(X)$ and $\psi \in \mathcal{B}^b(Y)$, then we let T_φ and $\widetilde{T}_\psi$ be the corresponding diagonal operators on $L^2(X * \mathfrak{H}, \mu)$ and $L^2(Y * \mathfrak{K}, \tau_*\mu)$, respectively. Chasing through the various maps shows that

$$W\widetilde{T}_{\varphi \circ \tau^{-1}} W^{-1} = UT_\varphi U^{-1}.$$

Then $V := W^{-1}U$ is an isomorphism of $L^2(X * \mathfrak{H}, \mu)$ onto $L^2(Y * \mathfrak{K}, \tau_*\mu)$ satisfying

$$VT_\varphi = \widetilde{T}_{\varphi \circ \tau^{-1}} V. \tag{F.12}$$

We need to see that V "decomposes" as a τ-isomorphism.[8] (The $V(x)$ will be essentially unique since V is determined by U and W.) Let $\{ f_n \}$ be a fundamental sequence for $X * \mathfrak{H}$ in $\mathcal{L}^2(X * \mathfrak{H}, \mu)$. Choose a representative $g_i \in \mathcal{L}^2(Y * \mathfrak{K}, \nu)$ for

[8]Notice that if $X = Y$ and $\tau = \text{id}$, then we recover the result — at least for unitary operators — that the Decomposable operators coincide with the commutant of the diagonal operators.

Vf_i. If $r_1, \dots, r_n$ are complex rational numbers and if $\varphi \in \mathcal{B}^b(X)$, then we can compute that

$$\begin{aligned}\int_{X \setminus N} |\varphi(x)| \Big\| \sum_{i=1}^n r_i g_i\big(\tau(x)\big) \Big\| \, d\mu(x) \\ = \int_{X \setminus N} \Big\| \sum_{i=1}^n r_i \widetilde{T}_{\varphi \circ \tau^{-1}} V f_i\big(\tau(x)\big) \Big\| \, d\mu(x)\end{aligned}$$

which, using (F.12), is

$$\begin{aligned}&= \int_{X \setminus N} \Big\| \sum_{i=1}^n r_i V T_\varphi f_i\big(\tau(x)\big) \Big\| \, d\mu(x) \\ &= \int_{Y \setminus M} \Big\| \sum_{i=1}^n r_i V T_\varphi f_i(y) \Big\| \, d\tau_*\mu(y) \\ &= \Big\| V\Big(\sum_{i=1}^n r_i T_\varphi f_i \Big) \Big\|\end{aligned}$$

which, since $\|V\| = 1$, is

$$\begin{aligned}&\le \Big\| \sum_{i=1}^n r_i T_\varphi f_i \Big\| \\ &= \int_{X \setminus N} |\varphi(x)| \Big\| \sum_{i=1}^n r_i f_i(x) \Big\| \, d\mu(x).\end{aligned}$$

Since φ is arbitrary, there is a μ-null set $N(r_1, \dots, r_n) \supset N$ such that

$$\Big\| \sum_{i=1}^n r_i g\big(\tau(x)\big) \Big\| \le \Big\| \sum_{i=1}^n r_i f_i(x) \Big\| \tag{F.13}$$

provided $x \notin N(r_1, \dots, r_n)$. Since the set of finite sequences of complex rational numbers is countable, there is a null set $N' \supset N$ such that $x \notin N'$ implies that (F.13) holds for any finite sequence of rationals. Thus if $x \in X \setminus N'$, then there is a well defined operator $V(x) : \mathcal{H}(x) \to \mathcal{K}\big(\tau(x)\big)$ such that $\|V(x)\| \le 1$ and such that for all n

$$V(x) f_n(x) = g_n\big(\tau(x)\big) \quad (x \notin N').$$

As described above, there is a $\tau_*\mu$-null set M' such that $\tau : X \setminus N' \to Y \setminus M'$ is a Borel isomorphism.

Let $\{ h_m \}$ be a fundamental sequence for $Y * \mathfrak{K}$ in $L^2(Y * \mathfrak{K}, \tau_*\mu)$. Then

$$y \mapsto \big(g_n(y) \mid h_m(y) \big)$$

is Borel (on $Y \setminus M'$) for all n and m. Equivalently,

$$x \mapsto \big(g_n\big(\tau(x)\big) \mid h_m\big(\tau(x)\big) \big)$$

is Borel (on $X \setminus N'$) for all n and m. Therefore

$$x \mapsto \big(V(x)f_n(x) \mid h_m\big(\tau(x)\big)\big) = \big(f_n(x) \mid V(x)^* h_m\big(\tau(x)\big)\big)$$

is Borel for all n. It follows that

$$x \mapsto V(x)^* h_m\big(\tau(x)\big)$$

is in $B(X * \mathfrak{H}|_{X \setminus N'})$. Thus if $f \in L^2(X * \mathfrak{H}, \mu)$,

$$x \mapsto \big(f(x) \mid V(x)^* h_m\big(\tau(x)\big)\big) = \big(V(x)f(x) \mid h_m\big(\tau(x)\big)\big)$$

is Borel for all m. Thus

$$h\big(\tau(x)\big) := V(x)f(x) \quad (x \notin N')$$

defines an element $h \in B(Y * \mathfrak{K}|_{Y \setminus M'})$ Furthermore,

$$\begin{aligned} \|h\|^2 &:= \int_Y \|h(y)\|^2 \, d\tau_*\mu(y) \\ &= \int_X \|h\big(\tau(x)\big)\|^2 \, d\mu(x) \\ &= \int_X \|V(x)\big(f(x)\big)\| \, d\mu(x) \\ &\le \int_X \|f(x)\| \, d\mu(x) \\ &= \|f\|^2. \end{aligned}$$

Thus we get a bounded operator V' from $L^2(X * \mathfrak{H}, \mu)$ to $L^2(Y * \mathfrak{K}, \tau_*\mu)$ defined by $V'f\big(\tau(x)\big) := V(x)\big(f(x)\big)$. In particular, $V'f_i = Vf_i$ by construction, and if $\varphi \in L^\infty(\mu)$, then using (F.12) we have

$$\begin{aligned} V'T_\varphi f_i &= \widetilde{T}_{\varphi \circ \tau^{-1}} V' f_i \\ &= \widetilde{T}_{\varphi \circ \tau^{-1}} V f_i \\ &= V T_\varphi f_i. \end{aligned}$$

Since the $T_\varphi f_i$ span a dense subset (Lemma F.17 on page 415), we have $V' = V$.

To complete the proof of part (c), we need to see that $V(x)$ is a unitary for almost all x. It follows from the above that if $g \in B(Y * \mathfrak{K})$, then $f(x) := V(x)^* g\big(\tau(x)\big)$ defines an element in $B(X * \mathfrak{H})$. Since $\|V(x)^*\| = \|V(x)\| \le 1$ for all x, we have

$$\begin{aligned} \|f\|^2 &= \int_X \|V(x)^* g\big(\tau(x)\big)\| \, d\mu(x) \\ &\le \int_X \|g\big(\tau(x)\big)\| \, d\mu(x) \\ &= \int_Y \|g(y)\|^2 \, d\tau_*\mu(y) \\ &= \|g\|^2. \end{aligned}$$

Thus there is a bounded operator $V'' : L^2(Y * \mathfrak{K}, \tau_*\mu) \to L^2(X * \mathfrak{H}, \mu)$ given by $V''g(x) := V(x)^* g\big(\tau(x)\big)$, and we have

$$\begin{aligned}(V''f \mid g) &= \int_X \big(V(x)^* f\big(\tau(x)\big) \mid g(x)\big)\, d\mu(x) \\ &= \int_X \big(f\big(\tau(x)\big) \mid V(x)g(x)\big)\, d\mu(x) \\ &= \int_Y \big(f(y) \mid Vg(y)\big)\, d\tau_*\mu(y) \\ &= (f \mid Vg) = (V^*f \mid g).\end{aligned}$$

Thus $V'' = V^*$. Since V is unitary, both

$$V^*V = \int_X^\oplus V(x)^*V(x)\, d\mu(x) \quad \text{and} \quad VV^* = \int_X^\oplus V(x)V(x)^*\, d\mu(x)$$

are the identity, and $V(x)$ is unitary for almost all x as required.

Define a representation π' of A on $L^2(Y * \mathfrak{K}, \tau_*\mu)$ by $\pi'(a) := V\pi(a)V^{-1}$, and observe that

$$\pi' = \int_Y^\oplus \pi'_y\, d(\tau_*\mu)(y),$$

where

$$\pi'_{\tau(x)}(a) = V(x)\pi_x(a)V(x)^{-1} \quad (x \notin N).$$

Furthermore, π' is equal to $W^{-1}\rho W$ which is equal to

$$\int_Y^\oplus \rho_y\, d(\tau_*\mu)(y).$$

Since $\pi'(a) = W^{-1}\rho(a)W$, there is a null set $M(a) \supset M$ such that

$$\pi'_y(a) = \rho_y(a) \quad (y \notin M(a)). \tag{F.14}$$

Since A is separable, there is a null set $M_A \supset M$ such that (F.14) holds for all a when $y \notin M_A$. Thus there is a μ-null set N_A such that $V(x)$ implements an equivalence between π_x and $\rho_{\tau(x)}$ for all $x \notin N_A$. This proves part (e). The last assertion is (F.12). □

An easy application of Proposition F.33 arises when we have bundles $X * \mathfrak{H}$ and $X * \mathfrak{K}$ in which the intertwining unitary commutes with actions of $L^\infty(X)$ as diagonal operators.

Corollary F.34. *Let A be a separable C^*-algebra and suppose that*

$$\pi = \int_X^\oplus \pi_x\, d\mu(x) \quad \textit{and} \quad \rho = \int_X^\oplus \rho_x\, d\mu(x)$$

are direct integral decompositions on $L^2(X * \mathfrak{H}, \mu)$ *and* $L^2(X * \mathfrak{K}, \mu)$*, respectively. Suppose that* $U : L^2(X * \mathfrak{H}, \mu) \to L^2(X * \mathfrak{K}, \mu)$ *is a unitary implementing an equivalence between* π *and* ρ *which commutes with the diagonal operators; more precisely, if* φ *is bounded Borel function on* X *and if* T_φ *and* T'_φ *are the corresponding diagonal operators on* $L^2(X * \mathfrak{H}, \mu)$ *and* $L^2(X * \mathfrak{K}, \mu)$*, respectively, then* $UT_\varphi = T'_\varphi U$*. Then* π_x *is equivalent to* ρ_x *for* μ*-almost all* x*.*

Proof. An examination of the beginning of the proof of Proposition F.33 on page 423 shows that we can apply that result with $\tau = \mathrm{id}_X$ and $W = I$. □

F.6 Isomorphism Bundle

Definition F.35. Let $X * \mathfrak{H}$ be a Borel Hilbert bundle with Borel sections $B(X, \mathfrak{H})$. The *isomorphism bundle* of $X * \mathfrak{H}$ is the set

$$\mathrm{Iso}(X * \mathfrak{H}) := \{\, (x, V, y) : V : \mathcal{H}(y) \to \mathcal{H}(x) \text{ is a unitary} \,\}$$

endowed with the smallest Borel structure such that, for all $f, g \in B(X, \mathfrak{H})$,

$$\psi_{f,g}(x, V, y) := \big(Vf(y) \mid g(x)\big)$$

define Borel functions from $\mathrm{Iso}(X * \mathfrak{H})$ to $\mathbf{C}$.

The isomorphism bundle admits a partially defined multiplication making it into a "groupoid". Two elements (x, V, y) and (y', U, z) are composable if $y = y'$, then their product is $(x, V, y)(y, U, z) := (x, VU, z)$. Elements of the form $(x, I_{\mathcal{H}(x)}, x)$ act as identities or units, and each element has an inverse: $(x, V, y)^{-1} := (y, V^*, x)$. Groupoids and their relevance to our interests here are discussed briefly in Remark G.26 on page 450.

Remark F.36. If X is a standard Borel space, then so is $\mathrm{Iso}(X * \mathfrak{H})$. Using Corollary F.12 on page 414, $X * \mathfrak{H}$ is Borel isomorphic to $\coprod_n X_n \times \mathcal{H}_n$ for Hilbert spaces $\mathcal{H}_n$ and $\{ X_n \}$ is a Borel partition of X. Then it is not hard to check that $\mathrm{Iso}(X * \mathfrak{H})$ is Borel isomorphic to $\coprod_n X_n \times U(\mathcal{H}_n) \times X_n$ where $U(\mathcal{H}_n)$ is the unitary group of $\mathcal{H}_n$ endowed with its standard Borel structure coming from the (Polish) weak operator topology and $X_n \times U(\mathcal{H}_n) \times X_n$ has the product Borel structure.

Remark F.37. Note that a function $U : Z \to \mathrm{Iso}(X * \mathfrak{H})$ is Borel if and only if $\psi_{f,g} \circ U$ is Borel for all f and g. In fact, it suffices to take f and g in fundamental sequences.

F.7 Product Borel Structure

Let G be a second countable locally compact group. Let $X * \mathfrak{H}$ be a Borel Hilbert bundle with fundamental sequence $\{ f_n \}$. Let $r : G \times X \to X$ be the projection onto the second factor.[9] We will have special need for the pull-back of $X * \mathfrak{H}$ via r

[9] Although we'll normally keep groupoid structure and terminology in the background, we use r for this map as it is the "range" map for the relevant groupoid structure on $G \times X$. See Remark G.26 on page 450.

and will use the notation $r^*(X * \mathfrak{H})$ in place of $X * \mathfrak{H}_r$. Note if $g \in B(r^*(X * \mathfrak{H}))$, then $g(s,x) \in \mathcal{H}(x)$. Since $\{\, (s,x) \mapsto f_n(x) \,\}_{n=1}^{\infty}$ is a fundamental sequence for $r^*(X * \mathfrak{H})$, $f \circ r \in B\big(r^*(X * \mathfrak{H})\big)$ for all $f \in B(X * \mathfrak{H})$ and

$$(s,x) \mapsto \big(f(x) \mid g(s,x)\big)$$

Borel for any $f \in B(X * \mathfrak{H})$.

Normally, we treat elements of $L^2(X)$ or even $L^2(X * \mathfrak{H}, \mu)$ as functions. Occasionally we have to admit that these elements are formally equivalence classes, and even more infrequently, it makes a difference. The next proposition will allow us to deal with such an instance. When such distinctions are necessary, will write $[f]$ to denote the almost everywhere equivalence class of a function f.

Proposition F.38. *Suppose that $f \in \mathcal{L}^2(X * \mathfrak{H}, \mu)$. Let $[f]$ be the class of f in $L^2(X * \mathfrak{H}, \mu)$. If V is a representation of G on $L^2(X * \mathfrak{H}, \mu)$ and if μ_G is Haar measure on G, then there is a $g \in B(r^*(X * \mathfrak{H}))$ such that for μ_G-almost all s,*

$$V(s)[f] = [g(s,\cdot)].$$

Proof. Since G is second countable, μ_G is σ-finite and there is a finite Borel measure ν on G which is equivalent to μ_G. Thus if $f \in \mathcal{L}^2(X * \mathfrak{H}, \mu)$, then $(s,x) \mapsto f(x)$ is in $\mathcal{L}^2\big(r^*(X * \mathfrak{H}), \nu \times \mu\big)$. Fubini's Theorem implies that

$$s \mapsto \int_X \big(h(s,x) \mid f(x)\big)\, d\mu(x) = \big(h(s,\cdot) \mid f\big)$$

is Borel and ν-integrable for all $h \in \mathcal{L}^2\big(r^*(X * \mathfrak{H}), \nu \times \mu\big)$. Thus, after modification on a null set, $s \mapsto h(s,\cdot)$ is a weakly Borel square integrable function of G into $L^2(X * \mathfrak{H}, \mu)$ with norm $\|h\|$. Since $s \mapsto V(s)f$ is continuous,

$$s \mapsto \big(h(s,\cdot) \mid V(s)f\big)$$

is Borel and we can define a bounded linear functional on $L^2\big(r^*(X * \mathfrak{H}), \nu \times \mu\big)$ by

$$\Phi(h) := \int_G \big(h(s,\cdot) \mid V(s)f\big)\, d\nu(s).$$

The Riesz Representation Theorem implies that there is a $g \in \mathcal{L}^2\big(r^*(X * \mathfrak{H}), \nu \times \mu\big)$ such that $\Phi(h) = (h \mid g)$.

Let $\{\, e_n \,\}$ be an orthonormal basis for $L^2(X * \mathfrak{H}, \mu)$. Then for all $\varphi \in C_c(G)$, $(s,x) \mapsto \varphi(s)e_n(x)$ is in $\mathcal{L}^2\big(r^*(X * \mathfrak{H}), \nu \times \mu\big)$ and

$$\int_G \varphi(s)\big(e_n \mid V(s)f - g(s,\cdot)\big)\, d\nu(s) = 0.$$

Thus there is a ν-null set M_n such that

$$(e_n \mid V(s)f - g(s,\cdot)) = 0 \quad \text{if } s \notin M_n.$$

If $M := \bigcup M_n$, then provided $s \notin M$, we have

$$V(s)[f] = [g(s,\cdot)]$$

as required. □

F.8 Measurable Sections

Unlike the discussion in Appendix B, in this Chapter we are working almost exclusively with Borel functions and Borel sections in particular. Of course, one advantage of Borel functions is that they are intrinsic to the space and do not depend on a choice of measure. It also seems appropriate to work in the Borel category when we are working with analytic and Standard Borel spaces where it is important that our σ-algebras actually be the Borel sets coming from a Polish topology somewhere. However, once we fix a measure μ, we could, as in [56] and [29], work with measurable sections of Borel Hilbert bundles. That is, we could define a section to be measurable if

$$x \mapsto \big(f(x) \mid f_n(x)\big) \tag{F.15}$$

is μ-measurable (that is, Borel with respect to the σ-algebra of sets in X for the completion of μ). Then Proposition F.6 on page 411 is still valid and the theory is essentially unchanged. That one arrives at the same direct integrals $L^2(X * \mathfrak{H}, \mu)$ follows from the observation that, as in the scalar case, every measurable section is almost everywhere equal to a Borel section.

Remark F.39. If f is measurable section of $X * \mathfrak{H}$, and if $\{ f_n \}$ is a fundamental sequence, then there is a Borel function b_n such that

$$b_n(x) = \big(f(x) \mid f_n(x)\big)$$

off a Borel μ-null set N_n. Thus

$$x \mapsto \big(f(x) \mid f_n(x)\big)$$

is Borel off $N := \bigcup N_n$. Let

$$g(x) := \begin{cases} 0_x & \text{if } x \in N, \text{ and} \\ f(x) & \text{if } x \notin N. \end{cases}$$

Then $g \in B(X * \mathfrak{H})$, and $f = g$ μ-almost everywhere.

Appendix G

Effros's Ideal Center Decomposition

The approach to proving the Effros-Hahn conjecture used by Gootman, Rosenberg and Sauvageot requires that we decompose representations into representations with primitive kernels. The normal means of decomposing arbitrary representations into irreducibles, or even factor representations, can't be done in manner that is sufficiently unique to suit our methods. Sauvageot's idea was to use *homogeneous* representations, and to invoke a structure theory, due to Effros, that allows a very useful decomposition of arbitrary representations into a direct integral of homogeneous representations.

G.1 Effros's Ideal Center Decomposition

Before we launch into the details of Effros's theory, we should define homogeneous representations and investigate a few of their properties. Recall that if E is a projection in the commutant $\pi(A)'$ of π, then π^E denotes the subrepresentation of π acting on $E\mathcal{H}_\pi$.

Definition G.1. A (possibly degenerate) representation of a C^*-algebra A is called *homogeneous* if

$$\ker \pi^E = \ker \pi$$

for every nonzero projection $E \in \pi(A)'$.

Notice that π is homogeneous exactly when every (possibly degenerate) nontrivial subrepresentation has the same kernel as π. Hence if π is nonzero and homogeneous, then π must be nondegenerate. Also, if π is homogeneous and if $E \in \pi(A)'$ is a nonzero projection, then $\pi(a) \mapsto \pi^E(a)$ is a well-defined, injective $*$-homomorphism of $\pi(A)$ onto $\pi(A)E$. Since $\pi(A)$ and $\pi(A)E$ are C^*-subalgebras of $B(\mathcal{H}_\pi)$, it follows that $\pi(a) \mapsto \pi(a)E$ is isometric for all $a \in A$. This is the harder half of the following useful characterization of homogeneous representations.

Lemma G.2. *If π is a (possibly degenerate) representation of A on $\mathcal{H}$, then π is homogeneous if and only if*

$$\pi(a) \mapsto \pi^E(a)$$

is isometric on $\pi(A)$ for every nonzero projection $E \in \pi(A)'$.

Suppose that P is a projection in $\pi(A)'$. Let $\mathcal{M} = P\mathcal{H}$ be the space of P, and define

$$[\pi(A)'\mathcal{M}] := \overline{\operatorname{span}}\{\, Th : T \in \pi(A)' \text{ and } h \in \mathcal{M}\,\}.$$

Let Q be the projection onto $[\pi(A)'\mathcal{M}]$. Then Q is central — that is, $Q \in \pi(A)' \cap \pi(A)''$. The corresponding subrepresentation of π, π^Q is called the *central cover* of π^E. Central covers play a big role in the general decomposition theory of type I representations (cf., e.g., [2, §2.1]). Our interest in central covers stems from a corollary to the next observation.

Lemma G.3. *Suppose that π is a (possibly degenerate) representation of a C^*-algebra A on $\mathcal{H}$. If P is a projection in $\pi(A)'$ and if π^Q is the central cover of π^P, then*

$$\ker \pi^Q = \ker \pi^P.$$

Proof. Let $\mathcal{M} := P\mathcal{H}$. Note that $a \in \ker \pi^P$ if and only if $\pi(a)h = 0$ for all $h \in \mathcal{M}$. Since $\mathcal{M} \subset [\pi(A)'\mathcal{M}]$, $a \in \ker \pi^Q$ implies that $a \in \ker \pi^P$. To prove the reverse inclusion, let $a \in \ker \pi^P$. To see that $a \in \ker \pi^Q$, it will suffice to see that $\pi(a)Th = 0$ for all $T \in \pi(A)'$ and $h \in \mathcal{M}$. But $\pi(a)Th = T\pi(a)h = 0$ if $a \in \ker \pi^P$. □

Corollary G.4. *Suppose that π is a factor representation of a C^*-algebra. Then π is homogeneous.*

Proof. Suppose that P is a nonzero projection in $\pi(A)'$. Since the center of $\pi(A)''$ is trivial, the central cover of π^P is π. Therefore $\ker \pi^P = \ker \pi$ by Lemma G.3. □

The converse of Corollary G.4 is false. If A is separable and not type I, then A has inequivalent irreducible representations π and ρ with the same kernel [126, Theorem 6.8.7]. Since π and ρ are irreducible, and therefore have trivial commutants, the commutant of $\pi \oplus \rho$ consists of operators of the form $\alpha 1_{\mathcal{H}_\pi} \oplus \beta 1_{\mathcal{H}_\rho}$, where α and β are complex constants. It follows easily that $\pi \oplus \rho$ is homogeneous and not factorial.

Recall that if π is a (possibly degenerate) representation of A on $\mathcal{H}$, then $\operatorname{ess}\pi$ is the subrepresentation of π on the essential subspace of $\mathcal{H}$ given by

$$\overline{\operatorname{span}}\{\, \pi(a)h : a \in A \text{ and } h \in \mathcal{H}\,\}.$$

Thus $\pi = \operatorname{ess}\pi \oplus 0$, where 0 denotes the zero representation on the orthogonal complement of the essential subspace for π. If $I \in \mathcal{I}(A)$ is an ideal in A, then we will write $e(I)$, or $e_\pi(I)$ if there is some ambiguity about which representation we're talking about, for the orthogonal projection onto the essential subspace of the restriction $\pi|_I$ of π to I.

Lemma G.5. *Suppose that π is a (possibly degenerate) representation of A on $\mathcal{H}$. If $I \in \mathcal{I}(A)$ and if $\{ u_n \}$ is an approximate identity for I, then $\pi(u_n) \to e(I)$ in the strong operator topology on $B(\mathcal{H})$. In particular, $e(I)$ belongs to the center $\pi(A)' \cap \pi(A)''$ of the von Neumann algebra generated by $\pi(A)$.*

Proof. By definition, $e(I)$ is the projection onto the subspace

$$\mathcal{M} = \overline{\operatorname{span}}\{ \pi(a)h : a \in I \text{ and } h \in \mathcal{H} \}.$$

But if $a \in I$, then $\pi(u_n)\pi(a)h = \pi(u_n a)h \to \pi(a)h$. It follows that $\pi(u_n)h' \to h'$ for all $h' \in \mathcal{M}$. But if $h'' \in \mathcal{M}^\perp$, then $\pi(a)h'' = 0$ for all $a \in I$. If $h \in \mathcal{H}$, then $h = h' + h''$ with $h' \in \mathcal{M}$ and $h'' \in \mathcal{M}^\perp$. Thus $\pi(u_n)h = \pi(u_n)h' \to h' = e(I)h$.

This proves the first assertion and that $e(I) \in \pi(A)''$ (which is the strong operator closure of $\pi(A)$). Since $e(I)$ is the projection onto an invariant subspace, we certainly also have $e(I) \in \pi(A)'$. □

In [45], Effros calls $\mathcal{Q}(I) := 1_{\mathcal{H}} - e(I)$ the *ideal center projection* associated to I.

Definition G.6. Suppose that π is a (possibly degenerate) representation of a C^*-algebra A on $\mathcal{H}$. The von Neumann algebra $\mathcal{IC}(\pi)$ generated by the ideal center projections $\mathcal{Q}(I)$, for $I \in \mathcal{I}(A)$, is called the *ideal center* for π.

Remark G.7. Notice that $\mathcal{IC}(\pi)$ is a subalgebra of the center $\pi(A)' \cap \pi(A)''$. In particular, $\mathcal{IC}(\pi)$ is an abelian von Neumann algebra. We can also view $\mathcal{IC}(\pi)$ as the von Neumann algebra generated by the $e(I)$'s.

Proposition G.8. *Let π be a (possibly degenerate) representation of A on $\mathcal{H}$. Then π is homogeneous if and only if $\mathcal{IC}(\pi) = \mathbf{C}1_{\mathcal{H}}$.*

Proof. If $\mathcal{IC}(\pi) \neq \mathbf{C}1_{\mathcal{H}}$, then there is an ideal center projection $Q = \mathcal{Q}(I)$ which equals neither 0 nor $1_{\mathcal{H}}$. Since $\pi|_I$ is the zero representation on $Q\mathcal{H}$, we must have $I \subset \ker \pi^Q$. Since $Q \neq 1_{\mathcal{H}}$, $I \neq \{0\}$. Let $\{ u_n \}$ be an approximate identity for I. If $h \in (Q\mathcal{H})^\perp$, then $\pi(u_n)h \to h$ by Lemma G.5. Thus for large n, $u_n \notin \ker \pi$. Thus $\ker \pi^Q \neq \ker \pi$, and π is not homogeneous (since $Q \neq 0$).

If π is not homogeneous, then there is a nonzero projection $P \in \pi(A)'$ such that

$$\ker \pi^P \neq \ker \pi. \tag{G.1}$$

Let $Q = \mathcal{Q}(\ker \pi^P)$. If $\{ e_n \}$ is an approximate identity for $\ker \pi^P$, then $\pi(e_n)P = 0$ for all n. Since $\pi(e_n) \to 1_{\mathcal{H}} - Q$, we have $(1_{\mathcal{H}} - Q)P = 0$. Hence $0 \neq P \leq Q$. But if $Q = 1_{\mathcal{H}}$, then $\operatorname{ess}(\pi|_{\ker \pi^P}) = 0$. This implies that $\ker \pi^P \subset \ker \pi$. Since we trivially have $\ker \pi \subset \ker \pi^P$, this contradicts (G.1). Thus Q is a nontrivial ideal center projection, and $\mathcal{IC}(\pi) \neq \mathbf{C}1_{\mathcal{H}}$. □

Recall that an ideal $I \in \mathcal{I}(A)$ is *prime* if $I \neq A$ and if whenever $J, K \in \mathcal{I}(A)$ satisfy $JK \subset I$, then either $J \subset I$ or $K \subset I$. All primitive ideals are prime and the converse holds for separable C^*-algebras (cf., [139, Proposition A.17 and Theorems A.49 and A.50]). Now we can establish the connection between homogeneous representations and primitive ideals.

Corollary G.9. *Suppose that π is a nonzero homogeneous representation of a C^*-algebra A on $\mathcal{H}$. Then $\ker \pi$ is a prime ideal in A. In particular, if A is separable, then $\ker \pi$ is a primitive ideal.*

Proof. Since $\ker \pi \neq A$, it suffices to show that if $J, K \in \mathcal{I}(A)$ satisfies $JK \subset \ker \pi$, then either $J \subset \ker \pi$ or $K \subset \ker \pi$. Since π is homogeneous, $\mathcal{Q}(K)$ is either 0 or $1_{\mathcal{H}}$. Equivalently, $e(K)$ is either 0 or $1_{\mathcal{H}}$. Recall that $e(K)$ is the projection onto

$$\mathcal{M} := \overline{\operatorname{span}}\{\, \pi(b)h : b \in K \text{ and } h \in \mathcal{H} \,\}.$$

If $a \in J$ and $b \in K$, then $\pi(a)\pi(b)h = \pi(ab)h = 0$. Thus, $\pi(J)\mathcal{M} = \{\,0\,\}$. If $e(K) = 1_{\mathcal{H}}$, then $\mathcal{M} = \mathcal{H}$ and $J \subset \ker \pi$.

On the other hand, if $e(K) = 0$, then $K \subset \ker \pi$. Thus $\ker \pi$ is prime. □

Lemma G.10. *Suppose that $X * \mathfrak{H}$ is an analytic Borel Hilbert bundle over an analytic space X.*[1] *Let*

$$\pi = \int_X^{\oplus} \pi_x \, d\mu(x)$$

be a direct integral of representations of a separable C^*-algebra on $L^2(X * \mathfrak{H}, \mu)$. If I is an ideal in A, then

$$e_\pi(I) = \int_X^{\oplus} e_{\pi_x}(I) \, d\mu(x).$$

Proof. Let $\{\, u_i \,\}$ be an approximate identity for I. Then

$$e_\pi(I) = \lim_i \pi(u_i) \quad \text{and} \quad e_{\pi_x}(I) = \lim_i \pi_x(u_i) \tag{G.2}$$

in the strong operator topology. Since $e_\pi(I) \in \pi(A)''$ and $\Delta(X * \mathfrak{H}, \mu) \subset \pi(A)'$, we have $e_\pi(I) \in \Delta(X * \mathfrak{H}, \mu)'$ and $e_\pi(I)$ is decomposable:

$$e_\pi(I) = \int_X^{\oplus} e_\pi(I)(x) \, d\mu(x).$$

It follows from the left-hand side of (G.2) and Proposition F.29 on page 421 that, after passing to a subsequence and relabeling, $\pi_x(u_i) \to e_\pi(I)(x)$ in the strong operator topology for μ-almost every x. Hence, the result follows from the right-hand side of (G.2). □

Remark G.11. If $1 \leq n \leq \aleph_0$ and if π is a homogeneous representation of A, then $n \cdot \pi$ is homogeneous. To see this, notice that π is homogeneous if and only if given an ideal $J \in \mathcal{I}(A)$ and a approximate identity $\{\, u_i \,\}$ for J, then the strong operator limit of $\pi(u_i)$ is either 0 of $I_{\mathcal{H}_\pi}$. Thus π is homogeneous if and only if $n \cdot \pi$ is.

[1] With the sole exception of Appendix F.1, we will use "Borel Hilbert bundle" to signify an analytic Borel Hilbert Bundle over an analytic Borel space X.

To prove Effros's decomposition theorem, we'll need a Borel selection result for representations. To discuss this, it's probably best to start with the treatment in section 4.1 of [2]. Let A be a separable C^*-algebra and $\mathcal{H}$ a separable complex Hilbert space. We let $\operatorname{rep}(A,\mathcal{H})$ be the set of $*$-homomorphisms of A into $B(\mathcal{H})$. (Thus $\pi \in \operatorname{rep}(A,\mathcal{H})$ is a representation exactly when it is nondegenerate.[2]) We give $\operatorname{rep}(A,\mathcal{H})$ the smallest topology for which

$$\pi \mapsto \bigl(\pi(a)h \mid k\bigr)$$

is continuous for all $a \in A$ and all $h, k \in \mathcal{H}$. Since

$$\|\pi(a)h - k\|^2 = \bigl(\pi(a^*a)h \mid h\bigr) - 2\operatorname{Re}\bigl(\pi(a)h \mid k\bigr) + \|k\|^2,$$

$\pi \mapsto \pi(a)h$ is continuous for all $a \in A$ and $h \in \mathcal{H}$.[3] Let $\{a_n\}$ and $\{h_n\}$ be dense sequences in the unit balls of A and $\mathcal{H}$, respectively. It is not hard to check that

$$d(\pi,\rho) := \sum_{n,m=1}^{\infty} 2^{-n-m}\|\pi(a_n)h_m - \rho(a_n)h_m\|$$

is a metric on $\operatorname{rep}(A,\mathcal{H})$ which is compatible with the given topology. Some more work is required to see that d is a complete metric and that $\operatorname{rep}(A,\mathcal{H})$ is second countable (cf. [26, Lemme 1]). Thus $\operatorname{rep}(A,\mathcal{H})$ is a Polish space.

We let $\operatorname{rep}_{\mathrm{hom}}(A,\mathcal{H})$ be the set of homogeneous representations of A (including the zero representation), and we let $\operatorname{rep}_{\mathrm{nh}}(A,\mathcal{H})$ be its complement: the set of nonhomogeneous $*$-homomorphisms of A into $B(\mathcal{H})$. We let $P(\mathcal{H})$ be the set of projections in $B(\mathcal{H})$, and let $P_0(\mathcal{H})$ the set $P(\mathcal{H}) \smallsetminus \{0\}$ of nonzero projections. We give $P(\mathcal{H})$ and $P_0(\mathcal{H})$ the relative Borel structure as subsets of the unit ball

$$B(\mathcal{H})_1 := \{\, T \in B(\mathcal{H}) : \|T\| \le 1 \,\}$$

equipped with the weak operator topology.

Theorem G.12. *The set* $\operatorname{rep}_{\mathrm{nh}}(A,\mathcal{H})$ *of nonhomogeneous $*$-homomorphisms of A into $B(\mathcal{H})$ is an analytic Borel space in its relative Borel structure. Furthermore*

$$\begin{aligned} E := \{\, (\pi, P) \in \operatorname{rep}_{\mathrm{nh}}(A,\mathcal{H}) \times P_0(\mathcal{H}) :& \\ P \in \pi(A)' \textit{ and } \pi(a) \mapsto \pi(a)P &\textit{ is not an isometry on } \pi(A). \,\} \end{aligned} \tag{G.3}$$

is a Borel subset of the standard Borel space $\operatorname{rep}(A,\mathcal{H}) \times P_0(\mathcal{H})$ *(and hence a standard Borel space in its relative Borel structure).*

Remark G.13. In [45, Theorem 1.7], Effros proves that both $\operatorname{rep}_{\mathrm{hom}}(A,\mathcal{H})$ and $\operatorname{rep}_{\mathrm{nh}}(A,\mathcal{H})$ are Borel subsets of $\operatorname{rep}(A,\mathcal{H})$, and hence that both are standard Borel spaces. However, for our purposes (Theorem G.14 on page 437), it is sufficient to have only that $\operatorname{rep}_{\mathrm{nh}}(A,\mathcal{H})$ is an analytic Borel space. Since this is a bit easier to establish, we have settled for the weaker assertion.

---

[2] Recall that we have reserved the term "representation" for nondegenerate homomorphisms of a C^*-algebra into $B(\mathcal{H})$.

[3] As pointed out in [2], the topology on $\operatorname{rep}(A,\mathcal{H})$ is also generated by the functions $\pi \mapsto \pi(a)h$.

Proof. By Lemma D.37 on page 391, $B(\mathcal{H})_1$ is a Polish space in the weak operator topology. Since $T \mapsto T^*$ is continuous in the weak operator topology,

$$B(\mathcal{H})_1^{\text{s.a.}} := \{\, T \in B(\mathcal{H})_1 : T = T^* \,\}$$

is also Polish. Let $\{\, e_n \,\}$ be an orthonormal basis for $\mathcal{H}$. If $T = T^*$, then

$$(T^2 h \mid k) = (Th \mid Tk) = \sum_n (Th \mid e_n)(e_n \mid Tk).$$

Since the pointwise limit of Borel functions is Borel,

$$T \mapsto (T^2 h \mid k)$$

is Borel on $B(\mathcal{H})_1^{\text{s.a.}}$. If

$$B(n,m) := \{\, T \in B(\mathcal{H})_1^{\text{s.a.}} : |(T^2 e_n \mid e_m) - (Te_n \mid e_m)| = 0 \,\},$$

then $B(n,m)$ is Borel and

$$P(\mathcal{H}) = \bigcap_{n,m} B(n,m)$$

is Borel and equals the set of orthogonal projections in $B(\mathcal{H})$. Thus the set $P_0(\mathcal{H}) = P(\mathcal{H}) \setminus \{\, 0 \,\}$ of nonzero projections is a standard Borel space. Since

$$\begin{aligned}\big(\pi(a)Th \mid k\big) &= \sum \big(Th \mid e_n\big)\big(\pi(a)e_n \mid k\big), \text{ and}\\ \big(T\pi(a)h \mid k\big) &= \sum \big(\pi(a)h \mid e_n\big)\big(Te_n \mid k\big),\end{aligned}$$

it follows that

$$(\pi, T) \mapsto \big(\big(\pi(a)T - T\pi(a)\big)h \mid k\big)$$

is Borel on $\operatorname{rep}(A,\mathcal{H}) \times B(\mathcal{H})_1^{\text{s.a.}}$ for all $a \in A$ and $h,k \in \mathcal{H}$. Let

$$P(a,n,m) := \{\, (\pi, P) \in \operatorname{rep}(A,\mathcal{H}) \times P_0(\mathcal{H}) : \big(\big(\pi(a)P - P\pi(a)\big)e_n \mid e_m\big) = 0 \,\}.$$

Then each $P(a,n,m)$ is Borel. Therefore if $\{\, a_k \,\}$ is dense in A, then

$$C := \{\, (\pi, P) \in \operatorname{rep}(A,\mathcal{H}) \times P_0(\mathcal{H}) : P \in \pi(A)' \,\} = \bigcap_{k,n,m} P(a_k, n, m) \tag{G.4}$$

is a Borel subset of the standard Borel space $\operatorname{rep}(A,\mathcal{H}) \times P_0(\mathcal{H})$. If $(\pi, P) \in C$, then

$$\begin{aligned}\|\pi(a)Ph\|^2 &= \big(\pi(a^*a)h \mid Ph\big)\\ &= \sum \big(\pi(a^*a)h \mid e_k\big)\big(e_k \mid Ph\big).\end{aligned}$$

Therefore both

$$(\pi, P) \mapsto \|\pi(a)Ph\| \quad \text{and} \quad (\pi, P) \mapsto \|\pi(a)h\|$$

are Borel functions on C for all $a \in A$ and $h \in \mathcal{H}$. If $\{ h_n \}$ is dense in the unit sphere of $\mathcal{H}$, then

$$\|\pi(a)P\| = \|\pi^P(a)\| = \sup_n \|\pi(a)Ph_n\|.$$

Therefore both

$$(\pi, P) \mapsto \|\pi(a)P\| \quad \text{and} \quad (\pi, P) \mapsto \|\pi(a)\|$$

are Borel. Thus

$$D(a) := \{ (\pi, P) \in C : \|\pi^P(a)\| = \|\pi(a)\| \}$$

is a Borel set. Since A is separable, and since π is homogeneous exactly when $\|\pi^P(a)\| = \|\pi(a)\|$ for all $a \in A$ and $P \in \pi(A)' \cap P_0(\mathcal{H})$,

$$D := \{ (\pi, P) \in C : \pi(a) \mapsto \pi(a)P \text{ is isometric on } \pi(A) \} \tag{G.5}$$

is Borel.

Let $E = C \setminus D$. Then E a Borel subset of the standard Borel space $\operatorname{rep}(A, \mathcal{H}) \times P_0(\mathcal{H})$ (in the product Borel structure), and coincides with the set described by (G.3). On the other hand, $\sigma : E \to \operatorname{rep}(A, \mathcal{H})$ given by $\sigma(\pi, P) = \pi$ is a Borel map into a Polish space with image exactly $\operatorname{rep}_{\mathrm{nh}}(A, \mathcal{H})$. Since E is a standard Borel space, $\sigma(E) = \operatorname{rep}_{\mathrm{nh}}(A, \mathcal{H})$ is analytic by [2, Theorem 3.3.4]. □

Theorem G.14 ([45, Theorem 1.8]). *Suppose that μ is a finite Borel measure on a second countable locally compact space X and that $X * \mathfrak{H}$ is a Borel Hilbert bundle over X. Let*

$$\pi := \int_X^\oplus \pi_x \, d\mu(x)$$

be a direct integral decomposition of a representation π of a separable C^*-algebra A on $L^2(X * \mathfrak{H}, \mu)$. Let $\mathcal{IC}(\pi)$ be the ideal center of π and $\Delta(X * \mathfrak{H}, \mu)$ the diagonal operators on $L^2(X * \mathfrak{H}, \mu)$. Then $\mathcal{IC}(\pi) \subset \Delta(X * \mathfrak{H}, \mu)$ if and only if π_x is homogeneous for μ-almost all x.

For the proof, we need some preliminaries on things measurable (as compared with Borel).

Lemma G.15. *Suppose that μ is a finite measure on X and that $\tau : X \to P(\mathcal{H})$ is μ-measurable. Then there is a Borel null set N such that $f : X \setminus N \to P(\mathcal{H})$ is Borel.*

Proof. Let $\{ h_n \}$ be a countable dense subset of the unit sphere in $\mathcal{H}$. Since $P(\mathcal{H})$ inherits its standard Borel structure from the weak operator topology, $\tau : X \to P(\mathcal{H})$ is Borel (resp. μ-measurable) if and only if $x \mapsto \big(\tau(x)h_n \mid h_m\big)$ is Borel (resp. μ-measurable) for all n and m (Lemma D.20 on page 375). Thus if $\tau : X \to P(\mathcal{H})$ is measurable, then there are null sets $N_{n,m}$ such that $x \mapsto \big(\tau(x)h_n \mid h_m\big)$ is Borel off $N_{n,m}$. Let

$$N = \bigcup_{n,m} N_{n,m}.$$

Then τ is Borel on $X \setminus N$. □

Remark G.16 (Absolutely measurable). If $(X, \mathscr{B})$ is a Borel space, then a subset $A \subset X$ is called *absolutely measurable* if it is μ-measurable for every finite Borel measure μ on $(X, \mathscr{B})$. This means that given μ, then there are Borel sets E_μ and F_μ such that $E_\mu \subset A \subset F_\mu$ and such that $\mu(F_\mu \setminus E_\mu) = 0$. Then collection of absolutely measurable subsets of X is a σ-algebra $\mathscr{A}$ in X which contains $\mathscr{B}$. Since analytic subsets of a Polish space are absolutely measurable [2, Theorem 3.2.4], it is often the case that $\mathscr{A}$ strictly contains $\mathscr{B}$. A function $f : (X, \mathscr{B}) \to (Y, \mathscr{M})$ is *absolutely measurable* if it is Borel from $(X, \mathscr{A})$ to $(Y, \mathscr{M})$; that is, f is absolutely measurable if the inverse image of every Borel set in Y is absolutely measurable in X. Alternatively, $f : X \to Y$ is absolutely measurable if its μ-measurable for all finite Borel measures on X.

Proof of Theorem G.14. Using Corollary F.12 on page 414, it will suffice to treat the case where $X * \mathfrak{H}$ is a trivial bundle $X \times \mathcal{H}$ for a fixed Hilbert space $\mathcal{H}$. Assume that $\mathcal{IC}(\pi) \subset \Delta(X * \mathfrak{H}, \mu) = \mathscr{L} \otimes 1_{\mathcal{H}}$. If the conclusion of the theorem were false, there would be a set $S \subset X$ such that $\mu(S) > 0$ and such that π_x was nonhomogeneous for all $x \in S$. After replacing X by S and μ by $\mu|_S$, we may as well assume that π_x is not homogeneous for all x in X. Therefore, for each $x \in X$, there is a nonzero projection $P_x \in \pi_x(A)'$ such that $\pi_x(a) \mapsto \pi_x(a)P_x$ is not isometric on $\pi_x(A)$. The problem is that we want to choose the P_x so that $x \mapsto P_x$ is a Borel function. The gymnastics required for this are provided largely by Theorem G.12 on page 435. Let E be as in (G.3) and define

$$\sigma : E \to \operatorname{rep}_{\mathrm{nh}}(A, \mathcal{H}) \quad \text{by } \sigma(\pi, P) = \pi.$$

Since Theorem G.12 implies that E and $\operatorname{rep}_{\mathrm{nh}}(A, \mathcal{H})$ are both analytic Borel spaces, there is an absolutely measurable cross-section

$$c : \operatorname{rep}_{\mathrm{nh}}(A, \mathcal{H}) \to E$$

for σ [2, Theorem 3.4.3]. Since, by assumption,

$$x \mapsto \big(\pi_x(a)h \mid k\big)$$

is Borel for all $a \in A$ and $h, k \in \mathcal{H}$, it follows that $\tau(x) := \pi_x$ defines a Borel map $\tau : X \to \operatorname{rep}(A, \mathcal{H})$. Let $\nu := \tau_* \mu$ be the push-forward measure on $\operatorname{rep}(A, \mathcal{H})$. Then c is ν-measurable, and it is not hard to see that $c \circ \tau$ is a μ-measurable function from X to E. Since the projection on the second factor, pr_2, is certainly Borel from E to $P_0(\mathcal{H})$, there is a μ-measurable map $x \to P_x$ (given by $\mathrm{pr}_2 \circ c \circ \tau$) of X into $P_0(\mathcal{H})$ such that $(\pi_x, P_x) \in E$ for all $x \in X$. Lemma G.15 on the previous page implies that there is a Borel null set $N \subset X$ such that $x \mapsto P_x$ is Borel on $X \setminus N$.

We claim there is a Borel set $T \subset X \setminus N$, an $a \in A$ and a $\epsilon > 0$ such that $\mu(T) > 0$ and such that

$$\|\pi_x(a)P_x\| \le \|\pi_x(a)\| - \epsilon \quad \text{for all } x \in T. \tag{G.6}$$

Suppose the claim is false. Let $\{ a_k \}$ be a dense sequence in A. Then

$$T(n, m) := \{ x \in X : \|\pi_x(a_n)P_x\| \le \|\pi_x(a_n)\| - \frac{1}{m} \}$$

has measure zero. Then

$$X \setminus \bigcup_{n,m} T(n,m) \tag{G.7}$$

is conull. If x belongs to (G.7), then

$$\|\pi_x(a_n)P_x\| = \|\pi_x(a_n)\| \quad \text{for all } n.$$

Thus $\pi_x(a) \mapsto \|\pi_x(a)P_x\|$ is an isometry which contradicts our choice of P_x. This proves the claim and there is a set T such that (G.6) holds on T for a given $a \in A$ and $\epsilon > 0$. Define $F \in \mathcal{B}^b(X, \mathcal{H})$ by

$$F(x) = \begin{cases} P_x & \text{if } x \in T, \text{ and} \\ 0 & \text{otherwise.} \end{cases}$$

Let $P := L_F$ be the corresponding decomposable operator, and let $E = L_{\mathbb{1}_T}$ be the diagonal operator corresponding to T. Then $P \in \pi(A)'$ and

$$\begin{aligned} \|\pi(a)P\| &= \operatorname{ess\,sup}_{x\in T} \|\pi_x(a)P_x\| \\ &\leq \operatorname{ess\,sup}_{x\in T} \|\pi_x(a)\| - \epsilon \\ &= \|\pi(a)E\| - \epsilon. \end{aligned} \tag{G.8}$$

Let $I := \ker \pi^P$, and let $Q = Q(I)$ be the ideal center projection onto the complement of the essential subspace of $\pi|_I$. If $\{u_n\}$ is an approximate identity for I, then $\pi(u_n)$ converges to $1_{\mathcal{H}} - Q$ in the strong operator topology. But $\pi(u_n)P = \pi^P(u_n) = 0$ for all n. Thus $(1_{\mathcal{H}} - Q)P = 0$, and we have $Q \geq P$. Thus $\ker \pi^Q \subset \ker \pi^P$. On the other hand, if $a \in \ker \pi^P$, then

$$\pi(a) = \lim_n \pi(au_n) = \pi(a)(1_{\mathcal{H}} - Q).$$

Thus $\pi(a)Q = 0$, and $a \in \ker \pi^Q$. Therefore

$$\ker \pi^Q = \ker \pi^P. \tag{G.9}$$

But, by definition, $Q \in \mathcal{IC}(\pi)$. Thus by assumption, $Q = L_{\mathbb{1}_W}$ for a Borel set $W \subset X$. Since $P_x \neq 0$ for all $x \in T$ and since $Q \geq P$, we must have $W \supset T$ μ-almost everywhere. Thus

$$E = L_{\mathbb{1}_T} \leq Q = L_{\mathbb{1}_W}. \tag{G.10}$$

But (G.8) and (G.9) imply that

$$\|\pi(a)Q\| = \|\pi(a)P\| < \|\pi(a)E\|.$$

But this contradicts (G.10). Therefore we conclude that π_x is almost everywhere homogeneous if $\mathcal{IC}(\pi) \subset \Delta(X * \mathfrak{H}, \mu)$.

To prove the converse, suppose that π_x is homogeneous for all $x \in X$. Let I be an ideal in A and let E be the projection onto the essential subspace of

$\pi|_I$. Since projections of this form generate $\mathcal{IC}(\pi)$, it suffices to show show that $E \in \Delta(X * \mathfrak{H}, \mu)$. Let $\{ u_n \}$ be an approximate identity for I. Then $\pi(u_n)$ converge to E in the strong operator topology. Since $E \in \mathcal{IC}(\pi) \subset \pi(A)' \cap \pi(A)'' \subset \pi(A)'' \subset \Delta(X * \mathfrak{H}, \mu)'$, E is decomposable. Let $E = L_F$ for some $F \in \mathcal{B}^b(X, \mathcal{H})$. We can assume that $F(x) \in P(\mathcal{H})$ for all x. Using Proposition F.29 on page 421, we can pass to a subsequence, relabel, and assume that $\pi_x(u_n) \to F(x)$ in the strong operator topology for μ-almost all x. Thus $F(x)$ is almost everywhere a projection in $\mathcal{IC}(\pi_x)$. Since each π_x is homogeneous, we can assume that $F(x)$ is either 0 or $1_{\mathcal{H}}$ for all $x \in X$. Thus $E = L_F \in \Delta(X * \mathfrak{H}, \mu)$ as required. □

Corollary G.17. *Suppose A is separable and that*

$$\pi = \int_X^{\oplus} \pi_x \, d\mu(x)$$

is a direct integral representation on $L^2(X * \mathfrak{H}, \mu)$. If π is homogeneous, then for almost all x, π_x is homogeneous with kernel equal to $\ker \pi$.

Proof. Fix $a \in A$. Then $\|\pi(a)\| = \operatorname{ess\,sup}_x \|\pi_x(a)\|$. Thus there is a null set N such that $x \notin N$ implies $\|\pi_x(a)\| \le \|\pi(a)\|$. If $\pi(a) = 0$, then $\|\pi_x(a)\| = \|\pi(a)\|$ for μ-almost all x. If $\|\pi(a)\| > 0$, let

$$S_n = \{ x : \|\pi_x(a)\| < \|\pi(a)\| - \frac{1}{n} \}.$$

Let E_n be the corresponding projection in $\pi(A)'$. Then

$$\|E_n \pi(a)\| < \|\pi(a)\|.$$

Since π is homogeneous, we must have $E_n = 0$ and $\mu(S_n) = 0$. Thus $\|\pi_x(a)\| = \|\pi(a)\|$ for μ-almost all x. Since A is separable, there is a μ-null set M such that $x \notin M$ implies $\|\pi_x(a)\| = \|\pi(a)\|$ for all $a \in A$. Thus $\ker \pi_x = \ker \pi$ for all $x \notin M$.

On the other hand, $\mathcal{IC}(\pi) = \mathbf{C}I_{\mathcal{H}_\pi}$ by Proposition G.8 on page 433. Thus $\mathcal{IC}(\pi)$ is a subalgebra of the diagonal operators $\Delta(H * \mathfrak{H}, \mu)$, and Theorem G.14 on page 437 implies that π_x is homogeneous for μ-almost all x. □

It should be kept in mind that if A is a separable C^*-algebra, then $\operatorname{Prim} A$ is a standard Borel space (Theorem H.40 on page 477).

Definition G.18. Let ρ be a separable representation of a separable C^*-algebra A. We say that ρ has an *ideal center decomposition* if there is a (standard) Borel Hilbert Bundle $\operatorname{Prim} A * \mathfrak{H}$ and a finite Borel measure μ on $\operatorname{Prim} A$ such that ρ is equivalent to a direct integral decomposition

$$\pi = \int_{\operatorname{Prim} A}^{\oplus} \pi_P \, d\mu(P)$$

on $L^2(\operatorname{Prim} A * \mathfrak{H}, \mu)$ where each π_P is homogeneous with kernel P.

The terminology in the definition is at least partially justified by the following result.

Lemma G.19. *Suppose that* $\operatorname{Prim} A * \mathfrak{H}$ *is a Borel Hilbert bundle,* μ *a finite measure on* $\operatorname{Prim} A$ *and*

$$\pi = \int_{\operatorname{Prim} A}^{\oplus} \pi_P \, d\mu(P)$$

an ideal center decomposition on $L^2(\operatorname{Prim} A * \mathfrak{H}, \mu)$. *Then the ideal center* $\mathcal{IC}(\pi)$ *of* π *is equal the algebra* $\Delta(\operatorname{Prim} A * \mathfrak{H}, \mu)$ *of diagonal operators on* $L^2(\operatorname{Prim} A * \mathfrak{H}, \mu)$.

Proof. Let $U \subset \operatorname{Prim} A$ be an open subset in the hull-kernel topology on $\operatorname{Prim} A$. Let E_U be the corresponding diagonal operator. Let

$$I_U := \bigcap \{ J \in \operatorname{Prim} A : J \notin U \}$$

be the ideal in A corresponding to U. Thus if P is a primitive ideal, then $I_U \subset P$ if and only if $P \notin U$. For any ideal I in A, let $e_P(I)$ be the essential projection for $\pi_P|_I$. Since π_P is homogeneous, $e_P(I)$ is either 0 or $1_{\mathcal{H}(P)}$. In particular,

$$e_P(I_U) = \begin{cases} 1_{\mathcal{H}(P)} & \text{if } P \in U, \text{ and} \\ 0 & \text{if } P \notin U. \end{cases}$$

If $e(I)$ is the essential projection for $\pi|_I$, then Lemma G.10 on page 434 implies that

$$e(I_U) = \int_{\operatorname{Prim} A}^{\oplus} e_P(I_U) \, d\mu(P) = E_U.$$

But every ideal in A is of the form I_U for some open set U. Thus the $e(I_U)$ generate $\mathcal{IC}(\pi)$. On the other hand, the E_U generate $\Delta(\operatorname{Prim} A * \mathfrak{H}, \mu) \cong L^\infty(\operatorname{Prim} A, \mu)$. □

To show that every representation has an ideal center decomposition, we have to sharpen Theorem G.14 on page 437 slightly. Our proof is taken from [45, Theorem 1.10] and [160, Lemme 1.10].

Proposition G.20. *Suppose that* μ *is a finite Borel measure on a second countable locally compact space* X *and that* $X * \mathfrak{H}$ *is a Borel Hilbert bundle over* X. *Let*

$$\pi := \int_X^{\oplus} \pi_x \, d\mu(x) \tag{G.11}$$

be a direct integral decomposition of a representation π *of a separable* C^**-algebra* A *on* $L^2(X * \mathfrak{H}, \mu)$. *Then* $\Delta(X * \mathfrak{H}, \mu) = \mathcal{IC}(\pi)$ *if and only if there is conull set* $Y \subset X$ *such that* π_x *is homogeneous for all* $x \in Y$ *and such that* $\ker \pi_x \neq \ker \pi_y$ *for all distinct points* $x, y \in Y$.

Proof. Suppose that $\Delta(X * \mathfrak{H}, \mu) = \mathcal{IC}(\pi)$. If I is an ideal in A, then let $Q(I)$ be the corresponding ideal center projection. Thus if $\{ u_n \}$ is an approximate identity for I, then $1_{\mathcal{H}} - Q(I)$ is the strong operator limit of $\pi(u_n)$. Recall that $\mathcal{IC}(\pi)$ is generated by the $Q(I)$. On the other hand, we can identify the lattice of projections $\mathcal{P}$ in $\mathcal{IC}(\pi)$ with the Boolean σ-algebra $\mathscr{B}/\mathscr{N}$ of Borel sets in X modulo the μ-null sets. Clearly, $\mathcal{P}$ is countably generated — say by $\{ E_i \}$. As $\mathcal{IC}(\pi)$ is generated by the ideal center projections, and since each E_i belongs to a sublattice generated by countably many ideal center projections (cf., e.g., [70, Chap. I §5 Theorem D]), there is a sequence $\{ Q_i \}$ of ideal center projections which generate $\mathcal{P}$ as a Boolean σ-algebra. Let S_i be a Borel set in X such that $Q_i = L_{\mathbb{1}_{S_i}}$. Define an equivalence relation on X by

$$x \sim y \quad \text{if and only if} \quad \mathbb{1}_{S_i}(x) = \mathbb{1}_{S_i}(y) \quad \text{for all } i.$$

Then $X/{\sim}$ is countably generated and is an analytic Borel space with the quotient Borel structure [2, Corollary 2 to Theorem 3.3.5]. Let $q : X \to X/{\sim}$ be the quotient map, and let $\nu = q_*\mu$ be the push-forward measure on $X/{\sim}$. Let $\mathscr{A}/\mathscr{M}$ be the Boolean σ-algebra of Borel sets in $X/{\sim}$ modulo the ν-null sets, and let Φ_q be the corresponding σ-homomorphism of $\mathscr{A}/\mathscr{M}$ into $\mathscr{B}/\mathscr{N}$. The definition of ν guarantees that Φ_q is injective. If $y \in q^{-1}\big(q(S_i)\big)$, then $y \sim x$ for some $x \in S_i$. Thus $\mathbb{1}_{S_i}(y) = 1$ and $y \in S_i$. This means that $q(S_i)$ is a Borel set in $X/{\sim}$, and that $\Phi_q([q(S_i)])$ is the class of S_i in $\mathscr{B}/\mathscr{N}$. Thus, as we're identifying $\mathcal{P}$ and $\mathscr{B}/\mathscr{N}$, we've shown that the image of Φ_q contains the Q_i. Since the Q_i generate, Φ_q is a bijection. It follows from Theorem I.37 on page 501 (and Remark I.35) that there is a Borel map $\varphi : X \to X/{\sim}$ which implements Φ_q and a null set X_0 such that $\varphi|_{X \smallsetminus X_0}$ is injective. It follows from Lemma I.36 on page 501 that $q = \varphi$ μ-almost everywhere. Thus, after enlarging X_0 if necessary, we can assume that q is one-to-one on $X \smallsetminus X_0$. Then the sets $T_i := S_i \smallsetminus X_0$ separate points of $X \smallsetminus X_0$. Let $J_i = \ker \pi^{Q_i}$. Since Q_i is an ideal center projection, $Q_i = Q(K_i) = 1_{\mathcal{H}} - e(K_i)$ for some ideal $K_i \in \mathcal{I}(A)$. If $a \in K_i$, then $\pi(a)Q_i = 0$, so $a \in J_i$, and $K_i \subset J_i$. Thus

$$Q_i = Q(K_i) \geq Q(J_i).$$

But if $\{ u_n^i \}$ is an approximate identity for J_i, then

$$\big(1_{\mathcal{H}} - Q(J_i)\big)Q_i = \lim_n \pi(u_n^i)Q_i = 0.$$

Thus, $Q_i \leq Q(J_i)$. Therefore $Q_i = Q(J_i)$ and $\pi(u_n^i)$ converges to $1 - Q_i$ in the strong operator topology. Using Proposition F.29 on page 421, passing to a subsequences and enlarging X_0 by a Borel null set for each i, we can assume that for all i we have

$$\pi_x(u_n^i) \to \big(1 - \mathbb{1}_{T_i}(x)\big)1_{\mathcal{H}(x)} \quad \text{for all } x \in X \smallsetminus X_0.$$

Therefore the projection onto the essential part of $\pi_x|_{J_i}$ is $\big(1 - \mathbb{1}_{T_i}(x)\big)1_{\mathcal{H}(x)}$. In particular, $x \in T_i$ if and only if $\pi_x(J_i) = \{ 0 \}$. Therefore if $x, y \in X \smallsetminus X_0$ and if $\ker \pi_x = \ker \pi_y$, then $x \in T_i$ if and only if $y \in T_i$. Thus $x = y$.

This shows that the kernels of the π_x are distinct on $X \smallsetminus X_0$. This suffices as the π_x are almost everywhere homogeneous by Theorem G.14 on page 437.

Now suppose that there is a μ-conull set Y as in the statement of the theorem. Then we may as well assume that $Y = X$. Then $x \mapsto \ker \pi_x$ is an injective Borel map f of X into $\operatorname{Prim} A$ (Lemma F.28 and Corollary G.9). On the other hand, in view of Theorem H.39, we can assume that $\operatorname{Prim} A$ is a Polish space. Then [2, Theorem 3.3.2] implies that f is a Borel isomorphism of X onto a Borel subset of $\operatorname{Prim} A$. Thus we can replace X by $\operatorname{Prim} A$, μ by the push-forward $f^*\mu$, π_x by $\pi_{f(x)}$ and (G.11) by

$$\pi = \int_{\operatorname{Prim} A}^{\oplus} \pi_P \, d\mu(P).$$

Now Lemma G.19 on page 441 implies that $\Delta(\operatorname{Prim} A * \mathfrak{H}, \mu) = \mathcal{IC}(\pi)$ as desired. □

One of the essential features of ideal center decompositions is that they have a strong uniqueness property. In general, to compare two direct integral decompositions of a given representation class, it is first necessary to assume that the associated diagonal operators are intertwined by the equivalence. Then we can apply Proposition F.33 on page 423. The next proposition, taken from [160, Lemme 1.7], gives a much stronger result for ideal center decompositions which says that the measure class of μ and the equivalence classes of the π_P are uniquely determined. Notice that there is no explicit mention of the diagonal operators.

Proposition G.21. *Suppose that*

$$\pi = \int_{\operatorname{Prim} A}^{\oplus} \pi_P \, d\mu(P) \quad \text{and} \quad \rho = \int_{\operatorname{Prim} A}^{\oplus} \rho_P \, d\nu(P)$$

are ideal center decompositions on $L^2(\operatorname{Prim} A * \mathfrak{H}, \mu)$ and $L^2(\operatorname{Prim} A * \mathfrak{K}, \nu)$, respectively. Suppose that $W : L^2(\operatorname{Prim} A * \mathfrak{H}, \mu) \to L^2(\operatorname{Prim} A * \mathfrak{K}, \nu)$ is a unitary implementing an equivalence between π and ρ. Then

(a) *μ and ν are equivalent measures,*

(b) *there is an essentially unique $\mathrm{id}_{\operatorname{Prim} A}$-isomorphism*[4] *$W' = \{W(P)\}$ of $\operatorname{Prim} A * \mathfrak{H}$ onto $\operatorname{Prim} A * \mathfrak{K}$ implemented by an almost everywhere family of unitaries $W(P) : \mathcal{H}(P) \to \mathcal{K}(P)$ such that $W(P)$ implements an equivalence between π_P and ρ_P for almost all P, and*

(c) *$W = UW'$ where U is the natural isomorphism of $L^2(\operatorname{Prim} A * \mathfrak{K}, \mu)$ onto $L^2(\operatorname{Prim} A * \mathfrak{K}, \nu)$ induced by the equivalence of μ and ν.*

(d) *If φ is a bounded Borel function on $\operatorname{Prim} A$ and if T_φ and $\widetilde{T}_\varphi$ are the respective diagonal operators on $L^2(\operatorname{Prim} A * \mathfrak{H}, \mu)$ and $L^2(\operatorname{Prim} A * \mathfrak{K}, \nu)$, then $W'T_\varphi = \widetilde{T}_\varphi W'$.*

Proof. Clearly, W must intertwine $\mathcal{IC}(\pi)$ and $\mathcal{IC}(\rho)$. Lemma G.19 on page 441 implies that W must therefore intertwine the respective diagonal operators. Hence Proposition F.33 on page 423 applies. Thus there are null sets N and M together with a Borel isomorphism $\tau : \operatorname{Prim} A \smallsetminus N \to \operatorname{Prim} A \smallsetminus M$ such that $\tau_*\mu$ is equivalent to ν. Also there is a τ-isomorphism $\{W(P)\}$ consisting almost everywhere of unitaries $W(P) : \mathcal{H}(P) \to \mathcal{K}(\tau(P))$ implementing an equivalence of π_P with $\rho_{\tau(P)}$,

[4] That is, a τ-isomorphism with $\tau = \mathrm{id}_{\operatorname{Prim} A}$.

and such that $W'T_\varphi = \widetilde{T}_{\varphi\circ\tau}W'$. However, equivalent representations have the same kernels, so

$$P = \ker \pi_P = \ker \rho_{\tau(P)} = \tau(P).$$

Therefore, τ is the identity map. It follows that $\mu = \tau_*\mu$ is equivalent to ν and that $W'T_\varphi = \widetilde{T}_\varphi W'$. This completes the proof. □

Theorem G.22 (Effros's Ideal Center Decomposition). *Every separable representation ρ of a separable C^*-algebra A has an ideal center decomposition.*

Proof. We clearly have $\mathcal{IC}(\rho) \subset \rho(A)'$, so Proposition F.26 on page 419 implies that there is a second countable locally compact space X, a Borel Hilbert Bundle $X * \mathfrak{K}$ and a finite measure ν on X so that ρ has a direct integral decomposition

$$\pi = \int_X^\oplus \pi_x \, d\nu(x)$$

with $\mathcal{IC}(\pi) = \Delta(X * \mathfrak{K}, \nu)$. Proposition G.20 on page 441 implies that we may assume that the map $\tau : X \to \operatorname{Prim} A$ given by $\tau(x) := \ker \pi_x$ is injective. It is Borel by Lemma F.28 on page 420, and as both X and $\operatorname{Prim} A$ are Polish,[5] it is a Borel isomorphism onto its range [2, Theorem 3.3.2]. Let $\mu := \tau_*\nu$. Then there is a Borel Hilbert bundle $\operatorname{Prim} A * \mathfrak{H}$, with $\mathcal{H}\big(\tau(x)\big) = \mathcal{K}(x)$, and an isomorphism, induced by τ, of $L^2(X * \mathfrak{K}, \nu)$ onto $L^2(\operatorname{Prim} A * \mathfrak{H}, \mu)$ which intertwines π with an ideal center decomposition. □

Remark G.23. Suppose that π is a representation of a separable C^*-algebra A with ideal center decomposition

$$\int_{\operatorname{Prim} A}^\oplus \pi_P \, d\mu(P)$$

on $L^2(\operatorname{Prim} A * \mathfrak{H}, \mu)$. If $1 \le n \le \aleph_0$, then Remark G.11 on page 434 implies that $n \cdot \pi_P$ is homogeneous with kernel P. Let $\mathcal{K}(P) = \bigoplus_{j=1}^{j=n} \mathcal{H}(P)$ and form the obvious Borel Hilbert bundle $\operatorname{Prim} A * \mathfrak{K}$. Then

$$\int_{\operatorname{Prim} A}^\oplus n \cdot \pi_P \, d\mu(P)$$

is ideal center decomposition of $n \cdot \pi$ on $L^2(\operatorname{Prim} A * \mathfrak{K}, \mu)$. If $n = \aleph_0$, then each $\mathcal{K}(P)$ has infinite dimension and we can replace $\operatorname{Prim} A * \mathfrak{K}$ with a constant field. Therefore we can realize an ideal center decomposition of any representation π with infinite multiplicity on $L^2(\operatorname{Prim} A, \mu; \mathcal{H})$ for a fixed separable infinite-dimensional Hilbert space $\mathcal{H}$ (cf., Corollary F.11 on page 414).

[5] Second countable locally compact spaces are Polish by Lemma 6.5, and $\operatorname{Prim} A$ has the Borel structure coming from a Polish space by Theorem H.40 on page 477.

G.2 Ideal Center Decompositions for Covariant Representations

The next proposition is critical for forming the restriction of a covariant representation to the stability groups as required in our proof of the Effros-Hahn conjecture. Although we could restrict to representations of infinite multiplicity (cf, Remark 9.6 on page 271), and therefore do away with one level of direct integrals, it seems worthwhile to work out a more general result here for future reference.

Proposition G.24. *Suppose that* $R = (\pi, V)$ *is a covariant representation of* (A, G, α) *on* $L^2(\operatorname{Prim} A * \mathfrak{H}, \mu)$*, and that*

$$\pi = \int_{\operatorname{Prim} A}^{\oplus} \pi_P \, d\mu(P) \tag{G.12}$$

is an ideal center decomposition of π *on* $L^2(\operatorname{Prim} A * \mathfrak{H}, \mu)$*. Let* $\operatorname{Prim} A * \mathfrak{H}_s$ *be the pull-back of* $\operatorname{Prim} A * \mathfrak{H}$ *via* lt_s^{-1} *as in Example F.18 on page 415. Then the following statements hold.*

(a) *The measure* μ *is quasi-invariant for the natural* G*-action on* $\operatorname{Prim} A$*.*

(b) *Let* $d : G \times \operatorname{Prim} A \to (0, \infty)$ *be a Borel choice of Radon-Nikodym derivatives as in Corollary D.34 on page 389. Then for each* $s \in G$*, there is a unitary*

$$W(s) : L^2(\operatorname{Prim} A * \mathfrak{H}, \mu) \to L^2(\operatorname{Prim} A * \mathfrak{H}_s, \mu)$$

given by $W(s)f(P) := d(s, P)^{\frac{1}{2}} f(s^{-1} \cdot P)$*.*

(c) *The operator* $U(s) := V(s)W(s^{-1})$ *is decomposable. Moreover, there is a* μ*-conull set* $X \subset \operatorname{Prim} A$ *such that if*

$$E := \{ (s, P) \in G \times \operatorname{Prim} A : P \in X \text{ and } s^{-1} \cdot P \in X \},$$

then there are unitaries $U(s, P) : \mathcal{H}(s^{-1} \cdot P) \to \mathcal{H}(P)$ *for all* $(s, P) \in E$ *such that*

$$(s, P) \mapsto \big(U(s, P) f(s^{-1} \cdot P) \mid g(P)\big)$$

is Borel on E *for all Borel sections* $f, g \in B(X, \mathfrak{H})$*, and such that for* all $s \in G$*,*

$$U(s) = \int_{\operatorname{Prim} A}^{\oplus} U(s, P) \, d\mu(P).$$

(d) *For all* $s, r \in G$ *and all* $P \in X$ *such that* $s^{-1} \cdot P \in X$ *and* $r^{-1}s^{-1} \cdot P \in X$*, we have*

$$U(sr, P) = U(s, P)U(r, s^{-1} \cdot P).$$

(e) *If* $(s, P) \in E$*, then*

$$\pi_P(a) = U(s, P)\pi_{s^{-1} \cdot P}\big(\alpha_s^{-1}(a)\big)U(s, P)^* \quad \text{for all } a \in A.$$

(f) *If $P \in X$, then $\sigma_P(t) := U(t, P)$ defines a unitary representation of G_P and (π_P, σ_P) is a covariant representation of (A, G_P, α).*

Proof. Recall that $s \cdot \pi := \pi \circ \alpha_s^{-1}$. Therefore

$$s \cdot \pi = \int_{\mathrm{Prim}\, A}^{\oplus} s \cdot \pi_P \, d\mu(P) \tag{G.13}$$

on $L^2(\mathrm{Prim}\, A * \mathfrak{H}, \mu)$. As in Example F.18 on page 415, we can define a unitary

$$T(s) : L^2(\mathrm{Prim}\, A * \mathfrak{H}, \mu) \to L^2(\mathrm{Prim}\, A * \mathfrak{H}_s, s \cdot \mu)$$

by $T(s)f(P) = f(s^{-1} \cdot P)$. Let

$$\pi'_P := s \cdot \pi_{s^{-1} \cdot P} = \pi_{s^{-1} \cdot P} \circ \alpha_s^{-1}.$$

Notice that π'_P is homogeneous with kernel P, and that $\{\, \pi'_P \,\}$ is a Borel field of representations on $\mathrm{Prim}\, A * \mathfrak{H}_s$. Thus

$$\pi' := \int_{\mathrm{Prim}\, A}^{\oplus} \pi'_P \, d(s \cdot \mu)(P) \tag{G.14}$$

is an ideal center decomposition on $L^2(\mathrm{Prim}\, A * \mathfrak{H}_s, s \cdot \mu)$. Since

$$\begin{aligned} T(s)s \cdot \pi(a)f(P) &= s \cdot \pi(a)f(s^{-1} \cdot P) \\ &= s \cdot \pi_{s^{-1} \cdot P}(a)\big(f(s^{-1} \cdot P)\big) \\ &= \pi'_P(a)\big(f(s^{-1} \cdot P)\big) \\ &= \pi'_P(a)\big(T(s)f(P)\big) \\ &= \pi'(a)T(s)f(P), \end{aligned}$$

$T(s)$ intertwines $s \cdot \pi$ an π'. Since (π, V) is covariant,

$$\begin{aligned} \pi(a) &= V(s)s \cdot \pi(a)V(s)^* \\ &= V(s)T(s^{-1})\pi'(a)\big(V(s)T(s^{-1})\big)^*. \end{aligned}$$

Thus π and π' are equivalent ideal center decompositions, and it follows from Proposition G.21 on page 443 that that μ an $s \cdot \mu$ are equivalent for all s. This proves part (a).

Let $d : G \times \mathrm{Prim}\, A \to (0, \infty)$ be a Borel choice of Radon-Nikodym derivatives as in Corollary D.34 on page 389. Then $W(s)f(P) := d(s, P)^{\frac{1}{2}} f(s^{-1} \cdot P)$ defines a unitary as required for part (b). Since $d(s, P)^{-1} = d(s^{-1}, s^{-1} \cdot P)$ for almost all P, we see that $W(s)^* = W(s)^{-1} = W(s^{-1})$. If

$$\pi'' := \int_{\mathrm{Prim}\, A}^{\oplus} \pi'_P \, d\mu(P) \tag{G.15}$$

is the ideal center decomposition on $L^2(\operatorname{Prim} A * \mathfrak{H}_s, \mu)$, then using (G.13), we compute that

$$\begin{aligned} W(s)V(s^{-1})\pi(a)f(P) &= W(s)\pi\big(\alpha_s^{-1}(a)\big)V(s^{-1})f(P) \\ &= d(s,p)^{\frac{1}{2}} s \cdot \pi(a)V(s^{-1})f(s^{-1}\cdot P) \\ &= d(s,P)^{\frac{1}{2}} s \cdot \pi_{s^{-1}\cdot P}(a)\big(V(s^{-1})f(s^{-1}\cdot P)\big) \\ &= d(s,P)^{\frac{1}{2}} \pi'_P(a)\big(V(s^{-1})f(s^{-1}\cdot P)\big) \\ &= \pi''(a)W(s)V(s^{-1})f(P). \end{aligned}$$

Therefore $U(s) := V(s)W(s^{-1})$ implements an equivalence between the ideal center decompositions for π and π'' on $L^2(\operatorname{Prim} A * \mathfrak{H}, \mu)$ and $L^2(\operatorname{Prim} A * \mathfrak{H}_s, \mu)$, respectively. Then we can apply Proposition G.21 on page 443 with $\mathfrak{K} = \mathfrak{H}_s$ and $\nu = \mu$. In particular, the map U appearing in that proposition is the identity, and for each s, there is an essentially unique id-isomorphism $\{ W_0(s,P) \}$ consisting of unitaries $W_0(s,P) : \mathcal{H}(s^{-1}\cdot P) \to \mathcal{H}(P)$ (for μ-almost all P) such that

$$U(s) = \int_{\operatorname{Prim} A}^{\oplus} W_0(s,P)\, d\mu(P),$$

and such that for μ-almost all P,

$$\pi_P(a) = W_0(s,P)\pi'_P(a)W_0(s,P)^* = W_0(s,P)\pi_{s^{-1}\cdot P}\big(\alpha_s^{-1}(a)\big)W_0(s,P)^*$$

for all $a \in A$. Since

$$U(sr) = U(s)W(s)U(r)W(s^{-1}) \tag{G.16}$$

and since the $W_0(s,P)$ are essentially unique, we have for each $(s,r) \in G \times G$,

$$W_0(sr,P) = W_0(s,P)W_0(r, s^{-1}\cdot P)$$

off an μ-null set $N(s,r)$. However, there is no reason that $\{ W_0(s,P) \}$ should be jointly Borel in s and P as required in condition part (c). The easiest way to address this defect is to repeat parts of the proofs of Proposition F.33 on page 423 and Proposition G.21 on page 443 while keeping track of the dependence on the G variable.

Let $\{ f_n \}$ be a fundamental sequence in $\mathcal{L}^2(\operatorname{Prim} A * \mathfrak{H}, \mu)$. Since $d(s,P) > 0$ everywhere,

$$f_n^s := W(s)f_n$$

defines a fundamental sequence in $L^2(\operatorname{Prim} A * \mathfrak{H}_s, \mu)$ for all s. Let $p : G \times \operatorname{Prim} A \to \operatorname{Prim} A$ be the projection onto the second factor, and let $p^*(\operatorname{Prim} A * \mathfrak{H}) = \operatorname{Prim} A * \mathfrak{H}_p$ be the pull-back by p. Then Proposition F.38 on page 428 implies that for each n, there is a $g_n \in B(p^*(\operatorname{Prim} A * \mathfrak{H}))$ and a μ_G-null set M_n such that

$$V(s)[f_n] = [g_n(s,\cdot)] \quad \text{if } s \notin M_n.$$

Then $M := \bigcup M_n$ is a μ_G-null set, and if $s \notin M$, we have

$$U(s)[f_n^s] = V(s)[f_n] = [g_n(s,\cdot)].$$

A computation similar to that establishing (F.13) on 424 shows that for all $s \notin M$ and any finite set $r_1, \dots, r_n$ of complex rational numbers there is a μ-null set $N(r_1, \dots, r_n)$ such that $P \notin N(r_1, \dots, r_n)$ implies that

$$\Bigl\|\sum_i r_i g_i(s,P)\Bigr\| \le \Bigl\|\sum_i r_i f_i^s(P)\Bigr\|. \tag{G.17}$$

Since the set of finite sequences of complex rational numbers is countable, there is a μ-null set N such that (G.17) holds for any finite sequence of rationals. Thus if $s \notin M$ and $P \notin N$, there is a well-defined contraction $U_0(s,P) : \mathcal{H}(s^{-1}\cdot P) \to \mathcal{H}(P)$ such that

$$U_0(s,P)\bigl(f_n^s(P)\bigr) = g_n(s,P).$$

Let Y' be the conull set $(G \setminus M) \times (X \setminus N)$. Since $g_n \in B\bigl(p^*(\operatorname{Prim} A * \mathfrak{H})\bigr)$,

$$(s,P) \mapsto d(s,P)^{-\frac12}\bigl(g_n(s,P) \mid f_m(P)\bigr) = \bigl(U_0(s,P)f_n(s^{-1}\cdot P) \mid f_m(P)\bigr)$$

is Borel on Y'.

Let $U'(s) : L^2(\operatorname{Prim} A * \mathfrak{H}_s, \mu) \to L^2(\operatorname{Prim} A * \mathfrak{H}, \mu)$ be the contraction given by the direct integral

$$U'(s) := \int_{\operatorname{Prim} A}^{\oplus} U_0(s,P)\, d\mu(P).$$

Since $U'(s)f_n = U(s)f_n$ for all n, $U'(s) = U(s)$. By Lemma F.20 on page 417, for all $s \notin M$, we have

$$U_0(s,P) = W_0(s,P) \quad \text{for } \mu\text{-almost all } P.$$

Therefore there is a $\mu_G \times \mu$-conull set $Y \subset G \times \operatorname{Prim} A$ such that $U_0(s,P)$ is unitary for all $(s,P) \in Y$. Also we claim that

$$U_0(sr,P) = U_0(s,P)U_0(r,s^{-1}\cdot P)$$

off a $\mu_G \times \mu_G \times \mu$-null set. To see this, let $N(s)$ be a μ-null set such that that $U_0(s,P) = W_0(s,P)$ provided $P \notin N(s)$. Then, since μ is quasi-invariant, $N := N(s) \cup N(sr) \cup s\cdot N(r)$ is a μ-null set, and if $P \notin N$, we have $U_0(sr,P) = W_0(sr,P)$, $U_0(s,P) = W_0(s,P)$ and $U_0(r,s^{-1}\cdot P) = W_0(r,s^{-1}\cdot P)$. Now the claim follows as the corresponding equality for W_0 holds $\mu_G \times \mu_G \times \mu$-almost everywhere.

We now invoke Ramsay's selection theorem via Lemma G.25 on the next page. Therefore there is a μ-conull set $Z \subset \operatorname{Prim} A$ and unitaries $U(s,P) : \mathcal{H}(s^{-1}\cdot P) \to \mathcal{H}(P)$ for all $(s,P) \in F := \{\,(s,P) : P \in Z \text{ and } s^{-1}\cdot P \in Z\,\}$ such that

(i) $U(s,P) = U_0(s,P)$ for $\mu_G \times \mu$-almost all (s,P),

(ii) $(s,P) \mapsto \bigl(U(s,P)f(s^{-1}\cdot P) \mid g(P)\bigr)$ is Borel on F for all $f, g \in B(X,\mathfrak{H})$,

(iii) if P, $s^{-1}\cdot P$ and $r^{-1}s^{-1}\cdot P$ are in Z, then $U(sr,P) = U(s,P)U(r,s^{-1}\cdot P)$.

Let E be the set of $(s,P) \in F$ such that

$$\pi_P(a) = U(s,P)\pi_{s^{-1}\cdot P}\bigl(\alpha_s^{-1}(a)\bigr)U(s,P)^* \quad \text{for all } a \in A.$$

Since $U(s,P) = W_0(s,P)$ for $\mu_G \times \mu$-almost all (s,P) and since F is $\mu_G \times \mu$-conull, it follows that E contains a Borel $\mu_G \times \mu$-conull set. Furthermore, if (s,P) and $(r, s^{-1}\cdot P)$ both belong to E, then since $E \subset F$, part (iii) implies that

$$\begin{aligned}\pi_P(a) &= U(s,P)\pi_{s^{-1}\cdot P}\big(\alpha_s^{-1}(a)\big)U(s,P)^* \\ &= U(s,P)U(r,s^{-1}\cdot P)\pi_{r^{-1}s^{-1}\cdot P}\big(\alpha_r^{-1}\big(\alpha_s^{-1}(a)\big)\big)U(r,s^{-1}\cdot P)^*U(s,P)^* \\ &= U(sr,P)\pi_{r^{-1}s^{-1}\cdot P}\big(\alpha_{sr}^{-1}(a)\big)U(sr,P)^*.\end{aligned}$$

Therefore E is closed under multiplication and we can apply Lemma G.27 on page 451, and there is a $X \subset Z$ such that parts (d) and (e) hold (as well as most of part (c)).

To get the rest of part (c), let Σ be the set of $s \in G$ such that

$$U(s) = \int_{\operatorname{Prim} A}^{\oplus} U(s,P)\,d\mu(P).$$

Using (G.16), it is not hard to see that Σ is a conull semigroup. Hence Lemma D.36 on page 391 implies that $\Sigma' = G$ and part (c) follows.

If $P \in Z$, then $t \mapsto U(t,P)$ is a Borel homomorphism of G_P into $U\big(\mathcal{H}(P)\big)$. Hence it is continuous by Theorem D.3 on page 370, and $\sigma_P(t) := U(t,P)$ defines a representation σ_P of G_P. If $P \in X$, then part (e) implies that

$$\pi_P(a) = \sigma_P(t)\pi_P\big(\alpha_t^{-1}(a)\big)\sigma_P(t)^* \quad \text{for all } a \in A. \tag{G.18}$$

Thus (π_P, σ_P) is a covariant representation of (A, G_P, α), and part (f) holds. This completes the proof of the proposition. □

G.2.1 Ramsay's Selection Theorem

We need the following technical lemma, which is an application of a beautiful Borel selection result due to Ramsay [140, Theorem 5.1].

Lemma G.25. *Let G be a second countable locally compact group and μ a quasi-invariant measure on an analytic G-space X. Suppose that $X * \mathfrak{H}$ is a Borel Hilbert bundle and that $Y \subset G \times X$ is a $\mu_G \times \mu$-conull set such that for each $(s,x) \in Y$ there is a unitary $U(s,x) : \mathcal{H}(s^{-1}\cdot x) \to \mathcal{H}(x)$ such that*

$$(s,x) \mapsto \big(U(s,x)f(s^{-1}\cdot x) \mid g(x)\big)$$

is Borel on Y for all $f,g \in B(X,\mathfrak{H})$, and such that

$$U(sr,x) = U(s,x)U(r,s^{-1}\cdot x)$$

holds off a $\mu_G \times \mu_G \times \mu$-null set.

Then there is a μ-conull set $Z \subset X$ such that

(a) *for each $(s,x) \in E = \{\,(s,x) : x \in Z \text{ and } s^{-1}\cdot x \in Z\,\}$ there is a unitary*

$$V(s,x) : \mathcal{H}(s^{-1}\cdot x) \to \mathcal{H}(x)$$

such that $V(s,x) = U(s,x)$ for $\mu_G \times \mu$-almost all (s,x) and

(b) *such that*

$$(s,x) \mapsto \big(V(s,x)f(s^{-1}\cdot x) \mid g(x)\big)$$

is Borel on E *for all* $f, g \in B(X, \mathfrak{H})$ *and such that*

(c) *if* $x, s^{-1}\cdot x, r^{-1}s^{-1}\cdot x \in Z$, *then*

$$V(sr,x) = V(s,x)V(r,s^{-1}\cdot x).$$

Remark G.26 (Ramsay's Selection Theorem). The proof of Lemma G.25 is essentially an invocation of Ramsay's selection theorem [140, Theorem 5.1]. It is a more complex application than we needed in the proof of Corollary D.34 on page 389, and we outline the set-up here. Recall that if X is an analytic Borel G-space and if μ is a quasi-invariant finite Borel measure, then $(G \times X, \mu_G \times \mu)$ is what Ramsay calls a virtual group in [140].[6] It is a special case of what he calls a measured groupoid in [142]. Without getting too involved in technicalities here, a groupoid $\mathcal{A}$ is a an analytic space with a partially defined multiplication satisfying certain axioms.[7] The set of composable pairs is denoted by $\mathcal{A}^{(2)}$. If, for example, $\mathcal{G} = (G \times X)$, then (s,x) and (r,y) are composable when $y = s^{-1}\cdot x$, and then we define their product by $(s,x)(r,s^{-1}\cdot x) := (sr,x)$, and each element has an "inverse" given by $(s,x)^{-1} = (s^{-1}, s^{-1}\cdot x)$. If $Y \subset X$, then

$$\mathcal{G}|_Y = \{\, (s,x) \in \mathcal{G} : x \in Y \text{ and } s^{-1}\cdot x \in Y \,\}$$

is a subgroupoid of $\mathcal{G}$ (in the sense that it is closed under the partially defined multiplication) called the *restriction of* $\mathcal{G}$ *to* Y. If Y is conull in X, then since μ is quasi-invariant, $\mathcal{G}|_Y$ is $\mu_G \times \mu$-conull in G, and $\mathcal{G}|_Y$ is called an essential reduction.[8] A map $\varphi : \mathcal{G} \to \mathcal{A}$ is called a (groupoid) homomorphism if whenever γ and η are composable in $\mathcal{G}$, then $\varphi(\gamma)$ and $\varphi(\eta)$ are composable in $\mathcal{A}$ and $\varphi(\gamma\eta) = \varphi(\gamma)\varphi(\eta)$. Ramsay's result says that if φ_0 is a Borel map of $\mathcal{G} = (G \times X, \mu_G \times \mu)$ into an analytic groupoid $\mathcal{A}$ which is almost everywhere a homomorphism in the sense that

$$\varphi(s,x)\varphi(r,s^{-1}\cdot x) = \varphi(sr,x) \quad \text{for } \mu_G \times \mu_G \times \mu\text{-almost all } (s,r,x),$$

then there is a Borel map $\varphi : \mathcal{G} \to \mathcal{A}$ such that $\varphi = \varphi_0$ $\mu_G \times \mu$-almost everywhere and such that there is a μ-conull set $Y \subset X$ such that the restriction of φ to $\mathcal{G}|_Y$ is a homomorphism.[9]

[6] A careful look at [140] will reveal that Ramsay requires that μ also be ergodic. This hypothesis is not used in the results we reference and is not required in the definition of a measured groupoid in [142] which uses the same results from [140] as we do. An excellent discussion, as well as proofs of Ramsay's results, can be found in Muhly's as yet unpublished notes on groupoids [108, Chap. 4].

[7] Since we are only interested in two concrete examples of groupoids here, the precise axioms are not necessary to follow the argument. For the curious, they are given in [140, §1]. A more up to date and more complete source for this discussion is [108].

[8] To see that $\mathcal{G}|_Y$ is conull, note that for each $y \in Y$, the quasi-invariance of μ implies that $\{\, s \in G : (s,y) \in \mathcal{G}|_Y \,\}$ is μ_G-conull.

[9] In heuristic terms, we would like to see that a Borel map $\varphi_0 : \mathcal{G} \to \mathcal{A}$ which almost everywhere a homomorphism is equal almost everywhere to a homomorphism. If $\mathcal{A}$ is an analytic groupoid and if $\mathcal{G}$ is σ-compact Polish groupoid, then Ramsay proves such a result in [142]. Here, although Prim A is Polish, it need not be σ-compact in the regularized topology. So we settle for Ramsay's result from [140] in which he shows φ_0 is equal almost everywhere to a bona fide homomorphism on an essential reduction of $\mathcal{G}$.

Proof of Lemma G.25. We want to apply Ramsay's Theorem to a map φ_0 from $\mathcal{G} := G \times X$ into the isomorphism bundle $\mathcal{A} := \operatorname{Iso}(X * \mathfrak{H})$ as defined in Definition F.35 on page 427. (In view of Remark F.36 on page 427, $\mathcal{A}$ is an analytic groupoid.) For each $(s,x) \in Y$, define $\varphi_0(s,x) := (x, U(s,x), s^{-1} \cdot x)$. Then φ_0 is Borel on Y, and we can extend φ_0 to a Borel function on all of $\mathcal{G}$ by letting φ_0 take a constant value on the complement of Y. We have arranged our hypotheses so that φ_0 is an almost everywhere homomorphism of $\mathcal{G}$ into $\mathcal{A}$. Thus, Ramsay's Theorem [140, Theorem 5.1] implies that there is a Borel map $\varphi : \mathcal{G} \to \operatorname{Iso}(X * \mathfrak{H})$ and a μ-conull set $X_0 \subset X$ such that $\varphi = \varphi_0$ $\mu_G \times \mu$-almost everywhere and such that φ restricted to $A_0 = \mathcal{G}|_{X_0}$ is a homomorphism.

Let $x \in X_0$. Then $(e,x) \in A_0$ and we can define $a(x) \in X$ by $\varphi(e,x) = \big(a(x), I, a(x)\big)$. If $(s,x) \in A_0$, then $(s,x) = (e,x)(s,x)(e,s^{-1}\cdot x)$. Therefore, for all $(s,x) \in A_0$ we have

$$\varphi(s,x) = \big(a(x), V(s,x), a(s^{-1}\cdot x)\big),$$

where $V(s,x)$ is a unitary mapping $\mathcal{H}\big(a(s^{-1}\cdot x)\big)$ onto $\mathcal{H}\big(a(x)\big)$. Since $\varphi(s,x) = \varphi_0(s,x)$ $\mu_G \times \mu$-almost everywhere and since there is a $s \in G$, such that $(s,x) \in A_0$ for μ-almost all x, there is a μ-conull set $Z \subset X_0$ such that $a(x) = x$ for all $x \in Z$. Let $A = \mathcal{G}|_Z = \{\,(s,x) : x \in Z \text{ and } s^{-1}\cdot x \in Z\,\}$. Then $\{\,V(s,x)\,\}_{(s,x)\in A}$ satisfies the requirements of the lemma. □

We will also need the following generalization of Lemma D.36 on page 391. (It can be proved in a general groupoid setting — cf., [140, Lemma 5.2] or [108, Lemma 4.9] — and is a critical component of the proof of Ramsay's selection theorem.)

Lemma G.27. *Let G be a second countable locally compact group and μ a quasi-invariant finite measure on an analytic Borel G-space X. Let $Y \subset G \times X$ be a $\mu_G \times \mu$-conull (measurable) set which is closed under multiplication; that is, Y is such that $(s,x), (r, s^{-1}\cdot x) \in Y$ implies $(s,x)(r,s^{-1}\cdot x) = (sr, x) \in Y$. Then there is a μ-conull Borel set $X_0 \subset X$ such that*

$$E := \{\,(s,x) : x \in X_0 \text{ and } s^{-1}\cdot x \in X_0\,\} \subset Y.$$

Proof. Let

$$Y^{-1} := \{\,(s^{-1}, s^{-1}\cdot x) : (s,x) \in Y\,\}.$$

In view of Remark D.33 on page 388, Y^{-1} is $\mu_G \times \mu$-conull (because, in the notation of Remark D.33, the measure $\kappa_*(\mu_G \times \mu)$ is equivalent to $\mu_G \times \mu$). Therefore $Y' := Y \cap Y^{-1}$ is also conull. Since μ_G and μ are σ-finite, Fubini's Theorem implies there is a μ-conull set X_0 such that $x \in X_0$ implies there is a μ_G-conull set $G(x) \subset G$ such that $G(x) \times \{\,x\,\} \subset Y'$. Let E be as in the lemma, and suppose that $(s,x) \in E$. Note that $s^{-1}G(x) \cap G(s^{-1}\cdot x)$ is μ_G-conull, and hence nonempty. Let $r \in s^{-1}G(x) \cap G(s^{-1}\cdot x)$. Then $(r, s^{-1}\cdot x)^{-1} = (r^{-1}, r^{-1}s^{-1}\cdot x) \in Y'$, and $(sr, x) \in Y'$. Since Y is closed under multiplication,

$$(s,x) = (sr,x)(r^{-1}, r^{-1}s^{-1}\cdot x) \in Y.$$

This completes the proof. □

The above lemma together with Ramsay's Selection Theorem (see Remark G.26 on page 450) provide the tools needed for the proof of Lemma 9.2 on page 267. Of course, we adopt the notations and conventions of Lemmas 9.1 and 9.2.

Proof of Lemma 9.2. We can view $\mathcal{G} := \operatorname{Prim} A \times \operatorname{Prim} A$ as a standard Borel groupoid with $\mathcal{G}^{(2)} = \{\, \big((P,Q),(Q',R)\big) : Q = Q' \,\}$ so that $(P,Q)(Q,R) = (P,R)$ and $(P,Q)^{-1} = (Q,P)$. Part (a) of Lemma 9.1 on page 264 together with (9.1) imply that $(\mathcal{G},\gamma)$ is a measured groupoid. Part (b) of Lemma 9.1 implies that $(P,Q) \mapsto D(P,Q)$ is almost everywhere a homomorphism of $\mathcal{G}$ into the multiplicative positive reals $\mathbf{R}^+_\times$. Therefore Ramsay's Theorem [140, Theorem 5.1] implies that there is a Borel map $D_0 : \mathcal{G} \to \mathbf{R}^+_\times$ such that $D = D_0$ holds γ-almost everywhere, and such that there is a μ-null set $X \subset \operatorname{Prim} A$ such that

$$D_0(P,Q)D_0(Q,R) = D_0(P,R) \quad \text{provided } P,\ Q \text{ and } R \text{ are in } X.$$

Using Lemma H.14 on page 463 and the fact that $D = D_0$ almost everywhere, it follows that (9.3) holds with D replaced by D_0. Furthermore, examining the proof of Corollary D.34 on page 389 (see (D.22) on page 390), there is a μ-null set $X_{00} \subset X \subset \operatorname{Prim} A$ such that

$$d(s,P)d(r,s^{-1}\cdot P) = d(sr,P) \quad \text{if } P,\ s^{-1}\cdot P \text{ and } r^{-1}s^{-1}\cdot P \text{ are in } X_{00}.$$

Let

$$Y_0 := \{\, (s,P) \in G \times \operatorname{Prim} A : P \in X_{00} \text{ and } s^{-1}\cdot P \in X_{00} \,\}.$$

Then the quasi-invariance of μ together with Fubini's Theorem implies that Y_0 is $\mu_G \times \mu$-conull. Let

$$Y = \{\, (s,P) \in Y_0 : D_0(P,s^{-1}\cdot P) = d(s,P)\rho(s^{-1},G_P)^{-1} \,\}.$$

In view of (9.3) (with D_0 in place of D), it follows that Y_0 is also conull. Suppose that (s,P) and $(r,s^{-1}\cdot P)$ are in Y. Since $Y \subset Y_0$, we must have $P, s^{-1}\cdot P, r^{-1}s^{-1}\cdot P \in X_{00} \subset X$. Therefore

$$d(s,P)d(r,s^{-1}\cdot P) = d(sr,P) \quad \text{and}$$
$$D_0(P,s^{-1}\cdot P)D_0(s^{-1}\cdot P, r^{-1}s^{-1}\cdot P) = D(P,r^{-1}s^{-1}\cdot P).$$

Since we also have $\rho(r^{-1}s^{-1},G_P) = \rho(s^{-1},G_P)\rho(r^{-1},G_{s^{-1}\cdot P})$, it follows that $(sr,P) \in Y$. Therefore Lemma G.27 on the previous page implies that there is a μ-conull set $X_0 \subset \operatorname{Prim} A$ such that $P \in X_0$ and $s^{-1}\cdot P \in X_0$ implies that

$$D_0(P,s^{-1}) = d(s,P)\rho(s^{-1},G_P)^{-1}.$$

Thus if $P \in X_0$ and if $t \in G_P$, then $D_0(P,P) = 1$ and

$$d(t,P) = \rho(t^{-1},G_P) = \gamma_{G_P}(t)^2\rho(e,G_P) = \gamma_{G_P}(t)^2. \qquad \square$$

Appendix H

The Fell Topology

The closed subsets $\mathscr{C}(X)$ of any topological space can given a topology, called the Fell topology, in which $\mathscr{C}(X)$ is compact. If X is a (not necessarily Hausdorff) locally compact space, then $\mathscr{C}(X)$ is Hausdorff. If X is a locally compact group, then the subset of closed subgroups is closed in $\mathscr{C}(X)$, and the set of closed subgroups inherits its own compact Hausdorff topology which is both natural and very useful. We develop some of the important properties of this topology in this section. If (A, G, α) is a dynamical system, then simple examples show that the stability map $P \mapsto G_P$ is often not continuous, where G_P is the stability group at P for the natural action of G on $\operatorname{Prim} A$. We show here that the stability map is at least a Borel map, and this will be of importance in our proof of the GRS-Theorem in Chapter 9. We also give Dixmier's proof of Effros's result that $\operatorname{Prim} A$ is a standard Borel space in the Borel structure coming from the usual hull-kernel topology on $\operatorname{Prim} A$. In the process, we see that $\operatorname{Prim} A$ has a possibly finer topology, called the regularized topology, which is a Polish topology and which generates the same Borel structure as the hull-kernel topology.

H.1 The Fell Topology

Let X be an arbitrary topological space. Let $\mathscr{C}(X)$ be the collection of all closed subsets of X (including the empty set). If $\mathscr{F}$ is a finite collection of open subsets of X and if K is a compact subset of X, then we define

$$\mathcal{U}(K;\mathscr{F}) := \{\, F \in \mathscr{C}(X) : F \cap K = \emptyset \text{ and } F \cap U \neq \emptyset \text{ for all } U \in \mathscr{F} \,\}.$$

It is not hard to check that the $\mathcal{U}(K;\mathscr{F})$ form a basis for a topology on $\mathscr{C}(X)$ called the *Fell Topology*. The first thing to prove about $\mathscr{C}(X)$ is that it is compact. This was proved by Fell in [51]. The key to his proof is the concept of a *universal net.* A net $\{\, x_\lambda \,\}_{\lambda\in\Lambda}$ in X is called universal if given any subset $F \subset X$, then either x_λ is eventually in F or it is eventually in $X \smallsetminus F$. Every net has a universal subnet [127, Theorem 1.3.8], and a space X is compact if and only if every universal net in X is convergent [127, Theorem 1.6.2].

Lemma H.1. *If X is any topological space, then $\mathscr{C}(X)$ is compact in the Fell topology.*

Proof. Let $\{ F_i \}$ be a universal net in $\mathscr{C}(X)$. Let

$$Z := \{ x \in X : \text{given any neighborhood } U \text{ of } x, \text{ we eventually have } F_i \cap U \neq \emptyset \}.$$

It is not hard to see that Z is closed. Thus it will suffice to see that $F_i \to Z$ in $\mathscr{C}(X)$.

Suppose that $Z \in \mathcal{U}(K; \mathscr{F})$. It will suffice to see that F_i is eventually in $\mathcal{U}(K; \mathscr{F})$. But if $U \in \mathscr{F}$, then there exists a $z \in Z \cap U$. Since U is a neighborhood of z, we eventually have $F_i \cap U \neq \emptyset$. So we just need to verify that we eventually have $F_i \cap K = \emptyset$. However, if this fails, then we can pass to a subnet, $\{ F_{i_j} \}$, so that we can find $x_{i_j} \in K \cap F_{i_j}$. Since K is compact, we may as well assume that $x_{i_j} \to z \in K$. Let V be a neighborhood of z. We eventually have $F_{i_j} \cap V \neq \emptyset$. Since F_i is universal, we eventually have $F_i \cap V \neq \emptyset$. Thus $z \in Z$. But this is a contradiction as $Z \in \mathcal{U}(K; \mathscr{F})$ implies $Z \cap K = \emptyset$. □

The following characterization of the Fell topology when X is locally compact not only gives an intuitive feel for convergence in the Fell topology, but it is also a useful technical tool.

Lemma H.2. *Suppose that X is a (not necessarily Hausdorff) locally compact space. Let $\{ F_i \}_{i \in I}$ be a net in $\mathscr{C}(X)$. Then $F_i \to F$ in $\mathscr{C}(X)$ if and only if*

(a) *given $t_i \in F_i$ such that $t_i \to t$, then $t \in F$, and*

(b) *if $t \in F$, then there is a subnet $\{ F_{i_j} \}$ and $t_{i_j} \in F_{i_j}$ such that $t_{i_j} \to t$.*

Proof. Assume that $F_i \to F$. Suppose that $t_i \in F_i$ and that $t_i \to t$. If $t \notin F$, then there is a compact neighborhood K of t disjoint from F. Thus we eventually must have $t_i \in K$. But $\mathcal{U}(K; \emptyset)$ is a neighborhood of F which implies that we eventually have $F_i \cap K = \emptyset$. This is a contradiction, so that (a) holds.

If $t \in F$, let

$$\Gamma := \{ (i, U) : i \in I,\ U \text{ is a neighborhood of } t \text{ and } F_i \cap U \neq \emptyset \}.$$

Given (i_1, U_1) and (i_2, U_2) in Γ, $U_3 := U_1 \cap U_2$ is a neighborhood of t, and $\mathcal{U}(\emptyset; \{ U_3 \})$ is a neighborhood of F. Hence, there is an i, dominating both i_1 and i_2, such that $F_i \cap U \neq \emptyset$. It follows that Γ is directed by $(i, U) \geq (i', U')$ if $i \geq i'$ and $U' \subset U$. For each $(i, U) \in \Gamma$, choose $s_{(i,U)} \in F_i \cap U$. Then $\{ s_{(i,U)} \}$ is a net converging to t. Let $t_{i_{(j,U)}} = s_{(j,U)}$ and $F_{i_{(j,U)}} = F_j$. Then $\{ t_{i_\gamma} \}_{\gamma \in \Gamma}$ is the subnet required in (b). Thus (a) and (b) both hold if $F_i \to F$.

Now suppose that (a) and (b) both hold. Suppose that $F \in \mathcal{U}(K; \mathscr{F})$. If we don't eventually have $F_i \cap K = \emptyset$, then we can pass to a subnet, relabel, and find $t_i \in F_i \cap K$ such that $t_i \to t \in K$. But then (a) implies $t \in F \cap K$, which is a contradiction. If $U \in \mathscr{F}$ and if we don't eventually have $F_i \cap U \neq \emptyset$, then we can pass to a subnet, relabel, and assume that $U \cap F_i = \emptyset$ for all i. But if $t \in F \cap U$, then (b) implies that we may also assume that there are $t_i \in F_i$ such that $t_i \to t$. But the t_i must eventually be in U. This is a contradiction and completes the proof. □

Lemma H.2 on the facing page implies that the limit of any net in $\mathscr{C}(X)$ is unique. Hence we obtain the following important corollary.

Proposition H.3. *If X is a (not necessarily Hausdorff) locally compact space, then $\mathscr{C}(X)$ is a compact Hausdorff space.*

An important application of the Fell topology is to the collection of closed subgroups of a locally compact group.

Corollary H.4. *Suppose that G is a locally compact group. Let*

$$\Sigma_G := \{ H \in \mathscr{C}(G) : H \textit{ is a closed subgroup of } G \}.$$

Then Σ_G is closed in $\mathscr{C}(G)$. In particular, Σ_G is a compact Hausdorff space.

Proof. Suppose that $H_i \to H$ with each H_i a closed subgroup of G. It will suffice to see that H is a subgroup. Since $e \in H_i$ for all i, item (a) of Lemma H.2 on the preceding page implies that $e \in H$. If $s, t \in H$, then using item (b) of Lemma H.2 on the facing page, we can pass to a subnet, relabel, and assume that there are $s_i, t_i \in H_i$ such that $s_i \to s$ and $t_i \to t$. But then $s_i t_i \in H_i$ and $s_i t_i \to st$. Thus $st \in H$. A similar argument shows that if $s \in H$, then $s^{-1} \in H$. □

The Urysohn Metrization Theorem implies that a compact Hausdorff space is Polish exactly when it is second countable. Therefore the following theorem is not surprising, but nevertheless, the proof is just a bit subtle.

Theorem H.5. *Suppose that X is a second countable (not necessarily Hausdorff) locally compact space. Then $\mathscr{C}(X)$ is Polish. In particular, Σ_G is Polish in the Fell topology for any second countable locally compact group G.*

To prove the theorem, we just need to show that $\mathscr{C}(X)$ is second countable whenever X is. For this, we need the following observation about locally compact spaces. Some care is needed as we are not assuming the space to be Hausdorff.

Lemma H.6. *Suppose that X is a second countable (not necessarily Hausdorff) locally compact space. Then there is a countable family $\{ K_i \}$ of compact sets in X such that given a neighborhood U of x in X, there is an i such that*

$$x \in \operatorname{int} K_i \subset K_i \subset U,$$

where $\operatorname{int} K_i$ denotes the interior of K_i.

Proof. Let $\{ U_j \}_{j \in J}$ be a countable basis for the topology on X. Let

$$A = \{ (i, j) \in J \times J : \text{there is a compact set } K \text{ such that } U_i \subset K \subset U_j \}.$$

For each $(i, j) \in A$, fix $K_{(i,j)}$ such that $U_i \subset K_{(i,j)} \subset U_j$. If $x \in U$, then since $\{ U_j \}$ is a basis, there is a j such that $x \in U_j \subset U$. Since X is locally compact, there is a compact neighborhood K of x such that $x \in K \subset U_j$. Since $\{ U_i \}$ is a basis, there is an i such that $x \in U_i \subset K$. Then $(i, j) \in A$ and

$$x \in \operatorname{int} K_{(i,j)} \subset K_{(i,j)} \subset U.$$

Thus $\{ K_{(i,j)} \}_{(i,j)\in A}$ suffices. □

Theorem H.5 on the previous page is an immediate consequence of the next lemma.

Lemma H.7. *If X is a second countable (not necessarily Hausdorff) locally compact space, then $\mathscr{C}(X)$ is second countable in the Fell topology.*

Proof. Let $\{ K_a \}_{a\in A}$ be a family of compact sets as in Lemma H.6 on the preceding page, and let $\{ V_i \}_{i\in I}$ be a countable basis for the topology. It will suffice to see that the neighborhoods

$$\mathcal{U}\Big(\bigcup_{a\in A_f} K_a ; \{ V_i \}_{i\in I_f} \Big)$$

form a basis for the Fell topology on $\mathscr{C}(X)$ as A_f and I_f range over finite subsets of A and I, respectively.

Suppose that $F \in \mathcal{U}(K; \mathscr{F})$. Since $F \cap K = \emptyset$ and since F is closed, each point in the compact set K has neighborhood disjoint from F. Then there is an $a_x \in A$ such that K_{a_x} is a neighborhood of x disjoint from F. Since K is compact, there is a finite set $A_f \subset A$ such that

$$K \subset \bigcup_{a\in A_f} K_a$$

and such that the above union is disjoint from F. Furthermore, if $\mathscr{F} = \{ U_1, \dots, U_n \}$ then there are i_j such that $V_{i_j} \subset U_j$ and $V_{i_j} \cap F \neq \emptyset$. Then

$$F \in \mathcal{U}\Big(\bigcup_{a\in F_f} K_a ; \{ V_{i_1}, \dots V_{i_n} \} \Big) \subset \mathcal{U}(K; \mathscr{F}).$$

This completes the proof. □

H.2 Basic Results

Let Σ be the set of closed subgroups of a locally compact group G with Fell's compact Hausdorff topology. A *continuous choice of left Haar measures on* Σ is a family of left Haar measures $\{ \mu_H \}_{H\in\Sigma}$ such that

$$H \mapsto \int_H f(s) \, d\mu_H(s)$$

is continuous for all $f \in C_c(G)$. Glimm showed such families exist in [61, p. 908]. In order to give his proof, we need to recall the notion of a *generalized limit*. Let D be a directed set and let c_D be the set of *bounded* D-convergent nets in $\mathbf{C}$. That is, D is the set of bounded nets $x = \{x_d\}_{d\in D}$ indexed by D such that $\lim_d x_d$ exists. For example, if $D = \mathbf{N}$, then $c_{\mathbf{N}}$ is the set of convergent sequences and is usually denoted by c.[1] Note that c_D is a subspace of the Banach space $\ell^\infty(D)$ of bounded functions on D with the sup norm: $\|x\| := \sup_{d\in D} |x_d|$. The linear functional γ sending $x \in c_D$ to $\lim_d x_d$ is of norm 1. Any norm 1 extension Γ of γ to $\ell^\infty(D)$ such that $\Gamma(x) \geq 0$ if $x_d \geq 0$ for all d is called a generalized limit over D. To see that such things exist, we proceed as follows (cf. [18, Theorem III.7.1]). Let $\ell^\infty_{\mathbf{R}}(D)$ be the Banach space of bounded real-valued functions on D. Define γ_1 and c^1_D in analogy with γ and c_D above. Let Γ_1 be any norm-one extension of γ_1 to $\ell^\infty_{\mathbf{R}}(D)$. Suppose that $x_d \geq 0$ for all d. If $\Gamma_1(x) < 0$, then let $y_d := \|x\|_\infty$ for all d. Then $\|y - x\|_\infty \leq \|x\|_\infty$ and $|\Gamma_1(y - x)| = \|x\|_\infty - \Gamma_1(x) > \|x\|_\infty$. This contradicts $\|\Gamma_1\| = 1$. Thus we must have $\Gamma_1(x) \geq 0$.

If $x \in \ell^\infty(X)$, then $x = x_1 + ix_2$ with each $x_k \in \ell^\infty_{\mathbf{R}}(D)$. It is not hard to see that

$$\Gamma(x) := \Gamma_1(x_1) + i\Gamma_1(x_2)$$

defines a linear functional on $\ell^\infty(D)$ extending γ. Clearly $\Gamma(x) \geq 0$ if $x_d \geq 0$ for all d, and $\|\Gamma\| \leq 2$. To see that $\|\Gamma\| = 1$, so that Γ is a generalized limit, we need a bit of fussing. Let x be an element in the unit ball of $\ell^\infty(D)$ taking only finitely many values. Then there is a partition $E_1, \dots, E_m$ of D and $\alpha_k \in \mathbf{C}$ such that $|\alpha_k| \leq 1$ and such that

$$x = \sum_{k=1}^m \alpha_k \mathbb{1}_{E_k}.$$

Then $\Gamma(x) = \sum_k \alpha_k \Gamma(\mathbb{1}_{E_k}) = \sum_k \alpha_k \Gamma_1(\mathbb{1}_{E_k})$. But $\Gamma_1(\mathbb{1}_{E_k}) \geq 0$, and $\sum_k \Gamma_1(\mathbb{1}_{E_k}) = \Gamma_1(\mathbb{1}_D) \leq 1$. Since $|\alpha_k| \leq 1$ for all k,

$$|\Gamma(x)| \leq \sum_{k=1}^m \Gamma_1(\mathbb{1}_{E_k}) \leq 1.$$

However, if x is any element in the unit ball of $\ell^\infty(D)$, then there is a sequence $x_n \to x$ in $\ell^\infty(D)$ such that each x_n is an element in the unit ball of $\ell^\infty(D)$ taking only finitely many values (cf. [57, Theorem 6.8(e)]). Since Γ is bounded, $\Gamma(x_n) \to \Gamma(x)$. Since $|\Gamma(x_n)| \leq 1$ by the above, we must have $|\Gamma(x)| \leq 1$. Thus $\|\Gamma\| = 1$, and we've shown that generalized limits exist.

Notice that if x and y are nets over D and if $x_d = y_d$ for sufficiently large d, then $x - y$ is eventually 0 and $\Gamma(x) = \Gamma(y)$ for any generalized limit Γ. It follows that for all nets x,

$$|\Gamma(x)| \leq \limsup_d |x_d| := \inf_c \Bigl\{ \sup_{d\geq c} |x_d| \Bigr\}.$$

[1] Unlike a sequence, a convergent net need not be bounded. This is the reason for the hypothesis that c_D consist of bounded convergent nets.

Now suppose that M is a directed set and that there is a function $N : M \to D$ which satisfies condition (b) in the definition of a subnet on page 4. Then given $x \in \ell^\infty(D)$, $m \mapsto x_{N_m}$ is a subnet of $\{ x_d \}_{d \in D}$. Let

$$c_D(M) := \{ x \in \ell^\infty(D) : \lim_m x_{N_m} \text{ exists} \}.$$

Then $c_D(M)$ is a subspace of $\ell^\infty(D)$ containing c_D. Let γ_M be the norm one linear functional on $c_D(M)$ given by $\gamma_M(x) := \lim_m x_{N_m}$. Repeating the above gymnastics shows that there is a generalized limit Γ which restricts to γ_M on $c_D(M)$. In particular, if $x \in \ell^\infty(D)$ has a subnet converging to c, then there is a generalized limit Γ such that $\Gamma(x) = c$. It follows that $a = \lim_d x_d$ if and only if $a = \Gamma(x)$ for all generalized limits over D.

Lemma H.8 (Glimm). *Suppose that $f_0 \in C_c^+(G)$ and that $f_0(e) > 0$. For each $H \in \Sigma$, let μ_H be the left Haar measure on H such that such that*

$$\int_H f_0(s) \, d\mu_H(s) = 1.$$

Then $\{ \mu_H \}_{H \in \Sigma}$ is a continuous choice of Haar measures on Σ.

Proof. We claim that for each compact set K in G, there is a number $a(K)$ such that

$$\mu_H(K \cap H) \le a(K) \quad \text{for all } H \in \Sigma. \tag{H.1}$$

To prove the claim, notice that there is an $\epsilon > 0$ and a neighborhood U of e in G such that $f_0(s) > \epsilon$ provided $s \in U$. Since K is compact, there are open sets $U_1, \dots, U_n$ such that

$$K \subset \bigcup_{i=1}^n U_i \quad \text{and} \quad U_i^{-1} U_i \subset U \quad \text{for all } i.$$

If $H \in \Sigma$, let $J = \{ j : U_j \cap H \ne \emptyset \}$. For each $j \in J$, pick $h_j \in U_j \cap H$. Then

$$g(s) := \sum_{j \in J} f_0(h_j^{-1} s)$$

defines an element of $C_c(G)$ such that $g(s) > \epsilon$ for all $s \in H \cap K$. Thus

$$\begin{aligned} \mu_H(H \cap K) &\le \frac{1}{\epsilon} \int_H g(s) \, d\mu_H(s) \\ &\le \frac{1}{\epsilon} \sum_{j \in J} \int_H f_0(h_j^{-1} s) \, d\mu_H(s) \\ &\le \frac{n}{\epsilon} \end{aligned}$$

Therefore we can set $a(K) = n/\epsilon$. This proves the claim.

Now suppose that $H_i \to H$ in Σ and that $f \in C_c(G)$. The first part of the proof implies that

$$i \mapsto \int_{H_i} f(s)\, d\mu_{H_i}(s)$$

is a bounded net. Let Γ be a generalized limit and define

$$\Phi(f) := \Gamma\Big(i \mapsto \int_{H_i} f(s)\, d\mu_{H_i}(s)\Big).$$

Since Γ is arbitrary, it will suffice to show that

$$\Phi(f) = \int_H f(s)\, d\mu_H(s).$$

The positivity condition on Γ implies that Φ is a positive linear functional on $C_c(G)$. If $f, g \in C_c(G)$ and if $f(s) = g(s)$ for all $s \in H$, then given $\epsilon > 0$,

$$C := \{\, s \in G : |f(s) - g(s)| \geq \epsilon \,\}$$

is a compact set disjoint from H. Using the definition of the topology on Σ, we must eventually have $H_i \cap C = \emptyset$. Thus

$$\Big|\int_{H_i} \big(f(s) - g(s)\big)\, d\mu_{H_i}(s)\Big| \leq \epsilon\big(a\big(\operatorname{supp}(f) \cup \operatorname{supp}(g)\big)\big)$$

Since ϵ was arbitrary, we must have $\Phi(f) = \Phi(g)$. Therefore

$$\varphi(f|_H) := \Phi(f)$$

is a positive linear functional on $C_c(H)$.

To see that φ is left-invariant, and therefore a Haar measure, fix $t \in H$ and a compact neighborhood C of t. Then if $t' \in C$,

$$s \mapsto f(ts) - f(t's)$$

has support in $K := C^{-1} \operatorname{supp}(f)$. The uniform continuity of f (Lemma 1.62 on page 19) implies that there is a neighborhood $U \subset C$ of t such that $t' \in U$ implies that

$$\big|f(ts) - f(t's)\big| < \frac{\epsilon}{a(K) + 1}.$$

Thus for all $L \in \Sigma$,

$$\Big|\int_L f(ts)\, d\mu_L(s) - \int_L f(t's)\, d\mu_L(s)\Big| < \epsilon.$$

Since $H \cap U \neq \emptyset$, we eventually have $H_i \cap U \neq \emptyset$. Thus for large i we can pick

$t_i \in H_i \cap U$. Then we compute that

$$\begin{aligned}\bigl|\varphi(\mathrm{lt}_t^{-1}(f)|_H) - \varphi(f|_H)\bigr| &= \Bigl|\Gamma\Bigl(i \mapsto \int_{H_i} f(ts)\,d\mu_{H_i}(s) - \int_{H_i} f(s)\,d\mu_{H_i}(s)\Bigr)\Bigr| \\ &\le \limsup_i \Bigl(\Bigl|\int_{H_i} f(ts)\,d\mu_{H_i}(s) - \int_{H_i} f(t_i s)\,d\mu_{H_i}(s)\Bigr| \\ &\qquad + \Bigl|\int_{H_i} f(t_i s)\,d\mu_{H_i}(s) - \int_{H_i} f(s)\,d\mu_{H_i}(s)\Bigr|\Bigr) \\ &\le \epsilon + 0 = \epsilon.\end{aligned}$$

Since ϵ is arbitrary, φ is left invariant and a Haar measure on H. Since

$$\varphi(f_0|_H) = \Gamma\Bigl(i \mapsto \int_{H_i} f_0(s)\,d\mu_{H_i}(s)\Bigr) = 1,$$

the uniqueness of Haar measure guarantees that φ is given by μ_H. Thus

$$\Phi(f) = \int_H f(s)\,d\mu_H(s).$$

This completes the proof. □

The following observations will be useful.

Lemma H.9 ([169, Lemma 2.12]). *Suppose that $\{\mu_H\}_{H\in\Sigma}$ is a continuous choice of Haar measures on Σ.*

(a) *If $K \subset G$ is compact, then there is a constant $a(K) \in \mathbf{R}^+$ such that $\mu_H(K \cap H) \le a(K)$ for all $H \in \Sigma$.*

(b) *If $\{f_i\}$ is a net of functions in $C_c(G)$ converging to $f \in C_c(G)$ in the inductive limit topology and if $\{H_i\}$ is a net in Σ converging to $H \in \Sigma$ (with the same index set), then*

$$\int_{H_i} f_i(s)\,d\mu_{H_i}(s) \to \int_H f(s)\,d\mu_H(s).$$

(c) *If Δ_H is the modular function on H, then $(s, H) \mapsto \Delta_H(s)$ is continuous on $G * \Sigma$.*

(d) *If X is a locally compact space and if $F \in C_c(X \times G \times \Sigma)$, then*

$$(x, H) \mapsto \int_H F(x, s, H)\,d\mu_H(s)$$

is continuous on $X \times \Sigma$.

Proof. Let $f \in C_c^+(G)$ be such that $f(s) = 1$ for all $s \in K$. By assumption,

$$H \mapsto \int_H f(s)\,d\mu_H(s)$$

defines a continuous function F on the compact set Σ. Part (a) follows with $a(K) = \|F\|_\infty$.

Using part (a), it follows that for any $f \in C_c(G)$,

$$\Bigl| \int_H f(s)\, d\mu_H(s) \Bigr| \le \|f\|_\infty a(\operatorname{supp} f).$$

Part (b) is an immediate consequence.

Let $f_0 \in C_c^+(G)$ be such that

$$\varphi(H) := \int_H f_0(s)\, d\mu_H(s) > 1 \quad \text{for all } H \in \Sigma. \tag{H.2}$$

Then φ is continuous and

$$\Delta_H(s) = \varphi(H)^{-1} \int_H f_0(sr^{-1})\, d\mu_H(s).$$

Let $f^r(s) := f(sr)$. The map $r \mapsto f^{r^{-1}}$ is continuous from G into $C_c(G)$ with the inductive limit topology. Therefore part (c) follows from part (b).

Part (d) is proved similarly: $(x, H) \mapsto F(x, \cdot, H)$ is continuous from $X \times \Sigma$ into $C_c(G)$ with the inductive limit topology. □

Lemma H.10. *There is a continuous function $\omega : G \times \Sigma \to (0, \infty)$ such that for all $g \in C_c(G)$ we have*

$$\int_H g(sts^{-1})\, d\mu_H(t) = \omega(s, H) \int_{s\cdot H} g(t)\, d\mu_{s\cdot H}(t),$$

where $s \cdot H := sHs^{-1}$. Furthermore, for all $s, r \in G$ and all $H \in \Sigma$, we have

$$\omega(sr, H) = \omega(r, H)\omega(s, r \cdot H). \tag{H.3}$$

Proof. The existence of $\omega(s, H)$ for each s and H follows from the uniqueness of Haar measure. Let $f_0 \in C_c^+(G)$ and $\varphi \in C^+(\Sigma)$ be as in (H.2). Then

$$\omega(s, H) = \varphi(s \cdot H)^{-1} \int_H f_0(sts^{-1})\, d\mu_H(t),$$

But $s \mapsto f(s \cdot s^{-1})$ is continuous from G to $C_c(G)$ with the inductive limit topology, and $(s, H) \mapsto s \cdot H$ is easily seen to be continuous. Therefore ω is continuous by part (b) of Lemma H.9 on the facing page. Equation (H.3) is a straightforward computation. □

We define

$$\rho(s, H) := \Delta_G(s^{-1})\omega(s, H). \tag{H.4}$$

Then ρ is continuous and satisfies

$$\rho(st, H) = \frac{\Delta_H(t)}{\Delta_G(t)} \rho(s, H) \quad \text{for all } s \in G \text{ and } t \in H \in \Sigma,$$

and in view of (H.3),

$$\rho(sr,H)=\rho(r,H)\rho(s,r\cdot H)\quad\text{for all } r,s\in G.$$

Given a function ρ as above, it is well-known (for example, see [139, Lemma C.2] or [56, Theorem 2.56]) that there is a quasi-invariant measure β^H on G/H such that

$$\int_G f(s)\rho(s,H)\,d\mu_G(s)=\int_{G/H}\int_H f(st)\,d\mu_H(t)\,d\beta^H(\dot s) \tag{H.5}$$

for all $f\in C_c(G)^2$, and such that, if $r\cdot\beta^H$ is defined by

$$\int_{G/H} g(\dot s)\,d(r\cdot\beta^H)(\dot s)=\int_{G/H} g(r\cdot\dot s)\,d\beta^H(\dot s),$$

then

$$\frac{d(r\cdot\beta^H)}{d\beta^H}(\dot s)=\frac{\rho(r^{-1}s,H)}{\rho(s,H)}=\rho(r^{-1},s\cdot H).$$

Since this material is primarily destined to be used in proving the Effros-Hahn conjecture, where separability is assumed, we'll often assume that G is second countable. For many of these results, more general versions are available, but usually at the expense of considerably more intricate arguments.

We want to see that the decomposition of Haar measure on G given by (H.5) holds for Borel functions. When G is second countable (so that all the measures involved are σ-finite), this is a straightforward consequence of Fubini's Theorem. What we want is summarized in the following.

Proposition H.11. *Suppose that G is second countable and $H\in\Sigma$. Let β^H be the quasi-invariant measure on G/H associated to ρ as above. If f is a Borel function on G such that $s\mapsto f(s)\rho(s,H)$ is μ_G-integrable, then*

(a) *$t\mapsto f(st)$ is Borel on H for all $s\in G$,*

(b) *there is a β^H-null set D such that $t\mapsto f(st)$ is μ_H-integrable for all $\dot s\notin D$,*

(c) *if*

$$F_f(\dot s):=\begin{cases}\int_H f(st)\,d\mu_H(t) & \text{is } \dot s\notin D\\ 0 & \text{if } \dot s\in D,\end{cases}$$

then F_f is Borel and β^H-integrable, and

(d) *we have*

$$\int_G f(s)\rho(s,H)\,d\mu_G(s)=\int_{G/H} F_f(\dot s)\,d\beta^H(\dot s). \tag{H.6}$$

Remark H.12. In practice, we will never introduce the function F_f. Instead, we will write (H.5) in place of (H.6), and assume that parts (a)–(d) of Proposition H.11 are understood.

We will need the following observation.

[2] As elsewhere, we have written $\dot s$ in place of sH to make the notation a bit easier to read.

Lemma H.13. *Suppose that $\tau : X \to Y$ is a Borel map and that μ is a measure on X. Define the push-forward measure on Y by $\tau_*\mu(E) := \mu\big(\tau^{-1}(E)\big)$. If f is a nonnegative Borel function on Y, then*

$$\int_Y f(y)\, d(\tau_*\mu)(y) = \int_X f\big(\tau(x)\big)\, d\mu(x). \tag{H.7}$$

Thus a Borel function f on Y is $\tau_*\mu$-integrable if and only if $f \circ \tau$ is μ-integrable.

Proof. If $E \subset Y$ is Borel and $f = \mathbb{1}_E$, then (H.7) holds by definition. If f is a nonnegative Borel function on Y, then there are nonnegative simple Borel functions $f_n \nearrow f$ and (H.7) holds by the Monotone Convergence Theorem. (Note that both sides may be infinite.) The rest is straightforward. □

Proof of Proposition H.11. Since $(s,t) \mapsto f(st)$ is Borel on $G \times H$, the first assertion follows from Fubini's Theorem applied to $\mu_G \times \mu_H$. Let $c : G/H \to G$ be a Borel cross section to the quotient map ([2, Theorem 3.4.1] or [104, Lemma 1.1]). Then $\sigma : G/H \times H \to G$, given by $\sigma(\dot{s},t) := c(\dot{s})t$ is a Borel isomorphism, and $\sigma_*(\beta^H \times \mu_H)$ is a Borel measure on G. If $f \in C_c(G)$ is nonnegative, then, using the left invariance of μ_H,

$$\begin{aligned}\int_G f(s)\, d\sigma_*(\beta^H \times \mu_H)(s) &= \int_{G/H}\int_H f\big(c(\dot{s})t\big)\, d\mu_H(t)\, d\beta^H(\dot{s}) \\ &= \int_{G/H}\int_H f(st)\, d\mu_H(t)\, d\beta^H(\dot{s}) \\ &= \int_G f(s)\rho(s,H)\, d\mu_G(s).\end{aligned} \tag{H.8}$$

Since (H.8) holds for all $f \in C_c(G)$, we have $\sigma_*(\beta^H \times \mu_H) = \rho(\cdot, H)\mu_G$. The rest follows from Fubini's Theorem applied to $\beta^H \times \mu_H$. □

Lemma H.14. *Suppose that G is second countable.*[3] *Let D be a Borel subset of G/H and let $A = q^{-1}(D)$ be its saturation in G. Let β be any quasi-invariant measure on G/H. Then D is β-null if and only if A is μ_G-null. In particular, any two quasi-invariant measures on G/H are equivalent.*

Proof. Clearly, $(s,\dot{r}) \mapsto \mathbb{1}_D(s\cdot\dot{r})$ is Borel. Thus Tonelli's Theorem implies that

$$\int_{G/H}\int_G \mathbb{1}_D(s\cdot\dot{r})\, d\mu_G(s)\, d\beta(\dot{r}) = \int_G\int_{G/H} \mathbb{1}_D(s\cdot\dot{r})\, d\beta(\dot{r})\, d\mu_G(s). \tag{H.9}$$

[3] This hypothesis implies that Haar measure is σ-finite and allows us to invoke Fubini's Theorem at will in the proof. In the general case, a bit more care must be taken and the general result states that provided β is a Radon measure, a subset has a β-null image if and only if it is locally μ_G-null [54, Proposition III.14.8].

If $A = q^{-1}(D)$ is a μ_G-null set, then

$$\begin{aligned}\int_G \mathbb{1}_D(s\cdot\dot r)\,d\mu_G(s) &= \int_G \mathbb{1}_A(sr)\,d\mu_G(s)\\ &= \Delta(r)^{-1}\int_G \mathbb{1}_A(s)\,d\mu(s)\\ &= 0.\end{aligned}$$

Thus the left-hand side of (H.9) is 0. Since the right-hand side also vanishes,

$$\int_{G/H} \mathbb{1}_D(s\cdot\dot r)\,d\beta(\dot r) = \beta(s^{-1}\cdot D) = 0 \quad \text{for } \mu_G\text{-almost all } s.$$

Therefore, for some $s \in G$, $\beta(s^{-1}\cdot D) = 0$. Since β is quasi-invariant, $\beta(D) = 0$.

On the other hand, if D is β-null, then by quasi-invariance, $\beta(s^{-1}\cdot D) = 0$ for all s. Thus the right-hand side of (H.9) vanishes. Arguing as above, there is a $\dot r \in G/H$ such that

$$0 = \int_G \mathbb{1}_D(s\cdot\dot r)\,d\mu_G(s) = \int_G \mathbb{1}_A(sr)\,d\mu_G(s) = \Delta(r)^{-1}\int_G \mathbb{1}_A(s)\,d\mu_G(s).$$

Thus $\mu_G(A) = 0$. This completes the proof of the first assertion.

However, the first assertion implies that any two quasi-invariant measures on G/H have the same null sets. Hence they are equivalent. □

Lemma H.15. *Suppose that H is a closed subgroup of G and that φ is a Borel function on G whose modulus is constant on H-cosets, and is such that $\dot s \mapsto |\varphi(s)|$ is in $\mathcal{L}^2(G/H,\beta^H)$. If $K \subset G$ is compact, then $\varphi|_K$ is in $L^2(G,\mu_G)$.*[4] *In particular, the inner product $(\varphi \mid f)$ in $L^2(G)$ is finite for all $f \in C_c(G)$. If $A \subset C_c(G)$ is dense in the inductive limit topology on $C_c(G)$ and if $(\varphi \mid f) = 0$ for all $f \in A$, then $\varphi(s) = 0$ for μ_G-almost all s.*

Proof. Suppose that $K \subset G$ is compact and that $g \in C_c^+(G)$ is such that $g(s) = 1$ for all $s \in K$. Then

$$F_{g^2}(\dot s) := \int_H g(st)^2\rho(st,H)^{-1}\,d\mu_H(t)$$

defines an element of $C_c(G/H)$. We have

$$\begin{aligned}\|\varphi|_K\|^2_{L^2(G)} &\le \int_G g(s)^2|\varphi(\dot s)|^2\,d\mu_G(s)\\ &= \int_{G/H}|\varphi(\dot s)|^2\int_H g(st)^2\,d\mu_H(t)\,d\beta^H(\dot s)\\ &\le \|F_{g^2}\|_\infty \||\varphi|\|^2_{L^2(\beta^H)}.\end{aligned}$$

[4]One says that φ is *locally in* $L^2(G)$.

Thus φ is locally in $L^2(G)$ as claimed.

If $f \in C_c(G)$ and if $K = \operatorname{supp} f$, then $(\varphi \mid f) = (\varphi|_K \mid f)$, which is finite by the above. If $(\varphi \mid f) = 0$ for all $f \in C_c(G)$, then for all compact sets K, $\varphi|_K = 0$ μ_G-almost everywhere. Thus $\varphi = 0$ μ_G-almost everywhere.

If $A \subset C_c(G)$ is dense and $f \in C_c(G)$, then there is a compact neighborhood K of $\operatorname{supp} f$ and a sequence $\{ f_i \} \subset A$ such that $\operatorname{supp} f_i \subset K$ for all i and $f_i \to f$ uniformly. Thus $(\varphi \mid f_i) \to (\varphi \mid f)$. The final assertions follows. □

H.3 Bruhat Approximate Cross Sections

Let $G \times \Sigma/{\sim}$ be the quotient topological space obtained from $G \times \Sigma$ by identifying (s, H) and (st, H) for all $s \in G$, $H \in \Sigma$ and $t \in H$.

Lemma H.16. *The quotient map $q : G \times \Sigma \to G \times \Sigma/{\sim}$ is open. Furthermore, $G \times \Sigma/{\sim}$ is a locally compact Hausdorff space.*

Proof. Let $V \subset G$ and $U \subset \Sigma$ be open. To see that q is an open map, it will suffice to see that

$$\mathcal{O} := q^{-1}\bigl(q(V \times U)\bigr) = \{ (st, H) : s \in V,\ H \in U \text{ and } t \in H \}$$

is open in $G \times \Sigma$. Fix $s \in V$, $H \in U$ and $t \in H$. Let $\{ (r_i, H_i) \}$ be a net converging to (st, H). It will suffice to see that a subnet is eventually in $\mathcal{O}$. Passing to a subnet, and relabeling, we find $t_i \in H_i$ such that $t_i \to t$ (Lemma H.2 on page 454). Then $r_i t_i^{-1} \to s$, and $r_i t_i^{-1}$ is eventually in V. Since we certainly eventually have $H_i \in U$, $(r_i, H_i) = (r_i t_i^{-1} t_i, H_i)$ is eventually in $\mathcal{O}$ as required.

Since q is open by the above, to prove the final statement, we just need to see that $G \times \Sigma/{\sim}$ is Hausdorff. Suppose that $\{ q(s_i, H_i) \}$ is a net converging to both $q(s, H)$ and $q(s', H')$. Since q is open, we can pass to a subnet, relabel, and assume that there are $t_i \in H_i$ such that $(s_i, H_i) \to (s, H)$ and $(s_i t_i, H_i) \to (s', H')$ (Proposition 1.15 on page 4). Thus $H = H'$ and $t_i \to s^{-1}s'$. Lemma H.2 on page 454 implies that $s^{-1}s' \in H$. Thus $q(s, H) = q(s', H)$. □

A locally compact space is paracompact if and only if it is the topological disjoint union of σ-compact spaces [31, Theorem XI.7.3]. Thus any locally compact group G is paracompact (Lemma 1.38 on page 10) as is $G \times \Sigma$. However, it is conceivable that $G \times \Sigma/{\sim}$ could fail to be paracompact. (However, I don't know of any examples where it isn't.) Since q is continuous, $G \times \Sigma/{\sim}$ is σ-compact whenever G is, and therefore paracompact. Thus any example where $G \times \Sigma/{\sim}$ fails to be paracompact would be pretty obscure.

Suppose that H is a closed subgroup of G. Then a continuous nonnegative function b on G with the properties that

$$\int_H b(st)\, d\mu_H(t) = 1 \quad \text{for all } s \in G$$

and such that the support of b has compact intersection with the saturation of any compact subset of G is called a *Bruhat approximate cross section* for G over H.

Since G/H is easily seen to be paracompact,[5] the existence of such functions is a special case of the following proposition which is a minor variation on [11, App. I, Lemme 1].

Proposition H.17. *Suppose that $G \times \Sigma/{\sim}$ is paracompact. Then there is a continuous function $\mathfrak{b} : G \times \Sigma \to [0, \infty)$ such that*

(a) $\displaystyle\int_H \mathfrak{b}(st, H)\, d\mu_H(t) = 1$ *for all $H \in \Sigma$ and $s \in G$, and*

(b) *if $K \subset G$ is compact, and if $q : G \times \Sigma \to G \times \Sigma/{\sim}$ is the quotient map, then*

$$\operatorname{supp} \mathfrak{b} \cap q^{-1}\big(q(K \times \Sigma)\big)$$

is compact.

We call $\mathfrak{b}$ a generalized Bruhat approximate cross section.

Remark H.18. In order to dispense with the paracompactness assumption in certain applications of Proposition H.17, it will be convenient to prove a version of the proposition that applies only to saturated subsets of $G \times \Sigma$ with paracompact quotient. Specifically, let L be a subset of G for which $q(L \times \Sigma)$ is a paracompact locally compact subset of $G \times \Sigma/{\sim}$.[6] Define $\mathcal{S}(L)$ to be the saturation $q^{-1}\big(q(L \times \Sigma)\big)$ of $L \times \Sigma$ in $G \times \Sigma$. We will prove that there is a continuous function $\mathfrak{b} \in C^+\big(\mathcal{S}(L)\big)$ such that

(a)′ $\displaystyle\int_H \mathfrak{b}(st, H)\, d\mu_H(t) = 1$ for all $s \in L$ and $H \in \Sigma$, and

(b)′ if $K \subset L$ is compact, then

$$\operatorname{supp} \mathfrak{b} \cap \mathcal{S}(K)$$

is compact, where $\mathcal{S}(K) := q^{-1}\big(q(K \times \Sigma)\big)$.

Of course, we recover Proposition H.17 by taking $L = G$.

Proof. For each $a = q(s, H) \in q(L \times \Sigma)$, let $f_a \in C_c^+\big(\mathcal{S}(L)\big)$ be such that

$$\int_H f_a(st, H)\, d\mu_H(t) > 0.$$

Let

$$U(a) := \{\, (r, K) \in \mathcal{S}(L) : f_a(r, K) > 0 \,\}.$$

Since q is open, $\{\, q\big(U(a)\big) \,\}_{a \in q(L \times \Sigma)}$ is an open over of $q(L \times \Sigma)$. Since we've assumed that the latter is paracompact, there is a locally finite subcover $\{\, V_i \,\}_{i \in I}$ of $\{\, q\big(U(a)\big) \,\}$. Again using paracompactness, there is a partition of unity, $\{\, \varphi_i \,\}_{i \in I}$

[5] Let G_0 be a σ-compact open subgroup of G as in Lemma 1.38 on page 10. Then it is not hard to see that $G_0 sH$ is a clopen subset of G for each $s \in G$ (it is a union of clopen right G_0 cosets). Thus G/H is the disjoint union of the images of these sets, each of which is clopen and σ-compact. Thus G/H is paracompact.

[6] The application we need is when L is compact. However, L could be any open σ-compact subset such as the subgroup generated by a compact neighborhood.

subordinate to $\{V_i\}$ ([139, Lemma 4.34]). For each $i \in I$, there is an $a_i = q(s_i, H_i)$ such that

$$V_i \subset q\big(U(a_i)\big).$$

Define $F_i(s, H) := \varphi_i\big(q(s, H)\big) f_{a_i}(s, H)$. Then $F_i \in C_c^+\big(\mathcal{S}(L)\big)$ and $\operatorname{supp} F_i \subset q^{-1}(V_i)$. Since $\{V_i\}_{i\in I}$ is locally finite, every $(s, H) \in G \times \Sigma$ has a neighborhood such that all by finitely many of the F_i vanish off that neighborhood. Thus we obtain a continuous function on $\mathcal{S}(L)$ via

$$F(s, H) := \sum_{i\in I} F_i(s, H).$$

If $(s, H) \in \mathcal{S}(L)$, then there is an $i \in I$ such that $\varphi_i\big(q(s, H)\big) > 0$. Since $V_i \subset q\big(U(a_i)\big)$, there is a $t \in H$ such that $(st, H) \in U(a_i)$. This implies that $f_{a_i}(st, H) > 0$ and that $\varphi_i\big(q(st, H)\big) = \varphi_i\big(q(s, H)\big) > 0$. Therefore $F_i(st, H) > 0$. Since F is continuous and nonnegative,

$$\int_H F(st, H)\, d\mu_H(t) > 0.$$

Suppose that $K \subset L$ is compact. Since $q(K \times \Sigma)$ is compact and since $\{V_i\}$ is locally finite, there is a *finite* subset $J \subset I$ such that $q(K \times \Sigma) \subset \bigcup_{i\in J} V_j$ and such that $V_i \cap q(K\times\Sigma) = \emptyset$ if $i \notin J$. Thus $i \notin J$ implies that $\operatorname{supp} F_i \cap q^{-1}\big(q(K\times\Sigma)\big) = \emptyset$. Thus

$$\begin{aligned} q^{-1}\big(q(K\times\Sigma)\big) \cap \operatorname{supp} F &\subset q^{-1}\big(q(K\times\Sigma)\big) \cap \Big(\bigcup_{i\in I} \operatorname{supp} F_i\Big) \\ &= q^{-1}\big(q(K\times\Sigma)\big) \cap \Big(\bigcup_{j\in J} \operatorname{supp} F_j\Big), \end{aligned}$$

which is compact.

Now fix $s \in L$ and let K be a compact neighborhood of s. Since $\operatorname{supp} F \cap q^{-1}\big(q(K \times \Sigma)\big)$ is compact by the above, the function $(s, t, H) \mapsto F(st, H)$ has compact support on $K \times G * \Sigma$. Thus there is a $F' \in C_c(K \times G \times \Sigma)$ such that $F(st, H) = F'(s, t, H)$ for all $s \in K$ and $t \in H$. Thus we can apply Lemma H.9 on page 460 to F', and conclude that

$$(s, H) \mapsto \int_H F(st, H)\, d\mu_H(t)$$

is continuous and greater than zero on $\mathcal{S}(L)$. Therefore we can define a continuous function by

$$\mathfrak{b}(s, H) := F(s, H)\Big(\int_H F(st, H)\, d\mu_H(t)\Big)^{-1}.$$

This function clearly satisfies (a)$'$ and (b)$'$ in Remark H.18 on the facing page. □

Corollary H.19. *Suppose that $K \subset G$ is compact. There there is a $b \in C_c^+(G \times \Sigma)$, called a* cut-down generalized Bruhat approximate cross section, *such that*

$$\int_H b(st, H)\, d\mu_H(t) = 1 \quad \text{for all } s \in K \text{ and all } H \in \Sigma.$$

Proof. Let L be a compact neighborhood of K. In view of Remark H.18 on page 466, the proof of Proposition H.17 on page 466 implies that there is a $\mathfrak{b} \in C^+\bigl(\mathcal{S}(L)\bigr)$ satisfying (a)$'$ and (b)$'$ on page 466. If U is a an open set such that $K \subset U \subset L$, then

$$K'' := \operatorname{supp} \mathfrak{b} \cap \mathcal{S}(K)$$

is compact and there is a $\varphi \in C_c^+(G \times \Sigma)$ such that $\varphi(s, H) = 1$ for all $(s, H) \in K''$ and such that φ vanishes off of the open set $\mathcal{S}(U) = q^{-1}\bigl(q(U \times \Sigma)\bigr)$. Thus

$$b(s, H) = \begin{cases} \varphi(s, H)\mathfrak{b}(s, H) & \text{if } (s, H) \in \mathcal{S}(U), \text{ and} \\ 0 & \text{if } (s, H) \notin K'' \end{cases}$$

is continuous. If $s \in K$ and $H \in \Sigma$, then since $\mathfrak{b}$ vanishes off K'',

$$\int_H b(st, H)\, d\mu_H(t) = \int_H \mathfrak{b}(st, H)\, d\mu_H(t) = 1. \qquad \square$$

Lemma H.20. *Suppose that G is a locally compact group and that Σ is the compact Hausdorff space of closed subgroups of G. For each $H \in \Sigma$, let $q^H : G \to G/H$ be the natural map. Let $K \subset G$ be compact. Then there is a positive constant $m(K)$ such that for all $H \in \Sigma$ we have*

$$\beta^H\bigl(q^H(K)\bigr) \le m(K).$$

Proof. To find a $m(K)$, let b be a cut-down generalized Bruhat approximate cross section (as in Corollary H.19) such that

$$\int_H b(st, H)\, d\mu_H(t) = 1 \quad \text{for all } s \in K \text{ and } H \in \Sigma.$$

Then

$$H \mapsto \int_G b(st, H)\rho(s, H)\, d\mu_G(s)$$

is a continuous function F_1 on Σ. Furthermore,

$$\begin{aligned} \int_H b(st, H)\rho(s, H)\, d\mu_G(s) &= \int_{G/H}\int_H b(st, H)\, d\mu_H(t)\, d\beta^H(\dot{s}) \\ &\ge \beta^H\bigl(q^H(K)\bigr). \end{aligned}$$

Thus we can let $m(K) = \|F_1\|_\infty$. $\square$

H.4 The Stabilizer Map

An important application of the Fell topology on the space Σ of closed subgroups of G arises when considering a transformation group (G, X). If X is a T_0-topological space, so that each stability group G_x is closed by Lemma 3.23 on page 95, then we obtain a natural map σ of X into Σ defined by

$$\sigma(x) = G_x := \{ s \in G : s \cdot x = x \}.$$

We call σ the *stabilizer map*. The stabilizer map is rarely continuous — even in rather nice situations such as the circle group $\mathbf{T}$ action on the complex numbers $\mathbf{C}$ by multiplication. A more involved example is the following which is taken from [59, p. 134]. Working out the details is an interesting exercise.

Example H.21. Let $G = \prod_{i=1}^{\infty} \mathbf{Z}_2$ and let $X = \prod_{i=1}^{\infty} [-1, 1]$. Let the ith factor of G act on the ith factor of X by reflection about the origin. Given $\mathbf{x} = (x_i)$, then $G_{\mathbf{x}} = \{ e \}$ if and only if $x_i \neq 0$ for all i. The stabilizer map is continuous at $\mathbf{x}$ if and only if $\mathbf{x} \in P = \{ \mathbf{x} \in X : x_i \neq 0 \text{ for all } i \}$. Notice that P has no interior.

Remark H.22. If G is a Lie group, if X is a second countable almost Hausdorff locally compact space and if $G\backslash X$ is T_0 (see Theorem 6.2 on page 173), then Glimm has shown that there is a dense *open* subset of X on which the stabilizer map is continuous [59, Theorem 3]. We have already seen in Example H.21 that if G is not a Lie group, then it is possible for the set where the stabilizer map is continuous to have no interior. However, as we will show in Theorem H.24 on the following page, the situation in Example H.21 is about as bad as it gets.

Our next result shows that the stabilizer map is always a Borel map. This will important in our study of the Effros-Hahn conjecture. (See Proposition H.41 on page 477.)

Proposition H.23. *Suppose that (G, X) is a topological transformation group with G locally compact and second countable, and with X Hausdorff. Then the stabilizer map $\sigma : X \to \Sigma$ is a Borel map.*

Proof. It suffices to see that $\sigma^{-1}\big(\mathcal{U}(K; \{ U_1, \dots, U_n \})$ is Borel for all compact sets $K \subset G$ and open sets $U_i \subset G$. To begin with, for any subset $K \subset G$, let

$$W(K) := \sigma^{-1}\big(\{ H \in \Sigma : H \cap K \neq \emptyset \}\big).$$

We claim that $W(K)$ is closed when K is compact. Suppose that K is compact, that $\{ x_i \} \subset W(K)$ and that $x_i \to x$. We want to show that $G_x := \sigma(x)$ is such that $G_x \cap K \neq \emptyset$. Let $h_i \in G_{x_i} \cap K$. Since K is compact, we can pass to a subsequence, relabel, and assume that $h_i \to h \in K$. Then $x_i = h_i \cdot x_i \to h \cdot x$. Since X is Hausdorff, we must have $h \in G_x$. Thus $x \in W(K)$, and $W(K)$ is closed.

Now suppose that $U \subset G$ is open. Since G is second countable, there are compact sets K_n such that

$$U = \bigcup_{n=1}^{\infty} K_n.$$

Thus

$$W(U) = \sigma^{-1}\big(\{\, H \in \Sigma : H \cap U \neq \emptyset \,\}\big) = \bigcup_{n=1}^{\infty} W(K_n)$$

is a F_σ subset (i.e., a countable union of closed sets) of Σ. In particular, $W(U)$ is Borel whenever U is open.

If K is compact and $U_1, \dots, U_n$ are open, then

$$\sigma^{-1}\big(\mathcal{U}(K; \{\, U_1, \dots, U_n \,\})\big) = W(K)^c \cap \bigcap_{i=1}^{n} W(U_i),$$

which is certainly Borel. □

Although Proposition H.23 on the previous page (in the form of Proposition H.41) is sufficient for our needs in the study of the Effros-Hahn conjecture, it is interesting to consider just how well or badly behaved the stabilizer map can be. Even in Example H.21 on the preceding page, the stabilizer map is still continuous on a large subset. We include a proof which shows that the stabilizer map is always continuous on a dense set. Note that this is a much stronger statement than is provided by Lusin's Theorem (Theorem 9.25 on page 298) which provides a large subset on which *the restriction* of σ is continuous.

Theorem H.24. *Suppose that (G, X) is a Polish transformation group with G second countable and locally compact. Then there is a dense G_δ subset of X on which the stabilizer map σ is continuous.*

Our proof follows [4, Chap. II Propositions 2.1 and 2.3]. To obtain points of continuity, we rely on the fact that lower semicontinuous functions on Baire spaces must be continuous on large subsets. Recall that $f : X \to \mathbf{R}$ is *lower semicontinuous* if $f^{-1}\big((a, \infty)\big)$ is open for all $a \in \mathbf{R}$. We say that a topological space X is a Baire space if the countable intersection of open dense sets is dense. Of course all Polish spaces are Baire spaces, but even non-Hausdorff spaces, such as the primitive ideal space of a C^*-algebra, can be Baire spaces [139, Corollary A.47 and §A.48].

Proposition H.25. *Suppose that X is a Baire space, and that $f : X \to \mathbf{R}$ is lower semicontinuous. Then the set C of x such that f is continuous at x is a dense G_δ subset of X.*

If X is a metric space and if f is a nonnegative lower semicontinuous function, then f in the increasing pointwise limit of a a sequence $\{\, g_n \,\}$ of continuous functions (see, for example, [168, 7K(4)] or [156, p. 60 #22]). Then [109, Theorem 48.5] and its proof imply that the set of points were f is continuous is a dense G_δ subset of X. In the general case, we'll make use of the notion of the *oscillation* of f at a point. If $f : X \to \mathbf{R}$ is any bounded function, let

$$\omega(f, U) := \sup\{\, |f(x) - f(y)| : x, y \in U \,\}, \quad \text{and}$$
$$\omega(f, x) = \inf\{\, \omega(f, U) : U \text{ is a neighborhood of } x \,\}.$$

Clearly, if f is continuous at x, then $\omega(f,x)=0$. On the other hand, if f is not continuous at x, then there is a net $x_i \to x$ such that $|f(x_i)-f(x)|\geq \epsilon>0$ for some ϵ. Then $\omega(f,U)\geq\epsilon$ for all neighborhoods U of x. Thus $\omega(f,x)\geq\epsilon$. That is, f is continuous at x if and only if $\omega(f,x)=0$.

Proof of Proposition H.25. Since $x\mapsto x/(1+|x|)$ is increasing,

$$g(x):=\frac{f(x)}{1+|f(x)|}$$

is lower semicontinuous and it is continuous at exactly those points where f is. Therefore, we can replace f by g and assume that f is bounded. Suppose that $x_i\to x$ and that for each i, $\omega(f,x_i)\geq\epsilon$. If U is any neighborhood of x, then U is also a neighborhood of some x_i, and $\omega(f,U)\geq\epsilon$. Therefore $\omega(f,x)\geq\epsilon$, and $\omega(f,\cdot)$ is upper semicontinuous. In particular,

$$E_n:=\{\,x:\omega(f,x)\geq\frac{1}{n}\,\}$$

is closed. Suppose that E_n were to contain a nonempty open set W. Let $\alpha:=\sup\{\,f(x):x\in W\,\}$. Since f is lower semicontinuous,

$$U:=\{\,x:f(x)>\alpha-\frac{1}{2n}\,\}\cap W$$

is a neighborhood of some $x_0\in W$, and $\omega(f,x_0)\leq\omega(f,U)\leq\frac{1}{2n}<\frac{1}{n}$. This contradicts $x_0\in E_n$. Thus $G_n:=X\smallsetminus E_n$ is open and dense. But

$$\begin{aligned}C&:=\{\,x\in X:\omega(f,x)=0\,\}\\&=X\smallsetminus\bigcup_{n=1}^{\infty}E_n\\&=\bigcap_{n=1}^{\infty}G_n\end{aligned}$$

is a G_δ subset which is dense since X is assumed to be a Baire space. □

Since G is second countable, it has a countable neighborhood basis at e. Consequently G is metrizable and there is a compatible metric d on G which is left-invariant; that is, $d(rs,rt)=d(s,t)$ for all $r,s,t\in G$ ([71, Theorem II.8.4]). Let

$$B_\epsilon(s):=\{\,r\in G:d(r,s)<\epsilon\,\}$$

be the ϵ-ball centered at s in G. Define $b\in(0,\infty]$ by

$$b:=\sup\{\,\epsilon:\overline{B_\epsilon(e)}\text{ is compact}\,\},$$

and let $a=\min\{\,\frac{b}{2},1\,\}$. Since d is left-invariant, $B_\epsilon(s)$ is pre-compact for all $0<\epsilon\leq a$.

For each $s\in G$, define $d_s:\Sigma\to[0,a]$ by

$$d_s(H):=\min\{\,a,d(s,H)\,\}\quad\text{where}\quad d(s,H)=\inf\{\,d(s,t):t\in H\,\}.$$

Lemma H.26. *For each $s \in G$, d_s is continuous from Σ to $[0,\infty)$.*

Proof. Fix $\epsilon > 0$, $s \in G$ and $H \in \Sigma$. Choose $t \in H$ such that

$$d(s,t) \leq d(s,H) + \frac{\epsilon}{2},$$

and let $V = B_{\frac{\epsilon}{2}}(t)$.

If $s \in H$, then $d_s(H) = 0$ and $U = \mathcal{U}(\emptyset; \{V\})$ is a neighborhood of H such that $H' \in U$ implies that $d(s,H') < \epsilon$. Thus d_s is continuous at H in the event $s \in H$.

If $s \notin H$ and if $d(s,H) > a$, then $K = \overline{B_a(s)}$ is compact and $U = \mathcal{U}(K;\emptyset)$ is a neighborhood of H such that $H' \in U$ implies $d_s(H') = a = d_s(H)$. Thus d_s is continuous at H in this case.

If $s \notin H$ and if $0 < d(s,H) \leq a$, then there is no harm in assuming that $\epsilon < d(s,H)$. Pick δ such that $d(s,H) - \epsilon \leq \delta < a$. Then $K = \overline{B_\delta(s)}$ is compact and $H' \in U := \mathcal{U}(K; \{V\})$ must satisfy

$$d(s,H) - \epsilon \leq d(s,H') \leq d(s,H) + \epsilon.$$

Thus d_s is continuous at H in all cases. □

Let $D = \{ s_i \}_{i=1}^{\infty}$ be a countable dense subset of G.

Lemma H.27. *The functions $\{ d_{s_i} \}_{i=1}^{\infty}$ separate points in Σ. If Z is any topological space and if $f : Z \to \Sigma$, then f is continuous at $z \in Z$ if and only if $d_{s_i} \circ f$ is continuous at z for all i.*

Proof. Suppose that H and H' are distinct elements in Σ. Without loss of generality, we can assume that $H \setminus H'$ is nonempty and choose $s \in H \setminus H'$. Since D is dense, there is a sequence $s_{i_k} \to s$. Then $d_{s_{i_k}}(H) \to 0$ while $d_{s_{i_k}}(H')$ is eventually bounded away from 0 by $\min\{ d(s,H')/2, a \}$. This proves the first assertion.

It follows that

$$H \mapsto \big(d_{s_i}(H)\big) \tag{H.10}$$

is a continuous injection of Σ into $\prod_{i=1}^{\infty}[0,a]$. Since Σ is compact, (H.10) is a homeomorphism onto its range. The remaining assertion follows easily from this. □

Lemma H.28. *Let (G,X) be as in Theorem H.24, and let σ be the stabilizer map. For each i, $d_{s_i} \circ \sigma$ is lower semicontinuous.*

Proof. It suffices to show that for each $x \in X$,

$$c := \liminf_{y \to x} d_{s_i}\big(\sigma(y)\big) \geq d_{s_i}\big(\sigma(x)\big).$$

If $c = a$, then there is nothing to show. Assume that $c < a$, and let $y_n \to x$ be such that

$$c = \liminf_n d_{s_i}\big(\sigma(y_n)\big).$$

Passing to a subsequence and relabeling, we can assume that $d_{s_i}\big(\sigma(y_n)\big) < a$ for all n. Therefore $d(s_i, G_{y_n}) < a$, and after passing to another subsequence, there is a $t_n \in G_{y_n}$ such that $d(s_i, t_n) < a$ and such that $d(s_i, t_n) \to c$. Since $\{ t_n \}$ belong to the pre-compact set $B_a(s_i)$, we can pass to a subsequence, relabel, and assume that $t_n \to t$. Since each $t_n \in G_{y_n}$ and since $t_n \cdot y_n \to t \cdot x$, we must have $t \cdot x = x$ (since X is Hausdorff). That is, $t \in G_x$ and

$$d_{s_i}\big(\sigma(x)\big) \leq d(s_i, G_x) \leq d(s_i, t) = c.$$

Thus each d_{s_i} is lower semicontinuous as claimed. □

Proof of Theorem H.24 on page 470. Proposition H.25 on page 470 and Lemma H.28 imply that each $d_{s_i} \circ \sigma$ is continuous on a dense G_δ subset A_i of X. By Lemma H.27 on the facing page, σ is continuous on the intersection $A := \bigcap A_i$. However, A is a G_δ subset and it is dense since X is a Baire space. □

Remark H.29. The proof of Lemma H.27 on the preceding page shows that a map $f : Z \to \Sigma$ is Borel if and only if $f \circ d_{s_i}$ is Borel for all i. Consequently, Lemma H.28 on the facing page gives a proof that σ is Borel (with the additional assumption that X is a Polish space).

H.5 Borel Issues

Let $\{ \mu_H \}_{H \in \Sigma}$ be a continuous choice of Haar measures on Σ. Let $K \subset G$ be compact and let $\mathscr{C}_K$ be the set of bounded Borel functions F on $K * \Sigma = \{ (s, H) \in G * \Sigma : s \in K \}$ such that

$$\varphi(H) := \int_H F(s, H)\, d\mu_H(s) \tag{H.11}$$

defines a Borel function on Σ. Then $\mathscr{C}_K$ contains $C(K * \Sigma)$ and is closed under monotone sequential limits. Therefore [127, Proposition 6.2.9] implies that $\mathscr{C}_K$ contains all bounded Borel functions. More generally, we have the following.

Lemma H.30. *Suppose that X is a Polish space and that F is a bounded Borel function on $X \times G * \Sigma$ with compact support in the second variable; that is, there is a compact set $K \subset G$ such that $F(x, s, H) = 0$ if $s \notin K$. Then*

$$\varphi(x, H) := \int_H F(x, s, H)\, d\mu_H(s)$$

defines a Borel function on $X \times \Sigma$.

Remark H.31. Since X is a standard Borel space, we could evoke Kuratowski's Theorem I.40 on page 503 and replace X by a second countable locally compact space such as $[0, 1]$. Then we could simply repeat the argument above invoking [127, Proposition 6.2.9]. Instead we'll give a direct proof. Note that we only use the fact that the Borel structure on X comes from a second countable topology.

Proof of Lemma H.30 on the preceding page. It will suffice to consider nonnegative F and to replace $G * \Sigma$ with $K * \Sigma$ as above. Since there are Borel simple functions $F_i \nearrow F$, it suffices to take F of the form $\mathbb{1}_B$ for some Borel set $B \subset X \times K * \Sigma$. Since X and $K * \Sigma$ are second countable, the Borel sets of $X \times G * \Sigma$ are generated by Borel rectangles $B_1 \times B_2$ with $B_1 \subset X$ and $B_2 \subset K * \Sigma$ each Borel (see Lemma 4.44 on page 141). Since the collection of Borel sets B such that

$$\varphi_B(x, H) := \int_H \mathbb{1}_B(x, s, H)\, d\mu_H(s)$$

defines a Borel function is a monotone class of sets containing the algebra of Borel rectangles, it contains all Borel sets by the Monotone Class Lemma ([57, Lemma 2.35]). □

Lemma H.32. *Let A be a* separable *C^*-algebra. Then every weakly Borel function $f : X \to A$ is Borel. That is, if $\varphi \circ f : X \to \mathbf{C}$ is Borel for all $\varphi \in A^*$, then f is Borel.*

Remark H.33. Since each $\varphi \in A^*$ is continuous, it is obvious that Borel functions are weakly Borel. The proof of the converse is essentially that of Lemma B.7 on page 333.

Proof. Let V be open in A. It suffices to see that $f^{-1}(V)$ is Borel. Since A is separable, there are closed balls

$$A_n := \{\, a \in A : \|a - a_n\| \le \epsilon_n \,\}$$

such that

$$V = \bigcup_{n=1}^{\infty} A_n.$$

Thus it will suffice to see the $f^{-1}(A_n)$ is Borel. But Lemma B.8 on page 333 implies that there is a countable set $\{\, \varphi_m \,\}$ in A^* such that

$$\|a - a_n\| \le \epsilon_n \quad \text{if and only if} \quad |\varphi_m(a - a_n)| \le \epsilon_n \quad \text{for all } m.$$

Thus

$$f^{-1}(A_n) = \bigcap_m \{\, x \in X : |\varphi_m\big(f(x) - a_n\big)| \le \epsilon_n \,\}.$$

This suffices since the sets on the right-hand side are Borel by assumption. □

H.6 Prim A Standard

Here we want to take a close look at the space Prim A with its usual hull-kernel topology. We want to see that Prim A is a standard Borel space. This result goes back at least to Effros's paper [45]. The treatment here uses the Fell topology on the closed subsets of Prim A, and is due to Dixmier [27].

Recall that an nonempty closed subset of a topological space is called *irreducible* if it cannot be written as a nontrivial union of two closed sets. Thus F is irreducible if and only if whenever C_1 and C_2 are closed sets such that $F = C_1 \cup C_2$, then either $C_1 = F$ or $C_2 = F$. In any topological space X, the closure of a point is irreducible. Here we have in mind the space $X = \operatorname{Prim} A$, where A is a *separable* C^*-algebra. Then points in X need not be closed. But X is at least second countable,[7] and since every closed subset of the primitive ideal space is the primitive ideal space of a quotient of A, every closed subset has the Baire property—that is, the countable intersection of open dense sets is again dense [139, Corollary A.47].

Lemma H.34. *Suppose that X is a second countable (not necessarily Hausdorff) locally compact space in which every nonempty closed subset has the Baire property. Then every irreducible subset is a point closure.*

Proof. (The proof is essentially that of Proposition 6.21 on page 186.) Suppose that F is irreducible. Then every nonempty open subset of F is dense in F. Since the relative topology on F is second countable, the intersection of all nonempty open subsets of F is dense. Any point in the intersection of every open set is necessarily dense in F. □

Lemma H.35. *Let $\mathscr{C}(X)$ be the closed subsets of a (not necessarily Hausdorff) locally compact space X. Define $\Theta : \mathscr{C}(X) \times \mathscr{C}(X) \to \mathscr{C}(X)$ by $\Theta(E, F) = E \cup F$. Then Θ is continuous when $\mathscr{C}(X)$ has the Fell topology.*

Remark H.36. Easy examples show the map taking (E, F) to $E \cap F$ is *not* continuous in general.

Proof. This is a straightforward consequence of Lemma H.2 on page 454. Suppose that $E_i \to E$ and $F_i \to F$ in $\mathscr{C}(X)$. If $t \in E \cup F$, then we may as well assume that $t \in E$. Then passing to a subnet and relabeling, we can assume that there are $t_i \in E_i \subset E_i \cup F_i$ such that $t_i \to t$. Similarly, if $t_i \in E_i \cup F_i$ and if $t_i \to t$, then we can pass to a subnet, relabel, and assume that t_i is either always in E_i or always in F_i. In either case, we must have $t \in E \cup F$. Thus $\Theta(E_i, F_i) = E_i \cup F_i \to E \cup F = \Theta(E, F)$ as required. □

Proposition H.37. *Suppose that X is a second countable (not necessarily Hausdorff) locally compact space. Then*

$$\mathscr{I} := \{ F \in \mathscr{C}(X) : F \textit{ is irreducible} \}$$

is a G_δ subset of $\mathscr{C}(X)$. In particular, $\mathscr{I}$ is a Polish space in the relative topology.

Proof. Let

$$\mathscr{E} := \{ (E_1, E_2) \in \mathscr{C}(X) \times \mathscr{C}(X) : E_1 \subset E_2 \} \text{ and}$$
$$\mathscr{F} := \{ (E_1, E_2) \in \mathscr{C}(X) \times \mathscr{C}(X) : E_2 \subset E_1 \}.$$

[7]This follows from [139, Theorem A.38] and [139, Proposition A.46]

Using Lemma H.2 on page 454, it is not hard to see that $\mathscr{E}$ and $\mathscr{F}$ are closed. Furthermore, a closed set F fails to be irreducible if and only if there are closed sets E_1 and E_2 such that $F = E_1 \cup E_2$ while $E_1 \not\subset E_2$ and $E_2 \not\subset E_1$. Thus $\mathscr{I}$ is the complement of the image under the continuous map Θ of the complement of $\mathscr{E} \cup \mathscr{F}$. Since open subsets of second countable metric spaces are the countable union of closed sets, and since $\mathscr{C}(X) \times \mathscr{C}(X)$ is compact, there are compact sets $K_i \subset \mathscr{C}(X) \times \mathscr{C}(X)$ such that

$$\mathscr{C}(X) \times \mathscr{C}(X) \setminus (\mathscr{E} \cup \mathscr{F}) = \bigcup K_i.$$

Thus $\mathscr{I}$ is the complement of

$$\bigcup \Theta(K_i).$$

Since Θ is continuous, $\Theta(K_i)$ is compact, and since $\mathscr{C}(X)$ is Hausdorff, closed. Thus $\mathscr{I}$ is the countable intersection of open sets; that is, $\mathscr{I}$ is a G_δ subset of $\mathscr{C}(X)$. Since $\mathscr{C}(X)$ is Polish (Theorem H.5 on page 455), $\mathscr{I}$ is Polish since all G_δ subsets of a Polish space are Polish (cf, [2, Theorem 3.1.2; 139, Lemma A.44; 168, Theorem 24.12]). □

Now we specialize to the case where $X = \operatorname{Prim} A$ for a separable C^*-algebra A. Then $\operatorname{Prim} A$ is a second countable locally compact space which certainly need not be Hausdorff in general. Define

$$\lambda : \operatorname{Prim} A \to \mathscr{C}(\operatorname{Prim} A)$$

by $\lambda(P) = \overline{\{ P \}} = \{ J \in \operatorname{Prim} A : J \supset P \}$. Since $\operatorname{Prim} A$ is a T_0-topological space, λ is injective. If V is open in $\operatorname{Prim} A$, then

$$\lambda(V) = \mathcal{U}(\emptyset; \{ V \}) \cap \lambda(\operatorname{Prim} A).$$

Consequently, λ is an open bijection of $\operatorname{Prim} A$ onto $\lambda(\operatorname{Prim} A)$. The following terminology is from [64].

Definition H.38. Suppose that A is a separable C^*-algebra. The *regularized topology* on $\operatorname{Prim} A$ is that obtained by identifying $\operatorname{Prim} A$ with its image in $\mathscr{C}(\operatorname{Prim} A)$ via the map λ.

Theorem H.39. *Suppose that A is a separable C^*-algebra. The regularized topology on* $\operatorname{Prim} A$ *is finer than the hull-kernel topology and generates the same Borel structure as the hull-kernel topology. With the regularized topology,* $\operatorname{Prim} A$ *is a Polish space and* $(G, \operatorname{Prim} A)$ *is a topological transformation group.*

Proof. Since λ is an open map, it is immediate that the regularized topology is finer than the hull-kernel topology. Since closed subsets of $\operatorname{Prim} A$ have the Baire property, the irreducible subsets $\mathscr{I}$ of $\mathscr{C}(\operatorname{Prim} A)$ are exactly the point closures (Lemma H.34 on the preceding page). Thus $\lambda(\operatorname{Prim} A) = \mathscr{I}$ is a G_δ subset of $\mathscr{C}(\operatorname{Prim} A)$ (Proposition H.37 on the previous page), and therefore Polish.

On the other hand, since the hull-kernel topology on $\operatorname{Prim} A$ is T_0 and second countable, $\operatorname{Prim} A$ is a countably separated Borel space with the Borel structure generated by the hull-kernel topology. Since λ is open, $\lambda^{-1} : \lambda(\operatorname{Prim} A) \to \operatorname{Prim} A$ is Borel map from the standard Borel space $\lambda(\operatorname{Prim} A)$ onto the countably separated Borel space $\operatorname{Prim} A$. Therefore λ^{-1} is a Borel isomorphism (Corollary 1 of [2, Theorem 3.3.4] and Corollary 1 of [2, Theorem 3.3.5]).

It only remains to show that the map $(s, \overline{\{P\}}) \mapsto \overline{\{s \cdot P\}}$ is continuous from $G \times \lambda(\operatorname{Prim} A) \to \lambda(\operatorname{Prim} A)$. (The corresponding map from $G \times \operatorname{Prim} A \to \operatorname{Prim} A$ is continuous in the hull-kernel topology by [139, Lemma 7.1].) Suppose that $s_i \to s$ and $\overline{\{P_i\}} \to \overline{\{P\}}$. Note that

$$\overline{\{s \cdot P\}} = s \cdot \overline{\{P\}} = \{\, s \cdot J : J \supset P \,\}.$$

Note that if $Q_i \in \overline{\{s_i \cdot P_i\}}$ and if $Q_i \to Q$, then $Q_i = s_i \cdot J_i$ with $J_i \in \overline{\{P_i\}}$. Furthermore, $J_i \to s^{-1} \cdot Q$ in $\operatorname{Prim} A$. Lemma H.2 on page 454 implies that $s^{-1} \cdot Q \in \overline{\{P\}}$. That is, $Q \in \overline{\{s \cdot P\}}$. On the other hand, if $J \in \overline{\{s \cdot P\}}$, then $J = s \cdot Q$ for some $Q \in \overline{\{P\}}$. Using Lemma H.2, we can pass to a subnet, relabel, and assume that there are $J_i \in \overline{\{P_i\}}$ such that $J_i \to Q$. Thus $s_i \cdot J_i \to J$, and $s_i \cdot J_i \in \overline{\{s_i \cdot P_i\}}$. Thus $\overline{\{s_i \cdot P_i\}} \to \overline{\{s \cdot P\}}$ by Lemma H.2. □

Since Theorem H.39 on the preceding page implies that the Borel structure coming from the hull-kernel topology is the same as that coming from the stronger Polish topology, we get the following as an immediate corollary.

Theorem H.40. *If A is a separable C^*-algebra, then* $\operatorname{Prim} A$ *is a standard Borel space in the Borel structure coming from the hull-kernel topology.*

We also obtain a crucial corollary to Proposition H.23 on page 469.

Proposition H.41. *Suppose that (A, G, α) is a separable dynamical system. Then the stabilizer map $P \mapsto G_P$ is Borel from* $\operatorname{Prim} A$ *into the compact Hausdorff space Σ of closed subgroups of G.*

Proof. Using Theorem H.39 on the preceding page, we can replace the hull-kernel topology on $\operatorname{Prim} A$ with it regularized topology. Then the result follows from Proposition H.23 on page 469. □

Appendix I

Miscellany

I.1 The Internal Tensor Product

In [139], we only considered a special case of the internal tensor product which sufficed for the study of imprimitivity bimodules. Here we want to consider a more general situation. Let A and B be C^*-algebras and suppose that X is a right Hilbert A-module, Y is a right Hilbert B-module and that $\varphi : A \to \mathcal{L}(\mathsf{Y})$ is a homomorphism. We can view Y as a left A-module — $a \cdot y := \varphi(a)y$ — and form the A-balanced module tensor product $\mathsf{X} \odot_A \mathsf{Y}$; recall that this is simply the quotient of the vector space tensor product $\mathsf{X} \odot \mathsf{Y}$ by the subspace N generated by

$$\{\, x \cdot a \otimes y - x \otimes \varphi(a)y : x \in \mathsf{X},\ y \in \mathsf{Y}, \text{ and } a \in A \,\}.$$

The B-module structure is given by $(x \otimes_A y) \cdot b := x \otimes_A y \cdot b$. Our object here is to equip $\mathsf{X} \odot_A \mathsf{Y}$ with a B-valued inner product and use [139, Lemma 2.16] to pass to the completion $\mathsf{X} \otimes_\varphi \mathsf{Y}$ which is a Hilbert B-module called the internal tensor product. I'll give a minimal treatment here just sufficient for our purposes. A more complete treatment can be found in Lance [98]. Here we take a slight short-cut, and merely equip $\mathsf{X} \odot \mathsf{Y}$ with a B-valued *pre*-inner product. As it turns out, elements in N all have 0-length, and we can, and do, view $\mathsf{X} \otimes_\varphi \mathsf{Y}$ as a completion of $\mathsf{X} \odot_A \mathsf{Y}$.[1]

Proposition I.1. *Let* X *be a Hilbert* A*-module and* Y *and Hilbert* B*-module with a homomorphism* $\varphi : A \to \mathcal{L}(\mathsf{Y})$*. Then there is a unique* B*-valued pre-inner product on* $\mathsf{X} \odot \mathsf{Y}$ *such that*

$$\langle\!\langle x \otimes y \,,\, z \otimes w \rangle\!\rangle_B := \big\langle y \,,\, \varphi\big(\langle x \,,\, z\rangle_A\big) w \big\rangle_B . \tag{I.1}$$

The completion $\mathsf{X} \otimes_\varphi \mathsf{Y}$ *is a Hilbert* B*-module which satisfies*

$$x \cdot a \otimes_\varphi y = x \otimes_\varphi \varphi(a)y$$

for all $x \in \mathsf{X}$, $y \in \mathsf{Y}$, *and* $a \in A$.

[1] Lance goes the extra mile and proves that the pre-inner product is actually an inner product (that is, definite) on $\mathsf{X} \odot_A \mathsf{Y}$.

Remark I.2. When the map $\varphi : A \to \mathcal{L}(\mathsf{Y})$ is understood, it is common to use the notation $\mathsf{X} \otimes_A \mathsf{Y}$ in place of $\mathsf{X} \otimes_\varphi \mathsf{Y}$.

Proof. As in Propositions 2.64, 3.16, and 3.36 of [139], the universal property of the algebraic tensor product $\odot$ implies that (I.1) determines a unique sesquilinear form on $\mathsf{X} \odot \mathsf{Y}$. The only real issue is to prove that this form is positive.

So let $t := \sum x_i \otimes y_i$. Then

$$\langle\langle t , t \rangle\rangle_B = \sum_{i,j} \langle y_i , \varphi(\langle x_i , x_j \rangle_A) y_j \rangle_B. \tag{I.2}$$

But [139, Lemma 2.65] implies that $M := \big(\langle x_i , x_j \rangle_A\big)$ is a positive matrix in $M_n(A)$. Thus there is a matrix D such that $M = D^* D$, and there are $d_{kl} \in A$ such that

$$\langle x_i , x_j \rangle_A = \sum_k d_{ki}^* d_{kj}.$$

Thus (I.2) equals

$$\begin{aligned} \sum_{i,j,k} \langle y_i , \varphi(d_{ki}^* d_{kj}) y_j \rangle_B &= \sum_{i,j,k} \langle \varphi(d_{ki}) y_i , \varphi(d_{kj}) y_j \rangle_B \\ &= \sum_k \Big\langle \Big(\sum_i \varphi(d_{ki}) y_i \Big) , \Big(\sum_i \varphi(d_{ki}) y_i \Big) \Big\rangle_B \\ &\geq 0 \end{aligned}$$

□

Now we want to see that each $T \in \mathcal{L}(\mathsf{X})$ determines an operator $T \otimes_\varphi 1$ on the internal tensor product $\mathsf{X} \otimes_\varphi \mathsf{Y}$.

Lemma I.3. *Let A, B, X, Y, and φ be as above, and let Z be a Hilbert A-module. If $T \in \mathcal{L}(\mathsf{X}, \mathsf{Z})$, then there is unique operator $T \otimes_\varphi 1 \in \mathcal{L}(\mathsf{X} \otimes_\varphi \mathsf{Y}, \mathsf{Z} \otimes_\varphi \mathsf{Y})$ such that*

$$T \otimes_\varphi 1(x \otimes_\varphi y) = Tx \otimes_\varphi y.$$

Furthermore, if $\mathsf{Z} = \mathsf{X}$, then $T \mapsto T \otimes_\varphi 1$ is a homomorphism φ_* of $\mathcal{L}(\mathsf{X})$ into $\mathcal{L}(\mathsf{X} \otimes_\varphi \mathsf{Y})$.

Proof. We clearly have a well-defined operator $T \otimes 1$ on $\mathsf{X} \odot \mathsf{Y}$, and if $t \in \mathsf{X} \odot \mathsf{Y}$ and $s \in \mathsf{Z} \odot \mathsf{Y}$, then straightforward calculations reveal that

$$\langle\langle (T \otimes 1)t , s \rangle\rangle_B = \langle\langle t , (T^* \otimes 1)s \rangle\rangle_B.$$

Similarly, if $S \in \mathcal{L}(\mathsf{Z}, \mathsf{X})$, then on the algebraic tensor product, $(T \otimes 1) \circ (S \otimes 1) = (TS \otimes 1)$. Since there is a $R \in \mathcal{L}(X)$ such that $\|T\|^2 1_\mathsf{X} - T^*T = R^*R$, for all $t \in \mathsf{X} \odot \mathsf{Y}$,

$$\begin{aligned} \|T\|^2 \|t\|^2_{\mathsf{X} \otimes_\varphi \mathsf{Y}} - \|(T \otimes 1)t\|^2_{\mathsf{X} \otimes_\varphi \mathsf{Y}} &= \langle\langle \|T\|^2 t , t \rangle\rangle_B - \langle\langle (T^*T \otimes 1)t , t \rangle\rangle_B \\ &= \langle\langle ((\|T\|^2 1_\mathsf{X} - T^*T) \otimes 1)t , t \rangle\rangle_B \\ &= \|(R \otimes 1)t\|^2_{\mathsf{X} \otimes_\varphi \mathsf{Y}} \geq 0. \end{aligned}$$

It follows that $T \otimes 1$ is bounded, and extends to an operator $T \otimes_\varphi 1 \in \mathcal{L}(\mathsf{X} \otimes_\varphi \mathsf{Y}, \mathsf{Z} \otimes_\varphi \mathsf{Y})$ such that $(T \otimes_\varphi 1)^* = T^* \otimes_\varphi 1$ and $\|T \otimes_\varphi 1\| \leq \|T\|$. The last assertion is straightforward. □

Let X_A, Y_B and $\varphi : A \to \mathcal{L}(\mathsf{Y})$ be as above. Suppose also that Z is a right Hilbert C-module and that $\psi : B \to \mathcal{L}(\mathsf{Z})$. Then we can form the Hilbert C-modules $\mathsf{X} \otimes_{\psi_* \circ \varphi} (\mathsf{Y} \otimes_\psi \mathsf{Z})$ and $(\mathsf{X} \otimes_\varphi \mathsf{Y}) \otimes_\psi \mathsf{Z}$. We expect these modules to be naturally isomorphic and we want to give a quick sketch of why this is so. The universal properties of the algebraic tensor product give us natural maps

$$\begin{aligned} \sigma &: \mathsf{X} \odot (\mathsf{Y} \odot \mathsf{Z}) \to (\mathsf{X} \odot \mathsf{Y}) \odot \mathsf{Z} \\ \sigma_1 &: \mathsf{X} \odot (\mathsf{Y} \odot \mathsf{Z}) \to \mathsf{X} \odot (\mathsf{Y} \otimes_\psi \mathsf{Z}) \\ \sigma_2 &: (\mathsf{X} \odot \mathsf{Y}) \odot \mathsf{Z} \to (\mathsf{X} \otimes_\varphi \mathsf{Y}) \odot \mathsf{Z}. \end{aligned}$$

As in the proof of Proposition I.1, we can equip the algebraic tensor product $(\mathsf{X} \odot \mathsf{Y}) \odot \mathsf{Z}$ with a C-valued pre-inner product such that

$$\begin{aligned} \big\langle\!\big\langle (x_1 \otimes y_1) \otimes z_1 \,,\, (x_2 \otimes y_2) \otimes z_2 \big\rangle\!\big\rangle_C &= \big\langle z_1 \,,\, \psi\big(\langle\!\langle x_1 \otimes y_1 \,,\, x_2 \otimes y_2 \rangle\!\rangle_B\big)(z_2)\big\rangle_C \\ &= \big\langle z_1 \,,\, \psi\big(\big\langle y_1 \,,\, \varphi\big(\langle x_1 \,,\, x_2\rangle_A\big)(y_2)\big\rangle_B\big)(z_2)\big\rangle_C \end{aligned}$$

It follows that σ_2 extends to an isomorphism of the completion of $(\mathsf{X} \odot \mathsf{Y}) \odot \mathsf{Z}$ with $(\mathsf{X} \otimes_\varphi \mathsf{Y}) \otimes_\psi \mathsf{Z}$. Similarly, σ_1 will extend to an isomorphism of $\mathsf{X} \otimes_{\varphi \otimes 1} (\mathsf{Y} \otimes_\psi \mathsf{Z})$ with the completion of $\mathsf{X} \odot (\mathsf{Y} \odot \mathsf{Z})$ with respect to the C-valued pre-inner product

$$\begin{aligned} \big\langle\!\big\langle x_1 \otimes (y_1 \otimes z_1) \,,\, x_2 \otimes (y_2 \otimes z_2) \big\rangle\!\big\rangle_C &= \big\langle\!\big\langle y_1 \otimes z_1 \,,\, \psi_*\big(\varphi\big(\langle x_1 \,,\, x_2\rangle_A\big)\big)(y_2 \otimes z_2) \big\rangle\!\big\rangle_C \\ &= \big\langle\!\big\langle y_1 \otimes z_1 \,,\, \varphi\big(\langle x_1 \,,\, x_2\rangle_A\big)(y_2) \otimes z_2 \big\rangle\!\big\rangle_C \\ &= \big\langle z_1 \,,\, \psi\big(\big\langle y_1 \,,\, \varphi\big(\langle x_1 \,,\, x_2\rangle_A\big)(y_2)\big\rangle_B\big)(z_2)\big\rangle_C \end{aligned}$$

Comparing these two pre-inner products, it is now clear that σ extends to the required isomorphism.

We summarize the above discussion in the following lemma.

Lemma I.4. *Suppose that* X *is a right Hilbert* A*-module,* Y *is a right Hilbert* B*-module and that* Z *is a right Hilbert* C*-module. Let* $\varphi : A \to \mathcal{L}(\mathsf{Y})$ *and* $\psi : B \to \mathcal{L}(\mathsf{Z})$ *be homomorphisms. Then the right Hilbert* C*-modules* $\mathsf{X} \otimes_{\psi_* \circ \varphi} (\mathsf{Y} \otimes_\psi \mathsf{Z})$ *and* $(\mathsf{X} \otimes_\varphi \mathsf{Y}) \otimes_\psi \mathsf{Z}$ *can be viewed as completions of the algebraic tensor products* $\mathsf{X} \odot (\mathsf{Y} \odot \mathsf{Z})$ *and* $(\mathsf{X} \odot \mathsf{Y}) \odot \mathsf{Z}$*, respectively. Furthermore, the natural algebraic isomorphism* $\sigma : \mathsf{X} \odot (\mathsf{Y} \odot \mathsf{Z}) \to (\mathsf{X} \odot \mathsf{Y}) \odot \mathsf{Z}$ *extends to an isomorphism of* $\mathsf{X} \otimes_{\psi_* \circ \varphi} (\mathsf{Y} \otimes_\psi \mathsf{Z})$ *onto* $(\mathsf{X} \otimes_\varphi \mathsf{Y}) \otimes_\psi \mathsf{Z}$.

I.2 Disintegration of Measures

More general formulations and references can be found in [47, Lemma 4.4] and [1]. We will settle for a special case suitable to our needs in the proof of the Gootman-Rosenberg-Sauvageot result in Chapter 9.

Suppose that X and T are Borel spaces, that μ is a measure on X and that $q : X \to T$ is Borel.[2] Then the *image or push-forward of μ under q* is the measure $\nu := q_*(\mu)$ on T given by $\nu(E) := \mu\big(q^{-1}(E)\big)$. Thus for any nonnegative (complex-valued) Borel function f on T we have

$$\int_T f(t)\,d\nu(t) = \int_X f\big(q(x)\big)\,d\mu(x). \tag{I.3}$$

If μ, and hence ν, is finite, then (I.3) holds for all bounded Borel functions on T or even all Borel functions in $\mathcal{L}^1(\nu)$ (see Lemma H.13 on page 463).[3]

Theorem I.5 (Disintegration of Measures). *Suppose that μ is a finite Borel measure on the second countable locally compact space X, that $(T, \mathscr{M})$ is a countably separated Borel space and that $q : X \to T$ is a Borel surjection. Let $\nu := q_*\mu$ be the image of μ by q. Then there is a family of (positive) Borel measures $\{\,\mu_t\,\}_{t\in T}$ on X and a ν-null set N such that*

(a) *For all $t \notin N$, μ_t is a probability measure with $\operatorname{supp}\mu_t \subset q^{-1}(t)$,*[4]

(b) *$\mu_t = 0$ if $t \in N$,*

(c) *For all $h \in \mathcal{B}^b(X)$,*

$$t \mapsto \int_X h(x)\,d\mu_t(x)$$

is bounded and Borel and

(d)

$$\int_X h(x)\,d\mu(x) = \int_T \int_X h(x)\,d\mu_t(x)\,d\nu(t).$$

If $\{\,\mu'_t\,\}_{t\in T}$ is another family of finite measures satisfying $\operatorname{supp}\mu'_t \subset q^{-1}(t)$ for ν almost all t as well as (c) *and* (d), *then $\mu_t = \mu'_t$ for ν-almost every t.*

Remark I.6. Theorem I.5 is illustrated by the special case where X is the product $A \times B$ and q is the projection on the first factor. Then the measures μ_a can be viewed as measures on B, and for all $h \in \mathcal{B}^b(A \times B)$ we have

$$\int_{A\times B} h(a,b)\,d\mu(a,b) = \int_A \int_B h(a,b)\,d\mu_a(b)\,d\nu(a).$$

---

[2]We are using terminology consistent with Appendix D.2. Thus a Borel space is a pair $(T, \mathscr{M})$ where $\mathscr{M}$ is a σ-algebra of sets in T. A function $q : (X, \mathscr{B}) \to (T, \mathscr{M})$ is called Borel if $q^{-1}(E) \in \mathscr{B}$ for all $E \in \mathscr{M}$. In many texts, the pair $(T, \mathscr{M})$ is called simply a *measurable space*, and a function q as above is called a measurable function.

[3]When working with measures on the Borel sets of a topological space T, there is a distinction between real-valued Borel functions and ν-measurable functions. A ν-measurable function is measurable with respect to the completion of ν. That is f is ν-measurable if for every open set $V \subset R$ there are Borel sets B and C such that $B \subset f^{-1}(V) \subset C$ with $\nu(B \setminus C) = 0$. Since $q^{-1}(B)$ and $q^{-1}(C)$ are Borel sets in X which differ by a μ-null set, it follows that $f \circ q$ is μ-measurable if and only if f is ν-measurable. In particular, (I.3) holds for integrable ν-measurable f as well.

[4]The support of a measure is used informally here. It has a clear definition in a topological space as the complement of the union of open null sets. Here $\operatorname{supp}\mu \subset B$ just means $\mu(X \setminus B) = 0$.

This iterated integral should be compared with that resulting from Fubini's Theorem in the case where μ is the product of two probability measures ν on A and σ on B. Then $q_*\mu = \nu$ and each $\mu_a = \sigma$.

The proof of Theorem I.5 given here depends on a classical result of Dunford and Pettis. Suppose that $(T, \mathscr{M})$ is a Borel space and that B is a Banach space. The dual space of B is the space B^* of bounded linear functionals on B. The value of $\varphi \in B^*$ on $b \in B$ is often denoted by $\langle b\,,\varphi\rangle$. A map $\boldsymbol{F} : X \to B^*$ is called *weak-$*$ Borel* if $t \mapsto \big\langle b\,, \boldsymbol{F}(t)\big\rangle$ is Borel for all $b \in B$.[5]

Theorem I.7 (Dunford-Pettis). *Let $(T, \mathscr{M}, \nu)$ be a finite measure space and let B be a separable Banach space with dual B^*. If $\Phi : L^1(\nu) \to B^*$ is a bounded linear map, then there is a bounded weak-$*$ Borel function $\boldsymbol{F} : T \to B^*$ such that for all $b \in B$*

$$\big\langle b\,, \Phi(g)\big\rangle = \int_T g(t)\big\langle b\,, \boldsymbol{F}(t)\big\rangle\,d\nu(t).$$

The function $\boldsymbol{F}$ is determined up to a ν-null set.

The proof of this result basically requires that we make sense out of "evaluating an element of $L^\infty(\nu)$ at a point." As usual, if ν is a measure on $(T, \mathscr{M})$, then we'll write $\mathcal{B}^b(T)$ for the bounded Borel functions on T, and reserve $L^\infty(\nu)$ for the Banach space of equivalence classes of functions in $\mathcal{B}^b(T)$ which agree ν-almost everywhere and equipped with the essential supremum norm.[6] Usually we ignore the fact that elements of L^∞ are equivalence classes and treat them as functions. This usually causes no harm provided one realizes that certain operations — such as evaluation at a point — makes no sense on L^∞. Our next lemma says that provided we restrict attention to a *separable* subspace of L^∞, we can pick functions representing elements of this subspace in a linear way. (In particular, on such a subspace, point evaluations do define linear functionals.) Such a map is called a *linear lifting for L^∞* and there is a considerable industry in finding such maps — even multiplicative ones (cf., [24, Chap. II, §11]). The lemma here is based on [13, §2, No. 5, Lemme 2] and suffices for our needs.

Lemma I.8. *Suppose that ν is a finite measure on $(T, \mathscr{M})$ and that $\pi : \mathcal{B}^b(T) \to L^\infty(\nu)$ is the quotient map. If S is a* separable *subspace of $L^\infty(\nu)$, then there is a linear map*

$$\rho : S \to \mathcal{B}^b(T)$$

such that $\pi\big(\rho(f)\big) = f$ for all $f \in S$.

Proof. Let D be a countable dense subset in S. Let S_0 be the rational subspace generated by D. Since S_0 is countable, it must have a countable basis $\{\, h_n \,\}$ as a vector space over the rationals. Let $f_n \in \mathcal{B}^b(T)$ be any function such that

[5] Note that B^* may not be separable even though B is. In particular, a weak-$*$ Borel function on a locally compact space need not be measurable as defined in Appendix B.

[6] Alternatively, we could have worked with the set $\mathcal{L}^\infty(\nu)$ of $\mathscr{M}$-measurable functions on T which are ν-essentially bounded.

$\pi(f_n) = h_n$. Then there is a unique **Q**-linear map $\rho' : S_0 \to \mathcal{B}^b(T)$ such that $\rho'(h_n) = f_n$ for all n. By linearity,

$$\pi\big(\rho'(h)\big) = h \quad \text{for all } h \in S_0.$$

Therefore the essential supremum of $\rho'(h)$ equals $\|h\|_\infty$. In particular, there is a null set $N(h)$ such that

$$|\rho'(h)(t)| \le \|h\|_\infty \quad \text{for all } t \in T \setminus N(h).$$

Since S_0 is countable, there is a null set N such that

$$|\rho'(h)(t)| \le \|h\|_\infty \quad \text{for all } h \in S_0 \text{ and } t \notin N.$$

For each $h \in S_0$, define

$$\rho(h) := \begin{cases} \rho'(h)(t) & \text{if } t \notin N, \text{ and} \\ 0 & \text{otherwise.} \end{cases}$$

Then we still have $\pi \circ \rho(h) = h$ for all $h \in S_0$ and ρ is isometric from the subspace $S_0 \subset L^\infty(\nu)$ into $\mathcal{B}^b(T)$ equipped with the supremum norm. Since the latter is a complete, ρ has a unique extension to all of S. □

Proof of Theorem I.7. Since ν is finite, we can identify $L^1(\nu)^*$ with $L^\infty(\nu)$, and define $\varphi : B \to L^\infty(\nu)$ by

$$\big\langle g\,,\, \varphi(a)\big\rangle := \big\langle a\,,\, \Phi(g)\big\rangle \quad \text{for } g \in \mathcal{L}^1(\nu).$$

Notice that

$$\begin{aligned} \|\varphi(a)\| &= \sup_{\|g\|_1 = 1} \big\|\langle a\,,\, \Phi(g)\rangle\big\| \\ &\le \sup_{\|g\|_1=1} \|\Phi\|\|g\|_1\|a\| \\ &= \|\Phi\|\|a\|. \end{aligned}$$

Thus φ is continuous, and since B is separable, the image $\varphi(B)$ is a separable subspace of $L^\infty(\nu)$. By Lemma I.8, there is a linear cross section

$$\rho : \varphi(B) \to \mathcal{B}^b(T)$$

for the quotient map $\pi : \mathcal{B}^b(T) \to L^\infty(\nu)$. Then for each $t \in T$, we can define $\boldsymbol{F}(t) \in B^*$ by

$$\big\langle b\,,\, \boldsymbol{F}(t)\big\rangle = \rho\big(\varphi(b)\big)(t).$$

Since $\rho\big(\varphi(b)\big) \in \mathcal{B}^b(T)$, $t \mapsto \boldsymbol{F}(t)(b) = \big\langle b\,,\, F(t)\big\rangle$ is Borel and $\boldsymbol{F}$ is weak-$*$ Borel with $\|\boldsymbol{F}(t)\| \le \|\varphi\|$ for all t. Moreover for all $g \in \mathcal{L}^1(\nu)$ and all $b \in B$, we have

$$\begin{aligned} \big\langle b\,,\, \Phi(g)\big\rangle &= \varphi(b)(g) \\ &= \int_T \rho\big(\varphi(b)\big)(t) g(t)\, d\nu(t) \\ &= \int_T g(t)\big\langle b\,,\, \boldsymbol{F}(t)\big\rangle d\nu(t). \end{aligned} \tag{I.4}$$

To prove that $\boldsymbol{F}$ is determined up to null sets, it suffices to see that if Φ is the zero map, then $\boldsymbol{F}$ is zero ν-almost everywhere. But if Φ is the zero map, then given $b \in B$, (I.4) is zero for all $g \in \mathcal{L}^1(\nu)$. Therefore there is a ν-null set $N(b)$ such that $\big(b\,,\, \boldsymbol{F}(t)\big) = 0$ if $t \notin N(b)$. Since B is separable, there is a null set N such that $\big(b\,,\, \boldsymbol{F}(t)\big) = 0$ for all $b \in B$. This suffices. □

Proof of the Disintegration Theorem on page 482. Recall that the Banach space $M(X)$ of complex measures on X equipped with the total variation norm can be identified with the dual $C_0(X)^*$ of $C_0(X)$. If $g \in \mathcal{L}^1(\nu)$, then $g \circ q$ is in $\mathcal{L}^1(\mu)$, and $(g \circ q) \cdot \mu$ is a complex measure in $M(X)$ which depends only on the class of g in $L^1(\nu)$. Thus we get a linear map $\Phi : L^1(\nu) \to C_0(X)^*$ which is bounded since

$$\|(g \circ q) \cdot \mu\| = \|g \circ q\|_{L^1(\mu)} \|\mu\| = \|g\|_{L^1(\nu)} \| \|\mu\|.$$

The Dunford-Pettis Theorem on page 483 implies there is a bounded weak-$*$ Borel function $\boldsymbol{F} : X \to M(X) = C_0(X)^*$ such that for each $h \in C_0(X)$ we have

$$\int_X h(x) g\big(q(x)\big)\, d\mu(x) := \Phi(g)(h) = \int_T g(t) \big(h\,,\, \boldsymbol{F}(t) \big) d\nu(t). \tag{I.5}$$

Let $\boldsymbol{F}(t)$ be given by the complex measure μ_t. Since $\boldsymbol{F}$ is weak-$*$ Borel, for each $h \in C_0(X)$,

$$t \mapsto \int_X h(x)\, d\mu_t(x)$$

is a Borel function bounded by $\|h\|_\infty \|\boldsymbol{F}\|$, and the right-hand side of (I.5) is equal to

$$\int_T g(t) \int_X h(x)\, d\mu_t(x)\, d\nu(t). \tag{I.6}$$

Therefore we see that for all $g \in \mathcal{L}^1(\nu)$ and $h \in C_0(X)$, we have

$$\int_X h(x) g\big(q(x)\big)\, d\mu(x) = \int_T g(t) \int_X h(x)\, d\mu_t(x)\, d\nu(t). \tag{I.7}$$

If h and g are nonnegative, then the left-hand side of (I.7) is too. Thus for each nonnegative $h \in C_0(X)$, there is a ν-null set $N(h) \in \mathscr{M}$ such that

$$t \mapsto \int_X h(x)\, d\mu_t(x)$$

is nonnegative off $N(h)$. Since X is second countable, there is a countable dense subset $\{ h_n \}$ of the nonnegative functions in $C_0(X)$. Let

$$N := \bigcup N(h_n).$$

Then N is ν-null and for all $t \notin N$,

$$\int_X h(x)\, d\mu_t(x) \geq 0 \quad \text{for all nonnegative } h \in C_0(X).$$

It follows that μ_t is a positive measure for all $t \in T \setminus N$. Since N is a ν-null set in $\mathscr{M}$, we can replace μ_t by the zero measure for all $t \in N$ without changing the conclusions above. Thus we'll assume from now on that each μ_t is either the zero measure or a *finite* positive measure.

Let $\mathcal{A}$ be the collection of functions h in $\mathcal{B}^b(X)$ for which $t \mapsto \int_X h(x)\,d\mu_t(x)$ is Borel and (I.7) holds for all $g \in \mathcal{L}^1(\nu)$. Then $C_0(X) \subset \mathcal{A}$. Let $\{ h_n \}$ be a sequence in $\mathcal{A}$ converging monotonically to h. Then $|h_n(x)| \le |h(x)| + |h_1(x)|$. Since μ_t is finite, the Dominated Convergence Theorem implies that

$$\int_X h_n(x)\,d\mu_t(x) \to \int_X h(x)\,d\mu_t(x) \quad \text{for all } t. \tag{I.8}$$

Thus $t \mapsto \int_x h(x)\,d\mu_t(x)$ is Borel. Furthermore

$$\Bigl| \int_X h_n(x)\,d\mu_t(x) \Bigr| \le \bigl(\|h\|_\infty + \|h_1\|_\infty \bigr) \|\mu_t\| \le \bigl(\|h\|_\infty + \|h_1\|_\infty \bigr) \|\boldsymbol{F}\|.$$

Hence another application of the Dominated Convergence Theorem implies that

$$\int_T g(t) \int_X h_n(x)\,d\mu_t(x)\,d\nu(t) \to \int_T g(t) \int_X h(x)\,d\mu_t(x)\,d\nu(t). \tag{I.9}$$

Applying the Dominated Convergence Theorem on the left-hand side of (I.7) shows that $h \in \mathcal{A}$. Thus $\mathcal{A}$ is a monotone sequentially complete class in $\mathcal{B}^b(X)$ containing $C_c(X)$. Since X is second countable, this means $\mathcal{A} = \mathcal{B}^b(X)$ [127, Proposition 6.2.9]. In particular, we have established part (c) and, taking g identically equal to 1 in (I.7), part (d).

Now let h be identically one and let $g = \mathbb{1}_E$. Then (I.7) implies that

$$\nu(E) = \int_E \mu_t(X)\,d\nu(t).$$

Since this holds for all $E \in \mathscr{M}$, we must have $\mu_t(X) = 1$ for ν-almost every t. Thus μ_t is ν-almost everywhere a probability measure. Thus we can assume from here on that μ_t is either the zero measure or a probability measure. This completes the proof of the existence part of the theorem with the exception of the statement about supports of the μ_t.

Since $(T, \mathscr{M})$ is countably separated, there is a countable family $\mathcal{D} \subset \mathscr{M}$ which separate points. There is no reason not to assume that $\mathcal{D}$ is closed under complementation so that for each $t \in T$,

$$T \setminus \{ t \} = \bigcup_{\substack{t \notin E \\ E \in \mathcal{D}}} E. \tag{I.10}$$

If k is a bounded Borel function on T, then $k \circ q \in \mathcal{B}^b(X)$ and it follows from (I.7) (and (I.3)) that for any $g \in \mathcal{L}^1(\nu)$ we have

$$\begin{aligned} \int_T k(t) g(t)\,d\nu(t) &= \int_X k\bigl(q(x)\bigr) g\bigl(q(x)\bigr)\,d\mu(x) \\ &= \int_T g(t) \int_X k\bigl(q(x)\bigr)\,d\mu_t(x)\,d\nu(t). \end{aligned}$$

Thus there is a ν-null set $N(k)$ such that

$$k(t) = \int_X k\big(q(x)\big)\,d\mu_t(x) \quad \text{for all } t \in T \setminus N(k).$$

In particular, if $k = \mathbb{1}_E$, then

$$\mathbb{1}_E(t) = \mu_t\big(q^{-1}(E)\big) \quad \text{if } t \in T \setminus N(\mathbb{1}_E). \tag{I.11}$$

Now let

$$N := \bigcup_{E \in \mathcal{D}} N(\mathbb{1}_E)$$

Since $\mathcal{D}$ is countable, N is a ν-null set and if $t \notin N$ and $E \in \mathscr{M}$, then (I.11) implies that $q^{-1}(E)$ is a μ_t-null set whenever $E \in \mathcal{D}$ and if $t \notin E$. It follows from (I.10) that $q^{-1}(T \setminus \{t\}) = X \setminus q^{-1}(t)$ is a μ_t-null set for all $t \in T \setminus N$. Thus $\operatorname{supp} \mu_t \subset q^{-1}(t)$ if $t \notin N$.

To prove the uniqueness assertion, note that using (I.5) and that $\operatorname{supp} \mu'_t \subset q^{-1}(t)$ for ν-almost all t, it follows that for all $g \in \mathcal{L}^1(\nu)$ and $h \in C_0(X)$ we have

$$\begin{aligned}\big\langle h\,,\,\Phi(g)\big\rangle &= \int_X g(q(x))h(x)\,d\mu(x)\\ &= \int_T \int_X g\big(q(x)\big)h(x)\,d\mu'_t(x)\,d\nu(t)\\ &= \int_T g(t) \int_X h(x)\,d\mu'_t(x)\,d\nu(t).\end{aligned}$$

Thus,

$$\Phi(g)(h) = \int_T g(t)\big\langle h\,,\,\boldsymbol{F}'(t)\big\rangle\,d\nu(t),$$

where $\boldsymbol{F}'(t) = \mu'_t$. But $\boldsymbol{F} = \boldsymbol{F}'$ ν-almost everywhere in view of the uniqueness assertion in Theorem I.7 on page 483. □

We took $\nu = q_*\mu$ in the statement of Theorem I.5 on page 482 for convenience. If $\nu = q_*\mu$ is absolutely continuous with respect to ρ, then we can replace $\nu(t)$ by $\frac{d\nu}{d\rho}(t)d\rho(t)$ in part (d). Since we can choose $\frac{d\nu}{d\rho}$ to be a Borel function with values in $[0,\infty)$, we can define $\mu'_t(\cdot) = \frac{d\nu}{d\rho}(t)\mu_t(\cdot)$. Since μ'_t is the zero measure whenever $\frac{d\nu}{d\rho}(t) = 0$, it follows that the set where $\operatorname{supp} \mu_t \subset q^{-1}(t)$ is ρ-conull. Therefore we obtain the following corollary.

Corollary I.9. *Suppose that μ is a finite Borel measure on the second countable locally compact space X, that $(T, \mathscr{M})$ is a countably separated Borel space and that $q : X \to T$ is a Borel surjection. If $q_*\mu \ll \rho$, then there is a family $\{\mu'_t\}_{t \in T}$ of finite Borel measures on X and a ρ-null set N such that*

(a) *For all $t \notin N$, μ'_t is a finite measure with* $\operatorname{supp} \mu'_t \subset q^{-1}(t)$,

(b) *For all $h \in \mathcal{B}^b(X)$,*

$$t \mapsto \int_X h(x)\,d\mu'_t(x)$$

is Borel and

(c)

$$\int_X h(x)\,d\mu(x) = \int_T \int_X h(x)\,d\mu_t'(x)\,d\rho(t).$$

I.3 Abelian Von Neumann Algebras

If B is a C^*-subalgebra of $C_0(X)$, then it not hard to see that $B \cong C_0(Y)$ where Y is the locally compact quotient space $X/{\sim}$, where $x \sim y$ if and only if $f(x) = f(y)$ for all $f \in B$. Thus B is exactly the subalgebra of $f \in C_0(X)$ which are constant on equivalence classes. Here we want to investigate the analogue of this result when $C_0(X)$ is replaced by $L^\infty(X, \mu)$ for a finite Borel measure μ.

To begin with, suppose that $(X, \mathscr{B})$ is an analytic Borel space and that we are given an equivalence relation $R \subset X \times X$. Let $X/{\sim}$ be the quotient space and $q : X \to X/{\sim}$ be the quotient map. The *quotient Borel structure* on $X/{\sim}$ is the largest Borel structure $\mathscr{B}/{\sim}$ making q a Borel map and consists of those sets E such that $q^{-1}(E) \in \mathscr{B}$. The pair $(X/{\sim}, \mathscr{B}/{\sim})$ is called the *quotient Borel space*. The forward image of μ under q is the measure $\nu := q_*(\mu)$ given by $\nu(E) := \mu\big(q^{-1}(E)\big)$.

Every function $g : X/{\sim} \to \mathbf{C}$ has a lift $\tilde{g} := q \circ q$ to X such that

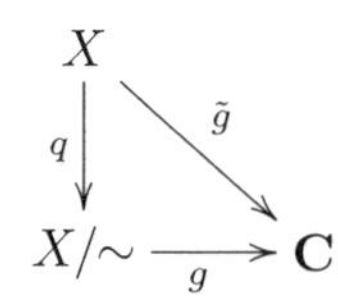

commutes, and g is Borel if and only if $\tilde{g}$ is. If we let $\widetilde{\mathcal{B}^b}(X)$ denote the bounded Borel functions on X which are constant on equivalence classes, then $g \mapsto g \circ q$ is a bijection between the bounded Borel functions on $X/\sim$ and $\widetilde{\mathcal{B}^b}(X)$ which induces a C^*-algebra isomorphism of $L^\infty(X/\sim, \nu)$ onto the image of $\widetilde{\mathcal{B}^b}(X)$ in $L^\infty(X, \mu)$. In this section, we will identify $L^\infty(X/\sim, \nu)$ with this subalgebra of $L^\infty(X, \mu)$. Note that in general, the equivalence classes in X need not be Borel (or even μ-measurable) so that points in $X/\sim$ need not be Borel. In particular, $\widetilde{\mathcal{B}^b}(X)$ need not be very large. However, if there is a *countable* family $f_1, f_2, \dots$ in $\widetilde{\mathcal{B}^b}(X)$ such that $x \sim y$ if and only if $f_n(x) = f_n(y)$ for all n, then $(X/{\sim}, \mathscr{B}/{\sim})$ is countably separated and therefore an analytic Borel space (by Corollary 2 to [2, Theorem 3.3.5]). (Thus $(X/{\sim}, \mathscr{B}/{\sim})$ is Borel isomorphic to an analytic subset of a Polish space by Corollary 1 to [2, Theorem 3.3.4].) It is standard terminology to call an equivalence relation for which $X/\sim$ is countably separated a *smooth equivalence relation.*

Theorem I.10. *Suppose that μ is a finite measure on a second countable locally compact space X. Let $\mathscr{L} = L^\infty(X, \mu)$ be the von Neumann algebra of diagonal operators on $L^2(X, \mu)$ and suppose that $\mathscr{A}$ is a von Neumann subalgebra of $\mathscr{L}$. Then there is a smooth equivalence relation on X such that $\mathscr{A}$ is equal to the* *image in $\mathscr{L}$ of the bounded Borel functions on X which are constant on equivalence classes. Thus, with the notation and identifications above, $\mathscr{A}$ can be identified with $L^\infty(Y,\nu)$ where $Y = X/\!\sim$ and ν is the image of μ via the quotient map.*

Proof. Since $\mathscr{A}$ is a von Neumann algebra, it is generated by its projections [110, Theorem 4.1.11]. Since $\mathcal{A}$ acts on a separable Hilbert space, its unit ball is a separable metric space in the weak (or strong) operator topology. Therefore, $\mathscr{A}$ is generated by a countable family $\mathbb{1}_{E_n}$, where each E_n is a Borel set in X. Then we can define a smooth equivalence relation on X by

$$x \sim y \quad \text{if and only if} \quad \mathbb{1}_{E_n}(x) = \mathbb{1}_{E_n}(y) \quad \text{for all } n.$$

Let $Y := X/\!\sim$ be the quotient space, $q : X \to Y$ the quotient map and $\nu := q_*(\mu)$. As above, we identify $L^\infty(Y,\nu)$ with the C^*-subalgebra of $L^\infty(X,\mu)$ which is the image of $\widetilde{\mathcal{B}^b}(X)$. Let $\mathscr{A}_0$ be the complex algebra generated by the $\mathbb{1}_{E_n}$. Then $\mathscr{A}_0 \subset L^\infty(Y,\nu)$, and since $L^2(X,\mu)$ is separable, the Kaplansky Density Theorem implies that every element of $\mathscr{A}$ is the strong limit of a *sequence* of elements in $\mathscr{A}_0$ ([2, Corollary to Theorem 1.2.2]). If $L_f \in \mathscr{A}$, then Proposition E.19 on page 406 implies that there is a μ-null set N and $\{ f_n \} \subset \mathscr{A}_0$ such that

$$f_n(x) \to f(x) \quad \text{for all } x \notin N.$$

Since

$$\{ x \in X : \lim_{n\to\infty} f_n(x) \text{ exists} \}$$

is both Borel and saturated with respect to our equivalence relation, there is a $g \in \widetilde{\mathcal{B}^b}(X)$ such that $f = g$ μ-almost everywhere. Thus $L_f = L_g$ and we've shown that

$$\mathscr{A} \subset L^\infty(Y,\nu). \tag{I.12}$$

On the other hand, since $(Y, \mathscr{B}/\!\sim)$ is analytic, the images in Y of the saturated sets E_n generate $\mathscr{B}/\!\sim$ as a σ-algebra [2, Theorem 3.3.5]. Therefore $\mathscr{A}_0$ is dense in $L^\infty(Y,\nu)$ in the strong operator topology coming from its natural representation on $L^2(Y,\nu)$. Suppose that g is a bounded Borel function on Y — which we view as a function in $\widetilde{\mathcal{B}^b}(X)$. Since $L^2(Y,\nu)$ is separable, we can invoke the Kaplansky Density Theorem as above to obtain a sequence in $\mathscr{A}_0$ converging to g in the strong operator topology on $L^\infty(Y,\nu)$ coming from its natural representation on $L^2(Y,\nu)$. Then Proposition E.19 implies that there are $f_n \in \mathscr{A}_0$ such that $f_n \to g$ ν-almost everywhere (as functions on Y). But then as functions on X, $f_n \to g$ μ-almost everywhere, and it follows that $L_{f_n} \to L_g$ in the strong operator topology. Thus $g \in \mathscr{A}$. This supplies the reverse containment for (I.12) and completes the proof. □

Lemma I.11. *Suppose that X and Y are standard Borel spaces and that μ and ν are finite Borel measures on X and Y, respectively. Suppose that $\varphi : L^\infty(Y,\nu) \to L^\infty(X,\mu)$ is a unital isomorphism onto a von Neumann subalgebra of $L^\infty(X,\mu)$. Then there is a Borel map $\tau : X \to Y$ such that $\varphi(f)(x) = f\big(\tau(x)\big)$.*

Proof. Let E be a Borel subset in Y. Then $\varphi(\mathbb{1}_E)$ is a projection and there is a Borel set $F \subset X$, determined up to a μ-null set, such that $\varphi(\mathbb{1}_E) = \mathbb{1}_F$. Thus we obtain a set map Φ from the Borel sets of Y into the boolean σ-algebra $\mathscr{B}/\mathscr{N}$ of Borel sets in X modulo the μ-null sets as in Appendix I.6. (We will adopt the notation and terminology of that appendix here.) Since φ is unital, $\Phi(Y) = [X]$. Since $\varphi(\mathbb{1}_{Y \setminus E}) = \mathbb{1}_X - \varphi(\mathbb{1}_E)$, we have $\Phi(Y \setminus E) = \varphi(E)'$. Since $\mathbb{1}_{E\cap F} = \mathbb{1}_E\mathbb{1}_F$, $\Phi(E \cap F) = \Phi(E) \wedge \Phi(F)$. It follows that $\Phi(E \cup F) = \Phi(E) \vee \Phi(F)$. To see that Φ is a σ-homomorphism, we need to see that

$$\Phi\bigl(\bigcup E_i\bigr) = \bigvee_i \Phi(E_i) \tag{I.13}$$

where $\{ E_i \}_{i=1}^{\infty}$ is a family of Borel sets in Y. But $\mathbb{1}_{\bigcup E_i}$ is the supremum in $L^\infty(Y, \nu)$ of the elements $\mathbb{1}_{\bigcup_{i=1}^n E_i}$ (cf., [29, Appendix II]). Since φ is an isomorphism of von Neumann algebras, it must be normal [29, I.4.3 Corollary 1]. In this context, the normality of φ implies that $\varphi\bigl(\mathbb{1}_{\bigcup E_i}\bigr)$ is the supremum in $L^\infty(X, \mu)$ of the elements $\varphi(\mathbb{1}_{\bigcup_{i=1}^n E_i})$. Therefore (I.13) holds, and Φ is a σ-homomorphism. Therefore Proposition I.31 on page 499 implies that there is a Borel map $\tau : X \to Y$ such that $\Phi(E) = [\tau^{-1}(E)]$. That is, in essence, $\varphi(\mathbb{1}_E)(x) = \mathbb{1}_E\bigl(\tau(x)\bigr)$. But $\varphi'(f)(x) := f\bigl(\tau(x)\bigr)$ is clearly a normal homomorphism from $L^\infty(Y, \nu)$ into $L^\infty(X, \mu)$ which agrees with φ on projections. Thus $\varphi = \varphi'$ [29, I.4.3 Corollary 2]. □

I.4 The Hilbert Space $L^2(G, H)$

If $\mathcal{H}$ is a complex Hilbert space and G a locally compact group with left-Haar measure μ, then it will be necessary to work with the Hilbert space $L^2(G, \mathcal{H})$. As we'll show shortly, we can think of $L^2(G, \mathcal{H})$ as the space of (equivalence classes) of certain $\mathcal{H}$-valued functions on G. To begin with, we simply *define* $L^2(G, \mathcal{H})$ to be the completion of $C_c(G, \mathcal{H})$ with respect to the norm $\|\cdot\|_2$ coming from the inner-product

$$(\xi \mid \eta) := \int_G \bigl(\xi(s) \mid \eta(s)\bigr)\,d\mu(s) \quad \text{for all } \xi, \eta \in C_c(G, \mathcal{H}). \tag{I.14}$$

This definition clearly gives us the usual space $L^2(G)$ when $\mathcal{H}$ is one-dimensional, and suffices for most purposes. We also will need to know that $L^2(G, \mathcal{H})$ is naturally isomorphic to the Hilbert space tensor product $L^2(G) \otimes \mathcal{H}$, which, to be consistent with the convention above, should be thought of as the completion of $C_c(G) \odot \mathcal{H}$.

Remark I.12. Instead of working with a locally compact group and Haar measure, we could equally well work with any locally compact space X with a Radon measure μ. Then the material below shows that $L^2(X, \mu) \otimes \mathcal{H}$ can be realized as the space $L^2(X, \mu, \mathcal{H})$ of equivalence classes of $\mathcal{H}$-valued functions on X.

Lemma I.13. *The natural identification of elements of $C_c(G) \odot \mathcal{H}$ with functions in $C_c(G, \mathcal{H})$ extends to a unitary isomorphism of the Hilbert space tensor product $L^2(G) \otimes \mathcal{H}$ with $L^2(G, \mathcal{H})$.*

Proof. It suffices to see that $C_c(G) \odot \mathcal{H}$ is $\|\cdot\|_2$-norm dense in $C_c(G, \mathcal{H})$. But Lemma 1.87 on page 29 implies that $C_c(G)\odot\mathcal{H}$ is dense in $C_c(G, \mathcal{H})$ in the inductive limit topology, and this clearly suffices. $\square$

Now we turn to the task of realizing $L^2(G, \mathcal{H})$ as a space of equivalence classes of $\mathcal{H}$-valued functions as in the scalar case.

Definition I.14. A function $\xi : G \to \mathcal{H}$ is said to be *essentially separately-valued* on $K \subset G$ if there is a null set $N \subset K$ and a separable subspace $\mathcal{H}_0 \subset \mathcal{H}$ such that $\xi(s) \in \mathcal{H}_0$ for all $s \in K \setminus N$. A function $\xi : G \to \mathcal{H}$ is said to be *weakly measurable* if

(a) The scalar-valued function $s \mapsto \big(\xi(s) \mid h\big)$ is measurable in the usual sense for all $h \in \mathcal{H}$, and

(b) ξ is essentially separately valued on every compact set $K \subset G$.

Although such functions are properly called *weakly measurable*, we will shorten this to just "measurable" here. Condition (b) may seem annoying and/or unmotivated at first, but its importance is illustrated by the proof of the next result. (The fact that this result is not immediate also illustrates why we've attempted to soft-pedal vector valued integration.)

Lemma I.15. *If $\xi : G \to \mathcal{H}$ is measurable, then $s \mapsto \|\xi(s)\|^2$ is measurable.*

Proof. Since Haar measure is complete, we can alter ξ on a null set without effecting its measurability. Since Haar measure is saturated, it suffices to show that the restriction of $s \mapsto \|\xi(s)\|^2$ to an arbitrary compact set is measurable. Hence we may assume that ξ takes values in a separable subspace $\mathcal{H}_0$ with countable orthonormal basis $\{ e_n \}_{n=1}^{\infty}$. Then Parseval's identity implies

$$\|\xi(s)\|^2 = \big(\xi(s) \mid \xi(s)\big) = \sum_{n=1}^{\infty} \big(\xi(s) \mid e_n\big)\big(e_n \mid \xi(s)\big),$$

where the sum is absolutely convergent. Since each $s \mapsto \big(\xi(s) \mid e_n\big)$ is measurable by assumption and since the sum is countable, the result follows. $\square$

Lemma I.16. *Every element of $C_c(G, \mathcal{H})$ is measurable. A linear combination of measurable $\mathcal{H}$-valued functions is measurable. If $\{ \xi_n \}$ is a sequence of measurable functions converging almost everywhere to a function $\xi : G \to \mathcal{H}$, then ξ is measurable.*

Proof. The first statement is routine once we note that image of $f \in C_c(G, \mathcal{H})$ is a compact metric space, and then recall that a compact metric space is second countable. Thus the range of f is separable and is contained in a separable subspace of $\mathcal{H}$. The second statement is straightforward. For the third, it is immediate that $s \mapsto \big(h \mid \xi(s)\big)$ is measurable for all $h \in \mathcal{H}$. We just have to see that ξ is essentially separately valued on each compact set K. But there are null sets N_n and separable subspaces $\mathcal{H}_n$ such that ξ_n takes values in $\mathcal{H}_n$ on $K \setminus N_n$. There is a null set N_0

such that $\xi_n(s) \to \xi(s)$ for all $s \in G \setminus N_0$. Let $N := \bigcup_{n=0}^{\infty} N_n$. Let $\bigvee \mathcal{H}_n$ be the subspace generated by the $\mathcal{H}_n$. Then if $s \in K \setminus N$,

$$\xi(s) \in \overline{\bigvee_{n=1}^{\infty} \mathcal{H}_n},$$

which is certainly a separable subspace. □

Definition I.17. A $\mathcal{H}$-valued function ξ on G is called *square integrable* if it is measurable and $s \mapsto \|\xi(s)\|^2$ is integrable. The set of square integrable functions on G is denoted by $\mathcal{L}^2(G, \mathcal{H})$. The set of equivalence classes in $\mathcal{L}^2(G, \mathcal{H})$ in which two functions are identified if they agree almost everywhere is denoted by $\underline{L}^2(G, \mathcal{H})$.[7]

Lemma I.18. *If $\xi \in \mathcal{L}^2(G, \mathcal{H})$, then ξ vanishes off a σ-compact set and is essentially separately valued on G.*

Proof. Since $\|\xi(s)\|^2$ is integrable, ξ clearly vanishes off a σ-finite set, and the first assertion follows from Lemma B.43 on page 349. Thus there are compact sets K_n such that ξ vanishes off $\bigcup K_n$. However, since ξ is measurable, it is essentially separately valued on each K_n. Thus ξ is essentially separately valued on the countable union $\bigcup K_n$. Since ξ vanishes off this set, this suffices. □

In view of Lemma I.16 on the previous page and the inequality

$$\|\xi(s) + \eta(s)\|^2 \leq 2(\|\xi(s)\|^2 + \|\eta(s)\|^2),$$

it is clear that $\underline{L}^2(G, \mathcal{H})$ is a vector space.

Lemma I.19. *If ξ and η are in $\mathcal{L}^2(G, \mathcal{H})$, then*

$$s \mapsto \big(\xi(s) \mid \eta(s)\big) \tag{I.15}$$

is integrable and

$$(\xi \mid \eta) := \int_G \big(\xi(s) \mid \eta(s)\big)\, d\mu(s)$$

defines an inner-product on $\underline{L}^2(G, \mathcal{H})$.

Proof. Since ξ and η are essentially separately valued by Lemma I.18, there is a null set N and a separable subspace $\mathcal{H}_0$ such that $\xi(s)$ and $\eta(s)$ belong to $\mathcal{H}_0$ for all $s \in G \setminus N$. Let $\{ e_n \}$ be an orthonormal basis for $\mathcal{H}_0$. Then if $s \in G \setminus N$,

$$\big(\xi(s) \mid \eta(s)\big) = \sum_{n=1}^{\infty} \big(\xi(s) \mid e_n\big)\big(e_n \mid \eta(s)\big).$$

Since each $s \mapsto \big(\xi(s) \mid e_n\big)$ and $s \mapsto \big(e_n \mid \eta(s)\big)$ are measurable and since the countable sum converges almost everywhere on G, it follows that (I.15) is measurable. Now integrability follows from the Cauchy-Schwarz and Hölder inequalities:

$$\int_G \big|\big(\xi(s) \mid \eta(s)\big)\big|\, d\mu(s) \leq \int_G \|\xi(s)\|\|\eta(s)\|\, d\mu(s) \leq \|\xi\|_2 \|\eta\|_2.$$

The rest follows as in the scalar case. □

[7] The notation $\underline{L}^2$ is a temporary artifact and will be dispensed with shortly.

Lemma I.20. *$\underline{L}^2(G, \mathcal{H})$ is a Hilbert space. In particular, if $\{\xi_n\}$ is a Cauchy sequence in $\mathcal{L}^2(G, \mathcal{H})$, then there is a $\xi \in \mathcal{L}^2(G, \mathcal{H})$ and a subsequence $\{\xi_{n_k}\}$ such that ξ_{n_k} converges to ξ almost everywhere and*

$$\lim_{n\to\infty} \|\xi_{n_k} - \xi\|_2 = 0.$$

Proof. At this point, to show that $\underline{L}^2(G, \mathcal{H})$ is a Hilbert space we need only show that it is complete. Since a Cauchy sequence with convergent subsequence is convergent, it suffices to prove the last statement. Therefore we can pass to a subsequence, relabel, and assume that

$$\|\xi_{n+1} - \xi_n\|_2 \leq \frac{1}{2^n} \quad \text{for } n \geq 1.$$

Now define

$$z_n(s) := \sum_{k=1}^{n} \|\xi_{k+1}(s) - \xi_k(s)\| \quad \text{and} \quad z(s) := \sum_{k=1}^{\infty} \|\xi_{k+1}(s) - \xi_k(s)\|,$$

with z taking values in $[0, \infty]$. Since the $\|\cdot\|_2$-norm is a norm, the triangle inequality implies

$$\|z_n\|_2 \leq \sum_{k=1}^{n} \|\xi_{k+1} - \xi_k\|_2 \leq 1.$$

In other words,

$$\|z_n\|_2^2 = \int_G z_n(s)^2 \, d\mu(s) \leq 1 \quad \text{for all } n,$$

and the Monotone Convergence Theorem implies

$$\int_G z(s)^2 \, d\mu(s) \leq 1 < \infty.$$

Therefore z is finite almost everywhere, and there is null set $N \subset G$ such that $\sum_{k=1}^{\infty} \xi_{k+1}(s) - \xi_k(s)$ is absolutely convergent in $\mathcal{H}$ for all $s \in G \setminus N$. Since $\mathcal{H}$ is complete, the series converges and there is a $\xi'(s)$ such that $\xi'(s) = \lim_{n\to\infty} \sum_{k=1}^{n} \xi_{k+1}(s) - \xi_k(s) = \lim_{n\to\infty} \xi_{n+1}(s) - \xi_1(s)$. Thus we can define ξ on $G \setminus N$ by

$$\lim_n \xi_n(s) = \xi'(s) + \xi_1(s) := \xi(s) \quad \text{for all } s \in G \setminus N.$$

We can define ξ to be identically 0 on N, and then ξ is measurable by Lemma I.16 on page 491.

We still have to see that $\xi \in \mathcal{L}^2(G, \mathcal{H})$ and that $\xi_n \to \xi$ in $\underline{L}^2(G, \mathcal{H})$. Let $\epsilon > 0$. By assumption, we can choose N so that $n, m \geq N$ implies

$$\|\xi_n - \xi_m\|_2 < \epsilon.$$

For all $s \in G \setminus N$, we have $\|\xi(s) - \xi_m(s)\| = \lim_n \|\xi_n(s) - \xi_m(s)\|$. Thus if $m \geq N$, Fatou's Lemma implies

$$\|\xi - \xi_m\|_2^2 \leq \liminf_n \|\xi_n - \xi_m\|_2^2 \leq \epsilon^2. \tag{I.16}$$

Since

$$\|\xi(s)\|^2 \leq \big(\|\xi(s) - \xi_m(s)\| + \|\xi_m(s)\|\big)^2 \leq 2\|\xi(s) - \xi_m(s)\|^2 + 2\|\xi_m(s)\|^2,$$

it follows from (I.16) that $\xi \in \mathcal{L}^2(G, \mathcal{H})$ and that $\xi_n \to \xi$. □

Proposition I.21. *Let G be locally compact group and $\mathcal{H}$ a complex Hilbert space. Then $C_c(G, \mathcal{H})$ is dense in $\underline{L}^2(G, \mathcal{H})$. In particular, we can identify $L^2(G, \mathcal{H})$ with $\underline{L}^2(G, \mathcal{H})$. Moreover, if $\xi \in \mathcal{L}^2(G, \mathcal{H})$, then there is a sequence $\{ \xi_n \} \subset C_c(G, \mathcal{H})$ such that $\xi_n \to \xi$ in $L^2(G, \mathcal{H})$ and $\xi_n(s) \to \xi(s)$ almost everywhere on G.*

Proof. Lemma I.20 on the preceding page implies that $\underline{L}^2(G, \mathcal{H})$ is a Hilbert space, so to identify $L^2(G, \mathcal{H})$ with $\underline{L}^2(G, \mathcal{H})$, it suffices to see that $C_c(G, \mathcal{H})$ is dense. The final assertion will then follow from Lemma I.20 on the previous page.

To see that $C_c(G, \mathcal{H})$ is dense, it will suffice to see that if $\xi \in \mathcal{L}^2(G, \mathcal{H})$ satisfies $(\eta \mid \xi) = 0$ for all $\eta \in C_c(G, \mathcal{H})$, then ξ is equal to 0 almost everywhere. Let N_0 be a null set and $\mathcal{H}_0$ a separable subspace with orthonormal basis $\{ e_n \}$ such that $\xi(s) \in \mathcal{H}_0$ for all $s \in G \setminus N_0$. If $z \in C_c(G)$, then by assumption the elementary tensor $z \otimes e_n$ satisfies $(z \otimes e_n \mid \xi) = 0$. That is,

$$\int_G z(s)\big(e_n \mid \xi(s)\big)\, d\mu(s) = 0 \quad \text{for all } z \in C_c(G).$$

It follows that there is a null set N_n such that

$$\big(e_n \mid \xi(s)\big) = 0 \quad \text{for all } s \in G \setminus N_n.$$

Thus $\xi(s) = 0$ if $s \notin N := \bigcup_{n=0}^{\infty} N_n$. Since N is a null set, we're done. □

For those who are dissatisfied with our "weak" definition of measurability in Definition I.14 on page 491, we include the following.

Definition I.22. A function $\xi : G \to \mathcal{H}$ is called *strongly measurable* if

(a) $\xi^{-1}(V)$ is measurable for each open set $V \subset \mathcal{H}$,[8] and

(b) ξ is essentially separately valued on each compact set $K \subset G$.

Our choice to call weakly measurable functions just "measurable" is, perhaps, justified by the following result.

Lemma I.23. *A function $\xi : G \to \mathcal{H}$ is strongly measurable if and only if ξ is (weakly) measurable.*

[8]Our definition of measurable set follows Rudin [156, Theorem 2.14] or [71, §III.11], and A is therefore measurable if and only if $A \cap K$ is measurable for all compact subsets K in G. In other treatments, such a set would be *locally measurable*.

Proof. Since $k \mapsto (k \mid h)$ is continuous on $\mathcal{H}$, strong measurability certainly implies measurability. So, assume that ξ is measurable. Since μ is saturated, it suffices to see that $\xi^{-1}(V) \cap K$ is measurable for all $V \in \mathcal{H}$ open and $K \subset G$ compact. Let N be a null set in G and $\mathcal{H}_0$ a separable subspace of $\mathcal{H}$ such that $\xi(s) \in \mathcal{H}_0$ for all $s \in K \setminus N$. Let

$$\xi_0(s) = \begin{cases} \xi(s) & \text{if } s \in G \setminus N, \text{ and} \\ 0 & \text{if } s \in N. \end{cases}$$

Since Haar measure is complete, it will suffice to show that $\xi_0^{-1}(V) \cap K$ is measurable:

$$\xi^{-1}(V) \cap K = (\xi_0^{-1}(V) \cap K \setminus N) \cup (\xi^{-1}(V) \cap N \cap K).$$

Since $\mathcal{H}_0$ is separable and since

$$\xi_0^{-1}(V) \cap K = \xi_0^{-1}(V \cap \mathcal{H}_0) \cap K,$$

there are countably many open balls $B_n := \{\, h \in \mathcal{H}_0 : \|h - h_n\| < \epsilon_n \,\}$ such that

$$\xi_0^{-1}(V) \cap K = \bigcup_{n=1}^{\infty} \xi_0^{-1}(B_n) \cap K.$$

So it will suffice to see that each $\xi_0^{-1}(B_n)$ is measurable. But since $\mathcal{H}_0$ is separable, there is a countable set $\{\, k_n \,\}$ of unit vectors in $\mathcal{H}_0$ such that $h \in \mathcal{H}_0$ satisfies

$$\|h - h_n\| < \epsilon_n \quad \text{if and only if} \quad |(h - h_n \mid k_m)| < \epsilon_n \quad \text{for all } m.$$

Since $\xi_0(s) \in \mathcal{H}_0$ for all s,

$$\xi_0^{-1}(B_n) = \bigcap_{m=1}^{\infty} \{\, s : |(\xi_0(s) - h_n \mid k_m)| < \epsilon_m \,\},$$

which is measurable since $s \mapsto \big(\xi_0(s) - h_n \mid k_m\big)$ agrees with $s \mapsto \big(\xi(s) - h_n \mid k_m\big)$ almost everywhere and the latter is measurable by assumption. □

I.5 The Dual of $L^1(G)$

We can view $L^1(G)$ as the completion of $C_c(G)$. We will need to know that the dual of $L^1(G)$ can be identified with $L^\infty(G)$ via the natural pairing: if $g \in L^\infty(G)$, then the corresponding functional φ should be given by

$$\varphi(f) := \int_G f(s)g(s)\,d\mu(s) \quad \text{for all } f \in C_c(G). \tag{I.17}$$

This is a straightforward result in functional analysis provided G is σ-compact. In fact, if (X, μ) is any σ-finite measure space, then the pairing in (I.17) identifies $L^1(X, \mu)^*$ isometrically with $L^\infty(X, \mu)$. On the other hand, it is also well-known that the identification of L^∞ with the dual of L^1 can fail if the σ-finite assumption on the measure space is dropped. Nevertheless, we can still recover this pairing for a locally compact group with Haar measure provided we *modify the definition of* $L^\infty(G)$ to account for locally null sets. (Recall a subset N is *locally null* if $N \cap K$ is null for every compact set $K \subset G$.)

Remark I.24. The material here applies equally well to $L^1(X, \mu)$ for any Radon measure on a *paracompact* locally compact space X. Then X is the disjoint union of clopen σ-compact sets [31, Theorem XI.7.3], and the proof proceeds exactly as below.[9]

Let $\mathcal{L}^\infty(G)$ denote the set of all bounded measurable functions on G equipped with the supremum norm. Since the pointwise limit of measurable functions is measurable, $\mathcal{L}^\infty(G)$ is a closed subspace of $\ell^\infty(G)$ and is therefore a Banach space. Let $S \subset \mathcal{L}^\infty(G)$ be the set of locally null functions on G; that is, functions which vanish off a locally null set. Then S is a closed subspace of $\mathcal{L}^\infty(G)$ and we can make the following definition.

Definition I.25. We define $L^\infty(G)$ to be the Banach space quotient $\mathcal{L}^\infty(G)/S$ with quotient norm

$$\|f + S\|_\infty = \inf_{g \in S} \Bigl(\sup_{s \in G} |f(s) + g(s)| \Bigr)$$

We also write $\|\cdot\|_\infty$ for the induced seminorm on $\mathcal{L}^\infty(G)$:

$$\|f\|_\infty := \inf_{g \in S} \Bigl(\sup_{s \in G} |f(s) + g(s)| \Bigr)$$

Lemma I.26. *Suppose that* $f \in \mathcal{L}^\infty(G)$.

(a) $\|f\|_\infty = \inf\{\, M : \{\, s : |f(s)| > M \,\}$ *is locally null*$\}$.

(b) $\|f\|_\infty = \inf\{\, M : K \subset G$ *compact implies* $\{\, s \in K : |f(s)| > M \,\}$ *is null*$\}$.

(c) $\{\, s : |f(s)| > \|f\|_\infty \,\}$ *is locally null.*

Sketch of the Proof. If $\{\, s : |f(s)| > M \,\}$ is locally null, then

$$g(s) := \begin{cases} f(s) & \text{if } |f(s)| > M, \textit{and} \\ 0 & \text{otherwise,} \end{cases}$$

defines a locally null function and $\sup_{s \in G} |f(s) - g(s)| \leq M$. Part (a) follows immediately, and part (b) follows from part (a). To prove part (c), notice that

$$\{\, s : |f(s)| > \|f\|_\infty \,\} = \bigcup_{n=1}^{\infty} \{\, s : |f(s)| > \|f\|_\infty + \frac{1}{n} \,\}.$$

The assertion follows as each set on the right-hand side is locally null, and the countable union of locally null sets is certainly locally null. $\square$

[9] In fact, we can *always* identify the dual of $L^1(X, \mu)$ with $L^\infty(X, \mu)$ provided X is locally compact, μ is a Radon measure and we define L^∞ as equivalence classes of functions which are equal locally almost everywhere. The proof is a bit involved: see [71, Theorem III.12.18] or [13, Chap. V.5.8 Theorem 4].

Now it is clear that $L^\infty(G)$ is the usual thing when G is σ-compact. In general, we have to think of $L^\infty(G)$ as equivalence classes of bounded measurable functions on G which can be identified if they agree *locally almost everywhere* — that is, if they agree off a locally null set.

If $g \in \mathcal{L}^\infty(G)$, then

$$\varphi(f) = \int_G f(s)g(s)\,d\mu(s)$$

defines a linear functional on $C_c(G) \subset L^1(G)$ which as norm $\|g\|_\infty$ in view of part (b) of Lemma I.26 on the preceding page. If $g' \in \mathcal{L}^\infty(G)$ induces the same functional, then g and g' must agree almost everywhere on each compact set $K \subset G$. Thus, g and g' define the same class in $L^\infty(G)$.

On the other hand, suppose that φ is a bounded linear functional with respect to the L^1-norm on $C_c(G)$; that is, $\varphi \in L^1(G)^*$. By Lemma 1.38 on page 10, G is the disjoint union

$$G = \bigcup_i G_i$$

of clopen σ-compact sets G_i. Certainly φ defines, by restriction, a bounded linear functional of norm at most $\|\varphi\|$ on $L^1(G_i)$. Since the restriction of Haar measure to G_i is σ-finite, there is a $g_i \in \mathcal{L}^\infty(G_i)$ such that for all $s \in G_i$, $|g_i(s)| \le \|\varphi\|$ and

$$\varphi(f) = \int_G f(s)g_i(s)\,d\mu(s) \quad \text{for all } f \in C_c(G_i).$$

Since the G_i are disjoint, we can define g by $g(s) := g_i(s)$ if $s \in G_i$. Since Haar measure is a Radon measure, to show that g is measurable it suffices to see that $g^{-1}(V) \cap K$ is measurable for all compact sets K. But this follows from the measurability of the g_i and the fact that the G_i are disjoint and open. Thus g is measurable and clearly bounded by $\|\varphi\|$. If $f \in C_c(G)$, then $f = f_1 + \cdots + f_n$ with each f_j supported in some G_{i_j}. Thus

$$\varphi(f) = \int_G f(s)g(s)\,d\mu(s) \quad \text{for all } f \in C_c(G).$$

Thus φ is given by a necessarily unique element $g \in L^\infty(G)$. We have proved the following result.

Proposition I.27. *If G is a locally compact group and if $L^\infty(G)$ is defined as in Definition I.25 on the preceding page, then we can identify the dual of $L^1(G)$ with $L^\infty(G)$ via the usual pairing given in* (I.17).

The following technicality will be needed in Appendix A.2. Recall that we use the notation

$$\mathcal{S}_c(G) = \{\, f \in C_c^+(G) : \|f\|_1 = 1 \,\}$$

to suggest that $\mathcal{S}_c(G)$ is a subspace of the state space of $L^\infty(G)$.

Lemma I.28. *$\mathcal{S}_c(G)$ is weak-$*$ dense in the state space of $L^\infty(G)$.*

Proof. Note that $\varphi \in \mathcal{L}^\infty(G)$ represents a positive element in $L^\infty(G)$ if and only if $\{ s \in G : \varphi(s) \not\geq 0 \}$ is locally null. Therefore it is not hard to see that φ represents a positive element in $L^\infty(G)$ if and only if

$$\int_G \varphi(s) f(s)\, d\mu(s) \geq 0 \quad \text{for all } \varphi \in \mathcal{S}_c(G).$$

Consequently, this result is an immediate consequence of the next lemma. □

Lemma I.29. *Suppose that A is a C^*-algebra with identity, and that C is a convex subset of the state space $\mathcal{S}(A)$ of A with the property that*

$$a = a^* \text{ and } \rho(a) \geq 0 \text{ for all } \rho \in C \text{ implies that } a \geq 0. \tag{I.18}$$

Then C is weak-$*$ dense in $\mathcal{S}(A)$.

Proof. Let $\overline{C}$ be the weak-$*$ closure of C. If C were not dense, then there would be a $\tau \in \mathcal{S}(A) \setminus \overline{C}$. Then τ has a convex open neighborhood D disjoint from the convex set $\overline{C}$. It follows from [139, Lemma A.40] that there would be an $a \in A$ and an $\alpha \in \mathbf{R}$ such that

$$\operatorname{Re} \tau(a) < \alpha \leq \operatorname{Re} \rho(a) \quad \text{for all } \rho \in C.$$

Since $\rho(a^*) = \overline{\rho(a)}$ for all $\rho \in \mathcal{S}(A)$, we can replace a by $a_0 := \frac{1}{2}(a + a^*)$ so that

$$\tau(a_0) < \alpha \leq \rho(a_0) \quad \text{for all } \rho \in C. \tag{I.19}$$

Then (I.18) implies that $a_0 - \alpha 1_A \geq 0$. Since τ is a state, and hence positive, we have $\tau(a_0) \geq \alpha$. This contradicts (I.19), and completes the proof. □

I.6 Point Maps

Under certain circumstances a map from a σ-algebra to itself may be induced by a map on the underlying space. Although there is considerable literature available (cf. [154, Chap. 15], [29, Appendix IV], [4, Proposition 2.14], [106, Theorem P.4]), we'll settle for a special case. Suppose that $(X, \mathscr{A})$ is a Borel space and that $\mathscr{M}$ is the collection of μ-null sets in $\mathscr{A}$ for some measure μ on $(X, \mathscr{A})$. We define $\mathscr{A}/\mathscr{M}$ to be the quotient $\mathscr{A}/\!\sim$ where $E \sim F$ if the symmetric difference $E \triangle F \in \mathscr{M}$. If $[F]$ denotes the equivalence class of F in $\mathscr{A}/\mathscr{M}$, then it is not hard to see that we can define a partial order on $\mathscr{A}/\mathscr{M}$ by $[E] \leq [F]$ if $E \setminus F \in \mathscr{M}$. Similar considerations show that

$$[E]' := [X \setminus E] \quad [E] \vee [F] := [E \cup F] \quad \text{and} \quad [E] \wedge [F] := [E \cap F] \tag{I.20}$$

are well-defined operations on $\mathscr{A}/\mathscr{M}$. Since the countable union of null sets is null, we can also define

$$\bigvee_{i=1}^{\infty} [E_i] := \Big[\bigcup_{i=1}^{\infty} E_i \Big]. \tag{I.21}$$

Let $(Y, \mathscr{B})$ be a Borel space and $\mathscr{N}$ the null sets in $\mathscr{B}$ for a measure ν on Y. A map $\Phi : \mathscr{B} \to \mathscr{A}/\mathscr{M}$ is called a σ-homomorphism if $\Phi(Y) = [X]$, $\Phi(Y \setminus E) = \Phi(E)'$, $\Phi(E \cup F) = \Phi(E) \vee \Phi(F)$ and $\Phi(\bigcup E_i) = \bigvee \Phi(E_i)$. Notice that if $\varphi : X \to Y$ is any Borel map, then $\Phi(E) := [\varphi^{-1}(E)]$ is a σ-homomorphism. If $\Phi(E) = \Phi(F)$ whenever $E \triangle F \in \mathscr{N}$, then Φ induces a map $\bar{\Phi} : \mathscr{B}/\mathscr{N} \to \mathscr{A}/\mathscr{M}$ which is also called a σ-homomorphism. We want to see that under mild hypotheses on Φ and $(Y, \mathscr{B})$, any σ-homomorphism is induced by a map φ. Since measures play no role in the next result, we'll introduce the terminology of a *σ-ideal*; we say that $\mathscr{N} \subset \mathscr{B}$ is a σ-ideal in $\mathscr{B}$ if $\mathscr{N}$ is closed under countable unions and if $\mathscr{N}$ has the property that any B in $\mathscr{B}$ which is contained in an element of $\mathscr{N}$ is itself in $\mathscr{N}$.

Remark I.30 (Boolean Algebras). In general, a set $\mathscr{B}$, together with operations $\vee$, $\wedge$ and $'$ modeled on (I.20), is called a *boolean algebra.* If there is a countable operation $\bigvee$ as in (I.21), then $\mathscr{B}$ is called a *boolean σ-algebra.* A subset $\mathscr{N}$ of $\mathscr{B}$ is called a σ-ideal if $\mathscr{N}$ is closed under $\bigvee$ and if $A \in \mathscr{N}$ and $B \subset A$ implies $B \in \mathscr{N}$. Then the quotient $\mathscr{B}/\mathscr{N}$ is again a boolean σ-algebra and the notion of σ-homomorphism is defined in the obvious way. A general discussion of boolean σ-algebras and their properties can be found in [154, §15.2].

Proposition I.31. *Suppose that $(X, \mathscr{A})$ is a Borel space and that $\mathscr{M}$ is a σ-ideal in $\mathscr{A}$. Suppose that $(Y, \mathscr{B})$ is a standard Borel space and that $\Phi : \mathscr{B} \to \mathscr{A}/\mathscr{M}$ is a σ-homomorphism. Then there is a Borel map $\varphi : X \to Y$ such that $\Phi(B) = [\varphi^{-1}(B)]$. If $\varphi_1 : X \to Y$ is another map implementing Φ, then $\varphi = \varphi_1$ off a set in $\mathscr{M}$.*

We won't give a complete proof here. Instead, we'll show that the proposition follows almost immediately from [154, §15.6 Proposition 19]. To do this properly, we need the following lemma.

Lemma I.32. *Suppose that $(X, \mathscr{B})$ is a standard Borel space. Then there is a Polish space Q such that $(Q, \mathscr{B}(Q))$ is Borel isomorphic to $(X, \mathscr{B})$.*

Proof. We may assume that X is a Borel subset of a Polish space P and that $\mathscr{B}$ is the relative Borel structure coming from $\mathscr{B}(P)$. Then [2, Theorem 3.2.1] implies that there is a Polish space Q and a one-to-one continuous map $f : Q \to P$ such that $f(Q) = X$. Then [2, Theorem 3.3.2] implies that f defines a Borel isomorphism of $(Q, \mathscr{B}(Q))$ onto $(X, \mathscr{B})$. □

Proof of Proposition I.31. Using Lemma I.32, we can assume that Y is a Polish space. If Y is uncountable, then the proposition is exactly [154, §15.6 Proposition 19]. If Y is countable, say $Y = \{ y_1, y_2, \dots \}$, then we can choose Borel sets $E_i \subset X$ such that $\Phi(\{ y_i \}) = [E_i]$. Since $\Phi(Y) = [X]$, we may as well assume that $\bigcup E_i = X$. Notice that if $i \neq j$, then $E_i \cap E_j \in \mathscr{M}$. We can define disjoint Borel sets by $B_1 = E_1$ and

$$B_n := E_n \setminus \bigcup_{i=1}^{n-1} E_i$$

such that $[E_n] = [B_n]$ and $\bigcup B_i = X$. (Some of the B_n could be empty.) Then we define a Borel function by

$$\varphi(x) = y_i \quad \text{if } x \in B_i.$$

If F is a Borel set in Y, then

$$\begin{aligned}[\varphi^{-1}(F)] &= \Big[\bigcup_{y_i \in F} B_i\Big] \\ &= \bigvee_{y_i \in F} [E_i] \\ &= \bigvee_{y_i \in F} \Phi(\{\, y_i \,\}) \\ &= \Phi(F),\end{aligned}$$

and φ implements Φ. If $\varphi_1 : X \to Y$ is another such map, then $i \neq j$ implies

$$\varphi^{-1}(\{\, y_i \,\}) \cap \varphi_1^{-1}(\{\, y_j \,\}) \in \mathscr{M}.$$

But φ and φ_i must agree off

$$\bigcup_{i \neq j} \varphi^{-1}(\{\, y_i \,\}) \cap \varphi_1^{-1}(\{\, y_j \,\}) \in \mathscr{M}.$$

This completes the proof. □

If the map Φ in Proposition I.31 on the previous page doesn't see null sets (more precisely, if Φ induces a σ-homomorphism on $\mathscr{B}/\mathscr{N}$), then we can significantly weaken the hypothesis that Y be a standard Borel space. The classical definition is as follows.

Definition I.33. We say that $(Y, \mathscr{B}, \nu)$ is a *standard measure space* if ν is σ-finite and if there is a ν-conull set $E \subset Y$ which is a standard Borel space in its relative Borel structure.

Standard measure spaces are more common than one might guess.

Lemma I.34. *Suppose that $(Y, \mathscr{B})$ is an analytic Borel space and that ν is a σ-finite measure on Y. Then $(Y, \mathscr{B}, \nu)$ is a standard measure space.*

Proof. Since ν is equivalent to a finite measure, we may as well assume that ν is finite. We can also assume that X is an analytic subset of a Polish space P ([2, Corollary 1 to Theorem 3.3.4]). Then we can extend ν to a measure $\bar{\nu}$ on P by $\bar{\nu}(E) := \nu(E \cap X)$. Since analytic subsets of Polish spaces are absolutely measurable ([2, Theorem 3.2.4]), there are Borel sets E and F in P such that

$$E \subset X \subset F \quad \text{and} \quad \bar{\nu}(F \setminus E) = 0 = \nu(X \setminus E).$$

Since E is, by definition, a standard Borel space, this suffices. □

Remark I.35. If $(Y, \mathscr{B})$ is an analytic Borel space and $\Phi : \mathscr{B} \to \mathscr{A}/\mathscr{M}$ is a σ-homomorphism, then Proposition I.31 on the preceding page does not apply (since it requires $(Y, \mathscr{B})$ to be standard). However, if ν is a σ-finite measure on $(Y, \mathscr{B})$ with null sets $\mathscr{N}$, then $(Y, \mathscr{B}, \nu)$ is a standard measure space by Lemma I.34. If $\Phi(F) = \Phi(E)$ whenever $F \mathbin{\triangle} E$ is a ν-null set, then Φ factors through $\mathscr{B}/\mathscr{N}$, and the next lemma is a useful substitute for Proposition I.31.

Lemma I.36. *Suppose that* $(X, \mathscr{A})$ *is a Borel space and that* $\mathscr{M}$ *is a* σ*-ideal in* $\mathscr{A}$*. Let* $(Y, \mathscr{B}, \nu)$ *be a standard measure space with null sets* $\mathscr{N}$*. If* $\Phi : \mathscr{B}/\mathscr{N} \to \mathscr{A}/\mathscr{M}$ *is a* σ*-homomorphism, then there is a Borel map* $\varphi : X \to Y$ *such that* $\Phi([F]) = [\varphi^{-1}(F)]$*. If* $\psi : X \to Y$ *is another such map, then* $\varphi = \psi$ *off a set in* $\mathscr{M}$*.*

Proof. Let N be a ν-null set in Y such that $Y \setminus N$ is a standard Borel space. Let $\mathscr{B}(Y \setminus N)$ be the Borel sets in $Y \setminus N$. Then we can define a σ-homomorphism $\Psi : \mathscr{B}(Y \setminus N) \to \mathscr{A}/\mathscr{M}$ by $\Psi(F) := \Phi([F])$. Proposition I.31 on page 499 implies that there is a Borel map $\varphi : X \to Y \setminus N \subset Y$ such that $\Psi(F) = [\varphi^{-1}(F)]$ for all Borel sets F in $Y \setminus N$. But if F is any Borel set in Y, then

$$\begin{aligned}\Phi\big([F]\big) &= \Phi\big([F \setminus N]\big)\\ &= \Psi(F \setminus N)\\ &= [\varphi^{-1}(F \setminus N)]\end{aligned}$$

which, since $\varphi(X) \subset Y \setminus N$, is

$$= [\varphi^{-1}(F)].$$

Thus φ implements Φ as required.

Suppose that ψ is another Borel map from $X \to Y$ such that $\Phi\big([F]\big) = [\psi^{-1}(F)]$. If $N \in \mathscr{N}$, then $M := \psi^{-1}(N) \in \mathscr{M}$. Fix $y_0 \in Y \setminus N$ and define $\psi_0 : X \to Y \setminus N$ by

$$\psi_0(x) = \begin{cases}\psi(x) & \text{if } x \in X \setminus M, \text{ and}\\ y_0 & \text{if } x \in M.\end{cases}$$

Then ψ_0 is Borel and equals ψ off a set in $\mathscr{M}$. Thus $\Psi\big([F]\big) = [\psi_0^{-1}(F)]$. Proposition I.31 on page 499 implies that $\varphi = \psi_0$ off a set in $\mathscr{M}$. Therefore $\varphi = \psi$ off a set in $\mathscr{M}$ as required. □

Let $(X, \mathscr{A}, \mu)$ and $(Y, \mathscr{B}, \nu)$ be standard measure spaces with null sets $\mathscr{M}$ and $\mathscr{N}$, respectively. Suppose that $\Phi : \mathscr{B}/\mathscr{N} \to \mathscr{A}/\mathscr{M}$ is a σ-homomorphism. Then Lemma I.36 implies that there is a Borel map $\varphi : X \to Y$ such that $\Phi\big([F]\big) = [\varphi^{-1}(F)]$. If Φ is a bijection, then Φ^{-1} is also a σ-homomorphism, and it makes sense to call Φ a σ-isomorphism. In this case, we could hope that we could choose φ to be a Borel isomorphism. However, this can fail to be the case (cf. [154, §15.6 Exercise 26]). But it is nearly the case.

Theorem I.37. *Suppose that* $(X, \mathscr{A}, \mu)$ *and* $(Y, \mathscr{B}, \nu)$ *are standard measure spaces with null sets* $\mathscr{M}$ *and* $\mathscr{N}$*, respectively. If* $\Phi : \mathscr{B}/\mathscr{N} \to \mathscr{A}/\mathscr{M}$ *is a* σ*-isomorphism, then there is a Borel map* $\varphi : X \to Y$ *and null sets* $X_0 \in \mathscr{M}$ *and* $Y_0 \in \mathscr{N}$ *such that* $\Phi([F]) = [\varphi^{-1}(F)]$ *and such that the restriction of* φ *to* $X \setminus X_0$ *is a Borel isomorphism of* $X \setminus X_0$ *onto* $Y \setminus Y_0$*.*

Proof. Lemma I.36 implies that there are Borel maps $\varphi : X \to Y$ and $\psi : Y \to X$ such that $\Phi\big([F]\big) = [\varphi^{-1}(F)]$ and $\Phi^{-1}\big([E]\big) = [\psi^{-1}(E)]$. Then $\psi \circ \varphi$ implements that identity map on $\mathscr{A}/\mathscr{M}$. The uniqueness assertion in Lemma I.36 implies that there is a null set $X_0' \in \mathscr{M}$ such that $\psi(\varphi(x)) = x$ if $x \notin X_0'$. Since $[\psi^{-1}(X \setminus X_0')] = \Phi^{-1}([X \setminus X_0']) = \Phi^{-1}([X])$, there is null set $Y_0' \in \mathscr{N}$ such that $\psi^{-1}(X \setminus X_0') = Y \setminus Y_0'$. However, we also have that $\varphi \circ \psi$ implements the identity on $\mathscr{B}/\mathscr{N}$. Thus there is a null set $Y_0 \in \mathscr{N}$ such that $\varphi(\psi(y)) = y$ for all $y \notin Y_0$. We can enlarge Y_0 if necessary so that $Y_0' \subset Y_0$. (This ensures that $\psi^{-1}(X \setminus X_0') \supset Y \setminus Y_0$.) As above, $\varphi^{-1}(Y \setminus Y_0)$ differs from X by a null set, and we can find $X_0 \in \mathscr{M}$ such that

$$X \setminus X_0 = \varphi^{-1}(Y \setminus Y_0) \cap (X \setminus X_0').$$

Of course, if $x \in X \setminus X_0$, then $\varphi(x) \in Y \setminus Y_0$. On the other hand, if $y \in Y \setminus Y_0$, then we have $\psi(y) \in X \setminus X_0'$ and $\varphi(\psi(y)) = y$. Thus $\psi(y) \in X \setminus X_0$. In particular, φ is a Borel isomorphism of $X \setminus X_0$ onto $Y \setminus Y_0$. □

As an application of these ideas, we consider isomorphisms of abelian von Neumann algebras. A collection $\mathcal{F}$ of positive elements in a von Neumann algebra $\mathcal{N}$ is said to be *filtering* if given $a, b \in \mathcal{F}$, then there is a $c \in \mathcal{F}$ such that $a \leq c$ and $b \leq c$. Naturally, we say $d = \operatorname{lub} \mathcal{F}$ if $d \in \mathcal{N}$, $a \leq d$ for all $a \in \mathcal{F}$ and if $c \in \mathcal{N}$ also satisfies $a \leq c$ for all $a \in \mathcal{F}$, then $d \leq c$. For example, if $\mathcal{N} = L^\infty(X, \mu)$ and if $\{ B_i \}$ are Borel sets in X, then $\operatorname{lub}\{ \mathbb{1}_{B_i} \} = \mathbb{1}_{\bigcup B_i}$. A filtering subset of $\mathcal{N}^+$ which is bounded above must have a lub as above [29, I.1.3 and Appendix II]. A positive linear map $L : \mathcal{N} \to \mathcal{M}$ is called *normal* if given a filtering subset $\mathcal{F}$ with $a = \operatorname{lub} \mathcal{F}$, then

$$L(a) = \operatorname{lub}\{ L(b) : b \in \mathcal{F} \}.$$

Since an isomorphism of one von Neumann algebra onto another must be an order isomorphism on the positive cones, it must be normal [29, I.4.3].

Corollary I.38 (von Neumann). *Let (X, μ) and (Y, ν) be standard measure spaces and suppose that $\Psi : L^\infty(Y, \nu) \to L^\infty(X, \mu)$ is an isomorphism. Then there are*

(a) *a μ-null set $N \subset X$,*
(b) *a ν-null set $M \subset Y$ and*
(c) *a Borel isomorphism $\tau : X \setminus N \to Y \setminus M$*

such that

$$\Psi(f)(x) = f(\tau(x)). \tag{I.22}$$

Remark I.39. Since elements of L^∞ are almost-everywhere equivalence classes of functions, (I.22) is a rather sloppy, but intuitive, way to state our conclusion. To be precise, we should say that if f is a bounded Borel function on Y and $[f]$ its class in $L^\infty(\nu)$, then the class of $\Psi([f])$ is given by the function $x \mapsto f(\tau(x))$. Obscuring this sort of detail is probably to everyone's advantage.

Proof. Let $\mathscr{A}$ be the Borel sets in X and $\mathscr{B}$ the Borel sets in Y. Let $\mathscr{M}$ and $\mathscr{N}$ be the ν- and μ-null sets, respectively. If E is a projection in $L^\infty(\nu)$, then $\Psi(E)$ is a projection in $L^\infty(\mu)$. Thus there are Borel sets $B \in \mathscr{B}$ and $A \in \mathscr{A}$ such that $E = \mathbb{1}_B$ and $\Psi(E) = \mathbb{1}_A$. Since A and B are determined up to null sets, we obtain a map

$$\Phi : \mathscr{B}/\mathscr{N} \to \mathscr{A}/\mathscr{M}.$$

We claim that Φ is a σ-homomorphism. Since Ψ is unital, we certainly have $\Phi([B])' = \Phi([B]')$. Since $\mathbb{1}_{B_1}\mathbb{1}_{B_2} = \mathbb{1}_{B_1\cap B_2}$, it follows that $\Phi([B_1]\wedge[B_2]) = \Phi([B_1]) \wedge \Psi([B_2])$. Since Ψ is normal,

$$\Phi\Bigl(\bigvee [B_i]\Bigr) = \bigvee \Phi([B_i]).$$

This proves the claim. Therefore Theorem I.37 on page 501 implies that there are null sets N and M as well as a Borel isomorphism τ as in (a), (b) and (c) such that

$$\Phi([B]) = [\tau^{-1}(B)].$$

Thus for μ-almost all x,

$$\Psi(\mathbb{1}_B)(x) = \mathbb{1}_B\bigl(\tau(x)\bigr).$$

Define $\Psi' : L^\infty(\nu) \to L^\infty(\mu)$ using τ: $\Psi'(f)(x) := f\bigl(\tau(x)\bigr)$. Then Ψ and Ψ' are isomorphisms which agree on the linear span of the projections in $L^\infty(\nu)$. Thus $\Psi = \Psi'$. □

Royden's proof of Proposition I.31 on page 499 for standard Borel spaces, requires Kuratowski's Theorem which says that any two uncountable standard Borel spaces are Borel isomorphic.

Theorem I.40 (Kuratowski's Theorem). *Every uncountable standard Borel space is Borel isomorphic to* $[0,1]$ *with the usual Borel structure.*

The usual reference for Kuratowski's Theorem is Remark (i) on page 451 of his classic treatise on Topology [96]. Since it is a formidable task to sort of the details from [96], it's nice to know that a more accessible treatment is given in [154, Theorem 15.10] where the result is proved for a standard Borel space of the form $\bigl(Q, \mathscr{B}(Q)\bigr)$ with Q an uncountable Polish space. To recover Kuratowski's Theorem as stated here, it suffices to apply Lemma I.32 on page 499.

I.7 Representations of $C_0(X)$

We need to revisit the situation in Appendix E.1, but in slightly greater generality so that we have an appropriate reference for our applications in Section 8.3. Let μ be a Radon measure on a locally compact space X. For each bounded Borel function $f \in \mathcal{B}^b(X)$ define an operator $L_f \in B\bigl(L^2(X,\mu)\bigr)$ by $L_f h(x) := f(x)h(x)$. As in Appendix E.1 we let $\mathscr{L}$ be the $*$-algebra of operators $\{\, L_f : f \in \mathcal{B}^b(X)\,\}$. As in Appendix I.5, we let $L^\infty(X,\mu)$ be the collection of *locally* μ-almost everywhere equivalence classes of functions in $\mathcal{B}^b(X)$.[10] If X is paracompact, then X is the disjoint union of clopen σ-compact sets [31, Theorem XI.7.3], and the proof of Proposition I.27 on page 497 shows that $L^\infty(X,\mu)$ is the dual of $L^1(X,\mu)$. For a Radon measure on an arbitrary locally compact space, the result still holds, but the proof seems quite difficult (see [71, Theorem III.12.18] or [13, Chap. V.5.8 Theorem 4]). We'll assume the general result here, but in fact, when we need to apply these

[10] That is, $f \sim g$ if f and g agree off a Borel locally null set.

results in Section 8.3 (in the form of Lemma I.42 on the facing page), X will be a locally compact abelian group and Proposition I.27 on page 497 suffices.

As in Appendix E.1, we let π_μ be the representation of $C_0(X)$ on $L^2(X,\mu)$ given by $\pi_\mu(f) := L_f$. Note that

$$\mathscr{L}' := \{\, T \in B\big(L^2(X,\mu)\big) : TL_f = L_f T \text{ for all } f \in \mathcal{B}^b(X)\,\}$$

is a von Neumann algebra. The following result is a variation on [2, Theorem 2.2.1].

Proposition I.41. *Suppose that μ is a Radon measure on a locally compact space X. Let π_μ, L_f and $\mathscr{L}$ be as above. Then*

(a) $\mathscr{L} = \mathscr{L}'$, *and*

(b) $\pi_\mu\big(C_0(X)\big)$ *is dense in $\mathscr{L}$ in the strong operator topology.*

Proof. Let $\mathscr{L}''$ be the von Neumann algebra generated by $\mathscr{L}$. The Kaplansky Density Theorem [110, Theorem 4.3.3] implies that the unit ball $(\mathscr{L})_1$ of $\mathscr{L}$ is dense in the unit ball $(\mathscr{L}'')_1$ of $\mathscr{L}''$ in the strong operator topology (and *a priori* in the weak operator topology).

Since each $h \in L^2(X,\mu)$ vanishes off a set which is the countable union of sets of finite measure, L_f depends only on the class of f in $L^\infty(X,\mu)$. Clearly $\|L_f\| \le \|f\|_\infty$.[11] A standard estimate shows that $\|L_f\| = \|f\|_\infty$. Since

$$(L_f h \mid k) = \int_X f(x)h(x)\overline{k(x)}\,d\mu(x),$$

it follows that $f \mapsto L_f$ defines an isometric map with is continuous from $L^\infty(X,\mu)$ with the weak-$*$ topology (as the dual of $L^1(X,\mu)$) into $B\big(L^2(X,\mu)\big)$ with the weak operator topology. Since the unit ball of $L^\infty(X,\mu)$ is weak-$*$ compact, its image, $(\mathscr{L})_1$ is compact, and hence closed, in the weak operator topology. Therefore it must equal $(\mathscr{L}'')_1$. Thus $\mathscr{L} = \mathscr{L}''$ and $\mathscr{L}$ is a von Neumann subalgebra of $B\big(L^2(X,\mu)\big)$.

Let $T \in \mathscr{L}'$. If $C \subset X$ is compact, then $\mathbb{1}_C \in L^2(X) \cap L^\infty(X)$. In particular, there is a Borel function $g_C \in \mathcal{L}^2(X)$ such that $T\mathbb{1}_C = g_C$. (Really we should say that the class of g_C is $T\mathbb{1}_C$, but this seems more distracting than helpful.) Since $T\mathbb{1}_C = T(\mathbb{1}_C)^2 = TL_{\mathbb{1}_C}\mathbb{1}_C = L_{\mathbb{1}_C}T\mathbb{1}_C$, we can assume that g_C vanishes off C. We view $\mathcal{L}^2(C)$ as the square integrable Borel functions on X that vanish off C. Then, if $h \in \mathcal{L}^2(C) \cap \mathcal{B}^b(X)$,

$$\begin{aligned}\Big|\int_X g_C(x)h(x)\overline{k(x)}\,d\mu(x)\Big| &= \big|(L_h T\mathbb{1}_C \mid k)\big| \\ &= \big|(TL_h\mathbb{1}_C \mid k)\big| \\ &= \big|(Th \mid k)\big| \\ &\le \|T\|\|h\|_2\|k\|_2.\end{aligned} \tag{I.23}$$

Using (I.23), it is not hard to show that $|g_C(x)| \le \|T\|$ for μ-almost all $x \in C$. Therefore we can assume that $g_C \in \mathcal{B}^b(X)$, that g_C vanishes off C and that $|g(x)| \le$

[11] Here, $\|\cdot\|_\infty$ is the essential sup norm on $L^\infty(X,\mu)$ as in Appendix I.5.

$\|T\|$ for all x. If $h \in \mathcal{B}^b(C)$ (viewed as bounded Borel function on X which vanishes off C), then $h \in \mathcal{L}^2(X)$ and

$$L_{g_C}h = g_C h = L_h g_C = L_h T\mathbb{1}_C = TL_h\mathbb{1}_C = Th$$

Since $\mathcal{B}^b(C)$ is dense in $L^2(C)$, $T = L_{g_C}$ on $L^2(C)$. Since $C_c(X)$ is dense in $L^2(X)$, it is not hard to see that $\{ L_{\mathbb{1}_C} \}$ converges to the identity on $L^2(X)$ as C increases. Then

$$Th = \lim_C TL_{\mathbb{1}_C}h = \lim_C T(\mathbb{1}_C h) = \lim_C L_{g_C}(\mathbb{1}_C h) = \lim_C L_{g_C}h.$$

Therefore $L_{g_C} \to T$ in the strong operator topology. Since $\mathscr{L}$ is a von Neumann algebra, $T \in \mathscr{L}$. This proves that $\mathscr{L}' \subset \mathscr{L}$. Since the other containment is clear, this proves part (a).

For part (b), let $f \in \mathcal{B}^b(X)$. If $C \subset X$ is compact, let V be a precompact open neighborhood of C. Given $\epsilon > 0$, Lusin's Theorem (cf., [156, Theorem 2.24] or [57, Theorem 7.10]) implies there is a $g_1 \in C(\overline{V})$ such that $\|g_1\|_\infty \le \|f\|_\infty$ and such that $\mu(\{ x \in \overline{V} : g_1(x) \ne f(x) \}) < \epsilon$. If $g_2 \in C_c^+(X)$ is such that $0 \le g_2 \le 1$, $g_2(x) = 1$ for all $x \in C$ and such that g_2 vanishes off V, then

$$g_{(C,\epsilon)}(x) := \begin{cases} g_1(x)g_2(x) & \text{if } x \in \overline{V} \\ 0 & \text{if } x \notin V, \end{cases}$$

defines an element of $C_c(X)$ such that $\|g_{(C,\epsilon)}\|_\infty \le \|f\|_\infty$ and such that

$$\mu(\{ x \in C : g_{(C,\epsilon)}(x) \ne f(x) \}) < \epsilon.$$

We can complete the proof by showing that $\pi_\mu(g_{(C,\epsilon)}) \to L_f$ in the strong operator topology as C increases and ϵ decreases. Since $\{ \pi_\mu(g_{(C,\epsilon)}) \}$ is bounded and since $C_c(X)$ is dense in $L^2(X)$, it will suffice to see that

$$\pi_\mu(g_{(C,\epsilon)})h \to L_f h \quad \text{in } L^2(X) \text{ for all } h \in C_c(X).$$

But if $C \supset \operatorname{supp} h$,

$$\|\pi_\mu(g_{(C,\epsilon)})h - L_f h\| = \|L_{(g(C,\epsilon)-f)}h\| \le \epsilon \|h\|_2.$$

This completes the proof. □

This rather technical result will be useful in understanding the type structure of transformation group C^*-algebras.

Lemma I.42. *Suppose that X, Y and Z are locally compact spaces and that μ is a Radon measure on Y and that ν is a Radon measure on Z. Let $i : Y \to X$ and $j : Z \to X$ be continuous injections, and let Π_1 and Π_2 be the corresponding representations of $C_0(X)$ on $L^2(Y,\mu)$ and $L^2(Z,\nu)$, respectively, given by*

$$\Pi_1(f)h(y) := f\big(i(y)\big)h(y) \quad \text{and} \quad \Pi_2(f)k(z) = f\big(j(z)\big)k(z).$$

If $i(Y) \cap j(Z) = \emptyset$, then Π_1 and Π_2 have no equivalent subrepresentations. In particular, Π_1 and Π_2 are not equivalent.

Proof. Let $\mathcal{H}$ be an invariant subspace for Π_1. We want to see that $\mathcal{H}$ must be of the form $L^2(E,\mu)$ for a Borel subset $E \subset Y$. Let $C(\mathcal{H})$ be the collection of Borel subsets F of Y such that $\mathbb{1}_F \in \mathcal{H}$. If $h \in \mathcal{H}$, let

$$E_n = \{\, y \in Y : |h(y)| \geq \frac{1}{n} \,\}.$$

Then E_n has finite measure and $y \mapsto h(y)^{-1}\mathbb{1}_{E_n}$ is a bounded Borel function on Y.

Claim 1. Let π_μ be the natural representation of $C_0(Y)$ on $L^2(Y,\mu)$. If g is a bounded Borel function on Y, then there is a net $\{\, f_i \,\}$ in $C_c(X)$ such that $\Pi_1(f_i) \to \pi_\mu(g)$ in the weak operator topology.

Proof of Claim 1. In view of Proposition I.41 on page 504, we can take $g \in C_0(Y)$. If $C \subset Y$ is compact, then i restricts to a *homeomorphism* of C onto its compact image in X. Thus, by the Tietze Extension Theorem, there is a $f_C \in C_c(X)$ such that $f_C\big(i(y)\big) = g(y)$ for all $y \in C$. Then, as in Proposition 8.27 on page 244, $\Pi_i(f_C) \to \pi_\mu(g)$ in the weak operator topology. This proves the claim. □

Claim 2. $\mathbb{1}_{E_n} \in \mathcal{H}$.

Proof of Claim 2. It suffices to see that $(\mathbb{1}_{E_n} \mid k) = 0$ for all $k \in \mathcal{H}^\perp$. Let g be the bounded Borel function $y \mapsto h(y)^{-1}\mathbb{1}_{E_n}(y)$. Using Claim 1, there is net $\{\, f_i \,\} \subset C_c(X)$ such that $\Pi_1(f_i) \to \pi_\mu(g)$ in the weak operator topology. Then since $\mathcal{H}$ is invariant for Π_1,

$$(\mathbb{1}_{E_n} \mid k) = \big(\pi_\mu(g)h \mid k\big) = \lim_i \big(\Pi_1(f_i)h \mid k\big) = 0.$$

This proves the claim. □

Notice that if $h_n := \mathbb{1}_{E_n}h$, then $h_n \to h$ in $L^2(Y)$ (by the Dominated Convergence Theorem). If $F \in C(\mathcal{H})$ and if $F \supset E_n$, then $\|\mathbb{1}_F h - h\|_2 \leq \|h_n - h\|_2$. Thus $\{\, \mathbb{1}_F \,\}_{F\in C(\mathcal{H})}$ converges strongly to the identity on $\mathcal{H}$ as F increases.

Claim 3. If $k \in \mathcal{H}^\perp$, then $\mathbb{1}_F k = 0$ for all $F \in C(\mathcal{H})$.

Proof of Claim 3. Note that $\mathcal{H}^\perp$ is also an invariant space. If $E \in C(\mathcal{H}^\perp)$, then $(\mathbb{1}_F \mid \mathbb{1}_E) = 0$. Since $\mathbb{1}_F$ and $\mathbb{1}_E$ are nonnegative functions, it follows that $\mathbb{1}_F\mathbb{1}_E = 0$.[12] But

$$\mathbb{1}_F k = \lim_{E\in C(\mathcal{H}^\perp)} \mathbb{1}_F\mathbb{1}_E k = 0.$$

This proves the claim. □

Now if $f \in L^2(Y)$, then we can write $f = f_1 + f_2$ with $f_1 \in \mathcal{H}$ and $f_2 \in \mathcal{H}^\perp$. Using Claim 3, we have

$$\lim_{F\in C(\mathcal{H})} \mathbb{1}_F f = \lim_{F\in C(\mathcal{H})} \mathbb{1}_F(f_1 + f_2) = \lim_{F\in C(\mathcal{H})} \mathbb{1}_F f_1 = f_1.$$

[12] If $\mathbb{1}_F(y)\mathbb{1}_E(y)$ is not zero for μ-almost every y, then $\int \mathbb{1}_F\mathbb{1}_E \, d\mu \neq 0$.

Thus if P is the orthogonal projection of $L^2(Y)$ onto $\mathcal{H}$, then $L_{\mathbb{1}_F} \to P$ in the strong operator topology. Since $\mathscr{L}_Y := \{ L_g : g \in \mathcal{B}^b(Y) \}$ is a von Neumann algebra (Proposition I.41 on page 504), $P \in \mathscr{L}_Y$ and $P = L_{\mathbb{1}_E}$ for some Borel set $E \subset Y$. Thus we have shown $\mathcal{H} = L^2(E)$ as claimed. Of course, a similar statement holds for Π_2 invariant subspaces of $L^2(Z, \nu)$.

Now suppose that Π_1 has a subrepresentation Π_1' equivalent to a subrepresentation Π_2' of Π_2. The above discussion proves that Π_1' must act on a subspace of the form $L^2(E)$ for some Borel set $E \subset Y$. Since $L^2(E)$ is not the zero subspace, E must contain a set of nonzero finite measure. By the regularity of μ, there must be a compact subset $C \subset E$ with $\mu(C) > 0$. Then $L^2(C)$ is an invariant subspace of Π_1 and determines a subrepresentation of Π_1 which is equivalent to a subrepresentation of Π_2 on a subspace of the form $L^2(F, \nu)$ for $F \subset Z$. As before, there is a compact subset $K \subset F$ such that $\nu(K) > 0$. $L^2(K)$ is invariant for Π_2, and the corresponding subrepresentation is equivalent to a subrepresentation of Π_1 on a subspace of $L^2(C)$. But $i(C)$ and $j(K)$ are disjoint compact sets in X. Then there is a $f \in C_c(X)$ such that f if identically zero on $i(C)$ and identically one on $j(K)$. But then $\Pi_1(f) = 0$ on $L^2(C)$ and $\Pi_2(f)$ is the identity on $L^2(K)$. This is a contradiction, and the completes the proof. □

I.8 Invariants of Morita Equivalent C^*-Algebras

Recall that a C^*-algebra is CCR if $\pi(A) = \mathcal{K}(\mathcal{H}_\pi)$ for all $\pi \in \hat{A}$. We say that A is GCR if $\pi(A) \supset \mathcal{K}(\mathcal{H}_\pi)$ for all $\pi \in \hat{A}$. A C^*-algebra is called *elementary* if it is isomorphic to the compact operators on some Hilbert space. Notice that a C^*-algebra is elementary if and only if it is Morita equivalent to the complex numbers $\mathbf{C}$.

The following observations probably should have been made in [139]. These results, and much more, are due to Zettl [174, 175]. The proofs here come from [81]. In the case of separable algebras, or more generally for σ-unital algebras, these results also follow from The Brown-Green-Rieffel Theorem [139, Theorem 5.55] which implies that Morita equivalent C^*-algebras are stably isomorphic.

Proposition I.43. *Suppose that* X *is an* A–B*-imprimitivity bimodule. If* B *is CCR, then* A *is CCR.*

Proof. Let $h_{\mathsf{X}} : \operatorname{Prim} A \to \operatorname{Prim} B$ be the Rieffel homeomorphism [139, Corollary 3.33(a)]. If $\pi \in \hat{A}$, then let $P = \ker \pi$. Then $h_{\mathsf{X}}(P) \in \operatorname{Prim} B$. Since B is CCR, $B/h_{\mathsf{X}}(P)$ is elementary. Since A/P and $B/h_{\mathsf{X}}(P)$ are Morita equivalent [139, Proposition 3.35], A/P is Morita equivalent to $\mathbf{C}$, and therefore elementary. Since π factors through A/P, π is onto the compacts on $\mathcal{H}_\pi$. Therefore, A is CCR as claimed. □

Proposition I.44. *Suppose that* X *is an* A–B*-imprimitivity bimodule. If* B *is GCR, then so is* A*.*

Proof. Let $h_X : \operatorname{Prim} A \to \operatorname{Prim} B$ be the Rieffel homeomorphism. Fix $\pi \in \hat{A}$ and let $P := \ker \pi$. [139, Proposition 3.25] shows that A/P is Morita equivalent to

$B/h_X(P)$. Since B is GCR, $B/h_X(P)$ has an elementary ideal which is therefore Morita equivalent to $\mathbf{C}$. The Rieffel correspondence implies that A/P must also have an elementary ideal I. Since $\pi(I) \neq \{0\}$, $\pi|_I$ is an irreducible representation on $\mathcal{H}_\pi$. Since I is elementary, $\pi(I) = \mathcal{K}(\mathcal{H}_\pi)$, and $\pi(A) \supset K(\mathcal{H}_\pi)$. Since π was arbitrary, A is GCR. $\square$

Notation and Symbol Index

Index

Bibliography

[1] William Arveson, *Operator algebras and invariant subspaces*, Ann. of Math. (2) **100** (1974), 433–532.

[2] ———, *An Invitation to C^*-algebras*, Springer-Verlag, New York, 1976. Graduate Texts in Mathematics, No. 39.

[3] ———, *Notes on measure and integration in locally compact spaces*, 1996. `http://www.math.berkeley.edu/~arveson`.

[4] Louis Auslander and Calvin C. Moore, *Unitary representations of solvable Lie groups*, Mem. Amer. Math. Soc. No. **62** (1966), 1–199.

[5] Larry Baggett and Adam Kleppner, *Multiplier representations of abelian groups*, J. Functional Analysis **14** (1973), 299–324.

[6] Teresa Bates, David Pask, Iain Raeburn, and Wojciech Szymański, *The C^*-algebras of row-finite graphs*, New York J. Math. **6** (2000), 307–324 (electronic).

[7] Sterling K. Berberian, *Counterexamples in Haar measure*, Amer. Math. Monthly **73** (1966), 135–140.

[8] Bruce Blackadar, *K-theory for operator algebras*, Second edition, Cambridge University Press, Cambridge, 1998.

[9] Étienne Blanchard, *Tensor products of $C(X)$-algebras over $C(X)$*, Astérisque (1995), 81–92. Recent advances in operator algebras (Orléans, 1992).

[10] Robert. J. Blattner, *On induced representations*, Amer. J. Math. **83** (1961), 79–98.

[11] Nicolas Bourbaki, *Éléments de mathématique. Fascicule XXIX. Livre VI: Intégration. Chapitre 7: Mesure de Haar. Chapitre 8: Convolution et représentations*, Hermann, Paris, 1963. Actualités Scientifiques et Industrielles, No. 1306.

[12] ———, *General topology, Part 1*, Addison-Wesley, Reading, 1966.

[13] ———, *Éléments de mathématique. Fasc. XXI. Livre VI: Intégration. Chapitre V: Intégration des mesures*, Hermann, Paris, 1967. Deuxième édition revue et corrigée. Actualités Scientifiques etIndustrielles, No. 1244.

[14] Richard Brauer, *Algebra der hyperkomplexen Zahlensysteme (Algebren)*, Math. J. Okayama Univ. **21** (1979), 53–89.

[15] Robert C. Busby and Harvey A. Smith, *Representations of twisted group algebras*, Trans. Amer. Math. Soc. **149** (1970), 503–537.

[16] François Combes, *Crossed products and Morita equivalence*, Proc. London Math. Soc. (3) **49** (1984), 289–306.

[17] Alain Connes, *Classification of injective factors*, Ann. of Math. **104** (1976), 73–115.

[18] John B. Conway, *A course in functional analysis*, Graduate texts in mathematics, vol. 96, Springer-Verlag, New York, 1985.

[19] David Crocker, Alexander Kumjian, Iain Raeburn, and Dana P. Williams, *An equivariant Brauer group and actions of groups on C^*-algebras*, J. Funct. Anal. **146** (1997), 151–184.

[20] Raúl E. Curto, Paul S. Muhly, and Dana P. Williams, *Cross products of strongly Morita equivalent C^*-algebras*, Proc. Amer. Math. Soc. **90** (1984), 528–530.

[21] Kenneth R. Davidson, *C^*-algebras by example*, Fields Institute Monographs, vol. 6, American Mathematical Society, Providence, RI, 1996.

[22] Leonard E. Dickson, *Linear associative algebras and abelian equations*, Trans. Amer. Math. Soc. **15** (1914), 31–46.

[23] Jan J. Dijkstra, *On homeomorphism groups and the compact-open topology*, Amer. Math. Monthly **112** (2005), 910–912.

[24] Nicolae Dinculeanu, *Vector measures*, International Series of Monographs in Pure and Applied Mathematics, Vol. 95, Pergamon Press, Oxford, 1967.

[25] Shaun Disney and Iain Raeburn, *Homogeneous C^*-algebras whose spectra are tori*, J. Austral. Math. Soc. Ser. A **38** (1985), 9–39.

[26] Jacques Dixmier, *Sur les structures boréliennes du spectre d'une C^*-algèbre*, Inst. Hautes Études Sci. Publ. Math. (1960), 5–11.

[27] ———, *Sur les espaces localement quasi-compacts*, Canad. J. Math. **20** (1968), 1093–1100.

[28] ———, *C^*-algebras*, North-Holland Mathematical Library, vol. 15, North-Holland, New York, 1977.

[29] ———, *von Neumann algebras*, North-Holland Mathematical Library, vol. 27, North-Holland Publishing Co., Amsterdam, 1981. With a preface by E. C. Lance, Translated from the second French edition by F. Jellett.

[30] Sergio Doplicher, Daniel Kastler, and Derek W. Robinson, *Covariance algebras in field theory and statistical mechanics*, Comm. Math. Phys. **3** (1966), 1–28.

[31] James Dugundji, *Topology*, Allyn and Bacon Inc., Boston, Mass., 1966.

[32] Nelson Dunford and Jacob T. Schwartz, *Linear Operators. I. General Theory*, Interscience Publishers, Inc., New York, 1958. (with the assistance of W. G. Bade and R. G. Bartle).

[33] ———, *Linear operators. Part II: Spectral theory. Self adjoint operators in Hilbert space*, Interscience Publishers John Wiley & Sons New York-London, 1963. (With the assistance of William G. Bade and Robert G. Bartle).

[34] Maurice J. Dupré, *The classification and structure of C^*-algebra bundles*, Mem. Amer. Math. Soc. **21** (1979), i–ix, 1–77.

[35] Maurice J. Dupré and Richard M. Gillette, *Banach bundles, Banach modules and automorphisms of C^*-algebras*, Vol. 92, Pitman (Advanced Publishing Program), Boston, MA, 1983.

[36] Siegfried Echterhoff, *On induced covariant systems*, Proc. Amer. Math. Soc. **108** (1990), 703–706.

[37] ———, *Erratum to: "On induced covariant systems" [Proc. Amer. Math. Soc.* **108** *(1990), no. 3, 703–706; MR 90f:46105]*, Proc. Amer. Math. Soc. **116** (1992), 581.

[38] ———, *Crossed products with continuous trace*, Mem. Amer. Math. Soc. **123** (1996), i–viii, 1–134.

[39] Siegfried Echterhoff, Steven Kaliszewski, John Quigg, and Iain Raeburn, *Naturality and induced representations*, Bull. Austral. Math. Soc. **61** (2000), 415–438.

[40] ———, *A categorical approach to imprimitivity theorems for C^*-dynamical systems*, Mem. Amer. Math. Soc. **180** (2006), i–viii, 1–169.

[41] Siegfried Echterhoff and Ryszard Nest, *The structure of the Brauer group and crossed products of $C_0(X)$-linear group actions on $C_0(X, \mathcal{K})$*, Trans. Amer. Math. Soc. **353** (2001), 3685–3712 (electronic).

[42] Siegfried Echterhoff and Jonathan Rosenberg, *Fine structure of the Mackey machine for actions of abelian groups with constant Mackey obstuction*, Pacific J. Math. **170** (1995), 17–52.

[43] Siegfried Echterhoff and Dana P. Williams, *Locally inner actions on $C_0(X)$-algebras*, J. Operator Theory **45** (2001), 131–160.

[44] ———, *Inducing primitive ideals*, preprint, 2005.

[45] Edward G. Effros, *A decomposition theory for representations of C^*-algebras*, Trans. Amer. Math. Soc. **107** (1963), 83–106.

[46] ———, *Transformation groups and C^*-algebras*, Ann. of Math. **81** (1965), 38–55.

[47] ———, *Global structure in von Neumann algebras*, Trans. Amer. Math. Soc. **121** (1966), 434–454.

[48] ———, *Polish transformation groups and classification problems*, General topology and modern analysis (Proc. Conf., Univ. California, Riverside, Calif., 1980), Academic Press, New York, 1981, pp. 217–227.

[49] Edward G. Effros and Frank Hahn, *Locally compact transformation groups and C^*-algebras*, Memoirs of the American Mathematical Society, No. 75, American Mathematical Society, Providence, R.I., 1967.

[50] James M. G. Fell, *The structure of algebras of operator fields*, Acta Math. **106** (1961), 233–280.

[51] ———, *A Hausdorff topology on the closed subsets of a locally compact non-Hausdorff space*, Proc. Amer. Math. Soc. **13** (1962), 472–476.

[52] ———, *Weak containment and induced representations of groups, II*, Trans. Amer. Math. Soc. **110** (1964), 424–447.

[53] ———, *Induced representations and Banach $*$-algebraic bundles*, Lecture Notes in Math., vol. 582, Springer-Verlag, New York, 1977.

[54] James M. G. Fell and Robert S. Doran, *Representations of $*$-algebras, locally compact groups, and Banach $*$-algebraic bundles. Vol. 1*, Pure and Applied Mathematics, vol. 125, Academic Press Inc., Boston, MA, 1988. Basic representation theory of groups and algebras.

[55] ———, *Representations of $*$-algebras, locally compact groups, and Banach $*$-algebraic bundles. Vol. 2*, Pure and Applied Mathematics, vol. 126, Academic Press Inc., Boston, MA, 1988. Banach $*$-algebraic bundles, induced representations, and the generalized Mackey analysis.

[56] Gerald B. Folland, *A course in abstract harmonic analysis*, Studies in Advanced Mathematics, CRC Press, Boca Raton, FL, 1995.

[57] ———, *Real analysis*, Second, John Wiley & Sons Inc., New York, 1999. Modern techniques and their applications, A Wiley-Interscience Publication.

[58] Georg Frobenius, *Über relationen zwischen den charakteren einer gruppe und denen ihrer untergruppen*, Sitzber. Preuss. Akad. Wiss. (1898), 501–515.

[59] James Glimm, *Locally compact transformation groups*, Trans. Amer. Math. Soc. **101** (1961), 124–128.

[60] ———, *Type I C^*-algebras*, Ann. of Math. (2) **73** (1961), 572–612.

[61] ———, *Families of induced representations*, Pacific J. Math. **12** (1962), 885–911.

[62] Elliot C. Gootman, *The type of some C^*- and W^*-algebras associated with transformation groups*, Pacific J. Math. **48** (1973), 93–106.

[63] Elliot C. Gootman and Dorte Olesen, *Spectra of actions on type I C^*-algebras*, Math. Scand. **47** (1980), 329–349.

[64] Elliot C. Gootman and Jonathan Rosenberg, *The structure of crossed product C^*-algebras: a proof of the generalized Effros-Hahn conjecture*, Invent. Math. **52** (1979), 283–298.

[65] Philip Green, *C^*-algebras of transformation groups with smooth orbit space*, Pacific J. Math. **72** (1977), 71–97.

[66] ———, *The local structure of twisted covariance algebras*, Acta Math. **140** (1978), 191–250.

[67] ———, *The structure of imprimitivity algebras*, J. Funct. Anal. **36** (1980), 88–104.

[68] Frederick P. Greenleaf, *Invariant means on topological groups and their applications*, Van Nostrand Reinhold Co., New York, 1969.

[69] Alain Guichardet, *Tensor products of C^*-algebras*, 1969. Math. Inst. Aarhus Univ. Lecture Notes No. 12–13.

[70] Paul R. Halmos, *Measure theory*, D. Van Nostrand Company, Inc., New York, N. Y., 1950.

[71] Edwin Hewitt and Kenneth A. Ross, *Abstract harmonic analysis. Vol. I: Structure of topological groups. Integration theory, group representations*, Die Grundlehren der mathematischen Wissenschaften, Band 115, Springer-Verlag, New York, 1963.

[72] ———, *Abstract harmonic analysis. Vol. II: Structure and analysis for compact groups. Analysis on locally compact Abelian groups*, Die Grundlehren der mathematischen Wissenschaften, Band 152, Springer-Verlag, New York, 1970.

[73] Einar Hille and Ralph S. Phillips, *Functional analysis and semi-groups*, American Mathematical Society, Providence, R. I., 1974. Third printing of the revised edition of 1957, American Mathematical Society Colloquium Publications, Vol. XXXI.

[74] Raphael Høegh-Krohn and Tor Skjelbred, *Classification of C^*-algebras admitting ergodic actions of the two-dimensional torus*, J. Reine Angew. Math. **328** (1981), 1–8.

[75] Karl Heinrich Hofmann, *Banach bundles*, 1974. Darmstadt Notes.

[76] ———, *Bundles and sheaves are equivalent in the category of Banach spaces*, K-theory and operator algebras (Proc. Conf., Univ. Georgia, Athens, Ga., 1975), Lecture Notes in Math, vol. 575, Springer, Berlin, 1977, pp. 53–69.

[77] Karl Heinrich Hofmann and Klaus Keimel, *Sheaf-theoretical concepts in analysis: bundles and sheaves of Banach spaces, Banach $C(X)$-modules*, Applications of sheaves (Proc. Res. Sympos. Appl. Sheaf Theory to Logic, Algebra and Anal., Univ. Durham, Durham, 1977), Lecture Notes in Math., vol. 753, Springer, Berlin, 1979, pp. 415–441.

[78] Astrid an Huef and Iain Raeburn, *Regularity of induced representations and a theorem of Quigg and Spielberg*, Math. Proc. Cambridge Philos. Soc. **133** (2002), 249–259.

[79] Astrid an Huef, Iain Raeburn, and Dana P. Williams, *An equivariant Brauer semigroup and the symmetric imprimitivity theorem*, Trans. Amer. Math. Soc. **352** (2000), 4759–4787.

[80] ———, *Proper actions on imprimitivity bimodules and decompositions of Morita equivalences*, J. Funct. Anal. **200** (2003), 401–428.

[81] ———, *Properties preserved under Morita equivalence of C^*-algebras*, Proc. Amer. Math. Soc (2006), (in press).

[82] Andrzej Hulanicki, *Means and Følner condition on locally compact groups*, Studia Math. **27** (1966), 87–104.

[83] Steven Hurder, Dorte Olesen, Iain Raeburn, and Jonathan Rosenberg, *The Connes spectrum for actions of abelian groups on continuous-trace algebras*, Ergod. Th. & Dynam. Sys. **6** (1986), 541–560.

[84] Thomas Jech, *Set theory*, Pure and Applied Mathematics, vol. 79, Academic Press, New York, 1978.

[85] Steven Kaliszewski and John Quigg, *Mansfield's imprimitivity theorem for full crossed products*, Trans. Amer. Math. Soc. **357** (2005), 2021–2042 (electronic).

[86] Irving Kaplansky, *The structure of certain operator algebras*, Trans. Amer. Math. Soc. **70** (1951), 219–255.

[87] Gennadi G. Kasparov, *Equivariant KK-theory and the Novikov conjecture*, Invent. Math. **91** (1988), 147–201.

[88] Esben T. Kehlet, *Cross sections for quotient maps of locally compact groups*, Math. Scand **55** (1984), 152–160.

[89] Eberhard Kirchberg and N. Christopher Phillips, *Embedding of continuous fields of C^*-algebras in the Cuntz algebra $\mathcal{O}_2$*, J. Reine Angew. Math. **525** (2000), 55–94.

[90] ———, *Embedding of exact C^*-algebras in the Cuntz algebra $\mathcal{O}_2$*, J. Reine Angew. Math. **525** (2000), 17–53.

[91] Eberhard Kirchberg and Simon Wassermann, *Exact groups and continuous bundles of C^*-algebras*, Math. Ann. **315** (1999), 169–203.

[92] ———, *Permanence properties of C^*-exact groups*, Doc. Math. **4** (1999), 513–558 (electronic).

[93] Adam Kleppner, *Continuity and measurability of multiplier and projective representations*, J. Functional Analysis **17** (1974), 214–226.

[94] ———, *Measurable homomorphisms of locally compact groups*, Proc. Amer. Math. Soc. **106** (1989), 391–395.

[95] Alexander Kumjian, Iain Raeburn, and Dana P. Williams, *The equivariant Brauer groups of commuting free and proper actions are isomorphic*, Proc. Amer. Math. Soc. **124** (1996), 809–817.

[96] Kasimierz Kuratowski, *Topology. Vol. I*, Academic Press, New York, 1966.

[97] E. Christopher Lance, *On nuclear C^*-algebras*, J. Functional Analysis **12** (1973), 157–176.

[98] ———, *Hilbert C^*-modules: A toolkit for operator algebraists*, London Math. Soc. Lecture Note Series, vol. 210, Cambridge Univ. Press, Cambridge, 1994.

[99] Ru Ying Lee, *On the C^*-algebras of operator fields*, Indiana Univ. Math. J. **25** (1976), 303–314.

[100] Horst Leptin, *Darstellungen verallgemeinerter L^1-Algebren*, Invent. Math. **5** (1968), 192–215.

[101] Lynn H. Loomis, *An introduction to abstract harmonic analysis*, Van Nostrand, New York, 1953.

[102] George W. Mackey, *Imprimitivity for representations of locally compact groups. I*, Proc. Nat. Acad. Sci. U. S. A. **35** (1949), 537–545.

[103] ———, *On induced representations of groups*, Amer. J. Math. **73** (1951), 576–592.

[104] ———, *Induced representations of locally compact groups. I*, Ann. of Math. **55** (1952), 101–139.

[105] ———, *Borel structures in groups and their duals*, Trans. Amer. Math. Soc. **85** (1957), 134–165.

[106] ———, *The theory of unitary group representations*, University of Chicago Press, Chicago, Ill., 1976. Based on notes by James M. G. Fell and David B. Lowdenslager of lectures given at the University of Chicago, Chicago, Ill., 1955, Chicago Lectures in Mathematics.

[107] Ralf Meyer, *Generalized fixed point algebras and square-integrable groups actions*, J. Funct. Anal. **186** (2001), 167–195.

[108] Paul S. Muhly, *Coordinates in operator algebra*, CMBS Conference Lecture Notes (Texas Tech 1990), 1999. In continuous preparation.

[109] James R. Munkres, *Topology: Second edition*, Prentice Hall, New Jersey, 2000.

[110] Gerard J. Murphy, *C^*-algebras and operator theory*, Academic Press Inc., Boston, MA, 1990.

[111] Francis J. Murray and John von Neumann, *On rings of operators*, Ann. of Math. (2) **37** (1936), 116–229.

[112] N. Dang Ngoc, *Produits croisés restreints et extensions of groupes*, 1977. Notes, Paris.

[113] May Nilsen, *C^*-bundles and $C_0(X)$-algebras*, Indiana Univ. Math. J. **45** (1996), 463–477.

[114] Dorte Olesen and Gert K. Pedersen, *Applications of the Connes spectrum to C^*-dynamical systems*, J. Funct. Anal. **30** (1978), 179–197.

[115] ———, *Applications of the Connes spectrum to C^*-dynamical systems. II*, J. Funct. Anal. **36** (1980), 18–32.

[116] ———, *Applications of the Connes spectrum to C^*-dynamical systems, III*, J. Funct. Anal. **45** (1982), 357–390.

[117] ———, *Partially inner C^*-dynamical systems*, J. Funct. Anal. **66** (1986), 263–281.

[118] Alexander Ju. Ol′šanskiĭ, *On the question of the existence of an invariant mean on a group*, Uspekhi Mat. Nauk **35** (1980), 199–200.

[119] Judith A. Packer and Iain Raeburn, *Twisted crossed products of C^*-algebras*, Math. Proc. Cambridge Philos. Soc. **106** (1989), 293–311.

[120] ———, *Twisted crossed products of C^*-algebras. II*, Math. Ann. **287** (1990), 595–612.

[121] ———, *On the structure of twisted group C^*-algebras*, Trans. Amer. Math. Soc. **334** (1992), 685–718.

[122] Judith A. Packer, Iain Raeburn, and Dana P. Williams, *The equivariant Brauer group of principal bundles*, J. Operator Theory **36** (1996), 73–105.

[123] Theodore W. Palmer, *Classes of nonabelian, noncompact, locally compact groups*, Rocky Mountain J. Math. **8** (1978), 683–741.

[124] Kalyanapuram Rangachari Parthasarathy, *Introduction to probability and measure*, Springer-Verlag New York Inc., New York, 1978.

[125] Alan L. T. Paterson, *Amenability*, Mathematical Surveys and Monographs, vol. 29, American Mathematical Society, Providence, RI, 1988.

[126] Gert K. Pedersen, *C^*-algebras and their automorphism groups*, London Mathematical Society Monographs, vol. 14, Academic Press Inc. [Harcourt Brace Jovanovich Publishers], London, 1979.

[127] ———, *Analysis now*, Graduate Texts in Mathematics, vol. 118, Springer-Verlag, New York, 1989.

[128] N. Christopher Phillips, *A classification theorem for nuclear purely infinite simple C^*-algebras*, Doc. Math. **5** (2000), 49–114 (electronic).

[129] Lev S. Pontrjagin, *Topological groups*, Princeton University Press, Princeton, 1946.

[130] John Quigg, *Duality for reduced twisted crossed products of C^*-algebras*, Indiana Univ. Math. J. **35** (1986), 549–571.

[131] John Quigg and Jack Spielberg, *Regularity and hyporegularity in C^*-dynamical systems*, Houston J. Math. **18** (1992), 139–152.

[132] Iain Raeburn, *Induced C^*-algebras and a symmetric imprimitivity theorem*, Math. Ann. **280** (1988), 369–387.

[133] ———, *On crossed products and Takai duality*, Proc. Edinburgh Math. Soc. (2) **31** (1988), 321–330.

[134] ———, *On crossed products by coactions and their representation theory*, Proc. London Math. Soc. **64** (1992), 625–652.

[135] ———, *Crossed products of C^*-algebras by coactions of locally compact groups*, Operator algebras and quantum field theory (Rome, 1996), Internat. Press, Cambridge, MA, 1997, pp. 74–84.

[136] ———, *Dynamical systems and operator algebras*, Proc. Centre Math. Appl. Austral. Nat. Univ. **36** (1999), 109–120.

[137] Iain Raeburn and Jonathan Rosenberg, *Crossed products of continuous-trace C^*-algebras by smooth actions*, Trans. Amer. Math. Soc. **305** (1988), 1–45.

[138] Iain Raeburn and Dana P. Williams, *Pull-backs of C^*-algebras and crossed products by certain diagonal actions*, Trans. Amer. Math. Soc. **287** (1985), 755–777.

[139] ———, *Morita equivalence and continuous-trace C^*-algebras*, Mathematical Surveys and Monographs, vol. 60, American Mathematical Society, Providence, RI, 1998.

[140] Arlan Ramsay, *Virtual groups and group actions*, Advances in Math. **6** (1971), 253–322 (1971).

[141] ———, *Nontransitive quasi-orbits in Mackey's analysis of group extensions*, Acta Math. **137** (1976), 17–48.

[142] ———, *Topologies on measured groupoids*, J. Funct. Anal. **47** (1982), 314–343.

[143] Jean Renault, *A groupoid approach to C^*-algebras*, Lecture Notes in Mathematics, vol. 793, Springer-Verlag, New York, 1980.

[144] ———, *The ideal structure of groupoid crossed product C^*-algebras*, J. Operator Theory **25** (1991), 3–36.

[145] Marc A. Rieffel, *Induced representations of C^*-algebras*, Advances in Math. **13** (1974), 176–257.

[146] ———, *Strong Morita equivalence of certain transformation group C^*-algebras*, Math. Ann. **222** (1976), 7–22.

[147] ———, *C^*-algebras associated with irrational rotations*, Pacific J. Math. **93** (1981), 415–429.

[148] ———, *Applications of strong Morita equivalence to transformation group C^*-algebras*, Operator algebras and applications, Part I (Kingston, Ont., 1980), Proc. Sympos. Pure Math., vol. 38, Amer. Math. Soc., Providence, R.I., 1982, pp. 299–310.

[149] ———, *The cancellation theorem for projective modules over irrational rotation C^*-algebras*, Proc. London Math. Soc. (3) **47** (1983), 285–302.

[150] ———, *Proper actions of groups on C^*-algebras*, Mappings of operator algebras (Philadelphia, PA, 1988), Progr. Math., vol. 84, Birkhäuser Boston, Boston, MA, 1990, pp. 141–182.

[151] ———, *Integrable and proper actions on C^*-algebras, and square-integrable representations of groups*, Expo. Math. **22** (2004), 1–53.

[152] Mikael Rørdam, *Classification of nuclear, simple C^*-algebras*, Classification of nuclear C^*-algebras. entropy in operator algebras, Encyclopaedia Math. Sci., vol. 126, Springer, Berlin, 2002, pp. 1–145.

[153] Jonathan Rosenberg, *A selective history of the Stone-von Neumann theorem*, Operator algebras, quantization, and noncommutative geometry, Contemp. Math., vol. 365, Amer. Math. Soc., Providence, RI, 2004, pp. 331–353.

[154] Halsey L. Royden, *Real analysis*, Third Edition, Macmillan Publishing Company, New York, 1988.

[155] Walter Rudin, *Functional analysis*, McGraw-Hill Book Co., New York, 1973. McGraw-Hill Series in Higher Mathematics.

[156] ———, *Real and complex analysis*, McGraw-Hill, New York, 1987.

[157] Shôichirô Sakai, *On a characterization of type I C^*-algebras*, Bull. Amer. Math. Soc. **72** (1966), 508–512.

[158] ———, *On type I C^*-algebras*, Proc. Amer. Math. Soc. **18** (1967), 861–863.

[159] Jean-Luc Sauvageot, *Idéaux primitifs de certains produits croisés*, Math. Ann. **231** (1977/78), 61–76.

[160] ———, *Idéaux primitifs induits dans les produits croisés*, J. Funct. Anal. **32** (1979), 381–392.

[161] Hiroshi Takai, *On a duality for crossed products of C^*-algebras*, J. Funct. Anal. **19** (1975), 25–39.

[162] Masamichi Takesaki, *Covariant representations of C^*-algebras and their locally compact automorphism groups*, Acta Math. **119** (1967), 273–303.

[163] Takasi Turumaru, *Crossed product of operator algebra*, Tôhoku Math. J. (2) **10** (1958), 355–365.

[164] Simon Wassermann, *Exact C^*-algebras and related topics*, Lecture Notes Series, vol. 19, Seoul National University Research Institute of Mathematics Global Analysis Research Center, Seoul, 1994.

[165] Nik Weaver, *A prime C^*-algebra that is not primitive*, J. Funct. Anal. **203** (2003), 356–361.

[166] Joseph H. M. Wedderburn, *A type of primitive algebra*, Trans. Amer. Math. Soc. **15** (1914), 162–166.

[167] André Weil, *L'intègration dans les groupes topologiques et ses applications*, Hermann et Cie., Paris, 1940. [This book has been republished by the author at Princeton, N. J., 1941.]

[168] Stephen Willard, *General topology*, Addison-Wesley, 1970.

[169] Dana P. Williams, *The topology on the primitive ideal space of transformation group C^*-algebras and C.C.R. transformation group C^*-algebras*, Trans. Amer. Math. Soc. **266** (1981), 335–359.

[170] ———, *Transformation group C^*-algebras with continuous trace*, J. Funct. Anal. **41** (1981), 40–76.

[171] ———, *The structure of crossed products by smooth actions*, J. Austral. Math. Soc. (Series A) **47** (1989), 226–235.

[172] ———, *From the Stone-von Neumann theorem to the equivariant Brauer group and beyond*, Operator algebras, quantization, and noncommutative geometry, Contemp. Math., vol. 365, Amer. Math. Soc., Providence, RI, 2004, pp. 401–422.

[173] Georges Zeller-Meier, *Produits croisés d'une C^*-algèbre par un groupe d'automorphismes*, J. Math. Pures Appl. (9) **47** (1968), 101–239.

[174] Heinrich H. Zettl, *Ideals in Hilbert modules and invariants under strong Morita equivalence of C^*-algebras*, Arch. Math. (Basel) **39** (1982), 69–77.

[175] ———, *Strong Morita equivalence of C^*-algebras preserves nuclearity*, Arch. Math. (Basel) **38** (1982), 448–452.

[176] Robert J. Zimmer, *Ergodic theory and semisimple groups*, Monographs in Mathematics, vol. 81, Birkhäuser Verlag, Basel, 1984.